Advances in Intelligent Systems and Computing

Volume 1397

The series "Advances in Intelligent Systems and Computing" contains publications on theory, applications, and design methods of Intelligent Systems and Intelligent Computing. Virtually all disciplines such as engineering, natural sciences, computer and information science, ICT, economics, business, e-commerce, environment, healthcare, life science are covered. The list of topics spans all the areas of modern intelligent systems and computing such as: computational intelligence, soft computing including neural networks, fuzzy systems, evolutionary computing and the fusion of these paradigms, social intelligence, ambient intelligence, computational neuroscience, artificial life, virtual worlds and society, cognitive science and systems, Perception and Vision, DNA and immune based systems, self-organizing and adaptive systems, e-Learning and teaching, human-centered and human-centric computing, recommender systems, intelligent control, robotics and mechatronics including human-machine teaming, knowledge-based paradigms, learning paradigms, machine ethics, intelligent data analysis, knowledge management, intelligent agents, intelligent decision making and support, intelligent network security, trust management, interactive entertainment, Web intelligence and multimedia.

The publications within "Advances in Intelligent Systems and Computing" are primarily proceedings of important conferences, symposia and congresses. They cover significant recent developments in the field, both of a foundational and applicable character. An important characteristic feature of the series is the short publication time and world-wide distribution. This permits a rapid and broad dissemination of research results.

Indexed by DBLP, INSPEC, WTI Frankfurt eG, zbMATH, Japanese Science and Technology Agency (JST).

All books published in the series are submitted for consideration in Web of Science.

More information about this series at http://www.springer.com/series/11156

G. Ranganathan · Xavier Fernando · Fuqian Shi ·
Youssouf El Allioui
Editors

Soft Computing for Security Applications

Proceedings of ICSCS 2021

Springer

Editors
G. Ranganathan
Department of Electronics
and Communication Engineering
Gnanamani College of Technology
Namakkal, India

Fuqian Shi
Rutgers Cancer Institute of New Jersey
New Brunswick, NJ, USA

Xavier Fernando
Ryerson Communications Lab
Toronto, ON, Canada

Youssouf El Allioui
Department of Computer Science
University of Sultan Moulay Slimane
Beni Mellal, Morocco

ISSN 2194-5357 ISSN 2194-5365 (electronic)
Advances in Intelligent Systems and Computing
ISBN 978-981-16-5300-1 ISBN 978-981-16-5301-8 (eBook)
https://doi.org/10.1007/978-981-16-5301-8

This Springer imprint is published by the registered company Springer Nature Singapore Pte Ltd.
The registered company address is: 152 Beach Road, #21-01/04 Gateway East, Singapore 189721, Singapore

The proceeding of ICSCS 2021 is dedicated to all the reviewers and authors for acknowledging their sincerity and dedication toward enhancing the computing and communication research community. Also, this proceeding is dedicated to all the researchers, who share their research knowledge and exclusive learning from experts and real-time research experiences to a wider audience.

Preface

It is our privilege to welcome you to the proceedings of the International Conference on Soft Computing for Security Applications [ICSCS 2021]. The theme for this year is "*Intelligent Soft Computing Technologies and Security Models*," a topic that is gaining significant research attention from both academic and industrial domains due to the relevance of advanced computing solutions to security challenges in real-time applications. This conference showcases how soft computing technologies feature prominently on how the advanced communication systems can be designed, implemented, and maintained, and this makes ICSCS an excellent platform to explore the advanced computing foundations for solving the emerging security issues.

ICSCS has totally received 274 papers, and based on technical review from the conference program committee, internal and external reviewers, 67 papers were selected for the conference. Also, each paper submitted to ICCSCS has received review decisions and comments from at least three reviewers. Moreover, the papers and presentations include in these proceedings and conference cover a wide range of topics including fuzzy sets and systems, neural networks, IoT, image processing, machine intelligence, web intelligence, intelligent security protocols, and so on. The conference event also provided the advanced technical research insights by providing valuable keynote sessions to the conference delegates and authors.

The aim of these proceedings is to bring together the knowledge and novel research data presented at each of the sessions at ICSCS 2021 and disseminate this valuable research ideas and information to a larger audience. Each session from the conference is covered by research articles from different conference tracks, and also, the Q&A session after each presentation has also helped the conference delegates to improve their future research scope and gain more useful research insights on computing technologies.

We would also especially like to thank the potential researchers, academicians, and student presenters for sharing their data and experiences and also the conference delegates for their continuous engagement.

Namakkal, India	Dr. G. Ranganathan
Toronto, Canada	Dr. Xavier Fernando
New Brunswick, USA	Dr. Fuqian Shi
Beni Mellal, Morocco	Dr. Youssouf El Allioui

Acknowledgements

The success of ICSCS 2021 depends completely on the continuous effort of researchers in the field of computing and communication systems, especially who have exclusively written and submitted a novel research articles on different computing topics. An equal appreciation is also deserved for the technical conference program committee members, internal and external reviewers, who have dedicated their efforts and time for the continuous assessment of multiple research articles submitted to the conference by maintaining the standards and quality of the conference.

We are pleased to acknowledge the keynote speaker, Dr. R. Kanthavel, King Khalid University, Kingdom of Saudi Arabia, for his outstanding efforts in producing a truly technical session that provides the right balance in all aspects of the state-of-the-art computing approaches and also their cooperation throughout the conference event. We are grateful for the support of our institution Dhirajlal Gandhi College of Technology for making this conference possible.

We would also thank all participants who have actively participated in the conference activities and enjoyed the opportunity to learn from one another. We would like to acknowledge all members of conference advisory committee and technical and non-technical program committee for providing excellent guidance.

On behalf of the organizing committee, we welcome you all to the proceedings of ICSCS 2021. We wish all the readers to make use of all the technical research experiences included in ICSCS 2021 proceedings.

Contents

About the Editors

Dr. G. Ranganathan working as a Professor and Head in Gnanamani college of Engineering and Technology, Namakal, India. He has done his Ph.D. in the Faculty of Information and Communication Engineering from Anna University, Chennai in the year 2013. His research thesis was in the area of Bio Medical Signal Processing. He has total of 29+ years of experience both in industry, teaching and research. He has guided several project works for many UG and PG Students in the areas of Bio Medical Signal Processing. He has published more than 35 research papers in International and National Journals and Conferences. He has also co-authored many books in electrical and electronics subjects. He has served as Referee for many reputed International Journals published by Elsevier, Springer, Taylor and Francis, etc. He has membership in various professional bodies like ISTE, IAENG etc., and has actively involved himself in organizing various international and national level conferences, symposiums, seminars etc.

Dr. Xavier Fernando is the Director of Ryerson Communications Lab that has received total research funding of $3,185,000.00 since 2008 from industry and government (including joint grants). He has (co-) authored over 200 research articles, three books (one translated to Mandarin), several book chapters and holds few patents. The present and past members and affiliates of this lab can be found at this LinkedIn group and Facebook page. He was an IEEE Communications Society Distinguished Lecturer and has delivered over 50 invited talks and keynote presentations all over the world. He was a member in the IEEE Communications Society (COMSOC) Education Board Working Group on Wireless Communications. He was the Chair of IEEE Canada Humanitarian Initiatives Committee 2017–18. He was also the Chair of the IEEE Toronto Section and IEEE Canada Central Area.

Dr. Fuqian Shi was graduated from College of Computer Science and Technology, Zhejiang University and got his Ph.D. on Engineering, and was a visiting Associate Professor at Department of Industrial Engineering and Management System, University of Central Florida, USA from 2012 to 2014. Dr. Shi is currently serving for Rutgers Cancer Institute of New Jersey, USA. Dr. Shi is a Senior Member of IEEE, Membership of ACM, and sever as over 30 committee board membership of international conferences; Dr. Shi also serve as associate editors of *International Journal of Ambient Computing and Intelligence* (IJACI), *International Journal of Rough Sets and Data Analysis* (IJRSDA), and special issue editors of IEEE Access. He published over 100 journal papers and conference proceedings, his research interests include fuzzy inference system, artificial neuro networks, and bio-mechanical engineering.

Dr. Youssouf El Allioui got his Ph.D. in 2012 from Sidi Mohamed Ben Abdellah University, Fez, Morocco. Currently he is a professor at the department of computer science, University of Sultan Moulay Slimane, Beni Mellal, Morocco. He works in a multidisciplinary environment involving machine learning, educational data mining (EDM), semantic web, knowledge management, Web Personalization and Recommender Systems, E-learning and semantic E-learning. His research focus is to study those research fields to come up with solutions to various real-world problems. He published several research papers at international journals and conferences. He served as a member of the technical committees of several international conferences and workshops.

Facial Image Denoising and Reconstruction

Sahil Vartak, Bhagyashree Wad, Pratik Varandekar, and Dnyaneshwar Kapse

Abstract Facial Image denoising removes the white Gaussian noise present in an image, retaining the facial features. White noise reduces the quality of images. The higher the noise present, the more difficult it is for facial recognition software to recognize the person, resulting in incorrect predictions. In this project, we aim to remove this image degrading noise from using a shallow CNN architecture. While denoising works on retaining a noisy face's features, the reconstruction model reconstructs a face's missing features, given a subset of the central portion of the facial image. We use a Generative Adversarial Network (GAN) to generate these missing portions of the face, where the Generator is involved in a Zero-Sum game with the Discriminator: both trying to minimize their loss the expense of the other.

Keywords Curriculum learning · Convolution neural network · Generative adversarial network · Image denoising · Image outpainting

1 Introduction

Images are used in a wide range of applications, from facial recognition to MRI to detection of cancer using AI. While Deep-Learning has made immense progress in Computer Vision, it cannot make accurate predictions on noisy images. In many cases, the models have confidently made predictions on plain white noise, assuming the noise to be an object or an animal [1]. Hence, it becomes of utmost importance that facial images be Noise Free before a trained model can make any predictions, as a could be catastrophic. Especially in security-sensitive applications such as biometric login methods, using Noise Free and clear images is a must as noisy images can lead to models misinterpreting a person for someone else.

S. Vartak (✉) · B. Wad · P. Varandekar · D. Kapse
Computer Engineering, Rajiv Gandhi Institute of Technology, Andheri (West), Mumbai, Maharashtra 400053, India

D. Kapse
e-mail: dnyaneshwar.kapse@mctrgit.ac.in

G. Ranganathan et al. (eds.), *Soft Computing for Security Applications*,
Advances in Intelligent Systems and Computing 1397,
https://doi.org/10.1007/978-981-16-5301-8_1

White noise disrupts an image's quality [2], rendering certain facial features to be altered or completely removed. Hence, for a Deep-Learning model to accurately make predictions on these images, the images need to be Noise Free.

To make the process of denoising more adaptable to a wide range of noise, we have trained a Convolution Neural Network: a learning-based approach. The paper discusses ways to make the training process more efficient by accommodating a new approach to train neural networks. To determine the quality of denoised images, we used Peak Signal to Noise Ratio (PSNR) as the metric for both noisy and actual images.

Image Outpainting [3] is a widely studied and ever-growing concept in Computer Vision. We apply the concept of Image Outpainting to reconstruct the missing portions of a face, such as edges of eyes, eyebrows, lips, and ears. A Generative Adversarial Network (GAN) is used where the Generator's main task is to learn the facial features of a face given a subset of its features. The Discriminator learns to differentiate between the generated and the original images. Hence, both Generator and Discriminator train themselves in competition: improvement leads to increased loss for the other and vice-versa. Our main objective is to reduce the Generator loss so that the Discriminator cannot differentiate between the generated and the real images. We again make use of PSNR as a metric to evaluate the quality of generated images.

2 Related Work

2.1 Image Denoising

Image Denoising methods can be broadly classified into two categories: Static filtering methods and learning-based methods. Although static methods are fast and provide good results at low noise levels, their performance declines as noise increases.

Average Filtering [4] is a static filtering method, wherein the weights of the filter are determined. This filter traverses each pixel in the image, and the subset of pixels that fall under the filter are averaged to produce the new pixel value. The method removes low-intensity noise, but as the filter size increases, the images become more blur.

Median Filtering [4, 5] removes the static filter weights used in Average Filtering; instead, it uses an array of a fixed size. This array traverses the entire Image, where each pixel value is defined by the median of the pixel values present in the array. In this technique, the images and essential facial features are lost due to some repeated pixels.

Weiner Filtering [6] is another static filtering method that proved effective against motion blur removal. Later, a variant of this method Weiner Filtering with Directionalet Domain was introduced [6], which showed its abilities in low-level denoising noise from images.

A modern approach to obtain a high-resolution image is discussed by [7–9]. The algorithm uses CNN: a learning-based technique, which returns better results than Average, Median, Gaussian, and Wiener filtering algorithms as mentioned in the paper. The algorithm uses adaptive weights, which prove to have an advantage on noise removal.

2.2 *Image Outpainting*

As explained in the paper [3], the concept updates the hidden portions of the "city/places365" images using a GAN architecture. Although the paper publishers obtain realistic-looking images of cities and nature, we try to achieve similar results to predict facial images' missing portions. Although realistic, the results presented could be improved by fine-tuning of hyperparameters of the Generator and Discriminator models to fit the model on our database consisting of facial images.

Some of the other methods used for Image Reconstruction are described by [10, 11]. These methods make use of a Convolution Neural network with feature extraction and pattern recognition algorithms to generate color saliency and direction maps respectively. However, both methods use feature extraction methods before feeding the image as input to their respective models, which we aim to overcome in this paper.

3 Dataset

3.1 *Description*

For both our Image Denoising and Image Reconstruction model, we have used the dataset [12]. From the dataset, we consider a total of 630 images of 105 different people. Each Image in our dataset has a dimension of 224×224, consisting primarily of people's faces (Fig. 1).

3.2 *Preprocessing*

First, we scale the images so that each pixel value lies between 0 and 1, such that I $\in [0, 1]$. Since each image pixel has a maximum value of 255, therefore, we divide each Image by 255.0

Image Denoising: The high-resolution images are used as targets. To generate noisy low-resolution images for input, we add Gaussian white noise [2] to all the high-resolution images.

Fig. 1 Example of noise-free images from the dataset

$$I_{\text{input}} = I_{\text{o}} + \frac{1}{2\sigma\sqrt{2\pi}} e^{\frac{-(x-\mu)^2}{2\sigma^2}} \tag{1}$$

where I_{input} = noisy image used as an input for the model, I_{o} = noise free image, σ = standard deviation $\in [0, 1]$ and $\mu = 0.0$ for white noise. Standard deviation can be varied depending upon the intensity of noise desired: the higher the intensity of noise, the higher the standard deviation value.

Image Reconstruction For images in the dataset, we first crop the 224 × 224 image to have dimensions 158 × 158. This is done to place the face in the center of the Image. We then define the mask to hide the boundary portions of the facial image.

$$M_{ij} = \begin{cases} 1 \; 30 < i < 126 \\ 0 \text{ else} \end{cases} \tag{2}$$

$$I_{\text{input}} = M_{ij} \times I_{\text{o}} + \mu x (1 - M_{\text{ij}}) \tag{3}$$

where M is the mask having dimension (158 × 158 × 1) and i, j is the number of rows and columns of pixels in the image. I_{o} is the cropped images having size (158 × 158 × 3), I_{input} is the masked image provided as the input to the generator model, and μ is the cropped image's mean. We then stack the Image I_{input} with ($M + \mu$ (1 − M)) to convert the (158 × 158 × 3) to (158 × 158 × 4).

4 Methodology

4.1 Image Denoising

To train CNN faster and make it more adaptive to different intensities of noise, we adopted the Curriculum Learning approach [13]. First, we define the bucket size, i.e., the total number of images in a bucket and the total number of buckets which are 300 and 6, respectively, in our case. Out of the 300 images in each bucket, we split the bucket data 270/30 images as the training/testing set. Each bucket has more complex images (higher noise intensity) than the images in the previous bucket, i.e., difficulty(B_i) < difficulty(B_j) where $i < j$ for all B_1, B_2, …, B_6. For the ith phase, we train the model on data from buckets having bucket number <= i. For example, in the first phase, we first train the model only on the first bucket then for the second phase, we again train the model using data from both 1st and 2nd bucket, and so on. To increase the difficulty of bucket data, we increase the standard deviation of the noise. For our model, each bucket data has standard deviation as follows: [0.01, 0.05, 0.10, 0.25, 0.35, 0.50] corresponding to each bucket and a fixed mean = 0.0 for all buckets.

For each phase, we train the neural network for 120 epochs using a Stochastic Gradient Descent optimizer. Using the grid search approach, we found a learning rate and momentum of 0.01 and 0.4 for the first 90 epochs respectively, which is changed to 0.0001 and 0.6 respectively for the last 30 epochs per iteration. We used a batch size of 64. The whole training correspond to a total of 720 epochs for all 6 phases combined. We used a sum of Mean Squared Error (MSE) and Cosine Similarity as our loss for the model. While MSE gives the difference in the pixel value of denoised and target Image, cosine similarity gives the contextual difference between the two images. The whole training schedule comprising of all 6 phases consisted of 10,920 iterations which corresponds to 72 minutes on a GPU provided by Google Colab.

$$\cos(\mathrm{X},\ \mathrm{Y}) = \frac{\sum_{\mathrm{i}} \sum_{\mathrm{j}} \mathrm{X}_{\mathrm{ij}} \mathrm{Y}_{\mathrm{ij}}}{\sqrt{\sum_{\mathrm{i}} \sum_{\mathrm{j}} \mathrm{X}_{\mathrm{ij}}^2} \sqrt{\sum_{\mathrm{i}} \sum_{\mathrm{j}} \mathrm{Y}_{\mathrm{ij}}^2}} \tag{4}$$

$$\mathrm{MSE}(X, Y) = \frac{1}{N} \sum_{i=1}^{n} (X_i - Y_i)^2 \tag{5}$$

$$\mathrm{Loss} = \mathrm{MSE}(X, Y) + (1 - \cos(X, Y)) \tag{6}$$

where X = denoised image generated by the model, Y = target/desired image. i, j is the number of rows and columns of pixels in an image.

4.2 Image Reconstruction

Due to bigger model size and memory constraints, we could not use curriculum learning to train the image reconstruction model. Instead, we used data augmentation techniques like Left-Right and Up-Down flip to increase dataset size from 630 to 1890 images, where we used 1800 images as the training set and 90 images as the test set. We ran the model for 60 epochs with a batch size of 8 in two phases The training was done using a GPU provided by Google Colab where the whole training phase corresponded to 13,500 iterations which took 108 minutes. In the first phase, we train the Generator for 15 epochs using MSE loss. In the next phase of 45 epochs, we train both Generator and Discriminator using $L_{\text{generator}}$ and L_{disc} [14] as the loss function, respectively.

$$L_{\text{generator}}(I_{\text{input}}, I_o) = \text{MSE}(G(I_{\text{input}}), I_o) - \alpha \log(D(G(I_{\text{input}}))) \tag{7}$$

$$L_{\text{disc}}(I_{\text{input}}, I_o) = -[\log(D(I_o))) + \log(1 - D(G(I_{\text{input}})))] \tag{8}$$

where D = Discriminator, G = generator, I_{input} = masked Image stacked with the mask, I_o = actual unmasked Image and $\alpha = 0.0004$, which controls the importance of discriminator output on generated images. Here, MSE is the Mean Squared Error as defined in Eq. 5.

We used an Adam optimizer to train the Generator and the Discriminator model with a learning rate = 0.0001, $\beta_1 = 0.4$ and $\beta_2 = 0.999$

5 Metrics

To determine the quality of denoised and generated images and to evaluate the model's performance on regenerated images, Peak Signal to Noise Ratio (PSNR) [15] is used as a metric. As the image quality improves, the PSNR for the image increases [16]. This means higher the noise present in an image, lower the PSNR and lower the noise present, higher the PSNR.

$$\text{PSNR} = 10 \log_{10}\left(\frac{\text{MAX}_I^2}{\text{MSE}}\right) \tag{9}$$

6 Model Architecture

6.1 Image Denoising

We use a three-layered architecture that performs the functions of extraction, mapping, and reconstruction of the Image. The first layer of our CNN has a kernel size of 9 × 9 and 128 filters. The subsequent layer has a kernel size of 1 × 1 with 64 filters. The final convolution layer consists of 5 × 5 kernel sizes with three filters corresponding to the RGB channel. A Rectified Linear Unit (ReLU) layer is used as an activation function. All layers are given a padded input, so the input and output size

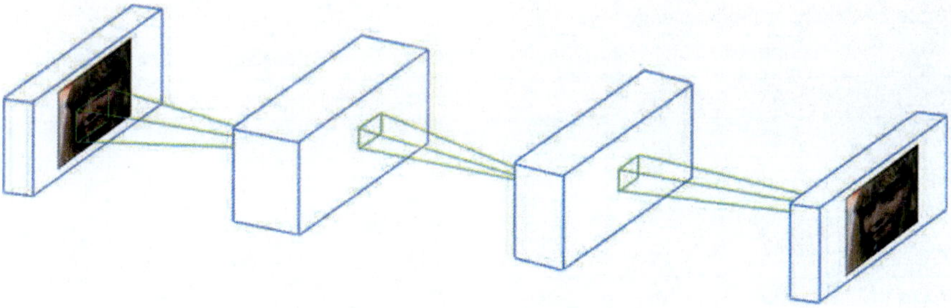

Fig. 2 Illustration for image denoising model's 3 layered architecture

is the same after each layer. We also trained the same model, with the same learning rate and momentum using a batch size of 16 and Stochastic Gradient Descent optimizer but without Curriculum Learning, for the same number of iterations as done for model with Curriculum learning using images from all the buckets at once. This was done to compare the model's performance when trained with and without Curriculum Learning (Fig. 2).

6.2 *Image Reconstruction*

The Image Reconstruction model consists of a Generator network and a Discriminator network. The model architecture is as follows (Tables 1 and 2).

Each layer of the Generator and Discriminator is followed by Rectified Linear Unit (ReLU), except the last layer of Discriminator is followed by sigmoid activation.

Table 1 Discriminator model architecture

Type	Filters/units	Strides	Kernel size
CONV	32	2	5
CONV	64	2	5
CONV	128	2	5
CONV	128	2	5
CONV	64	2	5
Flatten	–	–	–
Dense	512	–	–
Dense	1	–	–

Table 2 Generator model architecture

Type	Filters	Dilation rate	Strides	Kernel size	Dropout rate
CONV	64	1	1	5	0.1
CONV	128	1	2	3	0.1
CONV	256	1	1	3	0.1
CONV	256	2	1	3	0.1
CONV	256	4	1	3	0.1
CONV	256	8	1	3	0.1
CONV	256	1	1	3	0.1
DECONV	256	1	2	4	–
CONV	64	1	1	3	–
CONV	3	1	1	3	–

7 Results

7.1 Image Denoising

We tested our model against a wide range of noise intensity and compared our model's performance with static filtering methods (Fig. 3).

We compared our model's performance against another model with the same architecture but trained without a curriculum learning approach for the approximately same number of iterations (Figs. 4, 5, 6, 7, 8, 9; Table 3).

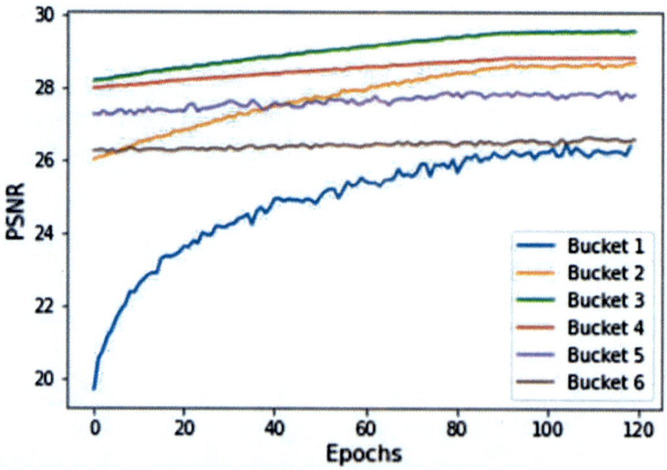

Fig. 3 Output PSNR when the model is trained with curriculum learning

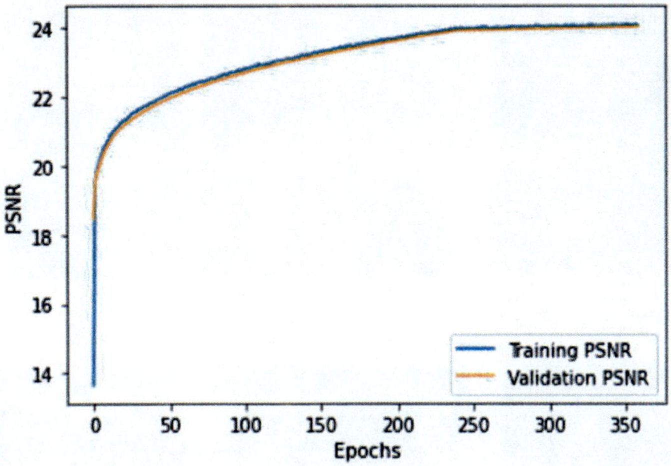

Fig. 4 Output training and testing PSNR when the model is trained without curriculum learning

Fig. 5 Actual clear images (224 × 224)

Fig. 6 Noisy images (224 × 224) with standard deviation = 0.15

Fig. 7 Denoised image (224 × 224) when input in Fig. 6

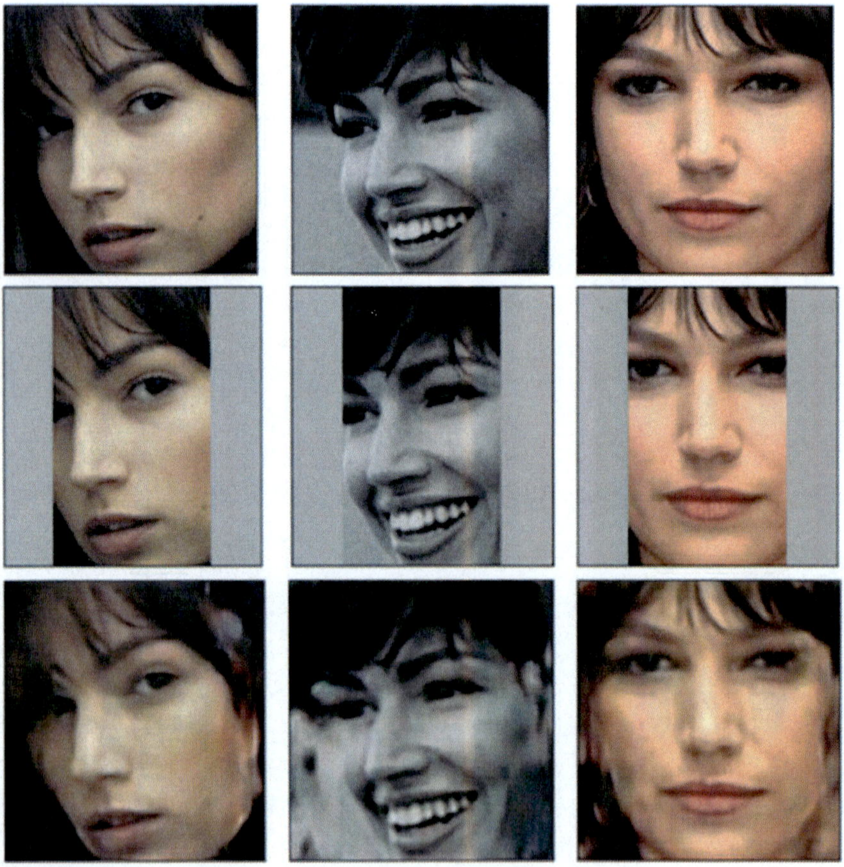

Fig. 8 Actual (top), masked (middle), and the output (bottom) image given by the model, output PSNR = 20.768

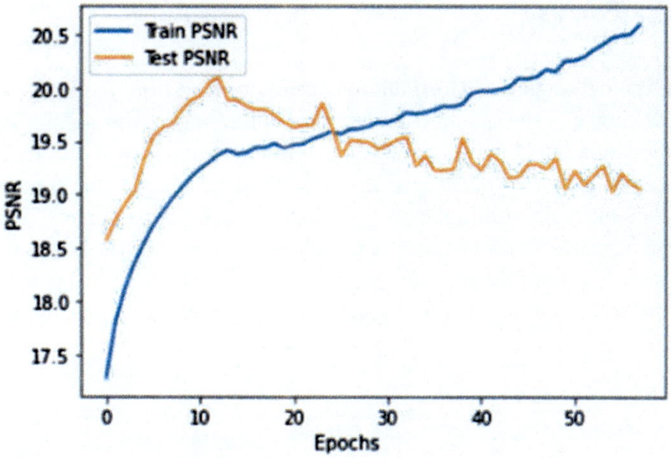

Fig. 9 Training versus testing PSNR of image reconstruction model

Table 3 Comparison of mean PSNR for different denoising methods on Fig. 5 for varying noise intensity

Standard deviation	Noisy image	Median filtering	Gaussian filtering	Model	Model (no curriculum learning)
0.07	23.509	29.370	30.050	29.398	26.993
0.15	17.368	26.132	24.759	28.101	25.995
0.20	15.175	24.448	22.524	27.047	24.969
0.30	12.227	21.829	19.235	24.614	22.591

8 Conclusion

In our project, Facial image denoising and reconstruction, we got rid of the unnecessary features (white noise) from the facial images that corrupted the images. The denoising process helped us focus on the data's key parameters and reduced the chances of making the images blur or unclear. All of the training done to the Image denoising model was done using the Curriculum Learning approach in which the difficulty level of training was increased gradually in a step-wise manner. This helped the model overcome Bottle-Neck capacity, which we showed by training the same model again without curriculum learning. This also helped us to focus on our second objective that was facial image reconstruction. The images were masked so that only some parts of the denoised images were given as input to the model for training purposes. After several epochs of the training process, the output images were compared with the data images before masking. The model was trained in such a manner so that it could predict the hidden parts of the images, thus making it look exactly like the target images

9 Future Works

One of the critical aspects of Curriculum learning used to train the Image denoising model is the choice of the standard deviation of noise that is to be added. We propose the same model architecture be trained in a similar fashion using a different standard deviation. Even though producing reasonably realistic images, the Reconstruction process needs more fine-tuning to reduce the high variance problem. The model, at present, can accurately reconstruct the image only when 60% of the face is visible. The results are less accurate for different values of percentage of image masked. The proposed model can be trained on a bigger dataset to reduce the high variance effect.

References

1. A. Nguyen, J. Yosinski, J. Clune et al., Deep neural networks are easily fooled: high confidence predictions for unrecognizable images, in *Computer Vision and Pattern Recognition (CVPR'15)* (IEEE, 2015)
2. D. Chowdhury, S. Gupta, D. Roy, D. Sarkar, C.C. Chattopadhyay, S.K. Das et al., *A Quantum Study on Digital Image Noises and their In-Depth Clusterization* (University of Engineering and Management Kolkata, India)
3. M. Sabini, G. Rusak et al., Painting Outside the Box: Image Outpainting with GANs
4. A. Susmitha, I. Mishra, D.S. Jain et al., Performance analysis of average and median filters for de noising of digital images. Int. Res. J. Eng. Technol. (IRJET), pp. 1435–1438 (2016)
5. A. Boyat, B.K. Joshi et al., Image denoising using wavelet transform and median filtering, in *2013 Nirma University International Conference on Engineering (NUiCONE)* (IEEE, 2013)
6. Z. Wang, C. Qu, L. Cui et al., Denoising images using Wiener filter in directionalet domain, in *International Conference on Computational Intelligence for Modelling Control and Automation,and International Conference on Intelligent Agents, Web Technologies and Internet Commerce* (2006)
7. Z. Liu, W.Q. Yan, M.L. Yang, Image denoising based on a CNN model, in *2018 4th International Conference on Control, Automation and Robotics*, pp. 389–393 (2018)
8. S. Ghose, N. Singh, P. Singh et al., *Image Denoising using Deep Learning: Convolutional Neural Network* (IEEE, 2020)
9. B. Garben, A. Grossman, S. Johnson-Yu, Image super resolution via convolution neural network
10. A. Sungheetha, R. Sharma, A novel capsNet based image reconstruction and regression analysis. J. Innov. Image Process. (JIIP) **2**(03), 156–164 (2020)
11. T. Vijayakumar, R. Vinothkanna, Retrieval of complex images using visual saliency guided cognitive classification. J. Innov. Image Process. (JIIP) **2**(02), 102–109 (2020)
12. Pins Face recognition, https://www.kaggle.com/hereisburak/pins-face-recognition
13. R. Palamuttam, N. Sohoni, H. Mohammad et al., Automated curriculum learning
14. S. Wang, Generative adversarial networks (GAN) a gentle introduction
15. Peak Signal-to-Noise Ration as an Image Quality Metric (2020). https://www.no.com/en-in/innovations/white-papers/11/peak-signal-to-noise-ratio-as-an-image-quality-metric.html
16. P. Kaur, J. Singh et al., A study on the effect of Gaussian noise on PSNR value for digital images. Int. J. Comput. Electric. Eng. **3**, 319–321 (2011)

One Method for RC5 Algorithm's Cryptographic Strength Improving

Lakhno Valeriy, Sahun Andrii, Khaidurov Vladyslav, Gusev Boris, Kravchuk Petro, and Kosenko Svitlana

Abstract Encryption of electronic data requires the use of crypto resistant functions. Especially relevant is the increase of the stability of block algorithms, in particular, RC5, which is part of various open cryptographic libraries. The article presents a method for improving the crypto stability of the classical RC5 algorithm with the help of nonlinear bit offset functions without increasing RC5 computational complexity. The modeling of the cryptographic system implemented based on RC5 crypto-algorithm modification is carried out in the article and obtained time and quality characteristics of the modified algorithm. The reliability of the cryptosystem obtained based on the improved RC5 algorithm is verified by the absence of statistical correlation between the input message blocks and the output blocks, as well as the absence of collisions, in which it is possible to obtain the same sequence of bits at the output with different messages at the input.

Keywords RC5 · Crypto algorithm · Nonlinear shift functions · Crypto resistance · Differential cryptanalysis

L. Valeriy
National University of Life and Environmental Sciences of Ukraine, Kiev, Ukraine
e-mail: lva964@nubip.edu.ua

S. Andrii (✉) · K. Vladyslav
Institute of Engineering, Thermophysics of NAS of Ukraine, Kiev, Ukraine
e-mail: avd29@ukr.net

G. Boris
National University of Life and Environmental Sciences of Ukraine, Kiev, Ukraine

K. Petro
Private Higher Educational Establishment "European University", Kiev, Ukraine
e-mail: p.kr@ukr.net

K. Svitlana
Head of the Laboratory of Educational and Scientific Work, State Scientific Research Forensic Center of the Ministry of Internal Affairs of Ukraine, Kyiv, Ukraine
e-mail: sveta_kosenko@bigmir.net

G. Ranganathan et al. (eds.), *Soft Computing for Security Applications*, Advances in Intelligent Systems and Computing 1397,
https://doi.org/10.1007/978-981-16-5301-8_2

1 Introduction

The OpenSSL package [1] is integrated in several of RSA Data Security Inc. The main products such as BSAFE, JSAFE, and S/MAIL, OpenVPN [2], are integrated for performing perspective hybrid secure routing, secured self-organizing network architecture, and others [3, 4] belong to packages that contain RC5. As a result, enhancing the cryptographic strength of RC5 in order to boost its cryptographic stability is important.

The cipher RC5 is characterized by a variable number of rounds, block length, key length [5, 6]. One way to make this algorithm "stronger" is to improve the basic RC5 algorithm by choosing nonlinear shift functions [7]. But there are some other methods to reach the ones [8, 9].

Some of the main parameters of the RC5 algorithm are changeable [10]. In addition to the secret key, algorithm has other parameters that can be changed [1, 7]:

- the size of the word w (in bits). RC5 algorithm encrypts two-words blocks (hereinafter referred to as A and B), and the valid values of w are natural numbers 16, 32, or 64, and 32 is the recommended value;
- number of rounds of the *R*—algorithm—t is permissible to use any integer from 0 to 255 inclusive as this parameter;
- secret key size in bytes *b*—any integer value from 0 to 255 inclusive.

The classical RC5 cryptographic algorithm before improving is given in Fig. 1.

For two blocks *A* and *B* in binary representation, the classical algorithm has the following strategy.

Before the first round, the operations of superimposing the extended *S*-key (detailed formation of the extended key *S* is given in [1]) on the data that is encrypted are performed:

$$A = (A + S_0) \bmod 2^w, \quad B = (B + S_1) \bmod 2^w. \tag{1}$$

The following actions are performed at each round [1, 3]:

$$A = ((A \oplus B) \ll B) + S_{2i}, B = ((B \oplus A) \ll A) + S_{2i+1}. \tag{2}$$

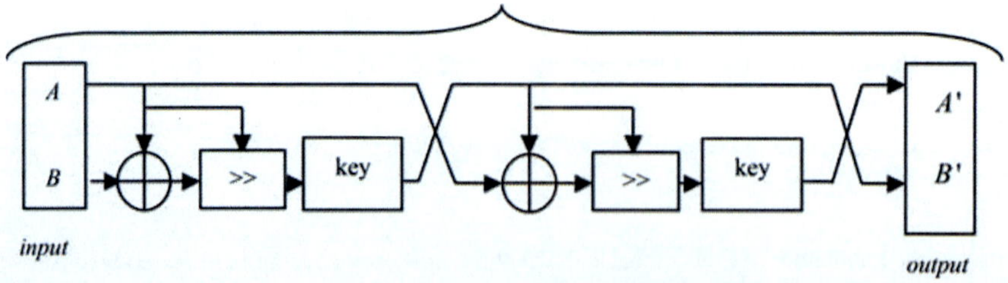

Fig. 1 One round of the classical RC5 cryptographic algorithm RC5 before improving

where *A*, *B*—binary message blocks (the classic size of each block is 32, 64, or 128 bits);

S_i—*S*-key extension table;

W—half block length (half the number of bits in the block—32, 64, or 128 bits).

Recurrent dependences (1) and (2) underlie the classical cryptosystem RC5. These formulas assume the use of the Feistel cipher, in which, after modulo addition operations and the *XORs* (eXclusive OR) operation, the positions of blocks A and B are exchanged. Decryption is performed in reverse order. At the time of submission of this work, the necessary crypto resistance of this algorithm is achievable. Because even after 6 rounds of linear cryptanalysis, and differential cryptanalysis, starting with 15 rounds, no longer makes sense, because there are only 2^{64} possible plain texts [11], due to the rapid development of cloud computing systems, the computational capabilities of cryptanalysis are growing.

2 Priming the Modification to the RC5 Algorithm

Increase in cryptographic stability of block diagrams and crypto algorithms can be achieved by different methods. Therefore, the task of improving the cryptographic algorithm RC5 in order to increase its cryptographic stability and speed is relevant.

Novelty of work is to modify the known cryptographic RC5 algorithm based on the use of bit offset functions during encryption of the binary representation of arbitrary electronic information.

This modification can improve the cryptographic stability of cryptosystem without increasing the computational complexity of the RC5 algorithm.

One round of RC5 algorithm's schema to be modified marked as ">>>" and *f* ()—functions shown in Fig. 2. In fact, basis parameters of rounds are nonlinear shift functions, k_{r_i}—round in number for *i* key implementation.

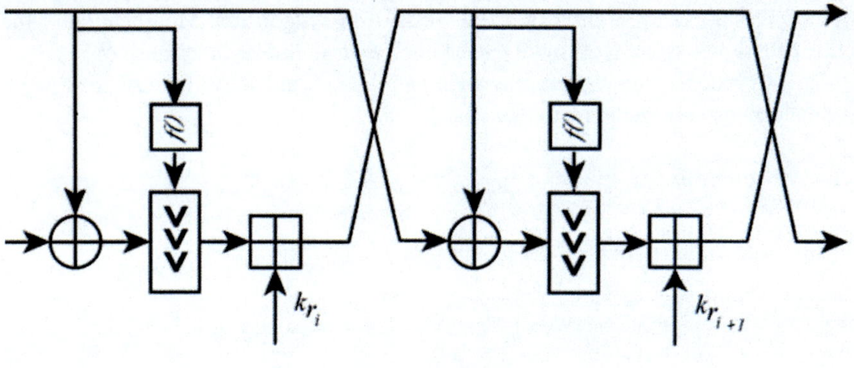

Fig. 2 One round of RC5 algorithm, to be modified using the bit offset function *f* ()

As a bit offset function used nonlinear functions. This principle of classical RC5 block exchanging will not be rejected. On platforms where the operation of cyclic shift on a variable number of bits is performed for a different number of clock cycles of the processor, the runtime attack on the RC5 algorithm is possible. Two variants of such an attack were formulated by cryptographic analysts Howard Hayes and Helena Handshuch [12]. They found that the key can be calculated after performing about 2^{20} encryption operations with high-precision measurements of execution time and then from 2^{28} to 2^{40} trial encryption operations. The simplest method of combating such attacks is to force shifts for a constant number of cycles (for example, during the slowest shift).

As is known, for the classic RC5 algorithm, approximately 2^{45} open texts are required for a successful attack [13]. When using more than 14 rounds, instead of the classic 12, it is almost impossible to break the algorithm using differential cryptanalysis [12]. Thus, the most realistic method of breaking the RC5 algorithm (not including the options with a small number of rounds and with a short key) is a full enumeration of possible options for the encryption key [12–14]. This is the basis that the RC5 algorithm is cryptographically resistant to differential analysis.

The choice of nonlinear functions for improving RC5 makes it possible to reduce the number of rounds to 10 so that this cryptosystem cannot be opened in a reasonable time using differential cryptanalysis. At the same time, the computational complexity of the implementation of the algorithm remains comparable to the classical version of RC5.

The maximum size of RC5's key that could be generated for improved RC5 (also for a classical RC5) algorithm is 2040 bits.

2.1 Nonlinear Functions to Improve RC5 Cryptocurrency

The proposed approach to improvement does not propose a modification of the algorithm, but an increase in its cryptocurrency by choosing nonlinear functions that use the Feistel cipher. It, in turn, is the basis of this algorithm. Modification of the classical cryptosystem RC5 involves the choice of nonlinear functions $f(K, r)$.

We can propose one way for improving an RC5 crypto algorithm crypto strength for data encryption in the form by using:

The first function:

$$f(K, r) = r + \left[w \sin\left(wr \sum_{s=1}^{w} L_{\text{bits}_{m_s}} \right) \right], \tag{3}$$

where $m = r^2 \bmod w$,

r—is the number of the round;

w—is the length of half of the encoded block;

$[x]$—is an integer part of the number x.

L_{bit}—is the binary representation of the symbol L_m coded word K, which has length w in binary presentation.

The second function:

$$f(K,r) = rw \exp\left(1 + \frac{r}{\left|w \sin\left(wr \sum_{s=1}^{w} L_{\text{bits}_{m_s}}\right)\right|}\right). \tag{4}$$

The third function:

$$f(K,r) = rth\left(w \sin\left(wr \sum_{s=1}^{w} L_{\text{bits}_{m_s}}\right)\right). \tag{5}$$

The essence of these functions lies in the nonlinearity dependence of the output bit sequences of the encryption blocks on the input bit sequences in the blocks during the formation of cyclic bit shifts. All three functions (3), (4), (5) have a range of valid values throughout their range of values, namely: $def(F_n(x)) \in (-\infty; +\infty)$.

Figure 3 shows graphs of all three nonlinear shift functions are proposed for use in the improved algorithm RC5. These functions (marked as "Func1", "Func2" and "Func3" in Fig. 3) meet the condition of formation of coefficients in the round.

In the study, we relied on the fact that the classic RC5 has a fixed number of rounds (cycles). Modification of RC5 by increasing rounds was not performed.

In order to test the theoretical assumptions about increasing the cryptographic strength of encryption, it is necessary to make simulation of algorithm operation. It is provided that for encryption you can upload text (txt) and graphic files (jpg) to the program, enter the keyword (K), the parameters of the round and the length of the encoding.

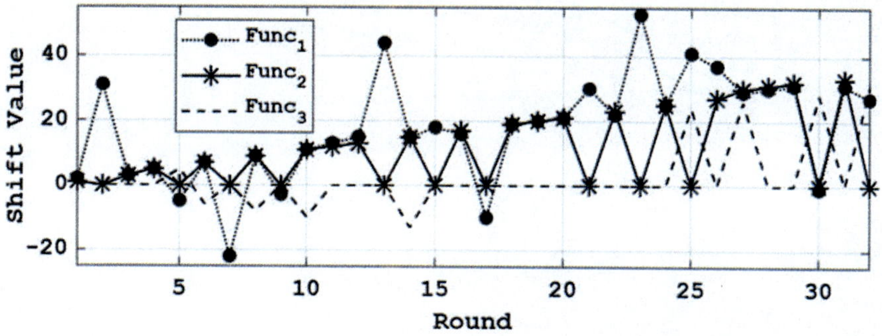

Fig. 3 Different shift functions values, depends of function number at RC5's rounds

2.2 Obtained Results of Improved RC5 Algorithm Operation

In the course of the work, a program was developed in the MatLab environment, which has the ability to protect (encrypt and decrypt) textual and graphical information.

The main module of the program consists of basic blocks: user interface, subroutine for complex protection of the graphic file, subroutine for formation of the protected image and its record in the current folder, subroutine for extraction, and decryption of the graphic container information.

The obtained results for the modified algorithm using the shift functions (3), (4), (5) are shown in Figs. 4 and 5. Figures 4 and 5 show the results of the program for the encryption and decryption for text and graphic documents of electronic data. The corresponding numerical results are shown in Tables 1 and 2. These tables show

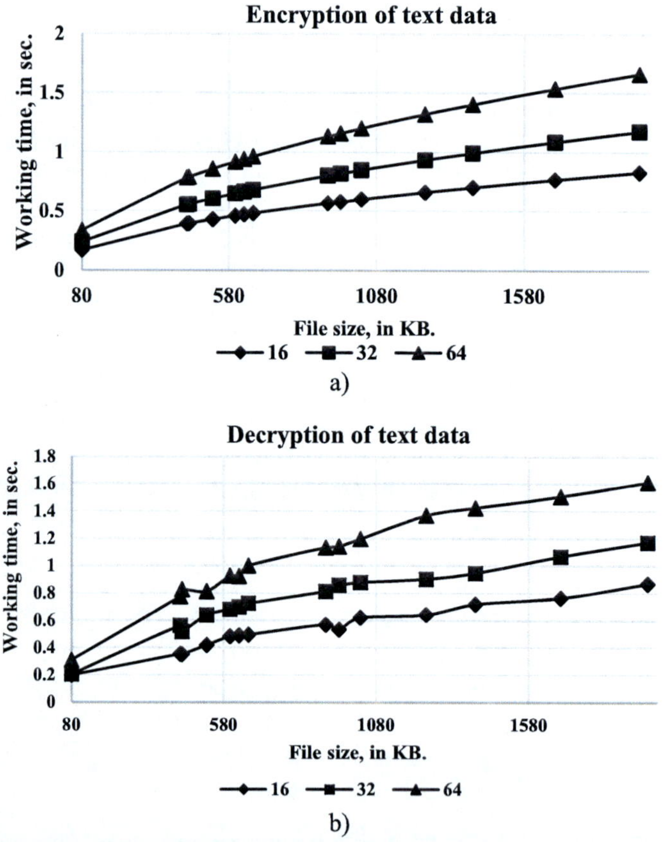

Fig. 4 The results of the program for encryption (**a**) and decryption (**b**) text information at $r = 16$

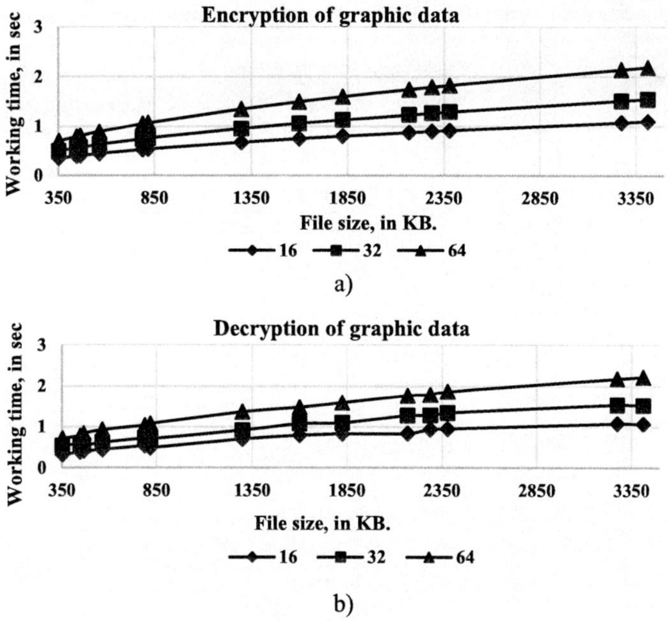

Fig. 5 The results of the program for encryption (**a**) and decryption (**b**) of graphic information when $r = 16$

Table 1 Operating time of encryption and decryption of graphic information with the number of rounds $r = 16$

File size, Kbytes	Program run time, seconds: encryption/decryption $w = 16/w = 16$	Program run time, seconds: encryption/decryption $w = 32/w = 32$	Program run time, seconds: encryption/decryption $w = 64/w = 64$
81	0.17/0.20	0.24/0.20	0.34/0.31
439	0.39/0.35	0.55/0.56	0.78/0.78
445	0.40/0.35	0.56/0.52	0.79/0.83
525	0.43/0.42	0.61/0.64	0.86/0.81
601	0.46/0.48	0.65/0.68	0.92/0.93
631	0.47/0.49	0.67/0.70	0.94/0.92
662	0.48/0.49	0.68/0.73	0.96/1.00
916	0.57/0.57	0.80/0.81	1.13/1.13
959	0.58/0.53	0.82/0.86	1.16/1.14
1029	0.60/0.62	0.85/0.88	1.20/1.20
1244	0.66/0.64	0.93/0.90	1.32/1.37
1406	0.70/0.72	0.99/0.95	1.40/1.42
1684	0.77/0.77	1.09/1.07	1.54/1.51
1968	0.83/0.87	1.17/1.17	1.66/1.61

Table 2 Operating time of encryption and decryption of graphic information modules at the number of rounds $r = 16$

File size, Kbytes	Program run time, seconds: encryption/decryption $w = 16/w = 16$	Program run time, seconds: encryption/decryption $w = 32/w = 32$	Program run time, seconds: encryption/decryption $w = 64/w = 64$
351	0.35/0.32	0.50/0.53	0.70/0.72
446	0.40/0.39	0.56/0.60	0.79/0.78
465	0.40/0.41	0.57/0.54	0.81/0.84
565	0.45/0.44	0.63/0.62	0.89/0.93
789	0.53/0.54	0.74/0.73	1.05/1.03
817	0.54/0.49	0.76/0.71	1.07/1.08
1298	0.67/0.70	0.95/0.91	1.35/1.37
1600	0.75/0.79	1.06/1.08	1.50/1.47
1825	0.80/0.83	1.13/1.10	1.60/1.59
2169	0.87/0.84	1.23/1.27	1.74/1.77
2288	0.90/0.94	1.27/1.28	1.79/1.79
2379	0.91/0.95	1.29/1.34	1.83/1.86
3273	1.07/1.08	1.52/1.54	2.14/2.18
3412	1.09/1.06	1.55/1.52	2.19/2.22

operating time of encryption and decryption of graphic information with the number of rounds $r = 16$ and various values of W-parameter.

Software implementation of the improved RC5 crypto algorithm includes the following modules: (1) text file download module; (2) graphic file download module; (3) module for converting bit information into text and graphics; (4) encryption/decryption results saving module; (5) bit information encryption module.

To test the algorithm, we took text data, which was presented to the binary code. Any other data for encryption can be converted to binary code and data encryption and decryption procedures can be used. Data sets taken for experimentation (for first/second/third offset nonlinear function) are shown in Table 1.

Any input data for testing the improved algorithm are converted to binary code; text data are converted according to the code system of ASCII, Windows, etc. If this is an image—the intensity of any pixel of any channel is converted to binary code. The number of bits per channel is usually 8, and the intensity of the pixel component is from 0 to 255.

Based on the results, it can be argued that the improved RC5 algorithm works fairly quickly with files of fairly large volumes. It should also be noted that the choice of the bit offset function does not significantly affect the encryption and decryption time of textual and graphical information.

3 Cryptoresistance Evaluation of the RC5 Algorithm Improving

3.1 Cryptanalytic Stability Evaluation

In experimental process have been used linear and differential cryptanalysis in experimental process. Plain texts were used for cryptanalysis.

We chose an attack based on data on the sequence of input and output bits in order to determine the relationship between them. This attack is the most effective method of differential cryptanalysis in relation to the RC5 family cipher.

In Table 3 shows a table of the number of input (plain) texts required to break the improved RC5 algorithm with a different number of rounds (from 10 to 14) compared to classical one.

In Fig. 6 shows number of input (plain) texts required to break the improved RC5 algorithm with a different number of rounds according to Table 3 results.

The number of open texts for breaking the classic RC5 by the same differential analysis method for 10 rounds is approximately equal to 2^{45}, for 14 rounds—approximately 2^{76}.

Studies have shown that improved RC5 has a high level of attack protection for attack for pairs of input and output bit sequences when $r \leq 12$. For $r > 12$ improved RC5 is resistant to this type of attack. Also is known that 12-round RC5 (with 64-bit blocks) is susceptible to a differential attack using 2^{44} chosen plain texts, but at 12 rounds.

Table 3 A table of the number of input (plain) texts required to break the improved RC5 algorithm with a different number of rounds

Number of rounds/number of nonlinear function	$r = 10$	$r = 11$	$r = 12$	$r = 13$	$r = 14$
Classical RC5	2^{45}	2^{47}	2^{49}	2^{53}	2^{56}
The first function	2^{46}	2^{50}	2^{56}	2^{62}	2^{80}
The second function	2^{46}	2^{52}	2^{58}	2^{66}	2^{88}
The third function	2^{46}	2^{54}	2^{60}	2^{66}	2^{90}

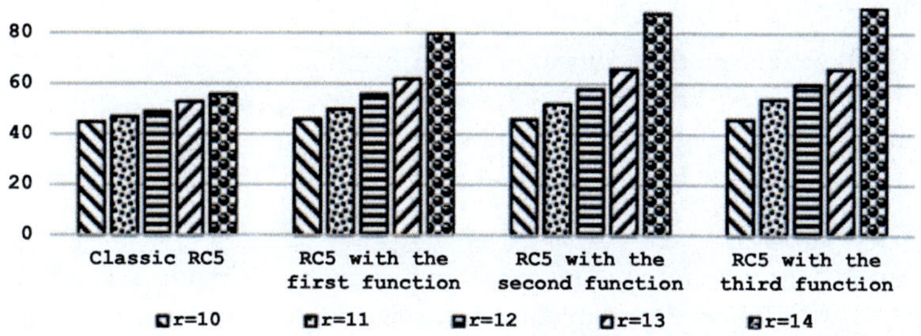

Fig. 6 Number of input (plain) texts required to break the improved RC5 algorithm with a different number of rounds

3.2 Determination of Additional Bits When Encoding in a Crypto Algorithm

Due to the fact, that the modified cryptographic algorithm RC5 is a block algorithm, we will determine the number of additional bits, that should be added to the message length of 106 characters if the encoding uses Unicode characters with codes $U +$ 0000 to $U +$ 007F.

Solution. Let's determine that the character encoding in the case of UTF-8 takes one byte (including zero in the high bit) and completely coincides with the 7-bit US-ASCII encoding. Therefore, encoding one character will require 8 bits. Thus, the encoded message of 106 characters contains 848 bits.

In the implemented test model, the cryptographic algorithm works with blocks of 64 bits length, i.e., the original text must be divided without remainder by 64. If we denote by the letters M and m length, respectively, the useful message and the length of the complementary bits, then:

$$M + m \equiv 0 \bmod 64 \Rightarrow m \equiv -848 \bmod 64 \equiv 48 \bmod 64$$

Therefore, 16 additional bits should be added at the end of the plaintext. The total length of the message will be 896 bits or 13 blocks of 64 bits, which corresponds to 14 rounds of the algorithm to create 14 blocks of encrypted text.

3.3 Simple Search Cryptanalysis (Brute Force)

Studies of data encrypted using the RC 5 algorithm have shown that the algorithm is reliable even after round 6 (against linear cryptanalysis).

While differential cryptographic analysis requires 2^{24} selected plain texts for a 5 round, 2^{45}—for 10 rounds, 2^{53}—for 12 rounds, 2^{68}—for 15 rounds. Due to the fact that there are 2^{64} possible different plain texts [8, 13], the cryptographic analysis is almost impossible for 15 or more rounds.

Let's determine the cryptographically strong of the modified crypto algorithm (the ability to obtain plaintext corresponding to the intercepted encrypted text) under the conditions that:

(1) it works in the mode of substitution/ permutation;
(2) block size $n = 64$ bits;
(3) encrypted text contains 12 units of information capacity;
(4) the speed of search of blocks by the cryptographic analyst of 108 blocks per second.

Indeed, for the substitution mode, each bit of plain text can be replaced by two values: "0" or "1"—the original plaintext and the encrypted text can have a different number of units.

Under the given conditions it is not known for sure, how many units are in plain text, therefore it is necessary to check all possible blocks, the number of which is equal to 2^{64}. Therefore, for a given search speed, the execution time of cryptanalysis will be: $t = 2^{64}/10^8$.

In order to avoid the register of the computing device in operations with large exponents, we use the properties of the natural logarithm:

$$\ln t = 64 \ln 2 - 8 \ln 10 \approx 25,94,$$

where:

$$t \approx e^{26} \approx 195729609426\,\mathrm{s} \approx 6206\,\mathrm{years}.$$

If the cryptanalyst has the best computing power, the cryptanalysis time can be significantly reduced.

For the permutation mode, cryptanalyst knows that there are exactly 10 units in the plaintext. When carrying out an attack by the method of complete search (brut—force) only those 64-bit blocks are used that have exactly 10 units. The total number of such blocks can be estimated by:

$$C_{64}^{10} \approx 151 \cdot 10^9.$$

When using the genetic search algorithm for cryptanalysis of this block cipher, there is a problem of forming the initial population. After all, with a variable key length in the RC5 algorithm, the population will also be variable, which significantly complicates the implementation of the cryptanalysis algorithm [6–9, 12, 13]. Although theoretically, the genetic algorithm, due to evolutionary parallelism can accelerate cryptanalysis, in practice the time of cryptanalysis (obtaining a key) has run out after 38 h of continuous operation of the test hardware platform based on CPU AMD Ryzen 5 2600 (4 cores). In practical conditions, the use of such cryptographic algorithm is sufficient. However, crypto resistance of improved by nonlinear function RC5 algorithm and classical one against search cryptanalysis (brute force) is the same.

4 Conclusion

Among all the existing methods to improve the crypto resistance of a symmetric cryptosystem based on RC5, such as: enlarging length of the key in bytes; enlarging number of round subkeys, and using linear cyclic shift operations were described the last one. After all, the aim of the work was to improve the crypto resistance of the RC5 even with a standard number of rounds ($r = 10$). It was assumed that the computational complexity of the implementation of the improved system should remain the same as in the classical algorithm RC5 ($k = 10$). That was achieved.

It should be noted that taking into account the improvements of the basic algorithm RC5 by choosing nonlinear shift functions (3), (4) and (5), we obtained a much more stable algorithm with the same computational capabilities compared to the classical one. We determine the additional bits when encoding in the crypto algorithm and calculate the cryptographic stability of the algorithm under certain conditions.

The obtained results of the improved RC5 crypto algorithm turn out to be more cryptographic strength than the traditional one, which makes it possible to increase the cryptographic stability of existing applied cryptosystems. Even at the standard value of the parameter $r = 10$, the improved version of the algorithm's crypto resistance increased from 2^{45} (for the classic RC5) to 2^{46} with third function (5) (with the applied in this paper method of cryptanalysis).

Thus, it can be argued that at the time of this study, the most likely method of compromising the RC5 algorithm (except for a few proven cases with a small number of rounds and a short key) can only be a complete search of possible encryption key options. This confirms the fact that the RC5 algorithm has almost no shortcomings due to its crypto resistance, even against described in [15, 16].

To check the cryptographic strength of this system, an attack was also performed based on the data on the sequence of input and output bits with the determination of the relationship between them. The reliability of improved cryptosystem is verified by the absence of statistical correlation between input message blocks and output blocks.

On future works, it makes sense to search for nonlinear functions that satisfy the condition of formation of better coefficients in the round. There are no limitations in applications of the proposed work provided that the improved algorithm is used in the same applications as the classic RC5.

References

1. OpenSSL, Cryptography and SSL/TLS Toolkit, https://www.openssl.org
2. RSA® BSAFE® Crypto-J JSAFE and JCE Software Module 6.2.4 Security Policy Level 1 (PDF). August 2020 Copyright © 2020 Dell Inc. or its subsidiaries. http://csrc.nist.gov/CSRC/media/projects/cryptographic-module-validation-program/documents/security-policies/140sp3172.pdf
3. P. Jyothirmai, J.S. Raj, S. Smys, Secured self organizing network architecture in wireless personal networks. Wirel. Pers. Commun. **96**(4), 5603–5620 (2017)
4. J.S. Raj, A novel hybrid secure routing for flying ad-hoc networks. J. Trends Comput. Sci. Smart Technol. (TCSST) **2**(03), 155–164 (2020)
5. Encyclopedia of Physical Science and Technology. by Robert A. Meyers (Editor). Academic Press; 3rd edition (October 22, 2001), vol 1, 15453 (2001)
6. E.B. Villanueva, R.P. Medina, B.D. Gerardo, An enhanced RC5 (ERC5) algorithm based on simple random number key expansion technique, in *2018 IEEE Symposium on Computer Applications & Industrial Electronics (ISCAIE)*, Penang, 2018, pp. 134–138. https://doi.org/10.1109/ISCAIE.2018.8405458

7. Tetiana Anatoliivna Zhovnovach, Andriy Viktorovych Sagun, Vladyslav Khaidurov, Hanna Martyniuk, Tetiana Leonidivna Scherbak.: "Modification of RC5 Cryptoalgorythm for electronic data Encryption Systems". December 2019. Ukrainian Scientific Journal of Information Security 25(3), 138–143 (2019). doi:https://doi.org/10.18372/2225-5036.25.14458
8. B.S. Kaliski, Y.L. Yin, On differential and linear cryptanalysis of the RC5 encryption algorithm, in Lecture Notes in Computer Science 963, *Advances in Cryptology—CRYPTO'95* (Springer, 1995), pp 171–184
9. Y.L. Yin, The RC5 encryption algorithm: two years on, RSA laboratories' CryptoBytes **2**(3), 14–16 (1997)
10. C. Ellison, B. Frantz, B. Lampson, R. Rivest, B. Thomas, T. Ylönen, SPKI certificate theory. RFC **2693**, 1–43 (1999)
11. Rivest, R, The RC5 encryption algorithm. FSE (1994)
12. Y. Sasaki, Y. Todo, New differential bounds and division property of Lilliput: block cipher with extended generalized Feistel network, in ed. by R. Avanzi, H.M. Heys, *SAC 2016: 23rd Annual International Workshop on Selected Areas in Cryptography*, volume 10532 of Lecture Notes in Computer Science, August 10–12, 2016 (St. John's, NL, Canada, Springer), pp. 264–283
13. A. Biryukov, E. Kushilevitz, Improved cryptanalysis of RC5 (2006). https://doi.org/10.1007/BFb0054119
14. L.R. Knudsen, W. Meier, Improved differential attacks on RC5. in ed. by N. Koblitz, *Advances in Cryptology—CRYPTO '96. CRYPTO 1996*. Lecture Notes in Computer Science, vol 1109 (Springer, Berlin, Heidelberg, 1996). https://doi.org/10.1007/3-540-68697-5_17/
15. G.C. Kessler, An overview of cryptography. https://www.cs.princeton.edu/~chazelle/courses/BIB/overview-crypto.pdf
16. RC5–72/Overall Project Stats, Aggregate Statistics, https://stats.distributed.net/projects.php?project_id=8

Multiclass Classification of Firewall Log Files Using Shallow Neural Network for Network Security Applications

Qasem Abu Al-Haija and Abdelraouf Ishtaiwi

Abstract Firewalls are essential devices to protect the communication networks by means of filtering out all incoming (and sometimes outgoing) traffic packets. The filtration process is performed by matching the traffic packets against predefined rules aiming to preclude cyber-threats from getting into the network. Accordingly, the firewall system proceeds with either to "allow," "deny," or "drop/reset" the incoming packet. Thus, an automated smart actions' classification process is essential for improved firewall operations. In this paper, we propose an intelligent classification model that can be employed in the firewall systems to produce proper action for every communicated packet by analyzing packet attributes using a shallow neural network (SNN). Specifically, the proposed model employs SNN with 150-neurons at the hidden layer to train and classify the Internet Firewall-2019 (IFW-2019) dataset into three classes, including: "allow, "deny," and "drop/reset." The experimental results exhibited our classification model's superiority, scoring an overall accuracy of 98.5% with a cross-entropy loss of 0.022 attained after 381 epochs for the 3-class classifier. Also, the proposed model was evaluated using several other evaluation metrics, including confusion matrix parameters, positive predictive value, true positive rate, harmonic mean, and false positive/negative rates. Eventually, the proposed model outperformed many other recent firewall classification systems in the same area of study.

Keywords Artificial intelligence (AI) · Shallow neural networks (SNN) · Network security · Supervised learning · Firewall device · Firewall logs · Classification

Q. A. Al-Haija (✉) · A. Ishtaiwi
Department of Data Science & Artificial Intelligence, University of Petra, Amman 1196, Jordan
e-mail: qasem.abualhaija@uop.edu.jo

A. Ishtaiwi
e-mail: aishtaiwi@uop.edu.jo

G. Ranganathan et al. (eds.), *Soft Computing for Security Applications*,
Advances in Intelligent Systems and Computing 1397,
https://doi.org/10.1007/978-981-16-5301-8_3

1 Introduction

Cyber attacks, intrusions, and cybercrimes are instigating an enormous influence on different networking systems communicating essential and private information such as financial, military, and medical systems [1]. Generally, data communication over the Internet is vulnerable to a wide range of potential cyber attacks and intrusions. Once the network infrastructure is breached, hackers could distribute the data to unauthorized parties and manipulate the network data's accuracy and consistency over its entire life cycle. Consequently, various safety methods have been used in the different stages of defense to address security concerns, such as Internet Firewalls [2], Intrusion Detection/Prevention Systems (IDS/IPS) [3–5], and reliable connectivity for several cyber physical and IoT network systems [6–8].

Firewall devices/servers are crucial security systems to defend communication networks from external cyber attacks [9]. They are usually installed at the networks' edges to monitor the traffic flow and protect the communication by filtering out all incoming (and sometimes the outgoing) data packets. The filtration process is typically performed by matching the network packets against predefined instructions and rules aiming to preclude cyber-threats from getting into the network. Hence, the firewall system proceeds with either to "allow," "deny," or "drop/reset" the incoming packet. Once the firewall decides what to do with the received traffic, it records all these decisions in the form of logs. Further analysis for the firewall log files will help improve the cyber-defense by integrating machine learning methods to provide automated and early classification and prediction for the traffic that bypasses the firewall system. Indeed, the recent coupling of cyber security and machine learning fields produced a robust and efficient security solution [10].

Classification task typically employs machine learning techniques to learn several class labels of examples from the problem space [11]. For example, classifying the emails as "inbox" or "junk." Several machine learning techniques can be employed to perform the classification/prediction tasks for the data records in the problem space [12]. Examples of classification-based machine learning techniques used to provide cyber security solutions: Artificial Neural Networks (ANN) [13], Shallow Neural Network (SNN) [14], Convolutional Neural Network (CNN) [15], K-Nearest Neighbors (KNN) [16], Decision Tree Method (DTM) [17], Majority Voting Method (MVM) [18], Support Vector Machine (SVM) [19], and others. The use of a proper algorithm is heavily based on several factors related to the dataset nature and complexity, such as the size of a dataset, the number of features, the types of features, the structure of data, the data labeling/clustering, and the data distribution [20].

In this paper, we employ a feedforward shallow neural network (SNN) to train and classify the communication traffic records provided by Internet Firewall (IFW-2019) dataset. SNN is a term used to describe neural networks that usually have only one hidden layer. IFW-2019 dataset [21] comprises 65,532 records from firewall logs files divided into four different categorical categories, namely, "allow," "deny," "drop," and "reset-both. Based on the collected records, our SNN-IFW network is trained for 387 epochs using scaled conjugate gradient back propagation, aiming to minimize the cross-entropy loss function. We show that the overall accuracy of the model is superior. In particular, the core contributions of the proposed work can be listed as follows:

- We provide a firewall prediction system that employs SNN architecture for classifying multiclass firewall log action records in communication networks.
- We evaluate our IDS' performance on recent and important datasets for Internet firewall log files (IFW-2019 [21]), scoring a 98.50% classification accuracy.
- We provide a detailed description of our implementation in conjunction with an extensive comparison with state-of-the-art solutions.

The rest of this paper is as follows: Sect. 2 introduces and justifies the IFW-2019 dataset used in our system. Section 3 provides details of the proposed system design and architecture. Section 4 presents the system development environment. Section 5 discusses the details about experimental evaluation and comparison. Finally, Sect. 6 concludes the paper.

2 Dataset of Internet Firewall

Data is the vital element of any intelligent system since it allows systems and stakeholders to build the required decisions based on definite facts and records. Data (categorical, numerical, images, etc.) are usually collected into organized records in a systematic dataset [5]. Dataset can be analyzed and used to address research investigations, formulate problem statements and validate theories and outcomes. In this work, our system concerns providing automated and intelligent classification for the security actions performed by firewall devices on the network traffic, and thus, we have used dataset [21].

IFW-2019 [21] is a recently composed dataset from the internet traffic records on a university's firewall devices (i.e., Specifically, Firat University, Turkey) and used for the automated prediction purposes of firewall actions in response to the network traffic data. IFW-2019 accumulates 65,532 firewall log files records with four different categorical labels listed for the output filed of each sample record including: "allow", "deny", "drop", and "reset-both". The data distribution among the different classes of the files are presented in Table 1.

Table 1 Statistics of traffic distribution of IFW-2019 dataset [22]

Actions	Allow	Deny	Drop	Reset-both
No. of records	37,640	14,987	12,851	54
Description	Permit the data packet	Block the data packet	Drop the data packet	Send TCP reset to both the client and server devices

IFW-2019 records are developed with 11-features and one class label using firewall log files. Features are carefully selected as numerical datatype to be efficiently applied to the machine learning techniques. The selected features include source port, destination port, network address translation (NAT) source port, NAT destination port, elapsed time for flow (in seconds), total bytes, bytes sent, bytes received, total packets, packets sent, and packets received [22]. Indeed, IFW-2019 dataset is nominated for evaluation in this research since it is publicly available as .CSV filetype and comprises a sensible number of distinctive samples which prevent classifier from being influenced by more recurrent class. Also, this dataset covers all common security actions for firewall devices/servers on the network traffic. Moreover, it can be powerfully preprocessed and programmed to generate multiclass classification for the firewall actions in the communication networks. Finally, IFW-2019 can be tailored, extended, updated, and stimulated.

3 Action Classification System Internet Firewall

Data classification is an extensive computational intelligence task concerns with organizing the data by means of categorizing the data records (instances) into pertinent classes (groups) for improved data processing, analysis, mining, allocation, manipulation, and retrieval [23]. Data classification tasks turn into crucial tasks when it comes to risk management and data security. Therefore, in this paper, we are interested specifically in experimentally investigating and implementing an automated and smart classification system of an essential data security application for improved communication processes. Figure 1 shows the workflow diagram to develop the proposed classification model.

Typically, data preprocessing subsystem is included almost in every machine learning based system as an essential phase for detection, recognition, and classification tasks. It is usually composed of several consecutive operations that are implemented on the original instances of the dataset in order to formulate the dataset instances in the adequate form that can be served and processed through machine learning based inference models such as the shallow neural network (SNN) model employed in this work. Consequently, our accumulated dataset records (i.e., instances) have undergone the following consecutive preprocessing operation:

- Dataset Importation, which is accountable to host and allocate the records of collected dataset from the .CSV filetype stored on local hard-disk drive within the MATLAB tables.
- Data Reshaping, which is accountable to customize the instances, unify the sizes and features for all records in the dataset. Also, this operation is essential to remove unneeded features from the input data and leave only the distinguishing features to improve the computational complexity of the neural network.
- Data Encoding, which is accountable for the one-hot-encoding for the categorical labels in order to be converted to unique numerical classes that can be processed

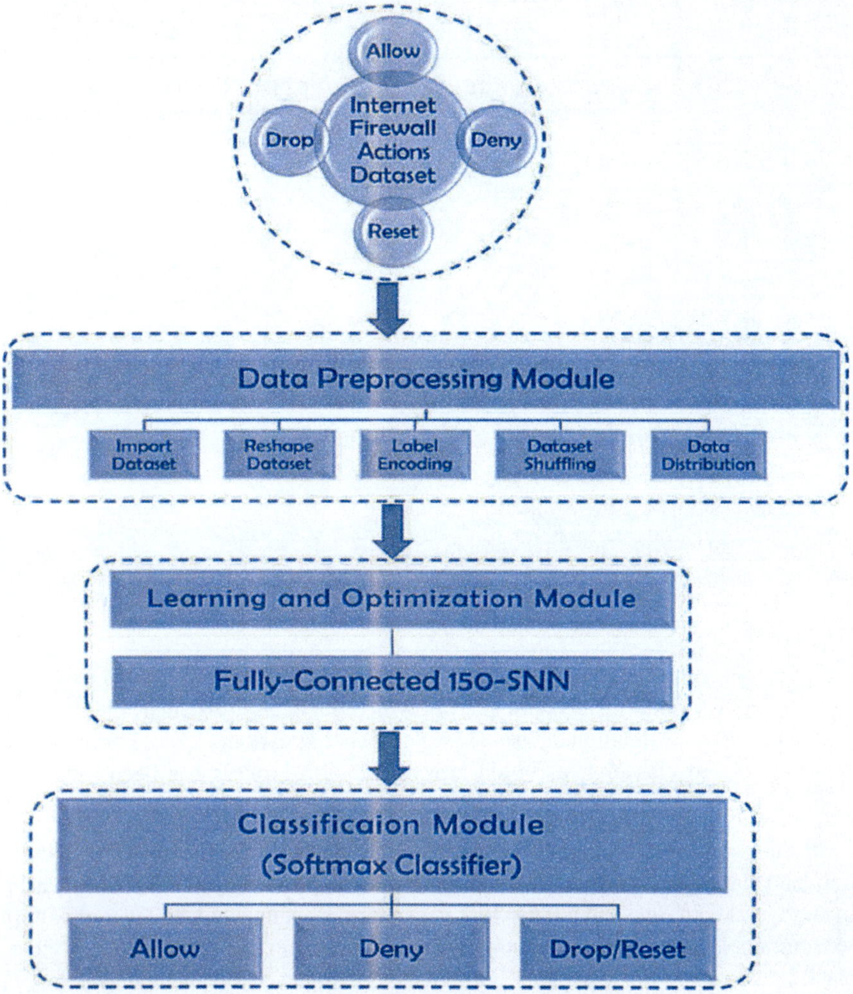

Fig. 1 Workflow diagram to develop the proposed classification model

through the inference based neural network module. Please note that the two categories "drop" and "reset" have been combined for better classification and due to the lack of instances of the "reset" class which contains a non-representative sample space with only 54 instances of firewall actions. Therefore, we have used the one-hot-encoding method to come up with three classes encoded as follows: allow (100), deny (010), and drop/reset (001).

- Dataset Shuffling, which is accountable for reordering the instances of provided dataset in an unsystematic fashion (i.e., randomized) to ensure a non-biased classification process and thus proving precise classification.

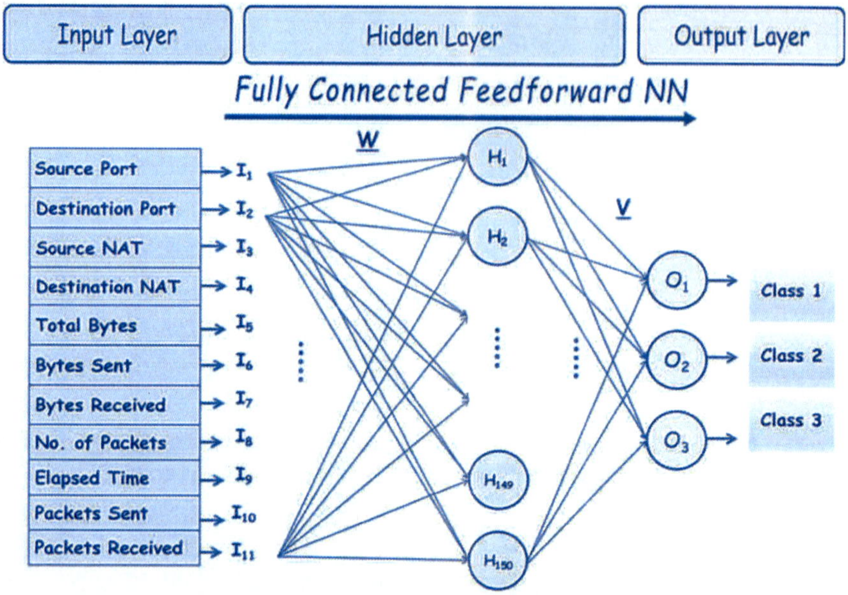

Fig. 2 Architecture of the proposed classification model: SNN-IFW

- Dataset Distribution is accountable of divided the dataset instances into training dataset (70%), validation dataset (15%), and testing dataset (15%). For improved and unbiased classifier, a 6-Fold cross validation process has been randomly performed to ensure different distributions of dataset at every run.

In this work, we developed our inference system using a feedforward SNN to train and classify the communication traffic records provided by IFW-2019 dataset. In short, data introduced to the network goes through a single hidden layer of pattern recognition. Figure 2 shows our inference system architecture developed to perform classification tasks in this paper. Our SNN is composed of an input vector $(\underline{I})$ with 11-inputs (I_1; I_2; I_3; ...; I_{11}) that connects the 11-features of IFW-2019 to the 150-neurons at the hidden layer (H_1; H_2; H_3; ...; H_{150}) in a fully connected fashion. Also, every single neuron is fully connected to the 3-neurons at the output layer (O_1; O_2; O_3) producing Softmax probabilities for the corresponding classes (Class_1 = allow; Class_2 = deny; Class_3 = drop/reset;). In the third class, we have combined both "reset-both" and "drop" actions in one class since "reset-both" has a small number of samples (i.e., only 54 samples). Finally, the trainable weight vectors corresponding to each of the parametrized layers are $\underline{W}$ and $\underline{V}$ from left to right respectively.

Moreover, Fig. 3 demonstrates the symbolic representation for the individual neurons with an input vector $\underline{I}$ of n elements and single output H. For every neuron, all elements of input vector $\underline{I}$ are multiplied by the corresponding weights in the weight vector $\underline{W}$ and subsequently supplied to the intersection of summation operation

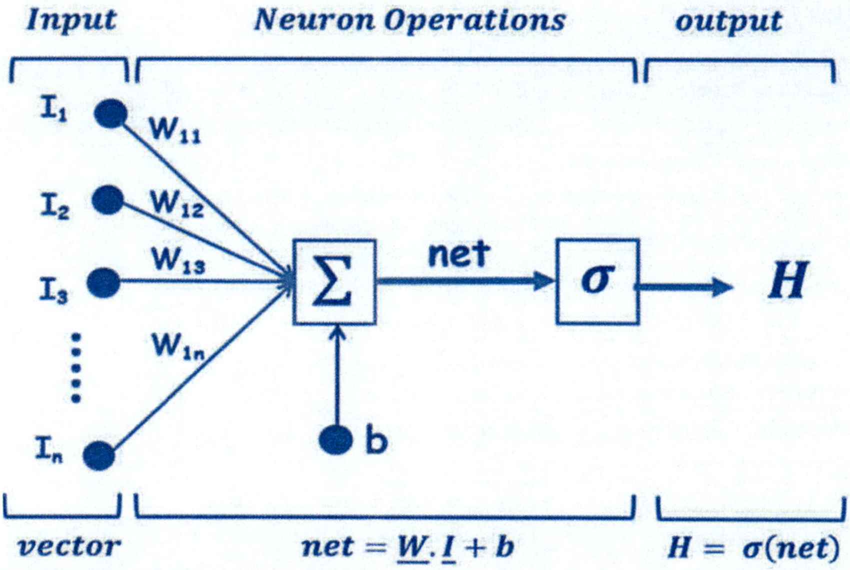

Fig. 3 A single-neuron module

production the dot product of weighs and inputs ($\underline{W}^T \cdot \underline{I}$). After All, the bias b is added to the dot product forming the *net* value. The final neuron output is activated using Sigmoid function $\sigma(net)$ as follows:

$$\text{net}^{[1]} = \left(\sum_{j=1}^{n} W_{1j} \cdot I_j \right) \underset{\text{then}}{\Rightarrow} H = \sigma\left(\text{net}^{[1]}\right)$$
$$= \frac{1}{1 + e^{-\text{net}^{[1]}}} \quad \text{for } i = 1, 2, 3, \ldots, n \ldots \tag{1}$$

Furthermore, to calculate the probabilities for the output classes, we have used the Softmax activation function (Multiclass classifier). Softmax is a normalized exponential formula that normalizes a vector of K real numbers ($\mathbb{R}^k$) into a probability distribution comprising of K real number-probabilities ($\mathbb{R}^k$) that are proportional to the exponentials of the input numbers [15]. The output layer computations via Softmax σ: $\mathbb{R}^k \longmapsto \mathbb{R}^k$ is defined as:

$$\text{net}^{[2]} = \left(\sum_{j=1}^{n} V_{1j} \cdot H_j \right) \underset{\text{then}}{\Rightarrow} \underline{O} = \sigma\left(\text{net}^{[2]}\right)_i$$
$$= \frac{e^{\text{net}_i^{[2]}}}{\sum_{j=1}^{K} e^{\text{net}_i^{[2]}}} \quad \text{for } i = 1, 2, 3, \ldots, K \ldots \tag{2}$$

Table 2 Summary of system development/verification environment

Specifications	Description
Computing platform	HP Pavilion × 360 Convertible (Model 14-ba100nx)
Processing unit (CPU)	Intel(R) Core (TM) i7-8550U CPU @ 1.80 GHz 1.99 GHz
Graphics (GPU)/memory (RAM)	4.00 GB NVIDIA GeForce 940MX/ 8.00 GB DDR4
Dataset preprocessing operations	Importing .CSV data into MATLAB tables One-hot encoding of categorical features ($1 - N$) [24] Converting tables to a double matrix
One-hot codes for class feature	Allow (100), deny (010), drop/reset (001)
Training/testing model	Shallow neural network with 150-hidden neurons
Dataset distribution proportions	Training: 70%, validating: 15%, testing: 15%
Dataset distribution numbers	Training: 45,872, validating: 9830, testing: 9830
Dataset distribution policy	DivideRand: divide targets into 3-sets using random indices
Optimization technique	Scaled conjugate gradient backpropagation (SCGB)
NN performance analysis	Minimization of cross-entropy loss (L_{CE}) function
Classification learner	Linear learner algorithm/initial learning rate ($\alpha = 0.0001$)
Validation frequency	6-fold cross validation/randomly performed at every run
Max number of epochs	387 epochs/shuffling performed at every epoch

4 Development and Experimental Environment

In order to develop and evaluate the proposed SNN-IFW classification system, the training and testing phases were carried out using IFW-2020 dataset. The predictive model is specified to differentiate between three classes; namely, 'allow', 'deny', 'and 'drop/reset' for the network packets. The proposed predictive model is implemented using MATLAB 2020b on a commodity laptop. Also, to optimize neural network training speed and memory, MEX calculation (MATLAB executable) has been used to train and simulate the network as well as for gradient calculations. Besides, the original dataset has undergone a preprocessing stage prior to the use of the neural network. The preprocessing module is responsible for the conversion of raw traffic records of IFW-2019 into a matrix of labeled features that can be trained by neural network's part of the classification system. To sum-up, the specifications and configurations of the test-bench environment are shown in Tables 2 and 3.

5 Results and Discussion

To measure the system performance, we have evaluated the developed classification model in terms of several evaluation metrics [28] including the following:

Table 3 Comparison with existing related work methods employing the firewall log files

Research method	Year	Classification technique	Accuracy	IM %
Ertam et al. [21]	2019	Support vector machine (SVM)	79.40	24.00%
Pîrîu et al. [25]	2019	Deep neural network (DNN)	92.82	06.10%
Banjongkan et al. [26]	2020	Convolutional neural network (CNN)	76.50	28.70%
Allagi et al. [27]	2020	Self-organizing feature map (SOFM)	97.20	01.34%
This model	2021	Shallow neural network (SNN)	98.50	–

- Multiclass Confusion Matrix to describe the performance of classifier model in terms of True Positives (**TP**), True Negatives (**TN**), False Positives (**FP**), and False Negatives (**FN**).
- Classification Accuracy (**ACC**) is the ratio of correct classifications to the total number of classifications. Conversely, Error Percent (**ERP**) is the fraction of misclassified samples to the total number of classifications.

$$\mathbf{ACC} = \frac{\mathbf{TP} + \mathbf{TN}}{\mathbf{TP} + \mathbf{TN} + \mathbf{FP} + \mathbf{FN}} \mathbf{ERP} = 1 - \mathbf{ACC} \ldots \tag{3}$$

- Positive Predictive Value (**PPV**)/Precision is the ratio of correctly positive labeled by the classifier to the total number of all positive labels. Conversely, False Discovery Rate (**FDR**) is calculated as the ratio of correctly positive labeled by the classifier.

$$\mathbf{PPV} = \frac{\mathbf{TP}}{\mathbf{TP} + \mathbf{FP}}, \ \mathbf{FDR} = 1 - \mathbf{PPV} \ldots. \tag{4}$$

- True Positive Rate (***TPR***)/Sensitivity is the ratio of all positive samples were correctly predicted as positive by the classifier. Conversely, False Negative Rate (**FNR**) is calculated as the ratio of all positive samples were incorrectly predicted.

$$\mathbf{TPR} = \frac{\mathbf{TP}}{\mathbf{TP} + \mathbf{FN}}, \ \mathbf{FNR} = 1 - \mathbf{TPR} \ldots \tag{5}$$

- True Negative Rate (**TNR**)/Specificity, is the ratio of all negative samples were correctly predicted as positive by the classifier. Conversely, False Positive Rate (**FPR**) is calculated as the ratio of all negative samples were incorrectly predicted.

$$\mathbf{TNR} = \frac{\mathbf{TN}}{\mathbf{TP} + \mathbf{FN}}, \ \mathbf{FPR} = 1 - \mathbf{TNR} \ldots \tag{6}$$

- Balanced Accuracy (**BAC**) is calculated as the average metric involving the classification sensitivity and precision.

$$\mathbf{BAC} = \frac{\mathbf{PPV} + \mathbf{TPR}}{2} \ldots \tag{7}$$

- Harmonic Mean (**HME**) is the first order weighted average ($\boldsymbol{F}1$) involving the 32 classification precision and recall.

$$\mathbf{HME} = \frac{\mathbf{2TP}}{\mathbf{2TP} + \mathbf{FP} + \mathbf{FN}} \tag{8}$$

Hence, Fig. 4 shows the confusion matrix results for our three-class classifier using training, validation, testing, and overall datasets. Based on the values obtained

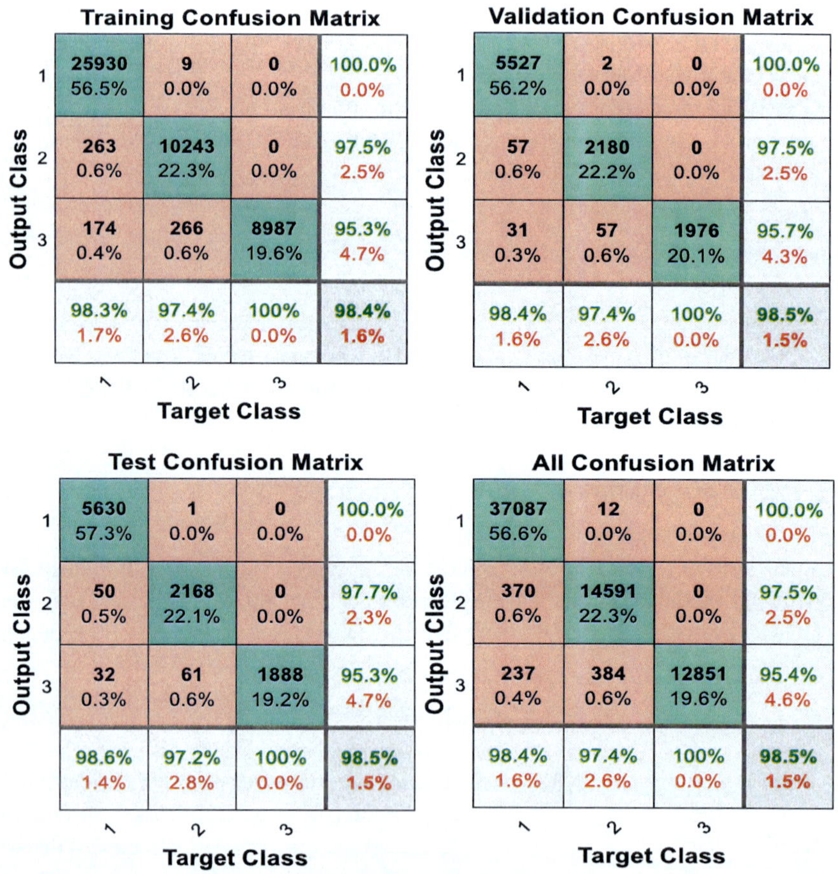

ACC	PPV	TPR	TNR	BAC	HME	ERP	FDR	FNR	FPR
98.50%	98.51%	98.47%	99.35%	98.49%	98.48%	1.50%	1.49%	1.53%	0.65%

Fig. 4 System evaluation: confusion matric results and the overall evaluation metrics

for ⟨TP, TN, FP, FN⟩ , we have computed the aforesaid evaluation metrics for our classification model.

Also, since the objective of the classification model is to produce output values as close as possible to the true values, thus the trainable weights of the model are iteratively adjusted aiming to minimize the Cross-Entropy Loss (L_{CE}) value. Hence, Softmax probability (p_i) for each predicted class (i) is compared to the true class label (t_i) and the loss (L_{CE}) is calculated that penalizes the probability based on how far it is from the true value [29]. A perfect model has a cross-entropy loss of 0. Cross-Entropy Loss (L_{CE}) is defined as:

$$\text{Cross Entropy Loss}(L_{\text{CE}}) = -\sum_{i=1}^{n} t_i . \log(p_i) \ldots \tag{9}$$

The learning optimization plots using Cross-Entropy Loss function versus taring epochs are illustrated in Fig. 5. The figure compares the optimization trajectories for training loss, validation loss, and testing loss functions attained throughout the training operation. According to the figure, the three loss curves are steadily declining with ongoing model training epochs. Afterward, the cross-entropy loss has saturated at 0.022 attained after 381 epochs for the 3-class classifier and thus the validation process stopped. The obtained loss value threshold indicated the robustness of our classification model scoring a significant classification accuracy and evading any under-fitting or over-fitting issues in the training process.

Furthermore, we have investigated the receiver operating characteristic (ROC) curve for our 3-class classifier. ROC curve represents the relative trade-offs between the true positive rate (benefits) at the y-axis against the false positive rate (costs)

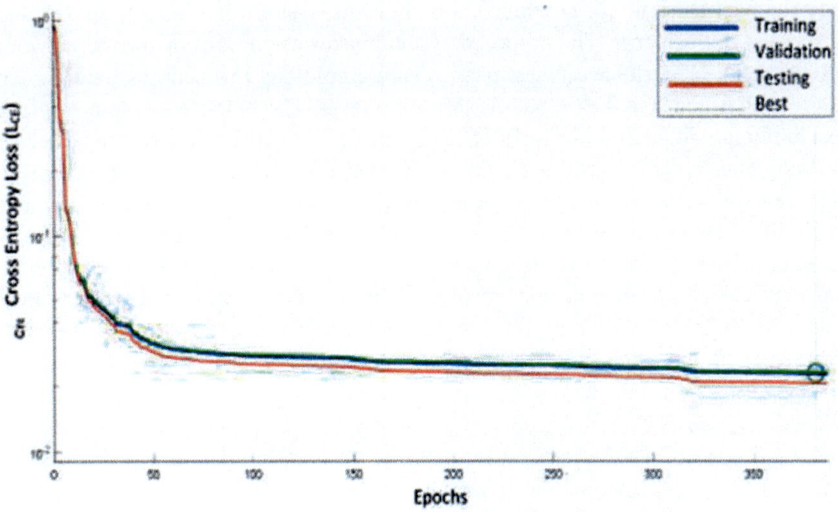

Fig. 5 SNN training performance using cross-entropy loss (epoch 381, validation stop)

at the x-axis for at various threshold settings. Typically, the classification model is established based on a continuous random variable ($\boldsymbol{X}$) which is compared with a predefined threshold ($\boldsymbol{T}$), therefore, the instance is classified as "positive" if $\boldsymbol{X} > \boldsymbol{T}$, and "negative" otherwise. $\boldsymbol{X}$ follows a probability density as follows:

$$f(x) = \begin{cases} f_1(x), \text{ For instance of class positive} \\ f_0(x), \text{ For instance of class negative} \end{cases} \tag{10}$$

Accordingly, the true positive rate (TPR) and false positive rate (FPR) for a given threshold ($\boldsymbol{T}$), can be computed as follows:

$$\mathrm{TPR}(T) = \int_T^\infty f_1(x)\mathrm{d}x \quad \text{and } \mathrm{FPR}(T) = \int_T^\infty f_0(x)\mathrm{d}x \ldots \tag{11}$$

Eventually, ROC curve plots parametrically $\mathrm{TPR}(T)$ versus $\mathrm{FPR}(T)$ with T as the varying parameter. The ROC curves of our three-classes classifier for the training, the validation, the testing, and the overall datasets are illustrated in Fig. 6. Since almost all experiments yield a point in the upper left corner (0, 1) of the ROC space, the classifier is approaching the perfect classification case recording 98.47% and 99.35% for classification sensitivity and classification specificity, respectively.

6 Conclusions

A dependable automated machine learning based internet firewall model to classify the packet traffic for communication network systems has been proposed and discussed in this paper. The proposed system makes use of shallow neural network (SNN) using 11-attributes at the input layer and employing 150-neurons at the hidden layer with 3-nodes at the output classification layer. The proposed system employs the multiclass internet firewall (IFW-2019) dataset with 70% of the records used for training dataset, 15% used for validation data set, and 15% used for testing dataset. To evaluate the performance of the model, the model was trained for 387 epochs recording a maximum accuracy of 98.5% achieved at epoch number 381 for the 3-class classifier. Besides, other machine learning metrics were evaluated to gain more insights into the system trajectory such as, positive predictive value, true positive rate, and others. Finally, based on the comparison with existing state-of-art in the field, the achieved outcomes surpassed the existing automated classification models for the firewall actions which is considered a contribution to this area of study. In future, we will consider the use of deeper neural network by incorporation the convolutional neural network (CNNs) to process the classification task for the firewall devices and software. Moreover, we will also seek to retain diversity of datasets and log files

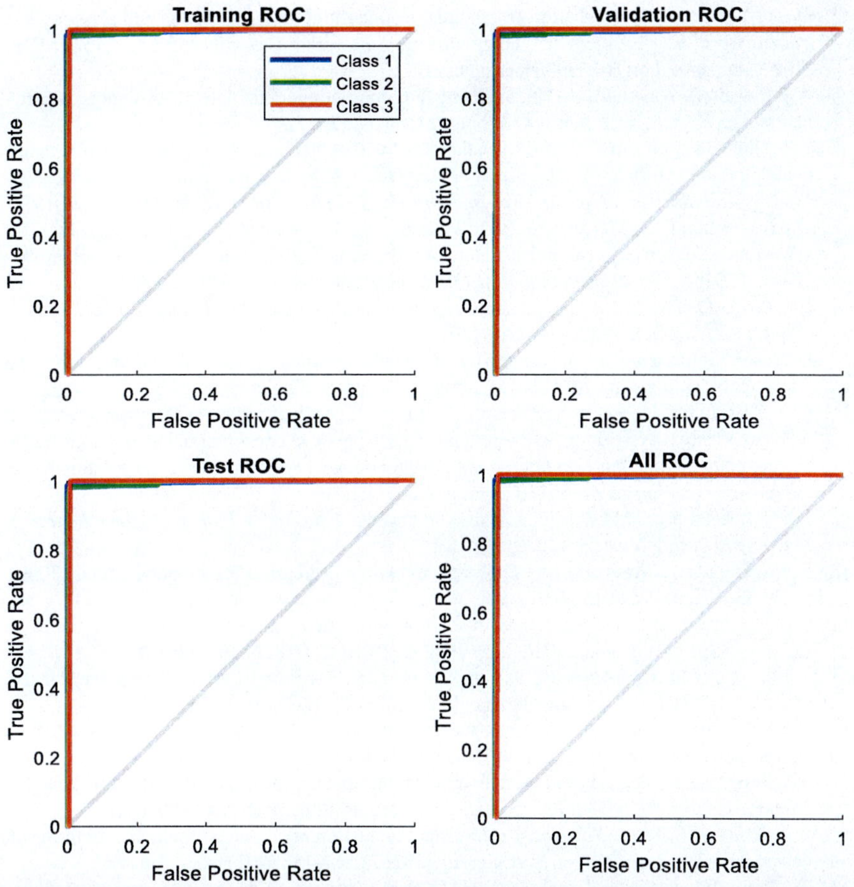

Fig. 6 Analyzing ROC Curve behavior for training, validation, testing, and the overall datasets

collected from different firewall system sources to further enhance firewall system conditionality under malicious communications.

References

1. Q.A. Al-Haija, L. Tawalbeh, Autoregressive modeling and prediction of annual worldwide cybercrimes for cloud environments, in *IEEE 10th International Conference on Information and Communication Systems (ICICS)*, pp. 47–51, 2019
2. W. Noonan, I. Dubrawsky, Firewall Fundamentals (Pearson Education, 2006)
3. G. Ranganathan, Real time anomaly detection techniques using pyspark frame work. J. Artif. Intell. **2**(01), 20–30 (2020)
4. P. Karunakaran, Deep Learning Approach to DGA Classification for Effective Cyber Security. J. Ubiquitous Comput. Commun. Technol. (UCCT) **2**(04), 203–213 (2020)

5. Q.A. Al-Haija, S. Zein-Sabatto, An efficient deep-learning-based detection and classification system for cyber-attacks in IoT communication networks. Electronics **9**(12), 2152. MDPI (2020). https://doi.org/10.3390/electronics9122152
6. A. Alrawais, A. Alhothaily, C. Hu, X. Cheng, Fog computing for the Internet of Things: security and privacy issues. IEEE Internet Comput. **21**, 34–42 (2017)
7. R. Mahmoud, T. Yousuf, F. Aloul, I. Zualkernan, Internet of things (IoT) security: current status, challenges, and prospective measures, in *Proceedings of the 10th International Conference for Internet Technology and Secured Transactions (ICITST)*, London, UK; pp. 336–341, 2015
8. Q.A. Al-Haija, C.D. McCurry, S. Zein-Sabatto, A real time node connectivity algorithm for synchronous cyber physical and IoT network systems, in *2020 SoutheastCon*, Raleigh, NC, USA, 2020, pp. 1–8. https://doi.org/10.1109/SoutheastCon44009.2020.9249730
9. E. Ucar, E. Ozhan, The analysis of firewall policy through machine learning and data mining. Wirel. Pers. Commun. **96**, 2891–2909 (2017)
10. G. Caspi, "Introducing deep learning: boosting cybersecurity with an artificial brain. Informa Tech" dark reading. Analytics (2016). http://www.darkreading.com/analytics
11. Q.A. Al-Haija, C.D. McCurry, S. Zein-Sabatto, Intelligent self-reliant cyber-attacks detection and classification system for IoT communication using deep convolutional neural network, in *Selected Papers from the 12th International Networking Conference. INC 2020*. Lecture Notes in Networks and Systems, vol 180 (Springer, 2021)
12. J. Brownlee, 4 types of classification tasks in machine learning, in *Python Machine Learning, Machine Learning Mastery*. (2020)
13. S. Haykin, *Neural Networks and Learning Machines*, 3rd edn. (Pearson publications, 2009). ISBN-13: 978-0-13-147139-9
14. C.C. Aggarwal, Machine learning with shallow neural networks. *Neural Networks and Deep Learning* (Springer, Cham, 2019). https://doi.org/10.1007/978-3-319-94463-0_2
15. Fei-Fei. *CS231n: Convolutional Neural Networks for Visual Recognition*. Computer Science, Stanford University (2019). Available online: http://cs231n.stanford.edu
16. J.S. Meneses, Z.R. Chavez, J.G. Rodriguez, Compressed kNN: K-nearest neighbors with data compression. Entropy **21**(3), 234 (2019). https://doi.org/10.3390/e21030234
17. Y.Y. Song, Y. Lu, Decision tree methods: applications for classification and prediction. Shanghai Arch Psychiatry **27**(2), 130–135. PMID: 26120265; PMCID: PMC4466856. (2015)
18. B.A. Tama, K.H. Rhee, An extensive empirical evaluation of classifier ensembles for intrusion detection task. Int. J. Comput. Syst. Sci. Eng. **32**(2), 149–158 (2017)
19. A. Ghose, Support vector machine (SVM) tutorial: learning SVMs from examples. Medium: Towards Data Sci. (2017)
20. R. Garg, Types of classification algorithms. Analytics India Mag. (2018)
21. F. Ertam, M. Kaya, Classification of firewall log files with multiclass support vector machine, in *6th International Symposium on Digital Forensic and Security (ISDFS)*, Antalya, pp. 1–4 (2019)
22. UCI. *Machine Learning Repository, Internet Firewall Data Set*. Center for Machine Learning and Intelligent Systems (2019)
23. Q.A. Al-Haija, A. Adebanjo, Breast cancer diagnosis in histopathological images using ResNet-50 convolutional neural network, in *2020 IEEE International IOT, Electronics and Mechatronics Conference (IEMTRONICS)*, Vancouver, BC, Canada, 2020, pp. 1–7. https://doi.org/10.1109/IEMTRONICS51293.2020.9216455
24. A. Wang, Encode smarter: how to easily integrate categorical encoding into your machine learning pipeline. Feature Labs (2019). https://blog.featurelabs.com
25. A.I. Pîrîu, M. Leonte, N. Postolachi, D.T. Gavrilut, Optimizing Cleanset Growth by Using Multi-Class Neural Networks, in *2018 20th International Symposium on Symbolic and Numeric Algorithms for Scientific Computing (SYNASC)* (Timisoara, Romania, 2018), pp. 425–429
26. A. Banjongkan et al., A comparative study of learning techniques with convolutional neural network based on HPC-workload dataset. Int. J. Mach. Learn. Comput. **10**(1) (2020)
27. S. Allagi, R. Rachh, Analysis of network log data using machine learning," in *IEEE 5th International Conference for Convergence in Technology*, India, pp. 1–3 (2019)

28. Q.A. Al-Haija, M. Smadi, S. Zein-Sabatto, Multi-class weather classification using ResNet-18 CNN for autonomous IoT and CPS applications, in *IEEE 7th Annual Conference on Computational Science & Computational Intelligence (CSCI'20)*, Las Vegas, USA (2020)
29. K.E. Koech, Cross-entropy loss function. Medium: Towards Data Sci. (2020)

On the Possibility of Evasion Attacks with Macro Malware

Risa Yamamoto and Mamoru Mimura

Abstract To detect unknown malware, several methods with machine learning techniques have been proposed. These methods extract the features of malware, and thereby classify samples as malicious or benign. The previous method extracts words from the source code and detects macro malware. This method constructs a language model latent semantic indexing (LSI) to extract the features. On the other hand, several methods to avoid the detection have been proposed. These methods add benign features to the source code for Android malware. These methods can be applied to the macro malware. In this study, we discuss the risk of macro malware that evades detection. This paper attempts to imitate benign macros by adding benign features to macromalware. Our method extracts the variables from macro malware and replaces them to frequent words of benign macros. Furthermore, our method inserts the frequent words that appear in benign macros or the LSI topics. These words are inserted as the arguments of some functions, which do not vary the main behavior. The target detection model is our previous method, which detects macro-malware with language models: Bag of words (BoW) and LSI. The detection rate is evaluated under two conditions that the attacker can access inside the model or not. As a result, the detection rate with BoW decreases to 1.5% under the situation that the attacker can access inside the model. Even if the attacker cannot access inside the model, the detection rate with LSI decreases by 73%.

Keywords VBA macro · Adversarial attack · Malware · Evasion

R. Yamamoto
Japan Ground Self-Defense Force, Tokyo, Japan

M. Mimura (✉)
National Defense Academy, Yokosuka, Japan
e-mail: mim@nda.ac.jp

G. Ranganathan et al. (eds.), *Soft Computing for Security Applications*,
Advances in Intelligent Systems and Computing 1397,
https://doi.org/10.1007/978-981-16-5301-8_4

1 Introduction

In recent years, information technologies have been developing and become more important. However, as well as this development, the risk of cyberattacks became a more serious problem. Automated cyberattacks would be the next big threat [15]. According to Symantec's report in 2019,[1] the Office macro downloaders trended upward toward the end of 2018. The number of detected macro downloaders is about 50,000 per a month at least, and about 270,000 at most. Moreover, EMOTET, one of the most popular malware in 2019, uses a macro to call a PowerShell script. As has been noted, the risk of attacks with macro malware is one of the serious threats. Macro is written in Visual Basic Application (VBA). VBA is an implementation of Microsoft's programming language Visual Basic 6 and is built into most Microsoft Office applications.

Current antivirus programs mainly adopt a pattern matching approach to detect malware [3]. However, detecting unknown malware is a challenging task.

To overcome this difficulty, detection methods with machine learning are studied [2, 10, 12, 13, 22]. In contrast, some studies have suggested vulnerabilities of machine learning-based detection, such as adversarial attacks [1, 4–8, 11, 16]. To improve the resilience of these detection methods, we should evaluate the risk of these attacks and consider the countermeasures. In fact, several studies have evaluated the risk [1, 4–8, 11, 16]. To evaluate the risk appropriately, the attacks should be feasible to maintain the behavior of malware. Considering a real situation of the attacks, the purpose of attackers is to send the malware to victims and exploit their computers by remote control. To achieve the goal, the behavior of malware should be remained. Therefore, maintaining the behavior of malware has to be considered to evaluate the actual risk of the attacks. While some studies focus on avoiding the detection, they hardly consider maintaining the behavior of malware [7, 11]. Moreover, resent studies about VBA macros have not revealed the risk of adversarial attacks [17, 18]. In this study, we discuss the risk of macro malware that evades machine learning-based detection, which maintains the actual behavior. This paper attempts two approaches. First, our method extracts the variables from the source code of macro malware and replaces them to the frequent words of benign macros. Second, our method inserts the words to the source code, which appear in benign macros or the LSI topics. The target is the detection model that was proposed in the previous study [19, 20]. This method uses bag of words (BoW) or latent semantic indexing (LSI) as natural language processing (NLP) to detect macro malware. This paper evaluates our method by measuring the detection rate.

The contributions of this paper are as follows.

1. Adding benign features to macro malware is effective for machine learning-based detection and it can be a risk.
2. The detection rate with BoW is decreased to 1.5% by inserting words that appear in only benign macros.

[1] https://docs.broadcom.com/docs/istr-24-2019-en.

3. The detection rate with LSI is decreased to 27% by inserting words that appear in only benign LSI topics.

This paper consists of seven sections. Section 2 describes the related works and Sect. 3 describes the background. Section 4 produces the detail of our method. Section 5 demonstrates the experimental result. Section 6 discusses the risk of VBA malware, the resilience of the detection model, the countermeasure, the limitation, and the ethics. Section 7 concludes this paper and describes the future work.

2 Related Work

2.1 Detection with Machine Learning

Jeoung et al. proposed a detection method of PDF malware that extracts the byte sequence and classifies by a convolutional neural network (CNN) [10]. This study evaluated not only CNN, but also support vector machine (SVM), Naive Bayes, decision tree, and random forest as the detection models. This study concluded that the CNN model is the best in accuracy and processing speed. The SVM model was the second best in accuracy.

Amin et al. proposed a detection method of Android malware that extracts the byte code from APK and converts into feature vectors [2]. This method adopted deep neural networks with restricted Boltzmann machine as the classifier.

Vidyarthi et al. proposed a detection method of PE file malware that adopts static analysis and dynamic analysis [22]. This study converts the source code into feature vectors for the static analysis and uses text mining technique for the dynamic analysis. This study concludes that this method can detect not only known malware, but also unknown malware that even avoids a sandbox. Kozachok et al. proposed a detection method of PE file malware that extracts not only the source code but also file size, encryption, API calls, and digital certificates as features [13].

Thus, malware detection with machine learning is widely studied. Furthermore, various types of malware are studied.

In regard to VBA malware, Kim et al. proposed a method that classifies malware by obfuscation techniques [12]. This study focuses that VBA malware is often obfuscated. However, detecting not-obfuscated malware is difficult for this method. Therefore, we focus on VBA malware detection that can detect not only obfuscated malware but also malware that is not obfuscated.

2.2 Evasion for Machine Learning-Based Detection

Ehteshamifar et al. proposed a method that avoids the PDF malware detection by static analysis and dynamic analysis [6]. This study executed the malicious code in only runtime to avoid the detection. Wang et al. proposed a method that uses generative adversarial network (GAN) to avoid the PDF malware detection [16]. This study concluded that the method considerably decreases the detection rate of the classifier.

Khormali et al. proposed a method that avoids the detection of the Win32 or IoT malware by using adversarial learning [11]. In this method, malware maintains its behavior. Huang et al. proposed a method that avoids the detection of the PE file malware by inserting the benign API calls [8]. Chen et al. proposed a method that avoids the detection of the PE file malware by inserting the benign source code by using CNN [4]. This study suggests that retraining or changing the method to create the training dataset can improve the detection rate.

Chen et al. proposed a method that avoids the detection of Android malware by inserting the adversarial examples to the classifier [5]. Abaid et al. proposed a method that avoids the detection of Android malware with adversarial learning [1]. Grosse et al. proposed a method that avoids the Android malware detection with dynamic analysis by inserting the benign source code [7]. Furthermore, this method can maintain the behavior of malware.

Thus, the risk of attacks with PDF, PE, and Android malware is suggested. However, the risk of attacks with VBA malware has not been suggested. We assume that inserting benign code to VBA malware can imitate the benign features likewise Android malware [4]. Therefore, this paper evaluates the risk of attacks with VBA malware that avoids malware detection.

2.3 Evasion for Natural Languages

Quiring et al. proposed an evasion method for authorship attribution of source code [21]. This method evades the attribution by rewriting source code that have the same functions as the original source code. This study concludes that the method is effective if attackers cannot access the classifier model.

Jain et al. proposed a method that evades Google's Perspective API [9] . This method evades the classifier by inserting a few words to texts. This study concludes that the method can decrease the harmful score of the texts.

Jifeng et al. proposed a method that evades the classifiers with deep learning by replacing characters or words [14]. This study concludes that method decreases the success rate of the classifiers efficiently.

Thus, some studies attempted to evade classification for natural languages. These studies do not use these techniques for malware detection. This paper focuses on macro malware detection with some natural language processing techniques.

3 Background

3.1 Bag of Words (BoW)

BoW is a basic method to convert a document into a feature vector. This method converts frequency of unique words in a document into elements of a feature vector. Because the length of a feature vector depends on the number of unique words, the dimensionality of a feature vector becomes considerably long, if the document contains many unique words. For this reason, the process speed of BoW is decreased, if the dimensionality of a feature vector is long. Therefore, BoW requires the optimum dimensionality of a feature vector.

3.2 Latent Semantic Indexing (LSI)

LSI is another method that solves the problem of BoW. LSI is a basic method of dimension reduction by singular value decomposition, and this method creates topics that are related with words in a document. Because LSI can set the dimensionality of a feature vector arbitrarily, this method can be used as a dimension reduction method and reduces the computational complexity.

We will introduce how LSI converts a document into a feature vector. First, this method calculates term frequency-inverse document frequency (TF-IDF).

$$\mathrm{TF} = \frac{\text{frecuancy of word } x \text{ in document } A}{\text{sum of frequency of all words in document } A} \tag{1}$$

TF is frequency of a word in document A and defined as (1).

$$\mathrm{IDF} = \log\left(\frac{\text{the number of all documents}}{\text{the number of documents that contain word } x}\right) \tag{2}$$

IDF is inverse document frequency and defined as (2).

$$\mathrm{TFIDF} = \mathrm{TF} \times \mathrm{IDF} \tag{3}$$

TF-IDF is defined as (3). In this model, if a word appears rarely in an entire corpus and appears frequently in a document, the TF-IDF value increases.

$$T = \begin{bmatrix} t_{1,1} & \cdots & t_{1,m} \\ \vdots & \ddots & \vdots \\ t_{n,1} & \cdots & t_{n,m} \end{bmatrix} \tag{4}$$

Thereafter, the matrix T is created by using TF-IDF. Matrix T is defined as (4). n is the number of documents that is contained a document group, and m is the number of unique in a document group.

Next, we apply singular value decomposition for matrix T.

$$T = USV^{\mathrm{T}} \tag{5}$$

$$T = \begin{bmatrix} u_{1,1} & \cdots & u_{1,m} \\ \vdots & \ddots & \vdots \\ u_{m,1} & \cdots & u_{m,m} \end{bmatrix} \begin{bmatrix} \sigma_1 & & 0 \\ & \ddots & \\ 0 & & \sigma_r \end{bmatrix} \begin{bmatrix} v_{1,1} & \cdots & v_{1,n} \\ \vdots & \ddots & \vdots \\ v_{n,1} & \cdots & v_{n,n} \end{bmatrix} \tag{6}$$

Matrix U is a square matrix of m rows and columns, and it is eigenvector that each row is orthogonal to matrix TT^{T}. Matrix V is a square matrix of n rows and columns, and it is eigenvector that each row is orthogonal to matrix TT^{T}. Matrix S is a diagonal matrix of n rows and m columns, and diagonal component σ is singular value. Matrix U stands for topic weight of each words, and matrix V stands for topic weight of each documents.

$$T' = \begin{bmatrix} u_{1,1} & \cdots & u_{1,k} \\ \vdots & \ddots & \vdots \\ u_{m,1} & \cdots & u_{m,k} \end{bmatrix} \begin{bmatrix} \sigma_1 & & 0 \\ & \ddots & \\ 0 & & \sigma_k \end{bmatrix} \begin{bmatrix} v_{1,1} & \cdots & v_{1,n} \\ \vdots & \ddots & \vdots \\ v_{k,1} & \cdots & v_{k,n} \end{bmatrix} \tag{7}$$

Because matrix S is a diagonal matrix in (5), component σ that is small value less affect to matrix T. Therefore, if we define the dimensionality k, the component σ except above to k is approximated to 0. This accomplishes to reduce dimensionality, and square errors become smallest by approximating singular value that is small value to 0. In other words, this method can reduce dimensionality with maintaining the feature of documents.

4 Proposed Method

4.1 Outline

The previous study [19, 20] proposed a detection method for macro malware, which converts the source code into features. In this method, at first, the source code is divided into words. Next, the method converts the words into a feature vector with BoW or LSI and classifies the macro into benign or malicious by SVM.

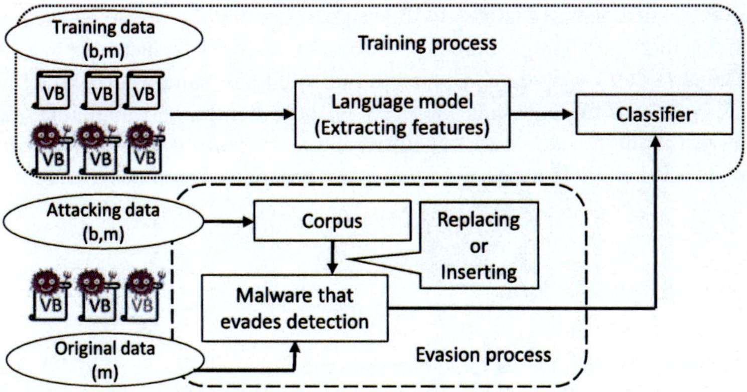

Fig. 1 Process of the proposed method

This method uses the words as the feature, which are not obfuscated. Hence, it can also classify macros that are not obfuscated. Therefore, this paper evaluates the risk to avoid this method by modifying the source code. Besides, we focus on maintaining the behavior of malware to modify the source code. In this paper, we modify the source code by replacing variables or inserting words. Figure 1 shows the process of the proposed method.

Our method is mainly divided into the training process and evasion process.

4.2 Training Process

In the training process, our method trains the detection model. This process was proposed in the previous study [19, 20]. First, our method extracts the words from the source code of a training dataset and converts them into the language models. The previous study used an SVM classifier and two language models: BoW and LSI. In this experiment, the detection models that use BoW or LSI for the language model are described as "BoW model" or "LSI model". Our method converts source code into the language model and uses the model to extract the feature vectors. The default number of LSI topics is 400, which was revealed as the optimal value in the previous study. Thereafter, our method trains the SVM classifier with the feature vectors.

4.3 Evasion Process

In this process, our method executes the evasion process to malware, which is to avoid detection. The term "original data" is used to indicate these malware samples to avoid detection. Hence, original data contain only malicious macros (malware).

This process adds benign features to the original data by the replacing method and inserting method with benign words. These benign words are selected from attacking data. The term "attacking data" is used to indicate these samples to extract benign features. Note that the attacking data contain both benign and malicious macros. This suggests benign features are not always the words simply extracted from benign macros. To add benign features, we attempt two approaches: replacing and inserting.

4.4 Replacing Method

This method replaces some variable names in the source code of original data. First, this method extracts the source code from original data. Thereafter, this method extracts the variable names that users can define arbitrary and replaces them into the benign words. These benign words are extracted from attacking data and used one by one. Hence, the maximum number of the replaced words depends on the variable. In other words, this method depends on the original data.

4.5 Inserting Method

This method inserts benign words to original data. First, this method extracts the source code from original data. Thereafter, this method inserts the words that appear in only benign macro or benign LSI topics of original data. To maintain the behavior of malware, the benign words are inserted as argument of following functions: "Asc", "AscB", "AscW", "LCase", "LTrim", "Trim", "StrReverse", "Ucase", or "Split". These functions do not prevent the behavior, because they only change order or case of strings that is assigned as argument. These benign words are extracted from attacking data and used one by one. In this method, we can insert these benign words without regard to the original data. Note that the attacking data contain a plenty of words to insert.

4.6 Implementation

We implemented our method with Python 3.6. We used gensim 3.6.0 [2] and sckit-learn 0.22.1[3] to implement the LSI and SVM model respectively.

[2] https://radimrehurek.com/gensim/.

[3] https://scikit-learn.org/stable/.

Table 1 Numbers of samples

2016		2017	
Benign	Malicious	Benign	Malicious
1200	1150	2220	719

5 Experiment

5.1 Dataset

To evaluate our method, we use actual benign and malicious macros. Table 1 shows the numbers of samples. These samples are obtained from Virus Total.[4] We selected the samples that were uploaded for the first time from 2016 to 2017 and the extension is doc, docs, xls, xlsx, ppt, or pptx. We defined malicious macro as the macros that are detected as malware by more than half of 58 antivirus programs. There is no duplication of these samples.

5.2 Outline

To evaluate our method, we conduct four experiments. Table 2 shows the outline of the experiments.

In this experiment, we use the same dataset. The training and attacking data consist of benign (b) and malicious (m) samples in 2016. The original data consist of malicious samples in 2017. The training and attacking data are selected as described in Table 2, respectively.

5.3 The Effect of the Replacing Method

To evaluate the effect of the replacing method, we conduct the following experiment. The replacing method extracts the variables of the source code and replaces them into benign words. In this experiment, we do not consider practicality. We use benign and malicious macros in 2016 as training data and attacking data, and malicious macros in 2017 as original data. The targets are a BoW model and a LSI model. Tables 3 and 4 show the performance of the replacing method. As a result, this simple method decreases the detection rate of the BoW model by only 3%. Moreover, this method hardly decreases the detection rate of the LSI model. Therefore, the replacing method is not effective for these models.

[4] https://www.virustotal.com/.

Table 2 Outline of the experiments

No.	Purpose	Method	Language model	Training data	Attacking data	data	How to select words
1	To evaluate the effect of the replacing method	Replacing	BoW	2016b,m	2016b,m	2017m	Frequent words in the benign macros
			LSI				
2	To evaluate the effect of the simple inserting method	Inserting	BoW				Only appear in the benign macros
			LSI				
3	To evaluate the effect of the modified inserting method						Only appear in the benign LSI topics
4	To optimize the number of the LSI topics						

Table 3 Performance of the replacing method for the BoW model

	Before the process	After the process
Training data	2016	2016
Attacking data	–	2016
Original data	2017m	2017m
Accuracy	0.78	0.76
Precision	1.00	1.00
Recall	0.78	0.76
F1	0.88	0.86

Table 4 Performance of the replacing method for the LSI model

	Before the process	After the process
Training data	2016	2016
Attacking data	–	2016
Original data	2017m	2017m
Accuracy	0.85	0.85
Precision	1.00	1.00
Recall	0.85	0.85
F1	0.92	0.92

5.4 *The Effect of the Simple Inserting Method*

Next, we evaluate the effect of the inserting method. In this experiment, we simply select the words that appear in only benign macros, but not in malicious macros of attacking data. We use benign and malicious macros in 2016 as training data and attacking data, and malicious macros in 2017 as original data. The targets are a BoW model and a LSI model. Figures 2 and 3 show the transition of the accuracy by the inserting method. The vertical axis indicates the accuracy, precision, recall, or F1 score. The horizontal axis indicates the number of the inserted words. The recall score of the BoW model is decreased to 1.5%. In contrast, the detection rate of the LSI model is hardly decreased. Therefore, this method is effective for the model that adopts simple algorithm such as the BoW model. However, this method is not effective for the LSI model.

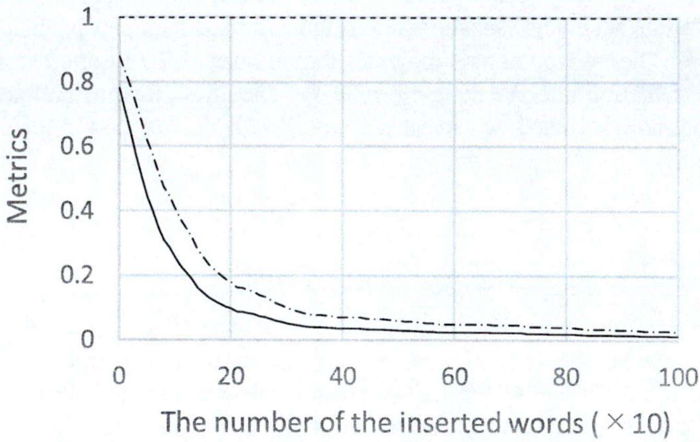

Fig. 2 Transition of the accuracy by the inserting method for the BoW model

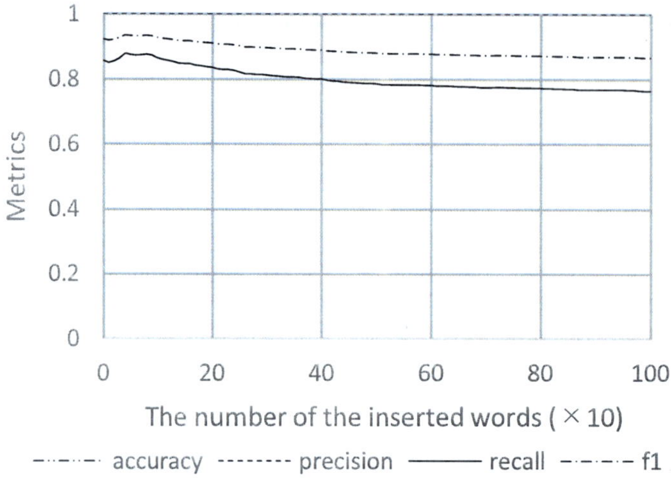

Fig. 3 Transition of the accuracy by the inserting method for the LSI model

5.5 The Effect of the Modified Inserting Method

To decrease the accuracy of the LSI model, we focus on the selection method of the words, which are inserted to original data. The modified method converts benign and malicious macros into each LSI model, respectively. These macros are attacking data. Thereafter, the modified method compares the benign LSI topics with the malicious LSI topics and selects the words that appear in only the benign LSI topics. To evaluate the modified method, we use benign and malicious macros in 2016 as training data and attacking data, and malicious macros in 2017 as original data. Figure 4 shows the transition of the accuracy. The vertical axis indicates the accuracy, precision, recall, or F1 score. The horizontal axis indicates the number of the inserted words. As a result, the detection rate is decreased to 27 %. Therefore, this method is effective for the LSI model. Hence, we select the words with this method in the following experiments.

5.6 Optimizing the Number of LSI Topics

To optimize the number of LSI topics, we vary the number of LSI topics. The number of the LSI topics is 100–900. We use benign and malicious macros in 2016 as training data and attacking data, and malicious macros in 2017 as original data. Figure 5

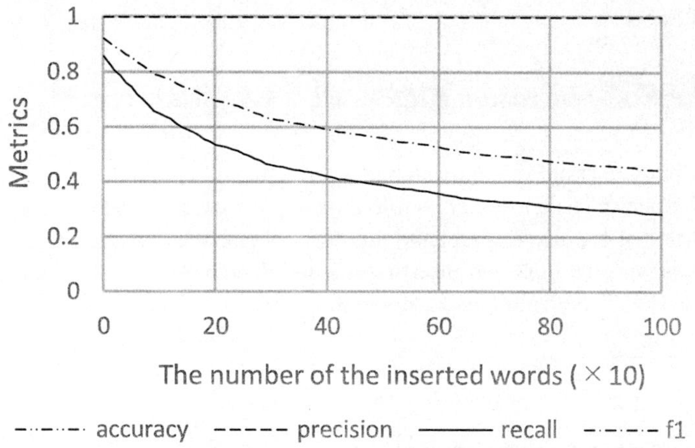

Fig. 4 Transition of the accuracy with the inserting method that focuses on LSI topics to select words

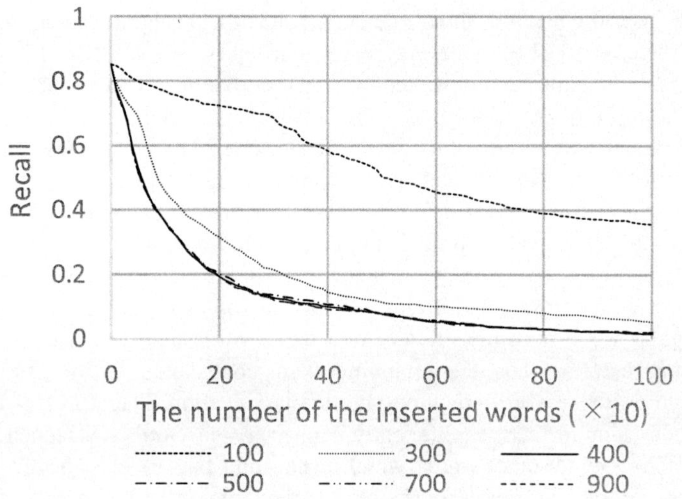

Fig. 5 Transition of the recall with the various number of LSI topics

shows the transition of the accuracy. The vertical axis indicates the recall score. The horizontal axis indicates the number of the inserted words. As a result, this method with 100 topics decreases recall to 36%. On the other hand, this method with 900 topics decreases recall to 22%. Therefore, there seems to be no significant correlation between the number of LSI topics and performance.

6 Discussion

6.1 The Risk of Evasion Attacks with VBA Malware

The experimental result demonstrated that the replacing method is not effective for the BoW model and LSI model. While the simple inserting method decreased the detection rate of the BoW model and did not decrease the detection rate of the LSI model. However, the modified inserting method decreased the detection rate of the LSI model. Hence, both models are vulnerable for the attacks that add benign features to malicious macros.

6.2 Resilience to the Attacks

The experimental result demonstrated that the detection rate of BoW is decreased by the simple inserting method, and the detection rate of LSI is hardly decreased by this method. Therefore, the LSI model is more resilient than the BoW model. However, the experimental result also demonstrated that the modified inserting method decreased the detection rate of the LSI model. As a consequence, our method decreases the detection rate of both the BoW model and LSI model. Therefore, these models are vulnerable for the attacks that add benign features to malware.

Next, we discuss why the simple inserting method does not decrease the detection rate of the LSI model. The words that appear in only benign macros include 2914 unique words. In contrast, the words that appear in only benign LSI topics include only 164 unique words. This difference is due to singular value decomposition of LSI. Singular value decomposition removes the words that do not contribute for determining features. Thus, the remaining words contribute a lot for classification. In other words, these words are important for classification. Therefore, the modified method adds benign features efficiently. However, the words that appear in only benign macros include many words, which do not contribute for classification. Hence, the simple method does not decrease the detection rate of the LSI model.

6.3 Countermeasure Against Evasion Attacks

This section discusses countermeasures against the evasion attacks with VBA malware. One approach to mitigating these attacks is that excluding some words that users can define arbitrarily. The target models of this study extract features from the whole source code. Hence, our method can insert arbitrary words to vary the features. These inserted words do not vary the behavior. In other words, these arbitrary words are not important for classification. Therefore, excluding these words can be a countermeasure. Another approach is retraining the model with evasion attacks. However, this can be applied after evasion attacks. Hence, both countermeasures should be simultaneously applied.

6.4 Limitation of This Study

The purpose of this study is to evaluate the risk of macro malware that evades machine learning-based detection, which maintains the actual behavior. In theory, our method can maintain the behavior of malware. However, we did not confirm the behavior by dynamic analysis. Moreover, we suggested two countermeasures. However, this paper did not evaluate these countermeasures to be effective for our method.

6.5 Ethics of This Study

This paper demonstrated that adding benign features to macro malware is effective and the attacks can be a risk. We paid attention to ethics in this study. Our method does not require particular resources and can be implemented easily. Therefore, attackers may imitate our method to evade detection. However, our method requires a dataset for attacks. Moreover, our method does not consider the conventional detection methods with pattern matching. For these reasons, the attackers cannot use our method briefly. However, the possibility of this cannot be completely denied. Therefore, we determined to not release the source code of our method and dataset for now. Our main purpose is to warn the risk of evasion attacks with VBA malware. We believe that this paper contributes development of detection methods that are resilient against evasion attacks.

7 Conclusion

We conclude that adding benign features to macro malware is effective for machine learning-based detection and it can be a risk. The experimental results show that our method decreases the detection rate to 1.5% for the BoW model and to 27% for the LSI model. Our method might be a risk for malware detection systems with machine learning. Finally, we suggest two countermeasures against evasion attacks. Further studies are required in order to implement the countermeasures and evaluate them. Evasion attacks with adversarial learning should be also evaluated.

References

1. Z. Abaid, M.A. Kâafar, S. Jha, Quantifying the impact of adversarial evasion attacks on machine learning based android malware classifiers, in *16th IEEE International Symposium on Network Computing and Applications, NCA 2017*, Cambridge, MA, USA, October 30–November 1, 2017 (2017), . pp. 375–384. https://doi.org/10.1109/NCA.2017.8171381
2. M. Amin, T.A. Tanveer, M. Tehseen, M. Khan, F.A. Khan, S. Anwar, Static malware detection and attribution in android byte-code through an end-to-end deep system. Future Gener. Comp. Syst. **102**, 112–126 (2020). https://doi.org/10.1016/j.future.2019.07.070
3. F. Biondi, T. Given-Wilson, A. Legay, C. Puodzius, J. Quilbeuf, Tutorial: an overview of malware detection and evasion techniques, in *Leveraging Applications of Formal Methods, Verification and Validation. Modeling—8th International Symposium, ISoLA 2018*, Limassol, Cyprus, November 5–9, 2018, Proceedings, Part I, pp. 565–586 (2018). https://doi.org/10.1007/978-3-030-03418-4_34
4. B. Chen, Z. Ren, C. Yu, I. Hussain, J. Liu, Adversarial examples for cnn-based malware detectors. IEEE Access **7**, 54360–54371 (2019). https://doi.org/10.1109/ACCESS.2019.2913439
5. S. Chen, M. Xue, L. Fan, S. Hao, L. Xu, H. Zhu, Automated poisoning attacks and defenses in malware detection systems: an adversarial machine learning approach. Comput. Secur. **73**(2017). https://doi.org/10.1016/j.cose.2017.11.007
6. S. Ehteshamifar, A. Barresi, Gross, T.R., Pradel, M, Easy to fool? Testing the anti-evasion capabilities of PDF malware scanners. CoRR abs/1901.05674 (2019), http://arxiv.org/abs/1901.05674
7. K. Grosse, N. Papernot, P. Manoharan, M. Backes, P. McDaniel, Adversarial Examples for Malware Detection, in *ESORICS 2017*, ed. by S.N. Foley, D. Gollmann, E. Snekkenes. LNCS, vol. 10493 (Springer, Cham, 2017), pp. 62–79. https://doi.org/10.1007/978-3-319-66399-9_4
8. Y. Huang, U. Verma, C. Fralick, G. Infantec-Lopez, B. Kumar, C. Woodward, Malware evasion attack and defense, in *2019 49th Annual IEEE/IFIP International Conference on Dependable Systems and Networks Workshops (DSN-W)* (2019), pp. 34–38. https://doi.org/10.1109/DSN-W.2019.00014
9. Jain, E., Brown, S., Chen, J., Neaton, E., Baidas, M., Dong, Z., Gu, H., Artan, N.S.: Adversarial text generation for google's perspective api, in *2018 International Conference on Computational Science and Computational Intelligence (CSCI)* (2018), pp. 1136–1141. https://doi.org/10.1109/CSCI46756.2018.00220
10. Y. Jeong, J. Woo, A.R. Kang, Malware detection on byte streams of PDF files using convolutional neural networks. Security and Communication Networks 2019, 8485365:1–8485365:9 (2019). https://doi.org/10.1155/2019/8485365
11. A. Khormali, A. Abusnaina, S. Chen, D. Nyang, A. Mohaisen, Copycat: practical adversarial attacks on visualization-based malware detection (2019)
12. S. Kim, S. Hong, J. Oh, H. Lee, Obfuscated vba macro detection using machine learning, in *2018 48th Annual IEEE/IFIP International Conference on Dependable Systems and Networks (DSN)* (2018), pp. 490–501. https://doi.org/10.1109/DSN.2018.00057
13. A.V. Kozachok, V.I. Kozachok, Construction and evaluation of the new heuristic malware detection mechanism based on executable files static analysis. J. Comput. Virol. Hacking Techn. **14**(3), 225–231 (2017). https://doi.org/10.1007/s11416-017-0309-3
14. J. Li, S. Ji, T. Du, B. Li, T. Wang, Textbugger: generating adversarial text against real-world applications, in: *26th Annual Network and Distributed System Security Symposium, NDSS 2019*, San Diego, California, USA, February 24–27, 2019. The Internet Society (2019), https://www.ndss-symposium.org/ndss-paper/textbugger-generating-adversarial-text-against-real-world-applications/
15. R. Maeda, M. Mimura, Automating post-exploitation with deep reinforcement learning. Comput. Secur. **100**, 102108 (2021)
16. D. Maiorca, B. Biggio, G. Giacinto, Towards adversarial malware detection: Lessons learned from pdf-based attacks. ACM Comput. Surv. **52**(4), 78:1–78:36 (2019). https://doi.org/10.1145/3332184

17. M. Mimura, An improved method of detecting macro malware on an imbalanced dataset. IEEE Access **8**, 204709–204717 (2020). https://doi.org/10.1109/ACCESS.2020.3037330
18. M. Mimura, Using fake text vectors to improve the sensitivity of minority class for macro malware detection. J. Inf. Secur. Appl. **54**, 102600 (2020)
19. M. Mimura, T. Ohminami, Towards efficient detection of malicious VBA macros with LSI, in *Advances in Information and Computer Security—14th International Workshop on Security, IWSEC* 2019, Tokyo, Japan, August 28–30, 2019, Proceedings (2019), pp. 168–185. https://doi.org/10.1007/978-3-030-26834-3_10
20. M. Mimura, T. Ohminami, Using LSI to detect unknown malicious VBA macros. J. Inf. Process. **28**, 493–501 (2020). https://doi.org/10.2197/ipsjjip.28.493
21. E. Quiring, A. Maier, K. Rieck, Misleading authorship attribution of source code using adversarial learning, in *28th USENIX Security Symposium (USENIX Security 19)*. USENIX Association, Santa Clara, CA (2019), pp. 479–496. https://www.usenix.org/conference/usenixsecurity19/presentation/quiring
22. D. Vidyarthi, S.P. Choudhary, S. Rakshit, C.R.S. Kumar, Malware detection by static checking and dynamic analysis of executables. IJISP **11**(3), 29–41 (2017). https://doi.org/10.4018/IJISP.2017070103

IoT-Based Smart Crop Field Monitoring and Protection System from Heavy Rainfall Utilizing Raspberry Pi

G. Dhanalakshmi, M. Anil, and P. Madhavi

Abstract This paper proposes an IoT-based intelligent and secured crop field monitoring framework and also Raspberry Pi architecture is included to prevent the crop fields from heavy rainfall. The proposed overall design is made up of various sensors that are coupled to the raspberry pi. Further, the proposed raspberry pi architecture gets linked to the private cloud. The private cloud provides data storage, which can be examined at any moment by the user. The sound, temperature, and moisture sensors are implemented to collect data that differs from one plant to another. In addition to the sensors, if the moisture limit is reached or the temperature rises, the siphon, which commences the flow of liquids via tubes will be blocked. Nevertheless, the proposed framework has also coordinated the downpour sensor, soil dampness sensor, temperature and pressure sensor for checking the temperature, and soil status along with the raspberry pi and Internet of Things [IoT] architecture. Downpour sensor recognizes the downpour and based on that the dc engine works and shields the downpour caused by heavy rainfall and cautions all the people surrounding the crop field with an alert.

Keywords Raspberry Pi · Crop field monitoring · RFID · Rainfall · Internet of Things · Python · Downpour sensor · Soil dampness sensor · Temperature and pressure sensor

1 Introduction

In this study, an Arduino-based crop field monitoring framework is proposed to not only sense soil moisture content and automatically run solar power motors, but also to protect crops from diseases. The proposed application will help in the detection of the diseases in the initial stage by carefully examining the affected region of plant leaves. In this way, the proposed system will help farmers to take preventive actions to maintain the full growth of plant with high productivity. The proposed system will provide

G. Dhanalakshmi · M. Anil · P. Madhavi (✉)
Department of ECE, Siddartha Institute of Technology and Sciences, Hyderabad, Telangana, India

G. Ranganathan et al. (eds.), *Soft Computing for Security Applications*,
Advances in Intelligent Systems and Computing 1397,
https://doi.org/10.1007/978-981-16-5301-8_5

great help to farmers whose crops have been lost due to under- or over-irrigation, as well as leaf diseases, and will ultimately benefit Indian agriculture and the Indian economy. According to "IoT Based Smart Crop Field Monitoring and Automation Irrigation System", Horticulture plays a vital part in the growth of our agricultural based country, as like an Asian country, Concerns about horticulture are always a hindrance to the country's prosperity. Smart farming, which entails modernizing the conventional horticulture practices as the only way to solve this problem. As a result, the planned concept aims to improve agribusiness by utilizing robotization and IoT advancements. The Internet of Things (IoT) leverages various applications for crop improvement review and selection, water support, and to select the most significant preferred position to enhance the agriculture with distributed computing to streamline the use of water composts in increasing the crop yield and assist in breaking down the barometrical state of the area. As indicated by "Rural Field Observing Utilizing IoT", modernizing the current methodologies can offer a response to the overall issues. As a result, ARDUINO-based with the computerization and Internet of Things upgrades the traditional agricultural activities. This permits field monitoring, decision support, and water system call support to automate the traditional agricultural processes. The execution of precision agriculture can advance the water utilization and uncommon choices that centers on the wellbeing systems based on the field exploitation and distributed computing advances. As indicated by "smart farm monitoring using lora enabled IoT", food scarcity will occur as a result of the progressive growth in population. To decrease the shortage of nourishment, we need to proficiently utilize the water. There will be such a large number of uncertain climatic fluctuations which will make farmers to face difficulties in avoiding natural calamities; nevertheless, the use of IoT applications will reduce the misfortunes. IoT helps to transmit the sensor information from the transmitter hub to collector hub. The hub includes Wi-Fi and MQTT to screen the information and store it in cloud database for administration. As indicated by "Farming Field Monitoring Using IoT" [1], we have many existing issues in horticulture and it requires discovering the modernizing ways with a goal for viable utilization. An ARDUINO-based smart farming targets utilizing the robotization and Internet of Things. Precision agriculture will enhance the agriculture field's water consumption and provide distinctive highlights that revolve around the field's security systems and distributed computing advances. The ideal temperature run is also expected to produce a higher yield.

2 Literature Survey

This part presents the advances of predominant crop management frameworks. In the previous few years, G. Sushant et al. suggested an intelligent agriculture monitoring framework that has been incorporated in a bunch of various sensors. It helped in administering the field climate by considering moistness, temperature, daylight, and elements, which could perceive the living creatures, the assistance in preventing crop destruction. A Wi-Fi module is used to disperse the data as information to the cloud. Fuse is one of the methods considered to execute accurate agribusiness, where customers can only recover the important information to set objectives for increasing the fruitfulness and productivity of their firm. This arrangement has been designed to be tolerable during the entire harvest cycle, starting with adventure illustration, planting, and finally purchasing and storage. Kaa is an open-source IoT platform that is used as a solution for asset planning, failure forecasting frameworks, animal tracking frameworks, and smart metering devices. The farmer can surely use fundamental changes, dilatations, or mixes in this area. From a technical perspective, the Internet of Things (IoT) is a vast collection of connected devices, gadgets, and actuators that operate on the Infobahn and are mostly reliant on various governments and farming procedures. Recently, the farmers have automated the crop monitoring process with IoT equipment. Each of these advancements has its own set of issues, and the farmers are being roused by some of the experts, who have successfully implemented the autonomous testing and identify the plant leaf contamination. At that point, when experts are finished by the nursery observatory framework by utilizing the IoT works of new sorts, Jyotirmayee Dash has proposed to monitor the plants by utilizing the NDVI (Normalized Vegetation Index) for standard sound and non-strong [2]. The utilization of the leaf region record (LAI) by Bauer for yield advancement. [3]. Rao, R. Nageswara proposed a water monitoring framework in the nursing partner, which includes low water volumes [1]. Magnificent utilization of IoT is delivered like Kothiya, Rathinkumar H. farmland mechanical technology framework for the location of the gate crasher and the prevailing water siphon [4]. Mazon-Olivo in Bertha utilized a CEP processor by perception mode that has been zeroed on this framework and informed the top purchaser with customized decisions [5]. Dr. N Suma utilizes GPS designing and IoT to screen water framework insurance, which has been unmatched along with the grass [6]. K. Sreeram enhanced the GSM and DTMF advancements, which are utilized in the agriculture domain [7]. Reza, ZarreenNaowal with auto-application and similarly jute plant with SVM stem disturbance are perceived [8]. Observing and attracting the nature of a faraway nursery using adaptive applications and locations has been proposed by Supun, Athukorala [7, 9]. In initial stages of setting up a nursery, it will be difficult for the new farmers to analyze any plant contamination issue Throughout this work, we appear to be separated throughout the execution of IoT using picture handling to control and view far away nursery structures. This technique conserves energy and provides benefits throughout both the radiant and blustery seasons.

3 Existing System

Due to the increasing complexity of labor and lack of automation in the present crop yield observation framework, the traditional farmers screen all agriculture parameters manually. The crop yield processing was also delayed due to the lack of automation. Due to the lack of protection, farmers will face crop loss during the period of excessive rains. We introduced the smart crop monitoring solution and establish autonomous protection from heavy rainfall by utilizing IoT and Raspberry Pi to eliminate all the complications and make the crop smarter.

4 Proposed System

The proposed system has integrated the rain sensor, soil moisture sensor, temperature, and dampness sensor for observing the temperature and soil status with raspberry and Internet of Things [IoT]. Rain sensor detects rain and activates the DC motor to shield the rain from heavy downpours while also sounding a warning to the people around (Fig. 1).

All of the sensors, including soil moisture, humidity, temperature, and rain sensor, are incorporated into the processing unit, which runs on a 5 volt power source. Crop protection, dc motor mechanism, LCD module, and internet of things module are all incorporated into the raspberry pi module for operation, and the suggested module was developed by using Python programming language. When the rain sensor detects water, it instantly activates the crop protection system and a buzzer sounds to inform the surrounding area.

A. **Raspberry Pi**

The Raspberry Pi is a little single board PC, planned especially for teaching purposes. It's compatible with operating systems like Android, Linux, net BSD, RISCOS,

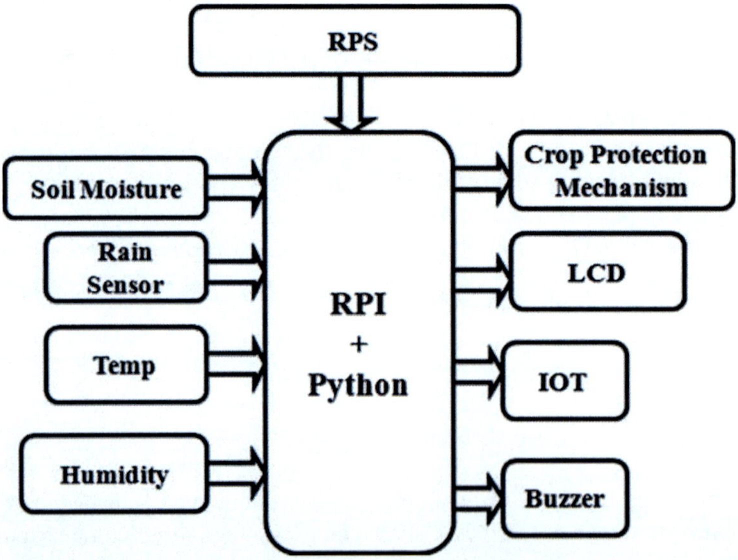

Fig. 1 Block diagram

Window-10, ARM 64, etc. It has a USB power port of 5V, 3A, which can deliver full power to USB devices. The CPU is 1.5 GHz 64/32bit quad core ARM cortex A 72, provided with four RAM slots, two with inbuilt Wi-Fi and other two without Wi-Fi. An SD card is also used to run the Raspbian Buster OS. The input power is rated at 5V DC, 3A (minimum), a power-bank with a 5V DC, 2.4A rating can be used, if no extra peripherals are connected to the raspberry pi, this has to be provided through a USB-C connector or GPIO header pin (Fig. 2).

B. IOT Module

Internet of Things [IoT] model is used for controlling any device or monitoring the device status through internet. This proposed system uses IoT module for considering all data parameters and post it into the cloud called as server. ESP8266 module is used as IoT module, which can operate through Wi-Fi frequency concept (Fig. 3).

The ESP8266 is a Wi-Fi module, which is generally utilized in IoT applications. It comprises a few forms like ESP-01 to ESP-11 ESP-12E and WeMos D1 Mini. This module may be used to create a web worker, send HTTP requests, control yield, read information, and interfere.

C. LCD Display

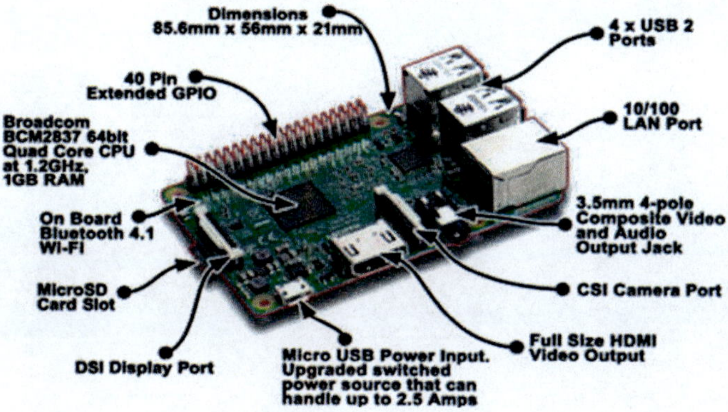

Fig. 2 Raspberry Pi

Fig. 3 ESP 8266

An electronic presentation device is used to deliver an image in a fluid gem show. It is a common showcase module that is used in DIYs and circuits, and it can also read a 16 × 2 presentation with 16 characters per line. It likewise comprises two lines. Each character is shown in 5×7-pixel grid design. Fluid Crystal Display [FCD] is used to show the boundaries for reporting the status of proposed framework. This can show 32 characters with 2 sections. When each sensor is activated, the corresponding message will be displayed in 16 * 2 LCD modules. In this, four data pins are used to transfer the data from micro preprocessor to LCD (Fig. 4).

D. **Buzzer**

A buzzer is a compact speaker, which also serves as an audio signaling device. The effect of piezoelectricity, in which crystals change shape when electricity is applied to them, is used here. Now when the electricity is applied with the correct frequency, the buzzer will make a sound. The buzzers are used in various applications such as alarm devices, timers, etc. The buzzer works as follows: a frequency of 1 KHZ is sent to a certain pin, a delay is used to stop it for a certain number of seconds, and then the buzzer interrupts the signal. This is repeated until the buzzer emits a short beep sound (Fig. 5).

E. **Soil Moisture Sensor Module**

Fig. 4 16X2 LCD

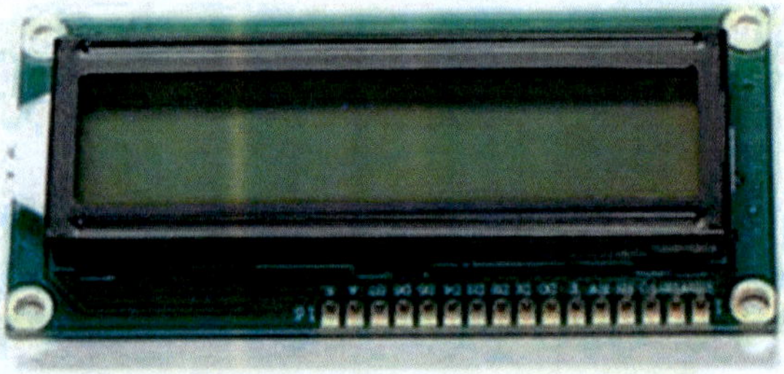

Fig. 5 Buzzer

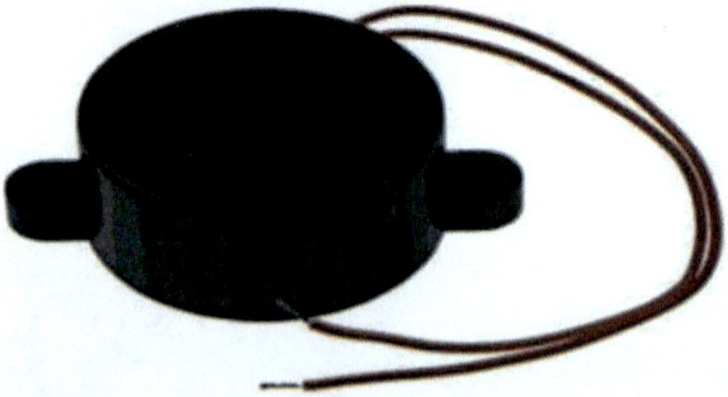

Fig. 6 Soil moisture module

It estimates the dampness content present in the dirt. The dampness sensor instructs the client to water the plants/crops in the same manner and displays this information on the LCD screen. As shown in Fig. 2, the sensor is connected to aluminum tests that are covered in dirt. It calculates the volumetric material in water by concentrating on some of the qualities of soil, such as dielectric constant, electrical opposition, and neutron connection. The sensor is aligned and a certain edge worth of soil's dampness content is selected once the sensor is coated in dirt and the IOT regulator administration is connected to the water system framework. This fully activates the dirt sensor (Fig. 6).

F. **Rain Sensor Module**

This module helps to measure the water content. The rain sensor alerts the user to water the plants/crops as required and displays this information on the LCD screen. The sensor is coupled to aluminum probes that are buried and covered under the earth, as illustrated. It calculates the volumetric material present in water by concentrating on some of the soil attributes, such as dielectric constant, electrical blockage, and neutron connection. The sensor is calibrated and a certain threshold value of soil moisture content is set when the sensor is buried in the soil and the IoT controller service is connected to the irrigation system. This completes the rain sensor's operation.

G. **Temperature Sensor**

In the crop field monitoring system, it is used to calculate the ambient temperature. This temperature sensor has three pins: one for activating the sensor, second pin for power, and third pin for delivering an analog output to provide temperature measurements to the microprocessor (Fig. 7).

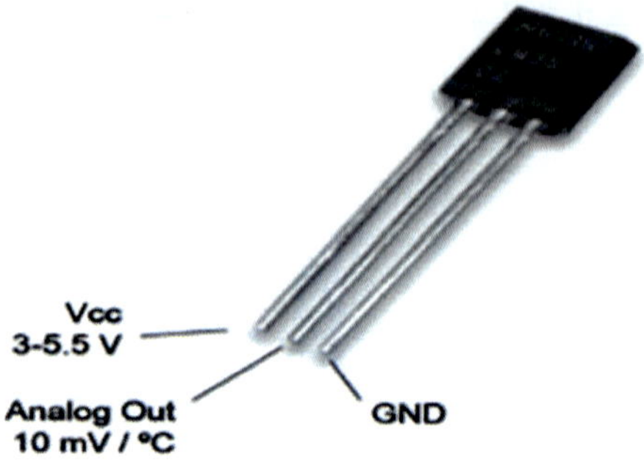

Fig. 7 LM35 sensor

Fig. 8 Motor

H. **Protection Mechanism**

The vibrating motor is a compact coreless DC engine that has utilized admittance to the clients, whenever it is obtained by face acknowledgment approval. A DC engine appeared in Fig. 6 is used to convert electric energy into mechanical energy. The DC motor's principle is that torque is experienced when a current flowing conductor is exposed in a magnetic field and it gains the possibility to move (Fig. 8).

I. **Software**

Python and machine learning based algorithms are used for developing the proposed framework. Python IDE is used to write and compile code. Raspbian operating system will be used for developing the desired model.

5 Results and Discussion

The proposed research work has designed and implemented a smart crop monitoring as well as protection system by utilizing the temperature, humidity, and rain sensor. The proposed system will automatically prevent heavy rainfall by using Raspberry pi as the processor. The hardware implementation is integrated with a temperature sensor, humidity, rainfall sensor, and a protection mechanism (Figs. 9, 10 and 11).

Fig. 9 Yield protecting hardware module

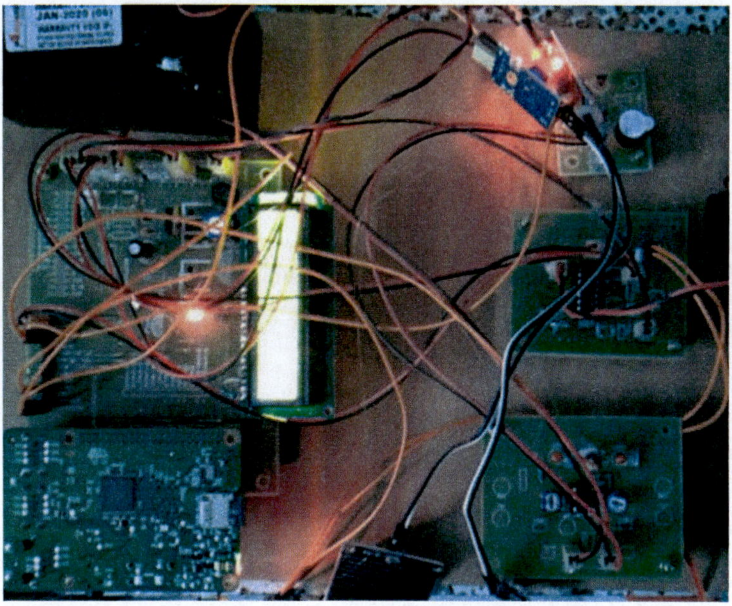

Fig. 10 Yield hardware model

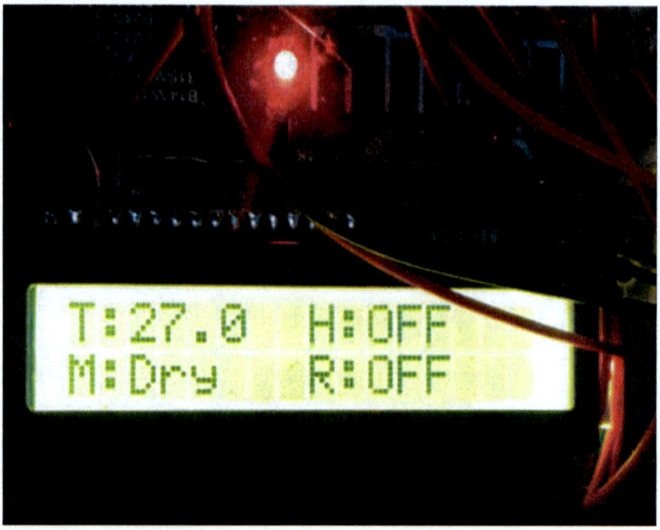

Fig. 11 Yield information on LCD

6 Conclusion

The proposed research work has designed and implemented a smart crop monitoring as well as a protection system with temperature, humidity, and rain sensors. Using the Raspberry Pi as the processor, the proposed system prevents the water from falling down prosperously. The proposed framework has incorporated a downpour sensor, soil dampness sensor, temperature and moistness sensor for checking the temperature and soil status by extensively utilizing the raspberry pi and Internet of Things [IoT] models. The proposed rain sensor detects the rain and depending on that, the dc motor operates for protecting the crops from heavy rainfall and successfully alerts the surrounding people with alarm.

References

1. A. Torres, Collaborative-smart-watering-repository (2015)
2. S. Khot, M.S. Gaikwad, Green house parameters monitoring system using Raspberry Pi and web server. Int. J. Innov. Res. Sci., Eng., Technol. **5**(5) (2016)
3. K. Ganesan, U. Walele, N. Hambire, P. Choughule, D. Oommen, Raspberry-Pi based automated greenhouse. Int. J.
4. Gartner, Gartner says the Internet of Things installed base will grow to 26 billion units by 2020 (2013)
5. M. Baskar, T. Gnansekaran, Multi-model network analysis for improved intrusion tracing towards mitigating DDoS attack. Asian J. Res. Soc. Sci. Humanit., **7**(3), 1343–1353 (2017)

6. A. Goap, D. Sharma, A.K. Shukla, C. Rama Krishna, An IoT based smart irrigation management system using machine learning and open source technologies. https://doi.org/10.1016/j.compag.2018.09.040. **155**, 41–49 (2018)
7. A. Muhammad, B. Haider, Z. Ahmad, IoT enabled analysis of irrigation rosters in the Indus basin irrigation system. https://doi.org/10.1016/j.proeng.2016.07.457 **154**, 229–235 (2016)
8. J. Gutiérrez, J. Francisco Villa-Medina, A. NietoGaribay, M.Á. Porta-Gándara, Automated irrigation system using a wireless sensor network and GPRS module. IEEE Trans. Instrum. Measur.
9. A. Ramadan, A. Mohamed, B.A. Bakry, M.K.H. Elbegawy, Effect of Sprinkler irrigation systems and irrigation frequency on water use efficiency and economical parameters for wheat production. Int. J. Sci. Res. Agric. Sci. **1**(4), 56–66 (2014)

A Systematic Review of Adoption of Blockchain and Machine Learning Technology and Its Application

Sharda Tiwari, Namrata Dhanda, and Harsh Dev

Abstract This research work provides a systematic review on blockchain-based applications present across multiple domains. Also, a comprehensive classification of technology applications across diverse sectors such as healthcare, supply chain, education, automobile industry, etc. has been presented. The twenty-first century has introduced significant innovations and discoveries that have completely changed our lives. In this way, Computers were first introduced followed by the internet, internet of things, and this has recently resulted in the introduction of "Blockchain" technology. This technology is considered as the most important innovation since the internet, and it is also known as the future internet since it creates a decentralized network chain. It has huge potential to impact, disrupt, and change human activities. Also, blockchain becomes an ideal tool for creating a trust-based solution. Nowadays several private, public sector firms, and government organizations are initiating new research work in the blockchain domain. This paper discusses innovative ideas, technologies, and its real-time applications by considering different industries, where blockchain can be applied to create new business opportunities.

Keywords Blockchain · IoT · Machine learning · Security · Data protection

1 Introduction

In the present health care system, patient data is stored in a centralized database, and its content stored in physical memory, allowing anybody with access to the system to modify it. Many challenges may be solved by using a decentralized approach in

S. Tiwari (✉) · N. Dhanda
Department of Computer Science and Engineering, Amity University, Noida, Uttar Pradesh, India

N. Dhanda
e-mail: ndhanda@lko.amity.edu

H. Dev
Department of Computer Science and Engineering, Pranveer Singh Institute of Technology, Kanpur, India

G. Ranganathan et al. (eds.), *Soft Computing for Security Applications*, Advances in Intelligent Systems and Computing 1397,
https://doi.org/10.1007/978-981-16-5301-8_6

IoT networks. Decentralized blockchain technology may be used to build trust-based solutions by increasing data availability, efficiency, and healthcare trust. Adopting a peer-to-peer model can handle billions of transactions between the devices and it also decreases the cost of installation and also maintains the data center and server. Blockchain and machine learning are two new technologies. The first improves the security of information, while the second improves the capacity to predict illness and provide recommendations. Blockchain is recognized for its ability to develop trust-based solutions. Blockchain is distributed ledgers that are managed by a network of users known as miners. Due to their decentralized nature, it has a large amount of computational power [1]. The two basic properties of blockchain are its decentralized peer-to-peer network and security established through a cryptographic function. In 1991 Stuart Haber and W. Scott Stornetta introduced the concept of linking digital timestamp of documents to help users in sending the backdated document or with a later date. Later on, Bayer, Stuart Haber, and W. Scott Stornetta used the Merkle tree to make the model efficient. In 2008, Satoshi Nakamoto improved the concept with Proof of Work (PoW), linking blocks without the requirement of a trusted third-party signature called Blockchain [2]. Blockchain is a growing list of blocks, which are cryptographically connected to ensure that the data are immutable. Each block in the blockchain contains a cryptographic hash on its parent block, timestamp, and other data related to transactions and distributed consensus [3]. The first popular application of blockchain is Bitcoin cryptocurrency, and later it is considered as one of the integrity tools for distributed applications and applied in many applications such as healthcare, Internet of Things, supply chain management, etc.

History of Blockchain.
1990: The concept of hyperledger.
2009: Santoshi Nakamoto created bitcoin and introduced blockchain.
2011–2012: Cryptocurrency deployment in an application.
2012–2013: Current transfer and digital payment system.
2013–2014: Financial market using blockchain beyond cash transaction.
2014–2015: Evolution of smart contract.
2015–2016: Permission blockchain network solution.
2016–2017: Market development and exploration across the industry.
2018–2020: Creating a Blockchain platform for different industries.

Block: Blocks are files, where the data about the application is stored. Data in a block are some or recent transactions made in the network and meta-data such as parent block hash, some or recent transactions made in the network, timestamp, etc. Blocks have header and data parts. Blocks header part is the shaded one and the data part is considered as the tree. Each block is connected to the previous block through the hash pointer.

Genesis Block: The first block of the blockchain is called the genesis block, which does not have any previous or parent block. All users of the respective network should start with the same genesis block to ensure the correctness of the blockchain.

Figure 1 shows how blocks are linked to form a blockchain, it is a simplified architecture of blockchain. each block contains the header and several transactions.

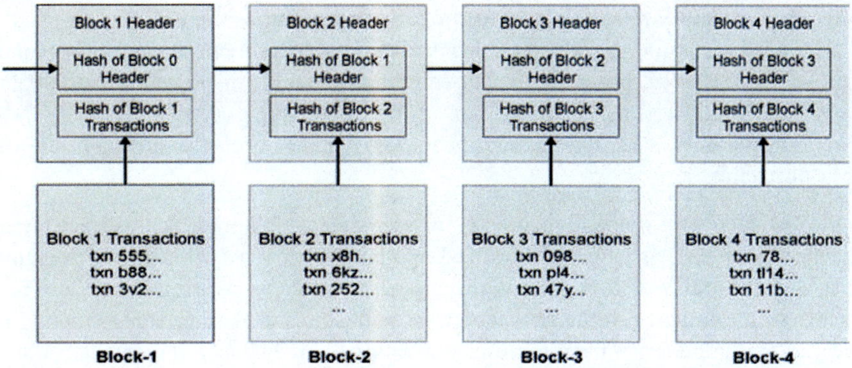

Fig. 1 Blockchain architecture

this transaction is hashed to generate fixed-length output which is added to the next block header, this sequence is continued till the last transaction every subsequent block contains the hash of the previous block. This hash forms a chain so that the linking of one block to another block will form a chain called blockchain.

Type of Blockchain: Every application has its requirement as well as infrastructure. The generic blockchain implementation may not fit into the application. Considering its usage and different application or environment, the following blockchain variants were introduced.

Public or Permissionless Blockchain: This allows any individual to join the network and participate in blockchain activities such as mining, block validation, etc.

Private Blockchain: It is private because of the genesis block they use. Their genesis blocks do not match with any other blockchain. The important note is that any node can connect to a private blockchain if it knows a bootstrap node address to sync, the network id, and genesis file. This node can perform any action on the private network such as mining, make transactions, deploy contracts, etc.

Permissioned Blockchain: It allows only the authenticated users into the network.

Public Permissioned Blockchain: This is similar to private blockchain but has restrictions. For example, a user is added to the network after authentication verification if he/she performs a write operation that is making transactions and mining blocks but anyone can have the read operation that is validation and further have a view on the transaction.

Private Permissioned Blockchain: It has more restrictions than the above. It allows the user only after performing an authentication verification for both read and write operations. The present health care system faces many issues such as

(a) In the present health care system, patient data gets stored in a centralized database and its content gets stored in the physical memory, so anyone with access to the system can modify the data.

(b) IoT devices have a risk of exposure to various security threats.
(c) IoT device processes the recorded data to provide future health care assistants.
(d) Present system has a high cost of installation and it maintains both the data center and server.
(e) In present system due to intermediaries, healthcare cost is too high due to brokerage.

Transparency of medical records is missing due to this prediction system faces many challenges. Using a decentralized approach to IoT network, it solves many issues. Decentralized blockchain is very useful to create trust-based solutions. This brings data availability, efficiency, and trust in healthcare. Furthermore, adopting a peer-to-peer model can handle billions of transactions between the devices and it also decreases the installation cost and maintains the data center and server.

Blockchain includes many properties such as transparency, integrity, redundancy, privacy, trust, and public verifiability. Nowadays many sectors use blockchain as a trust-based solution, wherein the detailed examples are presented in the upcoming sections. The proposed survey is divided into 5 segments. (a) Introduction (b) Literature Survey (c) Potential impact of blockchain on various sectors (d) Ongoing projects on blockchain (e) Future scope and conclusion.

Literature Review

Topic	Model/technology used	Findings	Author conclusion	Future scope
1. A framework for maintaining citizenship record on blockchain	Ethereum and smart contract is used to design framework for decentralized citizenship record	A framework for the problem is established that ensures immutability and public verifiability of all citizen	Ethereum is costly and healthcare requires a permission blockchain so it is required to develop a separate platform for health care in future	Ethereum to a custom blockchain dedicated for the different domain applications
2. Blockchain-based access control model to preserve privacy for personal health record systems	In this proposed work, a blockchain hyperledger fabric model is developed to preserve the privacy for PHR system	Privacy and security of model has been analyzed with 4 threat models such as temper attack, collusion attack, reply attack, and malicious attack	A comprehensive study on the effect on hyperledger fabric configurable parameters may be performed	hyperledger Fabric provides various configurable parameters. In future work. A comprehensive study on the effect of Hyperledger Fabric configurable parameters must be performed
3. Blockchain explorer based on RPC-based monitoring system	Show architecture for blockchain monitoring system	This system monitor blocks and transaction generated in bitcoin and ethereum networks	This system helps to detect illegal transactions in the future and also this system design expands to monitor additional data such as node and network information	In future, the design will be expanded to monitor additional data such as node and network information
4. A systematic literature review of blockchain-based applications: current status, classification, and open issues	This work provides a systematic literature review of blockchain-based applications on multiple domains	This adopt process suggested by Brier and Denyer (2012) and some features of PRISM statement	This describes classification of blockchain-based application, suitability of this technology, and guiding researchers to provide a roadmap of promising research have use	In future author facilitate proper blockchain and the corresponding mechanisms to tailor the blockchain to the actual need of the applications

(continued)

(continued)

Topic	Model/technology used	Findings	Author conclusion	Future scope
5. Blockchain technology and its potential for the construction industry	Smart contracts are used to track the purchase and verify items in construction, where supply chain reduces the counterparty risk	In construction industry two major sectors, where blockchain could have immediate impact to build information modeling and supply chain management	Blockchain technology solves many problems in construction industry including trust issues among parties, asset maintenance, lack of transparency in construction supply chain, reduce administrative paperwork, compliance, and collaboration issue and others	In future, construction industry will soon be able to solve several long-standing problems with high efficiency, transparency, agility, and low cost
6. Singapore blockchain ecosystem 2019	Model of Singapore blockchain ecosystem	This highlights the confluence of blockchain activity in Singapore and combined effort of MAS, IMDA has advanced the use of blockchain across a wide range of use cases	In Singapore, many sectors use blockchain like smart cities, financial services, transport mobility, banking logistics, and supply chain, academia, etc.	N.A
7. A blockchain-based framework for deleting malicious mobile applications in App stores	B2MDF framework	The framework consists of two internal and external private Blockchain forming a dual private blockchain as well as a consortium blockchain for the final decision	B2MF also provides useful features for third parties to develop their antimalware solutions	N.A

(continued)

(continued)

Topic	Model/technology used	Findings	Author conclusion	Future scope
8. Blockchain: opportunities for health care	Deloitte blockchain decision framework	Blockchain solution could address the HIPAA privacy rule by separating and encrypting identity PHI	Deloitte blockchain framework serves as a guide for organizations interest in utilizing blockchain technology	N.A
9. Blockchain technology: is it a good candidate for securing IoT sensitive medical data?	Secure IoT architecture for medical data based on blockchain technology	This system is based on contracts between the providers and consumers of data, which help to create public blockchain and conduct large scale evaluations	Adoption of blockchain model can handle billions of transactions between IoT interconnected devices and results in decreases maintenance and installation cost of data center and servers	In future, author aims to extend the system to work on a public blockchain and conduct large scale evaluations
10. A blockchain-based platform for consent management of personal data processing in the IoT ecosystem	ADVOCA TE Platform, a user-centric solution that allows data subjects to easily manage consents regarding access to their personal data in the IoT ecosystem	Prototype implementation of the proposed platform that supports the main consent's management functionality	This platform helps user to better consent management for incoming requests and current policy of providers	In future, author intends to explore further issues that can significantly enhance the quality of services provided by ADVOCATE such as the development of an intelligence component. It can provide recommendation to data subjects and further protect them from unwittingly exposing their personal data

(continued)

(continued)

Topic	Model/technology used	Findings	Author conclusion	Future scope
11. Nut BaaS: a blockchain-as-a-service platform	BaaS platform called NutBaaS provides blockchain service over cloud computing environments, such as network deployment and system monitoring, smart contracts analysis, and testing	This platform provides transparency of blockchain deployment in runtime and security to smart contracts	Buss platform has 3 advantages it improves security of smart contracts, transparency of blockchain deployment in runtime, reliability of services	In future work can be extended in different aspects: transparency of blockchain deployment and runtime, reliability of BaaS infrastructure (service), security of smart contracts
12. A systematic review of the use of blockchain in healthcare	33 publication was inspected by three reviewers. We have collected data as prompted by our research questions and assessed the publications using the predefined assessment criteria	blockchain technology research and its employment in healthcare are increasing at an unprecedented rate. Current trends of blockchain research in healthcare indicate that it is mostly used for data sharing, health records, and access control	Author concludes current trends of blockchain research in healthcare indicate that it is mostly used for data sharing, health records, and access control, but rarely for other scenarios, such as supply chain management or drug prescription management	In future author want to extend work in field of blockchain to improve medical claims management and supply chain in healthcare medicine apart from securing the medical record
13. A systematic review of blockchain	756 articles related to blockchain on the web of science core collection	Blockchain applications have increased from virtual currencies to financial applications to the entire social realm. Based on its applications, blockchain is delimited to blockchain 1.0, 2.0, and 3.0	Internet of Value can be achieved with blockchain through 5 areas such as corporate and market efficiency protection and privacy issue, Integration of blockchain Fintech, cross section technology implementation	N.A

(continued)

(continued)

Topic	Model/technology used	Findings	Author conclusion	Future scope
14. A survey of how to use blockchain to secure Internet of Things and the stalker attack	Business model for IoT using blockchain in this author uses Markow decision process	Combination of blockchain and IoT is powerful as blockchain provide resilience to attack and the ability to interact with peer in reliable and auditable way	Blockchain provides many opportunities of future IoT system	N.A
15. Secure and efficient data accessibility in blockchain-based health care system	Design a secure efficient data accessibility mechanism	System fulfills the requirements of confidentiality, integrity, and authentication that enhance security to current healthcare	Smart contract management improve healthcare scenario	N.A
16. ChainSplitter: towards blockchain-based industrial IoT architecture for supporting hierarchical storage	Paper presents a blockchain-based IIoT architecture to support verifiable and immutable services	In this hierarchical structure majority of blockchain stores in clouds. Most recent generated block in blockchain can overlay the network	Author plan to work on proposed blockchain-based IoT architecture in more real IoT applications	In future author wants work on the implementation of the proposed blockchain-based IoT architecture in more real IoT applications and thoroughly evaluates its performances such as latency and throughput

(continued)

(continued)

Topic	Model/technology used	Findings	Author conclusion	Future scope
17. BlendMAS: A blockchain enabled decentralized microservices architecture or smart public safety	The proposed platform follows the divide-and-conquer principle to decouple the SPS and security functionality into multiple containerized microservices that are computationally affordable to each individual computing platform	Proposed blend MAS gains the ability to offer a decentralized, scalable, and secured data sharing and access control to the distributed IoT based SPS system	BlendMAS framework is a promising approach to provide a decentralized, scalable, fine-grained architectural security mechanism for SPS system	N.A
18. Blockchain-based charging coordination mechanism for smart grid energy storage units	In this paper blockchain and smart contracts to build a decentralized charging coordination mechanism without the need for a centralized charging coordinator	Knapsack problem to maximize the power delivered to the ESUs	Author found that the costs for deploying and executing the implemented smart contract in Ethereum reasonable for a small scale community of ESUs	N.A
19. Developing a smart storage container for a blockchain-based supply chain application	Smart storage container is developed and connected to an Ethereum-based smart contract	Smart contract is based on multi signature wallet of three parties to process the payment and arbitrate disagreements	Author found many applications on blockchain in logistics	N.A

(continued)

(continued)

Topic	Model/technology used	Findings	Author conclusion	Future scope
20. Investigating quality requirements for blockchain-based healthcare system	Analysis of quality requirements blockchain in healthcare related system	Five quality requirements discussed in this paper such as availability, transparency, security usability, and performance	A consensual and crucial need is a sufficient number of nodes online to meet the requirements for providing the threshold number of validation signatures and maintaining the integrity of the consensus protocol	In future, authors extend the result and provide a deeper analysis on the proposed technical implementation in the literature for blockchain-based platforms for the discussed five scenarios
21. Blockchain enabled E-voting	The blockchain-based distributed–computing platform used by Moscow's active citizen program	Votes are recorded accurately permanently, securely, and transparently so no one can modify or manipulate votes	Blockchain requires much energy to perform authentication and validation and they are slow by using them for national e-voting night and not be practical	N.A
22. Blockchain application for healthcare data management	Author performed a preliminary survey to fill the gap that exists between purely technically focused manuscripts about blockchain	The finding shows that new digital platforms based on blockchain are emerging to enabling fast, simple and seamless interaction between data providers, including patients themselves	Provide a conceptual understanding of the technical foundations of the potential for blockchain technology in healthcare	N.A
23. Blockchain, the perfect data protection tool	Functional component of distributed ledger system	Blockchain is not only secure but also very efficient and faster to scale	Many multinational organizations believe that blockchain is the back bone of IoT	N.A

(continued)

(continued)

Topic	Model/technology used	Findings	Author conclusion	Future scope
24. Blockchain and data protection in the European Union	Paper explains detailed analysis of general data protection regulations (GDPR) in the European Union	This examine implication of distributed ledger technology for GDPR	The study has concluded GDPR can be easier for private and permissioned blockchain to comply with these legal requirements	N.A
25. Blockchain-based big data model for BIM modification audit and provenance in mobile cloud	Designed blockchain-based BIM model for data aggregation including data structure and basic computation consensus also analyzed system parameters such as security length block size, hashing time cost	Proposed method of BIM data organization based on private and public blockchain guarantees to trace, authenticate and prevent tempering with BIM historical data	BcBIM model helps to design applications in BIM ICT system such as accessing BIM cloud securely by engineering machine, constructions robots and wearable helmets in construction area	N.A
26. Machine learning adoption in blockchain-based smart applications: the challenges, and a way forward	In this paper author presents architecture for adoption of machine earning in blockchain based smart applications	Support vector machines (SVM), clustering, bagging, and deep learning (DL) algorithm neural network (CNN), and long short-term memory (LSTM) used to analyze the attacks on blockchain-based network	Author finds combination of these technologies can be applied in several smart applications such as Unmanned Aerial Vehicle (UAV), Smart Grid (SG), healthcare, and smart cities	In future research can be done on several issues related with combination of ML and BT like infrastructure availability, quantum resilience, and privacy issues
27. Improved heart disease diagnostic IoT model using machine learning techniques	The classification algorithm used for training the system are KNN, Naïve Bayes, and decision trees	In this paper, a novel pre-processing technique using machine learning algorithms was proposed to improve the classification of ECG data	This method can be used for identification of normality and abnormality of ECG signals and assist in making early and accurate diagnosis	This work will help in monitoring elderly people and help to take accurate treatment

(continued)

(continued)

Topic	Model/technology used	Findings	Author conclusion	Future scope
28. Converging blockchain and machine learning for healthcare	This paper gives an overview of combining blockchain and machine learning technologies improve healthcare sectors	Blockchain increases the accuracy and efficiency of machine learning models and improve their usability	This mode helps patients as well as doctors	This model can be further extended for inventory to prevent fraud

2 Research Methodology

To provide a clear, reproducible, and scientific literature review of blockchain-based applications, the method instructed and as some options of the PRISMA statement also Guidelines for systematic review in software engineering are adopted.

The methodological approach includes the subsequent steps:

1. Establish the requirement for the review, prepare a proposal for the review, and develop the review protocol.
2. Establish the analysis, choose the studies, assess the standard, take notes and extract information, combine the info.
3. Report the results of the review.

The first step is the systematic process of literature which is used to get an overview of the research area. The systematic mapping also describes the use cases of blockchain-based application literature. it helps to provide an overview of several literatures searched in which how many are useful in this literature review (Fig. 2).

2.1 Definition of Research Questions

It is the first step of the systematic mapping process this includes following

(a) Define use cases of blockchain in the industry.
(b) Blockchain application is developed for these use cases.
(c) Challenges and limitations of blockchain-based applications.
(d) How are these challenges and limitations being addressed.

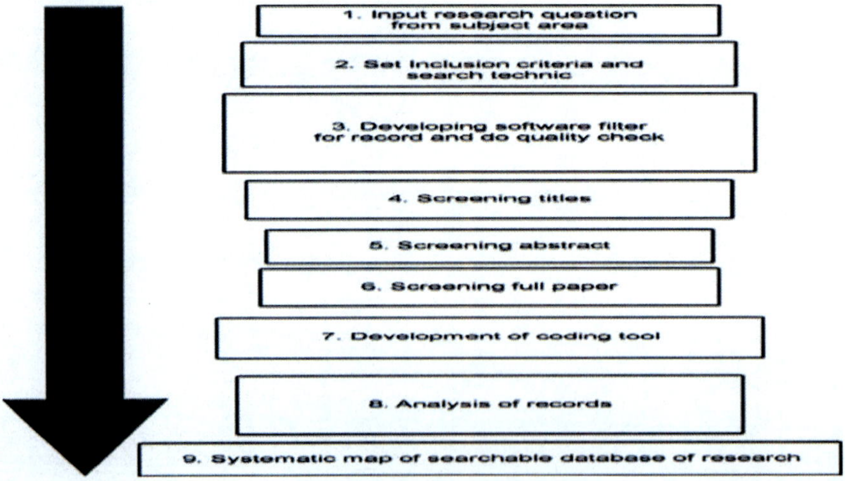

Fig. 2 The process of systematic mapping

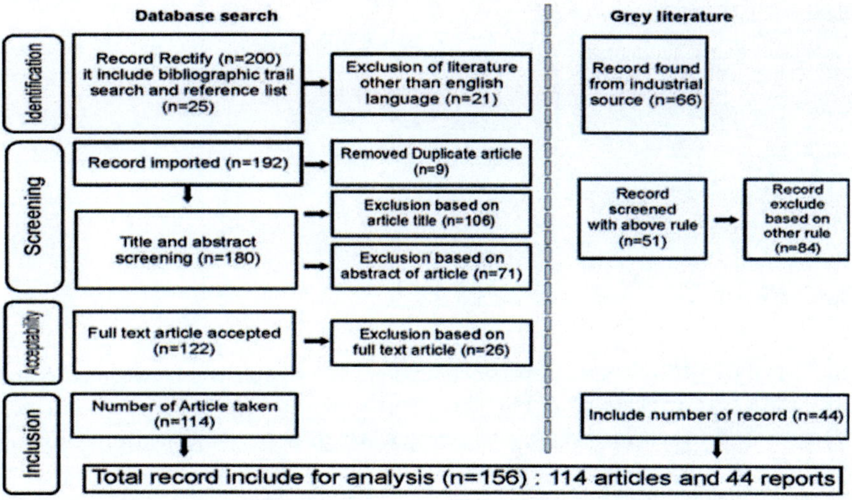

Fig. 3 Flow chart of search strategy

(e) The open research possibilities and future scope.

2.2 *Descriptive Analysis*

To address these primary questions a systematic literature search is done during July 2020. Firstly in the database, all the has term "blockchain" was searched in figure flow chart of the search strategy is shown in which some rules are set by the authors so that inclusion and exclusion criteria is performed. Some papers fulfill the inclusion criteria in which screening is done at various levels first based on the title after that screening is done based on the abstract at the last full-text screening is performed. In the flow chart of search strategy, Gray literature is also rectified through database search. Exclusion criteria are based on the subject area and type of document (Fig. 3).

Inclusion and exclusion Criteria

Selection criteria	Database		Gray literature
Inclusion	Research articles including review papers, journals, book chapters, industrial reports, conference proceedings, survey papers, etc.		English report
Exclusion	Title screening	Article related to blockchain	Non-conventional report to blockchain technology with general applications
	Abstract screening	The article belongs to software needed to the blockchain application	

(continued)

(continued)

Selection criteria	Database		Gray literature
	Full-text screening	Describe technical aspects of blockchain technology	

2.3 Result

Study analyzed 156 research papers and reports papers published between 2015 and 2020.in this we exclude Gray literature. This study provides the current research scenario in Blockchain, majority of selected papers were published in 2018 and 2019. This shows year-wise analysis of literature selected for review it includes Journals, conference papers, book chapters, and Industrial reports analysis shows upward trends in researcher that shows blockchain is very popular in researcher. Firstly blockchain is introduced in Bitcoin and it took several years to show its potential from past 4, 5 years researcher shows its possible application domain rather than finance such as Healthcare, Automobile industry, Governance, Identity management, etc. (Fig. 4).

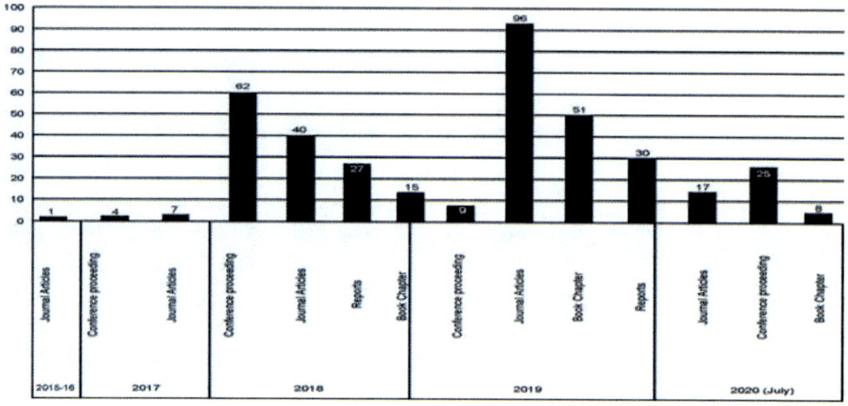

Fig. 4 Literature selected year-wise analysis and type of publication

3 Blockchain Application

3.1 Blockchain Impact on Various Sectors

Banking—The first implementation of this technology in banking is also referred to as the internet of money. first product of blockchain technology is bitcoin.it is a preeminent cryptocurrency it allows people to store their money securely and make a payment without an intermediary such as banks. So that vendors and purchases can transfer money much simple, cheaper, faster and more efficiently fashion means blockchain provides a great infrastructure to the monitory system [4].

Finance—The finance sector is first disrupted by blockchain. last major invention in the finance sector in 1950–60 such as ATM, the credit card that reached massive adoption after this many services are introduced like online banking, P2P lending, PayPal these are innovation just on the surface but core financial infrastructure and all related business and large process remains unchanged over a decade. thus finance sector has largely been stagnant and lagging in terms of innovation. for ex-in capital market transaction takes 3 days to settle. we need big technological revolution such as blockchain is the internet of money, internet of value is brightening innovation to create financial infrastructure will transform global economy it can bring new efficient solution not only for a money transfer but for global trade finance, clearing and settlement, insurance and security trade and many more services and product [5]. Blockchain improves not just the client-facing side of finance but also streamline compliance and regulatory functions like anti-money laundering and know your customer so we can say it has enormous potential to completely reshape the finance sector [6].

Supply chain—Another great implementation of blockchain in supply chain management, in this we use the authentication verification feature of blockchain. Many companies are already actively developing blockchain solution, so monitor and logistics in real-time to track and verify the original goods entering in their supply chain and the condition they kept in this can help the certified the origin and life cycle of the product this build trust in the system, prevent fraud and improved outcomes for supplier and customer [7–9].

Intellectual property—We can use blockchain to verify the authenticity of any assets it can resist any kind of copyright, patent, or trademark.

Financial Assets—This can be extended to all kinds of financial and real assets such as stocks, bonds, derivatives, commodities, and real estate all can be managed and transacted much more efficiently on blockchain in comparison to the traditional system [10, 11].

Digital identity—This is another huge application of blockchain it can be used in the authenticity of people and their personal information [12, 13]. Passport is one of the first areas where identity verification is useful, it is also used in the financial sector such as know your customer, anti-money laundering. Blockchain is good to manage physical and intangible assets and valuable information.

Luxury Goods—Luxury goods such as diamonds and other items are first where blockchain projects are innovating to provide transparency, security, and trust, bringing such efficiency in the luxury good market can unblock a lot of value by reducing counterfeiting and promoting higher ethical standards this is a very exciting innovation [14–16].

Healthcare—Blockchain can transform the health care industry to improve services such as information sharing with doctors, payers, partners, and patients, it also helps to change the way to store clinical information and movement of this within the organization. Decentralized blockchain brings data availability, efficiency, and trust in the healthcare system.

(a) **Patient consent and health data exchange**—Blockchain can bring a revolutionary change in healthcare. In the current healthcare system, patient records are stored in the database and its content is stored in physical memory, anyone who has access to the system can corrupt or modified it easily, this issue resolves with this technology it has lack of central administrator, it provides a complete longitudinal record for everyone better control for the patient over their information through verification consent, it reflects best known medical facts from genomics data to diagnostic medical imaging and can be reliably transferred over the system [17–19].
(b) **Clinical trial management**—Blockchain with EDC (electronic data transfer) produced data that automatically aggregated, replicated, and distributed to help researcher and practitioner for control, provenance tracking with greater audibility [20].
(c) **Outcome-Based Contracts**—Quantifying the worth of medical care is difficult. Accounting for the real costs of all aspects of healthcare can frustrate providers and patients alike, Blockchain-backed, outcome-based contracts tie clinical outcomes to costs. Providers ranging from pharmacies to hospitals can offer healthcare to consumers with a fee-for-value model instead of fee-for-services, and reduce the complexity of records [21, 22].
(d) **Claims adjudication**—There are lots of problems associated with the present healthcare management system such as fraud medical claims. there is a number of fraud related to medicals claims because in today's there is no way to find out how old diseases are, how long patients is suffered from a particular disease in this claimant attempts to misrepresent facts to wrongly gain insurance benefits that they are not entitled to. This does not only mean a deliberate attempt of providing an inappropriate claim but can also include misrepresentations and omissions of critical information at the time of a claim for financial gain. Some more common examples of medical insurance frauds are buying policies in the name of a dead person. non-disclosure of some form of critical illness while applying for insurance or forging documents, Securing policy in the name of a non-existent life assured by forging documents, stating false reason of death to claim additional benefit options. Second is an immediate payment-related problem in the case of bill claims management system.

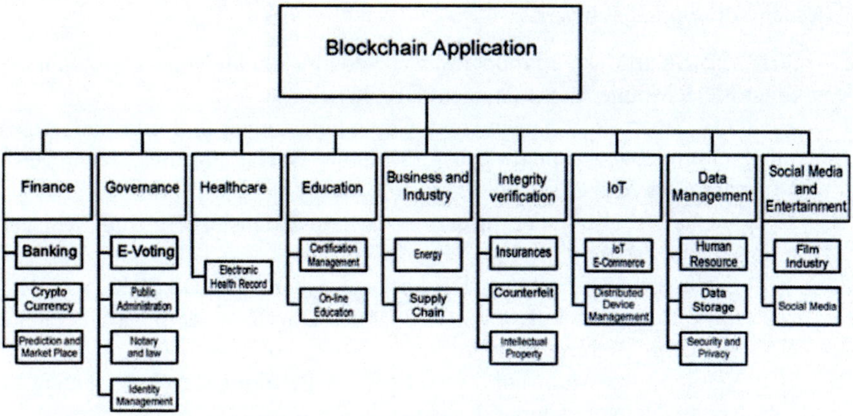

Fig. 5 Abstraction level of different types of blockchain application

Social Media—Steemit is a decentralized blockchain-based social network that uses the native cryptographic token to reward for contributing content and participating on the platform. in this way users are to incentivized and post high-quality content and build in an active community, this helps create strong network key facts, which then boasts the value of the whole network and its native tokens, in this way blockchain create a self-sustainable economic system, which distributes value fairly, i.e., this aspect of blockchain also known as crypto economics [23, 24].

Initial coin offering (ICO)—It can issue digital assets, directly on blockchain and distribute them among global community assets so it is a very controversial aspect, it is a real game-changer for start-ups and venture capital, along with rising capital this alternative provides the important benefits to establish a broad community of user and network effects, which create additional value for start-ups (Fig. 5).

3.2 Ongoing Projects on Blockchain Technology

(a) **Smartmesh**—It opens up a world parallel to the internet. It is an innovative blockchain company that develops internet-free communication and payment services, enabled by the cooperation of its user base. Smart mesh can connect poor billion unconnected people, who still lived in the pre-internet age and under information access, education, and opportunities to improve the standard of living, with smart mesh cell phones atomically connect to form parallel internet. Its foundation is IoT and IoT extends internet access to areas without providing privacy, saving energy, and bring e-commerce to areas without internet access to banks [25].

A major development in smash

1. 2018 Q2 SmartMesh Mobile Raiden Network Light Node Development, SmartMesh Mobile Micro-Payment Development
2. 2018 Q3 SmartMesh Off-internet Communication 2.0, Multi-hop Algorithm Optimization, SmartMesh Off-internet Social Features, SmartMesh Off-internet Payment Protocol Extension
3. 2018 Q4 SmartMesh Token Incentives, Swarm Extension Protocol, Whisper Extension Protocol
4. 2019 Q1 SmartMesh Android SDK Package, SmartMeshiOS SDK Package
5. 2019 Q2 Optimize Mobile Phone P2P Driver, Development of Mobile Phone Network Chip P2P New Standard
6. 2019 Q3 Optimize Multi-hop Mesh Network Performance, Development of New Mobile Network Chip P2P Standard
7. 2019 Q4 Construct SmartMesh Ecosystem.

(b) **Odem Network—**Odem launches 17th February 2018, it stands for the on-demand education marketplace. It is a blockchain-based platform. They directly connect educators with students and try to solve many of the issues in the current education system. The global market of education is fundamentally flawed. Long-standing educational institutions and intermediaries are inflating the cost of education and ensure the global elite access the top educators, By the time most students completed four years degree the world has drastically changed, and the skill they needed to implement also changes. Odem is creating a global marketplace for students, educators, and professionals interested in continuing education to come together and access the impersonal educational course that is relevant for their career scenario. Odem education is no longer a chain of the institution and comes instead of their curriculum with the modern marketplace and receives direct compensation for the course that hosts in the real world. Modern educational marketplace is powered by smart contracts and the Odem token is to make education more affordable, removes expensive intermediaries and application program processes, which are known to be a barrier in higher education, and also facilitate a prompt payment for Odem verified educators and issued an educational certificate to students, who complete the course and get certificate added to the public blockchain, which helps to build professors' reputation and career visibility for the students.

(c) **Cinemadrom—**The International blockchain network of the independent filmmaker and its mission is to create an independent global film industry system, where every first film returns on investment and every second brings profit to creators and investors. Authors, film workers, actors, producers, spectators, investors, miners, traders, and partners are the users of cinemadrom. Its ecosystem contains film distributions, cinema studios, film steaming, regional offices, social networks, cinema agency, and online film market. There are several features of this projects where mainly blockchain is used for film production and distributions, cryptocurrency for film producers, PoW, PoS,

PoT mining algorithm for users, multicurrency wallet for crypto users and filmmakers, multifunction smart contract for financing and film production, allocation of budget for production and marketing, distributions of profit. The main advantage of cinemadrom is the protection of transactions of all participants of film production, the introduction of electronic documentation in the film industry, instead of paper, confirmation of property and copyright for cinema products, an effective insurance system for movie projects, quick and cheap transfer, commission less transboundary transfer, and reduced film production cost [26].

(d) **Goeureka**—Goeureka is to create transparency anomalies to the online booking marketplace by utilizing blockchain technology to directly connect consumers with the hotel and driving commission tend down to zero. Currently, currency in the market is dominated by online travel agents like Booking.com, hotels.com, etc., where they charge up to 30% commission for booking, indirectly overtime increase the price of hotels around the world. There are several benefits to this project such as commission less model for the hotel leading to the best possible booking rate for consumers, automated rebooking features that ensure users obtain the best-guaranteed price, driving interoperability of loyalty points across different hotels groups, provide the ability to users to transfer Fiat currency and cryptocurrency.

(e) **Uhive**—It revolutionizes the social networking concept to deliver the ultimate user experience and new areas of user empowerment. it developed with advanced hybrid blockchain technology. This is a new social network that physical dimension design to mimic real world living satisfies human emotions and needs. Uhive networks consist of 2 worlds civilized; gray world users can choose the group according to their interest. It offers a new interactive user experience among social networks with interphase easy for anyone, to use. Uhive creates a new currency for social networks called Uhive coin, this can be used in the social network explicitly where business marketing trades and consumers purchase for goods and all services will be done. Uhive shares a huge portion of adds revenue with the user, so active user surely gets rewards [27] (Fig. 6).

4 Conclusion

The proposed research aims to present literature reviews on blockchain applications in the current scenario. The special concern of this research work is considered as a combination of blockchain with IoT systems in the healthcare industry to provide fully automated environment for patient monitoring and provide assistance and care to patients in real-time. The further aim of this research work is to design a healthcare framework system in which the data collection is done by different sensors. The IoT system after that by using pubic permissioned blockchain stored patient data. It provides immutable, secure and timeline database also enhances data protection of

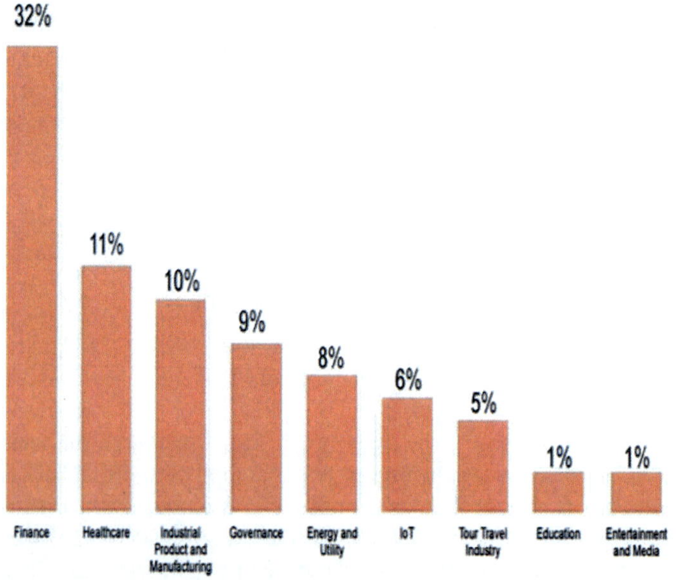

Fig. 6 Development of blockchain in different sectors till July 2020

patient health records. This is reliable data so this can be used by any other external agency, it may be a government agency, insurance company, or any other company. In order to enhance the ability of the system, machine learning has been incorporate for taking quick decisions, which in turn helps in the real-time monitoring of elderly people. This platform fastens the treatment time taken by the hospitals due to the lack of details on the medical history of the patients. We already know that IoT systems are more prone to attack and it cannot be protected easily. Nevertheless, Blockchain enhances security features of patient data and machine learning enhances the performance of the system.

References

1. A. Jha, R.K. Bhattacharjee, A framework for maintaining citizenship record on blockchain, in *BSCI'19*, July 8, 2019
2. S. Vasupongayya, Blockchain-based access control model to preserve privacy for personal health record systems. Hindawi Secur. Commun. Netw. **2019** (2019)
3. C. Lee, H. Kim, S. Maharjan, K. Ko, J. Won-Ki Hong, Blockchain explorer based on RPC-based monitoring system, in *IEEE 2019*
4. F. Casinoa, T.K. Dasaklisb, C. Patsakisa, A systematic literature review of blockchain-based applications: current status, classification and open issues (2018)
5. S. Nanayakkara, S. Peerera, D. Bandara, T. weerasuriya, Blockchain technology and its potential for the construction industry, in *IEEE 2019*
6. Singapore Blockchain Ecosystem 2019

7. S. Homayoun, A. Dehghantanha, R.M. Parizi, K.-K.R. Choo, A blockchain-based framework for detecting malicious mobile application in app stores, in *2019 IEEE Canadian Conference of Electrical and Computer Engineering (CCECE)*
8. R.J. Krawiec, D. Housman, M. White, M. Filipova, F. Quarre, D. Barr, A. Nesbitt, K. Fedosova, J. Killmeyer, A. Israel, L. Tsai, Blockchain: opportunities for health care, in *2016 Deloitee Report*
9. N. Rifi, N. Agoulmine, N.C. Taher, E. Rachkidi, Blockchain technology: is it a good candidate for securing IoT sensitive medical data? Hindawi Wirel. Commun. Mobile Comput. **2018** (2018)
10. K. Rantos, G. Drosatos, A. Kritsas, C. Ilioudis, A. Papanikolaou, A.P. Filippidis A blockchain-based platform for consent management of personal data processing in the IoT ecosystem. Hindawi Secur. Commun. Netw. **2019** (2019)
11. W. Zheng, Z. Zheng, X. Chen, K. Dai, P. Li, R. Chen, NutBaaS: a blockchain-as-a-service platform, in *IEEE 2019*
12. M. Hölbl, M. Kompara, A. Kamišali, L.N. Zlatolas, A systematic review of the use of blockchain in healthcare. Symmetry MDPI J. (2018)
13. M. Xu, X. Chen, G. Kou, A systematic review of blockchain, in *Financial Innovation 2019 South Western University of Finance and Economics*, Chengdu, China
14. E.F. Jesus, V.R.L. Chicarino, C.V.N. de Albuquerque, A.A. de A. Rocha, A survey of how to use blockchain to secure Internet of Things and the stalker attack. Hindawi Secur. Commun. Netw. **2018** (2018)
15. V. Ramani, T. Kumar, A. Braeken, M. Liyanage, M. Ylianttila, Secure and efficient data accessibility in blockchain based healthcare system, *in IEEE 2018*
16. G. Wang, Z.J. Shi, M. Nixon, S. Han, ChainSplitter: towards blockchain-based industrial IoT architecture for supporting hierarchical storage, in *2019 IEEE International Conference on Blockchain*
17. R. Xu, S.Y. Nikouei, Y. Chen, E. Blasch, A. Aved, BlendMAS: a blockchain enabled decentralized microservices architecture for smart public safety, in *2019 IEEE International Conference on Blockchain*
18. M. Baza, M. Nabil, M. Ismail, M. Mahmoud, E. Serpedin, M.A. Rahman, Blockchain-based charging coordination mechanism for smart grid energy storage units, in *2019 IEEE International Conference on Blockchin*
19. J. Hinckeldeyn, J. Kreutzfeldt Developing a smart storage container for a blockchain-based supply chain application, in *2018 Crypto Valley Conference on Blockchain Technology*
20. M. Kassab, J. DeFranco, T. Malas, G. Destefanis, V.V.G. Neto, Investigating quality requirements for blockchain-based healthcare systems, in *2019 IEEE/ACM 2nd International Workshop on Emerging Trends in Software Engineering for Blockchain (WETSEB)*
21. G. Antonial, P. Laptante, S. Counsill, Blockchain-enabled E-voting, in *IEEE 2018*
22. H. Guo, W. Li, M. Nejad, C.-C. Shen, Access control for electronic health records with hybrid blockchain-edge architecture, in *IEEE International Conference on Blockchain 2019*
23. D.V. Dimitrov, Blockchain application for healthcare data management. Healthc. Inform. Res. (2019)
24. A. Nayak, K. Datta Blockchain: the perfect data protection tool, in *International Conference on Intelligent Computing and Control 2017*
25. J. Wang, S. Wang, J. Guo, Y. Duo, S. Cheng, X. Li, A summary of research on bockchain in the field of intellectual property, in *ELSEVIER 2018 International Conference on Identification, Information and Knowledge in Internet of Things, IIKI 218*
26. P.J. Taylor, T. Dargahi , A. Dehghantanha, R.M. Parizi, K.-K. Raymond Choo, A systematic literature review of blockchain cyber security. J. Digital Commun. Netw. (2020)
27. G. Foroglou, A.-L. Tsilidou, Further application on blockchain, in *IEEE International Conference on Blockchain 2015*

Review of Malicious URL Detection Using Machine Learning

G. Pradeepa and R. Devi

Abstract Web URLs are the base for Internet to locate resources uniquely on the Internet. Recent report from poloalto networks states that more 86 thousand malicious URLs were registered during the Covid period between March and April 2020. Cyber-attacks through malicious URL causes lose more than billion in every year. Attacks through malicious URL are the handy way for the cyber-criminals. Systematic approaches are required to detect the malicious URL to prevent the cyber-attacks. Researchers proposed several techniques to detect the malicious URL. But it requires continuous efforts to block newly generated attacks. This paper presents overview of malicious URL detection techniques and recent research works and the issues. Also highlights the research challenges in malicious URL detection, which can help for the future researchers to bring out new solutions.

Keywords Malicious URL · Blacklist technique · Machine learning technique · Deep learning technique · Phishing detection

1 Introduction

URL is the base to locate the resources on the Internet. Cyber-criminals use the URL to attack the victims to steal the user credentials and sensitive information or install the malware in the victim's system. Such specially generated URLs are called malicious URLs. Since March to April 2020, nearly 86 thousand malicious URLs were registered [1], and it is increased year by year. Therefore, it has become a serious threat to Internet users. Examples of attacks are phishing, spamming, drive-by download, etc. Awareness and proper techniques are needed to detect the presence of malicious URLs to protect against such attacks. Most common technique to detect

G. Pradeepa (✉) · R. Devi
Department of Computer Applications, Vels Institute of Science, Technology & Advanced Studies, Chennai, India

R. Devi
e-mail: devi.scs@velsuniv.ac.in

G. Ranganathan et al. (eds.), *Soft Computing for Security Applications*,
Advances in Intelligent Systems and Computing 1397,
https://doi.org/10.1007/978-981-16-5301-8_7

malevolent URLs is blacklisting, rules-based technique and machine learning-based techniques.

Blacklist technique requires huge list of malicious URLs. URLs are blacklisted either by software or manually. Main issues in this technique are blacklist should be updated timely, requires time for search in the huge list of URLs and not able to detect newly generated malicious URLs [2]. Rules-based technique overcome the issues in blacklisting such away that extracting features from URLs both malicious and benign URLs and formularize the threshold values for each identified feature called rule. But assigning proper weight to the features becomes the serious issue in this technique [3]. To overcome all these issues, researchers are using machine learning techniques to classify the malicious URLs from the benign. Most common algorithms used in this binary classification problem are SVM (Support Vector Machine), LR (Logistic Regression), k-NN (k-Nearest Neighbors) and RF (Random Forest). Machine learning model automatically assign weight to the features and perform classification. Continuous research efforts have been devoted to this area to improve the accuracy of detection and speed.

The paper is organized as follows. Section 2 presented the overview of URL and features category. Related works are discussed in Sect. 3. Sections 4 and 5 highlight the research issues and challenges. Section 6 concludes the paper.

2 Background Study

The general structure of URL consists of various parts such as scheme, subdomain, second level domain (SLD), top level domain (TLD), port number, path of the file, parameters and anchors as shown in Table 1.

http://uniedu.example.uk:80/demo/test.jsp?id=333#main

Malicious and benign URLs are similar in structure. To do classification, feature extraction and feature selection are the two primary steps involved in the machine learning. From the large URL, dataset extracts the features and prepares the dataset for analysis. Usually, the large dataset includes both malicious and benign URLs.

Table 1 Various parts of the URL

Elements of URL	Description
http	Scheme which indicates the protocol to access the resource from the server
uniedu.example.uk	Domain name used to locate the resource on the Internet which is sub divided as subdomain (uniedu), secondary level domain (example) and top level domain (uk)
80	Port number
demo	Directory of the resource
test.jsp	Request webpage
id = 333	Parameters are key value pairs and are separated by &
#main	Anchor is the bookmark inside the webpage

Table 2 URL features category

Features	Sub-category	Examples
Lexical features [4]	Count feature Presence feature Ratio BoW (bag of words) features Length feature	Count the dots in the URL, present of numerical IP address or https in the URL, ratio special characters and numbers in the URL, length of the domain name, identifying the presence of malicious words in URL
Host-based features [4]	WHOIS features Domain name features Geo-location features Speed of connection features	Registration, update and expiry information, TTL value, existence of PTR record, location of IP, uplink speed
Content-based features [4]	HTML features JavaScript features Visual features Other features	Number of hyperlinks, images in the page, script length, count the suspicious javascript functions in the page sych as escape(), eval(), link(), etc., compute the visual similarity of the page with other page, count activeX objects
Popularity-based features [4]	Context features Page rank features	Context of the URL shared in the social media, Page rank value

Features which are extracted from the large dataset are grouped into different category as shown in Table 2.

Among the features, lexical features extraction is having no dependency with external servers so no risk on processing. Other features like host-based and popularity-based feature extraction require high dependency with external servers and risk also low. Content-based features extraction is having no dependency with external servers but in high processing risk [5].

Training and testing are second phase of the machine learning technique. Based on the dataset with features, the selected algorithm model will be trained and then will be tested as shown in Fig. 1.

Accuracy is the measure showcase the betterment of the classification model. Accuracy is calculated using Eq. 1.

- Accuracy is the measure of total correct predictions divided by the total of numbers of instances [6].

$$\text{accuracy} = \frac{\text{tp} + \text{tn}}{\text{tp} + \text{tn} + \text{fn} + \text{fp}} \tag{1}$$

where tp = true (correct) positives, tn = true (correct) negatives, fp = false(incorrect) positive and fn = false (incorrect) negatives.
Other measures are precision, recall and F1-Score as shown in Eqs. 2–4 respectively.

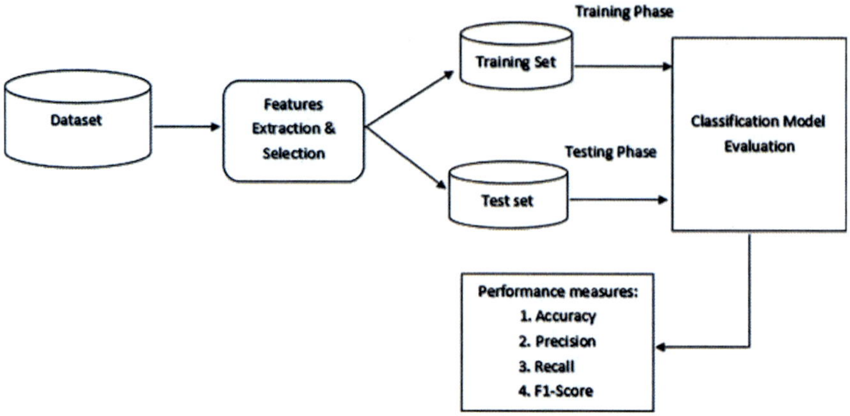

Fig. 1 Malicious URL detection process

- Precision (PR) is the measurement of number of true positive predictions divided by total number of positive predictions [5].

$$\text{Precision(PR)} = \frac{\text{tp}}{\text{tp} + \text{fp}} \tag{2}$$

- Recall (RC) is the measure of number of true positive predictions divided by total number of positives [5].

$$\text{Recall (RC)} = \frac{\text{tp}}{\text{tp} + \text{fn}} \tag{3}$$

- F1-score is a measure of harmonic means of precision and recall [5].

$$F1 - \text{Score} = \frac{2 * (\text{RC} \; * \text{PR})}{(\text{RC} + \text{PR})} \tag{4}$$

3 Related Works

Researchers contributed many approaches to detect malicious URL using machine learning techniques. Some of the considerable research work presented in this section. Harshal et al. [3] presented an approach to detect the phishing URL. Data were collected from phishtank and DOMZ. Phish List, BoW, word segmentation with n-grams, numerical and lexical features were extracted from the URLs and tested. Result shows highest accuracy in word segmenation+phisy list + numerical features in logistic regression algorithm.

Cho et al. [7] proposed a detection system which has 16 new features and adopt the big data technology for scalability of the system. URLs were collected from Phishtank, URLhaus and alexa. SVM and Random Forest algorithms were used for classification and the result shows accuracy of 96 % in Random Forest with 100 trees and 91% in SVM with 100 iteration.

Ripon et al. [8] presented a study report on malicious URL detection using machine learning. SVM and Random Forest algorithms were used for classification. Training and test data are divided into 60:40, 70:30 and 80:20 ratios. Test result shows accuracy of 92% in Random forest and 89% in SVM with 10 iterations.

Ozgur et al. [9] proposed phishing detection system using machine learning. Data are collected from different sources such as phish tank, Yandex search API, Ebbu phishing dataset 2017. Experimental result shows accuracy of 97.9% in Random Forest, 93% in adaboost, 94% in Kstar, 96% in k-NN, 95% in SMO and 96% in Naïve Bayes algorithms with only NLP-based features.

Kamel et al. [10] presented system for detecting phishing URL in twitter. UCI phishing dataset and tweeter. Initially the URLs are extracted from the tweets and then compared with blacklisted URL for detection. If it fails, then extract the features from the URL and apply machine learning algorithm to classify the URL either malicious or legitimate. Result shows the proposed system attain the accuracy of 90% in Logistic Regression, 93% in SVM and 96% in Random Forest.

Ammara et al. [11] proposed a system for phishing URL detection using machine learning algorithms. Data were collected from kaggle. 32 features were extracted from the dataset and grouped in two set such as N1 (weakest) and N2 (strongest). After normalization, classifier models were trained with features dataset. Begging, Naïve Bayes (NB), k-NN, SVM, RF, NN algorithms were used with PCA for feature reduction. 10-fold validation was performed to finalize the accuracy. The stacking with NN+RF+Bagging yields the accuracy of 97.4%.

Yogendra et al. [12] presented lightweight machine learning approach to detect phishing attacks. Data were collected from UNB dataset and Kaggle dataset. 24 lexical features extracted from the dataset and are converted as feature vector. Using feature vector, the classification models were trained and tested. Different classification algorithms were used (RF, NN, SVM, LR, k-NN) to classify the phishing URL. The result shows the attained accuracy of 95 and 99% in RF algorithm by using Kaggle dataset and UNB dataset respectively.

Gopinath et al. [13] proposed a method for malicious URL detection using machine learning. Dataset includes malicious and benign URLs collected from different sources and extracts the different features such as blacklisted IP addresses, DNS-based features (host-based), web-based features (content-based) and lexical features in addition to the blacklisted domain. Logistic regression (LR) algorithm was used for classification and achieved accuracy of 60%.

Ferhat et al. [14] presented a method to classify the malicious URL using machine learning algorithms. Data were collected from UCI Machine Learning Repository (https://archive.ics.uci.edu/ml/datasets/URL+Reputation). Lexical and host-based features were extracted and select the most suitable features by using PCA method. Random Forest and Gradient Boost classifiers were used for classification and achieved highest accuracy of 98.6% in Random Forest.

Patil et al. [15] presented methodology to detect malicious domain name and type of attacks based on multiclass classification. Nearly 42 new features were identified and used in this method. Dataset collected from alexa top sites, phishtank and jwspamspy. Different features sets were used which includes URL features (lexical), content-based features, host-based features and short URL features. Different SVM algorithms (OVA-SVM, OVO-SVM) and Multiclass online confidence weight learning were used for training and testing. The result shows that confidence weighted learning classifier gives accuracy of 98.44% in multiclass setting) and 99.86% in binary setting.

Mohammad et al. [16] presented a method for phishing URL detection using machine learning algorithms. Data were collected from Kaggle Web site. Lexical, host-based, content-based and popularity features were extracted from the dataset. It also uses the algorithm PCA for feature selection. Decision Tree (DT) and Random Forest (RF) algorithms were used for classification and achieved highest accuracy of 96.9% in Random Forest.

Arun et al. [17] proposed a method for phishing URL detection using machine learning algorithms. Data were collected from Irvine Machine Learning Repository which includes 1353 URLs (phishing and benign). Lexical, host-based features were extracted and used for classification. Decision Tree (DT), SVM, Naïve Bayes (NB) and Neural Network (NN) classifiers were used for classification and achieved accuracy of 91.5%, 86.69%, 86.14% and 84.87% respectively (Table 3).

4 Observed Issues

- Generating feature sets using different categories requires more processing time.
- Most of the research works includes large number of features to train and test the classifiers but only few researchers focus on the fitness of the features in the feature set to perform better classification.
- Based on the independent features extracted from dataset, classification accuracy may vary for different machine learning algorithms. Sometime few features are highly correlated which affects the prediction accuracy.

5 Research Challenges

Even though many techniques were proposed by the researchers, still some research problems are yet to address [4]. They are

- Handling huge volume of data in real time is the challenging task [4]
- Difficulty of extracting features some incur cost and time [4]

- Difficulty in feature selection, transformation, and reduction due to the high dimensional [4]
- Handling algorithmically generated URLs and short URLs [13]
- Detection of homographic domain names requires deep word processing [13].

Table 3 Summary of the existing research work

Authors and year	Features	Accuracy
Harshal et al. [3] and 2019	Lexical features BoW features Numerical features	
Cho et al. [7] & 2020	Lexical features Host-based features	RF: 96% SVM: 91%
Ripon et al. [8] & 2019	Lexical Features Host-based features	RF: 92% SVM: 89%
Ozgur et al. [9] & 2020	NLP features Word vector Hybrid	RF: 97.9% Adaboost: 93 Kstar: 94% k-NN: 96% SMO: 95% Naïve Bayes: 96%
Kamel et al. [10] & 2020	Lexical features Host-based features Content-based features Popularity-based features Lexical features Host-based features Content-based features Popularity-based features	LR: 90% SVM:93% RF:96%
Ammara et al. [11] & 2020		NB: 75%. k-NN: 95% SVM: 95%. RF: 97% Bagging: 97%, NN: 97% NN + RF + Bagging: 97.4% k-NN + RF + bagging: 97.2%
Yogendra et al. [12]	Lexical features	Kaggle dataset RF: 95%, LR:88% k-NN:94%, SVM:91% NN:93% UNB dataset RF: 99% , LR:92% k-NN:99%, SVM:98% NN:99%

(continued)

Table 3 (continued)

Authors and year	Features	Accuracy
Gopinath et al. [13] & 2020	Blacklisted IP addresses, DNS-based features (host-based), web-based features (content-based), and lexical features in addition to the blacklisted domain	LR: 60%
Ferhat et al. [14] & 2021	Lexical features, host-based features	RF: 98.6% Gradient Boosting: 96.5%
Patil et al. [15] & 2018	URL features (lexical), content-based features, host-based features, short URL features	Confidence weighted learning classifier 98.44% in multiclass setting 99.86% in binary setting
Mohammad et al. [16] & 2020	Lexical features, host-based features Content-based features, Popularity features	DT: 91.9% RF: 96.9%
Arun et al. [17] & 2019	Lexical features Host-based features	DT: 91.5% SVM: 86.69% NB: 86.14% NN: 84.87%

6 Conclusion

Malicious URL-based attacks are the serious threat to world. Attackers easily play with victims and steal the sensitive information. To detect these kinds of attacks requires awareness and systematic approach to detect and black the malicious URLs. This paper presents the details of URL and its features. Machine learning techniques requires proper feature sets extracted from large dataset and are categorized into different type. Review of recent research work highlights the data source, feature set used, functionality of different methods and achieved accuracies. Finally, Section 5 and 6 present the issues and research challenges which pave the path for new research works in future.

References

1. https://unit42.paloaltonetworks.com/how-cybercriminals-prey-on-the-covid-19-pandemic/
2. A. Joshi, L. Lloyd, P. Westin, S. Seethapathy, Using lexical features for malicious url detection—a machine learning approach, ArXiv (2019)
3. H. Tupsamudre, A.K. Singh, S. Lodha. Everything ıs in the name—a URL based approach for phishing detection, in *Cyber Security Cryptography and Machine Learning*, vol 11527. Lecture Notes in Computer Science (Springer, Cham, 2019)
4. S. Doyen, L. Chenghao, H. Steven, Malicious URL detection using machine learning: a survey. arXiv (2017)

5. Andrea, *Detecting Malicious URLs*, Master thesis, Masaryk University (2019)
6. H.M.J. Khan, Q. Niyaz, V. Devabhaktuni, S. Guo, U. Shaikh, Identifying generic features for malicious URL detection system, in *IEEE 10th Annual Ubiquitous Computing, Electronics & Mobile Communication Conference*, New York, 2019
7. C.D. Xuan, H.D. Nguyen, T.V. Nikolaevich, Malicious URL detection based on Machine learning. Int. J. Adv. Comput. Sci. Appl. **11**(1) (2020)
8. R. Pargiri, H. Katari, R. Kumar, D. Sharma, Empirical study on malicious URL detection using machine learning, Lecture Notes in Computer Science, in *International Conference on Distributed Computing and Internet Technology*, vol 11319 (Springer, Cham, 2019), pp. 380–388
9. O.K. Sahingoz, E. Buber, O. Demir, B. Diri, Machine learning based phishing detection from URLs, Expert Syst. Appl. **117**, 345–357 (2019)
10. K.A. Djaballah, K. Boukhalfa, Z. Ghalem, O. Boukerma, A new approach for the detection and analysis of phishing in social networks: the case of Twitter, in *IEEE Seventh International Conference on Social Networks Analysis, Management and Security*, pp. 1–8 (2020)
11. A. Zamir, H. Ullah Khan, T. Iqbal, N. Yousaf, F. Aslam, A. Anjum, M. Hamdani, Phishing web site detection using diverse machine learning algorithms. The Electron. Lib. **38**(1) (2020)
12. Y. Kumar, B. Subba, A lightweight machine learning based security framework for detecting phishing attacks, in *IEEE—International Conference on COMmunication Systems & NETworkS (COMSNETS)*, pp. 184–188 (2021)
13. G. Palaniappana, Sangeetha, B. Rajendrana, Sanjaya, S. Goyala, Bindhumadhava, Malicious domain detection using machine learning on domain name features, host-based features and web-based features. Procedia Comput. Sci. **171**, 654–661 (2020)
14. F.O. Catak, K. Sahinbas, V. Dörtkardeş, *Malicious URL Detection Using Machine Learning* (IGI Global, 2021)
15. D. Patil, J. Patil, Feature-based malicious URL and attack type detection using multi-class classification. The ISC Int. J. Inf. Secur. **10**(2), 141–162 (2018)
16. M.N. Alam, D. Sarma, F.F. Lima, I. Sah, Rubaiath-E-Ulfath, S. Hossain, Phishing attacks detection using machine learning approach, in *Third International Conference on Smart Systems and Inventive Technology (ICSSIT)*, Tirunelveli, India, pp. 1173–1179, 2020
17. A. Kulkarni, L. Brown, Phishing websites detection using machine learning. Int. J. Adv. Comput. Sci. Appl. **10**(7) (2019)

Securing IoT Using Artificial Intelligence and Feature Engineering

K. Geetha and S. H. Brahmananda

Abstract IoT is one of the promising technology which is changing the prospects into a new dimension. The adoption of new technology will uplift the contemplation. Every useful mechanization will have its own pros and cons. One of the major concerns with IoT is privacy and security. Designing a secure IoT ecosystem will increase the rate of embracing of new technology in day-to-day life. Securing the IoT systems against cyber-attack is a mandate. Artificial intelligence is creating a revolution in the technology by its ability. In this paper, we are proposing a methodology to design smart and secure IoT systems which can think and act efficiently when it senses something queer. The goal is to come up with secure IoT systems which can detect and mitigate cyber-attacks. Here, we propose machine learning model to detect botnet attack in IoT network using N-BaIOT benchmark dataset. Our model is able to classify network flow as benign or attack with an accuracy of 99%.

Keywords Artificial intelligence · Botnet · Cyber-attack · IoT · Security

1 Introduction

IoT is an emerging technology which is reforming world into a new era which involves technology everywhere. IoT is enabling non-living things to think and act according to the changing environment around them. This is possible since all these things are connected, controlled, and coordinated over the Internet. Advancements in machine learning and deep learning has enabled us to design smarter & efficient IoT systems which can sense good and bad happening around them and take decision as expected in order to preserve the privacy of user sensitive information. Cyber-attack is major concern of IoT systems since these tiny devices have ample of vulnerabilities [1]. Syed Rizvi listed IoT architecture and various devices connected based on domains. Author also listed IoT device vulnerabilities and major IoT security threats. The author proposed DDOS attack and Botnet attack are the major threat

K. Geetha (✉) · S. H. Brahmananda
Department of CSE, GITAM School of Technology, Bengaluru, India
e-mail: gkallesh@gitam.edu

G. Ranganathan et al. (eds.), *Soft Computing for Security Applications*,
Advances in Intelligent Systems and Computing 1397,
https://doi.org/10.1007/978-981-16-5301-8_8

to IoT devices [2]. In this, the author listed security issues at each IoT layer and also listed various deep learning algorithms which can be implemented to design intelligent security systems [3]. In this paper the author reviewed various ML/DL models can be implemented for astute smart IoT security systems.

Designing secure IoT systems involves designing capabilities for these IoT devices the ability to identify cyber-attack. This can be achieved through machine learning and deep learning algorithms. Botnet attack is a major concern in IoT devices. If the IoT systems are enabled with ability to detect known Botnet attacks using ML and DL algorithm, the systems can be designed to prevent such attacks. This paper is abiding list major botnet threats of IoT devices and possible solutions to detect and mitigate them.

2 Related Work

Much research has been carried out to detect and prevent botnet attacks [4]. Huy-Trung tried to detect IoT botnet on graph-based convolution neural network method. The author extracted Printable String Information (PSI) from ELF file of known malwares and generated Control Flow Graph (CFG). These CFGs are classified as benign or attack using Deep Graph Convolution Neural network (DGCNN). Since most of the malwares are encrypted and obfuscated the control flow graphs may not represent actual malware. Moreover, they implemented static analysis of malwares [5]. P Torres et al. tried to design the detection model based on Recurrent Neural Network (RNN) by behavior of connections. They tried to construct a binary vector of behavior model of known connections. Then implemented LSTM-based model to classify the connection as attack or normal. For their work, they considered dataset of network traffic captures taken from CVUT university campus networks [6]. A. Guerra-Manzanares et al. demonstrated machine learning-based models for IoT botnet detection. They implemented K-NN and Random Forest algorithm and demonstrated hybrid feature selection technique. The implemented various filter models for dimensionality reduction to reduce the computational load on machine learning models.

3 Botnet and Types of IoT botnet

Botnet is a network of bots which includes compromised devices. In the Botnet, the attacker, called the botmaster, controls the devices in the compromised network through the Command and Control (C&C) infrastructure through a bi-directional network communication channel. A botnet can launch thing, such as DDoS, Keylogging, Phishing, Spamming, Click Fraud, Identity Theft, and even the spread of other bot malware. A botnet can start several operations. The botmaster will have the complete control of all the compromised bots.

An IoT botnet is a group of compromised malwares infected IoT systems, such as cameras, switches, Webcams, wearables, etc. This malware enables a device-controlled attacker to perform tasks similar to a standard botnet. The attacker is targeting these IoT devices since the devices are having many vulnerabilities. Unlike conventional botnets, compromised IoT devices are continuously concentrating on some more devices to propagate their malware. While a conventional botnet can consist of tens or thousands of units, a larger IoT botnet with hundreds of thousands of affected units. Table 1 provides a list of different forms of IoT botnet.

4 Dataset

For demonstrating the credibility of ML and DL algorithm a well-structured dataset is very essential which includes the data of various attack scenarios which mimic the real attacks. In this, we are using N-BaIOT dataset [7]. This dataset focuses on two main IoT botnets Mirai and Bashlite and includes 10 classes of attacks (Ack, Scan, syn, UDP, UDPplain under Mirai attack and Combo, Junk, Scan, TCP, UDP under Bashlite attack). The dataset was recorded by using IoT devices such as doorbell, thermostat, baby monitor, security camera and webcams. The dataset includes categories and subcategories of various attacks to aid the detection as benign or attack class. Table 2 lists the dataset details.

5 Proposed Methodology

The proposed method is based on supervised machine learning technique. Machine learning and deep learning methods produce promising results for the prediction of zero-day risks. We designed linear Support Vector Machine (SVM) [8] based classifier for classifying the network communication as benign or attack. SVM is one of the most attention receiving machine learning algorithms currently. SVM can also be used for nonlinear data classification and can give better accuracy in analysis compared to other Machine Learning algorithms. SVMs are built on the notion of finding a hyperplane that best divides a dataset into two categories, as seen in Diagram 1.

The distance from each set is called the boundary between the hyperplane and the closest data point. The purpose is to pick a hyperplane with the highest margin between the hyperplane and any point in the training data, thus providing greater potential for new data to be correctly categorized. Maximizing the distances between the closest data point (both classes) and the hyperplane will help us determine the hyperplane correctly. The distance is called margin.

The dataset consists of 49,549 benign and 968,760 attack samples in the form of csv files [7, 9]. The attack sample includes the most discussed IoT botnets Mirai and Bashlite. The proposed methodology includes data pre-processing, model training

Table 1 Botnet types

Hydra (Chuck Norris and Tsunami)	Hydra was released in 2008 as an open-source botnet framework and infected routers. It also had DDoS and propagate features, which are the staples of potential IoT botnet malware families.. Several other malware variants of Hydra are Psyb0t discovered in 2009. It targeted routers and DSL modems, spreading to IoT devices through Telnet and SSH access
Aidra	Aidra was developed in 2012 and is defined by its own author as an IRC-based mass router scanner. It spreads by looking for free telnet ports that can be reached by default credentials
BASHLITE (also called gafgyt, Torlus, Lizkebab)	BASHLITE was first discovered in 2014 when it used ShellShock or the bash vulnerability to compromise Linux-based. IoT devices and launch DDoS attacks. Its source code was leaked in 2015. In 2016, it bad made a resurgence by infecting 1 million devices such as digital video recorders (DVRs) and cameras, routers, and Linux servers. BASHLITE is still active to date
LuaBot	T.naBot is tbs -fi-rsr malware variant written in the T.IIA prcigrafnmintT language that targeted Linux servers and IoT devices. It can also conduct application- let nel DDoS attacks
Remaiten (KTM-RM)	Remaiten was released in 2016. It combines the DDoS features of Tsunami and an improved version of the scanning abilities of BASHLITE. Once Remaiten has gained access to IoT applications, it installs the executable bot for several of the built-in architectures included in these devices, enabling it to initiate attacks tailored to particular architectures
Mirai	Mirai was discovered by MalwareMustDie researchers in August 2016 and had become the most herd IoT malware in existence. It gained attention by its massive DDoS attacks on well-known Web sites. Mirai's intrusion on Dyn, a controlled DNS service provider, triggered critical losses to Dyn's customers Web sites including Amazon. Twitter, Spotify, Reddit. Mirai is most powerful botnet malware and called as malware that "brought down the Internet"
Linux/IRCTelnet (NewAidra)	Its components are largely built from many IoT botnet predecessors. It takes parts from Aidra (root code), Tsunami (IRC protocol), BASHLITE (infection techniques), and Mirai (credential list). It targets DVRs and IP cameras

(continued)

Table 1 (continued)

Persirai	In 2017 Trend Micro discovered Persirai. It was one of the botnet malware families fighting for control over insecure IP cameras and was targeting over 1000 models. Using three vulnerabilities, the malware will gain, user credentials and execute order injections independent of strong password

Table 2 Dataset samples

Sample	Attack type	Attack class	No. of samples
Benign			49,549
Attack	Mirai attack	Ack	102,196
		Scan	107,686
		Syn	122,574
		Udp	237,666
		Udpplain	81,983
	Bashlite attack (gafgyt)	Combo	59,719
		Junk	29,069
		Scan	29,850
		Tcp	92,142
		Udp	105,875

Table 3 Results

Accuracy	0.993476
Precision	0.995853
Recall	0.933494
F1-score	0.962354

and attack prediction. During pre-processing the data is labeled as zero for benign data and one for attack data which is essential for supervised learning algorithms (Diagram 2).

The model is implemented by using scikit-learn kit. For support vector machine implementation the kernel used is linear. Every model should be evaluated for its ability to classify data into different classes. In this regard, the dataset is divided as 70% for training and 30% for testing. Once the model is trained using 70% of training data, its performance is evaluated by considering remaining 30% of the test data and the results are recorded. The trained model is able to classify the network flow as benign or attack with an accuracy of 99%. The results obtained are discussed in results and discussion section.

6 Results and Discussion

Following parameters are considered to evaluate the effectiveness of proposed model.

	Positive	Negative
True	True Positive: the number of data items correctly classified to the positive class	True Negative: the number of data items correctly classified to the negative class
False	False Positive: the number of data items wrongly classified to the positive class	False Negative: the number of data items wrongly classified to the negative class

Accuracy: The metric that offers the cumulative number of accurate forecasts made from all the models' observations.

$$\textbf{Accuracy} = \frac{\textbf{TP} + \textbf{TN}}{\textbf{TP} + \textbf{FP} + \textbf{TN} + \textbf{FN}}$$

Precision: The metric is the model's ability to provide optimistic output estimates the likelihood that the positive prediction of the model is positive, can be interpreted as.

$$\textbf{Precision} = \frac{\textbf{TP}}{\textbf{TP} + \textbf{FP}}$$

Recall: The ratio of the Tr-Pos (True Positive) divided by the total of Tr-Pos (True Positive) and Fl-Neg (False Negative) predictions.

$$\textbf{Recall} = \frac{(\textbf{Recall} \times \textbf{Precision})}{(\textbf{Recall} + \textbf{Precision})}$$

F1-score: Provided by the harmonic mean of precision and recall strides. If reference groups are unevenly distributed, this may be a safer estimate.

$$\textbf{F1} - \textbf{Score} = 2 \times \frac{(\textbf{Recall} \times \textbf{Precision})}{(\textbf{Recall} + \textbf{Precision})}$$

The results obtained by training the support vector machine model is as shown in Table 3.

The results obtained has been plotted in below confusion matrix. From the diagram, it has been noted that our model is able to correctly classify as benign or attack with an accuracy of 99%. For the proposed model we got precision is 99%, recall 93% and F1-measure is 96% (Diagram 3).

7 Conclusion

This paper proposes an IoT botnet detection mechanism using supervised machine learning technique. The proposed method is giving the prediction accuracy of 99%. IoT botnet threat are evolving drastically and most of them remain undetected for a longer time. We need specialized techniques to detect them and prevent them. Our study shows that using machine learning algorithm to classify network behavior as attack or benign is giving promising results compared to traditional detection techniques. Efficient implementation of artificial intelligence techniques for botnet detection and mitigation can systematically prevent the cyber-attacks on IoT networks. Further study may be considered by evaluating the performance of deep learning algorithm for getting better prediction accuracy of different categories of botnets as well as different benchmark dataset need to be considered for evaluating performance of proposed machine learning and deep learning algorithms.

Diagram 1 SVM classification

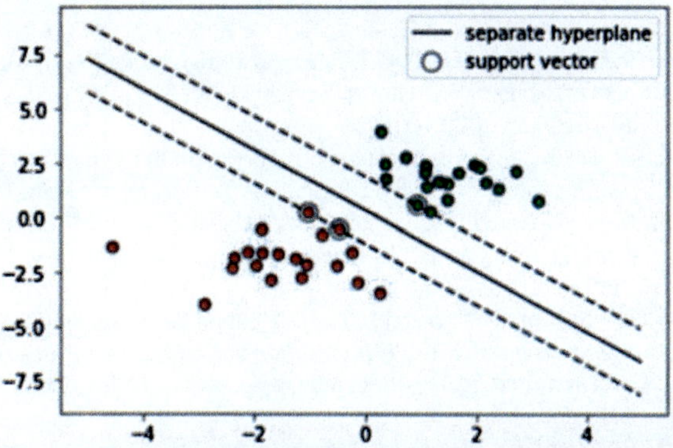

Diagram 2 Proposed model

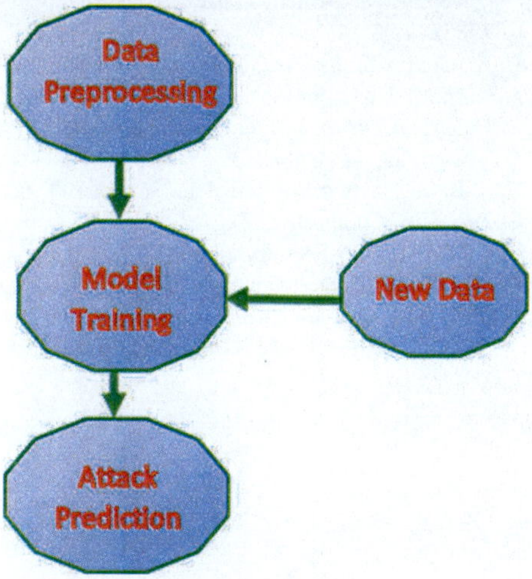

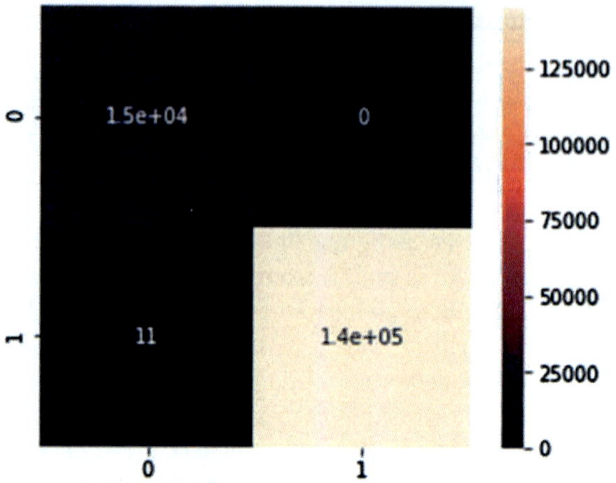

Diagram 3 Confusion matrix for SVM

References

1. R.J. Syed Rizvi, A.C. Orr, P. Ashokkumar, M.R. Rizvi, Identifying the attack surface for IoT network. Internet of Things, ScienceDirect **9**, 100162, ISSN 2542–6605 (2020). https://doi.org/10.1016/j.iot.2020.100162
2. K. Geetha, S.H. Brahmananda, A survey on IoT cyber-attacks and deep learning assistance that can be used to detect the attacks. J. Adv. Res. Dyn. Control Syst. **10**, 04-Special Issue
3. M.A. Al-Garadi, A. Mohamed, A. Al-Ali, X. Du, M. Guizani, A survey of machine and deep learning methods for Internet of Things (IoT) security. [Online]. https://arxiv.org/abs/1807.11023
4. H. Nguyen, Q. Ngo and V. Le, IoT botnet detection approach based on PSI graph and DGCNN classifier, in *2018 IEEE International Conference on Information Communication and Signal Processing (ICICSP)*, Singapore, pp. 118–122 (2018). https://doi.org/10.1109/ICICSP.2018.8549713
5. P. Torres, C. Catania, S. Garcia and C. G. Garino, 2016, An analysis of Recurrent Neural Networks for Botnet detection behavior, in *2016 IEEE Biennial Congress of Argentina (ARGENCON)*, Buenos Aires, Argentina, pp. 1–6. https://doi.org/10.1109/ARGENCON.2016.7585247.
6. A. Guerra-Manzanares, H. Bahsi, S. Nõmm, Hybrid feature selection models for machine learning based botnet detection in IoT networks, in *2019 International Conference on Cyberworlds (CW)*, Kyoto, Japan, pp. 324–327 (2019). https://doi.org/10.1109/CW.2019.00059.
7. Y. Meidan, M. Bohadana, Y. Mathov, Y. Mirsky, D. Breitenbacher, A. Shabtai, Y. Elovici, N-BaIoT: network-based detection of IoT botnet attacks using deep autoencoders, in *IEEE Pervasive Computing, Special Issue—Securing the IoT*
8. C.J.C. Burgesm, A tutorial on SVM for pattern recognition. Data Min. Knowl. Discov. **2** (1998)
9. Y. Mirsky, T. Doitshman, Y. Elovici & A. Shabtai, Kitsune: an ensemble of autoencoders for online network intrusion detection, in *Network and Distributed System Security (NDSS) Symposium*, San Diego, CA, USA (2018)

A Novel Framework for NIDS Using Stacked Ensemble Learning

H. Jagruthi and C. Kavitha

Abstract With growing information technology and increasing digitization, the data is increasing largely every year, data security becomes the largest concern and detecting abnormal behavior in the network is very much necessary and essential in almost all organizations like schools, industry, business, hospitals. In recent years, there is an increase in malicious activities in networks which causes the threat to safety of network. So, to make sure the data safety and security of their communication, this paper proposes a novel method of classifying the UNSWNB-15 dataset based on Convolution Neural Network (CNN) and Stacked ensemble learning. In this approach, CNN is used for primary classification to extract the featured data that avoids the deletion of key feature information. The second classifier is based on stacking ensemble method. Here the CNN has taken the benchmark dataset UNSWNB-15 to compare the performance with other machine learning algorithm like Decision tree classifier. The results say that secondary classifier has achieved high performance when compared to primary classifier. The better detection accuracy is achieved when compared to other Intrusion detection models.

Keywords Intrusion detection system · Stacking ensemble learning · UNSWNB-15 · Decision tree · Convolutional neural network · Machine learning

1 Introduction

From recent years, people are using Internet rapidly. Internet is used by many departments like education, hospitals, industries, banking, social media, business, marketing, and many others all are connected to World Wide Web and are reaching through Internet services. High usage of Internet services from above category leads

H. Jagruthi (✉)
Department of Information Science and Engineering, BNM Institute of Technology, Bengaluru, India

C. Kavitha
Department of Computer Science and Engineering, Dayanand Sagar Academy of Technology and Management, Bengaluru, India

G. Ranganathan et al. (eds.), *Soft Computing for Security Applications*,
Advances in Intelligent Systems and Computing 1397,
https://doi.org/10.1007/978-981-16-5301-8_9

to more enormous amount of data. The growth of Internet is leading to cyber-attacks and attacks on the data exchange between sender and receiver leads to misuse of integrity, availability, confidentiality of computer security policies.

To protect data, there are various teaching mechanism available and to know the usage of hacking mechanism do not high skills. On the other side of the coin, there are multiple software security tools such as antivirus, encryption mechanisms, firewalls, but these various security methods do not efficiently identify all the types of attacks. Attacking the user sensitive data manipulating user sensitive information leads to integrity of service provider with users. Since there is an increasing demand of security to the recent technologies design of security methods are very essential. One of the better mechanisms to identify the attacking is intrusion detection system.

IDS is a mechanism that monitors and filters the network activities that are affected by the attack. M. S Abhirami et al. in [1] divide IDS into two types: One is network intrusion detection system, and other one is Host Intrusion Detection System (HIDS). HIDS organizes the operating system files and NIDS (network-based intrusion detection system) is helpful in accessing traffic of incoming network. IDS can also be classified as signature-based and anomaly-based while signature-based is to detect bad pattern like malwares and anomaly-based is to detect good traffic. IDS majorly analyzes the incoming and outgoing traffic of all devices in the network. IDS conducts traffic analysis on the entire network and checks with the traffic that passes on to the network library of known attacks, once the attack is identified or seen with malicious behavior an alarm will be raised to admin. IDS based on deep learning & machine learning will be able to monitor large data and detects the malicious pattern in better way.

In previous work different machine learning algorithms like decision tree, naïve Bayes [2], logistic regression [3], Random forest, Gradient boosting [GB] fuzzy logic, Neural Networks, Support Vector Machine (SVM) [4], k-Nearest Neighbor (KNN) [5] are used for intrusion detection. Nowadays deep learning models like CNN, RNN are used for IDS. In this paper combination of ensemble stacking method is introduced so that efficiency of individual model can be enhanced. Ensemble learning is a machine learning approach which allows different algorithms to combine to improve prediction. Recent methods have proved that ensemble method is multi-skilled and will improve detection speed and boosts accuracy of prediction, moving on with the same opinion, proposed methodology improves the application of deep learning method CNN and supervised machine learning algorithms to give a novel classification framework called stacking ensemble learning.

Based on related works [6, 7], it is known that deep learning algorithms can get the better of feature engineering requiring better feature quality. With reference to earlier works, it can be told that classification using CNN is higher. In this paper, we are proposing IDS based on CNN and stacking ensemble learning, in the model the primary classifier is done using CNN where features are extracted using CNN to avoid missing important feature data, the secondary classifier used is stacking method to generate the final classification result.

2 Literature Review

Horng et al. [8] have given IDS that combines different methods called SVM and Clustering. The clustering type used is hierarchical that generates highly classified data which also improves time of detection. The experiment was done on KDD Cup99 dataset that was giving 95.72% accuracy and FPR of 0.7%. Many intrusion detection systems are proposed using ensemble learning [9–11].

Shrives and Dewangan [12] have proposed ensemble model based on two classifiers ANN and Bayesian network, these two are combined with gain ratio and feature selection technique, and proposed model is tested with KDD Cup 99 & NSL-KDD dataset. This method has used feature reduced system and has given better performance than a regular system. Here they have extracted data from original dataset and got better results with 98% TP model has used naïve Bayes, Bayesian Net & J48 [13]. Govindarajan [14] has given ensemble model using arcing classifier, with two methods RBF & SVM, experiments are conducted both on real-time dataset and benchmark dataset NSL-KDD. Here the base classifier has given better accuracy.

Smitha Rajgopal has proposed ensemble model, they have used random forest, logistic regression, K-nearest neighbor and SVM for UNSWNB-15 packet and flow data, and the model has achieved good performance with real-time dataset than emulated dataset [15]. In this, they have used anomaly network detection and method used is autoencoder, SVM & RF, the dataset used is NSL-KDD & UNSWNB-15 proposed methodology has given better classification accuracy. Data preprocessing has been applied to both NSL-KDD & UNSWNB-15 [16].

Olasehide Olayemi et al. [17] proposed the ensemble IDS using multiple model trees with tenfold cross-validation for binary & multiclass classification which has improved prediction rate for different machine learning with classification accuracy of 99.93% & low FAR 0.22%. Here most of them have used machine learning algorithm for ensemble approach [18–23]. In this paper, we have used deep learning technique to overcome the disadvantages that is required for better feature quality and many works have shown that the accuracy using CNN classifier is higher hence CNN is used as base classifier and Decision tree as secondary classifier which uses stacking method to give result by combining primary classifier CNN which improves further accuracy.

3 Methods

3.1 Framework

The proposed framework for IDS is shown in Fig. 1. Here the framework is divided as data processing and classifier. Data preprocessing involves feature conversion wherein all the dataset contains both numerical and character data that is there in networked data. In order to do machine learning step, the character data is converted

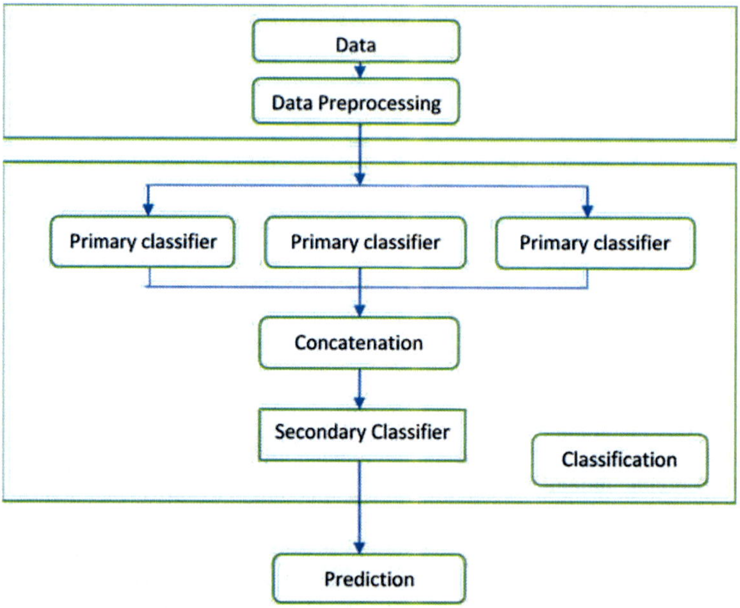

Fig. 1 Proposed methodology using stacking ensemble learning

into numeric data. In the next step, feature selection is done to remove unnecessary or redundant values without loss of data or without disturbing the accuracy of final model, and then the data normalization is applied to dataset to rescale the raw values to proper range. During data processing, the UNSWNB-15 dataset is normalized using standard scaler. The classifier is composed of both primary and secondary, and initially, primary classifier is based on convolutional neural network (CNN) which is used for extracting features. The secondary classifier is used to combine or concatenate primary classifier via stacking method, and then it is trained and tested. The final classified results are given as feature vectors to train the second classifier.

3.2 *Stacking Method*

Stacking ensemble model is a combined learning method that combines multiple classification models via meta-regressor or meta-classifier. Here, the primary model is trained on complete training dataset and the result will be given to new method training. Stacking involves two steps. In the first stage, the base classifier CNN involves in cross-validation process wherein one vector will be the final result. The result of CNN is given as input to the secondary classifier.

3.2.1 CNN as Primary Classifier

CNN (Convolutional Neural Network) is used for classifying the data as normal and malicious class like dos, worms, analysis, shellcode, backdoor, reconnaissance, fuzzers, exploits and generic. The accuracy of classification majorly depends on the CNN used to train the model. CNN is composed of convolutional, pooling and fully connected layers. Basically, CNN is used for categorizing images and speech recognition.

In this work CNN is used for classification and the input dataset used is UNSWNB-15 for conducting training by CNN. Once the training is preformed some quantity of testing data is used for doing classification. The various dense layers are considered along with ReLU (Rectified linear unit) activation function. The CNN is trained using training dataset and it is tested with testing dataset and the results of one layer is feeded as input to next layer, training dataset reduces the classification errors and helps in identifying attacks. CNN (Convolutional layers) use two dissimilar filters to handle 64bytes and 128bytes of data during mapping and filters the features in accordance with mapping outputs. Further the results of convolutional layer are forwarded to fully connected layer along with ReLU function, bias data and hidden value. Here fully connected layer calculates the feature score which helps to identify different attack types. The convolutional function is shown in Eq. 1.

$$H_j = f\left(h_{j-1} \otimes w_j + b_j\right) \tag{1}$$

Here w_j is kernel layer $f(y)$ is ReLU activation function $\otimes$ is convolution function h_j features used and b_j is bias value, after using CNN layers the value of h_j should be converted to vector. The decision is taken on input data of training and testing dataset using ReLU activation function. The result will be varying based on the function used in SoftMax layer of CNN. The final feature value is computed with the help of fully connected layer and results are classified as normal and malicious.

3.2.2 Decision Tree as Stacking Secondary Classifier

Decision tree classifier is a robust machine learning classifier, and it is very much helpful in decision-making process. Since the Decision tree dimension is independent of dataset it is built with huge capacity of dataset with many features and attributes. Decision tree consists of three categories leaves, edges, and nodes. Every node is attached with either one node and leaves or two nodes. Nodes consists of edges and are categorized based on attributes. Leaves are categorized according to decision value. Decision tree is starting with root followed by tree with branches till leaf node until it reaches the class. Initially decision tree algorithm constructs the tree and cut back for effective categorization. With trimming method part of tree is reduced or merged to reduce the complete size of decision tree. Depending on the volume of the dataset time complexity and space complexity differs.

Data is divided into training and testing set.

- Scaling all the variable using min-max scaler
- Import DT
- Use fit and score for function for classifier.

Algorithm: Hybrid CNN-DT ensemble model

- **Input:** UNSWNB-15 dataset
- K = 2 the number of models in ensemble
- Primary classifier CNN (Convolutional neural network) secondary classifier DT
- **Output**: Hybrid CNN-DT model M^1.

4 Experimental Setup

4.1 Dataset Description

The UNSWNB-15 dataset is used for proposed methodology. UNSWNB-15 dataset contains total 49 features out of which two are classes and 47 are features. Table 1 shows the training and testing samples used from UNSWNB-15 dataset.

One class talks about attack whether it is normal or malicious and another class denotes type of attack. Types of attack include normal, worm, shellcode, backdoor, analysis, reconnaissance, dos, fuzzers, exploits and generic. Either of the one label is used for classification. Finally, the whole dataset is divided into training and testing dataset. 42 features are used for training and testing dataset. The UNSWNB-15 dataset features are divided into 5 groups, namely basic features, content features, flow features, time features and additional features. Table 2 shows the types of attacks instances present in training and testing dataset.

Table 1 UNSWNB-15 dataset training and testing samples

Type of attacks	Training samples	Testing samples
Worms	130 samples	44 samples
Shellcode	1133 samples	378 samples
Backdoor	1746 samples	583 samples
Analysis	2000 samples	677 samples
Reconnaissance	10,491 samples	3496 samples
DOS	12,264 samples	4089 samples
Fuzzers	18,184 samples	6062 samples
Exploits	33,393 samples	11,132 samples
Generic	40,000 samples	18,871 samples
Normal	56,000 samples	37,000 samples
Total	1,75,341	82,332

Table 2 UNSWNB-15 dataset training and testing samples

S. No.	Attack	Description
1.	Fuzzers	Providing random data as input to the computer program
2.	Analysis	Set of instructions getting into application through email and other sources
3.	Backdoor	Trespassing the authentication and getting unauthorized access to system remotely
4.	DoS	Interrupting the network and not allowing the legitimate user to access the network
5.	Exploit	Set of instructions supporting bug to generate malicious behavior in the network
6.	Generic	It is an attack created against block cipher using hash function
7.	Reconnaissance	It is also called as probe; it sees network and tries to attack the system security
8.	shellcode	Attackers get into the network and take part of code and use it to make network vulnerable
9.	Worm	Malicious software infecting other computers while remaining active in current systems

4.2 *Performance Metrics*

There are standard performance metrics available to measure the performance of intrusion detection like accuracy, F1 score, false alarm rate (FAR), false positive rate (FPR), false negative (FN), precision and recall. True positive and true negatives classify the correct predictions, false positive and false negative are incorrect predictions. The performance metrics can be calculated using following Eqs. 2–7.

$$\text{Accuracy} = \frac{\text{TP} + \text{TN}}{\text{TP} + \text{TN} + \text{FP} + \text{FN}} \tag{2}$$

$$F1\ \text{score} = \frac{2 \times \text{precision} \times \text{recall}}{\text{precision} + \text{recall}} \tag{3}$$

$$\text{False positive rate} = \frac{\text{FP}}{\text{FP} + \text{TN}} \tag{4}$$

$$\text{False negative rate} = \frac{\text{FN}}{\text{FN} + \text{TP}} \tag{5}$$

$$\text{Precision} = \frac{\text{TP}}{\text{TP} + \text{FP}} \tag{6}$$

$$\text{Recall} = \frac{\text{TP}}{\text{TP} + \text{FN}} \tag{7}$$

5 Results and Discussion

For primary classifier, CNN and secondary classifier, decision tree, ROC curve is generated. Figures 2 and 3 show the ROC curve for CNN and decision tree. The ROC curve is called as receiver operating characteristic curve, and it gives the relationship between true positive and false positive classification at different levels. The *x*-axis gives the false positive classification and *y*-axis gives the true positive classification. The ROC curve for CNN shows the good training and testing accuracy.

The various performance metrics used are false positive rate (FPR), false negative rate (FNR), true positive rate (TPR) and true negative rate (TNR) meaning of these metrics are explained in Table 3.

Here, different authors have proposed stacking ensemble learning models using different machine learning classifiers like random forest, support vector machine, naive Bayes, logistic regression, autoencoder and KNN classifiers, when compared to other proposed methods our proposed methodology gives stacking ensemble learning model using deep learning techniques CNN and Decision tree and it gives better prediction accuracy compared to existing methods. Figure 4 shows the comparison of other ensemble models with proposed model.

Table 4 shows the comparison accuracy of binary classification with proposed ensemble learning and other methods. Here it is observed that all authors have used the UNSW-NB15 dataset in their works along with the real-time dataset used by Smitha Rajgopal and Ying-Feng. The authors have used ensemble learning methodology wherein different combinations of machine learning algorithms like random forest, support vector machine, naïve Bayes classifier, logistic regression, KNN, autoencoder is used for predicting accuracy for binary classification of attacks. The result shows that predicted ensemble methodology with CNN and decision tree achieves the better accuracy compared to existing methods.

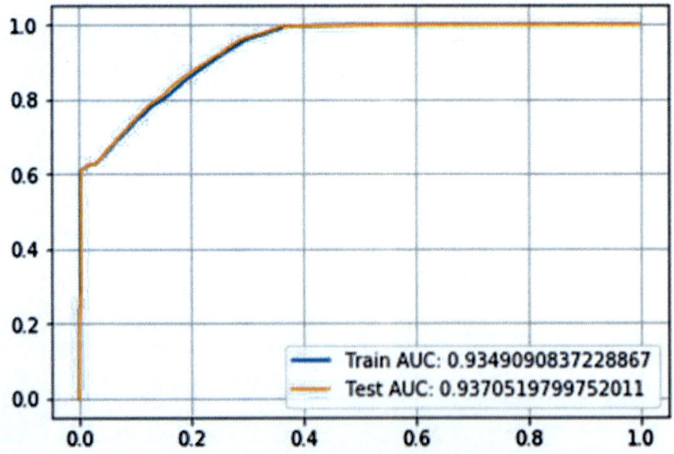

Fig. 2 ROC curve for CNN

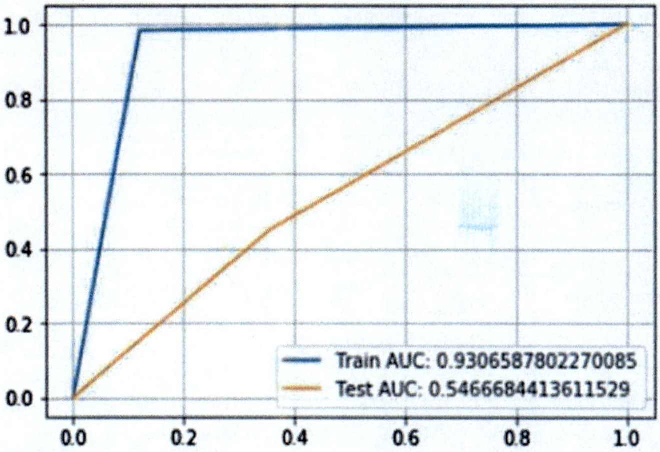

Fig. 3 ROC curve for DT

Table 3 Performance metrics

False positive (FP)	Normal behavior detected as normal and malicious
False negative (FN)	Contains attacks classified which is normal behavior
True positive (TP)	Attacks successfully detected by IDS
True negative (TN)	Normal and malicious both are identified as same

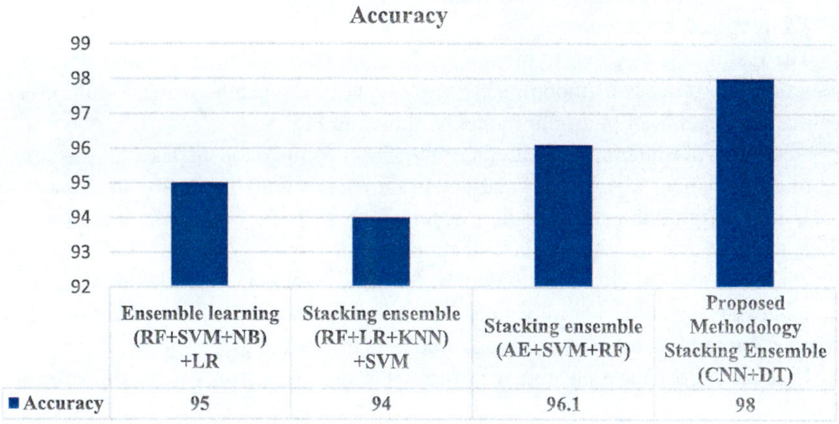

Fig. 4 Comparison of accuracy of different ensemble models with proposed stacking ensemble

Table 4 Comparison accuracy of proposed ensemble learning for binary classification

Author	Dataset	Method	Accuracy
M. S. Abirami, Umaretiya Yash and Sonal Singh	UNSW-NB15	Ensemble learning (RF + SVM + NB) LR	95
Smitha Rajagopal, Poornima Panduranga Kundapur, and Katiganere Siddaramappa Hareesha	UNSW-NB15	Stacking ensemble (RF + LR + KNN) + SVM	94
	UGR'16		97
Ying-Feng Hsu, ZhenYu He, Yuya Tarutani, Morito Matsuoka	UNSW-NB15	Stacking ensemble (AE + SVM + RF)	91.8
	NSL-KDD		91.7
	Palo Alto system log datasets		96.1
Proposed methodology	**UNSW-NB15**	**Stacking ensemble (CNN + DT)**	**98**

Table 5 shows the comparison accuracy of multiclass classification with proposed ensemble learning and other methods.

Basically, the stacking classifier at ground level involves in cross-validation training where one vector will be a final score. It will be in the form of Eq. 8.

$$\langle (X^1 0 \ldots X^1 M), XK \rangle \tag{8}$$

where k = Number of folds of cross-validation.

$X^1 M$ = output predicted for M classifier.

XK is the expected output.

The result will be given to meta-classifier that resolves the errors and optimizes the complete process of model. This process will be repeated using K-fold cross-validation to get final optimum model as shown in Fig. 5.

Accuracy of stacking classifier is 98%. The comparison of stacked ensemble learning, and other algorithms accuracy is shown in Fig. 6. Comparison shows that ensemble learning is getting highest accuracy of **98%** for binary classification and **85%** accuracy for multiclass classification.

Table 6 shows the classification report for binary stacking classifier. Results show that stacked ensemble learning gives the better accuracy for binary classification when compared to other ensemble methods given in literature survey. Here in the table, it can be seen the precision, recall and F1-score for Normal (0) classification and

Table 5 Comparison accuracy of proposed ensemble learning for multiclass classification

Method	Accuracy
CNN	66.04
Decision tree	75
Stacking ensemble	**85**

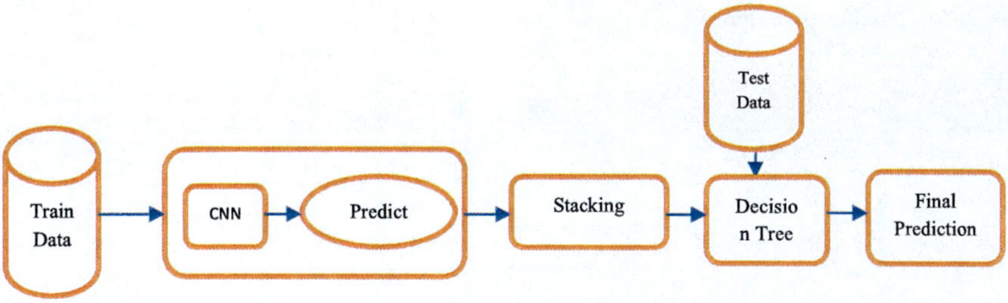

Fig. 5 Stacking ensemble

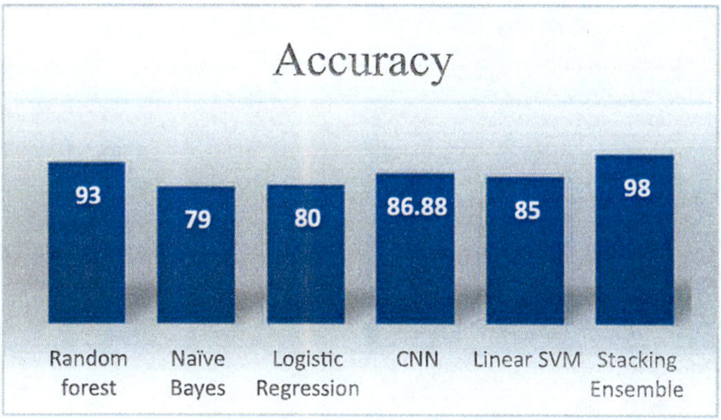

Fig. 6 Comparison of accuracy of different models with proposed stacking ensemble

Table 6 Classification report of binary stacking classifier

	Precision	Recall	F1-score
0	0.97	0.99	0.98
1	0.99	0.98	0.98
Accuracy			**0.98**
Macro-average	0.98	0.98	0.98
Weighted average	0.98	0.98	0.98

attack (1) classification. The precision and recall give the positive class predictions out of entire dataset. The accuracy of 98% gives the best score for binary stacking classifier.

Table 7 shows the classification report for multiclass stacking classifier. Results show that stacked ensemble learning gives the better accuracy for multiclass classification when compared to other ensemble methods given in literature survey. Here, it can be seen with precision, recall and F1-score of all nine different attack class types stating from normal class to nine different types of attack class. The distribution

Table 7 Classification report of multiclass stacking classifier

	Precision	Recall	F1-score
0	0.68	0.06	0.11
1	0.40	0.01	0.01
2	0.44	0.60	0.51
3	0.77	0.64	0.70
4	0.53	0.57	0.55
5	1.00	0.97	0.98
6	0.91	0.96	0.94
7	0.73	0.72	0.73
8	0.72	0.11	0.19
9	0.33	0.16	0.22
Accuracy			**0.85**
Macro-average	0.65	0.48	0.49
Weighted average	0.85	0.85	0.84

percentage of positive class prediction with respect to different attack categories can be seen. The accuracy of 85% gives the best score for multiclass stacking classifier.

6 Conclusion

Ensemble learning for UNSWNB-15 dataset has been proposed by many authors but ensemble learning using deep learning and machine learning algorithm Decision tree is proposed in our paper. Here, the fusing of CNN and stacked ensemble model is considered for prediction, primary classifier CNN is used to improve the performance of classification prediction secondary classifier Decision tree is trained using the values generated by primary classifier CNN. Results have shown that primary classifier and secondary classifier were trained and tested for UNSWNB-15 dataset. Experimental results show that stacked ensemble model using CNN and Decision tree have achieved better prediction accuracy compared to other machine learning algorithms like random forest and logistic regression.

References

1. M.S. Abirami, U. Yash, S. Singh, Building an ensemble learning based algorithm for improving intrusion detection system, in *Artificial Intelligence and Evolutionary Computations in Engineering Systems, Advances in Intelligent*
2. N. Moustafa, J. Slay, The signifcant features of the UNSW-NB15 and the KDD99 data sets for network intrusion detection systems, in *4th International Workshop on Building Analysis Datasets and Gathering Experience Returns for Security (BADGERS)* (IEEE, 2015), pp. 25–31

3. S. Bagui, E. Kalaimannan, S. Bagui, D. Nandi, A. Pinto, Using machine learning techniques to identify rare cyber-attacks on the unsw-nb15 dataset, Secur. Priv. **2**(6), e91 (2019)
4. S. Meftah, T. Rachidi, N. Assem, Network based intrusion detection using the unsw-nb15 dataset. Int. J. Comput. Digital Syst. **8**(5), 478–487 (2019)
5. M.H. Kamarudin, C. Maple, T. Watson, N.S. Safa, A logitboost-based algorithm for detecting known and unknown web attacks, IEEE Access **5**, 26190–26200 (2017)
6. S. Smys, B. Abul, W. Haoxiang, Hybrid Intrusion detection system for Internet of Things (IoT). J. ISMAC **2**(04), 190–199 (2020)
7. J.I.Z. Chen, S. Smys, Social multimedia security and suspicious activity detection in SDN using hybrid deep learning technique. J. Inf. Technol. **2**(02), 108–115 (2020)
8. S.J. Horng, M.Y. Su, Y.H. Chen, T.W. Kao, R.J. Chen, J.L. Lai, C.D. Perkasa, A novel intrusion detection system based on hierarchical clustering and support vector machines. Expert Syst. Appl. **38**(1), 306–313 (2011)
9. N.T. Pham, E. Foo, S. Suriadi, H. Jeffrey, H.F.M. Lahza, Improving performance of intrusion detection system using ensemble methods and feature selection, in *ACSW 2018: Australasian Computer Science Week (ACSW)*, January 29–February 2, 2018, Brisbane, QLD, Australia (2018)
10. Z.-H. Zhou, *Ensemble Methods: Foundations and Algorithms* (CRC Press, 2012)
11. M. Panda, M.R. Patra, Ensemble of classifiers for detecting network intrusion, in *Proceedings of the International Conference on Advances in Computing, Communication and Control* (ACM, 2009), pp. 510–515
12. A.K. Shrivas, A.M. Dewangan, An ensemble model for classification of attacks with feature selection based on KDD99 and NSL-KDD data set. Int. J. Comput. Appl. **15** (2014)
13. NSL-KDD datasey (2017). http://www.unb.ca/cic/research/datasets/nsl.html
14. M. Govindarajan, Hybrid intrusion detection using ensemble of classification methods. Int. J. Comput. Netw. Inf. Secur. **6**(2) (2014)
15. S. Rajagopal, P.P. Kundapur, K.S. Hareesha, A stacking ensemble for network intrusion detection using heterogeneous datasets. Secur. Commun. Netw. **2020**, 9p. Article ID 4586875. https://doi.org/10.1155/2020/4586875
16. Y.-F. Hsu, Z.Y. He, Y. Tarutani, M. Matsuoka, Toward an online network intrusion detection system based on ensemble learning, in 2159–6190/19/$31.00 ©2019 IEEE https://doi.org/10.1109/CLOUD.2019.00037
17. O. Olasehinde Olayemi, C. Olayemi Olufunke, B.K. Alese, Multiple model tree meta algorithms improvement of network intrusion detection predictions accuracy. Int. J. Inf. Secur. Res. (IJISR), 9(3), (2019)
18. F. Amiri, M.R. Yousefi, C. Lucas, A. Shakery, N. Yazdani, Mutual information-based feature selection for intrusion detection systems. J. Netw. Comput. Appl. **34**(4), 1184–1199 (2011)
19. Akashdeep, I. Manzoor, N. Kumar, A feature reduced intrusion detection system using ANN classifier. Expert Syst. Appl. **88**, 249–257 (2017)
20. S. Chaurasia, A. Jain, Ensemble neural network and KNN classifiers for intrusion detection. Int. J. Comput. Sci. Inf. Technol. **5**, 2481–2485 (2014)
21. N.F. Haq, A.R. Onik, F.M. Shah, An ensemble framework of anomaly detection using hybridized feature selection approach approach (HFSA), in *SAI Intelligent Systems Conference* (IEEE, 2015), pp. 989–995
22. N. Moustafa, J. Slay, UNSW-NB15: a comprehensive dataset for network intrusion detection systems, in *Military Communications and Information Systems Conference (MilCIS)* (IEEE, Australia, 2015)
23. B. Riyaz Sannasi Ganapathy, *A Seep Learning Approach for Effective Intrusion Detection in Wireless Networks using CNN*. School of Computing Science and Engineering, Vellore Institute of Technology, Chennai, Tamilnadu, India

Butterfly Algorithm Boosted Deep Random Vector Functional Link Network for Keystroke Dynamics

M. Rathi, A. V. Senthil Kumar, and Amit Dutta

Abstract In the advanced information technology domain, cyber-attacks are crucial which is spread throughout the world to hack confidential information such as intellectual property, secret of trade, banking details etc. Handling such security threats are very challenging, and many security schemes are in existence. Continuous Keystroke Dynamics Authentication (CKDA) is the most important and an interesting authentication technique which is non-intrusive, permanent, and inexpensive. It enhances the logical access security based on the typing rhythm. But CKDA needs huge number of samples to analyze and register typing characteristics of users. There are many existing works on continuous authentication using standard machine learning models to understand its pattern. Still, there is an absence of deep understanding on keystroke procedures depending on free text, mainly due to the adaptability and unpredictability of free text keystroke. It is very challenging to signify discrete keystroke using statistical constrained keystroke time features. Thus, this paper focuses on constructing a boosted continuous identity authentication model using Deep Random Vector Functional Link Network (DRVFLN). Unlike standard deep learning model which uses random values for parameters involved in classification, DRVFLN assigns the fine-tuned parameter value using the knowledge of Butterfly Optimization Algorithm (BOA). The experimental results proved that the Boosted Deep Random Vector Functional Link Network (BDRVFLN) prominently produces higher accuracy rate in classifying the users as genuine or imposter using continuous keystroke dynamic authentication.

M. Rathi (✉)
Department of Computer Technology, Dr. NGP Arts and Science College, Coimbatore, India

A. V. Senthil Kumar
PG & Research department of Computer Applications, Hindusthan College of Arts and Science, Coimbatore, India

A. Dutta
AICTE, Head Quarters, JNU Campus, Vasant Kunj, New Delhi, India

G. Ranganathan et al. (eds.), *Soft Computing for Security Applications*,
Advances in Intelligent Systems and Computing 1397,
https://doi.org/10.1007/978-981-16-5301-8_10

Keywords Security · Authentication · Continuous keystroke dynamic authentication · Deep random vector functional link network · Butterfly optimization algorithm · Classification

1 Introduction

Authentication is one of the important technique to ensure data integrity and user identity. Data integrity is secure data transmission between the network and the user within the optimal time assignment [1]. User identity is based on passwords, tokens, or PIN numbers. Among several authentication models, keystroke dynamics is one of the strongest authentication systems which recognize the person by their typing rhythm. This increases the strength of security because it not only checks the correct password value, but also the typing rhythm for entering the password. Still, after processing initial log-in session, the system does not know if there is any change of user during each session. Suppose if the user fails to log out or unattended after logged on, without locking the system, then there is a high chance of impostor who will gain the access of the system. In such situation, the impostors need not worry about the password and pretend as genuine users with the typing rhythm. They may access the file, delete, copy, or modify the content in the name of genuine user. This unsecured situation can be tackled by verifying that whether the user has changed after the log on process. This system is known as continuous authentication system [2]. Since, the user still uses keyboard after the procedure of log in, document typing, etc., keystroke dynamics can be used to accomplish continuous authentication. The major variance of Static Keystroke Dynamics and Continuous Keystroke Dynamics is in the former method typed information is fixed, while in later, it is dynamic and continuous verification is done.

The two foremost needs of a novel continuous authentication system are that the user should not be disturbed with their daily activities and every single keystroke has to be identified for discovering users' genuinity [3]. Many existing approaches are developed for continuous authentication based on verifying the genuineness of a user depending on fixed actions. Numerous datasets are used in this area of research to authenticate the users on static or continuous keystrokes. However, the performance of the imbalanced datasets can be boosted by using Navo Minority Over-sampling Technique (NMOTe) [4].

To overcome the challenges in the existing approaches because of the dynamic and continuous investigation of authentication in adversarial environment, this paper concentrates on observing in-depth knowledge about the continuous keystroke dynamic pattern in the users typing rhythm by developing a deep random vector functional link to predict the genuineness of the users. Unlike, other deep learning models which assign random values for the parameters involved, the proposed model fine tunes the parameter weights by assigning optimized value. This is done by inducing the knowledge of butterfly optimization algorithm, which selects the best values based on the fitness value obtained by them. Thus, the proposed model of authenticating the users using the continuous keystroke dynamics method produce more effective result.

2 Related Work

Lu et al. [5] developed a fixed length keystroke sequence by developing a deep learning model using RNN and CNN to understand the keystroke vectors of the specific user. They converted sequence of keystroke into vector sequence based on the time features of the keystroke. Thus, they authenticated the users based on their keystroke features. Ahmed and Traore [6] in their work designed a MLP which discovers the similarities of keystroke characteristics and individual keystroke behavior models. This model uses more than one hidden layer to perform the strong analysis of the pattern of keystroke sequence, but the weight assigned to the hidden units is random values. So, it uses backpropagation to reduce the error rate.

Mondal et al. [7] devised three policies to discover user identity by their typing rhythm on a keyboard. Different machine learning approaches are used in their work along with their developed model pair wise user coupling. They stated that by combining more approaches its performance gets increased. Ayotte et al. [8] investigated the user authentication model by using keystroke patterns. In their work, they reduced the keystroke data length which is needed for authentication of the users. A fusion based classifier is developed to accomplish this task which works based on the importance of features of keystroke data. Mohamed et al. [9] explored about how fast a user gets authenticated, and number of digraphs needed to effectively identify an impostor in a free text abandoned atmosphere. They used three different distance metrics to measure the similarity of the users and fused them with each other for achieving effective results.

Zhong et al. [10] presented a unique multi-objective binary bat algorithm for authentication of users using keystroke dynamics. The binary bat algorithm ranks and selects the features of keystroke dynamics simultaneously. V-shaped binarization function is used for extracting the important features of the dataset. Key down time attribute is considered to be more significant feature for classification of genuine users and imposters. Deng and Zhong [11] devised a new distance metric for classifying the authenticated and non-authenticated users using keystroke dataset. They stated that similarity measure introduced by them performs better in keystroke dynamics.

Alsultan et al. [12] developed a classification model for authentication of users using keystroke dynamics. The feature selection done using decision tree and support vector machine is integrated with ant colony optimization algorithm to effectively classify the genuine users and imposters based on the keyboard typing rhythm. Maheshwary and Pudi [13] investigated on user authentication based on the typing pattern and the timing of keystroke. In this work, nearest neighbor-based regression approach is used for discovering anomalies on DSN dataset. Mirjalili et al. [14] in their work developed a V-shaped binary bat algorithm for determining keystroke dynamic authentication. They compared this model with sigmoid binarization on various benchmark functions.

Parashar et al. [15] developed a fuzzy classification model integrated with bat algorithm for keystroke-based authentication. This fuzzy-based bat algorithm is evaluated with standard metaheuristic approaches. Rathi and Senthil Kumar [16] developed a neutrosophic fuzzy inference model to authenticate the users using continuous keystroke dynamics. The authors have proposed a boosted deep random vector functional link network whose weights are optimized using butterfly algorithm for continuous keystroke dynamic authentication which is discussed in the next section.

3 Background

3.1 Keystroke Vector

Keystroke sequences collected from the users are vectorized by splitting the overall keystroke data to fixed length and timing of each keystroke is converted to keystroke vector depending on the keystroke timing characteristics such as dwell time and flight time. Dwell time or hold time is the time interval between a key press and key release. Flight time is the time interval between a key press of one key and the key press of another [17]. Figure 1 represents the keystroke timing characteristics of two keys ‘A’ and ‘B’.

For example, the generic keystroke vector format for the first and second keys typed by the user is shown in Fig. 2, and the acronyms used in keystroke vector format are given in Table 1. It is characterized in Fig. 2 about the data format of the keystroke vector where H[*] signifies the key hold duration, KP[*] represents time stamp of the key press, KR[*] characterizes time stamp for key release, KP[*]KR[+] key press and key release of two different keys, like wise KP[*]KP[+] refers to two keys’ key press time. For example, the text “I love Music” is converted from keystroke text

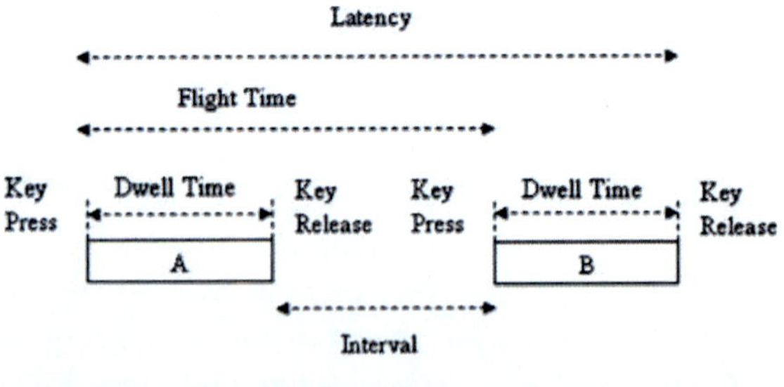

Fig. 1 Keystroke characteristics [17]

Fig. 2 Keystroke vector format

ID[1]	ID[2]	H[1]	H[2]	KP[1]KR[2]	KP[1]KP[2]

Table 1 Acronyms of keystroke vector format

Vector	Description
ID [1]	Identifier of the first key that is pressed
ID [2]	Identifier of the second key that is pressed
H [1]	Hold time or dwell time of first key
H [2]	Hold time or dwell time of second key
KP [1] KR [2]	Key press time of first key Key release time of second key
KP [1] KP [2]	Key press time of first and second keys

sequence to the keystroke vector as represented in Fig. 3. The keystroke vectors 1, 2, …, *m*, *n* are given in Table 2.

The keystroke vectors extracts the unique input mode of each individual and that is verified with the keystroke vectors stored in the database corresponding to the user. The methodology of predicting the user as a genuine user or an impostor is discussed in Sect. 4.

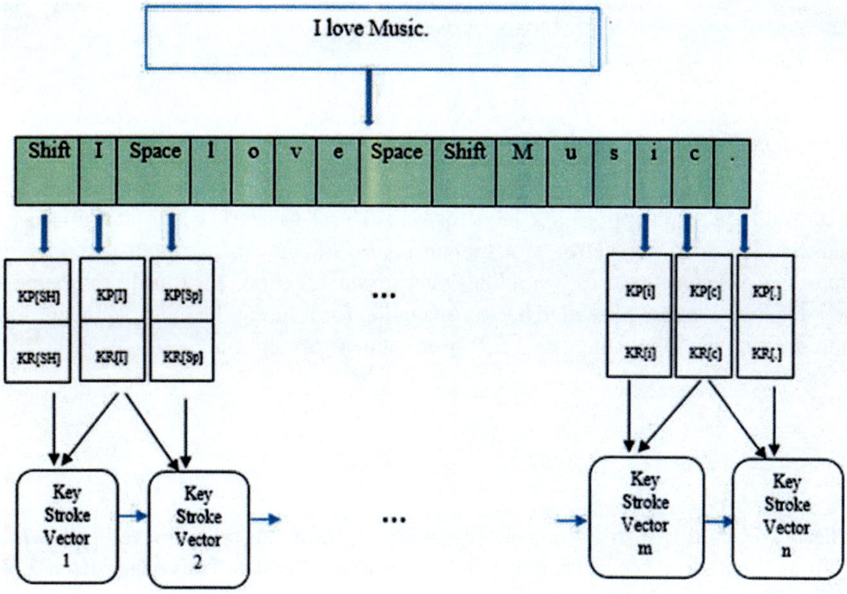

Fig. 3 Keystroke vector process

Table 2 Keystroke vectors of data sequence 'I love Music'

Keystroke vector	Description					
Keystroke vector 1	ID[SH]	ID[I]	H[SH]	H[I]	KP[SH] KR[I]	KP[SH] KP[I]
Keystroke vector 2	ID[I]	ID[Sp]	H[I]	H[Sp]	KP[I] KR[Sp]	KP[I] KP[Sp]
…	…					
Keystroke vector m	ID[i]	ID[c]	H[i]	H[c]	KP[i] KR[c]	KP[i] KP[c]
Keystroke vector n	ID[c]	ID[.]	H[c]	H[.]	KP[c] KR[.]	KP[c] KP[.]

3.2 *Artificial Neural Network*

The artificial neural network works with single input, hidden, and output layer for classification. The hidden layers comprised of the neurons which receives input from the input layer, and they are interconnected with links. Each link is assigned with weights and bias for activating the concern neuron nodes, and the activation function is used for integrating all the outputs generated by the hidden layers and the summation of it produce the binary classification.

3.3 *Multilayer Perceptron*

The multilayer perceptron is a feed forward neural network which comprised of multiple layers of perceptron's or hidden nodes. Except for the input nodes, each node is a neuron that uses a nonlinear activation function. MLP utilizes a supervised learning technique called backpropagation for training. Its multiple layers and nonlinear activation distinguish MLP from a linear perceptron.

3.4 *Deep Neural Network*

The deep neural network is a class of machine learning algorithm developed based on the convolutional neural network. It has dense structure, and it is constructed with greedy layer by layer method. It has the option of inbuilt feature extraction which is not presented in the ANN and MLP.

4 Methodology: Butterfly Optimization Boosted Deep Random Vector Functional Link Network for Prediction of Continuous Keystroke Dynamic Authentication

4.1 Overall Architecture

This proposed model performs prediction of user as authorized or unauthorized based on their continuous keystroke pattern. In this work, buffalo dataset [18] is used for authentication detection process. The input is fed into deep random vector functional link network (DRVFLN), and the number of keystroke vectors are determined by this network. It extracts n-gram features of the keystroke sequence. In DRVFLN to learn the sequence of keystroke, it uses the keystroke vector format as given in Fig. 3. When the inputs are transferred more depth to each hidden layer, its higher-level keystroke signature sequence are discovered by its internal operation. User's identity verification is related to binary classification issue. While authenticating the keystroke of users, it performs matching process with the database. If the sequence is matched with the corresponding user's entry, then it is decided as genuine user otherwise the user is an impostor. In one-time authentication process, the authenticity of the user is checked only during login process. However, in continuous keystroke dynamics authentication process, over a certain period of typing time, the user's identity is verified. The overall architecture of the proposed work to predict the authorized user is shown in Fig. 4.

The DRVFLN is enhanced by assigning the weights of the links which connects the layers by adapting the butterfly optimization algorithm. Instead of random assignment of weight values, this work utilizes the behavior of the butterfly to discover the potential weights which produce more accurate prediction results in continuous keystroke authentication.

4.2 Random Vector Functional Link Network (RVFLN)

A standard framework of random vector functional link network (RVFLN) is depicted in Fig. 5. It comprised of both nonlinear and original features from the input layer. While passing to hidden layer, it is converted to nonlinear features [19]. Assume that if 'DS' be the input data attributes with 'HN' be hidden nodes, then for each output node, there will be total of DS + HN inputs.

The parameters of hidden layers are generally assigned randomly during process of training, where β_r is the output weight that is to be calculated, and it is mathematically represented as

$$\min_{\beta_r} \|\mathrm{DA}\beta_r - \mathrm{OY}\|^2 + \vartheta \|\beta_r\|^2 \tag{1}$$

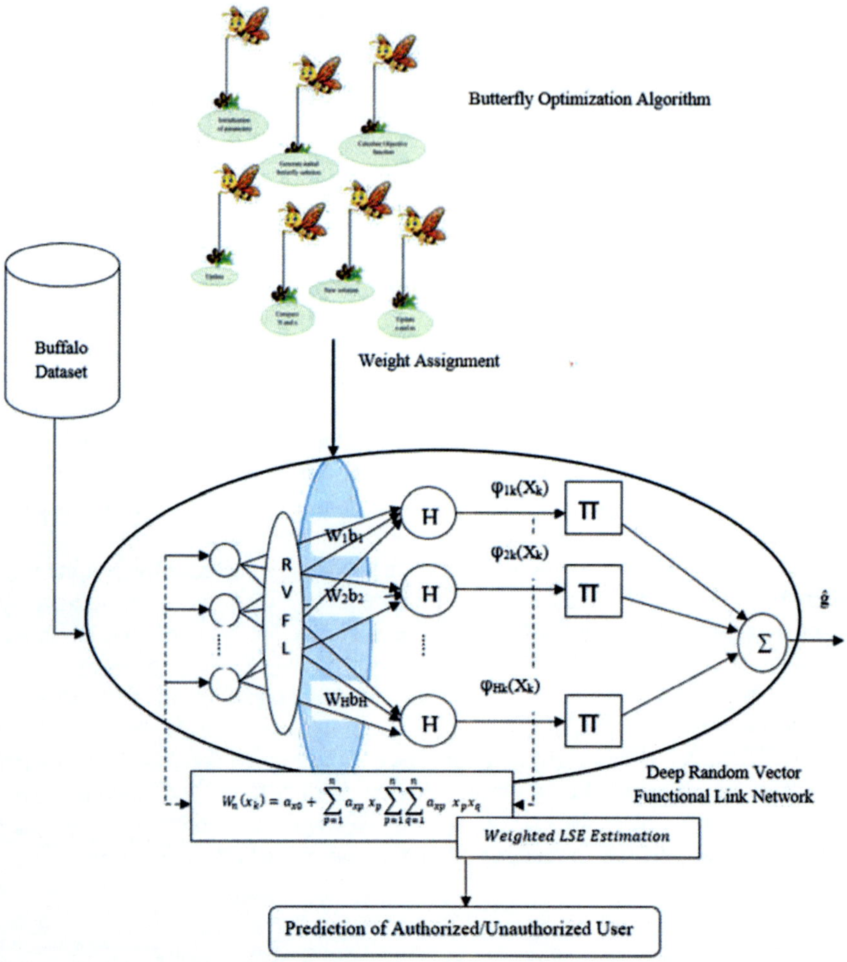

Fig. 4 Overall architecture for predicting the authorized user

where DA = [HF, OF] which refers to the addition of original and hidden features, ϑ signifies standardize parameters and OY refers to target vector. By assuming $\vartheta = 0$, then the equation can be closely solved by finding the primal and dual space as shown in Eqs. 2 and 3, respectively.

$$\beta_{r=}\left(\mathrm{DA}^{T}\mathrm{DA}+\vartheta I\right)^{-1}DA^{T}\mathrm{OY} \tag{2}$$

$$\beta_{r=}\mathrm{DA}^{T}\left(\mathrm{DADA}^{T}+\vartheta I\right)^{-1}\mathrm{OY} \tag{3}$$

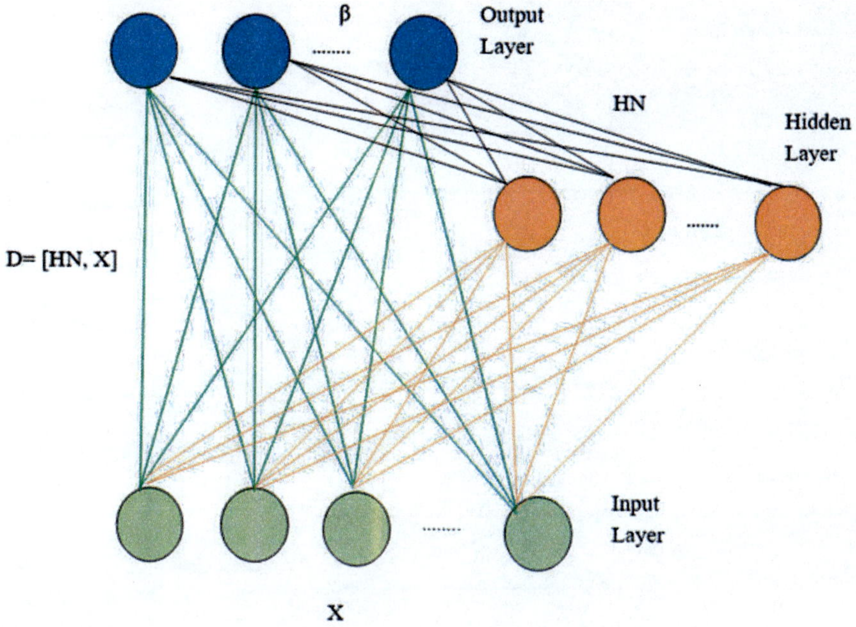

Fig. 5 General structure of random vector functional link network

4.3 Deep Random Vector Functional Link Network

The standard RVFL network is extended by applying deep learning concept, and it is termed as deep random vector functional link network (DRVFLN) [20]. The standard RVFL is a single layer feed forward neural network, in which the hidden neurons, the weights, and bias values are randomly generated and kept fixed. But, DRVFLN is a multilayer deep learning network, which comprised of stack of RVFL for processing the input patterns in which the weights and bias are optimized by adapting butterfly optimization algorithm.

The DRVFLN is used for understanding the deep knowledge about the patterns of continuous keystroke dynamic authentication to predict the authenticated or unauthenticated entries more precisely. DRVFLN is constructed using stacked hierarchy of hidden layers in which each layer's input is the output of the previous layer as illustrated in Fig. 6. The continuous input of the key buttons are received until it reaches the specific length $L = [10, 30, 50, 70, 100]$ to facilitate the network processing. Once it reaches the length L, then keystroke sequence vectorization is started which compares with the personal keystroke classification model learnt by the DRVFLN. It checks whether the number of abnormalities exceeds the threshold value p, which should be less than 50% for genuine users. If the threshold exceeds 50%, then the user is treated as imposter.

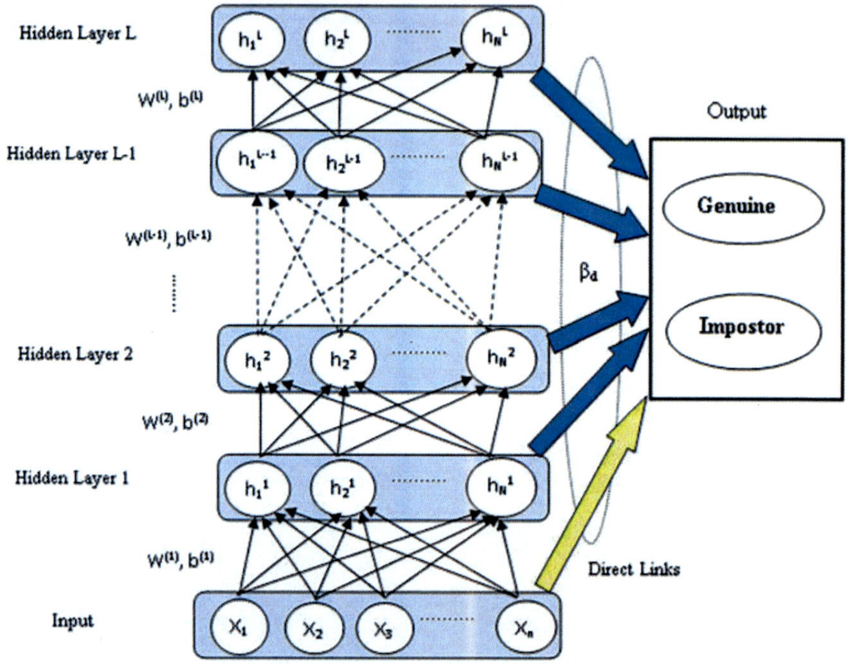

Fig. 6 Deep random vector functional link network

Once the data is fed into each layer, they convert into their internal representation. Though, DRVFLN has the ability to have varying network size, for simplicity HL hidden layers of stacks are used with same number of hidden nodes HN. The output produced by the first Hidden layer is formulated as

$$\mathrm{HL}^1 = g\left(\mathrm{XWt}^{(1)}\right) \tag{4}$$

For other hidden layers >1 it is represented as

$$\mathrm{HL}^{Lv} = g\left(\mathrm{HL}^{Lv-1}\mathrm{Wt}^{(Lv)}\right) \tag{5}$$

where $\mathrm{Wt}^{(1)} \in R^{(\mathrm{DS} \times \mathrm{HN})}$ signifies the weight matrix among input and first hidden layer and $\mathrm{Wt}^{(Lv)} \in R^{(\mathrm{HN} \times \mathrm{HN})}$ is the weight matrix between intermediate hidden layers. During training phase, the parameters of hidden neurons namely weights and biases are arbitrarily produced inside the appropriate series, and it is reserved static. The nonlinear activation function is represented as g(.) function. The entire process of DRVFLN from input to output layer is represented as

$$\mathrm{DR} = [\mathrm{HL}^{(1)}\mathrm{HL}^{(2)} \ldots \mathrm{HL}^{(Lv-1)}\mathrm{HL}^{(Lv)}X] \tag{6}$$

The DRVFL approach is alike to the traditional RVFL, in which former uses nonlinear features from stacked hidden layers with its received attributes. The outcome is denoted as

$$\mathrm{OY} = \mathrm{DR}\beta_{\mathrm{d}} \tag{7}$$

where $\beta_d \in R^{(\mathrm{NLvl+d})\,\mathrm{xk}}$, k refers to classes involved in continuous keystroke authentication.

The existence of the linear combination among features and output layer weight matrix β_d in DRVFL is clearly shown from Eqs. 6 and 7. During the training process, it directly enables to vary the weights depending on the attribute's contribution from different layers. Moreover, this model differentiates itself from the standard deep learning architecture based on threefold. They are:

- DRVFL network is developed based on the inspiration of the standard RVFL network and its outer layer comprised of both original features and nonlinear features using direct links, while the deep learning model in contrast does not consider the original features during weight calculation.
- In this work, DRVFL network is used for classification of genuine or imposter's keystrokes, whereas standard deep learning handles the time-series problems.
- DRVFL is common, and it can be deployed with any modified RVFL.

4.4 Butterfly Optimization Algorithm (BOA)

In this proposed model, the performance of the deep random vector functional network is boosted by adapting butterfly optimization algorithm (BOA) [21]. It is involved in assignment of parameter values namely weights and biases. This will boost the performance of the standard DRVFL, by assigning the optimized weight values selected by the BOA. Instead of assigning the values randomly, the butterflies choose the best values using its fitness function.

BOA is developed with inspiration of food seeking behavior of butterflies. With their sensing smell, sight, touch, taste, and hearing, they identify their food source, partner mating is helpful for their migration from one place to another as shown in Fig. 7, and pseudo code of the BOA is discussed in Algorithm 1.

This sensing helps to lay eggs in appropriate place and escape from predators. The smell is its most vital sense which assist butterfly to discover its food, i.e., nectar, even it is from a long distance. Its sense receptors are scattered throughout the parts of the body such as palps, antennas, and legs. The receptors are nerve cells of butterflies which is termed as chemoreceptors.

Butterflies have more accurate sense of location of fragrance source, they can find difference among fragrances and intensities. In BOA, butterflies are search agents, which generate fragrance with appropriate intensity which is related to its fitness. The physical intensity of stimulus is used for formulating fragrance which is defined as

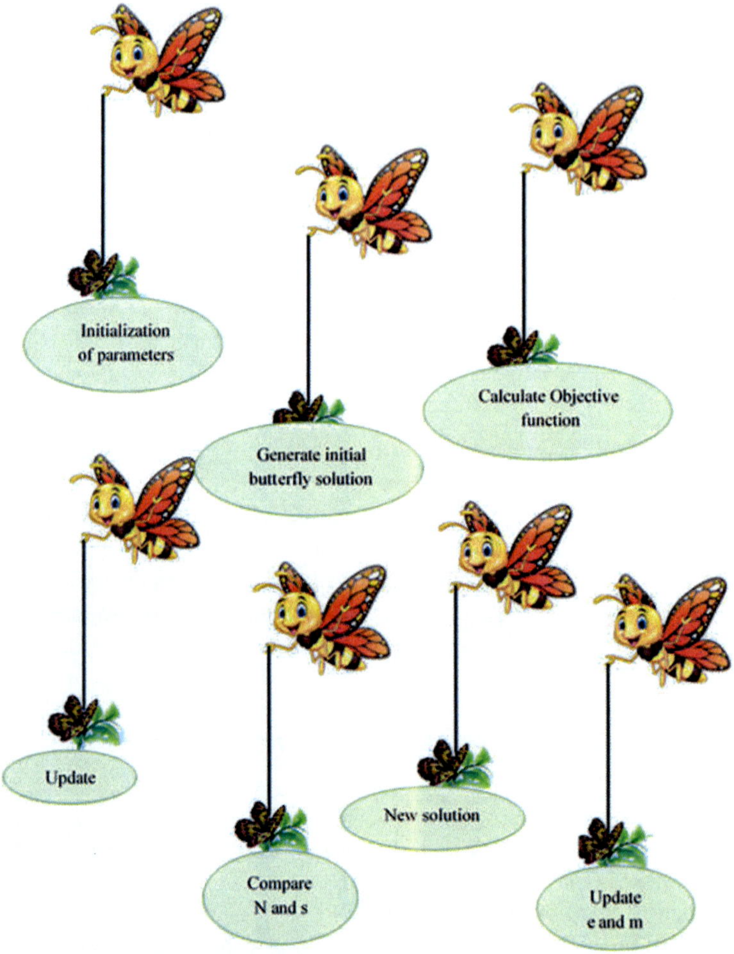

Fig. 7 Butterfly optimization algorithm

$$\mathrm{PFr}_j = mI^e \tag{8}$$

where PFr refers to the fragrance perceived by *j*th butterfly, *m* is the sensor modality, *I* refers to the stimulus intensity, and *e* refers to the power exponent.

Algorithm 1 Butterfly Optimization Algorithm

Initialize $\boldsymbol{objf(Y), Y = (y_1, y_2, \ldots, y_{ndm}), ndm}$ = no.of dimensions
Generates primary population of **N** Butterflies $\boldsymbol{y_i = (i = 1, 2, \ldots, N)}$
Stimulus Intensity $\boldsymbol{I_i}$ at $\boldsymbol{Y_i}$ is determined by $\boldsymbol{f(y_i)}$
Describe modality sensor *m*, exponent power *e* and probability switch *s*
while termination condition is not met
{
  for each *bfy* in population
    Compute *PFr* for *bfy* using equation 8
  Find the best *bfy*
  for each *bfy* in population
  {

   Produce an arbitrary *rn* from [0,1]
   if $rn<s$ **then**
     Transfer in the direction of best butterfly/solution
   else
     Transfer arbitrarily
   end if
  }
  Modify *e* value
}
Output: Optimized weight selection

5 Results and Discussions

This section discusses about performance analysis of the proposed butterfly optimization boosted deep random vector functional link network (BDRVFLN) for prediction of continuous keystroke dynamic authentication deployed using Python software. In this work, to perform prediction of continuous keystroke authentication, buffalo dataset [18] with 148 instances and three different sections are considered. Baseline and rotation are two different subsections involved in each section. There are 75 users in baseline subset with same type of keyboard in all three sections. In rotation subset, 73 users with different keyboards for all three sections are used. The proposed model BDRVFLN is compared with standard artificial neural network (ANN), multilayer perceptron (MLP), and deep neural network (DNN). The evaluation metric used are precision, recall, accuracy, and mean square error.

Figure 8 shows performance of four different network models to predict authentication using continuous keystroke dynamic based on precision. The performance of boosted deep random vector functional link network (BDRVFLN) produced highest

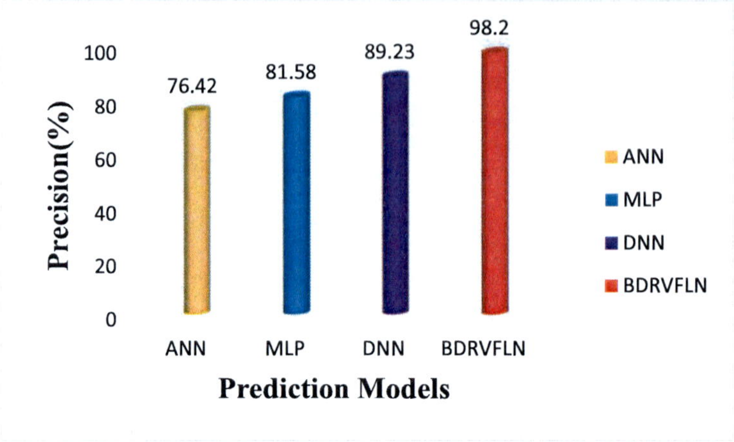

Fig. 8 Performance comparison based on precision

rate of precision while comparing with ANN, MLP, and DNN. This is because its outer layer comprised of both original features and nonlinear features using direct links. While the deep learning model, ANN and MLP in contrast does not consider the original features during weight calculation and it has higher chance of error. The overfitting problem is the main factor of other three models which resulted in less precision value.

Figure 9 illustrates the recall value obtained by four different classification models which are involved in prediction of continuous keystroke dynamic authentication. ANN, MLP, and DNN produce recall values of 75.9%, 80.63%, and 85.47%, respectively. While the proposed BDRVFLN achieves 97.58% which is the highest recall

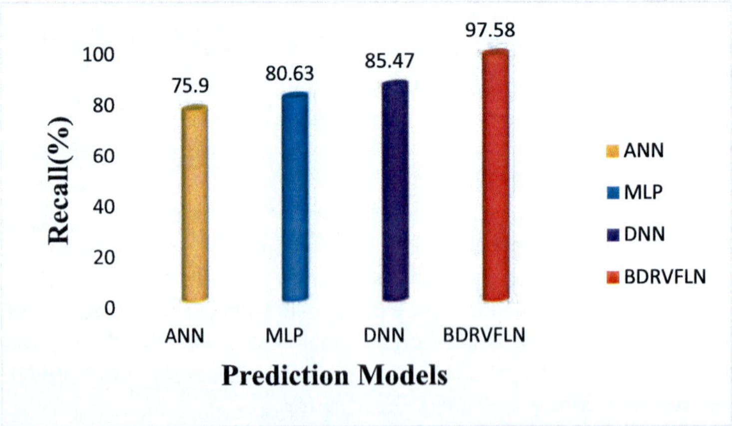

Fig. 9 Performance comparison based on recall

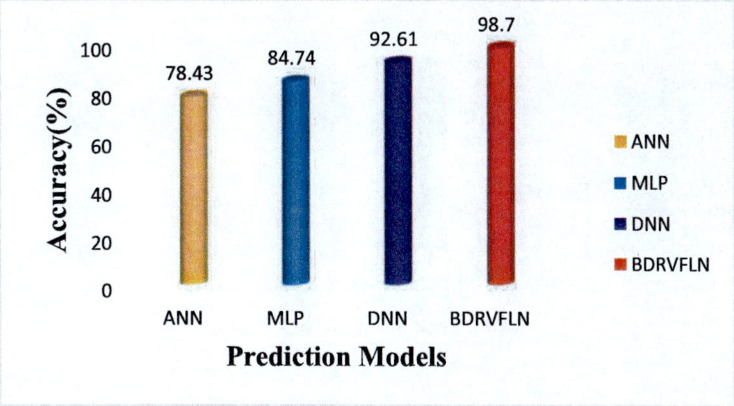

Fig. 10 Performance comparison based on accuracy

value compared to other three models. This is due to the weights assigned in the hidden units of all the three existing models which is assigned by random values. This may fail to produce optimized prediction. But BDRVFLN uses the knowledge of butterfly optimization algorithm to assign the weight values of the hidden nodes, and thus, by fine-tuning the parameters it produces high recall rate by predicting a greater number of authenticated users correctly.

Figure 10 reveals the accuracy produced by ANN, SVM, DNN, and BDRVLFN for predicting the genuine and imposters based on the continuous keystroke dynamic authentication. BDRVFL network is used for classification of genuine or imposter's keystrokes, whereas standard deep learning handles the time-series problems. The butterfly optimization model boost the performance of the BDRVLFN by fine-tuning the parameters involved in the model. The food searching behavior of the butterflies are used for assigning the best values to the parameters in RVFL network. Thus, BDRVLFN produces highest accuracy of genuine and imposter classification in buffalo dataset.

Figure 11 shows that mean square error rate produced by proposed BDRVFLN which is very less compared to other three models. This is because ANN, MLP, and DNN assign the weights during training phase with the help of backpropagation, and the weights are adjusted. But while using BDRVFLN, it assigns the weights using butterfly optimization algorithm, and all the input features are considered using direct link. Thus, the error rate is greatly reduced by BDRVFLN, and prediction of continuous authentication using keystroke dynamic is performed effectively.

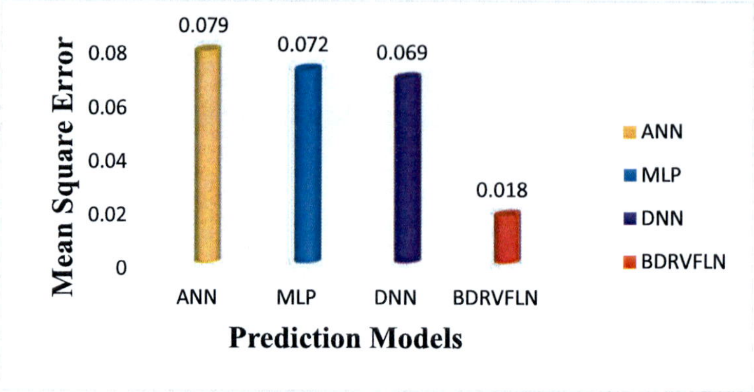

Fig. 11 Performance comparison based on MSE

6 Conclusion

One of the economical biometric authentication method is keystroke dynamics which can be integrated effortlessly into prevailing security systems with less user interference to improve the continuous authentication of users. The primary motive of this paper is to develop a deep learning model which is known as deep random vector functional link which is used for continuous keystroke dynamic authentication by extracting the significant features of dataset from various hidden layers along with original features. The standard DRVLFN is boosted by inducing the butterfly optimization algorithm, which boosts the performance by fine-tuning the parameters involved in this model. The simulation result proved that the proposed DRVLFN achieves better detection rate compared to the other standard classification models in continuous keystroke dynamic authentication. The ANN, MLP, and DNN use the concept of backpropagation to update its weight and bias parameters as they work in trial and error basis by comparing with the observed and expected results. The randomness in assignment of parameters may highly influence the accuracy of classification. In conventional DNN, only the extracted features are involved in classification, whereas in the proposed model DRVLFN, it uses both the extracted features and the original features to classify the instances with unknown labels. In future, various real-time keystroke dataset with variants of deep learning model will be developed.

References

1. J.I.Z. Chen, K.-L. Lai, Data conveyance maximization in bilateral relay system using optimal time assignment. J. Ubiquit. Comput. Commun. Technol. (UCCT) **2**(02), 109–117 (2020)
2. L. Ballard, D. Lopresti, F. Monrose, Forgery quality and its implication for behavioral biometrics security. IEEE Trans. Syst. Man Cybernet. Part B **37**(5), 1107–1118 (2007)

3. F. Alshanketi, I. Traore, A.A. Ahmed, Improving performance and usability in mobile keystroke dynamic biometric authentication. in *2016 IEEE Security and Privacy Workshops (SPW)* (2016), pp. 66–73
4. N. Chakrabarty, S. Biswas, Navo minority over-sampling technique (NMOTe): a consistent performance booster on imbalanced datasets. J. Electron. **2**(02), 96–136 (2020)
5. X. Lu, S. Zhang, P. Hui, P. Lio, Continuous authentication by free-text keystroke based on CNN and RNN. Comput. Secur. **96**(101861), 1–10 (2020)
6. A. Ahmed, I. Traore, Biometric recognition based on free-text keystroke dynamics. IEEE Trans. Cybernet. **44**(4), 458–472 (2014)
7. M. Soumik, B. Patrick, Person identification by keystroke dynamics using pairwise user coupling. Proc. IEEE Trans. Inf. Forensics Secur. **12**(6), 1319–1329 (2017)
8. B. Ayotte, J. Huang, M.K. Banavar, D. Hou, S. Schuckers, Fast continuous user authentication using distance metric fusion of free-text keystroke data. in *2019 IEEE/CVF Conference on Computer Vision and Pattern Recognition Workshops (CVPRW)* (2019), pp. 2380–2388
9. T.M. Mohamed, H.M. Moftah, Simultaneous ranking and selection of keystroke dynamics features through a novel multi-objective binary bat algorithm. Future Comput. Inf. J. **3**(1), 29–40 (2018)
10. Y. Zhong, Y. Deng, A.K. Jain, Keystroke dynamics for user authentication. in *2012 IEEE Computer Society Conference on Computer Vision and Pattern Recognition Workshops* (2012), pp. 117–123
11. Y. Deng, Y. Zhong, Keystroke dynamics user authentication based on gaussian mixture model and deep belief nets. Int. Sch. Res. Not. **2013**, Article ID 565183, 7 (2013)
12. A. Alsultan, K. Warwick, H. Wei, Non-conventional keystroke dynamics for user authentication. Pattern Recogn. Lett. **89**(1), 53–59 (2017)
13. S. Maheshwary, V. Pudi, Mining keystroke timing pattern for user authentication. in *5th International Workshop on New Frontiers in Mining Complex Patterns* (Springer, Cham, 2016)
14. S. Mirjalili, S.M. Mirjalili, X.S.Yang, Binary bat algorithm. Neural Comput. Appl. **25**(3), 663–681 (2014)
15. S. Parashar, J. Senthilnath, X.S. Yang, A novel bat algorithm fuzzy classifier approach for classification problems. Int. J. Artif. Intell. Soft Comput. **6**(2), 108–128 (2017)
16. M. Rathi, A.V. Senthil Kumar, A. Alrabea, A. Amine, A unified neutrosophical identification of authenticated users with continuous keystroke dynamics. Int. J. Adv. Sci. Technol. **29**(3), 14394–14406 (2020)
17. M. Rathi, A.V. Senthil Kumar, Exploration of keystroke dynamics based authentication on fixed-text and on free-text. Int. J. Comput. Sci. Eng. **7**(1), 807–812 (2019)
18. Y. Sun, H. Ceker, S. Upadhyaya. Shared keystroke dataset for continuous authentication. in *2016 IEEE International Workshop on Information Forensics and Security (WIFS)* (2016), pp. 1–6
19. Y. Ren, P.N. Suganthan, N. Srikanth, G. Amaratunga, Random vector functional link network for short-term electricity load demand forecasting. Inf. Sci. **367**, 1078–1093 (2016)
20. H. Cecotti, Deep random vector functional link network for handwritten character recognition. in *2016 International Joint Conference on Neural Networks (IJCNN)* (2016), pp. 3628–3633
21. S. Arora, S. Singh, An improved butterfly optimization algorithm for global optimization. Adv. Sci. Eng. Med. **8**(9), 711–717 (2017)

Hybrid Context-Aware Recommendation System Using Deep Autoencoder

Amartya Roy Chowdhury and Anita Pal

Abstract Nowadays, smart devices are revolutionizing mobile computing systems, and recommender system are a major driving force. There are a wide variety of recommender systems, but the most popular is collaborative filtering where recommendation is given based on user preference. But recently, deep learning-based models like RBM and autoencoder are extensively being used in recommender system mainly because of their ability to model complex nonlinear data with activation functions, and moreover, making a deep learning-based model is comparatively much easier as compared to other models as they are extremely flexible. Using deep autoencoder as a model in recommendation is very new and is gaining popularity nowadays, and so we propose a deep autoencoder-based hybrid recommender model using collaborative filtering. We will demonstrate that (a) this model performs better than shallow autoencoders and (b) adding contextual information improves the model performance.

Keywords Recommender system · Deep autoencoder · Collaborative filter · Deep learning · Context awareness · Hybrid recommendation

1 Introduction

With the increasing amount of irrelevant information in the Web, finding relevant information is becoming difficult day by day. As a result, recommender systems can significantly ease the task by predicting items based on user preference on that item. Collaborative filtering and content-based algorithm has been very popular in the past, but as the nonlinear problem becomes more complex, they do not do a very good task in giving relevant recommendation. Recently with the tremendous advancement of using deep learning in various fields, not a lot of work has been done to improve the

A. R. Chowdhury (✉) · A. Pal
Department of Mathematics, NIT Durgapur, Durgapur, India

A. Pal
e-mail: anita.pal@maths.nitdgp.ac.in

G. Ranganathan et al. (eds.), *Soft Computing for Security Applications*,
Advances in Intelligent Systems and Computing 1397,
https://doi.org/10.1007/978-981-16-5301-8_11

quality of recommendation, and furthermore, most of the model does not incorporate context that is present in real life [1]. In this paper, we are proposing a model which will be incorporating a deep learning-based autoencoder and will be using real-time contextual information.

Autoencoders [2] are feedforward neural network where the input is the same as output. They compress the input into a lower-dimensional code and then construct the output from this representation. An autoencoder consists of three parts encoder, hidden layer, and decoder.

Autoencoders are mainly dimensionality reduction technique with few advantages. It is worth mentioning here that RBMs [3] and classical autoencoders are very similar in their functional form although their interpretation and functionality are quite different. More specifically, the deterministic function that maps from input to mean hidden representation is the same in both models.

In this paper, a new hybrid approach using collaborative filtering is proposed for estimating top-N-recommendation using predicted rating from the deep autoencoder network as shown in Fig. 1. Context provides additional information about the real world. For example, [4] used user behavior as contextual information in social

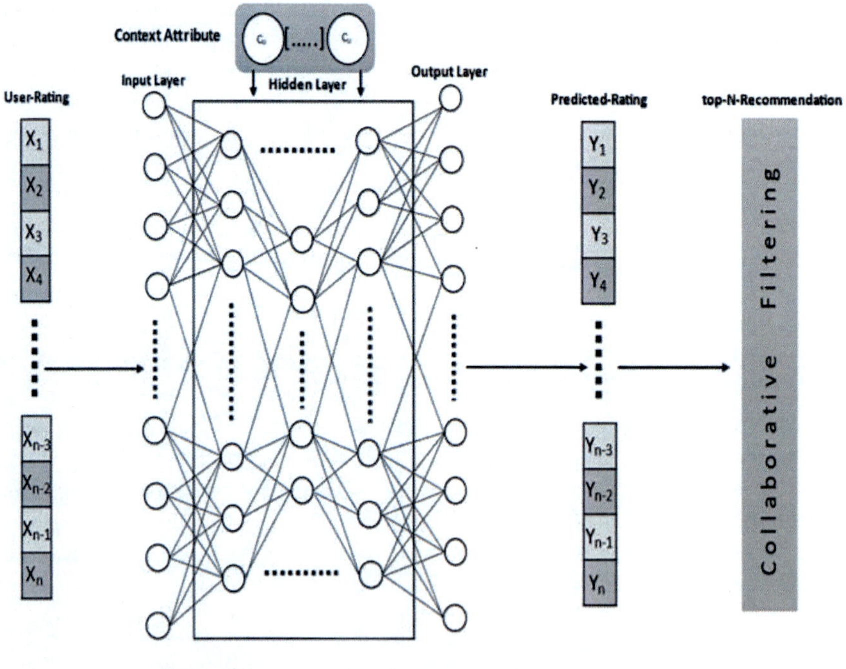

Fig. 1 Deep autoencoder network with two encoder networks and two decoder networks

network analysis. There is an existing work that provides contextual recommendation [5]. Compared to previous attempts, our model takes the matrix of ratings as an input, while the matrix of contextual information is applied to the hidden layers in a deep-autoencoder-based model and then provide further recommendation using collaborative filtering.

In this paper, Sect. 2 goes through the preliminaries. Then, in Sect. 4, we describe our model. Finally, in Sect. 5, experiments are discussed, and Sect. 6. discusses the results.

2 Preliminaries

2.1 *Collaborative Filtering*

Collaborative filtering is recommendation based on user preference. Given a matrix of users A having items B, collaborative filtering is used to find the correlation between *i*th user and *j*th item. In this paper, we will find the predicted ratings from the deep-autoencoder model and then feed those ratings into the collaborative filtering module to make top-N recommendations.

2.2 *Deep Autoencoders*

The following model is based upon autoencoders which was proposed by Hinton [2]. The loss is minimized by measuring the difference between the original input and reconstructed value and then backpropagating the loss back into the network. In this paper, both encoder and decoder are a part of our model. Given the training samples $X_i \in R$, the encoder maps the input vector X_i into a hidden representation $H_i \in R$ through a nonlinear mapping as follows:

$$e = \sigma(W_1 * x + b_1) \tag{1}$$

where σ is the nonlinear activation function. In this paper, we will be experimenting with various activation function, W_1 is the weight given to the input variable and b_1 is the bias variable. Then, the decoder maps the encoder representation back to its original representation through a series of hidden layers in a similar way as follows:

$$d = \sigma(W_2 * x + b_2) \tag{2}$$

To minimize reconstruction error between e and d, we will be using one of the most common used measure that is the mean square error (MSE).

$$E_{\mathrm{MSE}} = \frac{1}{m}\sum_{i=1}^{m}\left(\frac{1}{2}\|x_i - \hat{y}_i\|^2\right) \tag{3}$$

Here, m is the total no. of ratings, $\hat{y}_i$ is the predicted rating, and x_i is the real rating.

In addition, we will be using L2 regularization to avoid overfitting, so our loss function becomes:

$$\mathrm{Loss} = E_{\mathrm{MSE}} + \lambda \sum_{i=1}^{N} W_i^2 \tag{4}$$

where λ is a constant to adjust the weight of W_i.

It is practically impossible to rate every movie by the user, so the dataset is quite sparse; therefore, we will be imposing a sparsity constraint on the hidden layers to regularize the objective function, and mathematically, it is given by:

$$J_{\mathrm{sparse}}(\theta) = \beta \sum_{j}^{m} p \log \frac{m}{\sigma_i^{`}} + (1-p)p \log \frac{1-n}{1-\sigma_i^{`}} \tag{5}$$

where β is the weight adjustment parameter, m is the no of neurons of hidden layer, n is the activation value, and σ_i' is the activation function for the ith hidden layer over the whole set. Given a small σ, sparsity constraint is added in such a way, so that the hidden layers activation is close to 0. Here, in this experiment, σ is chosen as 0.01.

$$\begin{aligned} J_{\mathrm{DE}} &= \mathrm{Loss} + J_{\mathrm{sparse}}(\theta) \\ &= \frac{1}{m}\sum_{i=1}^{m}(\frac{1}{2}||x_i - \hat{y}_i||)2 + \lambda \sum_{i=1}^{N} w_i^2 + \beta \sum_{j}^{m} p \log \frac{\sigma}{\sigma_i^{`}} \\ &+ (1-p)p \log \frac{1-\sigma}{1-\sigma_i^{`}} \end{aligned} \tag{6}$$

3 Related Works

Recently the no. of deep learning-based recommender models has increased in the past few years. The earliest works in deep learning were done by Hinton, Bengio [6]. Then, Gao [7] tried to use generative [8] modeling for item recommendation. A recent survey [9] gives us a summary of all the work that has been done into recommender system.

In 2009, [10] BellKor's team used restricted Boltzmann machine and content-based recommender for the Netflix challenge. While deep is extensively used in computer vision, anomaly detection, and speech recognition [11], sparsity in data has always been a problem, and most of the work tend to force the network to put zero when encountering empty values, but this is not feasible as in real world almost 95% of the dataset will be sparse.

The earliest work on autoencoder as a recommendation system was done by Menon, Sanner [12]. Then, a lot of work followed to use autoencoder as a recommendation system like, the variational autoencoder-based model [1] or an even more recent work [13] on shallow autoencoder-based recommendation system where the author used extremely sparse data. Our model is different from the existing architecture, as we will be incorporating contextual information into every hidden layer into the deep autoencoder.

4 Model

Our model is inspired by *Deep Autoencoder* [14] with few changes (a) We will be experimenting with different activation function. (b) We will be incorporating contextual information to every hidden layer in the network.

The model, as shown in Fig. 1, is a typical five-layer deep autoencoder. The model takes input preference from the user which is in the form of a vector. If A and B are the matrices of users and items, the input vector is of the form $X_i \in \mathbb{R}^N$, where N is the total no of ratings. This vector of information is then fed into our feedforward network according to the equation $e = \sigma(W_{1*}\mathrm{x} + b_1)$, where σ is the activation function, $W_1 \in \mathbb{R}^{k\,\mathrm{X}\,N}$ and $b_1 \in \mathbb{R}^{\mathrm{k}}$ are the weight and bias vector where k is the no of units in the layer.

The encoder then tries to compress the data into the n-dimensional (in Fig: only three layers are shown for representational purposes) hidden layer, where the contextual attribute is multiplied to the hidden layer. In our model, we can apply multiple contextual information as desired. However, in this fig, we have shown only one contextual attribute C_U which is a user-context matrix. The mathematical equation after applying contextual information is given as:

$$h(x) = \sigma(W_i * x + W_u * Cu + b_1) \tag{7}$$

Here, $h(x)$ is the output which is then given to the decoder and W_u is the weight matrix of dimension $\mathbb{R}^{k\,\mathrm{X}\,u}$ multiplied to user-context matrix C_U. The decoder then mirrors this output back to n-dimensional vector X_i, which is represented by $\mathrm{Y_i}$ in our model (predicted ratings) in accordance to $d = \sigma(W_{2*}\mathrm{x} + b_2)$. In general, d is not an exact reconstruction of h thus giving a loss which is denoted by Eq. (4). Note that the input vector is very sparse while the decoder output $d \in \mathbb{R}^{\mathrm{N}}$ is very dense. Finally, we will be backpropagating the values of weight and bias vector to get new values in every step and to minimize the loss. The output of the decoder is then fed into the collaborative filtering to get relevant recommendations.

Algorithm 1: The Deep Autoencoder algorithm

Input: Dataset $X = \{X_1, X_2, \dots X_N\}$ with n ratings, no of hidden layers (L), activation function (σ)

Output: Predicted ratings $Y = \{Y_1, Y_2, \dots Y_N\}$

Method: Deep Autoencoder Algorithm

1) $L = \{L_1, L_2 \dots L_n\}$
2) Initialize σ, W_1, b_1, λ=0.01, n=0.01
3) **for** each (L_i in L) **do**
4)   define hidden layer input $e = \sigma(W_1 * x + b_1)$
5)   define hidden layer output $d = \sigma(W_2 * x + b_2)$
6)   **while** not stopping criterion **do**
7)   compute e
8)   compute cost $= \beta \sum_j^m p \log \frac{n}{\sigma_i} + (1\text{-}p) p \log \frac{1-n}{1-\sigma_i} + \lambda \sum_{i=1}^{N} w_i^2$
9)   compute d
10)   compute loss function (4)
11)   **end while**
12) **end for**
13) BP to get updated values of W_1, b_1, W_2, b_2

Algorithm 2: The Hybrid Algorithm

Input: Dataset containing the context attributes A = age matrix, G=gender matrix, O = occupation matrix

Output: recommended items

Method: Hybrid-context Algorithm

1) Conduct feature extraction of matrix A, G and O
2) $L = \{L_1, L_2 \dots L_n\}$
3) **for** each (L_i in L) **do**
4)   concatenate extracted value with *deepautoencoder(X_i)*
5)   **for** each i, j ∈ users, items **do**
6)   Compute sim(i, j) by Pearson correlation
7)   Compute predicted ratings $R_{ij} = R_i + \frac{\sum_{u_k \in N} Sim(i,j) \times (R_{k,j} - R_k)}{\sum_{u_k \in N} Sim(i,j)}$
8)   **end for**
9) **end for**

5 Experiments and Results

5.1 Dataset

We will be evaluating our model on real-world datasets. The MovieLens-1M, MovieLens-100K dataset provides 100K/1M ratings for 1K/6K users and 1.7K/4K movies. Contextual information for the 1M dataset is three separate matrices of age, sex, and gender. Contextual information for the Movielens-100k dataset is a matrix of tags. The dataset also consists of movie ratings by users on the scale of 1–5.

5.2 Experimental Setup

For evaluating purposes, we will be splitting the dataset into 90% training set and 10% test set. RMSE and MSE are related by Eq. (8)

$$\mathrm{RMSE} = \sqrt{\mathrm{MSE}} \tag{8}$$

We consider baseline method for comparison with our model like neighborhood-based methods (*User-KNN, Item-KNN*), deep learning-based methods (*AutoRec*), and model-based methods (*SVD++*).

- *User-KNN*: User-based k-nearest neighbor collaborative filtering is used to predict the similarity between users. We selected k = 10 for this experiment.
- *AutoRec*: This is the simplest autoencoder that accepts user preference as input.
- *SVD++*: *SVD++* is an improved version of SVD algorithm, which is a matrix factorization method and achieves better accuracy due to the addition of implicit feedback information.
- *Item-KNN*: Item-based k-nearest neighbor collaborative is used to predict the similarity between items. We selected k = 10 to for this experiment.

For most of our experiments, we will be using a batch size of 256 and $\lambda = 0.001$, and we will be running the algorithms for 500 epochs/network. Below table shows how our deep autoencoder performs with various baseline models.

From Table 1, we can see that our model (DeepAutoRec) performs much better when there is more information, i.e., on large datasets although loss in both the dataset is much less as compared to other models and the improvement of *SVD++* over DeepAutoRec in Movielens-1M dataset is very small.

Table 1 MSE and loss calculated for various models with train/test ratio taken as 90/10%

Algorithms	MovieLens-1M		Movielens-100K	
	MSE	Loss	MSE	Loss
User-KNN	0.771	1.502	0.757	1.439
Item-KNN	0.780	1.205	0.782	1.150
AutoRec	0.772	1.481	0.779	1.252
SVD++	**0.688**	1.502	**0.675**	1.149
DeepAutoRec	0.695	**0.997**	0.780	1.059

The bold values represent the lowest one in the column. The lower the value the better

5.3 *Side Information*

To incorporate side information, we perform one hot encode on the three contextual information *age*, *gender*, and *occupation* matrix. The overall matrix is given by $C_U = A*G*O$, where A is the age matrix, G is the gender matrix, and O is the occupation matrix. Binary representation of C_U is then concatenated with the training set. For this experiment, we have divided the entire training set in five clusters.

5.4 *Impact of Contextual Information*

At first sight, it may seem that the impact of contextual information has limited effect, but in the 1M dataset, the information pertaining to users and ratings are not uniform, the ratings dataset has a lot of information as compared to the no. of users, thereby adding extra contextual information has little effect. Users who have rated less movies or movies with less ratings should see a bit of an improvement as compared to others. We compute the RMSE of the training set, to understand the impact of contextual information after dividing the entire training set in five clusters. Whenever there is less rating for a movie, the side information becomes more important. A careful analysis of Table 2 shows that there is an improvement in RMSE in all the clusters depending on the amount of information in that cluster. The

Table 2 RMSE calculated on a cluster of movie ratings. The entire training set was divided into a set of five clusters. For instance, the first cluster contains 20% of the movie ratings. DeepAutoRec* is when contextual information is concatenated to the DeepAutoRec model

Clusters	DeepAutoRec	DeepAutoRec*
Cluster zero	0.867	0.858
Cluster one	0.857	0.840
Cluster two	0.865	0.846
Cluster three	0.863	0.824
Cluster four	0.864	0.842
Cluster avg	0.863	0.852

lesser the ratings, the more improvement we get, this is very desirable as effective use of contextual information can help us deal with new products.

According to Table 2, we get a maximum improvement of 4% according to cluster three.

5.5 *Impact of Different Activation Function and Increasing the No. of Layers*

In this experiment, we have used three activation functions scaled exponential linear units (SELU), hyperbolic tangent(tanh) and exponential linear unit (ELU). The activation functions are used in accordance with Eqs. (1) and (2), and basically, activation function acts as a filter that is applied to the input before sending it to the output. From the table, we can see that with deeper layers does bring down RMSE, with increasing the no of layers will increase the network's ability to generalize. In this experiment, we set the lowest layer as 128 and will go up to as high as 1024, and also to avoid overfitting, we will put the value of $\lambda = 0.01$ from 0.001.

5.6 *Impact of Different Activation Type and no. Of Layers*

Going deeper from one layer to another in encoder to decoder with any activation function provides good improvement in RMSE, and from Table 2, we get the maximum improvement when the no. of layers is increased from 2 to 4 and then the % of improvement decreases, in case of *Tanh, ELU* activation layer (from 1.105 to 0.957) after that if we increase the no. of layers, it does help but with diminishing returns.

6 General Results

Our deep learning-based model as explained in this paper is different to other attempts as our model incorporates contextual information into a unique neural network. Our model is an effective way to recommend items, although our model suffers from some problems (discussed in Sec. 7.), it offers various other advantages like modeling a complex nonlinear equation using activation functions. Tables 1, 2, and 3 summarize the various results that we get on the movielens-1 M and 100 k dataset. In Table 1. *Deep AutoRec* defeats state-of-the art model *SVD++* in terms of data loss in the algorithm but the MSE is a bit higher than *SVD++,* we are also getting a much better MSE for *Deep AutoRec* in MovieLens-1 M dataset. For the 100K dataset, the results are not so satisfying as *SVD++* has much better result although by a small

Table 3 Evaluation of RMSE for DeepAutoRec model with different no. of layers and different activation types

Activation type	No. of layers	RMSE
SELU, SELU	2	0.901
	4	0.899
	8	0.875
	12	0.866
ELU, ELU	2	0.953
	4	0.915
	8	0.902
	12	0.894
Tanh, ELU	2	1.105
	4	1.011
	8	0.957
	12	0.950

margin. In Table 2, we tried to compute the RMSE of the training set when contextual information is applied, and we find that we get a maximum improvement of 4% for cluster no. three. For Table 3, we compute RMSE of our model with different activation function, and we find that with increasing no of layers, the RMSE improves but at higher layers the improvement is very minimal. The results that we get are plotted and shown in Fig. 2 and Fig. 3 but only for only 4 and 12 layers.

One major problem of collaborative filtering is scalability since with increasing no of users and items, it becomes very difficult to give relevant recommendation. With increasing information, the nonlinear model becomes more complex and as a result training also becomes difficult. Making an efficient model which can be trained in reasonable amount of time is always desirable. We have trained our model on a GTX 1050Ti and on 8 GB memory, and it took around 15 min on an average to train around 500 epochs/ network.

7 Conclusion

In this paper, we introduced a new kind of recommendation system using deep autoencoders and showed how our deep learning model performs, and we successfully trained our model with various activation functions and with multiple hidden layers, thereby increasing the learning rate, and further improving the generalization performance of our model with more accurate result. The output ratings that we get from the deep autoencoder model is then fed into the collaborative filtering module whose output is the predicted ratings which can now be used to give relevant recommendations using kNN or any other technique. For this paper, we have taken movie ratings as an input, but our model can be trained with various other ratings.

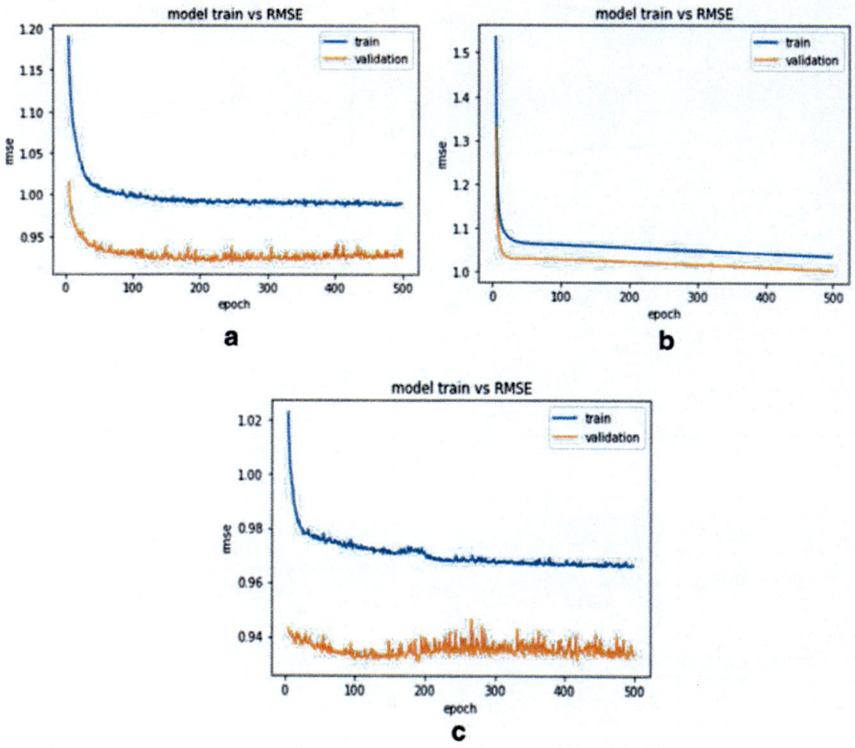

Fig. 2 Training the algorithm with four layers (two encoder layers and two decoder layers). Y-axis: RMSE and X-axis: epoch. Here **a** denotes tanh, ELU activation, **b** denotes SELU, SELU and **c** ELU, ELU. For **a** the graph clearly indicates that the graph overfits more easily as compared to **b** and **c** as the gap between train and validation is less

This work can be improved in many ways, and we plan to do so in the future. Like, adding more dropout noise (we have added only one dropout value that is 0.8), adding more contextual information and implementing cross-validation and tune hyperparameters for more effective results. Our model suffers from the same problem; as any other deep learning models do, it requires a lot of training and a lot of data. Now getting massive amount of data in a real-life scenario becomes time-consuming and training the complex model by incorporating the data becomes even more difficult as it requires serious amount of GPU and CPU power.

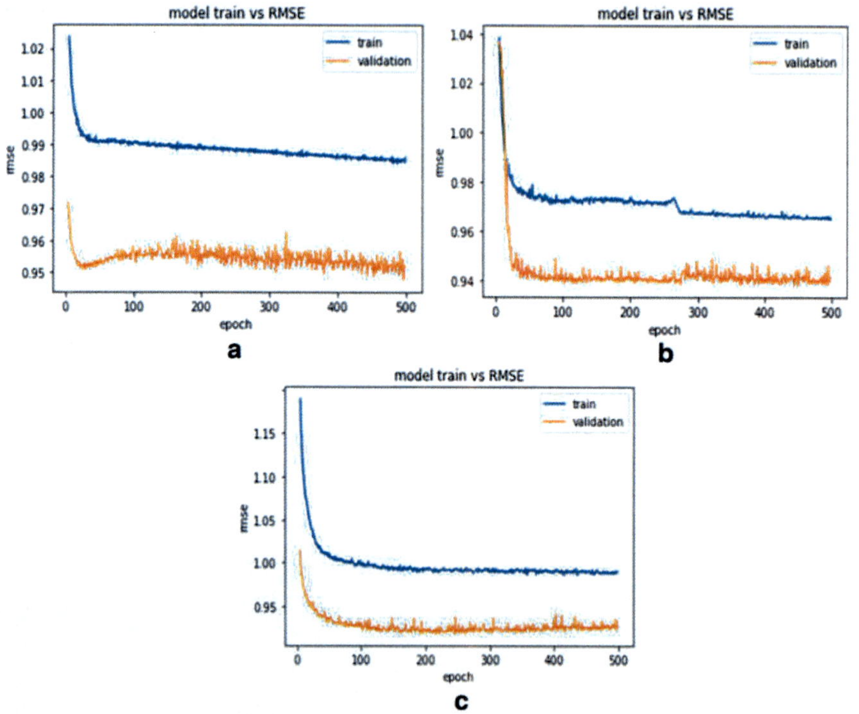

Fig. 3 Training the algorithm with 12 layers (six encoder layers and six decoder layers). Y-axis: RMSE and X-axis: epoch. Here **a** denotes tanh, ELU activation, **b** denotes SELU, SELU, and **c** ELU, ELU. Here, we have added a dropout of 0.8. From the graph, we can see that adding dropout and increase in no of layers helps as the graph overfits much slowly as compared to four layers, and we also see an improvement in RMSE

References

1. L. Dawen, K.G. Rahul, H.D. Matthew, J. Tony, *Variational Autoencoders for Collaborative Filtering* (2018)
2. D.E. Rumelhart, Hinton, *Distributed Processing and Exploration in the Microstructure of Cognition* (MIT Press, Cambridge, MA, USA 1986), pp. 318–362
3. S. Ruslan, M. Andriy, Hinton: *Restricted Machines for Collaborative Filtering* (2007)
4. R. Mina, A. Seyyed, A context-aware and user behavior-based recommender system with regarding social network analysis. in *14th International Conference on e-Business Engineering* (China, 2017)
5. H. Huang, Context-aware location recommendation using geotagged photos social media. Int. J. Geo-Inf. (2016)
6. Y. LeCun, Bengio, H. Geofrey, *Convolutional Networks for Images, Speech, and Time Series* (1995)
7. M. Gao, J. Zhang, J. Yu, J. Li, J. Wen, Q. Xiong, Recommender system based on generative adversarial network (2020). arXiv preprint arXiv:2003.02474
8. I. Goodfellow, Y. Bengio, M. Mirza, S. Ozair, B. Xu, A. Courville, Generative Adversarial Network (2014). arVix preprint. arXiv:1406.2661

9. G. Pooja, Context aware computing systems: a survey. in *IEEE 5th International Conference on IoT in Social, Mobile, Analytics and Cloud* (2018)
10. R. Bell, Y. Koren, C. Volinsky, *The BellKor Solution to Netflix Prize* (2007)
11. V. Athanasios, D. Nikolaos, D. Anastasios, Protopapadakis Eftychios: vol. 3. *Deep Learning for Computer Vision* (2018)
12. S. Suvash, M.K. Aditya, S. Scott, X. Lexing, *Autoencoders Meet Collaborative Filtering* (2015)
13. S. Harald, Embarrassingly shallow autoencoders for sparse data (2019). arVix preprint. arXiv: 1905.03375
14. O. Kuchaiev, B. Ginsburg, Training Deep Autoencoders for Collaborative Filtering (2017). arXiv preprint arXiv:1708.01715

Exploration and Implementation of RSA-KEM Algorithm

Manish Kumar and Seba Susan

Abstract Cryptography is used to protect information. Encryption and decryption can confirm the confidentiality, integrity of information and protect information from tampering, forgery and counterfeiting. RSA (Rivest, Shamir, Adleman) uses two keys: private key and public key. RSA-KEM (key encapsulation mechanism) is a hybrid encryption algorithm that uses RSA trapdoor permutation along with a key derivation function. RSA is more powerful than other symmetric key algorithms, and its keys are two large randomly generated prime numbers. It is hard to factor a large prime number which is a product of two prime numbers. This paper presents an implementation of a RSA-KEM encryption decryption based on RSA public key algorithm and KEM. RSA contains three functions: key generation, encryption and decryption. In RSA-KEM, password-based key derivation function is used for key stretching. Password, salt and iteration are used for generation of derived key. Advanced encryption standard is used for key-wrapping scheme. AES is a vigorous security algorithm and can be implemented in both hardware and software. It is resilient to hacking attempts due to long key lengths of 128,192 and 256 bits. RSA-KEM creates a random integer '*r*', manages symmetric encryption key by '*r*' along key derivation function (KDF) and then encodes '*r*' with RSA.

Keywords Encryption · Decryption · Key encapsulation mechanism · Advanced encryption standard · Primality testing

1 Introduction

The principles and methods of transforming plaintext into ciphertext, then converting ciphertext into plaintext is cryptography, and method and technique of converting ciphertext into plaintext without using key are known as cryptanalysis. In symmetric key, only one key is used for encryption, and the same key is again used for decryption. Both parties have the same key. In public key cryptography, one of the keys is

M. Kumar (✉) · S. Susan
Department of Information Technology, Delhi Technological University, Delhi, India

G. Ranganathan et al. (eds.), *Soft Computing for Security Applications*,
Advances in Intelligent Systems and Computing 1397,
https://doi.org/10.1007/978-981-16-5301-8_12

Table 1 Different types of cryptanalytic attacks

Cryptanalytic attacks	Description
Ciphertext only	Same ciphertext is available for decryption
Known plaintext	Only copy of ciphertext and its plaintext is available to the cryptanalyst for decryption
Chosen plaintext	Only use encryption algorithm/machine and can be used for many plaintext and corresponding ciphertext for trying to find key
Chosen ciphertext	Only use decryption algorithm/machine and can be used for many symbols or string of plaintext to find key

available in public domain, and other related key is kept secret. It is not possible to derive one key with use of another key and algorithm. In encryption, we use an algorithm and one of the keys. Using different key, we get different ciphertext using the same algorithm. After the ciphertext is sent to the receiver, using the other key and algorithm, the original plaintext is derived. Techniques like RSA PKCS#1 [1] use deterministic padding scheme. When server confirms the padding and, in some way, leaks the outcome, it may be a threat for system. In RSA-OAEP [2], keys are less efficient as compared to other schemes. It can leak information through timing attack. In RSA-KEM [3], the keys are two large randomly generated prime numbers. It is hard to factor a large prime number which is a product of two prime numbers. RSA cryptosystem is universally accepted. RSA-based system has two keys one private and other one is public. One key (public) is available for everyone, and another key (private) is with only one person. If '*X*' has to send some data to '*Y*', '*X*' does this by using public key of '*Y*'. After receiving data from '*X*', '*Y*' uses its private key to get the data. RSA-KEM is an accepted key encapsulation mechanism. It uses RSA trapdoor permutation along with a key derivation function (KDF). The RSA-KEM key transport algorithm uses store-and-forward concept for sending data to a beneficiary. This is done by using the beneficiary public key. The security depends on algorithm used and key length. In public key cryptography, two keys are required. Public key is available in public domain and can be used by everyone for sending messages and for signature verification. Private key is known by only one person and is important for decrypting messages and digital signature. Some of the attacks and security services are given in Table 1. The organization of the paper is as follows. Section 2 describes RSA, Sect. 3 describes RSA-KEM and its implementation, Sect. 4 analyses the results, and Sect. 5 concludes the findings of the paper.

2 RSA (Rivest, Shamir, Adleman) Algorithm

In early days, all cryptographic systems used permutation and substitution, and after that rotor encryption and decryption machine was used. Public key cryptography is the latest development which uses mathematical functions. Security for cryptography

Table 2 Different algorithms for public key cryptography

PKC algorithm	Encryption/Decryption	Digital signature	Key exchange
RSA	Used	Used	Used
Elliptic curve	Not used	Used	Used
Diffie–Hellman	Not used	Not used	Used
DSS	Not used	Used	Not used

system is based on key length and the effort involved in cracking the cipher. Nowadays, due to computational requirement, we use PKC in key management, digital signature. Some of the PKC algorithms are given in Table 2.

Asymmetric algorithm works on two keys, encryption by one key and decryption by another connected key. It is impossible to find the decryption key when only encryption key and cryptography algorithm is known. RSA algorithm is accepted as a general-purpose technique for public key cryptography [4–7]. RSA processing of multiple blocks is shown in Fig. 1.

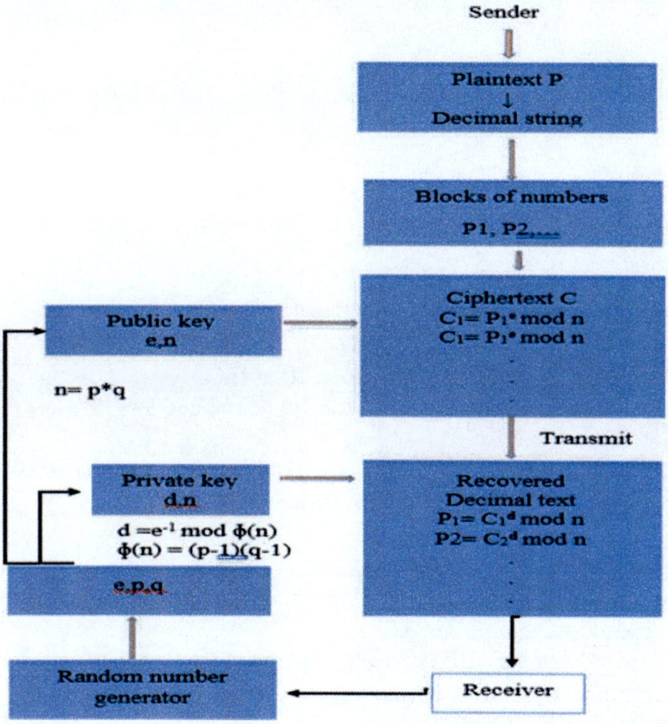

Fig. 1 RSA processing of multiple blocks

2.1 Algorithm

Key Generation

In RSA, each party who wishes to communicate using encryption required to make a pair of keys: One is public key, and other one is private key. We select two distinct primes p and q picking randomly and close in magnitude but varying in length by few numbers that makes factoring unbreakable. Using primality test, we find prime numbers p and q, and it is kept secret. Public key is available in public domain and can be used by everyone for sending messages and for signature verification. Private key is known by only one person, and it is important for decrypting messages and digital signature.

Key generation for RSA algorithm [8].

i. Take two prime numbers p and q.

Prime integers p and q be taken randomly and near same bit length.

ii. Calculate number $n = p \times q$

iii. Calculate Euler's totient function equal to $(q - 1)\ (p - 1)$

Same as $[n - (q + p - 1)]$.

iv. Take 'e' which is greater than 1 and less than Euler's totient function and gcd equal to 1.

'e' is public key and of small bit length.

v. Calculate d which is congruent to $[e - 1 \ (\text{mod}\ \varphi(n))]$.

Solving for an example:

Take two prime numbers $p = 17$ and $q = 11$.

Compute $n = \text{p} \times \text{q} = 17 \times 11 = 187$

Compute $\phi(n) = (p - 1)(q - 1) = 16 \times 10 = 160$.

Pick e such that e is relatively prime to $\phi(n) = 160$ and less than $\phi(n)$. Picking $e = 7$.

Decide d such that $d * e \equiv 1 \ (\text{mod}\ 160)$ and $d < 160$, value of $d = 23$.

Finally, public key = (7,187) and private key = (23,187).

Encryption

'X' keeps his public key open for all and keeps the private key confidential. 'Y' wants to send message 'm' to 'X'.

'Y' first uses public key of 'X' which is available for everyone.

Ciphertext that 'Y' sends is $m^e (\text{mod}\ n)$.

'Y' then sends ciphertext to 'X'.

Decryption

When 'X' gets the ciphertext sent by 'Y', it uses its private key to decrypt it.

This is done by calculating $c^d \pmod{n}$.
This is equal to the original message 'm'.
A can recover the original message 'm' by reversing the scheme.

2.2 *Primality Testing of a Number*

Miller Rabin primality testing [5, 9] is used to test the primality of a given number. For a given number 'n', we use this test to find whether this number is prime or not. This used the property of primes that is explained below.

Properties of Primes

Whenever p is prime, q is +ve odd integer and q less than p.
Subsequently, $a^2 \bmod p = 1$
only possible in two cases

(1) a mod p = 1
(2) a mod p = −1.

Let p be prime number > 2,
then $p - 1$ equal to $2^k\ q$, where $k > 0$,
if 'a' is integer such that $1 < a < p - 1$,
then any one of the two states is correct.
1. a^q Mod P = 1
or equal to $a^q = 1 \pmod{p}$
2. $a^q, a^{2q}, a^{4q}, \ldots, a^{2^{(k-1)}q}$ is congruent to −1 Mod P

Miller Rabin Algorithm

Input: $n > 3$, number that is to be tested for prime and it should be odd number.
Find k and q, k is positive number, and q is odd.
Using $n - 1 = 2^k.q$
Taking random 'a' such that 'a' is positive and less then n − 1.
If we get a^q mod n = 1, its outcome be
'undetermined' not able to tell it is not prime.
when $j = 0$ to $(k - 1)$.
if $a^{2j}.q$ mod $n = n-1$, then result confirmed that the number is not prime, but composite.

3 RSA-KEM (Rivest, Shamir, Adleman-Key Encapsulation Mechanism) Algorithm

RSA-KEM [3, 4, 10, 11] is a hybrid RSA algorithm incorporating key encapsulation mechanism. It uses RSA trapdoor permutation along with a key derivation function (KDF). The RSA-KEM key transport algorithm uses store-and-forward concept for sending data to a beneficiary. This is done by using the beneficiary RSA public key. Key encapsulation mechanism is a way to secure a symmetric key to send information from one place to another using public key algorithm. In symmetric key, the encryption and decryption keys are available to both parties: one is sender, and the other is receiver. The encryption key is known to all parties, and the decryption key is derived from it. In some cases, only one key is used for the encryption and decryption. In public key algorithm, encryption and decryption are slow, and it is not reliable for transmitting long messages. In public key cryptography, one key is available in public domain, and other related key is kept secret. It is not possible to derive one key with the use of the other key and algorithm. Public key, which is available in public domain, is known by everyone and can be used to encrypt messages and verify signatures. In private key system, only the receiver knows the private key that is used to decrypt messages and sign signatures. Public key algorithms are good for exchange of symmetric keys. After exchange of key, this key is further used for encryption of long messages. Public key system is used to exchange symmetric keys, which are relatively short, then this symmetric key is used to encrypt long messages. First a random symmetric key is generated, and then it is encrypted using public key algorithm. It is sent to the recipient, where it is decrypted to get the symmetric key. In RSA-KEM [3], password-based key derivation function (PBKDF) is used for key derivation. Advanced encryption standard (AES) is used for key wrapping. AES has ten or more rounds depending on the key size. Rounds consist of substitution, transposition and XOR operations, and the output of AES is 128 bits long which is given as input to RSA. Complete workflow of RSA-KEM is given in Fig. 2.

3.1 Algorithm

For sending a symmetric key with public key algorithm, first we.

- Generate a random symmetric key
- Encrypt it with public key algorithm
- Send it to recipient
- Recipient decrypts it, using public key and gets symmetric key.

Generally, symmetric key is small, we use padding for security purpose, but padding security is not secure. In KEM, we first generate random element using finite field. Then, we use hashing for deriving the symmetric key.

$M =$ symmetric key 128/256 bits length,

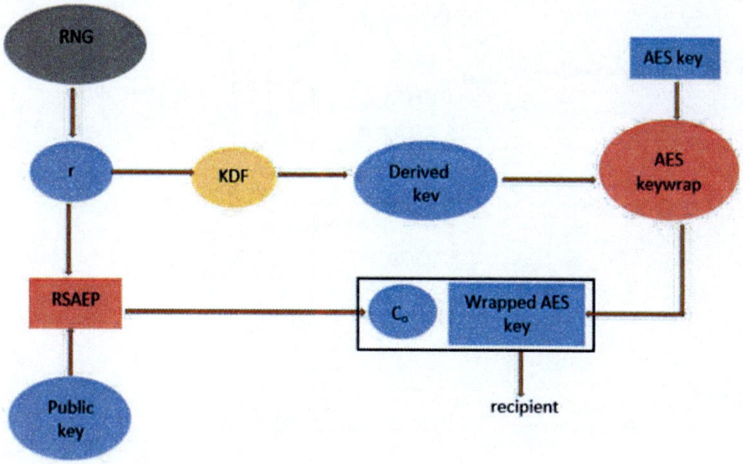

Fig. 2 RSA-KEM workflow

we make it larger m, $1 < m < n$, then
$c = m^e(\text{mod } n)$
Alice can get m using
$m = c^d \text{ (mod } n)$
the M can be derived by reversing padding scheme used
In place of generating random symmetric key M, Bob generates random m, $1 < m < n$
Deriving symmetric key $M = \text{KDF}(m)$
$c = m^e(\text{mod } n)$
$m = c^d \text{ (mod } n)$
then
symmetric key $M = \text{KDF}(m)$
M is calculated from m, but reverse is not possible. KDF is one-way function. If attacker somehow gets M, it would not be possible to get the plaintext m.

RSA-KEM creates a random integer 'r', manages symmetric encryption key by 'r' along key derivation function (KDF) and then encodes 'r' with RSA.

3.2 *Password-Based Key Derivation Function (PBKDF)*

Traditionally, the password is hashed and stored in database, which can be easily attacked by brute force or rainbow table attack. Hashed password can be used as a key in encryption and decryption in 3DES, AES and other algorithms. As CPU and GPU processors are getting more powerful and faster, cracking password is easy. We make password hash harder to break by using salt. Salt is a large sequence of

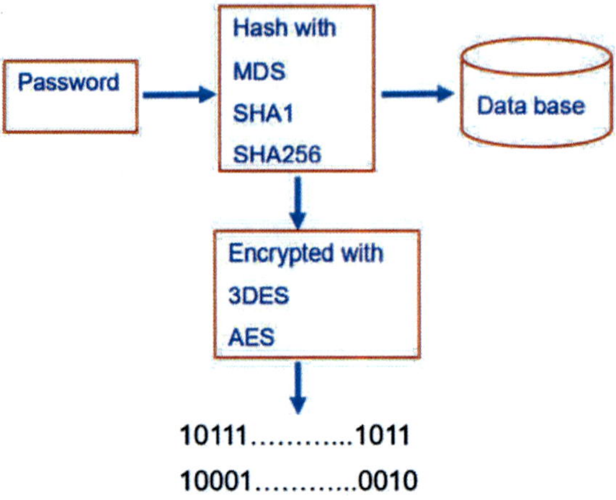

Fig. 3 Password hashing

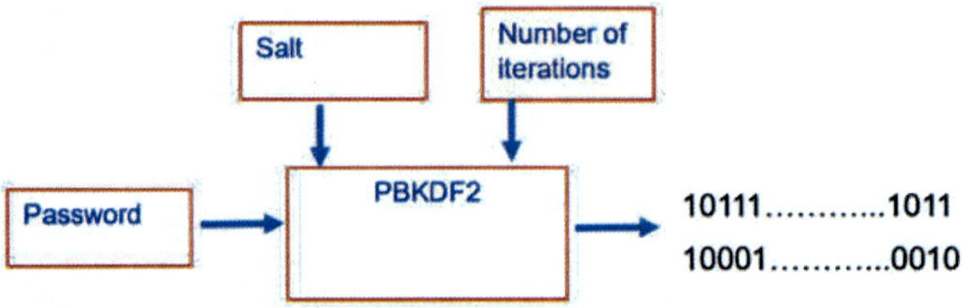

Fig. 4 Password-based key derivation function (PBKDF)

randomly generated data that is added with the password. We use password-based key derivation function. Block diagram of password hashing is given in Fig. 3.

Salt is a password before hashing. It protects from brute force attack but not from rainbow attack. It can be attacked by attacker using rainbow table. Only rainbow table size increases by adding salt as a permutation of password added into table. Password-based key derivation function [12] is used for security against brute force attack and rainbow table attack. In this method, time taken to test each possible case is increased. The number of iterations is added. It slows down the key generation and safeguards against rainbow table attack. For protection from rainbow attack, we use password-based key derivation function (key stretching algorithm). Block diagram of PBKDF is given in Fig. 4.

3.2.1 Algorithm

PBKDF has these input parameters:

DK = PBKDF (Password, salt, c, sha256, dklen)

Password = like manish, India, killer…
C is number of iteration > 1000
Salt is a sequence of bits, 32 bits
dklen is length of key we want to generate, 128 bits
DK is derived key
Derived key DK = K1 + K2 + K3 + … Kdklen/hlen
Ki = F (Password, Salt, c, i)
Function F is XOR of pseudorandom function
F(Password, Salt, c, i) = J1 ^ J2 ^… ^Jc
Ji = PRF (Jprevious).

3.3 *Advanced Encryption Standard (AES)*

Advanced encryption standard (AES) [11, 13] is used for key wrapping scheme. 128 bits/192 bits or 256 bits of block size and 128 bits of key as input to AES are required. AES has ten or more rounds depending on key size. Rounds consist of substitution, transposition and XOR operations, and output of AES is 128 bits, which is used as the input to RSA. Block diagram of AES and key size with rotations is given in Fig. 5 and Table 3, respectively.

Steps of AES [11, 13]

a. Substitution bytes
b. Shift rows

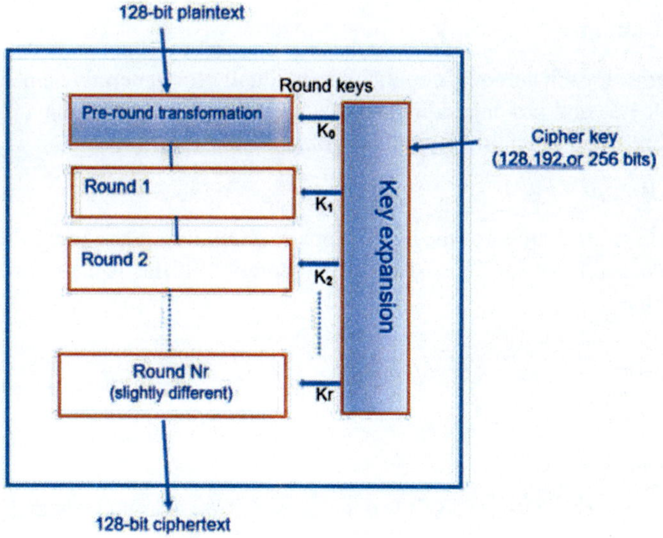

Fig. 5 Block diagram of advanced encryption standard

Table 3 Rotation and key size

r	Key size
10	128
12	192
14	256

Table 4 Rows and their rotations

First row	Rotated by 0 bytes
Second row	Rotated by 1 bytes
Third row	Rotated by 2 bytes
Fourth row	Rotated by 3 bytes

c. Mix columns
d. Add round key.

a. **Substitution Bytes**

In this each byte, ai,j is replaced by S(ai,j). Every byte is changed by a different value. S-box is used which is multiplicative inverse over Galois field GF(2^8) for changing the value that is replaced.

b. **Shift Rows**

In this, circular right shift is done. Each row is shifted by fixed number of bytes. For the first row (R_0), we do nothing. In second row (R_1), 1 bit is shifted, in the third row (R_2), 2 bits are shifted and so on. Rows and their rotations are given in Table 4.

c. **Mix Columns**

In this segment, each column is operated individually for generating a new column. Column's bytes and pre-defined matrix are used for output. Multiplication is done with column's bytes and pre-defined matrix.

d. **Add Round Key**

In this segment, the output of the mix column is XORed with the key. Output of this stage is given as input for the next round of operation. If this is the last round, then it is ciphertext.

3.3.1 Algorithm

Initial round

- AddRound key—Bitwise XOR is done on each byte of round key and states.

Main round

- SubBytes—Each byte ai,j is replaced by S(ai,j), using S-box.
- ShiftRows—Circular right shift is done.
- mixColumns—Multiplication with column's bytes and pre-defined matrix.
- AddRound Key—Output of the mix column is XORed with key.

Final round

- SubBytes
- ShiftfRows
- AddRound Key.

An example of AES, with the plaintext (in hexadecimal), key and resulting ciphertext, is shown below. Expansion of 16 byte key into ten rounds keys is given in Table 5, and progression of state through AES encryption process is in Table 6.

Table 5 Expansion of 16 byte key into ten round keys

Plaintext	0123456789abcdeffedcba9876543210
Key	0f1571c947d9e8590cb7add6af7f6798
Ciphertext	ff0b844a0853bf7c6934ab4364148fb9

Table 6 Progression of state through AES encryption process

Start of round	After Sub Bytes	After Shift Rows	After Mix Columns	Round Key
01 89 fe 76 23 ab dc 54 45 cd ba 32 67 ef 98 10				0f 47 0c af 15 d9 b7 7f 71 e8 ad 67 C9 59 d6 98
0e ce f2 d9 36 72 6b 2b 34 25 17 55 ae b6 4e 88	ab 8b 89 35 05 40 7f f1 18 3f f0 fc E4 4e 2f c4	ab 8b 89 35 40 7f f1 05 F0 fc 18 3f C4 e4 4e 2f	B9 94 57 75 E4 8e 16 51 47 20 9a 3f C5 d6 f5 3b	dc 9b 97 38 90 49 fe 81 37 df 72 15 B0 e9 3f a7
⋮	⋮	⋮	⋮	⋮
Cc 3e ff 3b A1 67 59 af 04 85 02 aa A1 00 5f 34	4b b2 16 e2 32 85 cb 79 F2 97 77 ac 32 63 cf 18	4b b2 16 e2 85 cb 79 32 77 ac f2 97 18 32 63 cf	4b 86 8a 36 B1 cb 27 5a Fb f2 f2 af Cc 5a 5b cf	B4 8e f3 52 Ba 98 13 4e 7f 4d 59 20 86 26 18 76
Ff 08 69 64 0b 53 34 14 84 bf ab 8f 4a 7c 43 b9				

Keywords	Auxiliary function
WO = 0f 15 71 c9 W1 = 47 d9 e8 59 W2 = 0c b7 ad d6 W3 = af 7f 67 98	RotWord (w3) - 7f 67 98 af SubWord (xl) = d2 85 46 79 Rcon (1) = 01 00 00 00 Y1 ffi Rcon (1) = d3 85 46 79 = zl
W4 = w0 ⊕ zl = dc 90 37 b0 W5 = W4 ⊕ wl = 9b 49 df e9 W6 = w5 ⊕ w2 = 97 fe. 72 3f W7 = w6 ⊕ w3 = 38 8115 a7	RotWord (w7) = 81 15 a7 38 = x2 SubWord (x4) = 0c 59 5c 07 = y2 Rcon (2) = 02 00 00 00 Y2 ⊕ Rcon (2) = 0e 59 5c 07 = z2
⋮	⋮
W36 = w32 ⊕ z9 = fd 0d 42 cd W37 = W36 ⊕ W33 = 0e 16 e0 lc W38 = w37 ⊕ w34 = c5 d5 4a 6e W39 = W38 ⊕ w35 = f9 6b 41 56	RotWord (w39) = 6b 41 56 f9 = x10 SubWord (×9) = 7f 83 bl 99 = y10 Rcon (10) = 36 00 00 00 YlO ⊕ Rcon (10) = 49 83 bl 99 = z10
W40 = w36 ⊕ z10 = b4 8e f3 52 W41 = w40 ⊕ w37 = ba 98 13 4e W42 = w41 ⊕ w38 = 7f 4d 59 20 W43 = w42 ⊕ w39 = 86 26 18 76	

4 Results

In this paper, we have implemented RSA and RSA-KEM algorithm using GNU multi-precision library in Linux platform. The application uses a 512 bit modulus RSA implementation, which is sufficient for non-critical applications. The libraries GNU MP arbitrary precision library (C/C++) and open SSL crypto library (C/C++) are used. The GMP library is a cross-platform library, implying that our application should work across platforms with least modifications.

In RSA algorithm, we take two prime numbers, $p = 53$ and $q = 59$ as the input. Then $n = 3127$. '*Y*' take $e = 3$, finally, '*Y*' chooses $d = 2011$, values $n = 3127$ and $e = 3$ is public (3, 3127) and value $d = 2011$ secret (2011, 3127). '*Y*' wants to send the letters '*m*', '*a*', '*n*', '*i*', '*s*', '*h*' to '*X*'. Putting letter as a number from 1 and 26 ('*a*' = 1 and '*z*' = 26), 'Y' and 'X' perform encryption and decryption as shown in Table 7 and Table 8. Input as 'manish kumar' is given in RSA, and its encryption and decryption are given in Figs. 6, 7 and 8.

In Miller–Rabin primality testing, we gave input of 20 digits, 24 digits and 25 digits numbers and check the primality of the numbers; test results are given in Table 9, and the screenshot of Miller Rabin test is given in Fig. 9.

In password-based key derivation function, we gave input plaintext as Delhi, technological, university, and the outputs are shown in Table 10. A number of iterations are taken as 1000, and salt is a 32 bit sequence.

Table 7 Y's encryption using $e = 3$, $n = 3127$

Message letter	Corresponding number 'm'	m^e	Encrypted message m^e mod n
m	13	2197	2197
a	1	1	1
n	14	2744	2744
i	9	729	729
s	19	6859	605
h	8	512	512

Table 8 X's decryption using, $d = 2011$, $n = 3127$

Encrypted message m^e mod n	c^d	c^d mod n	Original message letter
2197	2.63174900339550537925966745684471e + 6720	2197	m
1	1	1	a
2744	3.89438825533405629735806630461 77e + 6914	2744	n
729	8.81169481703714985949896268962 18e + 5756	729	i
605	1.28842287298233740726429966819 58e + 5594	605	s
512	2.19731096228713700992556454807 58e + 5448	512	h

In AES, we use key as 'manish dtu' and plaintext as 'info system', and when the ciphertext is decrypted we get the same plaintext. For different keys and plaintexts, the output is given in Table 11.

Output of AES is given as input for RSA module. Output of RSA is sent to recipient. Getting message from sender, recipient decrypts it using its private key and gets the symmetric key. Screenshot of AES is given in Fig. 10.

5 Conclusions

Previously used techniques like RSA PKCS#1 use deterministic padding scheme. When server confirms the padding and, in some way, leaks the outcome, it may be a threat for the system. It allows small amount of randomness 8 bits in range 1–255. In RSA-OAEP, keys are less efficient as compared to other schemes. It can leak information through timing attacks. In RSA-KEM, the keys are two large randomly generated prime numbers. It is hard to factor a large prime number which is a product of two prime numbers. For primality testing in RSA, Miller–Rabin algorithm is

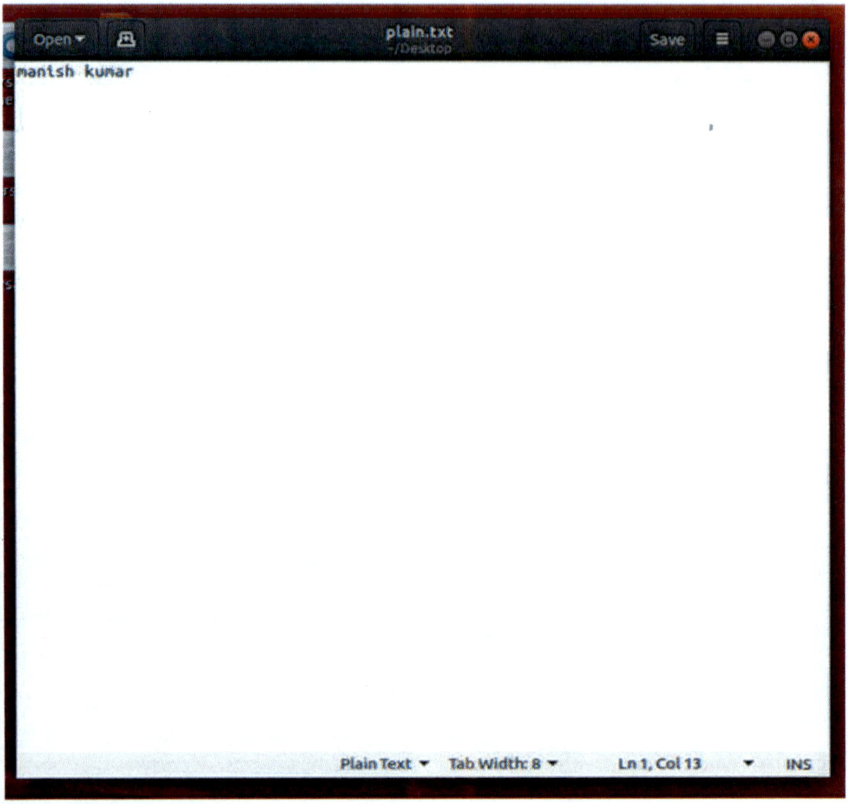

Fig.6 Plaintext for RSA algorithm

used. It takes the integer as an input and tests whether that number is prime or not. RSA contains three functions: key generation, encryption and decryption. It takes number of bits for n where (n = pq) and number of bits for e (public exponent) as a input. In RSA-KEM, password-based key derivation function is used for key stretching algorithm. Password, salt and iteration are used for generation of derived key. Advanced encryption standard is used for key wrapping scheme. AES has ten or more rounds depending on the key size. Rounds consist of substitution, transposition and XOR operations, and output of AES is 128 bits long, which is used as an input to RSA. Future work should be in quantum computers since the architecture of a quantum computer might be able to break RSA. Shor's algorithm is used for integer factorization which is polynomial time for quantum computer. This can be threat for RSA security. Classical cryptography can be broken by quantum computer. All current systems would be on threat.

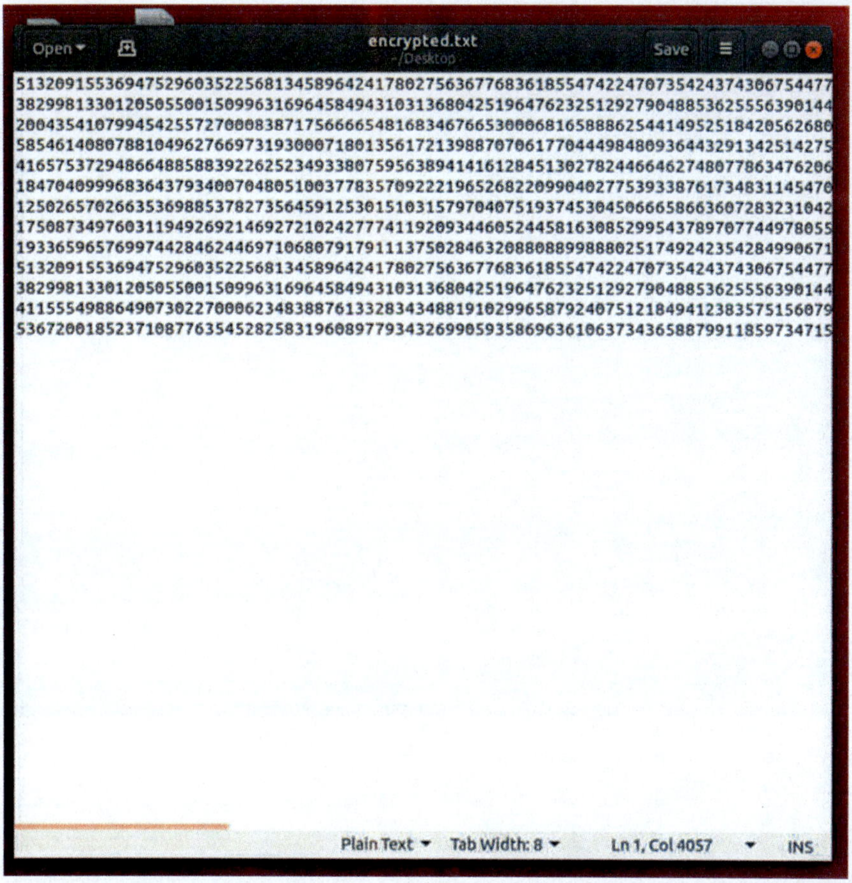

Fig.7 Encrypted output of plaintext for RSA algorithm

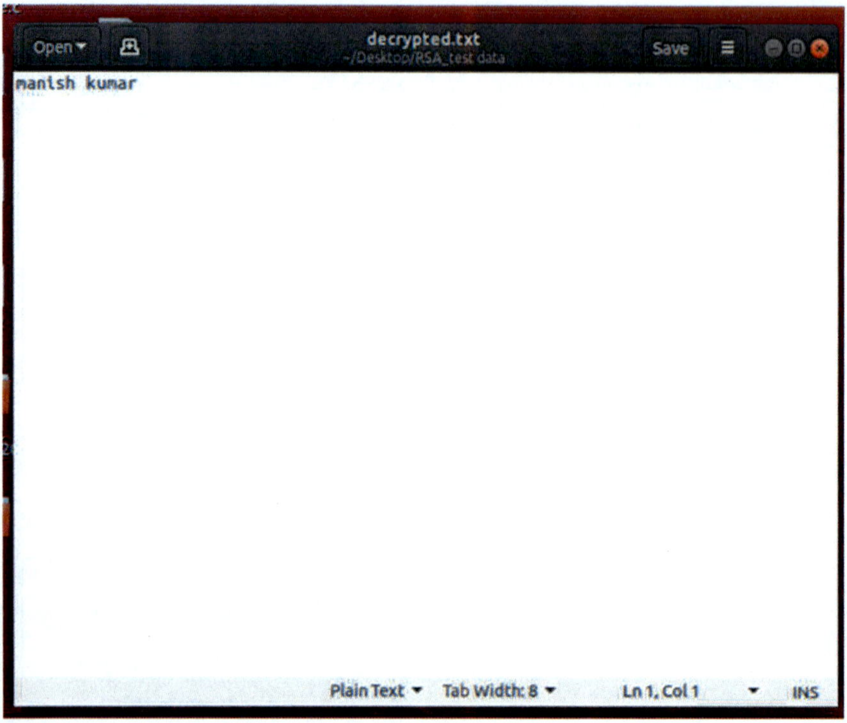

Fig. 8 Decrypted output of RSA algorithm

Table 9 Test results of Miller–Rabin algorithm

No. of digits	Number	Output
20	10,013,236,879,455,627,894	Composite
20	10,089,886,811,898,868,001	Prime
24	250,000,000,000,000,000,000,015	Composite
24	253,977,540,775,422,754,427,545	Prime
25	1,000,000,000,000,000,000,000,061	Composite
25	1,015,910,163,101,691,017,710,181	Prime

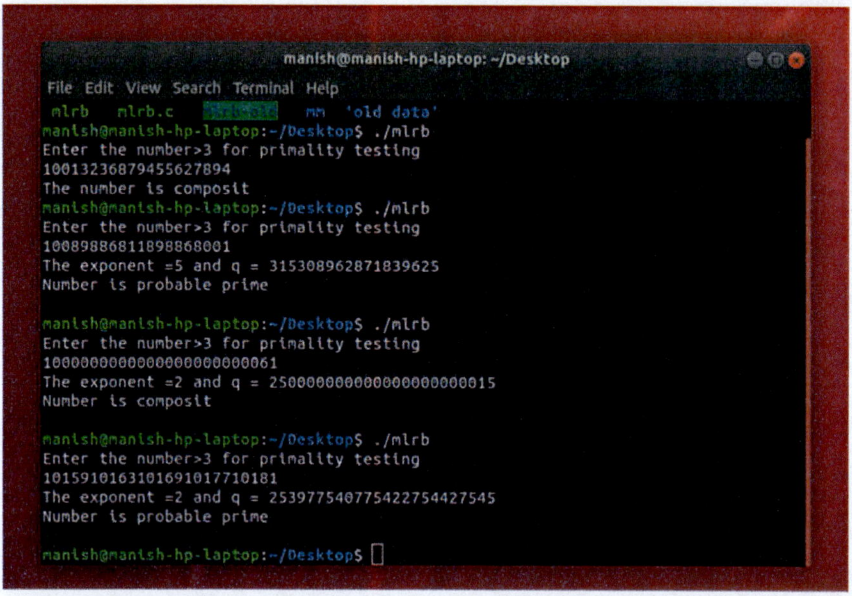

Fig. 9 Decrypted output of RSA algorithm

Table 10 Output of PBKDF

S. No.	Plaintext	Hash function output (PBKDF)
1	Delhi	2dc4518bfdeeb54a48136242b0df4470d9d2933bee1593e2d46912d0d69c1ede
2	Technological	2b0e27f4d53899137f00daacb0feeb9d3438483fc48a87daf6f81866ccefe629
3	University	362e2e0b74f956e3e595f76b2f897798ea66b5ca4cfa8d83d174f3354128f54c

Table 11 Results of AES

S. No.	Key	Plaintext	Ciphertext	Decrypted ciphertext
1	manish dtu	info system	B72243ff5024b7bcd8a83b90d5c47484	info system
2	manish kumar	this is secret	65eb480610e6ed25414f78707fdbbb	manish kumar
3	I am key	Database2	A222960c3cfc2dfa14aa8be681c7a2d	I am key

Fig. 10 Screenshot of output of AES

References

1. J. Jonsson, B. Kaliski, *Public-Key Cryptography Standards (Pkcs)#1: Rsa Cryptography Specifications Version 2.1 Rfc 3447* (February 2003)
2. A. Boldyreva, H. Imai, K. Kobara, How to strengthen the security of RSA-OAEP. IEEE Trans. Inf. Theory **56**(11), 5876–5886 (2010)
3. J. Randall, B. Kaliski, J. Brainard, S. Turner, Use of the RSA-KEM key transport algorithm in the cryptographic message syntax (CMS). Proposed Standard 5990 (2010)
4. R.L. Rivest, A. Shamir, L. Adleman, A method for obtaining digital signatures and public-key cryptosystems. Commun. ACM **21**(2), 120–126 (1978)
5. W. Stallings, *Cryptography and Network Security: Principles and Practice* (Upper Saddle River, Pearson, 2017)
6. K. Chitra, V. Prasanna Venkatesan, An antiquity to the contemporary of secret sharing scheme. J. Innov. Image Process. (JIIP) **2**(01), 1–13 (2020)
7. S. Shakya, Efficient security and privacy mechanism for block chain application. J. Inf. Technol. **1**(02), 58–67 (2019)
8. B. Kaliski, *The Mathematics of the RSA Public-Key Cryptosystem* (RSA Laboratories, 2006)
9. M. Agrawal, N. Kayal, N. Saxena, PRIMES is in P. Ann. Math. 781–793 (2004)
10. M. Abe, R. Gennaro, K. Kurosawa, V. Shoup, Tag-KEM/DEM: a new framework for hybrid encryption and a new analysis of Kurosawa-Desmedt KEM. in Annual International Conference on the Theory and Applications of Cryptographic Techniques (Springer, Berlin, Heidelberg, 2005), pp. 128–146
11. J. Jonsson, M.J.B. Robshaw, Securing RSA-KEM via the AES. in International Workshop on Public Key Cryptography (Springer, Berlin, Heidelberg, 2005), pp. 29–46

12. C. Percival, S. Josefsson, The scrypt password-based key derivation function. IETF Draft (2006). URL: http://tools.ietf.org/html/josefsson-scrypt-kdf-00.txt. Accessed 30 Nov 2012
13. P. Mahajan, A. Sachdeva, A study of encryption algorithms AES, DES and RSA for security. Glob. J. Comput. Sci. Technol. (2013)

Botnet Attack Detection Using Machine Learning Algorithm Integrated With Ensemble Algorithm

N. L. Lincy, Uma Devi, and E. R. Vimina

Abstract Botnet investigation helps in understanding the idea of attack and the usual way of doing things utilized by the aggressors. Botnet assaults are hard to follow given their quick speed, pestilence nature, and more modest size. AI functions as a only source to botnet assault issues. It encourages location as well as helps in avoidance from bot assault. The efficient learning framework can peruse the client's activities and conduct in the digital world. It can undoubtedly distinguish the conducting nature and part of each action via online media. Nonetheless, the dark cap local area works just in the personal responsibility and spotlights on proliferating vindictive exercises. The botnet is perhaps the most arising danger for the computerized society. The proposed probe model undertakings improved the nature of results by far-reaching botnet location and criminological examination. In this paper, the botnet attack has been detected by using machine learning techniques by ensemble classifier model to improve the accuracy rate by hybriding the model of classifier. The botnet dataset has been used in this approach of detecting the botnet attack. The performance has been calculated and compared with other related work.

Keywords Botnet · Machine learning algorithm · Ensemble classifier · SVM · KNN · Decision tree

1 Introduction

A botnet is a combination of bot-controlled Web mischief. Advanced criminals employ botnets to perform activities like unauthorized discharge, data theft, as well as DDoS assaults [1]. They can turn thousands of machines into botnets and use them

N. L. Lincy (✉) · U. Devi · E. R. Vimina
AmritaViswa Vidyapeetham, Cochin, India

U. Devi
e-mail: umadevi@asas.kh.amrita.edu

E. R. Vimina
e-mail: viminaer@asas.kh.amrita.edu

G. Ranganathan et al. (eds.), *Soft Computing for Security Applications*,
Advances in Intelligent Systems and Computing 1397,
https://doi.org/10.1007/978-981-16-5301-8_13

all at the same time to begin illegal activities. There are some serious security flaws in concoctions which are being targeted from the outset by hackers. Before this, the malicious activities have been executed, the PCs are usually infiltrated by using these vulnerabilities [2]. To utilize her organizational skills and talents, we ought to turn her administrative job down an official position and grant her a bonus of 20% of her sales-force performance royalties for her individual contributions. Mechanized and electrically energize whatever needed electrical assault [3].

In different cases, cybercriminals will offer admittance to the botnet network, once in a while known as a "zombie" organization, with the goal that other cybercriminals can utilize the organization for their own malignant exercises, for example, actuating a spam crusade [4]. The quantity of botnets will shift based on the capacity of the proprietor to contaminate unprotected ones. Concretely, for instance: According to our client estimates, a DDoS attack in August 2017 started with more than 75,000 bots as part of a botnet. An initial capability attack in December 2016 leveraged a botnet with nearly 13,000 people to send about 270,000 log in attempts every hour against the Akamaiases. That tells me nothing, as I do not know who sent it [5].

The impacts of a botnet assault can be destroying, from moderate gadget execution to tremendous Internet charges and taken individual information [6]. There are additionally legitimate ramifications to consider, for instance, if your PC is utilized as a component of a botnet assault, you might be lawfully liable for the results of any vindictive exercises that have begun from your gadget [7, 8].

2 Review of the Literature

In 2019, Wang et al. [9] manages BotMark that consequently identifies bots with half and half examination of stream-based and chart-based traffic practices. The stream-based identifier comprises of closeness-based part and solidness-based segment. The half and half of the two finders can describe the botnet practices more exhaustively than any individual indicator.

We used a bidirectional long short-term recurrent neural network (BLSTN) with embedded LSTM for botnet detection in 2018. For this dataset, we tested the amassing of context-oriented data from both the historical and future to determine if it returns exactness or misfortune measurements. The four mirai botnet assault models obtained high accuracy and low success [10].

In 2020, Vinaykumar et al. [11] deal with the botnet recognition framework dependent on a two-level profound learning structure for semantically segregating botnets and authentic practices at the application layer of the area name framework (DNS) administrations. In the main level of the structure, the comparability proportions of DNS questions are assessed utilizing siamese organizations dependent on a predefined limit for choosing the most successive DNS data across Ethernet associations. In the second level of the structure, a space age calculation dependent on profound learning designs is proposed for ordering ordinary and unusual area names. The system is profoundly versatile on a product equipment worker because of its expected plan of dissecting DNS information. The proposed system was assessed utilizing two datasets and was contrasted and ongoing profound learning models. Different perception techniques were additionally utilized to comprehend the attributes of the dataset and to imagine the installing highlights. The trial results uncovered significant enhancements regarding F1-score, speed of location, and bogus caution rate.

In 2019, Vinaykumar et al. [11] tried a recruitment approach to both supporting learning and our strategy to reduce traffic. An important feature of the expected approach is the organizational narrowing of the organization, which can currently reduce the amount of information by about half. The paper exhibits in the former segment that anticipated method discover pace of 98.30% what is more, a low FPR of 0.010% at the 60 s time-window size. The capture of bot identification using a neural organization is related to the size and size of the database, as the number of batches involved in the exploration is important. This is where our proposed network reduction strategy plays an important role. The use of such a method can reduce the preparation time required, and in addition increase the learning speed of the most recent highlights from the Web framework. In addition, the proposed bot detection method appears to meet the maximum specificity and can identify new bots.

The focus is on DNS query and response patterns in 2021 and aims to identify any legitimate DNS query and response patterns using the proposed DNS query and response rules. The results showed that the proposed method would detect botnets with 99.35% accuracy and a false-positive rate of 0.25%. The efficacy of the proposed method is evaluated by comparing it to well-known DNS-based approaches. The proposed approach outperforms other methods that can be introduced as part of any antivirus IDS product, according to the findings.

İn 2017, Chowdhury et al. [12] propose a graph-based botnet detection method for detecting evolving bot behaviors. This is novel since current methods primarily depend on flow-based features, which do not account for changes in network topology induced by bot activities. The in degree, out degree, in degree weight, out degree weight, clustering coefficient, node betweenness, and eigenvector centrality are seven graphed-based features that can be linked to bot activities. Filtering out inactive nodes, which are unlikely to be bots, improves the proposed algorithm even further. We test the proposed methods on the CTU-13 datasets, which is the largest dataset with bot-labeled nodes. Our proposed procedure can detect bots by searching a small number of nodes, according to numerical results (less than 0.1% of all nodes).

3 Proposed Botnet Attack Detectıon Method

To discover and examine a botnet attack, the testbed needs to be run in a physical test environment. It uses genuine vehicles for traffic. This dataset includes both preparation and testing for normal following and malware, too. Botnet analysis is performed to identify the make and type of attack. Using this AI architecture, non-conforming AI calculations with relatively better results are obtained for a wide range of outcomes.

3.1 Machine Learning Model

Fixing the location of a botnet with the SVM algorithm and the histogram vector [13], the printed spam breakdown could be performed using the KNN model [14]; the creator utilized the breakdown process of information extraction, separating out P2P2P botnet traffic by using AI methods [15]. The creators separated 17 highlights first and afterward eliminated five highlights from them due to the ostensible qualities. As a result, they are separated from the broadcast-based institution. The structure can be further enhanced with the help of the Hive and Mahout model to detect shared botnet attacks using the AI method [16]. Bot movement [17] is identified by both order and control and assault stage utilizing traffic conduct investigation and by applying AI arrangement. The creator has identified the bot action with the help of tree splitting as an AI program. Subsequent to changing over time area network correspondence to recurrence space organization, ref. [18] discusses how subsequent to the area network traffic organisation, it deals with P2P botnet traffic. Using signal making, they utilized an AI method that each pair of hubs employs and object includes P2O-precedencementation in ref. [19].

The work consolidated both P2P and online traffic lastly arranged the P2P information by applying ideal SVM model. Then again, reasonable DDoS discovery and moderation model planned by the various classifier [20] worked through numerous information pieces. The practical model also has real-world broadcast rather than manual information [21]. The creators have taken a blank and a limited amount of tagged data from a set prepared by KNN. Research findings have also shown that it is possible to improve outcomes by combining more than one count [22]. In ref. [23], experimented past work depended on the strategy for building one classifier for every lump. The creator further inferred that the different lumps with the numerous outfits the grouping method. Essentially,ref. [24] utilized twofold arrangement issue with the assistance of a gathering of a classifier. Did investigation through multisample train set analyzed the exhibition of different examination model. In ref. [25], executed system for imbalanced dataset is utilizing outfit method. The creators proposed scientific categorization of class awkwardness sorted by internal troupe strategy. In the show [26], they demonstrated their work using mindless backwood trees, KNN, and J48 AI. In ref. [27], deals with a self-learning botnet location framework in outfit classifier improved speculation ability.

3.2 Ensemble Classifier Model

Botnet investigation system gives the subtleties of botnet recognition and its examination through bit-by-bit measure. It is clear to break down the botnet when botnet distinguishing proof cycle gets finished. There is an alternate method to get the highlights separated and to dissect them. AI calculation is used to improve the accuracy of botnet detection. The function mainly controls a specific type of botnet database that enters the organization from within the denial management (DoS), shared distribution management (DDoS), and animal power information. Combined botnet traffic is the import of SSH, HTTP, and SMTP traffic that points to client behavior. It is further organized and defined by a set of structures that detects violent traffic from normal traffic. This database is then filtered for standard traffic and botnet traffic, and botnet traffic test is selected for additional testing. This cycle has yielded 42 credits, provided names for each event, and separated them in the preparation and testing of datasets. There has been analyzed eight different types of classifier with the boosting, bagging, and voting with KNN, DT, and SVM; the combinations are:

AdaBoost with decision tree
AdaBoost with K-nearest neighbor
Bagging with decision tree
Bagging with K-nearest neighbor
Bagging with support vector machine
Voting with K-nearest neighbor and decision tree
Voting with decision tree and support vector machine
Voting with support vector machine and K-nearest neighbor.

Some traits are identified with isolation-based and others with stream-based. The discovery of characteristics is grouped together; F2P2P and non-service streams come from these credits. NetMate separates 42 distinct credits that are mixed and deposited into NetOrange. Moreover, it is into conventional Internet traffic and peer to peer. The flow of typical of traffic is identical to the flow of genuine flow.

4 Result and Analysis

The botnet is an enormous organization of traded off PCs, which is told by botherder. The receptive methodology alludes to the proof that ought to be protected in one spot for after death of bot assaults. This proof is additionally applied for the examination of botnet traffic and to recover the pertinent data from it. As a result, it uncovers the examination with a single classifier.

Table 1 presents the performance metrics of a different classifier with accuracy, precision, recal and f1-score. the decision tree calculation shows 93, 92, 93% review, and 95%. On account of KNN calculation, 95, 95, 93.48 and 95% F1 score are noticed. In this way SVM shows 76, 81, 76 and 67% (Fig. 1).

Table 1 Performance analysis of different classifier model

Classifiers	Decision tree	KNN	SVM
Accuracy	94	95	76
Precision	92	95	82
Recall	93	95	76
F1 score	95	95	66

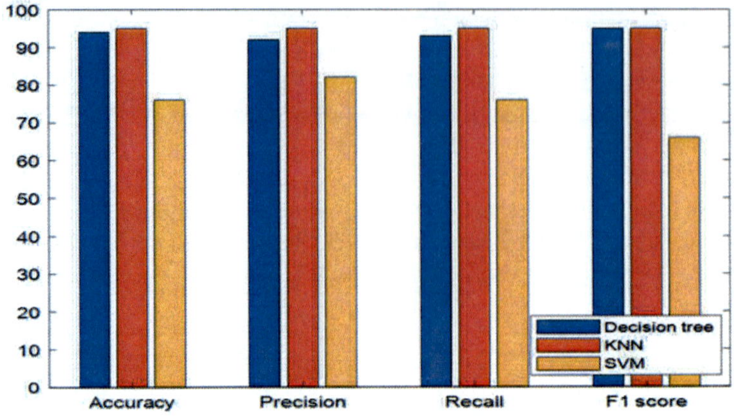

Fig. 1 Performance analysis of different classifier model

AdaBoost choice tree likewise builds exactness from 93 to 98%, improving learning cycle of a choice tree, and most elevated precision is accomplished by utilizing delicate democratic guideline since it combines the power of two calculations and gives more weight to the choice of a more efficient mathematics. The separator yield alone does not provide a surprising bot detection. A bot proof demonstration that uses a split group is better than a single split.

Table 2 shows the correlation graph of the distinctive group of classifiers. The AdaBoost with SVM diminishes the exhibition because SVM is a solid student, while AdaBoost is utilized primarily to improve powerless students. Also, AdaBoost gives examining to prepare the occurrence as per the intricacy of characterization; that is, more weight is given to the cases that are difficult to order (Fig. 2).

Table 2 Performance analysis of different ensemble classifier model

Classifier	AdaBoost with DT	AdaBoost with KNN	Bagging with DT	Bagging with KNN	Bagging with SVM	Voting with KNNand DT	Voting with DT and SVM	Voting with SVM and KNN
Accuracy	98	95	95	95	76	95	85	95
Precision	99	95	95	95	81	96	87	95
Recall	98	95	95	95	76	96	85	95
F1 score	99	95	96	94	67	95	83	95

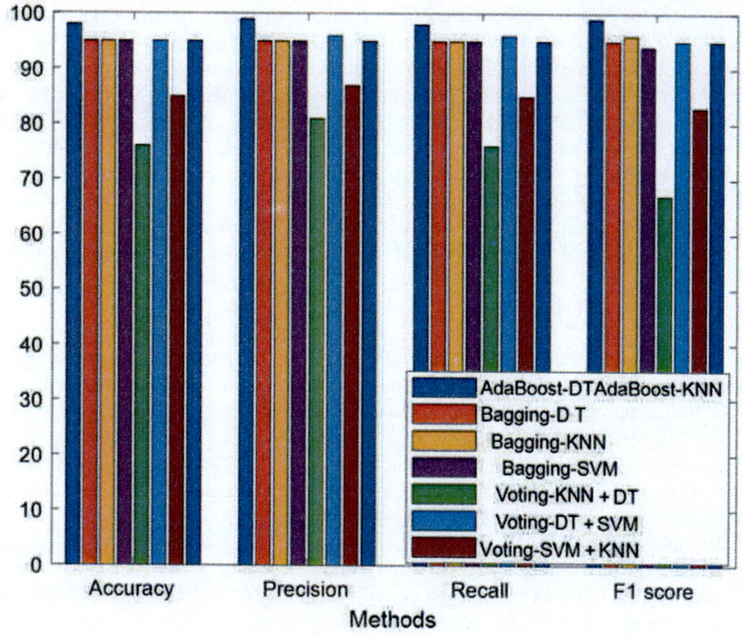

Fig. 2 Performance analysis of different ensemble classifier model

5 Conclusion

Existing classifiers are integrated with an integrated model that identifies botnet traffic. This combination of different dividers works best considering that it is done by combining forces with different numbers. From then on, using a data mining tool, a dividing calculator blanket was used. This result shows that the group model improved various parameters such as accuracy, precision, review, and F1 score in botnet traffic detection compared to the previous single collection calculations. In any case, an investigative model can be made to investigate botnet crimes later. The AI method can be used to test large-scale botnet attack information with costume segmentation combinations.

References

1. J. Vania, A. Meniya, H.B. Jethva, A review on botnet and detection technique. Int. J. Comput. Trends Technol. **4**(1), 23–29 (2013)
2. R.S. Abdullah, M.F. Abdollah, Z.A. Muhamad Noh, M.Z. Mas'ud, S.R. Selamat, R. Yusof, Revealing the criterion on botnet detection technique. Int. J. Comput. Sci. Issues (IJCSI) **10**(2), 208 (2013)

3. P. Amini, M.A. Araghizadeh, R. Azmi, A survey on Botnet: classification, detection and defense. in *2015 International Electronics Symposium (IES)*, (IEEE, 2015), pp 233–238
4. S. Anwar, J. Mohamad Zain, F. Zolkipli, Z. Inayat, A review paper on botnet and botnet detection techniques in cloud computing. in *Proceedings of the ISCI*, (2014), pp. 28–29
5. A. Karim, R. Bin Salleh, M. Shiraz, S.A.A. Shah, I. Awan, N.B. Anuar, Botnet detection techniques: review, future trends, and issues. J. Zhejiang Univ. Sci. C 15(11), 943–983 (2014)
6. M. Mahmoud, M. Nir, A. Matrawy, A survey on botnet architectures, detection and defences. IJ Netw. Secur. **17**(3), 264–281 (2015)
7. H. Dhayal, J. Kumar, Botnet and p2p botnet detection strategies: a review. in 2018 International Conference on Communication and Signal Processing (ICCSP), (IEEE, 2018), pp. 1077–1082
8. Z. Al-Othman, M. Alkasassbeh, S. Al-Haj Baddar, A state-of-the-art review on IoT botnet attack detection. arXiv preprint arXiv:2010.13852 (2020)
9. W. Wang, Y. Shang, Y. He, Y. Li, J. Liu, BotMark: Automated botnet detection with hybrid analysis of flow-based and graph-based traffic behaviors. Inf. Sci. **511**, 284–296 (2020)
10. C.D. McDermott, F. Majdani, A.V. Petrovski, Botnet detection in the internet of things using deep learning approaches. in *2018 International Joint Conference on Neural Networks (IJCNN)*, (IEEE, 2018), pp. 1–8
11. R. Vinayakumar, M. Alazab, S. Srinivasan, Q.-V. Pham, S.K. Padannayil, K. Simran, A visualized botnet detection system based deep learning for the internet of things networks of smart cities. IEEE Trans. Ind. Appl. **56**(4), 4436–4456 (2020)
12. S. Chowdhury, M. Khanzadeh, R. Akula, F. Zhang, S. Zhang, H. Medal, M. Marufuzzaman, L. Bian, Botnet detection using graph-based feature clustering. J. Big Data **4**(1), 1–23 (2017)
13. S. Kondo, N. Sato, Botnet traffic detection techniques by C & C session classification using SVM. in *Proceedings of the International Workshop on Security*, (Vienna, Austria, September 2007)
14. R.M. Alguliev, R.M. Aliguliyev, S.A. Nazirova, Classification of textual e-mail spam using data mining techniques. Appl. Comput. Intell. Soft Comput. **2011**, 8 (2011). Article ID 416308
15. S. Garg, A.K. Singh, A.K. Sarje, S.K. Peddoju, Behaviour analysis of machine learning algorithms for detecting P2P botnets. in *Proceedings of the 15th international conference on Advanced computing technologies (ICACT)*, (Rajampet, India, September 2013)
16. K. Singh, S.C. Guntuku, A. Thakur, C. Hota, Big data analytics framework for peer-to-peer botnet detection using random forests. Inf. Sci. **278**, 488–497 (2014)
17. D. Zhao, I. Traore, B. Sayed et al., Botnet detection based on traffic behavior analysis and flow intervals. Comput. Secur. **39**, 2–16 (2013)
18. P. Narang, V. Khurana, C. Hota, Machine-learning approaches for P2P botnet detection using signal-processing techniques. in *Proceedings of the 8th ACM International Conference on Distributed Event-Based Systems*, (Mumbai, India, May 2014), pp. 338–341
19. P. Barthakur, M. Dahal, M.K. Ghose, A framework for P2P botnet detection using SVM. in Proceedings of the International Conference on Cyber-Enabled Distributed Computing and Knowledge Discovery (CyberC), (Sanya, China, October 2012), p. 195_0
20. S. Bhatia, D. Schmidt, G. Mohay, Ensemble-based ddos detection and mitigation model. in Proceedings of the Fifth International Conference on Security of Information Networks, (Jaipur, India, October 2012), pp. 79–86
21. M.M. Masud, J. Gao, L. Khan, J. Han, B. Thuraisingham, A practical labeled approach to classify evolving data streams: training with limited amount of data. in *Proceedings of the Eighth IEEE International Conference on Data Mining ICDM'08*, (Pisa, Italy, December 2008), pp. 929–934
22. A. Bijalwan, N. Chand, E.S. Pilli, C. Rama Krishna, Botnet analysis using ensemble classifier. Perspect. Sci. **8**, 502–504 (2016)
23. M.M. Masud, J. Gao, L. Khan, J. Han, B. Thuraisingham, Mining concept-drifting data stream to detect peer to peer botnet traffic. Technical Reports, (University of Texas at Dallas, Richardson, Texas, 2008), Tech. Report UTDCS-05-08
24. X.Y. Liu, J. Wu, Z.-H. Zhou, Exploratory undersampling for class imbalance learning. IEEE Trans. Syst. Man Cybern. Part B (Cybern.) **39**(2), 539–550 (2009)

25. M. Galar, A. Fernandez, E. Barrenechea, H. Bustince, F. Herrera, A review on ensembles for the class imbalance problem: bagging-, boosting-, and hybrid-based approaches. IEEE Trans. Syst. Man Cybern. Part C **42**(4), 463–484 (2012)
26. R. McKay, B. Pendleton, J. Britt, B. Nakhavanit, Machine learning algorithms on botnet traffic: ensemble and simple algorithms. in *Proceedings of the 3rd International Conference on Compute and Data Analysis*, (Kahului, HI, USA, 2019), pp. 31–35
27. M. Nazemi Gelian, H. Mashayekhi, Y. Mashayekhi, A self-learning stream classifier for flow-based botnet detection. Int. J. Commun. Syst. **32**, 1–15 (2019)

Artificial Intelligence-Based Automation System for Health Care Applications: Medbot

R. Pradeep, S. Praveen Kumar, S. Sasikumar, P. Valarmathie, and P. V. Gopirajan

Abstract In this day and age of digitalization, every division of the industry is looking into implementing cutting-edge technologies into their products to meet the needs of the modern era. The use of Artificial Intelligence-based Automated Agents is very popular in some large-scale applications. In traditional health care system, most of the query regarding the functionalities is often not addressed rapidly around the clock. The proposed work focuses to develop an Artificial Intelligence-based automated system called Medbot, which uses Natural Language Processing (NLP) and Machine Learning (ML) to develop a personalized Virtual assistant for solving the queries related to medical devices. It stands in the place of technical support experts in comprehending the specific functionalities and features of the medical equipment which is oftentimes more intricate to handle. The Medbot responses to the user's query are quicker than a traditional system which is skimming an entire manual provided by the manufacturer. Based on the test constraints, this chatbot attained more than 90% accuracy for almost every intent provided. The bot is platform-independent and can be integrated with the web, mobile, and other most commonly used messaging applications and it can also handle multiple query requests by a large number of users at the same time and can be deployed securely. Based on the test constraints, this chatbot attained more than 90% accuracy for almost every intent provided.

Keywords Artificial intelligence · Machine learning · NLP · Medbot · Automated agents

1 Introduction

Modern healthcare is one such critical domain that needs advancement in technology like Artificial Intelligence (AI) which is one of the top technologies in the modern era to provide the best care for patients. Computer technology is now much smarter and

R. Pradeep (✉) · S. Praveen Kumar · S. Sasikumar · P. Valarmathie · P. V. Gopirajan
Department of Computer Science Engineering, Saveetha Engineering College, Chennai, India

G. Ranganathan et al. (eds.), *Soft Computing for Security Applications*,
Advances in Intelligent Systems and Computing 1397,
https://doi.org/10.1007/978-981-16-5301-8_14

more autonomous in solving the significant challenges faced in medical equipment [1]. The manual for a medical device is often time strenuous to refer to and it requires the technician skimming through 500+ pages to find the answer for a single query, instead of utilizing this time to improve the patient outcomes. A straightforward approach in solving this problem is a Medbot that uses NLP to address the user's query.

Chatbots are available in many domains including Healthcare that abate the complications faced by the end-user. There have been applications developed that aim to help the patients by providing them with relevant information about diseases and their overall health through Q&A chatbots. Cognitive chatbots that can perform sentiment analysis on the user to track their mood and promote better emotional health have been developed. Apart from these, chatbots have been developed with the expertise of doctors to track the symptoms of the users and arrive at a diagnosis. From our extensive research, we found that a fully developed application with an integrated chatbot that replaces a manual of a medical device for the technician's use is yet to be developed.

Technicians are required to work with sophisticated and advanced medical equipment daily [2]. A big step forward in improving customer service is the development of real-time virtual assistance to remote technicians through AR. But this still requires a human on the other side to walk the technician through the process [3]. A virtual assistant in the form of a chatbot [4] eliminates the need to contact the customer care personnel for trivial issues and focuses on bettering the patient care and diagnosis [5]. The proposed Medbot is developed using a Python framework called Rasa, a contextual assistant that employs Machine Learning [6]. It consists of two components, Rasa NLU for Natural Language Processing that performs intent classification and entity extraction, and Rasa Core for Dialogue Management through contextual decision making [7]. Through the utilization of these two components, Rasa is capable of holding conversations through which it executes interactive learning [8].

The Medbot is available in the form of both a web application and a mobile application to serve as a reference point for a technician while using a medical device [9]. Besides its ability to answer the end-user's query, the various features have incorporated in it are responses in form of text, videos, image, documents to ensure that the technician's query has been properly resolved [10], a speech to text conversion of the query asked by the user, full chat history with intents and confidence level logged into a folder among other [11]. Data collection and storage of information in cloud repository were explained in [12] for biomass applications which suggest the significance of the central repository. This present study was carried out with the development of a chatbot using RASA package and the bot was tested for its efficiency with multiple input test conditions and their results were also reported.

2 Experimental Workflow Analysis

The workflow of the Medbot application involves various stages, wherein at each phase the data is modulated to provide the best result for the user input. It is mainly composed of Rasa NLU (Natural Language Understanding) and Rasa Core. In the first step, the data is collected from various sources. Then, the intents and entities are labeled and classified from the data. In later stage, the processed data is given to the RASA NLU which understands the user message and figures out entities and labels the intents. Further, the RASA CORE handles the conversation flow, utterances and actions of the bot and decides what to do next based on the intents and entities. The RASA NLU and RASA CORE works interdependently until the desired response is obtained which satisfies the user query. The response is then delivered to the pipelines which can be configured to efficiently fine-tune dialog manager and response selection (Fig. 1).

The next phase involves in training and evaluating the data to help the model identify intents and entities correctly to give accurate results for the user input. Subsequently, the Medbot can be integrated with an existing Web site or application for user access. In later stage, the data and application are constantly regulated for improving the interaction with the users and giving maximum accuracy for the user messages. Furthermore, the history of chats conversions are stored as logs in the local device and all successful logs were identified after getting the feedback from the stakeholder and maintained in the distributed cloud environment for handling multiple query among different devices.

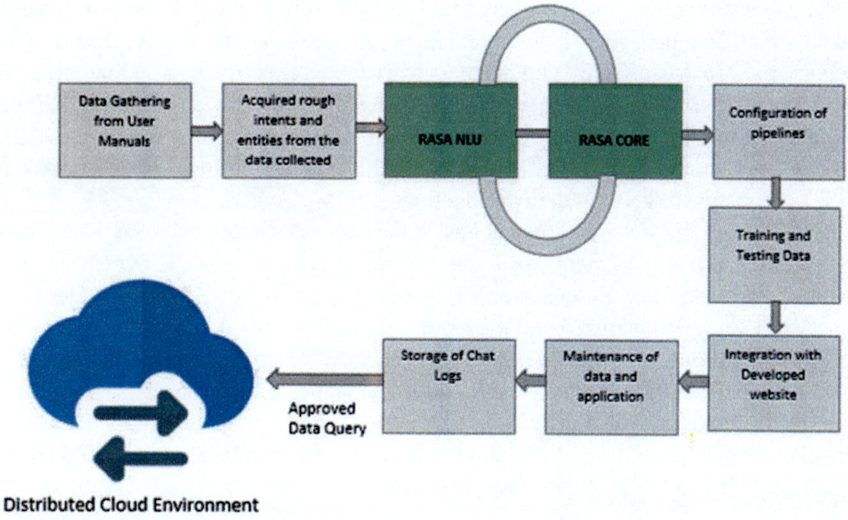

Fig. 1 Workflow analysis for Medbot

3 Module Description

3.1 Data Collection

Datasets related to fields like persistent warnings and failure, sounds of trouble with the tube, sliprings and brushes operation, were collected from previously published papers and they were validated with the information available from "Saveetha Multi-Speciality hospital". The dataset provided is a manual of a Computed tomography (CT) scanner consisting of 800+ pages. It contains 20 chapters which are further broken down into lessons in each chapter that comprises a detailed, step-by-step procedure concerning the CT scanner's functionalities. Keywords were used to narrow down the important paragraphs and procedures from the vast content available and this was done by identifying the actions that are performed, in other words the functionalities and some common troubleshooting questions. After narrowing down the paragraphs that are to be focused, the intents and entities are extracted from it. Intents provide the intention of the user's input and entities provide additional information to support the intent in order to provide faster and better responses. From here, the extracted intents and entities are used to form stories. Stories provide structure and provide dialog management.

3.2 Preprocessing

Once the data is ready, the preprocessing is done before training the Rasa model. It mainly contains three files namely the *nlu.md, domain.yml* and *stories.md* file. The *nlu.md* file consists of all the intents that are extracted from the dataset that are used by the Medbot to classify the user's query. For example, low disk space and hard drive failure are two intents that are looked upon when the user has any query regarding low disk space and hard drive failure, whether it is a straightforward question or a color or message regarding the two functions.

The domain file is a *yaml* file that has all the responses extracted from the manual in accordance with the intents. It is also where the type of response, whether it is a button or an image or a video is specified in terms of the payload given to it. Once the query of the user is analyzed and the intents are derived, the domain file is accessed to get the appropriate response to the query. Lastly, the *stories.md* file includes the flow of the conversation to make the Medbot seem more human-like. This is a special feature available in Rasa that helps us to leverage the full potential of an AI chatbot. Each set of stories contains different possible flows in which the conversation could be driven with an intent and its respective response.

3.3 Intent Identification

Each intent that is identified from the data can be matched with the user's input. To make sure that the matched intent is as accurate as possible, different examples of the questions that could be asked in relation to the intent is listed. There must be at least ten such examples for a reliable match. But it is not limited to it as the more unique examples that are provided, the more accurate the matches are and it will yield a higher probability of the accurate results. Many features like intent classification and preprocessing can be done with the help of pipelines. Rasa provides two pre-trained pipelines for a simplified model but the components can also be completely customizable according to the user's needs. They are pre-trained embeddings spacy and supervised embeddings. Pre-trained embeddings spacy is preferred when the developer has less training data and multiple intents whereas supervised embedding is preferred for the vice versa.

3.4 Response Selection

The response that the user gets is based on how efficient the stories are formed in Rasa core. Matching an intent with a response will turn the chatbot into an FAQ (Frequently Asked Question) bot instead of a conversational bot. Stories help retain a human-like conversation between the user and the bot. Each time a query by the user is matched with intent, the stories.md file is referred to check how the conversation must be driven and what could be the next potential query from the user is predicted. With this feature, the efficiency and the accuracy increase by a great percentage.

Figure 2 shows the architectural design view of the proposed Medbot. This contains the significant cloud repository block. This cloud repository stores the entire information relevant to the system developed. The responses that are given to the user can be in terms of text, links to images, video and also documents. These forms of responses can be mentioned in the domain.yml file. Config.yml file is used to

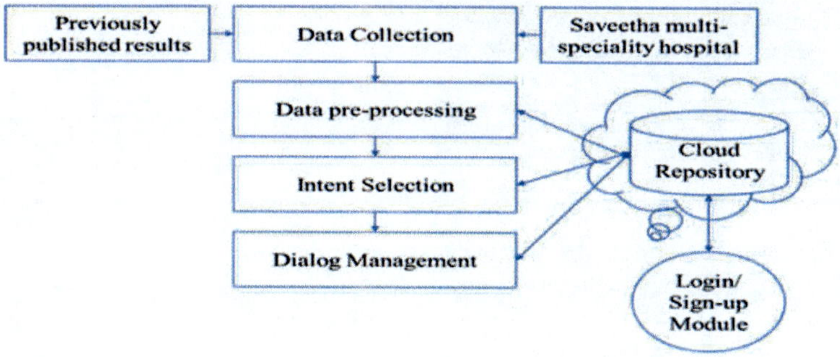

Fig. 2 Medbot architecture

specify policies according to the cahtbot's needs on deciding which action to take at each step in a conversation. There are Machine Learning-based policies and also rule-based policies to cater the developer's specific needs.

3.5 Suggestions and Dialogue Management

To make the user's experience more pleasant, we have incorporated suggestions when a query that has a broad context is handled. For example, if the user utters gantry, they might mean error in gantry or the components in gantry. When such a scenario occurs, we provide all the possible intents of the gantry to the user to help them narrow down their question. This has greatly helped the user to understand what they require and also establish a proper communication channel between the user and the chatbot. As mentioned earlier, stories are very crucial to develop an efficient chatbot. They are used to teach Rasa real conversation designs to learn from the user. Since dialogue management is a feature that is specific to Rasa, it is essential to build quality stories to leverage the maximum benefit that the platform can offer. Figure 3 shows the system Architecture of Artificial Intelligence Agent-Medbot.

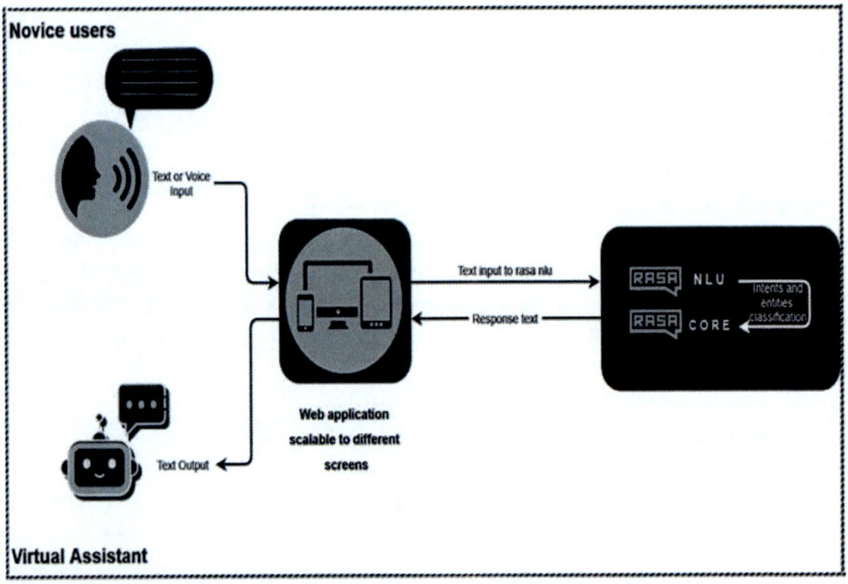

Fig. 3 System architecture of artificial intelligence agent-Medbot

4 Implementation and Result

The Medbot is a web application that is user-friendly and responsive to different devices of various sizes. It shows an effective and efficient response to various intents and entities. Figure 4 shows the Medbot Query Response with Text format. The bot was trained with 800 training set and test were performed with the ratio of 80:20 training set and test set. Their response time for multi-query was recorded and its efficiency was stored in the repository for future predictions. Further, a multi-inter-query model was developed based on the efficiency reported and the test was iterated till the positive progress in the efficiency percentage.

It can take inputs in the form of texts or speech and also the user queries are answered in various forms like images, audio, and video formats with a manual for additional references in any of the cases. The Rasa NLU converts the unstructured text messages into intents and entities to make the system understand the natural language query. Similarly speech to text conversion is handled with the SpeechRecognition PyDub package. Moreover, different suggestions were given by the chatbot to the end-user for achieving more accurate results. Figure 5 shows Medbot Query Response with Text and Speech type.

Security, which might be a matter of concern for some providers, has been resolved through the use of a Login system with password protection. To make the Medbot more portable, we have enabled the integration of it to other messaging platforms like Slack, Telegram, Messenger, and many others. Despite still being in its development phase, our Medbot has shown promising results bypassing many test cases. Figure 6 shows the Medbot User Manual for reference to the end-user. Multi-query inputs were given to the model developed and their query bounce rate, response retention rate, inter-query response time were calculated. A statistical correlation was calculated

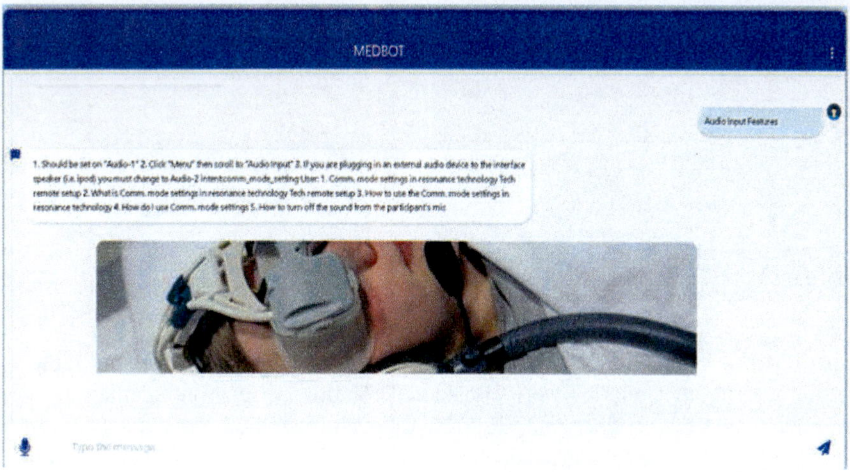

Fig. 4 Medbot query response with text

Fig. 5 Medbot query response with text and speech

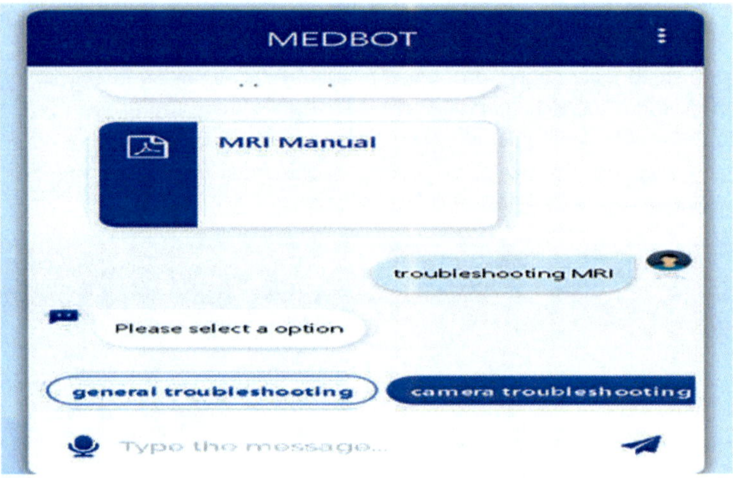

Fig. 6 Medbot user manual

with Pearson correlation coefficient and their results were displayed in Table 1. 19 number of inter-queries were considered for this correlation calculation their minimum and mean bounce rate, response time, and retention rate were analyzed.

Figure 7 shows the Medbot Launcher for the Sign In and Sign Up process. Overall, the chatbot attained more than 90% accuracy for almost every intent provided. Figure 8 shows the inter-query time graph. In this figure, bounce rate, retention rate,

Table 1 Inter-query correlation

Multi-query data	Variable	Observations	Obs. with missing data	Obs. without missing data	Minimum	Maximum	Mean	Std. deviation
Q1–Q7	0.02	19	0	19	0.040	0.400	0.220	0.113
Q8–Q12	0.92	19	0	19	0.226	0.820	0.257	0.311
Q13–Q19	0.83	19	0	19	0.431	0.690	0.029	0.296

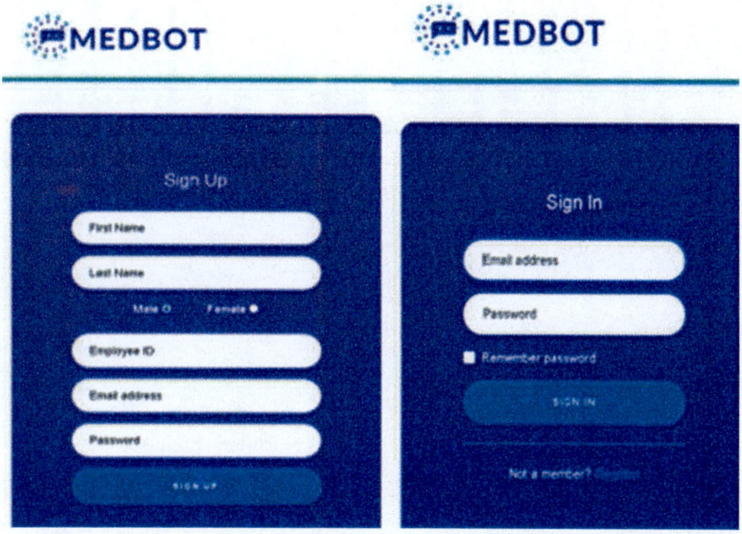

Fig. 7 Medbot launcher

and response time were reported and it is clear that the initial queries consume more time and the time gets reduced when the number of queries increases. This clearly shows that the developed model learns from the queries and the response time is directly proportional to the increase in the number of queries. Also, the bounce rate and retention rate were decreased, which means, bounce rate and retention rate are inversely proportional to the number of queries. In the future, this study was planned to enhance the accuracy results by collecting and storing different new intents asked by the users to train the chatbot.

Additionally, the results were compared with [13–15] for showing the impact of this present study. Efficiency graphs were generated for showing comparisons. Details of these [13–15] studies were displayed in Table 2.

Though these studies were carried out for different applications, their results help in formulating this proposed model. Each studies were given names as S1, S2 and S3 and the proposed study was named as P. Results of these four studies were compared based on the response time, bounce rate and retention rate.

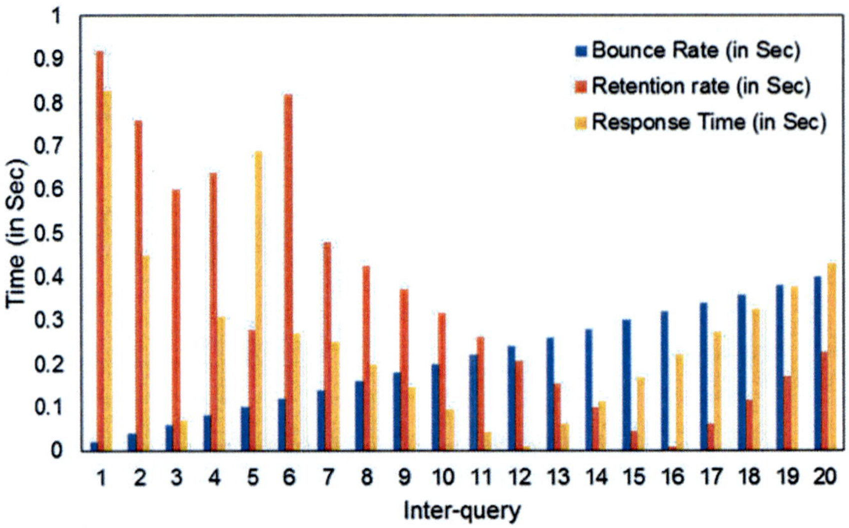

Fig. 8 Inter-query response graph

Table 2 Details of related studies

Study	Name	Methodology used	Application
[13]	S1	Vector chatbot	Learning management system
[14]	S2	Recommender system	Cultural heritage
[15]	S3	Deep strategic mediatization	Information management system

Figure 9 shows the efficiency comparison between the studies considered (S1, S2 and S3) Versus Proposed study (P). S1 attained 89% of efficiency was the moderate result compared with the other models. Whereas, models S2 and S3 achieved a comparatively better response time. Root mean square (R^2) was calculated based on the response time as the prime factor and the efficiency equation was generated as shown in the graph. Compared with the other models, it is clear that proposed study achieved higher efficiency. R^2 value of 0.8199 shows that the proposed study achieved a high level of correlation between the three parameters.

4.1 Validation Process

Well pre-processed data were considered for analysis. To validate the results, manual optimization was performed with the help of real-time requests obtained from the hospital patients and suitable responses given by Doctors. Manual form was given to the stackeholders and their suitable recorded results were compared with the

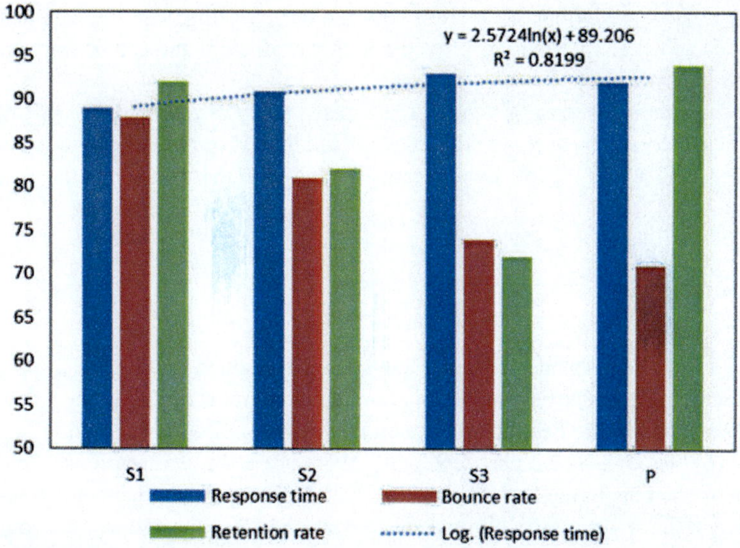

Fig. 9 Efficiency comparison

suggestions provided by this Medbot. 5 number of patients were chosen randomly and they were made to enter their query in a manual form prepared. Doctors answered their query by witnessing their medical report. Efficiency for this manual process was calculated after proper examination of request query and response query. Figure 10 denotes the comparison of manual form entry results efficiency with the results

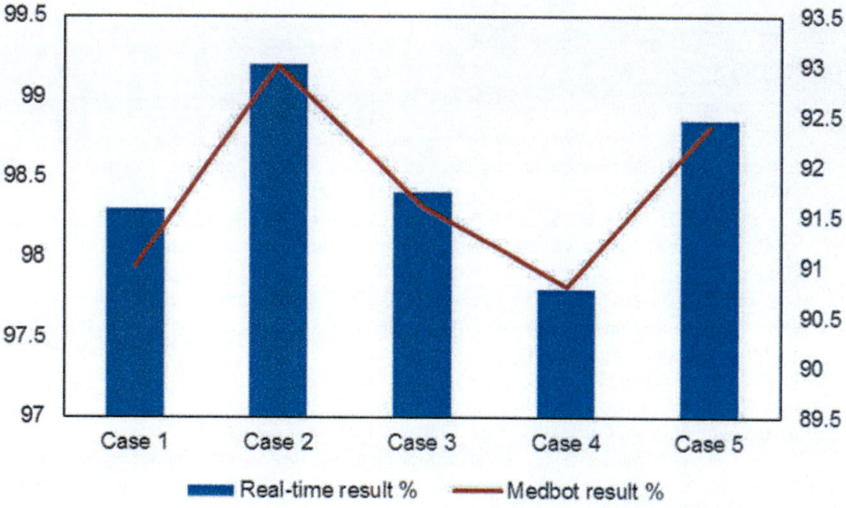

Fig. 10 Comparison of real-time results versus Medbot results

obtained with the Medbot. Validation between these results was analyzed and 90% efficiency achieved by Medbots proved that it was efficient and can be implemented in real-time systems.

For the 5 cases used in this study, form-based analysis reported 98.51% average efficiency and 91.76% average efficiency for Medbot. From the graph, it was proven that the efficiency of Medbots was relatively high with that of the manual process.

5 Conclusion

Artificial Assistant Agents are being increasingly attached to the lives of people in the present, that they've become a daily companion to most of them. The application of AI assistants in the healthcare domain is way beyond our imagination in recent times [16]. Though in the past few years, there are a lot of inventions to enhance the use of chatbots in the medical world, technological advancements for healthcare technicians have been very limited. However, "Medbot" is tailored to equip the technicians with quick access to handle a medical device without much effort and time and act as a handbook in times of critical need [17]. Based on the results reported in this study, it was evident that the Medbot can assist the technicians in the absence of instructions which will ultimately improve the efficiency and performance of the laboratory assistants. Real-time implementation of Medbots in this current Covid situation is highly recommended due to the unavailability of resources. And this present study aims at bringing the medical support technology reach to the end-user to access remotely. Further, this present work was planned to be enhanced to be accessed regional languages.

References

1. O.V. Bitkina, H.K. Kim, J. Park, Usability and user experience of medical devices: An overview of the current state, analysis methodologies, and future challenges. Int. J. Ind. Ergon. **76**, 102932 (2020)
2. J. Balsa, I. Félix, A.P. Cláudio, M.B. Carmo, I.C. e Silva, A. Guerreiro, ... M.P. Guerreiro, Usability of an intelligent virtual assistant for promoting behavior change and self-care in older people with type 2 diabetes. J. Med. Syst. **44**(7), 1–12 (2020)
3. M. Kaczmarek, A. Bujnowski, K. Osiński, E. Birrer, T. Neumann, B. Teunissen, Ella4Life virtual assistant-user-centered design strategy-evaluation following laboratory tests. in *2020 13th International Conference on Human System Interaction (HSI)* (IEEE, 2020 June), pp. 307–311
4. S. Sai Sowmya, N.V. Ravindhar, N. Bharathiraja, B. Rohith, N. Sathish Kumar, Virtual personal assistant with chatbot. Int. J. Pure Appl. Math. **119**(15), 2073–2080 (2018)
5. E. Cho, M.D. Molina, J. Wang, The effects of modality, device, and task differences on the perceived human likeness of voice-activated virtual assistants. Cyberpsychol. Behav. Soc. Netw. **22**(8), 515–520 (2019)

6. D. Sonntag, Medical and health systems. in *The Handbook of Multimodal-Multisensor Interfaces: Language Processing, Software, Commercialization, and Emerging Directions*, vol. 3, (2019), pp. 423–476
7. B. Sasikumar, D. Naveenraju, K. Anand, S. Hariharan, P. Sudhakaran, N. Bharathiraja, Diabetes prediction using sensors by analysing skin temperature. J. Eng. Sci. Technol., 15(2), 1357–1370
8. N. Ouerhani, A. Maalel, H.B. Ghézela, SPeCECA: a smart pervasive chatbot for emergency case assistance based on cloud computing. Clust. Comput. **23**(4), 2471–2482 (2020)
9. U. Bharti, D. Bajaj, H. Batra, S. Lalit, S. Lalit, A. Gangwani, Medbot: conversational artificial intelligence powered chatbot for delivering tele-health after covid-19. in *2020 5th International Conference on Communication and Electronics Systems (ICCES)* (IEEE, 2020 June), pp. 870–875
10. E.H. Almansor, F.K. Hussain, Survey on intelligent chatbots: state-of-the-art and future research directions. in *Conference on Complex, Intelligent, and Software Intensive Systems* (Springer, Cham, 2019), pp. 534–543
11. H. Kazi, B.S. Chowdhry, Z. Memon, MedChatBot: an UML based chatbot for medical students. Int. J. Comput. Appl. **55**, 1–5 (2012). https://doi.org/10.5120/8844-2886
12. P.V. Gopirajan, K.P. Gopinath, G. Sivaranjani, J. Arun, Optimization of hydrothermal liquefaction process through machine learning approach: process conditions and oil yield. Biomass Convers. Biorefinery (2020). https://doi.org/10.1007/s13399-020-01233-8
13. M. Casillo, F. Clarizia, G. D'Aniello, M. De Santo, M. Lombardi, D. Santaniello, CHAT-Bot: a cultural heritage aware teller-bot for supporting touristic experiences. Pattern Recogn. Lett. **131**, (2020). https://doi.org/10.1016/j.patrec.2020.01.003
14. E. Bezverhny, K. Dadteev, L. Barykin, S. Nemeshaev, V. Klimov, Use of chat bots in learning management systems. Procedia Comput. Sci. **169**, (2020). https://doi.org/10.1016/j.procs.2020.02.195
15. M. Wiesenberg, R. Tench, Deep strategic mediatization: organizational leaders' knowledge and usage of social bots in an era of disinformation. Int. J. Inf. Manage. **51**, (2020). https://doi.org/10.1016/j.ijinfomgt.2019.102042
16. N.M. Saravana Kumar, Implementation of artificial intelligence in imparting education and evaluating student performance. J. Artif. Intell. 1(1), 1–9 (2019)
17. V. Suma, Computer vision for human-machine interaction-review. J. Trends Comput. Sci. Smart Technol. (TCSST) **1**(02), 131–139 (2019)

An Approach for Maintaining Safety at Work Places During the COVID-19 Pandemic Using Deep Learning and Contact Tracing Applications

Deveshi Thanawala, Aditya Sarin, Mayank Kapil, and Priyanka Verma

Abstract The outbreak of "COVID_19" has affected our lives and more so, our economy. One question that comes to mind is how to prevent the spread of this virus, and one possible solution to this is to work from home. However, there are certain industries which cannot work from home, for example, manufacturing, health care and production. For such workplaces, in this paper, we have proposed a mask detection module that uses the YOLOv3 algorithm to classify whether an employee is wearing a mask or not. A training accuracy of 96% and testing accuracy of 93.36% were obtained using the YOLOv3 algorithm. Along with the designed algorithm, four standard convolutional neural network architectures—VGG-16, DenseNet, InceptionNet and MobileNet—are also tested on the given dataset. The highest test accuracy of 91.21% was obtained from the MobileNet classifier. The paper also proposes an additional stage consisting of a company Web site that can keep a daily track on the employee's health and sanitization using a multipurpose wearable as well as an application that can be used to contact tracing purpose.

Keywords Safety during COVID-19 · YOLOv3 algorithm · Convolutional neural networks · Contact tracing application using Bluetooth and GPS

1 Introduction

Due to the rapid spread of the virus, sanitation and hands-free technology will play a key role in reopening workplaces during the pandemic. In addition to existing office security guidelines, there are additional safety measures that businesses can take to reduce the spread of the coronavirus in the workplace. Office buildings are prime places for diseases to spread, since there are many people that have to pass through

D. Thanawala (✉) · A. Sarin · M. Kapil · P. Verma
Mukesh Patel School of Technology, Management and Engineering, Electronics and Telecommunication, Mumbai 400054, India

P. Verma
e-mail: priyanka.verma@nmims.edu

G. Ranganathan et al. (eds.), *Soft Computing for Security Applications*, Advances in Intelligent Systems and Computing 1397,
https://doi.org/10.1007/978-981-16-5301-8_15

common entry points like lobbies, security checkpoints and elevators. As we start to go back to work, there are additional safety protocol measures that organizations can take to provide a safer work environment for employees from the moment they enter the premises. Hence, it is crucial to use the latest technology to maintain sanitization principles at workplace. Due to the advancement in technology, we can try our best to alleviate the problems being faced by individuals and large corporations during and after the pandemic. Many industries have thoroughly suffered due this pandemic, and in order to resume work in batches on their campus, they need state-of-the-art software that satisfies their sanitization needs and to develop a healthy virus-free environment. This is only possible if these companies try to adopt certain measures such as to detect if the employee maintains a normal temperature daily, wears a mask and possibility of contact tracing in case someone gets the virus. For large corporations with thousands of employees, it is necessary for them to keep a track of contact; hence, we can make use of sensor technology to fight against the virus and try to maintain COVID-19-free offices.

In this paper, we have proposed a mask detection module that uses the YOLOv3 algorithm to classify whether an employee is wearing a mask or not. The paper also proposes an additional stage consisting of a company Web site that can keep a daily track on the employee's health and sanitization using a multipurpose wearable as well as an application that can be used for contact tracing.

The outline of the paper is as follows: Sect. 2 explains about the literature survey done in order to conduct this research. Section 3 describes the data collection process. Section 4 explains the results and analysis, and Sect. 5 states the conclusion drawn from this research. The last section lists down the references used for this paper.

2 Literature Survey

Yang et al. [1] have proposed using YOLO algorithm along with CNN for face detection. Feature pyramid network (FPN) is used for detecting targets at different scales, and K-means clustering is used for computation of anchor boxes. The average detection time was 0.027 s. This paper has displayed results in terms of detection time and not on accuracy. Blue et al. [2] have proposed an algorithm that takes the output of YOLOv3 as input and does preprocessing on this. Limitation of proposed work includes situations when some sharp object is present and when there is too much noise in the image. For improvisation, image processing techniques can be employed to reduce the noise. Accuracy of the YOLOv3 model was 99% for detecting a sheep. Kanimozhi et al. [3] used SSD algorithm along with MobileNet that is proposed for multiple real-time object detection. However, if the distance is more than 30 m, then the detection rate is reduced. It takes time of 3–5 s to move the objected detected box over the next video. In order to make the model that has better generalization ability and ensure the richness of data, Hu et al. [4] used the Internet search engine to obtain the helmet wearing auxiliary data. Compared with YOLOv3, the mAP reached 93.5% and detection rate reached 35 fps. Vallamreddy et al. [5] proposed these functionalities by relying on the GPS module already present on the smartphone, route mapping done on the Android application and a strong Bluetooth connection with the wearable. Limitation is that HC-05 Bluetooth module has a quite low range when compared to other sensors or devices. Improvisation is that we can use the HM-10 Bluetooth 4.0 sensors that provides better stability and also a wider range. Ahmed et al. [6] used the centralized, decentralized and hybrid approaches that combine features from both the centralized and the decentralized architectures. Improvisation is that we can use automated data entries that do not rely on users and can self-detect symptoms or use contact tracing to update automatically. Jin et al. [7] used a secure end-to-end key exchange mechanism between two remote parties by cooperation of multiple devices at each party using quick response (QR) codes. Amoh et al. [8] used neural networks to automatically identify cough segments, and preliminary results indicate that method can develop into a real-time cough identification technique in continuous cough monitoring systems. Limitation is that uses contact microphones placed on the chest wall for sound recording. Improvisation is that use minimal contact sensors so that it does not become a hindrance increase the size of dataset to provide holistic solution. CNN model accuracy is 92.7%, and recurrent attains an accuracy of 87.23%.

Tôn et al. [9] demonstrated how wearables can be programmed semi-automatically to enact a medical rule-based alert process that assists hospital doctors in taking care of outbreak patients. Tripathy et al. [10] proposed method of using easy band that had sensors and that can sense other such devices present in a radius of 1–4 m (6–13 feet). Easy Band for social distance aware safe mobility in public domain. Limitation is that the device stores these data only for the next 15 days. This can be improved by using good quality sensors and by storing all of the data from each user in a separate database. In the model proposed by Wang et al. [11], the model not only does face mask detection but also does social distancing violation detection. The three classes that are simultaneously detected are masked and unmasked faces, as well as people to determine social distancing. Hsu et al. [12] used YOLOv4 that is supplemented by CenterLoss and pruning algorithm, so that it can be better used in the actual environment. Using cross-entropy on center distance to compute center loss instead of just the squared Euclidean distance stabilizes the center loss network. Time for detection was 0.137 s. Abbas et al. [13] used Bluetooth module for proximity detection, but smartphones will not be accessible to migrant workers and laborers (Bluetooth low energy can be used to reduce the battery usage). Zeinalipour-Yazti et al. [14] proposed variety of protocols under development that aim to make MCTA more transparent and effective. Examples include the following: The World-level Open-Source Anonymous Protocol. In the app proposed by Kaiser et al. [15], the app hosts a fuzzy neural network model that integrates data of employees' health status from the industry's database, proximity and contact tracing data from the mobile devices and user-reported COVID-19 self-test data. For tracing proximity and tracking contact, we have employed K-nearest neighbor and threshold-based algorithms. However, an asymptomatic case cannot be detected. Kumar et al. [16] proposed using wearable sensors, IoT and telemedicine for remote monitoring of COVID-19 patients to assess their condition and also of non-COVID-19 ones to reduce risks of cross-infection. Gupta et al. [17] have proposed different technologies used in COVID-19 like information searching, chatbots, big data and robotics, and Sodar by Google Sodar makes an augmented reality two-meter range rings around one to follow social distancing rules.

After conducting an exhaustive literature survey, the gaps that we found out were: For the mask detection algorithm, the mode faces an issue in detecting objects of different aspect ratio. For improving this, we have used data augmentation techniques. Also, noise in the image is removed by using image processing. All images were resized to a size of 416 * 416 * 3 so that training is faster. Since detection time for these images was slower, we have used threading methodologies for processing the frames. In all the above papers, same kind of masks was used for testing the model. However, we have tested our model with different variations of masks. For the contact tracing application, we have relied on automated data entry instead of user entry so that no user using the app can go untraced. For the wearable, minimum sensors are used so that no hindrance is caused. For the Web site, we have also integrated a restful API which allows the users and authorities to keep a track on COVID-19 affected areas.

3 System Description

The objective of this research is:

- To design a mask detection algorithm that can accurately classify whether an employee is wearing a mask or not.
- To develop a Web site and an application interface along with a smart wearable for contact tracing purposes.

Figure 1 describes the block diagram of the proposed model.

3.1 Data Collection

The dataset that is used in our mask detection model at the entry gate was taken from Kaggle. It has two subdivisions. The first category comprises of images of people wearing masks, and the other category comprises images of people not wearing masks.

The total number of images in our dataset is 32,209 (Table 1).

The proportion in which we have divided the dataset is in an 80:20 ratio, wherein we have used 80% of the dataset images to train the dataset and the remaining 20% of the images for testing the model on new images. Figure 2 describes the number of images in the dataset separated into two categories, with mask and without mask.

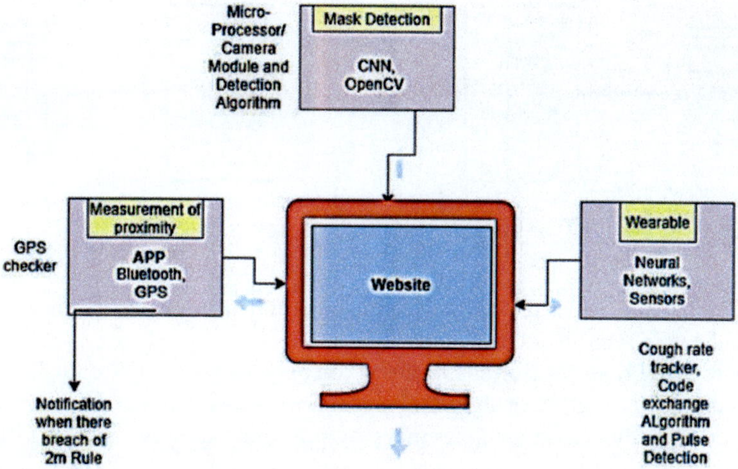

Fig. 1 Block diagram of the proposed model

Table 1 Distribution of images in dataset

Category type	Number of images
With mask	14,265
Without mask	17,944
Total	32,209

Fig. 2 Number of images in each class

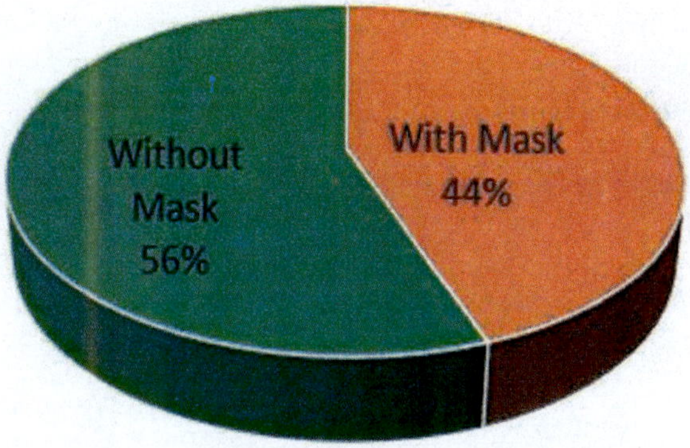

3.2 Classification Model: YOLOv3 Algorithm

The first part of our model would contain a mask detection model that could be placed at the entry gate of a workplace; in-built with a camera and would detect whether the person entering has worn a mask or not. If the person is not wearing a mask and tries to enter the premises, then he/she would not be granted access unless he wears a

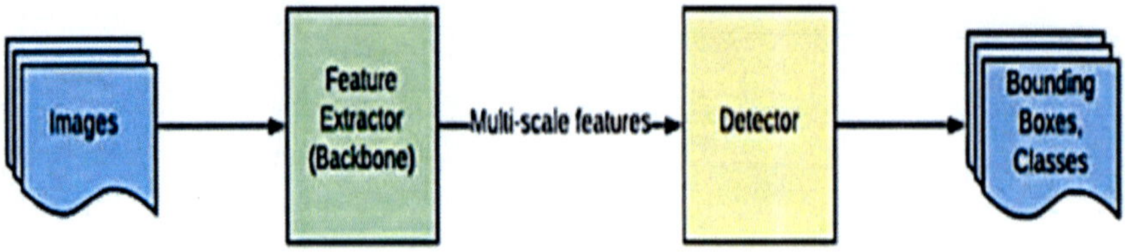

Fig. 3 Flowchart for the YOLOv3 algorithm [18]

mask. We have made use of object detection algorithm YOLOv3 and also compared our results with four other CNN models.

Figure 3 describes the flowchart for the You only learn once (YOLO) algorithm. YOLO is an object detection algorithm that internally uses CNN architecture. It can detect multiple objects in a single image by predicting classes and also identifies locations on the image. It applies a single neural network on the entire image, and this neural network then divides the image into grid cells and produces probabilities for each region, then finally it applied predict boxes for each region with the highest probability.

It contains 53 CNN layers using the DarkNet-53 model that is staked with more 53 layers thereby producing 106 layers for YOLOv3. In addition, some of the most important features added in the YOLOv3 algorithm are residual blocks, skip connections and upsampling. Each convolutional layer is followed by batch normalization layer and Leaky ReLU activation function. An activation function is used to introduce nonlinearity in the network so that the network can perform more complex operations.

The YOLOv3 algorithm does not contain any pooling layers, but additional convolutional layers with stride = 2 are used to downsample feature maps and thereby reducing the risk of losing low-level features in the data. YOLOv3 makes detections at layers 82, 94 and 106. These networks downsample the image by the factors of 32, 16 and 8 which are called the stride and show that how the outputs at these layers are smaller than the input of the network [19–21].

The following Eq. (1) computes the loss related to the predicted bounding box position (x, y).

$$\lambda_{\text{coord}} \sum_{i=0}^{S^2} \sum_{j=0}^{B} 1_{ij}^{\text{obj}} (x_i - \hat{x}_i)^2 + (y_i - \hat{y}_i) \tag{1}$$

Equation (1) describes loss for bounding box position.

Table 2 describes the hyperparameter that was used for tuning the YOLOv3 algorithm.

Table 2 Hyperparameters used for YOLOv3

Hyperparameter	Property
Batch size	64 images/step
Subdivisions	16
Max batches	4000
Filters	21
Learning rate	1e-3
Momentum	0.9
Delay	0.0005
Image width and height	416 × 416

3.3 Web site and Contact Tracing Application Along with Multipurpose Wearable

For this part of the model, we have designed an application interface that could help the employees to maintain the 2 m rule and stay safe. The application would use Bluetooth technology for this purpose and notify/alert the authorities as well as the employees when someone breaks this rule. In addition, this application can be used for contact tracing purposes. It uses GPS technology to keep a track on the zone being visited by the employee in the past week, in terms of red, yellow and green (according to COVID-19 affected areas). These data are available for one-month period, and hence, it could help the authorities if someone is tested positive. This data would also be stored onto the company Web site along with the other details of the employee. The figure below illustrates the application and Web site interface. The Web site for the company is created using ReactJS for the frontend with HTML EJS templating and NodeJS along with ExpressJS for the backend that is integrated with firebase. Figure 4 depicts the user interface of the application where it contains two pages; the first screen is for the employee login where he adds the number and the email-id, and the second page is the profile page that contains all the details of that employee with the image.

Figure 5 shows screens 3 and 4 of the application. Screen 3 shows the campus tracker screen where it displays Bluetooth data (hex code) from the current day. Screen 4 shows the in-state tracker screen that displays a history of the places visited by the user in past month by using the GPS data. The places are also color coded with respect to the intensity of the COVID-19 affected regions. Color code red is for regions that have high density of COVID-19 cases. Color orange is for regions with moderate number of cases. Color code green is for regions with less positive cases.

Each employee working in a particular company could be provided with a smart wearable that could keep a track on person coughs in a day and hence take significant actions if it reaches a certain limit, thereby notify others. Another use this of this wearable is that it uses code exchange technology. As soon as a person comes in contact with another person wearing the same band, a hex code is exchanged between the two, and hence, this would be useful in large companies that usually have more

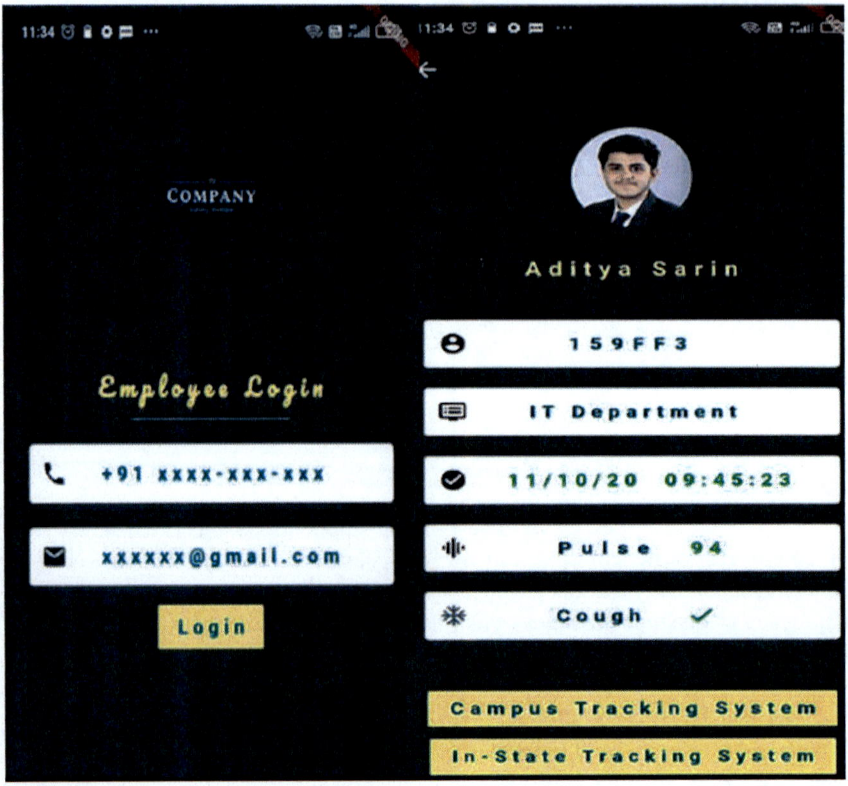

Fig. 4 Application log in page design

than 200–300 employees. The code exchange program would make use of barcode or basic encrypted keys and GPS technology.

4 Results and Analysis

The YOLOv3 algorithm was tested using different epochs, and each accuracy obtained has been depicted in Table 3.

Table 4 below provides the training as well as the test accuracy obtained after training various transfer learning algorithms and comparing them with YOLOv3 algorithm.

The mask detection algorithm was carried out using the best three algorithms DenseNet, MobileNet and You only learn once (YOLO) algorithm of Open CV.

The highest training and test accuracy were obtained from the YOLO algorithm. We had used a ratio of 80:20 for training and testing the model, respectively. The dataset contains approximately 4000 images. Figure 6 depicts a bar chart represen-

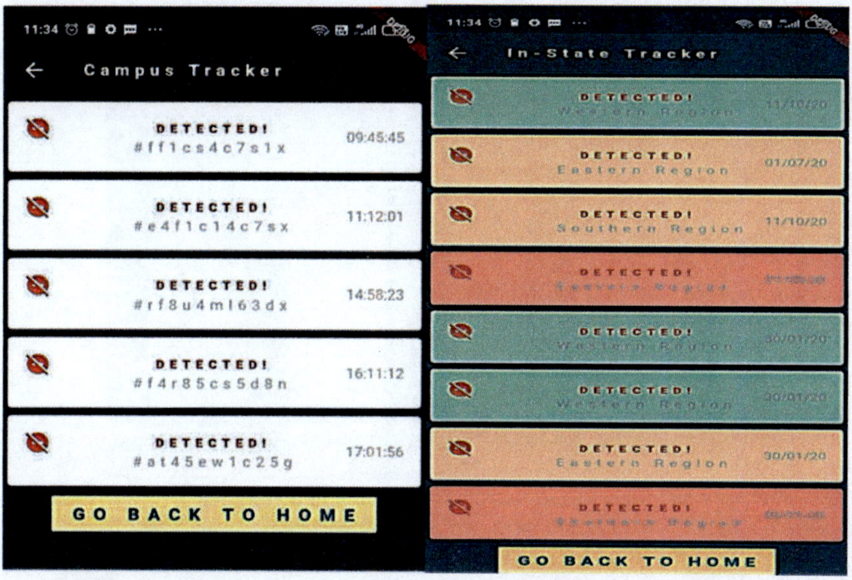

Fig. 5 Application interface

Table 3 Accuracy of YOLOv3 for different epochs

Sr No.	Epochs	Train accuracy (%)	Test accuracy (%)
1	5	84	81.11
2	15	92	89.28
3	20	95	91.01
4	25	94	84.25
5	10	96	93.63

Table 4 Accuracy of other CNN classifiers

Sr No	Model	Train accuracy (%)	Test accuracy (%)
1	ImageNet	84	81.29
2	DenseNet	92	89.89
3	MobileNet V2	95	91.21
4	VGG-16	94	84.27
5	YOLOv3	96	93.63

tation for the comparison of YOLO algorithm with other transfer learning models. Figure 7 shows the output one gets from the mask detection algorithm that we have applied in the stage 1 of our model. The model also provides the degree of accuracy with each and every frame that it detects for the mask.

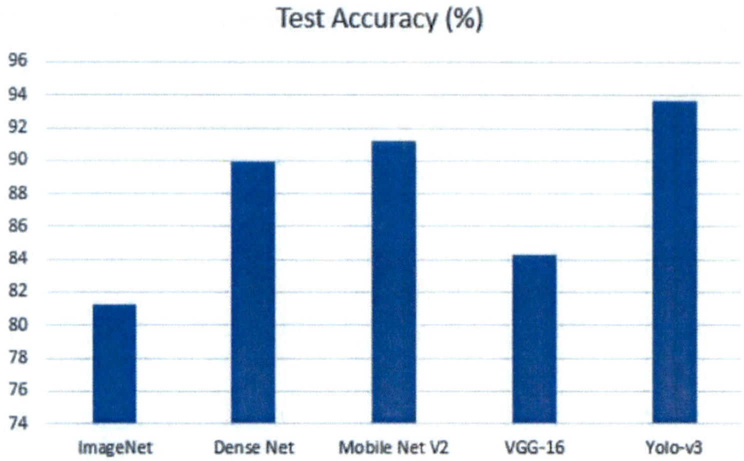

Fig. 6 Comparison of YOLO algorithm with other transfer learning models

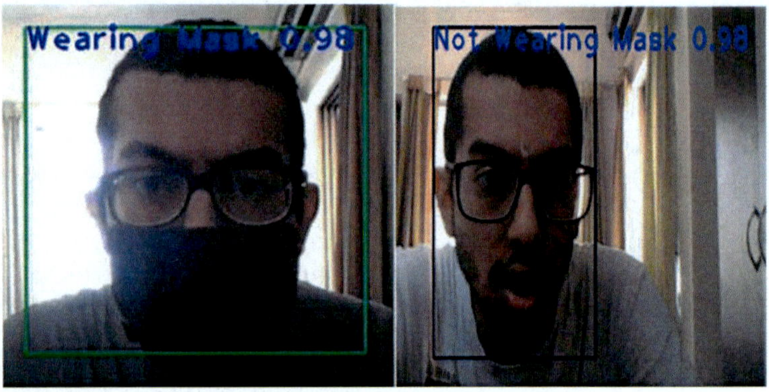

Fig. 7 Result of mask detection algorithm

5 Conclusion

The model has been created in the view of alleviating difficulties faced by companies to restart as well as continue smooth workflow during and after the unprecedented pandemic, COVID-19. The model contains a mask detection stage that uses YOLOv3 algorithm, and a test accuracy of 93.63% with 10 epochs and a batch size of 64 is obtained. The second stage contains the company Web site which acts as the employee database that stores employee data in real time so that helps officials to keep a track on each employee whether they are wearing masks, their pulse rate, temperature and other details. We have also created an application that contains an in-campus tracker and an in-state tracker that facilitates back tracing if an employee is exposed to

COVID-19. The smart wearable helps to keep a track on the employee's temperature and pulse rate.

We were not able to train the YOLOv3 algorithm after 25 epochs with the current batch size due to requirement for large computational power. The model does not classify whether the mask is worn properly or not. In the future, we plan to test the model with larger batch size and number of epochs as well as to devise a more efficient mechanism so that the same model can be used for public transport as well. The mask detection model cannot detect multiple bounding boxes with great accuracy.

References

1. W. Yang, Z. Jiachun, Real-time face detection based on YOLO. in *2018 1st IEEE International Conference on Knowledge Innovation and Invention (ICKII)* (Jeju, 2018), pp. 221–224. https://doi.org/10.1109/ICKII.2018.8569109.
2. S.T. Blue, M. Brindha, Edge detection based boundary box construction algorithm for improving the precision of object detection in YOLOv3. in *2019 10th International Conference on Computing, Communication and Networking Technologies (ICCCNT)* (Kanpur, India, 2019), pp. 1–5. https://doi.org/10.1109/ICCCNT45670.2019.8944852
3. S. Kanimozhi, G. Gayathri, T. Mala, Multiple real-time object identification using single shot multi-box detection. in *2019 International Conference on Computational Intelligence in Data Science (ICCIDS)* (Chennai, India, 2019), pp. 1–5. https://doi.org/10.1109/ICCIDS.2019.8862041
4. J. Hu, X. Gao, H. Wu, S. Gao, Detection of workers without the helments in videos based on YOLO V3. in 2019 12th International Congress on Image and Signal Processing, BioMedical Engineering and Informatics (CISP-BMEI) (Suzhou, China, 2019), pp. 1–4. https://doi.org/10.1109/CISP-BMEI48845.2019.8966045
5. S. Vallamreddy, S. Rajeev, S. Radhakrishna, Smart wearable for easy navigation. in *TENCON 2019—2019 IEEE Region 10 Conference (TENCON)* (Kochi, India, 2019), pp. 278–283. https://doi.org/10.1109/TENCON.2019.8929605
6. N. Ahmed et al., A survey of COVID-19 contact tracing apps. IEEE Access **8**, 134577–134601 (2020). https://doi.org/10.1109/ACCESS.2020.3010226
7. Y. Jin, M. Tomoishi, A secure end-to-end key exchange mechanism by cooperation of multiple devices using QR codes. in 2019 29th International Telecommunication Networks and Applications Conference (ITNAC) (Auckland, New Zealand, 2019), pp. 1–7. https://doi.org/10.1109/ITNAC46935.2019.9078020
8. J. Amoh, K. Odame, Deep neural networks for identifying cough sounds. IEEE Trans. Biomed. Circuits Syst. **10**(5), 1003–1011 (2016). https://doi.org/10.1109/TBCAS.2016.2598794
9. L. Tôn, L. Lê, Enacting a rule-based alert business process in smart healthcare using IoT wearables. in *2019 IEEE 23rd International Enterprise Distributed Object Computing Workshop (EDOCW)* (Paris, France, 2019), pp. 104–107. https://doi.org/10.1109/EDOCW.2019.00029
10. A.K. Tripathy, A.G. Mohapatra, S.P. Mohanty, E. Kougianos, A.M. Joshi, G. Das, EasyBand: a wearable for safety-aware mobility during pandemic outbreak. IEEE Consum. Electron. Mag. **9**(5), 57–61 (2020). https://doi.org/10.1109/MCE.2020.2992034
11. X. Wang, Z. Chen, B. Wei, M. Ling, Application of pruning Yolo-V4 with center loss in mask wearing recognition for gymnasiums and sports grounds of colleges and universities. in *2020 IEEE 6th International Conference on Computer and Communications (ICCC)* (Chengdu, China, 2020), pp. 1373–1377. https://doi.org/10.1109/ICCC51575.2020.9345257

12. H.-H. Hsu, N.-F. Huang, C.-H. Han, Collision analysis to motor Dashcam videos with YOLO and mask R-CNN for auto insurance. in 2020 International Conference on Intelligent Engineering and Management (ICIEM) (London, UK, 2020), pp. 311–315. https://doi.org/10.1109/ICIEM48762.2020.9160263
13. R. Abbas, K. Michael, COVID-19 contact trace app deployments: learnings from Australia and Singapore. IEEE Consum. Electron. Mag. **9**(5), 65–70 (2020). https://doi.org/10.1109/MCE.2020.3002490
14. D. Zeinalipour-Yazti, C. Claramunt, COVID-19 mobile contact tracing apps (MCTA): a digital vaccine or a privacy demolition? in *2020 21st IEEE International Conference on Mobile Data Management (MDM)* (Versailles, France, 2020), pp. 1–4. https://doi.org/10.1109/MDM48529.2020.00020
15. M.S. Kaiser et al., iWorksafe: towards healthy workplaces during COVID-19 with an intelligent Phealth app for industrial settings. IEEE Access **9**, 13814–13828 (2021). https://doi.org/10.1109/ACCESS.2021.3050193
16. S.S. Kumar, Emerging technologies and sensors that can be used during the COVID-19 pandemic. in *2020 International Conference on UK-China Emerging Technologies (UCET)* (Glasgow, UK, 2020), pp. 1–4. https://doi.org/10.1109/UCET51115.2020.9205424
17. R. Gupta, S.K. Pal, A. Khattar, K. Baliyan, Technology applications for health safety decision making under COVID-19 pandemic management. in *2020 International Conference on Decision Aid Sciences and Application (DASA)* (Sakheer, Bahrain, 2020), pp. 788–793. https://doi.org/10.1109/DASA51403.2020.9317205
18. https://towardsdatascience.com/dive-really-deep-into-yolo-v3-a-beginners-guide-9e3d2666280e
19. https://towardsdatascience.com/yolo-v3-object-detection-53fb7d3bfe6b
20. A. Sungheetha, R. Sharma, A novel capsnet based image reconstruction and regression analysis. J. Innov. Image Process. (JIIP) **2**(03), 156–216 (2020)
21. T. Vijayakumar, R. Vinothkanna, Retrieval of complex images using visual saliency guided cognitive classification. J. Innov. Image Process. (JIIP) **2**(02), 102–109 (2020)

Online Certificate Generation and Verification Using Blockchain Framework

Prasad Jadhav, Aseem Godambe, Rutwik Gaikwad, and Kiran Deshpande

Abstract There are many cases reported of certificate forgery every day, and many of them go undetected. The purpose of this study is to develop a system that would authenticate certificates to their real owners and could verify the same on a specific portal. To achieve this, in this paper we focus on Ethereum Blockchain as it is immutable, transparent, scalable, and also cost-effective. Universities would be able to generate their own certificates on the web portal itself by uploading their certificate template in Scalable Vector Graphics (SVG) format and the student data in comma-separated values (CSV) format. Each generated certificate would be allotted to the respective student within the system. Each certificate would be given a unique hash code so that it could be verified easily and there would be no scope of duplication. The authentication and verification would be done on the same web portal. To sum up, this method would save time and efforts that are required to verify a certificate manually and would result in an effective, secure way to generate certificates.

Keywords Blockchain · Ethereum · Distributed system · Security · Encryption · Hashing · Scalable vector graphics (SVG) · Comma-separated values (CSV) · Digital certificate · Authentication · Verification

---

P. Jadhav · A. Godambe (✉) · R. Gaikwad · K. Deshpande
A.P. Shah Institute of Technology, University of Mumbai, Mumbai 400615, India
e-mail: aseemgodambe@apsit.edu.in

P. Jadhav
e-mail: prasadjadhav@apsit.edu.in

R. Gaikwad
e-mail: gaikwadrutwik@apsit.edu.in

K. Deshpande
e-mail: kbdeshpande@apsit.edu.in

G. Ranganathan et al. (eds.), *Soft Computing for Security Applications*,
Advances in Intelligent Systems and Computing 1397,
https://doi.org/10.1007/978-981-16-5301-8_16

1 Introduction

Everyone has a particular talent or excels in a field, and certificates are a great way to reflect the achievements. It proves that an individual has acquired knowledge of a particular skill. Companies need skilled employees, and certificates prove to be a valid solution to depict ones capabilities. Displaying a certificate of a particular skill verifies that the certificate holder is competent in that skill and helps the company in the hiring process. Also due to the current pandemic situation, online learning and certifications through online courses have drastically increased. People have started to adapt to online learning, and hence, the certificates hold a greater value to validate the progress of an individual, and also it adds a great value to their profiles.

As certificates hold such higher values, people tend to misuse it and generate a fake copy of the certificate of the skill which they have not acquired. The companies or the organization where an individual is applying offers an advantage to the person having the necessary certificates. It is difficult for the organization to check individual certificates as they may have to contact the issuer, and it takes a lot of time and effort. The fake certificates can be easily generated using any online website or normal photo-editing software. As a result, it makes it more difficult for the deserving candidate to have all the genuine certificates. A survey conducted by UK's national qualifications agency UK NARIC across 17 countries in different universities found the common problem of difficulty in verifying documents. According to the survey, it has been found that 62% of them verify the documents by contacting the institutes awarding them, while 14% of them did not even bother to check the originality of the certificate. It stated that around 75% of the certificates submitted were fraud [1]. Similar research and investigation conducted by the BBC Radio 4's File on Four program found out about thousands of fake degree certificates in the UK are from "Diploma Mill" in Pakistan. This deteriorates the standard of employees being hired and unfair against the students who are hardworking and achieving the results [2].

The motive of this research is to create a system which is secure, easy to access, and verify certificates. The organization hiring an individual could verify the certificate online without the need to contact the issuer. Blockchain technology along with different hashing methods is used to maintain a secure record of the information. An Scalable Vector Graphics (SVG) format certificate template is used specifically to reduce the cost of storing multiple images on the blockchain network. This creates a scalable and economical system for an organization to implement.

2 Preliminary Concept

2.1 Blockchain

The first idea of collecting various certificates and documents in one block was led out by Stuart Haber and W. Scott Stornetta back in 1991 [3]. Their basic idea was to create a system in which the documents and certificates cannot be tampered by anyone. They decided to implement it using a cryptographically secured chain of blocks. This further led to the concept of blockchain which was introduced in 2008 by Satoshi Nakamto (or group of people). At the initial stage, blockchain was focused on crypto-currency. But later on, blockchain extended in various other fields. Blockchain could also be considered as an online ledger that provides decentralized and transparency in data sharing.

The concept of blockchain was to store data in a block along with a timestamp. Once the data is entered into the block and it is verified, then it becomes immutable and open in public to provide transparency. The block is provided with a hash value of the current block and also the hash value of the previous block in the chain. This ensures the trackability of the block from the chain.

2.2 *Ethereum*

The advancement in the field introduced Ethereum into the blockchain. Ethereum focused on another usage of blockchain. This led to the division as Bitcoin was for payment network Ethereum started its use in computing. The other advantage of Ethereum is that it is an open-source project.

Ethereum introduced Ethereum Virtual Machine (EVM). These are programmable blockchain. It works on a high-level language developed for this purpose like solidity. It also offers smart contracts. Smart contracts basically streamline multiple complex tasks or processes involved in completing an objective. Smart contracts basically work on a simple statement "if/when... then...". It is just the coding done to perform tasks automatically if one situation occurs then what would be its countermeasure. Smart contracts not only have the same time but also provide reliable security [4]. The various aspects of smart contracts are self-verifying and self-executing. It is also tampering resistant, involves fewer intermediaries, and also has low transaction cost.

2.3 *SVG—Scalar Vector Graphics*

SVG is an Extensible Markup Language-based vector image format for two-dimensional graphics [5]. It has additional features and supports interactivity and animation. SVG was developed by the World Wide Web Consortium as an open standard. It allows us to define vector-based graphics in XML format on the web. It is recommended by W3C (World Wide Web Consortium) and is also integrated with DOM and XSL.

SVG provides multiple advantages over other image formats such as it can be created in a text editor and also provides an option to edit it later on. As its written in XML format, it offers options to compress, index, search, and script the images. They do not lose quality when zoomed and therefore can have high resolution when printing. As SVG provides the ability to edit the image we can use it to store the template of the certificate separately and print multiple certificates of the same template without actually saving the whole certificate.

3 Literature Survey

3.1 *Shanmuga Priya R, Swetha N "Online Certificate Validation Using Blockchain" [6]*

This document summarizes the problems of forged certificates and how blockchain can solve the issue. NetBeans IDE and Android Studio are used to develop the server communication and the application to scan the QR codes. EthereumJS is used for faster Ethereum applications. The application involves the user uploading his/her certificate like 10th grade marksheet, college certificates, government certificates, and so on to the portal. After uploading the certificate, the data is then sent to the issuer for validation. The issuer (e.g., school which is responsible for validation of 10th grade marksheet) has to validate the data received by the application for verification. Once verification is successful, the data will be stored on the server, else it would be discarded. On the mobile application, a QR code will be generated based on the certificate number. The QR code can then be shared with anyone else for verification in case of necessity. When the QR code is scanned, an One-Time Password (OTP) will be sent to the registered mobile number for verification. After proper authentication, the user can view the certificate. If the number of scans goes beyond the permitted limit, the location of the scanner will be sent to the authorized user with a permission link. From that link, the authorized user can either allow or deny the person. The major drawback of this system is that verification can be delayed by the issuer as the certificate has to reach the organization for verification and the data is sent to blockchain only after verification. The OTP system can get tedious. This system can get costly as well.

3.2 *Nitin Kumavat, Swapnil Mengade, Dishant Desai, JesalVarolia "Certificate Verification System Using Blockchain" [7]*

This document summarizes that during the course of education, the students achieve many certificates. Students produce these certificates while applying for jobs in public or private sectors, where all these certificates are needed to be verified manually. There can be incidents where students may produce fake certificates, and it is difficult to identify them. The solution proposed in this system uses Ethereum and Interplanetary File System (IPFS). IPFS is a peer-to-peer, content address system. It is very similar to BitTorrent and Merkle Dag. Unlike HTTP which restricts or provides low latency on transfer of large amounts over the network which uses IP addressing, IPFS uses content addressing. As a result, it creates a distributed system of different nodes across the network. It returns us a hash value, and it uses this value to retrieve the data. They propose to add the data with the certificate on the IPFS network to generate the hash for it and later on add the data to the blockchain network with the help of EVM. This topology includes two distributed networks to provide additional security to the model. While retrieving the certificate they would compare the hash and pull the entire certificate which has been stored on the IPFS network. The advantage of this system was it is more secure and provides reliability. With two distributed networks, it is almost impossible to tamper the data. On the other hand, the disadvantage of the system is that IPFS is a tedious process to set up, and storing the certificate image on any distributed blockchain platform costs a lot of money and hinders the scalability of the system.

4 Proposed System Architecture

4.1 Methodology

This proposed system revolves around creating a secure and fast method to generate and verify the certificates using the features of SVG and blockchain. The certificate's data is hashed and stored on a blockchain network to provide security and immutability to the data. On the blockchain network, this data is stored on blocks. The block contains the hashed data, timestamp, and the ID of the next block. Then, the block is added to the network. The template of the certificate which is in SVG format is stored on a database server that is not a blockchain-based network as the certificate template will not be modified. The data which is hashed includes the name of the certificate holder, their email, issuing date, expiry date for the certificate, and information regarding other fields of the certificate. This data can only be submitted by the certificate issuer, and thus, ensuring the right data is provided. Anyone can check certificates of the students using the unique ID provided to each student for their certificates or from a student's profile (Fig. 1).

4.2 Certificate Generation

Certificates can be generated only by the organizations registered on the website. They need to upload the template of the certificate in the SVG format. The SVG template should have a proper ID for a particular field like the space for the name of the receiver should have a well-defined ID, for example, 's_name'. Once the SVG template is uploaded, the organization needs to upload a CSV file that has the record

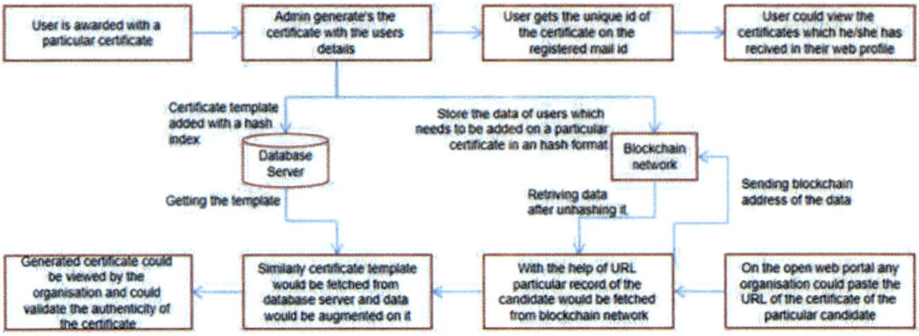

Fig. 1 Architecture of proposed system

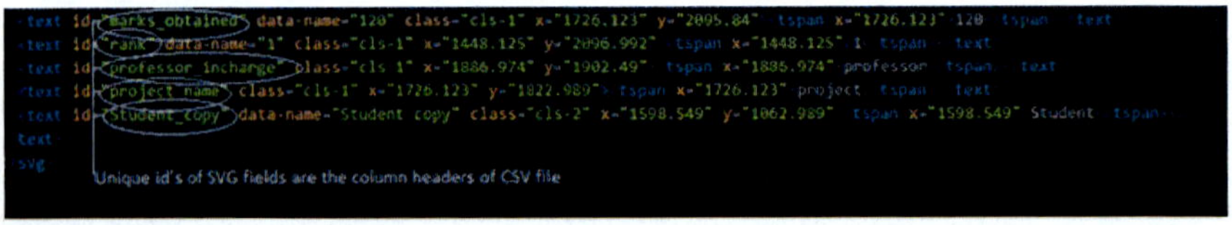

Fig. 2 Example of a certificate template in an SVG format indicating the IDs of the fields of the certificate

cert_id	email	Student_copy	marks_obtained	rank	professor_incharge	project_name
1	abc1@gmail.com	Rahul Sharma	70	2	Prof. Gopal Yadav	Water sanitisation
2	abc2@gmail.com	Vishal Pande	79	1	Prof. Gopal Yadav	Soil quality analysis

Fig. 3 CVS file with the data of students winning the certificate for the above given SVG template

of the students who have won the certificates. The CSV needs to have the first row as the IDs of the fields marked on the SVG template, along with it the CSV needs to have a compulsory email column that would contain the email of the receiver. Other columns of the CSV would be the remaining values that needed to be filled on the template like the receiver's name, issuing date and expiry date of the certificate, etc. The only condition is that the ID of the field written on the SVG template should be the column header of the CSV file. This method is done to ensure that the blanks on the template could be filled by the right data in the CSV file. Once the upload action of the template and CSV file is done, the certificates would be auto-generated for all the students receiving the certificates and would be available in their profile on the website. This would be done with the help of email IDs associated with each student present in the CSV file (Figs. 2 and 3).

4.3 *Certificate Validation*

In this process, the certificates are validated. As the data is stored on the blockchain network, there is no way to edit the data once it is appended into the blocks on the network. The process works as when the certificate data is uploaded by the organization in CSV format, the data is extracted from the CSV and uploaded on the blockchain network. Once the data is uploaded on the network, students who have won those certificates are informed through an email that they can view their certificates now in their profile and are provided a unique key or URL for the certificate. Students could use or provide this key URL to anyone to authenticate the certificates. When a particular key is used, the data of that candidate is retrieved from the blockchain and the corresponding template is fetched from the database server. The data from the blockchain is mapped on the SVG templates as the fields of the template are pre-recorded and thus generates a certificate in front of the user which is not tampered with by any means. Thus, an entity could verify that the certificate is genuine and possessed by the actual owner of the certificate (Fig. 4).

4.4 *Working of Application*

In our application on the landing page, any individual could see a certificate by just entering its unique ID on the search bar. There is an admin login; once logged in, the admin one could add certificate and CSV file to generate certificates. Once the data is uploaded by the admin, it would be stored temporarily in a database and would be uploaded on the blockchain network periodically using a CRON script every day at midnight. This is done to ensure every certificate's data is uploaded on

Fig. 4 Certificate uploaded on the web portal which is verified and genuine certificate

Fig. 5 Homepage of the web portal where anyone could search for any certificate to verify its authenticity

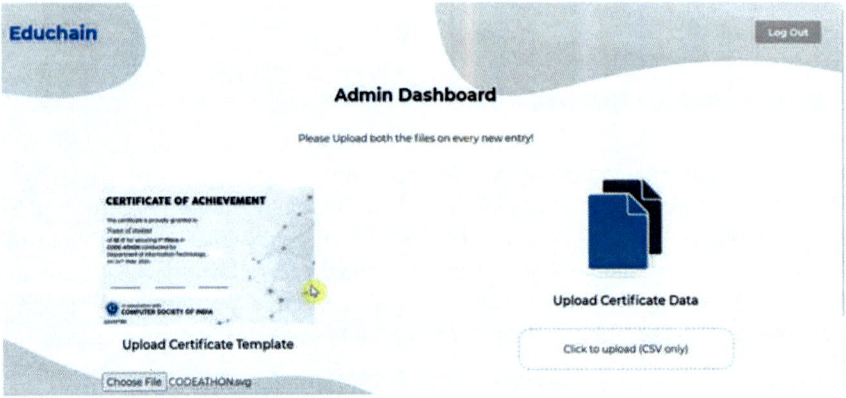

Fig. 6 Admin dashboard to upload SVG and CSV file

the blockchain network because if we fail to upload any data, we cannot rewrite the blocks on the network. Once the data is uploaded, students are informed that their certificate is uploaded. When logged in as students, they could see the certificates achieved by them in their profile and would see a green tick indicating that the certificate is validated. Students can share their certificate links with anyone to view their certificates (Figs. 5 and 6).

5 Conclusion

In this paper, we have successfully proposed a system where blockchain technology can be used to store and retrieve certificate data. The project will help companies issue certificates securely through Blockchain and can be verified by anyone with the unique link/code to each certificate. The system uses the concept of SVG templates for the certificates which would be stored on a local server and to store data over the blockchain for secure and reliable storage. This will minimize the cost of storing the entire certificate on the blockchain network. Storing only the data of the certificate will minimize cost and thereby turn out to be cost-efficient. The only drawback of this system is that the template of the certificate needs to be properly created with the SVG's text area ID to match with the header of the CSV file. Only a properly crafted SVG and CSV pair will result in proper certificate generation through Blockchain.

Acknowledgements The authors would like to sincerely thank A. P. Shah Institute of Technology for supporting this research by providing necessary resources and guidance.

References

1. UK: NARIC Survey: universities and education fraud—75% cannot spot a fake certificate—Feb 2020, https://uknaric.org/2020/02/26/survey-universities-and-education-fraud-75-cannot-spot-a-fake-certificate/
2. BBC File on 4 exposes a multi-million pound global trade in fake diplomas (2018). https://www.bbc.co.uk/programmes/b09ly731
3. Blockchain. https://en.wikipedia.org/wiki/Blockchain
4. Smart Contracts. https://www.ibm.com/blogs/blockchain/2018/07/what-are-smart-contracts-on-blockchain/
5. Scalable Vector Graphics. https://en.wikipedia.org/wiki/Scalable_Vector_Graphics
6. R. Shanmuga Priya, N. Swetha, Online Certificate Validation Using Blockchain. Published in International Journal Advanced Networking Applications (IJANA)
7. N. Kumavat, S. Mengade, D. Desai, J. Varolia, Certificate verification system using blockchain. Int. J. Res. Appl. Sci. Eng. Technol. (IJRASET) (2019)
8. O. Miquel, M. Joan, P. Gerson, B. Tez David, Using blockchain as a tool for tracking and verification of official degrees: business model, in *29th European Regional Conference of the International Telecommunications Society (ITS): "Towards a Digital Future: Turning Technology into Markets?"*, Trento, Italy, 1st–4th August 2018 (2018)
9. K. Kuvshinov, I. Nikiforov, J. Mostovoy, D. Mukhutdinov, *Disciplina: Blockchain for Education* (2018), pp. 1–17
10. D.T.T. Anh, M. Zhang, B.C. Ooi, G. Chen, Untangling blockchain: a data processing view of blockchain systems. IEEE Trans. Knowl. Data Eng. **4347**(c), 1–20 (2018)
11. Kumar, N.M. Saravana, Implementation of artificial intelligence in imparting education and evaluating student performance. J. Artif. Intell. **1**(01), 1–9 (2019)
12. S. Smys, J.I. Zong Chen, S. Shakya, Survey on neural network architectures with deep learning. J. Soft Comput. Paradigm (JSCP) **2**(03), 186–194 (2020)

A Survey on LiDAR-Based SLAM Technique for an Autonomous Model Using Particle Filters

Md. Shafiqul Islam, Md. Tarequl Islam, and Mohammad Gazi Golam Faruque

Abstract Simultaneous localization and mapping, otherwise known as SLAM, are the standard approach for autonomous model building and localization. A SLAM method is a real-time system where information about an unknown surrounding environment is collected. Based on this information, a model is generated online, and the position of the agent is estimated in parallel. There are many different types of SLAM techniques used over the years like EKF-based SLAM, Gridmap-based SLAM, and visual SLAM. This paper aims to review the most recent SLAM techniques from above and also an overview of 2D LiDAR-based SLAM implementation that uses particle filters. In particular, it focuses on the current approaches available for speeding up the process to deal with large-scale scenarios. Particle and Kalman filtering algorithms are the main areas of research that are reviewed as ways to minimize the amount of data needed. It is shown that particle filter-based methods are, in effect, an effective way to create large-scale consistent models of an indoor environment that may be used to successfully localize an autonomous vehicle using a 2D LiDAR scanner.

Keywords SLAM · Kalman filter · Particle filter · LiDAR · Survey

1 Introduction

SLAM signifies a family of algorithms in which a map of an unidentified environment is being constructed. At the same time, the position and orientation of the autonomous entity are calculated [1]. Several areas may benefit from the implementation of SLAM algorithms in autonomous vehicles. The mining industry, marine exploration, and planetary exploration should be examples [1]. SLAM is expressed with a probability density function,

$$P(x_t, m|z_{1:t}, u_{1:t}) \tag{1}$$

Md. Shafiqul Islam (✉) · Md. Tarequl Islam · M. G. G. Faruque
Department of Computer Science and Engineering, Khwaja Yunus Ali University, Sirajganj, Bangladesh

G. Ranganathan et al. (eds.), *Soft Computing for Security Applications*,
Advances in Intelligent Systems and Computing 1397,
https://doi.org/10.1007/978-981-16-5301-8_17

where x_t is denoted as the current position of the vehicle, m is called the map of surroundings, $z_{1:t}$ is denoted as the observation for different aspects of time, and lastly, $u_{1:t}$ is denoted as the control units for different units of time from the odometry [2]. The pose of the vehicle directly depends on the model or map, observation, and control units.

SLAM consists of different sub-problems; such as the extraction of landmarks, the association of data, estimate of state, update of state, and update of landmarks. Any of the smaller elements have several ways of solving them.

SLAM is a well-known problem that has been widely studied for autonomous vehicles in the last 40 years. The modeling of an unknown environment and localizing in the same is an essential requirement to build truly autonomous exploratory agents. Different types of sensors have been used in order to gather meaningful data about any environment. Laser, sonar, and vision systems are examples of the most commonly used sensors. Additionally, sensors such as compasses, gyroscopes, and the global positioning system are used to estimate the state of the robot [3]. However, all sensors provide imperfect data, often known as measurement noise, and also have other limitations that make it difficult to navigate through the environment. For example, light and sound ranging sensors most commonly have the problem that light and sound cannot penetrate solid objects such as walls.

At present, different types of cameras and light detection and ranging (LiDAR) sensors are the primary types of external sensors used in the SLAM technology solution. One can categorize environment sensing elements into active sensors such as LiDAR sensors and sonar sensors, where the environment is measured through sending out an active beam or pulse of energy and measuring the return and passive sensors such as cameras where the ambient energy is measured. Monocular and binocular (stereo) cameras are different subclasses of cameras.

Another interesting type of sensors is internal state sensors such as inertial measurement units (IMUs), odometry sensors, and GPS devices. These sensors are mainly used to estimate the current state of the mobile robot. The state may include the position, orientation, and velocity information, concerning some globally used coordinate systems.

Additionally, some new types of sensors that may be interesting for SLAM methods are event cameras [4], solid-state LiDARs, and RADAR sensors.

It is also interesting here to note the different problems that affect environment sensors. Cameras are deeply dependent on the light condition of the area. This light condition is limited by some factors. Sampling frequency, angle of view, and angular resolution are those factors. When the satellites are blocked, GPS is affected in the indoor environment. Odometry sensors such as wheel encoders are as a standard affected by quantization noise. LiDAR-based sensors operate using the Time of Flight (ToF) method. This is the way to measure the distance from the sensor center to an obstructing object. The measurement is largely unaffected by illumination conditions. Additionally, laser-based solutions have long ranges and exhibit higher precision in indoor and outdoor scenes compared to cameras or sonar sensors.

For the presented application, we choose to focus only on indoor environments [5–8]. In such an environment, a low-cost LiDAR sensor presents a cost-effective easy to use ranging solution.

2 Overview of Different Slam Techniques

At the 1986 IEEE Robotics and Automation Conference in San Francisco, California, the probabilistic SLAM problem was born. Probabilistic approaches were just starting to be incorporated into robotics and artificial intelligence at the time (AI). A variety of researchers were interested in using estimation-theoretic approaches to solve mapping and localization problems. Bailey and Durrant-Whyte [9] reviewed the history of SLAM approaches in 2006. Their work focused on mainly two methods: EKF-based SLAM and FastSLAM. Both of these approaches are probabilistic approaches; however, in the case of FastSLAM, particle filters and EKF are used. FastSLAM exploits the conditional independence property of environmental features concerning the trajectory estimation.

Since 2006, more SLAM methods have been proposed based on probabilistic approaches. Grisetti G presented a Rao-Blackwellized particle filter-based method [10]. In this method, the number of sampled particles has been reduced to manage computation requirements. Additionally, a gridmap-based representation of the model is adopted. A laser scanner is used to perform online synchronized localization and gridmap formation.

Most recently, graph-based techniques have been suggested, where multiple local sensor views of the entire model are limited by motion constraints.

2.1 EKF-Based SLAM

In an Extended Kalman Filter (EKF)-based method, the SLAM problem is solved using sensor data obtained from the robot's motion and rotation. For example, by using wheel encoders and gyroscopes, this can be done. Besides, it is often important to collect environmental information by using, for example, a laser range finder (LiDAR) or camera. With this knowledge, SLAM preserves track of where the robot is probable to be located on a map and keeps track of unique landmarks observed [11].

Because the robot odometry (which gives the location of the robots) is always inaccurate, we cannot rely on the odometry directly. To correct the location of the robot, we can use laser scans of the surroundings. This is done by removing and re-observing features from the environment as the robot travels some distance. Such attributes are generally called landmarks. The heart of the EKF-based SLAM method is an Extended Kalman Filter (EKF). It is responsible for adding new landmarks to the model, updating old landmarks, and also updating the current trajectory. The

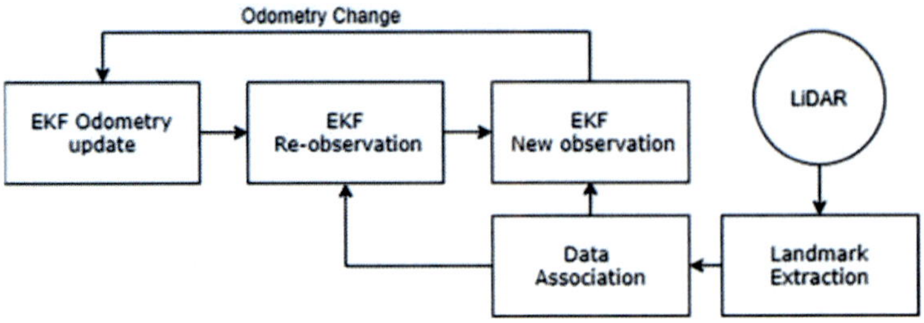

Fig. 1 EKF-based SLAM process

EKF keeps track of an estimation of the uncertainty in the position of the robot and also of the uncertainty in the environment of the landmark positions (Fig. 1).

First and foremost, the robot uses observed landmarks and information from the sensors to localize itself within the map being generated. Preferably, landmarks should be visible from various angles and not isolated from other landmarks. These prerequisites provide a good opportunity for the EKF algorithm to differentiate between landmarks at a later date. Besides, the landmarks used should also be stationary [11].

2.2 *GridMap-Based SLAM*

In a gridmap-based method, the model is expressed as a regular pattern of cells arranged in a grid-like manner. Each cell in the model corresponds to a unified physical space. A regular grid-like organization in 2D is known as a gridmap; while in 3D, the same is expressed as a voxel-map. Laser scans from a scanning laser range finder are collected and integrated online into a grid representation. During this integration, the continuous laser scans are discretized into the cell representation. The integration of new laser scans is only possible when a reliable estimate of the current pose (orientation and position) is available. Odometry as a second input is (generally) required. Grisetti G developed a factored particle filter-based approach and tested it on a wide range of data sets—providing an efficient and superior solution to the SLAM problem using gridmaps [10]. His methodology has also recently been integrated into the software suite of the robot operating system (ROS) [12], where a module based on the algorithm of Adaptive Monte Carlo Localization (AMCL) [13] uses the maps generated by RBPF-based SLAM. Within ROS, the interface is transparent and well documented. Although RBPF-based SLAM itself is in the research domain, we are aware of many groups that have successfully used it, and we believe it is the most realistic method for navigating indoor robots or similar applications (Fig. 2).

The laser rangefinder scans help to avoid obstacles and surround models in a way that it can be used for vehicle navigation, exploration, and path planning of the

Fig. 2 Gridmap-based SLAM

environment, for the map models obtained. The model is generally a 2D collection of cells regularly organized in a grid pattern. Each cell corresponds to a physical unit area in the real world. The grid representation automatically provides some robustness around sensor noise, as the cells act as a discretization of continuous range data.

2.3 *Graph-Based SLAM*

Another solution to addressing the SLAM problem is GraphSLAM. Using a graph, the idea behind GraphSLAM is to answer the SLAM problem. The graph includes nodes representing the robot $x_0 \ldots. x_t$ poses as well as landmarks on the map denoted as $m_0 \ldots. m_t$. However, to resolve the SLAM problem, this is not enough as there is also dependency between the robot's positions $x_t, x_{t-1}, x_{t-2} \ldots x_{t-n}$ and the landmarks $m_0 \ldots m_t$ [14]. The dependency is the distance between neighboring locations such as x_{t-1}, x_t, as well as the distance to the landmarks $m_0 \ldots m_t$ across locations. Due to data from the odometry source u_t, these dependencies can be constructed [14]. There is a possibility to convert the graph described earlier to a form of the sparse matrix, which can be used more efficiently, whereas dense matrices [15].

We can see how the landmarks $m_0 \ldots m_t$ and the locations that are $x_0 \ldots x_t$ and the constraints are linked together. The square circles are the robot's pose in different aspects of time. The squares are the landmarks in the environmental model. The dotted line represents the robot's motion based on the robot-specified commands. And lastly, the clear lines represent the estimation of the distance from the landmarks (Fig. 3).

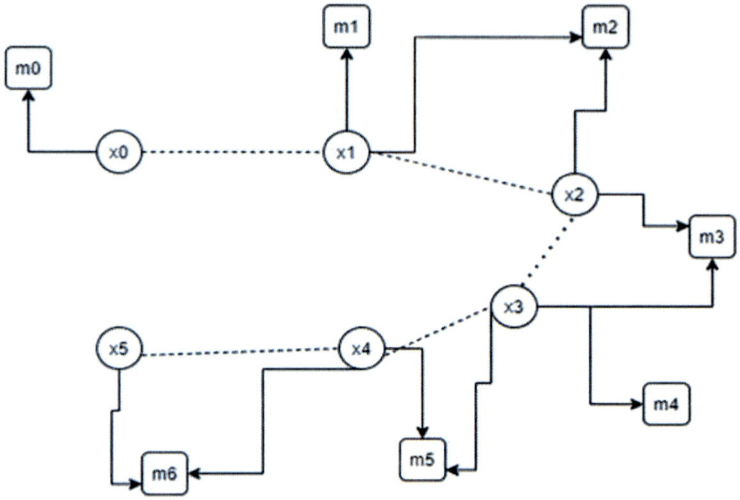

Fig. 3 Graph-based SLAM

2.4 Comparison Between Different SLAM Techniques

It is possible to compare SLAM algorithms using different qualitative and quantitative metrics. We decided to compare the interesting approaches based on applicative features such as robustness, computational requirement, and failure modes. The comparison is as follows,

EKF SLAM

- Robustness: Faulty data association results in incorrect results, so the EKF SLAM's robustness is poor [16].
- Computational Efficiency: The complexity of the algorithm is $O(N^2)$. Here N is the number of landmarks observed or contained in the map. In case of a single landmark being observed, the whole map needs to be updated. Therefore, it is suitable for minor maps with fewer landmarks due to substantial computation [17]. But localization and data association will be problematic when the landmarks are few in the environment [16].

Gridmap-based RBPF SLAM

- Robustness: Gridmap-based SLAM is significantly more robust than EKF because of the data association problem. Instead of landmarks, using a particle filter reduces the impact of erroneous data association.
- Computational Efficiency: The computational complexity is stated as $O(M(\log(N)))$. Here N is the number of landmarks and M is the number of particles used to compute the complexity [16].

Graph-based SLAM

- Robustness: The GraphSLAM algorithm is more robust than that of the EKF algorithm, partially because it is possible to re-examine older data associations if an incorrect data association has been created. This effectively decreases the likelihood of creating erroneous future data correlations, thus increasing the robustness of the data.
- Computational Efficiency: In contrast to the EKF algorithm that uses $O(N^2)$ where N is the number of landmarks, the use of memory is linearly dependent on the number of landmarks, $O(N)$ [15]. Besides this, the computational load is period-dependent; so, if the path of the surrounding is larger, the algorithm cost can be very expensive.

From the above comparison, we can see that the EKF SLAM is less robust in comparison to Gridmap-based RBPF SLAM and graph-based SLAM. We also see that the EKF SLAM algorithm is more costly in terms of efficiency. So, we have chosen to implement Gridmap-based SLAM with a 2D LiDAR sensor.

3 Implementation of RBPF-Based SLAM with LiDAR

The main idea behind the Rao-Blackwellized particle filter-based SLAM is to calculate the joint probability $P(x_{1t}, m|z_{1:t}, u_{1:t-1})$ of the environment map m and the dynamic position $x_{1:t} = x_1 \ldots x_t$ of the vehicle or the robots [10]. Here, $z_{1t} = z_1 \ldots z_t$. are the observations of different aspects of time. Measurements of the odometry in different aspects of time are obtained by $u_{1:t-1} = u_1 \ldots u_{t-1}$, by the vehicle. The following factorization is used in the RBPF-based SLAM.

$$P(x_{1:t}, m|z_{1:t}, u_{1:t-1}) = P(m|x_{1:t}, z_{1:t}) * P(x_{1:t}|z_{1:t}, u_{1:t-1}). \quad (2)$$

The factorization helps us to estimate the robot's trajectory first and after that, provided that trajectory, to generate maps of the environment. Since the generation of maps relies heavily on the robot's pose estimation, this approach provides an accurate computation. This factorization is often mentioned as Rao-Blackwellization.

RBPF algorithm

1. For each particle, do,
2. Using previous pose x_{t-1}, sampling a new pose x_t from the odometry data
3. Use the sampled pose as the beginning of the estimate, perform a local search in the map by maximizing the scan-matching score (match laser scan to map)
4. Update weight of the particle using the likelihood best-found pose
5. Calculate the pose x_{t-1} of the particle to best-found pose x_t
6. Find the number of effective particles N_{eff} from total weights.
7. If $N_{eff} < 0.5 * N_{total}$, do a weighted random resampling of all particles.

3.1 Hardware Used

To test the RBPF algorithm, we have used a 2D LiDAR sensor from SICK for sensing the surrounding environment, and for processing all the sensing data, we have used Jetson Xavier NX Devkit. The performance of the hardware was quite good to implantation the algorithm. This section briefly describes the implementation of the hardware.

LiDAR

- By the Time of Flight (ToF) process, LiDAR obtains the object distance from the middle of the laser scanner, in order to obtain the object coordinates in the LiDAR coordinate system, which has the benefits of high frequency, high accuracy, long detection range, and negligible effects of external illumination. In large-scale, light–dark environments with LiDARs, SLAM approaches can be used. The knowledge from the point cloud gathered by LiDAR may be used to approximate the pose of the vehicle in an iterative optimization process since the initial value of the pose transformation is known or guessed (Fig. 4).

We used sick TIM781s LiDAR [18] for the implementation. To determine the distance from the center point that is the scanning point to the LiDAR base, the LiDAR uses the ToF method. The ToF solution means that the LiDAR transmitter emits an ultra-short laser pulse, and the laser is transmitted onto the target to reflect sparsely. The LiDAR receiver absorbs the diffuse reflection light, and the scan point is determined as per the moment the laser beam flies in the atmosphere. The distance from the scanning point of the LiDAR is computed as,

$$d = c \times (\Delta t/2) \tag{3}$$

Fig. 4 Sick TIM781s LiDAR

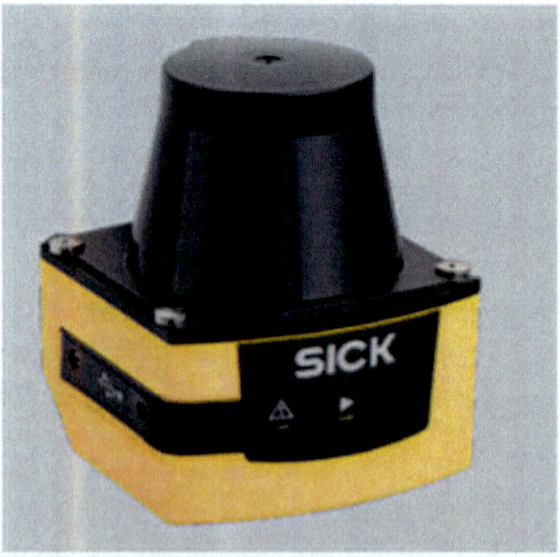

Table 1 TIM781s LiDAR working parameters

Parameter names	Numerical value
Light source	Infrared (850 nm)
Working angle (Horizontal)	270°
Frequency	15 Hz
Resolution	0.33°
Working range	0.05 m ... 25 m (>90% remission)
Blindspot	0 m ... 0.05 m
Maximum scanning distance	8 m

In LiDAR, light is emitted, and by the light, the distance from the center of the LiDAR to the object's scanning point is calculated, and this is achieved by the transmit time and receive time from the scanner.

Table 1 shows the different features of the LiDAR scanner we have used for our system.

Jetson Xavier NX Devkit

- We used the embedded Xavier NX System-on-Module (SoM), Development Kit. A CPU, DRAM, a GPU, and flash storage are included. SoM is an embedded single board that has only a row of different connector pins for different I/O. All the pins on the embedded board are attached to different types of ports like HDMI, USB, and Ethernet. A Jetson NX development kit is an identical board to a Raspberry Pi or any other single board computers which are paired with a development board.

Figure 5 shows the Jetson Xavier NX Devkit which is our computing unit. We get Nvidia's custom Carmel ARM-based cores with the board. We also get a 384-core volta-based GPU, and a hexa-core processor, and 8 GB LPDDR4x RAM. Display-Port, HDMI, 4x USB 3.1 ports, Gigabit Ethernet, Bluetooth, Wi-Fi, 2x camera

Fig. 5 Jetson Xavier NX Devkit

connectors, an SSD M.2 slot, and 40 GPIO pins added to the development board [19].

3.2 *Software Used*

From the software perspective, we used an open-source RBPF-based SLAM software package and developed the approach that is available as a library and easily used as a black box [20]. Making changes to the algorithm was done using C++.

4 Results

The results of the implementation of RBPF-based SLAM are shown in the below section. This section represents real-life experiments of the algorithm. Each figure from below shows 2D maps of the model or environments. All the maps generated are highly accurate.

4.1 *Maps–From the Simulation*

The section below shows the simulated maps of the algorithm. For this simulation, we used the recorded online data. We were able to see this simulation by the GUI of the QT creator.

Figure 6 shows 2D maps from the simulation from the recorded data with higher accuracy. This is a typical grid-based map. Each map consists of various particles. The black area is the obstacles in the environment that were detected by the scanner, and the white spaces are the area where the vehicle can move.

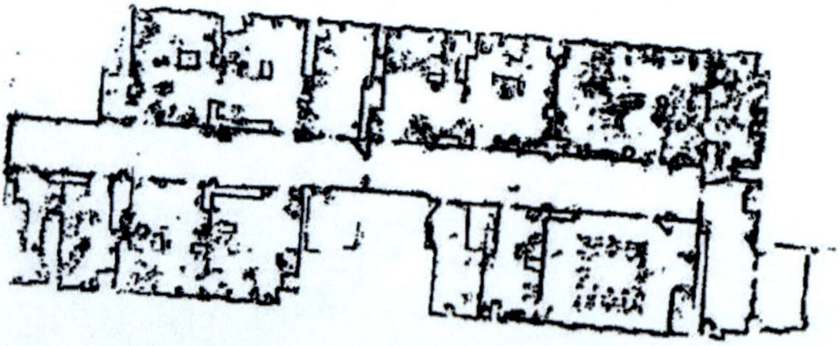

Fig. 6. 2D maps generated from simulation

4.2 Maps from 2D LiDAR Scan Data

The section below shows the generated maps from the implemented algorithm. For this, we used Jetson NX Devkit as the computation unit. For the data collection, we used a 2D LiDAR rangefinder.

Figure 7 shows 2D maps for the whole environment from the LiDAR data with higher accuracy. Here the red lines show the path of the vehicle moving throughout the environment. The white areas are the spaces to move the vehicle, and the blue ones are the obstacles in the environment.

Figure 8 shows 2D maps of a room from the whole environment which is generated from the LiDAR scanned data. In this section, we can see the black obstacles are the collection of particles.

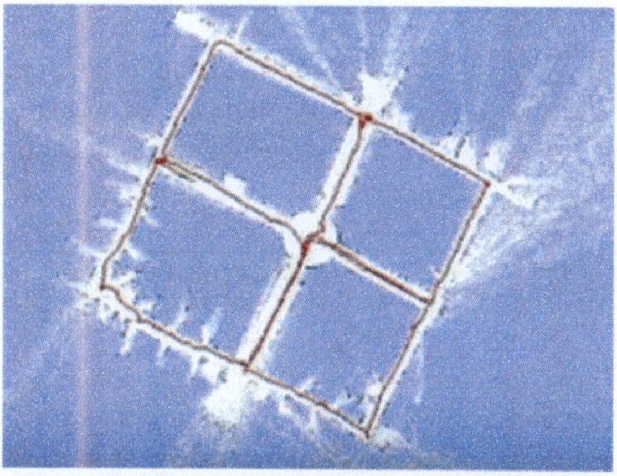

Fig. 7. 2D maps of the whole environment from LiDAR data

Fig. 8. 2D maps of a single room from the LiDAR data

Fig. 9 Detailed view of the portion of a map generated from the LiDAR data

Fig. 10 Position value of the scanner in the environment from the algorithm

```
m_count 25
Average Scan Matching Score = 316.927
Neff = 20.6614
Registering Scans: Done
Update frame 235
Update ld= 1.02217 ad=0.0463506
Laser Pose= 0.119441 -0.134954 3.12935
```

Figure 9 shows zoomed 2D maps of a small portion of the map from the whole environment which is generated from the LiDAR scanned data. These are the particles of a bigger portion of the maps.

Figure 10 shows the position and orientation of the scanner that is x, y, and theta. This is the real-time position of the vehicle in the environment.

The findings demonstrate that the RBPF algorithm is capable of solving the SLAM problem with great precision. In the ideal case, the generated maps from the algorithms are identical, and the real-time positions of the scanner are highly accurate.

5 Conclusion

The general SLAM approach is now a well-established and well-understood aspect of robotics. In this article, we have focused on discussing different types of SLAM techniques used in recent times and also which is popular. From the above review, we have chosen RBPF-based SLAM technique which is more robust and has low computational complexity as compared to EKF-based SLAM or graph-based SLAM. We then implement RBPF-based SLAM in real hardware like 2D LiDAR from sick for sensing the environment, and for computational processing, we used Jetson Xavier NX devkit. We conclude this article by providing some general recommendations. Indoor navigation is an interesting application area for SLAM. By using a laser scanner as the primary sensor, a map may be generated using the complete SLAM approach. For such maps, 2D Gridmap-based SLAM methods are suggested. Then, during autonomous navigation, the pre-calculated map may be used to perform online localization. 2D Gridmap-based SLAM would be more reliable if full interactive mapping, and localization is feasible but may require more features such as detection of dynamic objects and filtering them out from the map. Additionally, computational requirements for online mapping and localization for high-speed navigation may prove to be a bottleneck.

6 Limitations and Discussion

In general, SLAM technology has the following challenges: managing the system's real-time nature and accuracy relationship, causing real-time localization failure due to the highly dynamic nature of the surrounding environment, and significant usage of memory resources induced by a larger environment or model. Besides this, it will be more complex to perform this algorithm when it comes to a complex outdoor environment. Apart from that, because the thesis was about testing and revising SLAM algorithms, a practical vehicle motion model was not a priority. Since the model has been developed, it could be possible to implement other SLAM techniques in the future for a better review. A newer algorithm like HectorSLAM can be implanted in the future for a more accurate result as it uses no odometry.

References

1. M.W.M. Gamini Dissanayake, P. Newman, S. Clark, H.F. Durrant-Whyte, M. Csorba, A solution to the simultaneous localization and map building (slam) problem. IEEE Trans. Robot. Automat. **17**(3), 229–241, 2001
2. https://liu.diva-portal.org/smash/get/diva2:1218791/FULLTEXT01.pdf
3. S. Thrun, *Robotic Mapping: A Survey. Exploring Artificial Intelligence in the New Millennium*, ed. by G. Lakemeyer, B. Nebel. The Morgan Kaufmann Series in Artificial Intelligence (Hardcover) (2002). ISBN ISBN-10: 1558608117

4. E. Mueggler, H. Rebecq, G. Gallego, T. Delbruck, D. Scaramuzza, The event camera dataset and simulator: event-based data for pose estimation, visual odometry, and SLAM. Int. J. Rob. Res. **36**(2), 142–149 (2017)
5. A.J. Davison, D. Murray, Simultaneous localization and map-building using active vision. IEEE Trans. Pattern Anal. Mach. Intell. **24**(7), 865–880 (2002)
6. C. Estrada, J. Neira, J.D. Tardós, Hierarchical SLAM: real-time accurate mapping of large environments. (IEEE) Trans. Robot. **21**(4), 588–596 (2005)
7. P.M. Newman, J.J. Leonard, Consistent, convergent, and constant-time SLAM, in *International Joint Conference on Artificial Intelligence (IJCAI), Acapulco*, Mexico, pp. 1143-U1150 (2003)
8. P.M. Newman, On the structure and solution of the simultaneous localization and mapping problem. Ph.D. Thesis, University of Sydney (1999)
9. T. Bailey, H. Durrant-Whyte, Simultaneous localization and mapping (SLAM): part II. IEEE Robot. Autom. Mag. **13**(3), 108–117 (2006)
10. G. Grisetti, C. Stachniss, W. Burgard, Improved techniques for grid mapping with Rao-blackwellized particle filters. IEEE Trans. Robot. **23**(1), 34–46 (2007)
11. EKF SLAM for dummies, https://ocw.mit.edu/courses/aeronautics-and-astronautics/16-412j-cognitive-robotics-spring-2005/projects/1aslam_blas_repo.pdf, 2005. Accessed: 24-03-2017
12. M. Quigley et al., ROS: an open-source robot operating system, in *Proceedings of the ICRA-Workshop on Open-Source Robotics, software: ros.org* (2009)
13. S. Thrun, W. Burgard, D. Fox, *Probabilistic Robotics* (MIT Press, 2005)
14. R. Siegwart, R. Illah NourBakhsh, D. Scaramuzza, *Introduction to Autonomous Mobile Robots*, 2nd edn. (Springer, London Limited 2003) (2011)
15. C. Molina, F. Andreas, Graphslam algorithm implementation for solving simultaneous localization and mapping. Master's thesis, Universidad de Chile (2016)
16. S. Thrun, W. Burgard, D. Fox, *Probabilistic Robotics*, 1st edn. (Cambridge, MIT Press, 2005)
17. C. Stachniss, Robot mapping EKF SLAM. http://ais.informatik.uni-freiburg.de/teaching/ws12/mapping/pdf/slam04-ekf-slam.pdf (2012)
18. https://www.sick.com/us/en/c/products
19. https://developer.nvidia.com/embedded/jetson-xavier-nx-devkit
20. https://openslam-org.github.io/gmapping.html

TrafficNN: CNN-Based Road Traffic Conditions Classification

Monjurur Kader Shipu, Faisal Al Mamun, Shamim Hossen Razu, and Ms. Nishat Sultana

Abstract Traffic congestion can affect different socioeconomic aspects of a country. It is necessary to develop an automated system that can perform better than traditional human effort in traffic controlling. So, detection of traffic conditions is crucial for building a smart traffic control system to prevent traffic jam escalation. Modern deep learning approaches can help in detection of road traffic conditions especially convolutional neural network (CNN). We suggest a novel model in this paper that can classify road traffic conditions using CNN. Our proposed model 'TrafficNN' classifies five different road traffic conditions with an accuracy of 82%. To train and test our model, we use the traffic conditions images which are collected from the Internet. To prove the strength of our model, we compare it with several pre-trained models like VGG16, ResNet50, InceptionV3 and DenseNet121. The comparison result proves the significance of our model to extend its successful application for developing an automated traffic controlling system in the near future.

Keywords Convolutional neural network · Deep learning · Traffic congestion detection · Traffic control system

M. K. Shipu (✉) · F. A. Mamun · S. H. Razu · Ms. Nishat Sultana
Department of Computer Science and Engineering, Daffodil International University, Dhaka 1207, Bangladesh
e-mail: monjurur15-8644@diu.edu.bd

F. A. Mamun
e-mail: faisal15-8628@diu.edu.bd

S. H. Razu
e-mail: hossen15-7367@diu.edu.bd

Ms. Nishat Sultana
e-mail: nishat.cse@diu.edu.bd

G. Ranganathan et al. (eds.), *Soft Computing for Security Applications*,
Advances in Intelligent Systems and Computing 1397,
https://doi.org/10.1007/978-981-16-5301-8_18

1 Introduction

Traffic congestion management is an overhead for almost every big city in the world. Especially developing countries like Bangladesh where some roads are still under development and other roads are narrow compared to large volumes of vehicles. Research shows that for 240 working days, the cost of traffic congestion is 4495.6544 million USD in Dhaka, the capital of Bangladesh [1]. Traffic congestion in Dhaka consumes approximately 5 million working hours per day, according to a report by the Accident Research Institute of BUET [2]. Between January 2009 and April 2019, at least 25,526 people have died and nearly 20,000 people have been seriously injured in road accidents in Bangladesh [3]. So, we can see that traffic congestion has a significant effect on the various socioeconomic aspects of a country.

Nowadays, traffic congestion issues impact the transport system in cities and create severe problems in many countries. The optimization of heavy traffic congestion, particularly with multiple junction nodes, is a major issue that needs to be addressed. It is difficult to reduce road traffic congestion with existing traffic control systems, as it is operating manually most of the time. However, it is important to have a smart traffic management system in place to maintain a reliable transportation system. Many researches have already been done to reduce traffic congestion and ensure a reliable smart traffic management system. Many of them proposed to use sensor and IoT-based devices in every vehicle to know road traffic conditions, which is pretty expensive and complex to implement. Artificial intelligence (AI), machine Learning (ML) and deep learning (DL) have been used extensively in this field of study in recent years. But nobody tries to classify traffic conditions on the basis of real-life scenarios.

We are using convolutional neural network (CNN) in this study, which is a part of deep learning (DL). We proposed a CNN-based model named as 'TrafficNN' to dedicatedly classify road traffic conditions which can be used to automatically control road traffic congestion in near future. In order to develop an efficient, low-cost and real-time smart traffic management system, this paper proposes the use of CNN to resolve several defects and enhance traffic management based on the classification of road traffic conditions.

2 Related Works

Before moving toward our works, here we will discuss related works in the field of smart traffic control systems. Here we will discuss others research papers, their method, their strategies and problems about their works.

In 2020, S. R. Mugunthan proposed a model based on Li-Fi technology for communication between vehicle and traffic signals. In his model, every car has to be equipped with two Li-Fi devices for vehicle-to-vehicle interaction, and also the traffic signals have to be equipped with Li-Fi device in order to generate proper signal.

But the main problem here is Li-Fi technology can be affected by sunlight and other light on road [4]. Priyadharshini and Manikandan presented an automatic traffic control system in 2019, where PIR sensor is used to calculate vehicular density and Arduino is used to count the number of vehicles. And then that information is used to generate traffic lights [5]. In 2018, Tarun Kumar and Dharmender Singh Kushwaha proposed a model for creating green corridors for emergency vehicles by placing RFID modules in emergency vehicles. They create green corridors by finding least, moderate and heavy congestion using OpenCV based on surveillance camera images [6]. Ramkumar Eswaraprasad and Linesh Raja proposed an IoT-based traffic management where they use hybrid artificial neural network with hidden Markov model (HANN-HMM) to make better decisions about the traffic management [7]. All of these sensors and IoT-based models are expensive and complex to implement. It is also unrealistic to put sensors and devices in every vehicle on the road.

Lam et al. presented a traffic congestion detection method by counting the number of vehicles based on online images. To assess two degrees of congestion, namely normal and congested, the image correlation coefficient has been used along with a threshold for the number of vehicles detected [8]. Jiyong Chung and Keemin Sohn suggest a supervised learning method to calculate traffic density where the number of vehicles is collectively counted [9]. In 2018, Teresa Pamuła presented a model for road traffic condition classification based on video traffic data, where four levels of traffic conditions can be determined based on number of vehicles [10].

In 2020, Vinothkanna proposed to use deep learning in order to detect road conditions to plan optimal paths for self-driving cars [11]. In 2018, Wang et al. presented a CNN model named as 'TrafficNet' for the monitoring of unnecessary traffic congestion based on existing surveillance systems. This model classifies traffic conditions as congestion and non-congestion with an accuracy of 90% [12]. Manchanda et al. proposed a hybrid CNN model adding CNN and support vector machine (SVM) together. They classify traffic congestion as dense and sparse traffic and also add accident and fire class to classify risks for road safety [13]. In 2018, Chakraborty et al. proposed a model that classifies surveillance camera images as two classes named as congested and non-congested using YOLO, SVM and deep CNN [14]. For explaining non-recurring traffic congestion, Sun et al. presented a deep neural network. They classify non-recurring traffic congestion as three classes named as accident, sports game and adverse weather [15]. A deep neural network for traffic flow prediction based on the traffic performance index was proposed by Yi et al. in 2017 as an indicator of traffic flow conditions [16].

After analyzing those previous work, we observed that all models that count numbers of vehicles based on surveillance camera images are not classifying traffic conditions properly. Most of the CNN-based deep learning model just classifies traffic images as congestion and non-congestion categories. And they also do not classify traffic conditions based on real-life traffic situations. So, we proposed a model that can classify traffic conditions images based on real-life applicable scenarios. So that in near future, it can be used for a smart traffic management system.

3 Convolutional Neural Network (CNN)

Convolutional neural network, also known as CNN, is a deep learning algorithm which takes images as input, learns various important features in an image and is able to classify them between different classes [17]. A standard CNN consists of single or multiple blocks of convolutional and pooling layer and then one or more fully connected layers after that [18].

A digital image has a set of pixels arranged in a grid-like manner that contains pixel values to show how bright each pixel is and what color it contains [19]. CNN takes this grid-like topology as input and processes it through convolutional layers, then pooling layers extract specific features from it, and then finally the output of this layer works as input of a fully connected layer which finally classifies the images through various trained neurons. These layers are organized in such a way that they define simple patterns (lines, curves, etc.) first and then more complicated patterns (faces, objects, etc.).

3.1 Convolutional Layer

The convolution layer is CNN's key building block that does most of the computational heavy lifting. The main goal of this layer is to extract high-level features from the input dataset. These layers are composed of filters and feature maps. A filter is a small matrix used for feature extraction. The feature map is created after applying filters to the previous layer [20].

3.2 Pooling Layer

To build a new collection of the same number of pooled feature maps, the pooling layer operates separately on each feature map. This means that the pooling layer would often reduce the size of each feature map by a factor of 2, such as halving each dimension, reducing the number of pixels or values to one quarter of the size of each feature map [21]. In the pooling operation, two common functions used are as follows:

- Max Pooling: Calculate the maximum value the pooling window holds.
- Average Pooling: Calculate the average value the pooling window holds.

3.3 Flatten Layer

To create a single long feature vector, we flatten the output of the convolutional layers. In order to add it to the next layer, flattening alters the information into a one-dimensional array. This means that we place all the pixel data in one row and align it with the fully connected layer [22].

3.4 Fully Connected Layer

Fully connected layers are the final layers in the CNN structure, which could be one or more layers and positioned after a series of convolution and pooling layers. In this layer, neurons have complete contact with all neurons in the prior and subsequent layers.

3.5 Padding

Padding is a term which refers to the number of pixels that are added to the image as it is processed by the CNN filter. If we want the output to be the same size as the input, then we use padding as 'same' and if we want no padding, then we use padding as 'valid'.

3.6 Rectified Linear Units (ReLU)

The rectified linear unit (ReLU) acts as an activation function which returns 0 for any negative value, but returns the same value for any positive value. Therefore, it may be written as $f(x) = \max(0, x)$ [23]. It is a nonlinear activation function used to increase nonlinearity in convolutional neural networks.

3.7 Batch Normalization

We normalize data before passing it to a model, so it only works at the very first layer of the neural network. If the data becomes not normalized at some later layer, then that layer has to deal with un-normalized data which is not as good as normalized data. So, to make the data at every layer normalized, we use batch normalization which can help to reduce overfitting.

3.8 Dropout

Dropout is a technique of regularization which can assist to minimize overfitting. During training, randomly selected neurons are ignored where dropout technique is used. That means their participation to the activation of downstream neurons on the forward pass is temporarily omitted, and any weight adjustments are not added to the neuron on the backward pass [24].

3.9 Softmax

Softmax is an activation function that is used for multicategory classification in the output layer of neural network models. This function produces a vector describing the probability distributions of a list of possible outcomes [25].

4 Proposed Methodology

In this section, along with our proposed model, we will discuss our data collection method. It has been found that in order to achieve better results, we need to increase data quality and quantity. So, for all machine learning algorithms, data is an essential component.

4.1 Data Collection

In this study, we use a new dataset to train our proposed model. We create the new dataset by collecting images manually from the Internet and publicly available information along with it. This dataset has five classes of road traffic conditions such as low traffic, medium traffic, high traffic, emergency traffic and traffic accidents. This dataset has a total of 2500 images, among them 2000 images are used for training and 500 images are used for testing purposes. Here each class has 500 images; we considered 100 images for testing, and the remaining 400 images are for training.

4.2 Data Augmentation

Data augmentation is commonly used when training data is limited. We know training a CNN on a small dataset makes it vulnerable to overfitting. We gathered 100 images of each class as raw data. The dataset was artificially stretched to discourage overfitting. After collecting 500 images for five classes, we augmented all the main data using four different augmentation techniques, and these techniques are given below:

- Rotation Range (±30°)
- Horizontal Flip
- Width Shift Range (15%)
- Height Shift Range (15%)

So, we have a total of 2500 images after data augmentation to incorporate our proposed model, but before it was only 500 images.

4.3 Data Preprocessing

For our proposed model, it requires a fixed pixel for all data images. So, we resized all data with a fixed resolution of 256 × 256 pixels. To normalize our data, we reduce pixel values by dividing it with 255 and get the pixel values from 0.0 to 1.0. When passing images to our proposed model, we use RGB color channels which helps CNN to easily identify features.

4.4 TrafficNN Architecture

We are now going to discuss our proposed model for achieving the expected outcome. A simplified picture of our proposed TrafficNN model is shown (see Fig. 1).

We actually have six groups of convolution and pooling. And each of this group has two convolutions before max pooling. For every convolutional layer, we have a constant size filter of 3 × 3. We have increased the number of feature maps at each subsequent group of convolution and pooling layers. After every convolution layer, we added a batch normalization layer to make data normalize. Finally, before each fully connected (FC) layer, we added a dropout layer for regularization. We use softmax as an activation function in the final layer.

5 Performance Evaluation

In this section, we will discuss how we train our proposed 'TrafficNN' model and what performance we get from it. We will also compare our model's performance with some pre-trained models like VGG16, ResNet50, InceptionV3 and DenseNet121.

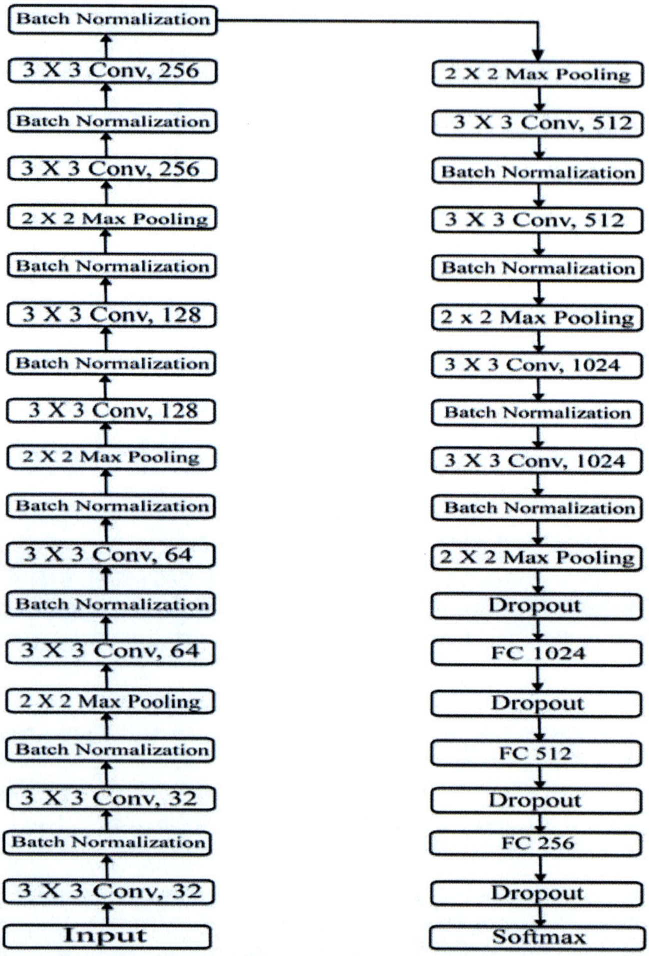

Fig. 1 TrafficNN architecture

5.1 Training the Model

Our 'TrafficNN' model was trained on our own traffic dataset where 80% images used as training set and other 20% as test set. So, we have 400 images in each class and a total of 2000 images for training. We have done all the coding and training of our model in Google Colab notebook using the free cloud GPU provided by google. We use Python-based TensorFlow 2.0 library along with Keras API to build, train and test our model.

We use Adam as an optimizer which can handle sparse gradients on noisy problems [26]. For 'TrafficNN', we used Adam optimizer with a learning rate of 0.000004. We

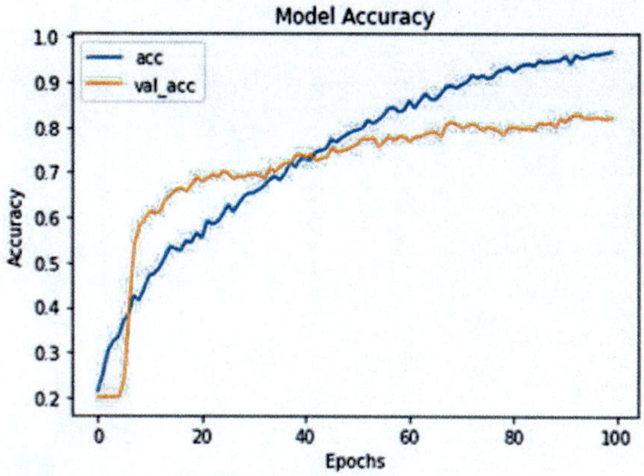

Fig. 2 TrafficNN training and validation accuracy

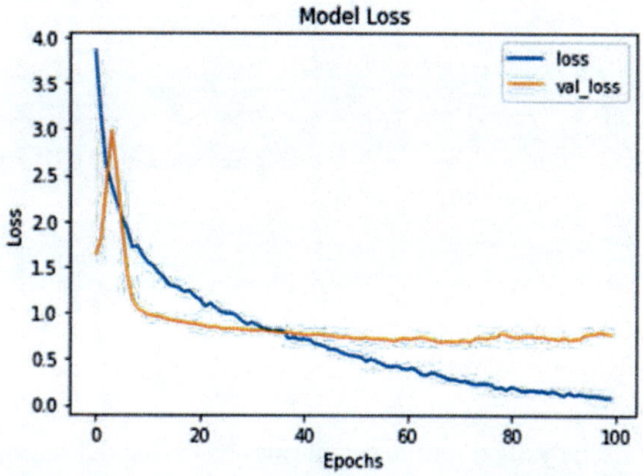

Fig. 3 TrafficNN training and validation loss

find this learning rate through hyperparameter tuning using TensorFlow callbacks methods (Figs. 2 and 3).

We have used sparse categorical cross-entropy as a loss function. We know our training's objective was to make the cross-entropy as small as possible. So, to train the model and observe the performance, we used batch size of 25 and 100 epochs. Our proposed 'TrafficNN' model gives 97% accuracy on the training set and 82% accuracy on the test set.

5.2 Result Discussion

We determined precision, recall and f1-score from working on a test dataset which contains 500 images, where each class has 100 images. After classification, only 89 images are misclassified and the rest of them give correct predictions.

Table 1 TrafficNN classification report

Class	Precision	Recall	F1-score	Support
Emergency traffic	0.83	0.90	0.87	100
High traffic	0.90	0.87	0.88	100
Low traffic	0.83	0.79	0.81	100
Medium traffic	0.81	0.68	0.74	100
Traffic accident	0.75	0.87	0.81	100

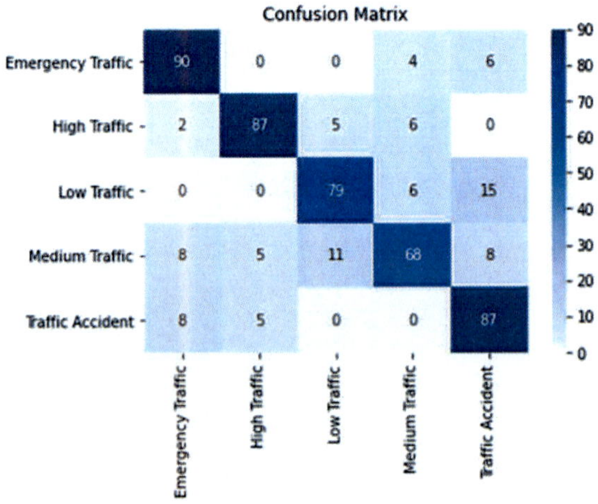

Fig. 4 Confusion matrix for TrafficNN

Many of our traffic conditions images looked so similar for vehicle count, lighting condition and similar type of vehicles. It actually happened due to less difference between features of our determined classes. We also assume that if we have more data, then this would not happen. Despite that this model also did pretty good for those cases (Table 1).

Indeed, we got 82% accuracy on the test dataset for our 'TrafficNN' model which was quite optimal for unseen data. We are showing the result with the confusion matrix within the given figure (see Fig. 4).

5.3 Comparison Using Transfer Learning

In this part, we compared different pre-trained models with our 'TrafficNN' model such as VGG16, ResNet50, InceptionV3 and DenseNet121. All of those existing models used in this experiment were pre-trained on the ImageNet dataset by default. We also use a transfer learning technique called fine-tuning, to re-train those models on our own traffic images dataset. We use the same number of epochs, batch size and learning rate for fine-tuning those pre-trained models. But this time, we use a

Table 2 Comparison of road traffic conditions classification

Model	Training accuracy (%)	Test accuracy (%)	Learning rate	Number of epochs
TrafficNN	97	82	0.000004	100
VGG16	90	65	0.001	10
ResNet50	94	76	0.001	10
InceptionV3	91	68	0.001	10
DenseNet121	90	75	0.001	10

different learning rate and number of epochs from 'TrafficNN' training. Because in this transfer learning approach, that learning rate and number of epochs do not work well and show a huge amount of overfitting.

From the table (see Table 2), we observed that all of those pre-trained models gave at least 90% accuracy on the training dataset. But they show different test accuracy, as different amounts of noise occur for the test dataset. Finally, we observed that under almost the same conditions, every pre-trained model's training accuracy is satisfied, but their test accuracy shows overfitting signs. We also observed that our 'TrafficNN' model performs much better with less overfitting and a test accuracy of 82%.

6 Future Work

In our dataset, all the images are of good resolution. But in the future, we will try to use blur images, low resolution images, distorted images and surveillance camera images as our traffic image dataset. We will also try to integrate more classes to accurately analyze road traffic conditions.

In the future, this work can be a part of a smart traffic control system, where our model will work as an artificial intelligence (AI) system to control traffic light smartly. It can also give special corridors for emergency vehicles like ambulance, fire service van and police car to move quickly. Then it can be used to detect accidents just after happening it, through the AI monitoring system with the help of our proposed model.

7 Conclusions

We know our existing traffic control system causes traffic accidents and traffic congestion that put heavy loads on businesses and works because of a bad traffic monitoring system. There is no doubt that a lot of research work has happened in this field. But nobody tries to build a CNN-based dedicated architecture to study road traffic conditions properly. We proposed a CNN-based 'TrafficNN' architecture to detect traffic conditions on the road. Our proposed model classifies road traffic conditions with an accuracy of 82%. We have compared our model with some pre-trained models like VGG16, InceptionV3, ResNet50 and DenseNet121 and found that our model gives the highest accuracy among them.

This model along with deep learning and AI can be incorporated into current surveillance systems to detect and report road traffic conditions automatically. In case, there is congestion on that road, then it can divert the upcoming traffic to another route to save the time of upcoming vehicles to prevent more congestion.

References

1. S. Chakraborty, Traffic congestion in Dhaka city and its economic impact. Traffic congestion in Dhaka city: its impact on business and some remedial measures (2010)
2. *Traffic Jam in Dhaka: Colossal Loss*, https://www.thedailystar.net/frontpage/colossal-loss-1553002. Last Accessed 31 Jan 2021
3. *526 people killed in road accident in the last 10 years*, https://www.dhakatribune.com/bangladesh/nation/2019/06/12/25-526-people-killed-in-road-accident-in-the-last-10-years. Last Accessed 31 Jan. 2021
4. S. Mugunthan, Concept of Li-Fi on smart communication between vehicles and traffic signals. J. Ubiquit. Comput. Commun. Technol. 9–69 (2020)
5. K. Priyadharshini, S. Manikandan, Automatic *Traffic Control System Based on the Vehicular Density* (2019)
6. T. Kumar, D.S. Kushwaha, *Information and Communication Technology for Competitive Strategies* (Springer, 2019), pp. 99–108
7. R. Eswaraprasad, L. Raja, Improved intelligent transport system for reliable traffic control management by adapting internet of things, in *2017 International Conference on Infocom Technologies and Unmanned Systems (Trends and Future Directions)(ICTUS)*, pp. 597–601 (2017)
8. C.-T. Lam, H. Gao, B. Ng, A real-time traffic congestion detection system using on-line images, in *2017 IEEE 17th International Conference on Communication Technology (ICCT)* (2017), pp. 1548–1552
9. J. Chung, K. Sohn, Image-based learning to measure traffic density using a deep convolutional neural network. IEEE Trans. Intell. Transp. Syst. **19**, 1670–1675 (2017)
10. T. Pamula, Road traffic conditions classification based on multilevel filtering of image content using convolutional neural networks. IEEE Intell. Transp. Syst. Mag. **10**, 11–21 (2018)
11. M.R. Vinothkanna, A survey on novel estimation approach of motion controllers for self-driving cars. J. Electron. **2**, 211–219 (2020)
12. P. Wang, L. Li, Y. Jin, G. Wang, Detection of unwanted traffic congestion based on existing surveillance system using in freeway via a CNN-architecture trafficnet, in *2018 13th IEEE Conference on Industrial Electronics and Applications (ICIEA)* (2018), 1134–1139
13. C. Manchanda, R. Rathi, N. Sharma, Traffic density investigation & road accident analysis in India using deep learning, in *2019 International Conference on Computing, Communication, and Intelligent Systems (ICCCIS)* (2019), pp. 501–506
14. P. Chakraborty et al., Traffic congestion detection from camera images using deep convolution neural networks. Transp. Res. Record **2672**, 222231 (2018)
15. F. Sun, A. Dubey, J. White, DxNAT|Deep neural networks for explaining non-recurring traffic congestion, in *2017 IEEE International Conference on Big Data (Big Data)* (2017), 21412150

16. H. Yi, H. Jung, S. Bae, Deep neural networks for traffic ow prediction, in *2017 IEEE International Conference on Big Data and Smart Computing (BigComp)* (2017), pp. 328–331
17. *A Comprehensive Guide to Convolutional Neural Networks | the ELI5 way*, https://towardsdatascience.com/a-comprehensive-guide-to-convolutional-neural-networks-the-eli5-way-3bd2b1164a53. Last Accessed 3 Feb. 2021
18. F. Sultana, A. Sufian, P. Dutta, Advancements in image classification using convolutional neural network, in *2018 Fourth International Conference on Research in Computational Intelligence and Communication Networks (ICRCICN)* (2018), 122–129
19. *Convolutional Neural Networks, Explained*, https://towardsdatascience.com/convolutional-neural-networks-explained-9cc5188c4939. Last Accessed 3 Feb. 2021
20. *Crash Course in Convolutional Neural Networks for Machine Learning*, https://machinelearningmastery.com/crash-course-convolutional-neural-networks/. Last Accessed 4 Feb. 2021
21. *A Gentle Introduction to Pooling Layers for Convolutional Neural Networks*, https://machinelearningmastery.com/pooling-layers-for-convolutional-neural-networks/. Last Accessed 4 Feb. 2021
22. *The Most Intuitive and Easiest Guide for Convolutional Neural Network*, https://towardsdatascience.com/the-most-intuitive-and-easiest-guide-for-convolutional-neural-network-3607be47480 Last Accessed 4 Feb. 2021
23. *Rectified Linear Units (ReLU) in Deep Learning*, https://www.kaggle.com/dansbecker/rectified-linear-units-relu-in-deep-learning Last Accessed 4 Feb. 2021
24. *Dropout Regularization in Deep Learning Models With Keras*, https://machinelearningmastery.com/dropout-regularization-deep-learning-models-keras/. Last Accessed 4 Feb. 2021
25. B. Yuan, Efficient hardware architecture of softmax layer in deep neural network, in *2016 29th IEEE International System-on-Chip Conference (SOCC)* (2016), 323–326
26. D.P. Kingma, Ba, J. Adam, *A Method for Stochastic Optimization*. arXiv preprint arXiv:1412.6980 (2014)

Review of Malware Detection Using Deep Learning

Tanvi Mane, Prachi Nimase, Prahalad Parihar, and Pragati Chandankhede

Abstract Malware is a program which is designed to harm or exploit any programmable device, service, or network by containing malicious code in it. Malware detection refers to the process of detecting the presence of malware on a system or whether a specific program is malicious or benign. Most common types of malware are rootkits, spyware, Trojan horses, viruses, and worms. Malware detection systems act as an early warning system for any organization or individual to prevent confidential information from getting disclosed. In this paper, deep neural network (DNN) is used to learn the features that optimally represent the given training data. In this, deep neural network learns the features from portable executable files. Thus, deep neural network-based systems are effective to detect existing and unknown malware with low false positive rates. The proposed system detects whether a given portable executable (PE) file is malware or not using deep neural network (DNN).

Keywords Malware · Deep neural network · Portable executable · Network security · Anti-malware · Malware detection system · Anti-malware system

1 Introduction

Malware is a software which may include viruses, trojans, worms, and other harmful programs whose purpose is to cause damage and access confidential information. As technology advances, new malicious software emerges everyday. Due to this, malware has become a serious threat to most companies and organizations. As malware can access confidential information, it can cause serious damage to the company like data theft or crippling of network security and the list goes on. Malware is a threat to the computer users regardless which operating systems and hardware

T. Mane · P. Nimase · P. Parihar (✉) · P. Chandankhede
Department of Computer Engineering, Excelsior Education Society's, K.C. College of Engineering and Management Studies and Research, Thane, India

P. Chandankhede
e-mail: pragati.chandankhede@kccemsr.edu.in

G. Ranganathan et al. (eds.), *Soft Computing for Security Applications*,
Advances in Intelligent Systems and Computing 1397,
https://doi.org/10.1007/978-981-16-5301-8_19

platforms that they are using. Microsoft Windows is the most popular operating system, and the popularity also makes it the most favorite platform to be attacked by the adversaries. Hence, early detection of such malware becomes necessary. There are various traditional algorithms such as signature-based, behavior-based, and heuristic-based which help in malware detection. But these traditional approaches have their own shortcomings which make them less efficient. Hence, there arises a need for an advanced approach. In this paper, we have introduced a deep neural network (DNN) approach for malware detection which is basically a neural network but contains more number of hidden layers which performs the binary classification operation of detecting the files into malicious or benign with low false positive rate. For detection purposes, our main focus is on portable executable (PE) files. For this purpose, we use Benign & Malicious PE Files which contains the data about portable executable files. This system is useful as it gives alert to the user before installing any such PE file into the system. The system can do a fast, static, and accurate analysis on any PE file, which keeps hackers out of the computer and protects the information from getting compromised.

2 Literature Review

Malware detection is very important for every modern organization or individual. Static and dynamic analyses are common techniques used for malware detection but due to emerging malware every day, and these techniques are not sufficient. Hence, authors Baptista et al. [1] proposed a novel malware detection system based on machine learning and binary visualization. This approach depends on binary visualization by converting a file's binary data in an image format. Important features which are capable of distinguishing malware are extracted from the image. It then builds a neural network for analyzing and detecting malicious files. This approach is robust and fast and hence appropriate for larger datasets.

Alternative approaches using supervised binary classifiers have been proposed by authors Shukla et al. [2] where robust binary classifiers are used to detect where portable executable (PE) files are malicious or not. But the limitation that if the PE file is encrypted then the model might not give satisfactory results.

As new malware emerges everyday, there arises a need for a detection system which can adapt to the constantly changing malware ecosystem. Authors Zhao et al. [3] have attempted to create this system wherein they have used dynamic and static feature-based detection using machine learning. They have extracted static and dynamic features like strings, DLLs, assembly instruction sequences, and frequency of API calls. Then feature selection is done using different algorithms like n-gram algorithm and information gain algorithm. These valuable features further help in building classifier modules like Naïve Bayesian and SVM for detecting malware programs.

The traditional signature-based algorithm is no longer robust as it cannot detect unknown malware. Authors Yewale and Singh [4] have introduced a malware detection system which is based on opcode frequency. Machine learning can be a suitable approach to complement traditional methods. Machine learning algorithms using opcode sequences can recognize malicious code more efficiently. In this approach, they have implemented four machine learning models BOOST, SVM, random forest, and decision tree, out of which accuracy of random forest was highest as compared to others. But the major issue is that models require extensive memory requirements.

The author Malik et al. [5] introduced a method where execution of a program is monitored while running when the program and its associated model is loaded. While running a program associated model will indicate a flag; if the program behavior is suspicious also, the data structure of the program is reflected into memory. This malware detection is done for two types Kernel rootkits and user-level malware. Three machine learning techniques, i.e., logistic regression, SVM, and random forest are used where feature selection is done using F-score for system call epochs. But these techniques have been proposed for certain names of malware only.

Malware detection based on Objective—Oriented Association Mining where authors Xiao et al. [6] compared the traditional association mining method with objective-oriented association mining (OOA). Traditional association mining cannot detect whether a PE file contains malware or not, whereas OOA will classify the file into malware or benign but as support and confidence value that decides the association rules for a given dataset. So if support value decreases, association rule set increases and this increased number in rules will slow down the OOA algorithm. To overcome this problem, authors proposed a same OOA algorithm but with improvement where selection criteria are set for association rules to remove no discrimination power rules using OOApriori_Gen algorithm and to speed up the mining which belongs to two objects, i.e., malware and benign.

To detect and categorize malware into their respective families, authors Vinaykumar et al. [7] designed a scalable framework, i.e., ScaleMalNet which performs the task into two phases. In the first phase, various machine learning and deep learning algorithms are developed for static and malware analysis. Also, detection of malware from images using deep learning is evaluated where the file is converted into image. In the second phase, learned and extracted features are passed to the next layer of model to classify the malware into their respective families. But deep learning techniques give better accuracy and detection rate than various machine learning algorithms.

Vyas et al. [8] worked with the 28 features which are extracted from the metadata of portable executable (PE) file to apply four different supervised learning algorithms, i.e., decision tree, SVM, KNN, and random forests to detect malware on the network. Out of these four algorithms, random forest gives better accuracy, in which the malware was detected with the average rate of 98% with 0.5 s as analysis speed. However, this detection is implemented with only four types of malware which are Trojan, backdoor, worm, and virus.

In study approach, authors Samantray et al. [9] observed that sometimes it is difficult to update a signature database frequently for the signature-based malware detection system. Thus, new aged malwares cannot be detected by these signature-based systems. To understand the hidden behavior of any malware, the feature extraction techniques used in deep learning are useful in order to detect the file as malware or benign with good accuracy.

In order to face the latest malware, the author [10] proposes a new way to classify the file as malicious or benign using convolutional neural network (CNN). In this technique, firstly the system generates images from the specified file and after that using a trained CNN model it predicts whether the file contains the malicious code or not with the accuracy of 96%. Since it produces images and then the predicting model will return the result, it causes a more time-consuming process.

The authors [11] Ke He and Dong Seong Kim have introduced a malware detection method with malware images using deep learning techniques. They have studied the efficiency of convolutional neural networks (CNN) against redundant API injection. Based on the analysis, they have designed a malware detection system that converts files into image representations and classifies the image representation with CNN. The CNN is implemented with spatial pyramid pooling layers (SPP) to deal with input of any size. The neural networks we deployed in the classifier are limited to plain three-layer network and a reasonably complex resnet50 network. Since the main goal is to investigate the effectiveness of the input color space and spatial pyramid pooling, the model we used was not optimized at all. The authors [12] Ben Athiwaratkun and Jack W. Stokes have proposed several deep learning architectures for classifying malware including LSTM- and GRU-based language models and a character-level CNN. The main motivation for this work is to find neural architectures that yield improved performance. In this work, they have shown that the features learned from an LSTM language model help improve the performance compared to random-weight architectures, with the LSTM model and temporal max pooling outperforming other competing models.

3 Deep Learning Over Machine Learning

To understand the difference, basic terminology of machine learning and deep learning need to be understood. Machine learning deals with increasing performance for any model application by improving or updating experiences without being explicitly programmed, while deep learning is acquainted with the use of neural networks which tries to learn the features from vast amount of data.

Major issue with machine learning algorithms is that they depend on feature engineering. But by using deep learning, the feature engineering phase is not required. Also, deep learning performs better than machine learning when data size and number of features are large.

Another major concern is that data processing needs to be done at a high level, and we need GPU for deep learning processing. Machine learning algorithms require more memory usage but predictive accuracy is low whereas deep learning algorithms offer more flexibility and also more accuracy for most of the applications.

Most importantly, machine learning algorithms try to extract features from given raw data, while deep learning tries to learn the feature from given data.

4 Discussion

The proposed system uses a deep learning model, i.e., deep neural network (DNN) to detect malware in any portable executable file. Deep neural networks are able to learn the features that optimally represent the given training data. The features are extracted from binaries and follow feature selection which minimizes dimension of the data. DNN is composed of interconnected layers, normalization layers, and dropout layers. Dropout and batch normalization are used in between connected layers, to reduce overfitting and speed optimize training models. The deep neural networks are capable of learning to differentiate benign and malicious files. For each of the classes benign and malware, we develop a model of neurons with activation function. We do not know which neurons contribute most to the benign or malware classification. But there is an issue when a file is an encrypted file or compiler file or installer file so the system will not be able to detect malware.

This methodology uses static analysis techniques as it is fast and requires a simple setup. So, using this methodology, users can check whether the file is malicious or not by giving any executable portable file as input to software. And software will give warning if file is malicious otherwise users can easily execute the file (Fig. 1).

The dataset "Benign & Malicious PE Files" is available on Kaggle [13] for research on malware detection with respect to Windows Portable Executable (PE) files. PE files start with a DOS header, followed by a PE header, and a number of sections. Each section has a header which describes its data and resources (Fig. 2).

These PE files will help us decide whether a given input file by the user is malicious or benign. The dataset consists of 19,611 unique files which has 79 features. These features include e_magic, e_cblp, e_cp, e_crlc, e_cparhdr,e_minalloc,e_maxlloc, and _e_csum which represent first few bytes of a file, number of bytes on last page, number of pages, number of relocation, header size, minimum and maximum paragraph needed, and checksum value, respectively.

In this system, after importing all the required libraries like pandas, NumPy, etc., main features from the training dataset are extracted. The DNN model contains an input layer, hidden layer, and an output layer. It predicts whether the file given by the user is benign or malicious. If the prediction is 1, then the given file is malicious, and if the prediction is 0, file is benign. This information can be helpful to the user and protect the user from executing malicious files.

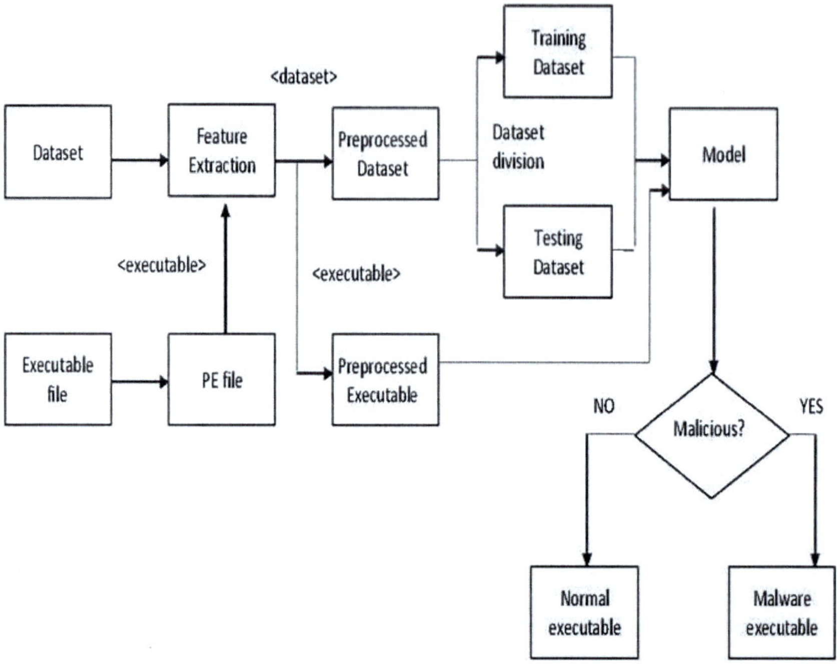

Fig. 1 System flowchart

5 Future Work

In the future, we are planning to implement the proposed approach on different datasets. For more accurate evaluation of the predictors, the models will be tested on all the possible types of malware: spyware, adware, rootkits, backdoor, banking malware, etc., so that a maximum number of possible families are known to the system before the launch of the project for real-world environments. In this paper, the DNN model will predict a given PE file as benign or malware but for the future work we can classify malware into different types like spyware, Trojan horse, backdoor, ransomware, worm, etc.

6 Conclusion

Malware is considered as a salient security threat on the Internet today. It is found that most of the problems for any organization or individual related to the Internet like denial of service have basically malware as their threat and expose the enterprise or individual. Malware detection system basically works on the signature present on any malicious program or behavioral parameters of a given code. Unlike these

Fig. 2 PE file format

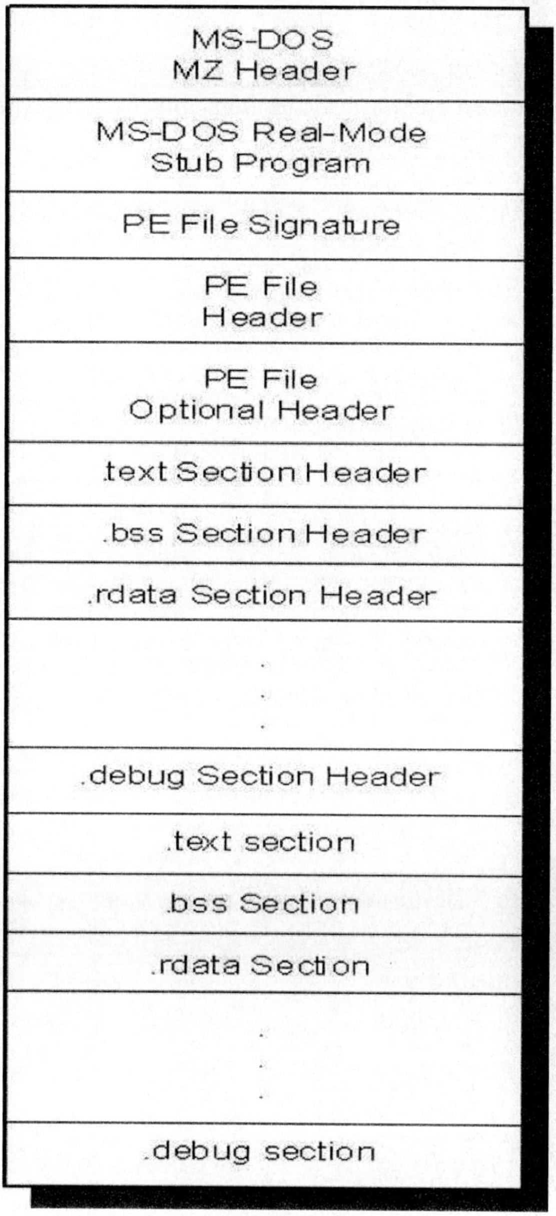

traditional methods, deep neural networks learn the features from a given PE file and extract the low-level features from a given PE file to detect malware or benign. Machine learning has a demerits of handling huge dataset, and this can be overruled by DNN.

References

1. I. Baptista, S. Shiaeles, N. Kolokotronis, A novel malware detection system based on machine learning and binary visualization, in *2019 IEEE International Conference on Communications Workshops (ICC Workshops)* (2019)
2. H. Shukla, S. Patil, D. Solanki, L. Singh, M. Swarnkar, H.K. Thakkar, On the design of supervised binary classifiers for malware detection using portable executable files, in *2019 IEEE 9th International Conference on Advanced Computing (IACC)* (2020)
3. J. Zhao, S. Zhang, B. Liu, B. Cui, Malware detection using machine learning based on the combination of dynamic and static features, in *2018 27th International Conference on Computer Communication and Networks (ICCCN)* (2018)
4. A. Yewale, M. Singh, Malware detection based on opcode frequency, in *2016 International Conference on Advanced Communication Control and Computing Technologies* (2017)
5. Z. Xu, S. Ray, P. Subramanyan, S. Malik, Malware detection using machine learning based analysis of virtual memory access patterns, in *Design, Automation & Test in Europe Conference & Exhibition (DATE)* (2017)
6. X. Xiao, D. Yuxin, Z. Yibin, T. Ke, D. Wei, Malware detection based on objective-oriented association mining, in *2013 International Conference on Machine Learning and Cybernetics* (2014)
7. R. Vinayakumar, M. Alazab, K.P. Soman, P. Poornachandran, S. Venkatraman, Robust intelligent malware detection using deep learning. J. IEEE Access **7** (2019)
8. R. Vyas, X. Luo, N. McFarland, C. Justice, Investigation of malicious portable executable file detection on the network using supervised learning techniques, in *2017 IFIP/IEEE Symposium on Integrated Network and Service Management (IM)*
9. O.P. Samantray, S. Narayan Tripathy, S.K. Das, A study to understand malware behavior through malware analysis, in *2019 IEEE International Conference on System, Computation, Automation and Networking*
10. S. Choi, S. Jang, Y. Kim, J. Kim, Malware detection using malware image and deep learning, in *2017 International Conference on Information and Communication Technology Convergence*
11. K. He, D.-S. Kim, Malware detection with malware images using deep learning techniques, in *2019 18th IEEE International Conference On Trust, Security and Privacy in Computing and Communications/13th IEEE International Conference On Big Data Science And Engineering (TrustCom/BigDataSE)*
12. B. Athiwaratkun, J.W. Stokes, Malware classification with LSTM and GRU language models and a character-level CNN, in *2017 IEEE International Conference on Acoustics, Speech and Signal Processing (ICASSP)*
13. Benign & Malicious PE Files, https://www.kaggle.com/amauricio/pe-files-malwares

ANN Based Hybrid RSSI-TDOA DOA Estimation and Beamforming

Mane Sunita Vijay and Bombale Uttam Laxman

Abstract Smart antenna system comprises of an antenna array interfaced with sophisticated electronic signal processing system that performs direction of arrival (DOA) estimation of the received signals and beamforming to form beams in the directions of the signals of interest (SOI) and nulls or no beams in the directions of the unwanted interferers. In this paper, the problem of DOA evaluation using uniform linear antenna arrays, with reduced computational complexity and cost is modeled by artificial neural network (ANN) and is followed by beamforming. The work consists of two main parts of DOA evaluation and beamforming for system of eight element Uniform Linear Array (ULA). This paper presents a hardware system incorporating three transmitters and eight element ULA interfaced with the hardware system. To perform DOA estimation artificial neural network optimized hybrid RSSI-TDOA model is devised to optimize the angles of arrival. Further, these DOA values are received and processed by RLS beamforming algorithm implemented in MATLAB. The MATLAB simulation results show that beams are formed in the directions of desired signals of interest and a perfect null is formed in the direction of the unwanted interferer and get steered dynamically as per the corresponding positions of the three transmitters in real time.

Keywords DOA (Direction of Arrival) · Uniform Linear Array (ULA) · Signals of Interest (SOI) · Beamforming · Artificial Neural Network (ANN) · RSSI (Received Signal Strength Indicator) · TDOA (Time Difference of Arrival) · RLS (Recursive Least Square)

M. S. Vijay (✉)
K. B. P. College of Engineering, Satara, Maharashtra, India
e-mail: sunita.mane@kbpcoes.edu.in

B. U. Laxman
Department of Technology, Kolhapur, Maharashtra, India

G. Ranganathan et al. (eds.), *Soft Computing for Security Applications*,
Advances in Intelligent Systems and Computing 1397,
https://doi.org/10.1007/978-981-16-5301-8_20

1 Introduction

DOA evaluation and beamforming are the two steps followed by the smart antenna for directing its signal only to the desired users while suppressing the signals to the undesired users or interferers. Several techniques are available for performing the tasks of DOA estimation [1–16] and beamforming [17–23]. Many powerful high-resolution algorithms are available for DOA estimation along with conventional classical subspace based methods and their variants that mainly employ the matrix operation on cross-spectral matrix and covariance matrix like Eigen-Value Decomposition and Eigen-Value Decomposition respectively. Computational complexity and overhead of such methods are quite high and have capped their significance prominently. Traditional methods are based on linear algebra, requiring complex matrix inversion and calibrated antennas, and are much affected by the manufacturing fault and other physical parameters and structure of the covariance matrix thus making them difficult to meet real-time requirements and expensive for implementation. On the other hand, the efficiency of major beamforming techniques reduces because of steering vector errors. These approaches limit to get correct DOA estimation which degrades the performance of the beamformer also.

Many types of beamforming non-blind algorithms like SMI, LMS, RLS, CGM, and blind algorithms like CMA, DDA are recognized [20] along with their variants and can be used as per user requirements. LMS, RLS, and CMA are the most commonly used beamforming algorithms and among these algorithms, RLS has an upper hand in terms of performance [21].

Sometimes it becomes necessary to reconfigure the uniform linear array into a system of sub arrays to overcome certain mismatch problems and errors in steering vector along with necessary software algorithms [22].

The ANN method is typically a two step method involving: a training phase and testing phase. The familiar input-output data patterns are initially applied to ANN for training. This training can be implemented offline. Once trained, it can directly replace the complex system dynamics. The ANN has the merits of massive parallelism because of which it is used for hardware implementation, adaptive learning capability, nonlinear property, as well as generalization capability [24, 25].

Finding direction of arrival of the radio signal is actually an inverse problem of receiving the signal from a known direction by mapping relationship of many parameters of signal with its arrival direction. Advancing fields of artificial intelligence and machine learning, lead many researchers to do the research on problem of DOA evaluation referred to ML techniques [26, 27].

With the recent advances in the fields of ANN, Internet of things (IOT), Machine Learning (ML) and Artificial Intelligence (AI) the wireless communication technology and optimization techniques along with smart antennas find diversified applications in many fields. Optimization based on Support Vector Machine (SVM) is used for performance analysis of wind turbines [28]. ANN is used in power management of street lights by adjusting its intensity as per weather conditions [29]. AI is put to work to deploy the chosen antennas in diversified situations with improved throughput, signal performance, and speed of data conveyance [30]. IOT based sensors along with wireless communication technology are put to work for optimization of productive agronomics [31]. ANN is used to identify vacant parking slots from live camera images using mobile net [32].

Acknowledging all this information and taking into account the aforementioned limitations, this paper explores the problem of DOA evaluation of mobile users using uniform linear array of antennas, with reduced computational complexity and cost by using a suitable ANN trained with input output pairs and is followed by beamforming. For DOA estimation RSSI based and TDOA based artificial neural network optimizations methods are used and a combination of the two methods is given to deliver more accurate results. Here, we have devised a method that performs DOA estimation using ANN giving angle optimization coefficients to produce a hybrid RSSI-TDOA model, and using the obtained information steps for the beamforming.

This paper presents a hardware system using three transmitters and eight elements Uniform Linear Array operating at 2.4 GHz frequency, interfaced with hardware system. To perform DOA estimation artificial neural network optimized hybrid RSSI-TDOA model is devised to optimize the angles of arrival based on RSSI and TDOA techniques. Estimated DOAs of the three transmitters are recorded by the master controller and the MATLAB program displays these values.

Further, these DOA values are received and processed by RLS beamforming algorithm implemented in MATLAB.

In this work, the focus is on,

(1) Designing antennas for three transmitters operating at 2.4 GHz frequency with suitable interface.
(2) Designing eight element uniform linear array operating at 2.4 GHz frequency with suitable interface.
(3) Devising easy to implement, cost effective, accurate, and computationally efficient DOA estimation using hardware platform.
(4) Devising a beamforming algorithm on MATLAB platform having a proper interface or link with 'DOA estimation hardware platform'.

This paper is outlined as follows. Section 2 delivers the problem formulation along with the research questions (RQs). Section 3 describes the proposed system model for DOA evaluation describing a novel, cost efficient, easy to implement, computationally efficient artificial neural network optimized hybrid DOA estimation algorithm. Section 4 discusses the beamforming model for the proposed system. Section 5 discusses the hardware system and explains the hardware items and systems implemented in the work. Section 6 discusses Results obtained and the conclusion of the paper is in Sect. 7.

2 Problem Formulation

As mentioned earlier the total work planned is a projection of the design problem formulated below:

Three mobile (moving) transmitters Tx1, Tx2, Tx3 operating at 2.4 GHz frequency with Tx1, Tx2 transmitting the desired signal of interest (SOI), and Tx3 transmitting the unwanted interfering signal is to be designed to transmit the signals. A uniform linear array (ULA) having eight antenna elements, operating at 2.4 GHz frequency is to be designed and is considered to operate at receiver end with a proper hardware system and interface. Easy to implement, cost effective, accurate, and computationally efficient DOA estimation algorithm has to be devised on a proper hardware platform with a suitable interface with the ULA. A MATLAB window should display the DOAs of the three transmitters in dynamic manner. A suitable beamforming algorithm is to be implemented on MATLAB platform with proper interface with 'DOA estimation hardware platform'. MATLAB beam spectrum should display the beams and null along with their DOAs and should show dynamic variations as per the changes in the transmitter angles in real environment. The beams and null should get steered dynamically as per the variations in the corresponding transmitter angles.

Looking forward to hardware based DOA estimation along with beamforming a system is proposed that is framed into two approaches to cover the required tasks.

Hardware approach:

In a broad view, following parts are used for system hardware with suitable interface and software. For this the hardware approach considers,

- Antenna for each of the three transmitters Tx1, Tx2, Tx3 operating at a frequency of 2.4 GHz.
- Selection and use of Node MCU ESP 8266 for each of the transmitters Tx1, Tx2, Tx3.
- Eight element ULA array operating at a frequency of 2.4 GHz.
- Selection and use of Node MCU ESP 8266 for each antenna element of the eight elements ULA that will be programmed to act in slave mode with suitable interface with the master controller.
- Arduino mega 2560 board that will act as a master controller and perform the DOA estimation with a suitable interface with slave Node MCUs and Laptops.
- Laptop interface for display of the results.

Software approach:

For performing the DOA estimation and beamforming following tasks are performed with proper hardware system and interface. For this the software approach considers,

- ANN optimized RSSI method to be used to calculate the angles of arrival of the signals from the three transmitters.
- ANN optimized TDOA method to be used to calculate angle of arrival of the signals from the three transmitters.

- ANN optimized hybrid of RSSI-TDOA to be devised to refine the angle of arrival measurement.
- Capturing the angles estimated for TX1, TX2, and TX3 by the DOA estimation algorithm, weights for each of the antenna element signals are to be calculated with RLS beamforming algorithm implemented on MATLAB platform that can form beams in the directions of TX1 and TX2 and a null or no beam in TX3 direction. The beams and null are required to get steered dynamically as per the variations in the corresponding transmitter angles.

Considering the intended work and projected approaches following research questions are framed to signify the proposed methodologies. The presented work should render optimal answers to these questions.

RQ1: Whether RSSI and TDOA techniques along with ANN can be used for the optimization of direction of arrival estimation on the designed hardware platform?

RQ2: Whether beamforming algorithm on MATLAB platform can be interfaced with the above mentioned 'DOA estimation hardware platform'?

RQ3: Whether the estimated DOA angles change dynamically as per the changes in the transmitter positions and beamforming algorithm output beam spectrum reflect the corresponding changes dynamically in the form of beams and null?

As a solution to the given problem and considering the intended hardware and software implementation a three transmitter, eight element ULA system is devised that implements an ANN optimized hybrid RSSI-TDOA DOA estimation algorithm with proper training and testing executions on a proper hardware platform with suitable interface with the ULA. A MATLAB window displays the DOAs of the three transmitters in dynamic manner. RLS beamforming algorithm is implemented on MATLAB platform with proper interface with 'DOA estimation hardware platform'. MATLAB beam spectrum displays the beams and null along with their DOAs and shows dynamic variations as per the changes in the transmitter angles in real environment.

3 System Model for DOA Estimation

For DOA estimation we have arranged the antenna elements as shown in Fig. 1. The angles of arrival are measured by treating the antenna elment1 in the ULA as the reference element and measuring all the angles with respect to this element in −90°–90° span.

In this work, RSSI and TDOA are the two methods used for DOA estimation and they are optimized with the use of artificial neural network. Finally, DOA is estimated with the help of a hybrid of the two methods to give more accurate results.

For RSSI method, let us assume three matrices, a matrix for received signal strength at eight elements of ULA, called received signal strength array and is denoted by RSSA; a matrix of angle adaptation coefficients for eight elements of ULA, denoted by AAC_{RSSI}; a matrix for the angle of arrival denoted by AOA_{RSSI}.

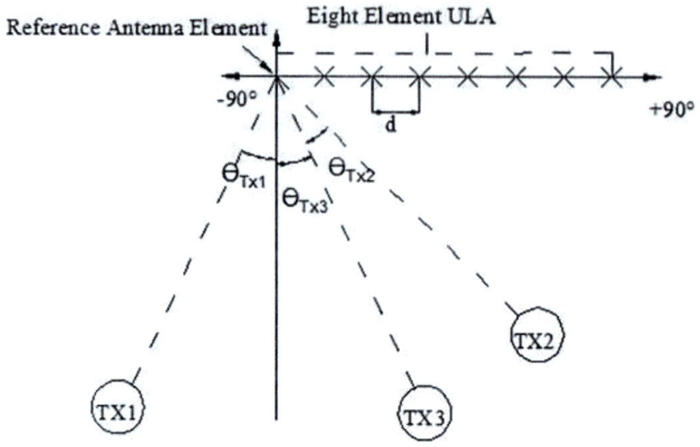

Fig. 1 Direction of arrival measurement setup

It is clear that RSSI values at ULA at certain angle bear some relation with it. Thus received signal strength multiplied by some unknown factor can yield the angle of arrival. We are using an ANN with this relation to find unknown factors which we term as angle optimization coefficients (AAC_{RSSI}). To obtain the angle optimization coefficient we are using two phases a training phase and testing phase for the ANN.

Now it is clear that the matrices mentioned, bear a relation as follows,

[Angle adaptation coefficients] [Received signal strength array] = [Angle of arrival]

$$[AAC_{RSSI}]\,[RSSA] = [AOA_{RSSI}]$$

where, angle adaptation coefficients are unknown for known angle of arrival.

It resolves to the relation given by,

$$\begin{bmatrix} AAC_1 & AAC_2 & \ldots & AAC_8 \end{bmatrix} \begin{bmatrix} RSSA_1 \\ RSSA_2 \\ \vdots \\ RSSA_8 \end{bmatrix} = [AOA] \tag{1}$$

Now the angle adaptation coefficients can be determined by,

$$[AAC_{RSSI}] = [AOA_{RSSI}][\ RSSA]^{-1} \tag{2}$$

For TDOA method, let us assume three matrices, a matrix for time delay of arrival related values at eight elements of ULA, called time delay array and is denoted by TDA; a matrix of angle adaptation coefficients for eight elements of ULA, denoted by AAC_{TDOA}; a matrix for the angle of arrival denoted by AOA_{TDOA}.

It is clear that TDOA values at certain angles bear some relation with it. Thus time delay of arrival multiplied by some unknown factor can yield the angle of arrival. We are using artificial neural network (ANN) with this relation to find the unknown factor which we term as angle optimization coefficients (AAC_{TDOA}). To obtain the angle optimization coefficient we are using two phases a training phase and testing phase for the ANN.

Now it is clear that the matrices mentioned, bear a relation as follows,

[Angle adaptation coefficients] [Time delay of arrival related array] = [Angle of arrival]

$$[AAC_{TDOA}]\,[TDA] = [AOA_{TDOA}]$$

Where, angle adaptation coefficients are unknown for known angle of arrival. It resolves to the relation given by,

$$\begin{bmatrix} AAC_1 & AAC_2 & \ldots & AAC_8 \end{bmatrix} \begin{bmatrix} TDA_1 \\ TDA_2 \\ \vdots \\ TDA_8 \end{bmatrix} = [AOA]. \tag{3}$$

Now the angle adaptation coefficients can be determined by,

$$[AAC_{TDOA}] = [AOA_{TDOA}][\ TDA]^{-1} \tag{4}$$

For determining the angle of arrival the ANN works in two phases called a training phase and a testing phase respectively.

Algorithm of ANN to find AAC_{RSSI}(/AAC_{TDOA}) involves following steps:

1. Take RSSI values (/TDOA related values) for ith iteration.
2. Adjust weights of the hidden layer.
3. Check confidence factor [CF]. If CF is less go to step 2, else go to step 4.
4. Increment value of i.
5. en iterations are not finished go to step 2 else stop.

Thus following relations comes true,

$$[AAC_{RSSI}][RSSA] = [AOA_{RSSI}]$$

$$[AAC_{TDOA}][TDA] = [AOA_{TDOA}]$$

where, AAC_{RSSI} (/AAC_{TDOA}) represents angle adaptation coefficients generated from artificial neural network, RSSA are real-time RSSI values (/TDA are real-time delay of arrival related values) received at the respective antenna elements of eight element ULA, AOA_{RSSI} (/AOA_{TDOA})gives the real-time angle value of the transmitter location.

For testing phase it is required to generate an RSSI value (/time delay related value) matrix that includes the actual respective values received at the receiver array and it should be fed to the trained neural network so that it will return actual angle of arrival of the signal from transmitter.

3.1 ANN Optimized Hybrid RSSI-TDOA Model Based DOA Estimation

To estimate the angle of arrival correctly we take into account the angles determined by both the RSSI based and TDOA based DOA estimation models using ANN.

Below mentioned steps are followed to determine the angle of arrival.

1. Read AOA_{RSSI}.
2. Read AOA_{TDOA}.
3. If,

$$|AOA_{RSSI} - AOA_{TDOA}| < \mu_{th} \tag{5}$$

Then,

$$AOA = \frac{AOA_{RSSI} + AOA_{TDOA}}{2} \tag{6}$$

Else,

Discard the angles.

Then steps 1 to 3 are repeated for next AOA values to get correct angle of arrival estimation. In this process, erroneous results are eliminated by applying a value μ_{th}. Where, μ_{th} is the step threshold value and is determined practically to get error free readings.

This method efficiently gives the DOA estimation in dynamic manner to show the direction of transmitters even though they change their directions in real time.

4 System Model for Beamforming

Figure 2 shows the required signal s(k) arriving at angle θ_0 with interference signals $i_1(k)$, $i_2(k)$,…$i_N(k)$ arriving on ULA of M antenna elements. The resulting input signals are $x_1(k)$, $x_2(k)$,…, $x_M(k)$ on respective M antenna elements. For strengthening the required signal and nullify the interferer's weights are estimated and applied for respective antenna element signal. Then y(k) is the combined received signal.

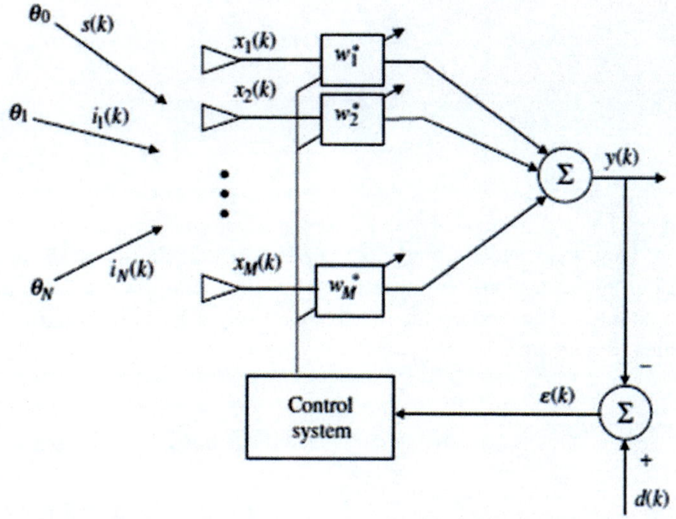

Fig. 2 RLS beamforming

Error, ε(k) is the difference between this received signal and the required signal, used as a controlling signal by adaptive algorithm to estimate the adaptive weights. Here d(k) is the reference signal required which has high correlation with the required signal. The adaptive algorithm attempt to reduce ε(k) by multiplying weights with the respective array element signals so as to make the output, combined received, signal equal to the required signal[20, 21].

4.1 Recursive Least Squares (RLS)

By recursively estimating the desired correlation vector and correlation matrix we can remove the computational load in this method.

Initially, the statistics of signal are unknown and we should estimate the array correlation matrix $(\overline{R}_{xx})$ and the signal correlation vector $(\overline{r})$. $(\overline{R}_{xx})$ and $(\overline{r})$ calculated each instant in time for a range of snapshots. There instant values are:

$$\hat{R}_{xx}(k) \approx \overline{x}(k)\overline{x}^{H}(k) \tag{7}$$

And

$$\hat{r}(k) \approx d^{*}(k)\overline{x}(k) \tag{8}$$

We know that,

$$\hat{R}_{xx} = \sum_{i=1}^{k} \overline{x}(i)\overline{x}^{H}(i) \tag{9}$$

$$\hat{r} = \sum_{i=1}^{k} d^{*}(i)\overline{x}(i) \tag{10}$$

where block length is k and $\hat{R}_{xx}$ and $\hat{r}$ are correlation calculations at the end of time sample k. For position of current source, earlier time samples are of no use so we have to forget them. Thus modifying the above two equations for forgetting the early time samples we can write,

$$\hat{R}_{xx}(k) = \sum_{i=1}^{k} \alpha^{k-i}\overline{x}(i)\overline{x}^{H}(i) \tag{11}$$

$$\hat{r}(k) = \sum_{i=1}^{k} \alpha^{k-i} d^{*}(i)\overline{x}(i) \tag{12}$$

Here exponential weighting factor is nothing but the forgetting factor α and $0 \leq \alpha \leq 1$. In the case where, $\alpha = 1$ the least squares algorithm is taken into consideration. This value considers infinite memory. Two terms can be derived by splitting the above two equations, as,

$$\begin{aligned}\hat{R}_{\text{xx}}(k) &= \alpha \sum_{i=1}^{k-1} \alpha^{k-1-i}\overline{x}(i)\overline{x}^{H}(i) + \overline{x}(k)\overline{x}^{H}(k) \\ &= \alpha \hat{R}_{xx}(k-1) + \overline{x}(k)\overline{x}^{H}(k)\end{aligned} \tag{13}$$

$$\begin{aligned}\hat{r}(k) &= \alpha \sum_{i=1}^{k-1} \alpha^{k-1-i} d^{*}(i)\overline{x}(i) + d^{*}(k)\overline{x}(k) \\ &= \alpha\hat{r}(k-1) + d^{*}(k)\overline{x}(k)\end{aligned} \tag{14}$$

If we define a gain vector $\overline{g}(k)$ as,

$$\overline{g}(k) = \frac{\alpha^{-1}\hat{R}_{xx}^{-1}(k-1)\overline{x}(k)}{1 + \alpha^{-1}\overline{x}^{H}(k)\hat{R}_{xx}^{-1}(k-1)\overline{x}(k)} \tag{15}$$

Therefore,

$$\hat{R}_{\text{XX}}^{-1}(k) = \alpha^{-1}\hat{R}_{\text{XX}}^{-1}(k-1) - \alpha^{-1}\overline{g}(k)\overline{x}^{H}(k)\hat{R}_{\text{XX}}^{-1}(k-1) \tag{16}$$

This equation is known as the Riccati equation applied to RLS method. Multiplying by denominator the Eq. (15) can be solved to,

$$\overline{g}(k) = \left[\alpha^{-1}\hat{R}_{\mathrm{xx}}^{-1}(k-1) - \alpha^{-1}\overline{g}(k)\overline{x}^{H}(k)\hat{R}_{\mathrm{xx}}^{-1}(k-1)\right]\overline{x}(k) \tag{17}$$

Thus,

$$\overline{g}(k) = \hat{R}_{\mathrm{xx}}^{-1}(k)\overline{x}(k) \tag{18}$$

For deriving recursive relation with an aim of updating weight vectors optimum Wiener solution is kept iterating for k iterations. By use of Eq. (14) then we get,

$$\overline{w}(k) = \hat{R}_{\mathrm{xx}}^{-1}(k)\hat{r}(k) = \alpha\hat{R}_{\mathrm{xx}}^{-1}(k)\hat{r}(k-1) + \hat{R}_{\mathrm{xx}}^{-1}(k)\overline{x}(k)d^{*}(k) \tag{19}$$

From Eqs. (16) and (19) we get,

$$\begin{aligned}\overline{w}(k) &= \hat{R}_{xx}^{-1}(k-1)\hat{r}(k-1) - \overline{g}(k)\overline{x}^{H}(k)\hat{R}_{xx}^{-1}(k-1)\hat{r}(k-1)\\ &\quad + \hat{R}_{xx}^{-1}(k)\overline{x}(k)d^{*}(k)\\ &= \overline{w}(k-1) - \overline{g}(k)\overline{x}^{H}(k)\overline{w}(k-1) + \hat{R}_{xx}^{-1}(k)\overline{x}(k)d^{*}(k)\end{aligned} \tag{20}$$

With the substitution of equation (18) in above equation we can derive,

$$\begin{aligned}\overline{w}(k) &= \overline{w}(k-1) - \overline{g}(k)\overline{x}^{H}(k)\overline{w}(k-1) + \overline{g}(k)d^{*}(k)\\ &= \overline{w}(k-1) + \overline{g}(k)\left[d^{*}(k) - \overline{x}^{H}(k)\overline{w}(k-1)\right]\end{aligned}$$

Thus with the use of previous values, array correlation and the vector correlation estimate can be acquired for future values called weight vectors ($\overline{w}$).

For the final equation of the weight vectors we get,

$$\begin{aligned}\overline{w}(k) &= \overline{w}(k-1) - \overline{g}(k)\overline{x}^{H}(k)\overline{w}(k-1) + \overline{g}(k)d^{*}(k)\\ &= \overline{w}(k-1) + \overline{g}(k)\left[d^{*}(k) - \overline{x}^{H}(k)\overline{w}(k-1)\right]\end{aligned} \tag{21}$$

In the case of RLS algorithm, the inversion of the large correlation matrix is not required as in the case of SMI algorithm and its convergence is much faster than the LMS algorithm. The recursive equations related to RLS algorithm quite easily update the inverse of the correlation matrix.

RLS algorithm converges much faster and provides good beam formation along with exact null placement. Hence RLS algorithm is selected as a potential candidate for beamforming algorithm.

The estimated DOAs of the three transmitters are recorded by the master controller and the MATLAB program reads these values and is given as input angles to the RLS beamforming algorithm.

Here transmitter TX1, TX2 signals are treated as desired signals and TX3 signal is treated as interferer. The algorithm records and displays these angles and forms beams in the required desired directions and a null or no beam is formed in the direction of unwanted interferer. Also, it is observed that the pattern changes dynamically in accordance with the changes in transmitter DOAs.

5 Hardware Platform

The system hardware consists of three transmitters including their antennas and their respective processors, an eight element uniform linear array of antennas connected with their respective processors having one independent processor for each antenna, a master controller processor that is connected to all the eight processors connected to ULA antennas and power supply and communication links between the master controller and the eight antenna element processors. The master controller operates in coordination with the eight clock synchronized slave processors receiving signals from respective eight antenna elements of ULA and helps in performing the tasks of DOA estimation and beamforming. A description of each of the hardware elements is given below.

5.1 *Antenna*

Using HFSS software three patch antennas are designed and then fabricated for transmitting signals from TX1, TX2, and TX3 transmitters with following specifications.

Dielectric constant, $\varepsilon_r = 4.3$, Substrate material thickness, $h = 1.6$ mm,
Frequency, $f = 2.45$ GHz, Input impedance $= 50\Omega$,
Patch width $= 37.6099$ mm, Patch length, $L = 29.1589$ mm,
Inset feed point, $F_i = 10.7915$ mm, Width of feed line, $W_F = 3.0389$ mm,
Gap of feed line, $G_{pf} = 0.2015$ mm, Substrate length, $L_g = 38.7589$ mm,
Substrate Width, $W_f = 47.2099$ mm,

The antenna design is as shown in Fig. 3 and the fabricated antenna appears as shown in Fig. 4.

The antenna response is shown in Fig. 5 and it indicates that the antenna is made to operate at 2.4 GHz frequency.

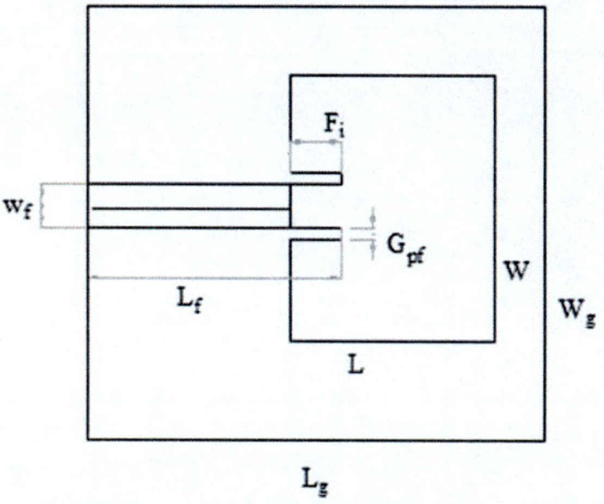

Fig. 3 Antenna designed for 2.4 GHz frequency

Fig. 4 Fabricated antenna for 2.4 GHz frequency

5.2 *Eight Element Uniform Linear Array (ULA)*

Using HFSS software uniform linear array consisting of eight antenna elements is designed and is fabricated as shown in Fig. 6. It has following specifications.

Frequency, f = 2.45 GHz

Inter-element distance, d = λ/2

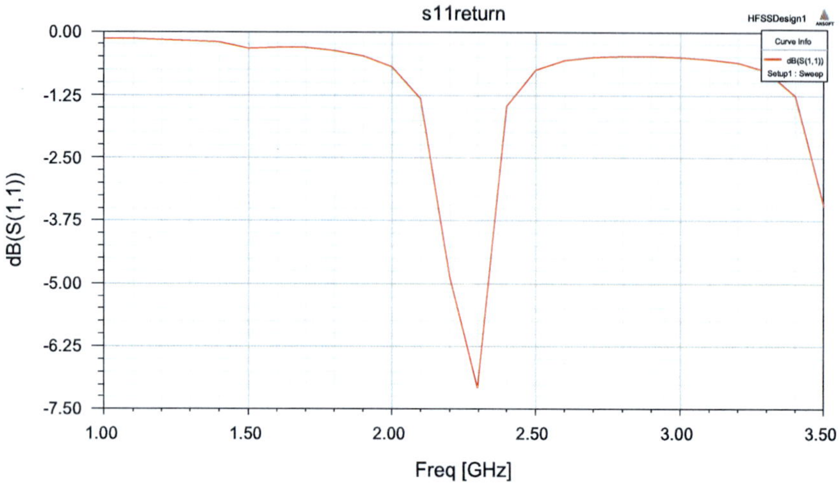

Fig. 5 Antenna response

Fig. 6 Antenna array design

5.3 System Configuration

The hardware system includes an Arduino mega 2560 board acting as master controller to which eight NodeMCU ESP8266 boards (one for each antenna element of the eight element uniform linear antenna array (ULA)) are connected as given in Fig. 7. To the antenna terminals of these node MCU boards, eight antenna elements in the uniform linear array are connected using SMA connectors. There are three transmitters named Tx1, Tx2, and Tx3. These transmitters are devised by using a Node MCU ESP8266 board with a separate patch antenna connected to its antenna terminals for each one of them. Tx1, Tx2 is supposed to transmit the desired signals of interest (SOI) whereas Tx3 is supposed to transmit the undesired interfering signal. So we can treat signals from Tx1 and Tx2 as desired signals and the one from Tx3 as interferer. The power (+5 V and 0 V) to this whole system is derived from a laptop by connecting a USB to TTL converter between the USB of laptop and power connector on the Arduino Mega board which also serves for the communication between laptop and Arduino Mega board. RX0 pin of Arduino Mega board is connected to TXD0 of each of the Node MCU ESP8266 boards to which each of the eight antenna elements from the ULA are connected. TX0 of Arduino Mega board is connected to RXD0 of each of the eight Node MCU ESP8266 boards. These connections are used for communication purposes. TX3, RX3 of Arduino Mega board are connected to D2, D3of each of the eight Node MCU ESP8266 boards respectively. Power supply to

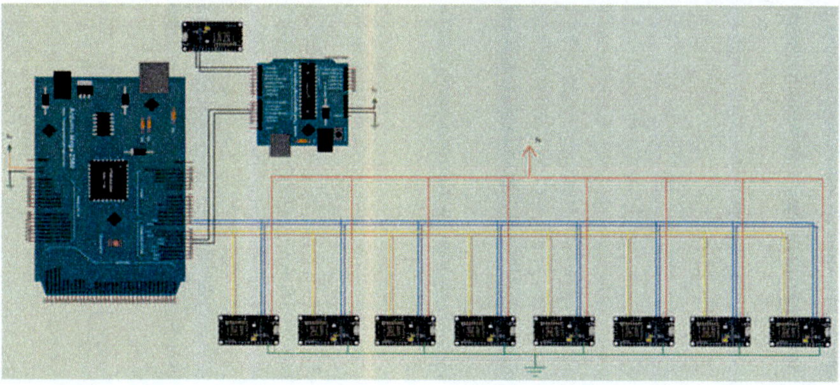

Fig. 7 Connection diagram of master Arduino Mega 2560 board with Arduino Uno and eight slave Node MCUs

all these boards is (+5 V and ground) common and is supplied from laptop USB port via a USB to TTL converter (Figs. 8 and 9).

The Arduino Mega board is programmed to work as a master controller. The eight Node MCUs ESP8266 used to receive data from the eight antenna elements are programmed as slave controllers with synchronized clocks.

Fig. 8 Photograph of three transmitters TX1, TX2 and TX3

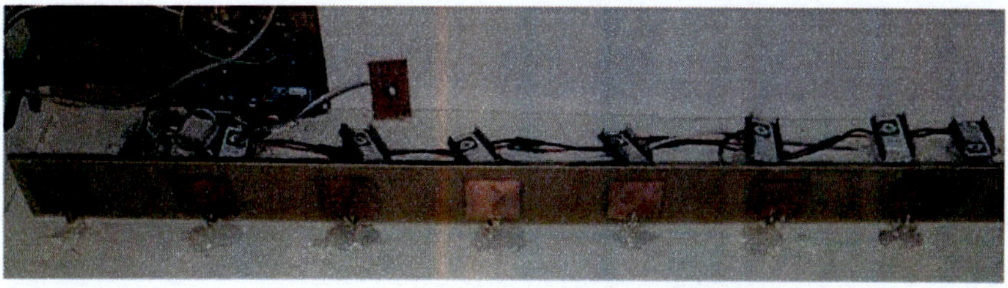

Fig. 9 Photograph of master Arduino Mega 2560 board with eight slave Node MCUs

The data transmission from the three transmitters can be differentiated at the receiver end due to the transmitter address identification. Thus, once the transmitter sends a data packet the eight elements of antenna array receive it in the corresponding antenna element processor (Node MCU ESP8266). All the eight Node MCUs receiving data from corresponding antenna elements in a ULA are connected to Arduino Mega 2560 master controller. Data transmission between Arduino Mega 2560 master controller and the eight antenna Node MCUs is done via serial communication.

For RSSI based DOA estimation method, these eight slave controllers receive the actual RSSI values of the signal transmitted from the transmitter at fixed known angle and are recorded as training dataset for that angle. Several such datasets constitute the training data for ANN. The trained ANN file containing the angle optimization coefficients [.h5] is loaded in the master controller. The setup is now ready for actual RSSI based DOA estimation and the eight slave controllers receive the actual RSSI values of the signal transmitted from the transmitters at unknown angles. Using the trained ANN file in the master controller the actual DOAs of the transmitters are then estimated.

For TDOA based DOA estimation method, these eight slave controllers receive the actual TDOA related values of the signal transmitted from the transmitter at fixed known angle and are recorded as training dataset for that angle. Several such datasets constitute the training data for ANN. The trained ANN file containing the angle optimization coefficients [.h5] is loaded in the master controller. The setup is now ready for actual TDOA based DOA estimation and the eight slave controllers receive the actual TDOA related values of the signal transmitted from the transmitters at unknown angles. Using the trained ANN file in the master controller the actual DOAs of the transmitters are then estimated.

Once the two values from the RSSI and TDOA based DOA estimation are received the master controller calculates the final DOA value from the hybrid RSSI-TDOA model code.

The estimated DOAs of the three transmitters are recorded by the master controller and the MATLAB program reads these values and is given as input angles to the RLS beamforming algorithm.

Here transmitter TX1, TX2 signals are the desired signals and TX3 signal is the undesired interferer. Figure 11 shows photograph of three transmitters TX1, TX2, and TX3 and Fig. 12 shows photograph of master Arduino Mega 2560 board with eight slave Node MCUs.

The algorithm in Arduino Mega master controller estimates and displays these angles and the beamforming algorithm in MATLAB captures these values and forms beams in the required desired directions and a null or no beam in the direction of undesired interferer. The array factor plot is displayed on the screen showing the beam directions. The null placement and the beams in the direction of desired users are found to be correct. Also, the beam pattern containing the desired beams and null get steered dynamically as per the variations in the transmitter positions.

6 Results and Discussion

In this paper, an easy to implement, cost effective, accurate, and computationally efficient DOA estimation model using hardware platform is devised using three transmitters, eight element ULA interfaced with proper hardware system incorporating a novel artificial neural network optimized hybrid RSSI-TDOA DOA estimation technique, which in turn is interfaced with a MATLAB platform where MATLAB based RLS beamforming algorithm captures the DOAs of the three transmitters to form the beams.

Table 1 presents the results of the estimated DOAs for the three transmitters when all the three transmitters are at the same angular position from the reference element.

Table 2 presents the results of the estimated DOAs for the three transmitters when all the three transmitters are at different angular positions from the reference element.

Table 1 Estimated DOAs for three transmitters at same angular position

Approximate Actual Transmitter Angle	Estimated DOA for Tx1	Estimated DOA for Tx2	Estimated DOA for Tx3
-40^0	-40^0	-40.7^0	-40^0
-30^0	-29.9^0	-30.8^0	-30^0
-20^0	-19.9^0	-20.7^0	-20.3^0
-10^0	-10^0	-10^0	-10.7^0
0^0	0.7^0	0^0	0^0
10^0	10.5^0	10^0	10^0
20^0	20^0	20.8^0	20.7^0
30^0	30^0	30.4^0	30.7^0
40^0	40^0	39.8^0	40.3^0
50^0	50.4^0	49.7^0	50.6^0
60^0	60^0	60.5^0	60^0

Table 2 Estimated DOAs for three transmitters at different angular positions

Approximate Actual Transmitter Angles			Estimated DOAs		
Tx1	Tx2	Tx3	Tx1	Tx2	Tx3
0^0	10^0	40^0	0.8^0	10^0	40^0
10^0	30^0	50^0	10^0	29.7^0	50^0
0^0	-10^0	-40^0	0^0	-10.7^0	-39.8^0
-10^0	-30^0	-40^0	-10^0	-30^0	-39.7^0
10^0	0^0	-40^0	10^0	0^0	-39.5^0
-30^0	10^0	-20^0	-30^0	10^0	-19.6^0
30^0	-10^0	20^0	30^0	-10^0	20^0

Observing the DOA estimation results obtained in Tables 1 and 2 for the approximate actual angle values of the three transmitters, it is observed that the algorithm implemented for DOA estimation efficiently and accurately determines the DOAs of the three transmitter signals.

The estimated DOAs of the three transmitters are recorded by the master controller and the MATLAB program reads these values and is given as input angles to the RLS beamforming algorithm. Here transmitter TX1, TX2 signals are treated as desired signals and TX3 signal is treated as interferer. The algorithm records and displays these angles and forms beams in the required desired directions and a null or no beam is formed in the direction of interferer.

The array factor plot displayed on the screen shows the beam directions and the null placement and is observed to be correct. Figures 10, 11 and 12 show that the angle positions of the three transmitters are determined accurately and accordingly the beams are directed in TX1, TX2 directions and a null is formed in the TX3 direction. Also, it is observed that the pattern changes dynamically in accordance with the changes in the transmitter DOAs.

Considering the research questions presented in section 2, it can be said that the results obtained in Tables 1 and 2 affirms the functional performance of the DOA estimation model- 'Artificial neural network optimized hybrid RSSI-TDOA DOA estimation' implemented on the hardware platform to deliver accurate results. This performance assessment with reference to the corresponding discussions covered in Sect. 3 and Sect. 5 justifies the RQ1 positively.

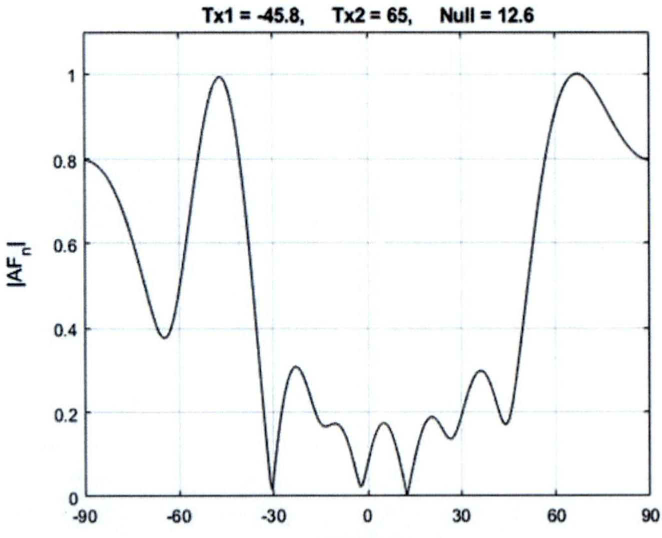

Fig. 10 Array factor plot for RLS algorithm with two desired users at approximate $-45^0, 65^0$ and one undesired interferer at 12^0 with $d = \lambda/2$, $N = 8$

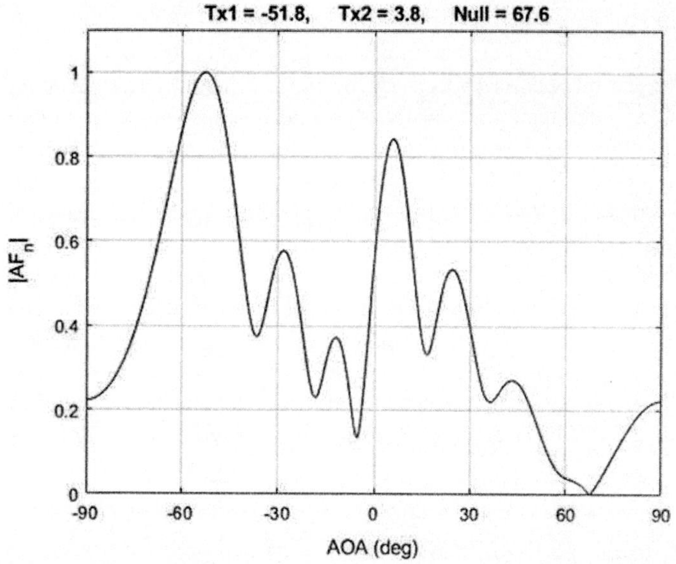

Fig. 11 Array factor plot for RLS algorithm with two desired users at approximate $-51^0, 3^0$ and one undesired interferer at 67^0 with $d = \lambda/2$, $N = 8$

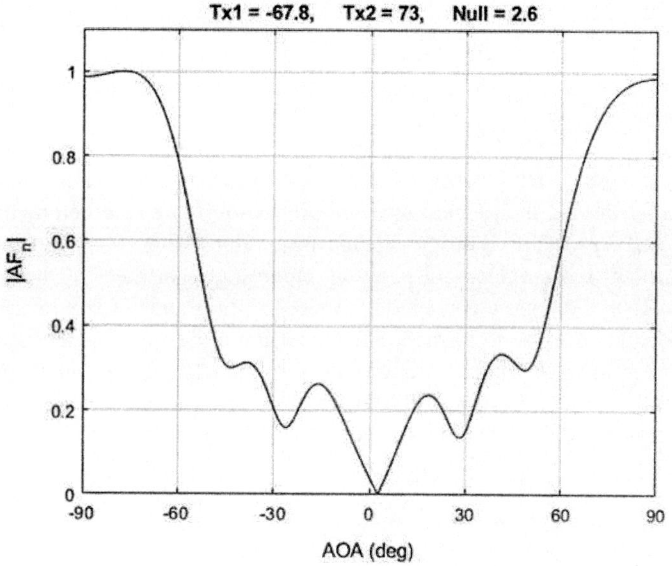

Fig. 12 Array factor plot for RLS algorithm with two desired users at approximate $-67^0, 73^0$ and one undesired interferer at 2^0 with $d = \lambda/2$, $N = 8$

Referring to Sect. 5 and the results obtained in Figs. 10, Fig. 11 and Fig. 12 it is confirmed that the RLS algorithm implemented on MATLAB platform and interfaced with hardware implemented DOA model performs well and the output beam spectrum results confirm the accuracy and the dynamic responsiveness of the system that dynamically alters the beam positions and null position as per the changes in the angular positions of the three transmitters. It is observed that the transmitter position-angles (estimated DOAs of the transmitter positions) steer the beams and null in the output spectrum in accordance with the moving transmitters dynamically. Thus RQ2, RQ3 is also positively justified. Considering the performance overview of the complete system it can be noted that it encompasses potential DOA evaluation as well as beamforming approach, which is easy to implement, cost effective, accurate, and computationally efficient. It assures the success of the practical implementation of the DOA and beamforming approaches in real time. This method is found satisfactory in terms of the results. Hence the acceptability related to RQ1, RQ2, and RQ3 is affirmed.

7 Conclusion

Many powerful high-resolution algorithms are available for DOA estimation along with conventional classical subspace based methods and their variants that mainly employ EVD associated with the cross-spectral matrix or SVD over the covariance matrix of the received signal. Computational complexity and overhead of such methods are quite high.

The proposed ANN based hybrid RSSI-TDOA model delivers accurate results for the eight element ULA system. Estimated DOAs of the three transmitters TX1, TX2, TX3 by varying their locations from -90^0 to + 90^0 are recorded by the master controller and the MATLAB program reads these values. These values show accurate results. Further, these values are given as input angles to the RLS beamforming algorithm. TX1, TX2 are considered as desired signals and TX3 as interferer. The results show that beams are formed in the direction of desired signals and a perfect null is formed in the direction of interferer. It is found that in accordance with the changes in the angle positions of the transmitters the beams and null in the output spectrum are steered dynamically to follow the real-time changes in transmitter angle positions.

References

1. R. Tanaka, Y. Haneda, "Direction-of-arrival estimation based on joint diagonalization of matrices in different direct-to-reverberation ratios", 2016 IEEE International Workshop on Acoustic Signal Enhancement (IWAENC), Xi'an, 2016, pp. 1-5.

2. Y. Fayad, C. Wang, Q. Cao and A. E. Hafez, "Improved ESPRIT algorithm used in spatial subspace for NULA," 2016 13th International Bhurban Conference on Applied Sciences and Technology (IBCAST), Islamabad, 2016, pp. 620–623.
3. M. Zohourian, G. Enzner, R. Martin, Binaural Speaker Localization Integrated Into an Adaptive Beamformer for Hearing Aids. IEEE/ACM Transactions on Audio, Speech, and Language Processing **26**(3), 515–528 (March 2018)
4. R. O. Schmidt, "A signal subspace approach to multiple emitter location and spectral estimation," Ph.D. dissertation, Dept. Elect. Eng., Stanford Univ., Stanford, CA, November 1981.
5. R.O. Schmidt, Multiple emitter location and signal parameter estimation. IEEE Trans. Antennas Propagation **34**, 276–280 (March 1986)
6. R. O. Schmidt, "Multiple emitter location and signal parameter estimation," Proc. RADC Spectrum Estimation Workshop (Rome, NY), pp. 243–258, 1979.
7. R. H. Roy, "ESPRIT-Estimation of signal parameters via rotational invariance techniques," Ph.D. dissertation, Dept. Elect. Eng., Stanford Univ., 1987.
8. A. Swindlehurst and T. Kailath, "Azimuth/elevation direction finding using regular array geometries," IEEE Trans. on Aerospace and Electronic Systems, Vol. 29, No.1, pp. 145–155, January 1993.
9. R. Roy, A. Paulraj, and T. Kailath, "Estimation of signal parameters via rotational invariance techniques – ESPRIT," in Proc. IEEE ICASSP, Vol. 4, Tokyo, Japan, pp. 2495–2498, 1986.
10. R. Roy, T. Kailath, ESPRIT –Estimation of signal parameters via rotational invariance techniques. Opt. Eng. **29**(4), 296–313 (April 1990)
11. R. Roy, A. Paulraj, T. Kailath, "ESPRIT-A subspace rotational approach to estimation of parameters of cissoids in noise", IEEE Trans. Acoustics, Speech, Signal Processing **34**(4), 1340–1342 (October 1986)
12. A. Paulraj, R. Roy, and T. Kailath, "Estimation of signal parameters via rotations invariance techniques –ESPRIT," in Proc. Nineteenth Asilomer Conf. Circuits, Syst., Comput. (Asilomar, CA), November 1985.
13. A. Paulraj, R. Roy, and T. Kailath, "A subspace rotation approach to signal parameters estimation," in Proc. IEEE, pp. 1044–1445, July 1986.
14. G. Xu, S. Silverstein, R. Roy, T. Kailath, Beamspace ESPRIT. IEEE Trans. Signal Processing **42**, 349–356 (February 1994)
15. Qinglong Li, Xueliang Zhang and Hao Li, "Online direction of arrival estimation based on deep learning," *Researchgate*, May 2018.
16. P. Gupta, K. Aditya and A. Datta, "Comparison of conventional and subspace based algorithms to estimate Direction of Arrival (DOA)", 2016 International Conference on Communication and Signal Processing (ICCSP), Melmaruvathur, 2016, pp. 0251–0255.
17. K.-L. Du, A.K.Y. Lai, K.K.M. Cheng, M.N.S. Swamy, Neural methods for antenna array signal processing: a review. Elsevier, Signal Processing **82**, 547–561 (2002)
18. S.A. Vorobyov, A.B. Gershman, Z.-Q. Luo, Robust adaptive beamforming using worst-case performance optimization: A solution to the signal mismatch problem. IEEE Trans. Signal Processing **51**, 313–324 (February 2003)
19. R. Michael and I. Psaromiligkos, "Robust minimum variance beamforming with dual response constraints", IEEE Transaction on Signal Processing, vol. 60, 2010.
20. Sunita V. Mane, Uttam L. Bombale, "Survey of Recent Beamforming Methods for Adaptive Antenna Array System," JETIR June 2019, Volume 6, Issue 6
21. Sunita V. Mane, Uttam L. Bombale, " Performance analysis of LMS, RLS, CMA algorithms for beamforming in Smart Antenna Systems," International Journal of Recent Technology and Engineering, ISSN: 2277–3878, Volume-8 Issue-4, November 2019.
22. S.V. Mane, U.L. Bombale, *Robust and Efficient DOA-Steered Adaptive MVDR-FROST Beamforming Model for Multi-Source Low SNR Environment* (Springer, IJWI, 2021)
23. D. Grenier, B. Elahian, A. Blanchard-Lapierre, Joint delay and direction of arrivals estimation in mobile communications. Springer, SIViP **10**, 45–54 (2016). https://doi.org/10.1007/s11760-014-0700-1

24. Konstantinos A. Gotsis, Katherine Siakavara and John N. Sahalos, "On the direction of arrival (DOA) estimation for a switched-beam antenna system using neural networks," *IEEE transactions on antennas and propagation*, vol. 57, no. 5, May 2009.
25. Ahmed H. El Zooghby, Christos G. Christodoulou and Michael Georgiopoulos, "Performance of Radial-Basis Function Networks for Direction of Arrival Estimation with Antenna Arrays," IEEE TRANSACTIONS ON ANTENNAS AND PROPAGATION, VOL. 45, NO. 11, NOVEMBER 1997.
26. M. Al-Bajari, J.M. Ahmed, M.B. Ayoob, "Performance Evaluation of an Artificial Neural Network-Based Adaptive Antenna Array System", Springer-Verlag, Berlin Heidelberg 2010, EvoApplications 2010. Part II, LNCS **6025**, 11–20 (2010)
27. SELÇUK ÇAYLAR, "A Novel Neural Network Based Approach For Direction Of Arrival Estimation," A Thesis Submitted To The Graduate School Of Natural And Applied Sciences Of Middle East Technical University, September 2007.
28. Shakya, Subarna, "Performance analysis of wind turbine monitoring mechanism using
29. integrated classification and optimization techniques," Journal of Artificial Intelligence 2, no. 01 (2020): 31–41.
30. Smys, S., Abul Basar, and Haoxiang Wang, "Artificial Neural Network Based Power
31. Management for Smart Street Lighting Systems," Journal of Artificial Intelligence 2, no.01 (2020): 42–52.
32. A. Bashar, Artificial Intelligence Based LTE MIMO Antenna for 5th Generation Mobile Networks. J. Artif. Intell. **2**(03), 155–162 (2020)
33. S. Manoharan, Supervised Learning for Microclimatic parameter Estimation in a Greenhouse environment for productive Agronomics. J. Artif. Intell. **2**(03), 170–176 (2020)
34. Kamel, Khaled, S. Smys, and Abul Bashar. "Tenancy Status Identification of Parking Slots Using Mobile Net Binary Classifier." Journal of Artificial Intelligence 2, no. 03 (2020): 146–154.

FPGA Implementation of Nested Binary Phase Codes Using DDS Approach for Radar Pulse Compression

Anitha Bhavani Chekka and Naga Jyothi Aggala

Abstract Binary Phase codes such as barker codes are widely used in radar pulse compression processing. Most of the radar signal processing applications use pulse compression codes to improve the range resolution. This can be achieved by phase codes such as barker codes and nested barker codes with reduced sidelobes. A field programmable gate array (FPGA) with low cost hardware architecture can be used effectively to implement these pulse compression codes. This paper proposes a novel approach to implement binary phase or barker codes and nested barker codes using a direct digital synthesizer and system generator which is a token of MATLAB Simulink Xilinx block set. The generated binary codes further tested and implemented on FPGA ZYNQ board (ZC7045-2ffg900) extensively used in radar signal processing.

Keywords Barker codes · Nested barker codes · Direct digital synthesizer · Nested barker code · Hardware co-simulation

1 Introduction

The range resolution of radar can be improved by a process called pulse compression [1] which is also known to be as intrapulse modulation. The name was given because the pulse which is to be transmitted will undergo time-dependent modulation. It also combines the dynamic benefit of a pulse with a long duration and also pulses with a short duration. The pulse with a shorter duration plays an important role to improve the range resolution but shows a limitation on transmitted power, which in turn deteriorates the signal-to-noise ratio value. Hence, there is an urgent requirement that is needed to improve transmitted power upon increasing the duration of the

A. B. Chekka (✉)
Department of Electronics and Communication Engineering, Vignan's Institute of Engineering for Women, JNTUK, Visakhapatnam, AP, India

N. J. Aggala
Department of Electronics and Communication Engineering, Vignan's Institute of Information Technology, Visakhapatnam, AP, India

G. Ranganathan et al. (eds.), *Soft Computing for Security Applications*,
Advances in Intelligent Systems and Computing 1397,
https://doi.org/10.1007/978-981-16-5301-8_21

transmitted pulse. The pulse compression allows us to produce better range resolution using long pulse as well as high transmitted power.

The pulse compression process can be done either in the analog domain or in the digital domain. Analog pulse compression technique [2, 3] is carried out by a correlation process which is mainly used for narrowband and some medium band of frequency applications. The process of digital pulse compression technique is performed using frequency, binary phase, and polyphase coding techniques, etc. The transmitted pulse is modulated using a carrier that can be varied either in frequency or phase.

The echo signal which is at the receiver is passed through a matched filter to improve max peak power with a narrow pulse. The modulation of input pulse allows the radar system to simultaneously achieve long pulse energy and with lesser pulse duration. As a result, the pulse modulation which is done in phase and frequency modulation provides the improved time-bandwidth product and increases the signal-to-noise ratio or with enhanced compression gain of the matched filter.

Recent advancements in the field of digital domain most commonly used waveform is a barker and nested barker codes in radar signal processing. Binary phased coding in analog or digital form has been widely using in radar applications for better range resolution. The binary phase coded signal produces mainly two-phase coded signal consists of In phase with a sign '−' produces a phase shift of 0^0 and out of phase signal with a sign ' + ' produces 180^0 phase shift with reference to an input signal. The autocorrelation value of N length code with a peak value of N whereas all sidelobes will have a value equal to 1. The longest possible binary barker codes of a length 13 with peak sidelobe ratio is -22.3 dB.

As long as the length of the barker code is increased range resolution can be improved which is possible by using nested barker codes. In most of the radar applications, the least PSLR to be obtained are −30 dB is possible only by raising the length of the barker code. Nested barker can codes can be framed by considering the Kronecker product between two barker codes. The length of nested barker code can be increased to a maximum value up to 65.further these nested codes can be compared based on autocorrelation, cross-correlation, bit error rate, and merit factor. A novel approach has been proposed in this paper generation and implementation of a 13-bit Barker code and 5X13 nested barker code using FPGA.

Generally, due to the cheap availability of DSP processors, they are widely used for many radar applications. The DSP processors allow us flexibility for programming purposes whereas architectural design is not flexible. The customized applications required by radar can be performed using FPGA which is flexible in programming as well as implementation. Peter J. Kajenski [2] shown that the best range resolution will be provided by nested barker codes as well as suggested the best technique reduce the sidelobes that occurred by generated code. Ahmed Sobhy [4] has proposed a novel technique to cancel the sidelobes observed in nested codes. Naga Jyothi [5] has proposed a VLSI architecture design of barker codes using FPGA also observed the merit factor values for generated codes. In continuation of this paper, we propose a novel architecture to generate and implement barker as well as nested barker codes using FPGA.

The field programmable gate array (FPGA) [6] is an integrated circuit that enables an integration for various applications and also eliminates the complexity of design for radar signal processing applications. FPGA can also be widely used for the implementation of various digital circuits to perform various mathematical operations. The feasibility of programmability in FPGA allows modifying complete design within the same device. FPGA's are also very helpful to design various digital integrated circuits and digital frequency synthesizers in various radar applications.

In the proposed work the FPGA digital integrated device is programmed using two different software tools MATLAB and Xilinx Vivado. The design functionality is tested and the programming is done in the MATLAB Simulink environment. The linkage between these two software packages is done by using a token called system generator [7]. This helps to convert the Simulink MATH code to the VHDL programming language recognized by the IDE software. The IDE is an environment that can be used for the implementation of the FPGA. The design and implementation of a 13-bit barker and Kronecker product of 5×13 barker code are done using a multi-DDS approach further tested by the FPGA kit.

2 Binary Phase Coded Waveforms

In the process of pulse compression, a pulse with a long time interval T is divided into N sub-pulses each of width τ. To improve the range resolution the phase of each sub pulse is changed. The phase of each sub pulse is chosen to be either 0° or 180° which are represented using a sign representation of ' + ' and '−' respectively. The generated waveform is ranging with a duration T to both sides of maximum peak response, or central peak. In the obtained waveform the portions that are observed except the central main peak are called time sidelobes.

A barker code [8, 9] can be defined as a sequence of binary with the reduced aperiodic autocorrelation value. The pulse compression process for any radar system allows us to improve better range and resolution with reduced radar pulse and increased power. The maximum possible length of barker code is 13 bit which is the ratio of the autocorrelation of the main lobe level to the autocorrelated sidelobe level. The binary sequence with phase shift 0° or 180° and with an equal magnitude of sidelobes passing into a matched filter known as a barker code. A barker sequence of N values of $+1$ and -1 for $j = 1$ to N such that.

$$\mathrm{x(n)} = \sum_{i=1}^{N-1} \mathrm{a(i)a(i+1)} \quad \text{for all } j = 1 \text{ to } N \tag{1}$$

Table 1 represents the known barker codes and the sequence representations.

Table 1 Barker coded sequences

Code length	Sequence
2	[1 −1], [1 1]
3	[1 1 −1]
4	[1 −1 1 1], [1 −1 −1 −1]
5	[1 1 1 −1 1]
7	[1 1 1 −1 −1 1 −1]
11	[1 1 1 −1 −1 −1 1 −1 −1 1 −1]
13	[1 1 1 1 1 −1 −1 1 1 −1 1 −1 1]

2.1 Nested Barker Codes

There are only two phases that can be produced by using the barker code with minimum achievable sidelobes. The longest length of barker code is 13 which is an odd code proven to be a major limitation. There is also a strong hypothesis is that the possible even coded lengths are 2 and 4 only. The 13-bit code length barker code's main lobe to peak sidelobe ratio can be achieved the value of 22.8 dB only which is lesser than the required practical value is of 30 dB. A mismatched filter can be used to improve this value and also nested codes provide a better solution to this problem that can be observed from literature. Nested binary codes can be acquired by using the Kronecker product [8] of two Barker codes whose response of the matched filter response is satisfied. If an N-bit Barker code is denoted by Bp, and another Bq, then a pq coded bit can be constructed as Bp ⊗ Bq. The Kronecker product is simply that the Bp barker code is repeated for q times, with each repetition multiplied by the corresponding element of the Bq code. For example, a 65-bit code can be obtained as the product B5 ⊗ B13. These codes can show the peak sidelobe level greater than 1.

3 Generation of a Barker and Nested Barker Codes Using FPGA

Radar mostly acting as a target detection device that uses electromagnetic waves to determine the direction, range, altitude, or speed of targets [8]. Digitally up-converted frequency is essential for a digital radar transmitter, which can be achieved by using either frequency and phase modulation.

A Binary phase coded waveform that can be generated is baker code whose sidelobe ratio can be reduced using nesting the codes. Generally, the binary sequence with N bits, a_k for $k = 0,1,2....N-1$, i.e., N values will be of 0's and 1's and the ACF of such binary sequence can be calculated is given in Eq. (1). If p ones and zeros are present in a given sequence then the duty cycle is given as equations should be centered and should be numbered with the number on the right-hand side.

$$C = \frac{p-1}{N-1} \tag{2}$$

3.1 *Direct Digital Synthesizers (DDS)*

Direct digital synthesizers (DDS) [10] are important elements in various communication systems. These elements can be used for implementing various modulation techniques in the digital domain, such as PSK (Phase shift keying), FSK (Frequency shift keying), and MSK (minimum shift keying) and can also be used to make demodulators, digital up and down converters. Generally, a DDS can be used to generate a sinusoid signal digitally as a complex or real value by a lookup table.

The Direct digital synthesizer compiler block consists of the sample values of a sine wave by a lookup table. A digital integrator is utilized to generate a suitable phase value which is matched by the sample value that is stored in the lookup table to the desired resultant waveform. A user-friendly interface is needed to accept system-level parameters such as the desired frequency content and to eliminate the noise in the resultant waveforms. The architecture of a single DDS block is presented in Fig. 1.

Direct digital synthesizers use an addressing mechanism consists of a suitable lookup table to represent arbitrarily generated frequency samples of the sine wave. These samples are taken from a lookup table by using a quantization process which includes both Phase and amplitude. The lookup table samples are phase slope values that are generated using integrator as shown in Fig. 1.

A digital to analog converter (DAC) is used, to obtain the required analog output, followed by a low-pass filter to obtain the desired frequency waveform in the analog

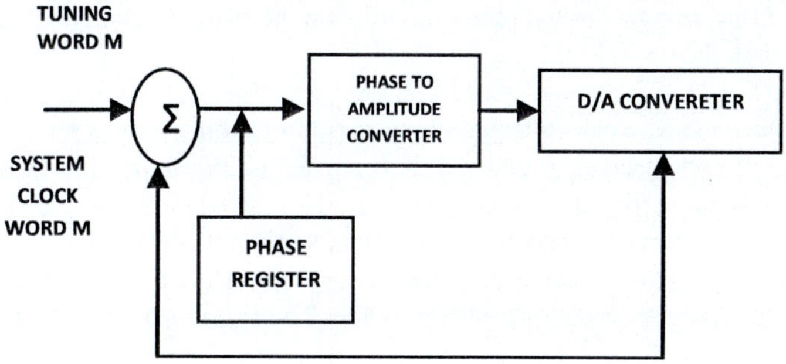

Fig. 1 The architecture of direct digital synthesizer

domain. The lookup table commonly consists of the amplitude levels of the generated signal of cosine or sinusoid.

The obtained waveform from DDS whose frequency is measured using the formula given below

$$f_{out} = \frac{f_{clk} \Delta\theta}{2^B} \quad (3)$$

where f_{clk} represents system clock frequency, $\Delta\Theta$ represents phase increment value and B represents no of bits in the phase accumulator.

3.2 The Proposed Method to Generate Barker Codes Using DDS

The objective of the proposed method is to generate binary phase coded waveform such as barker and nested barker codes to improve sidelobe ratio can be done in Multi-DDS [11] implementation in parallel processing. The FPGA is programmed using two distinctive software packages: MATLAB and Xilinx Vivado Design Suite [7]. MATLAB/Simulink is the environment where the design functionality is verified and programming is performed. IDE is one where the program will be arranged to implement on the FPGA. The main linkage between these two packages is done by using System Generator. It is a token used in MATLAB, which converts the Simulink math code into a VHDL code that is recognized by the IDE software.

3.3 Generation and Implementation of 13-Bit Barker Code Using FPGA

A 13-bit barker code consists of the binary sequence [1, 1, 1, 1, 1, -1, -1, 1, 1, -1, 1, -1, 1] where 1 indicates to generate a waveform with 0^0 phase shift whereas -1 indicates 180° phase shift to undergo the phase modulation the logic diagram to obtain 13-bit barker code is shown in Fig. 2.

There are 13 DDS compilers with 1 MHz frequency that are considered to generate 13-bit barker code whereas a phase shifter is used for the binary sequence positions of 6, 7, 10,12 the code needs to produce a change in a phase shift of 180°. The sampling interval of 0.01 samples per second is considered for the DDS compiler block design. By using a multiplexer, upon activation of selection inputs the individual waveform with corresponding phases can be transmitted for a certain interval so that the resultant signal can be observed as a 13-bit barker code. The simulated results are shown in Fig. 3.

Further, the design is implemented on the FPGA -ZYNQ board(ZC7045-2ffg900), which generates hardware co-simulation JTAG block shown in Fig. 4 and is giving

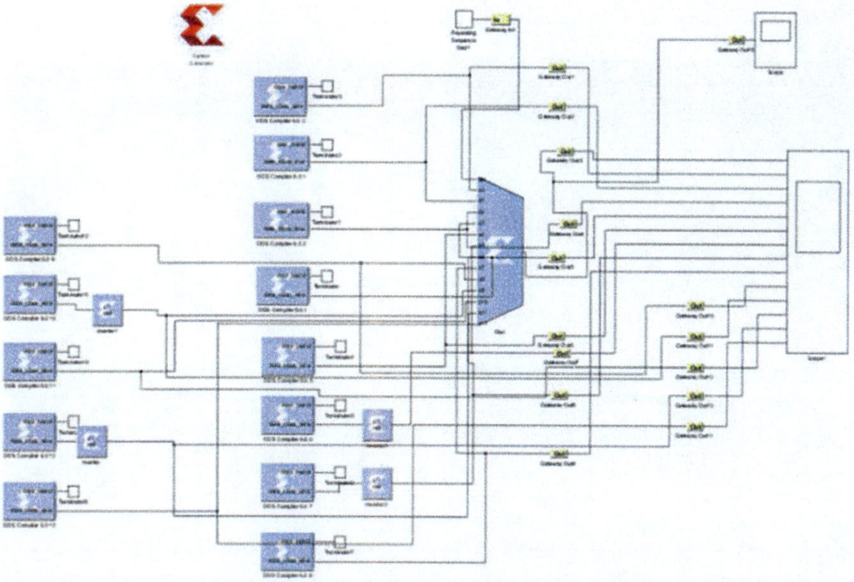

Fig. 2 Logic diagram for 13 bit barker code generation

similar results as the DDS compiler as shown in Fig. 5. The generated block allows us to debug the proposed design and can be processed for real-time applications.

3.4 Generation and Implementation of Nested Barker Code Using FPGA

A nested binary code that is used to improve range resolution in turn is also used to get better results in terms of PSLR compared with barker codes. By considering the Kronecker product of two barker codes nested barker codes can be generated. In this proposed work generation and implementation of B5 ⊗ B13 nested code is done to generated 65 binary codes using a multi-DDS approach.

A 5-bit barker code consists of the binary sequence is given by [1 1 1 -1 1] where the fourth bit of binary sequence will undergo with a phase shift of 180^0 and all the remaining binary sequences undergoes 0^0 phase shift. The initial binary sequence of 5 bit Barker code undergoes nesting 13 bit Barker code sequence whose representation is given by [1, 1, 1, 1, 1, -1, -1, 1, 1, -1, 1, -1, 1].Similarly, the remaining coding sequences also undergo nesting of 13-bit barker code the proposed model is shown in Fig. 6.

The proposed design is implemented on the FPGA -ZYNQ board(ZC7045-2ffg900), which is used to generate the JTAG block shown in Fig. 6 and is giving similar results as the DDS compiler as shown in Fig. 7. The generated block allows

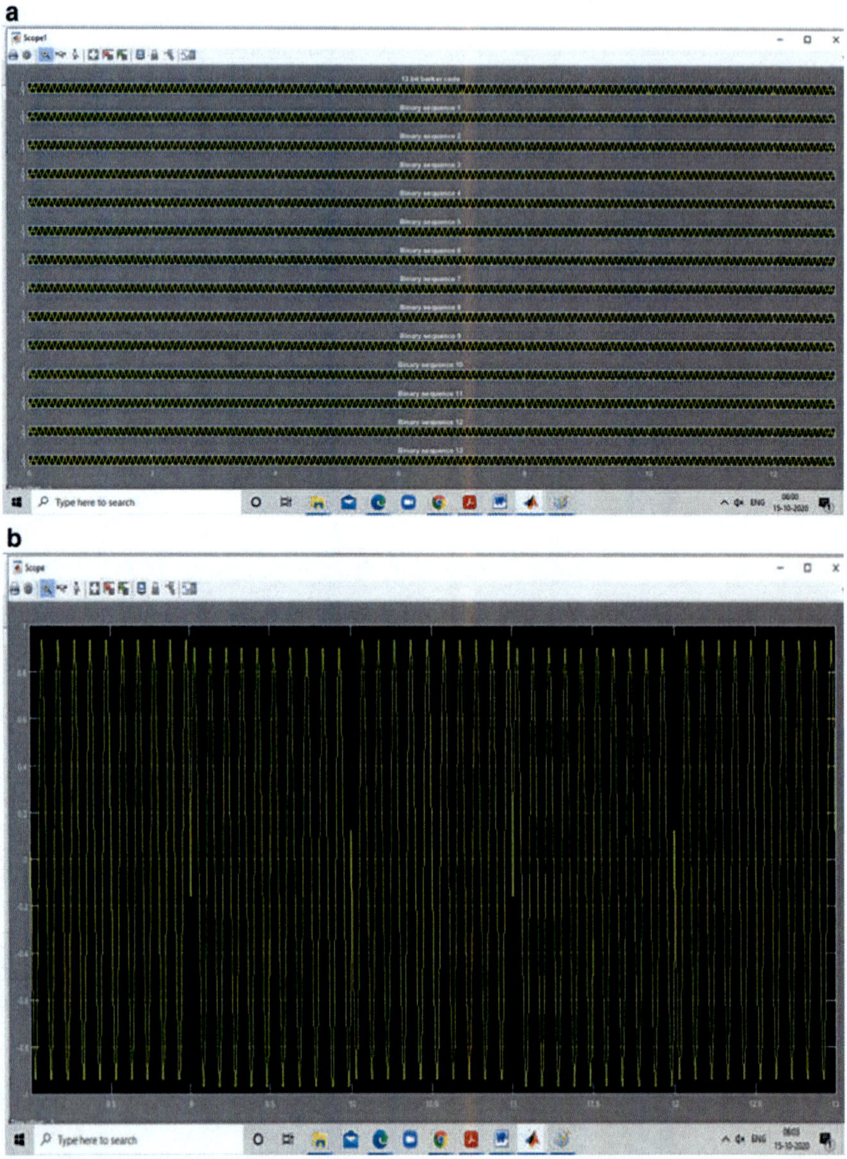

Fig. 3 **a** and **b** Simulated result for 13-bit barker code using multi-DDS approach

us to debug the proposed model for real-time applications. The RTL schematic of the proposed design is shown in Fig. 8. Table 3 gives the utilization report for the proposed architecture and static as well as dynamic power dissipation details are shown in Fig. 9 which gives the better results compared with traditional approach (Table 3).

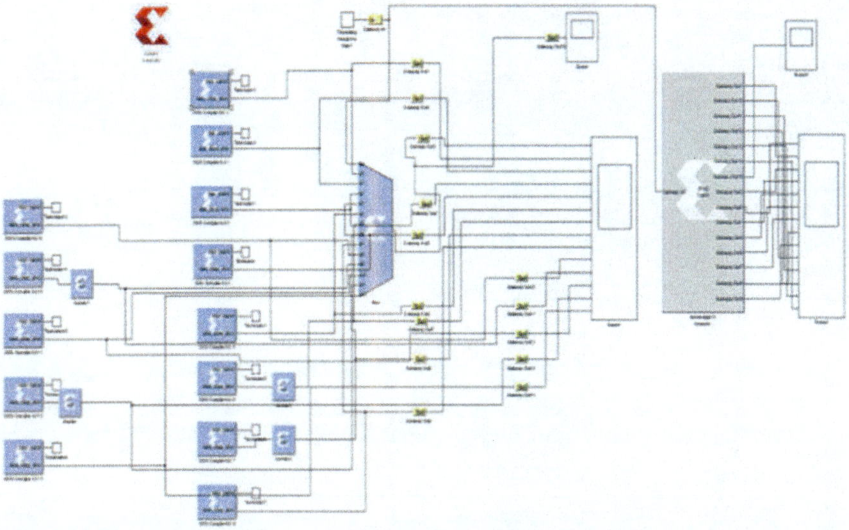

Fig. 4 Logic diagram for 13-bit barker code with JTAG block

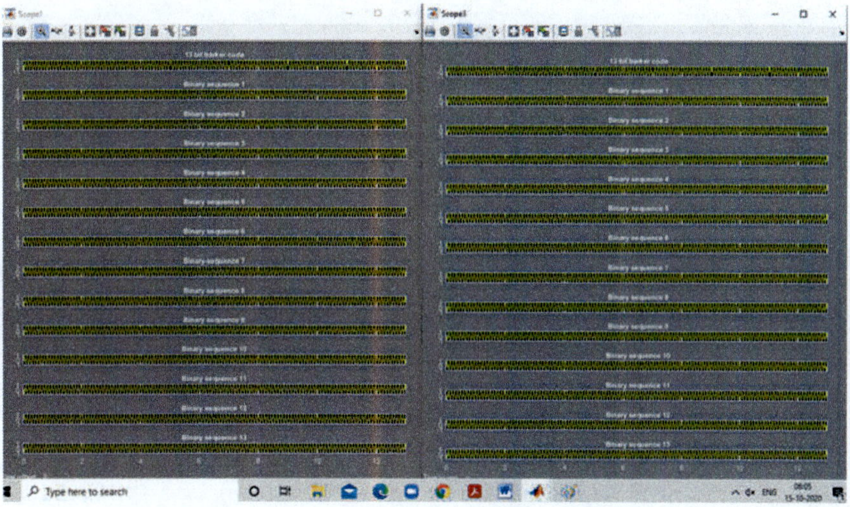

Fig. 5 Simulated results **a** simulated result from DDS **b** simulated result from hardware co-simulation JTAG block

4 Conclusion

In this paper, the generation of 13-bit barker code and nested code of B5 ⊗ B13 is carried out by a multi-DDS approach using MATLAB with the Xilinx block set. The

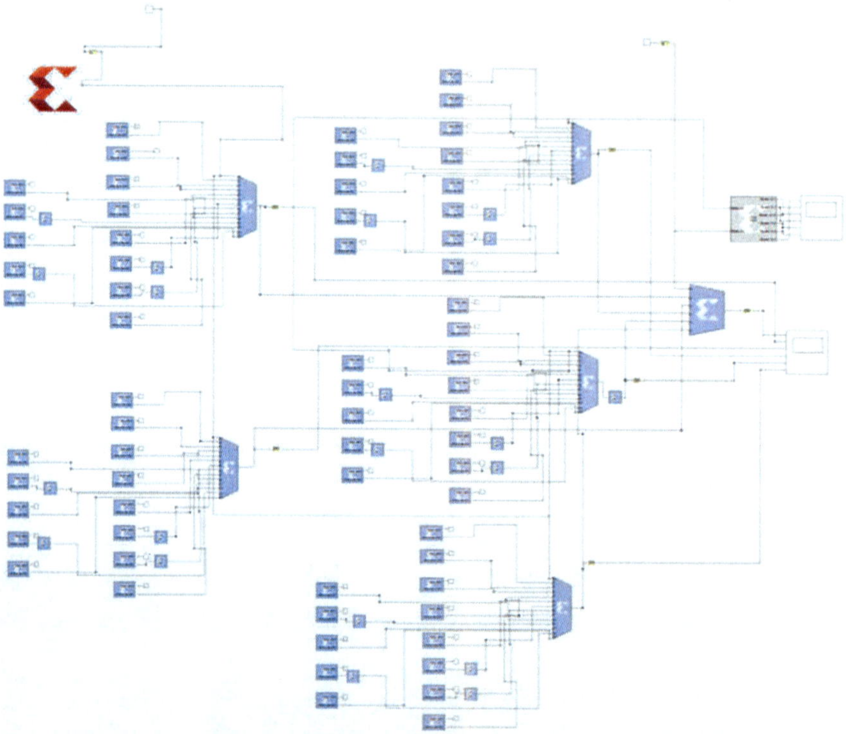

Fig. 6 Logic diagram for B5 ⊗ B13 Nested barker code with JTAG block

performance of the proposed design is verified on the FPGA -ZYNQ board (ZC7045-2ffg900) by generating hardware co-simulation JTAG block. The implemented hardware co simulated results are compared with the Simulink block results. The JTAG standards allow us to debug any design for real-time radar signal processing applications. In this proposed method the implementation of a barker and nested barker code using a multi-DDS approach allows lower complexity, flexibility, and better programmability. The design architecture can be further used to model an MIMO system for radar applications.

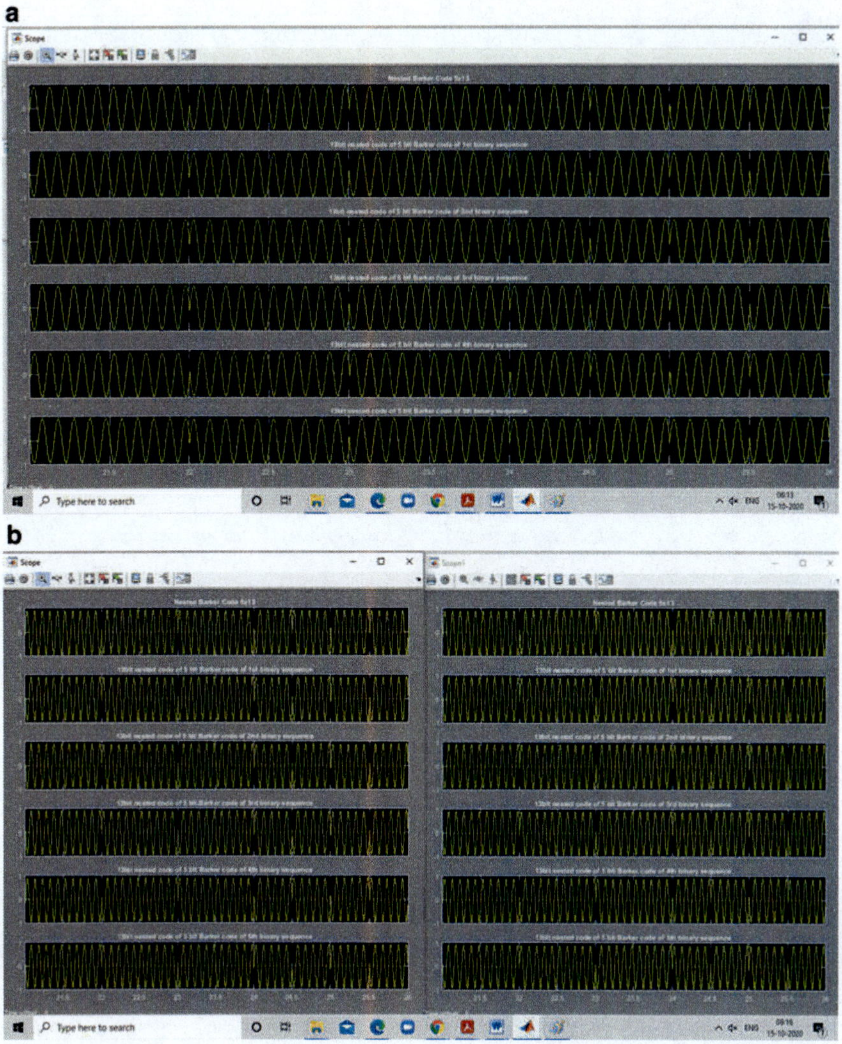

Fig. 7 **a** Simulation result of proposed design using a multi-DDS approach. **b** Comparison results of B5 ⊗ B13 nested barker code using DDS and hardware Co-sim block

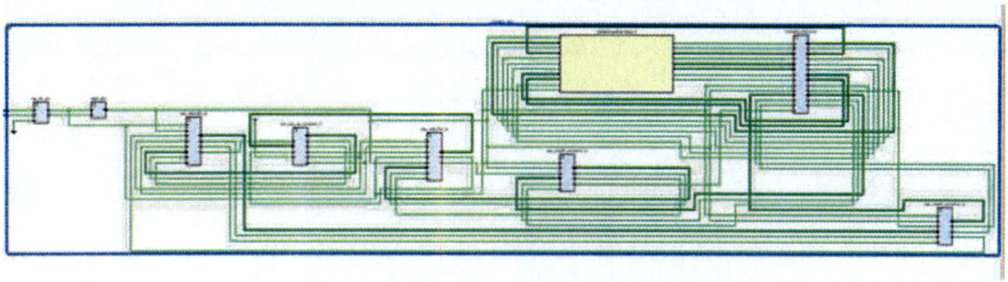

Fig. 8 RTL schematic of proposed B5 ⊗ B13 nested barker code

Table 2 Nested barker coded sequences

S. No	Nested code	Length of sequence
1	B7	7
2	B11	11
3	B13	13
4	B2 ⊗ B7	14
5	B4 ⊗ B4	16
6	B4 ⊗ B5	20
7	B4 ⊗ B7	28
8	B7 ⊗ B7	49
9	B5 ⊗ B11	55
10	B5 ⊗ B13	65

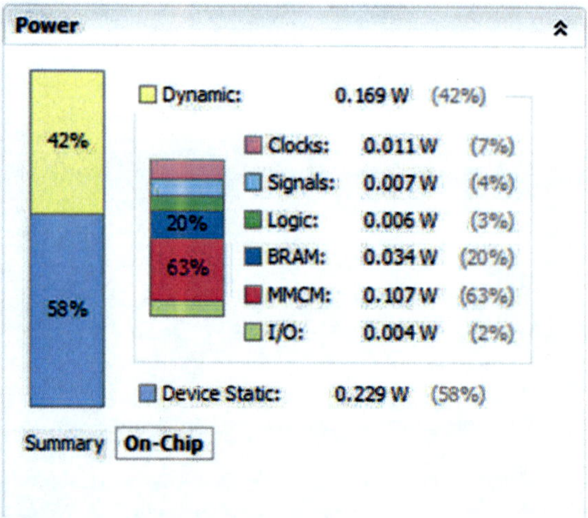

Fig. 9 Static and dynamic power dissipation details of the proposed architecture

Table 3 Utilization report for proposed architecture

Resource	Utilization	Available	Utilization%
LUT	1646	218,600	0.75
LUTRAM	1	70,400	0.01
FF	2876	437,200	0.66
BRAM	34.50	545	6.33
IO	2	362	0.55
BUFG	4	32	12.50
MMCM	1	8	12.50

References

1. E.C. Farnett, G.H. Stevens, M. Skolnik, Pulse compression radar. Radar Handbook **2**, 10–11 (1990)
2. P.J. Kajenski, Design of low-sidelobe phase-coded waveforms. IEEE Trans. Aerosp. Electron. Syst. **55**(6), 2891–2898 (2019)
3. M.C.A. Bhavani et al. "*Analog Pulse Compression Technique with Improved SNR and Reduced Sidelobes*"
4. H.E. Abd El-Rahman et al. "A Novel Sidelobe Cancellation Method for Binary Barker Code Pulse Compression." *International Conference on Aerospace Sciences and Aviation Technology*. Vol 17. No. Aerospace Sciences and Aviation Technology, ASAT-17–April 11–13, 2017. The Military Technical College, 2017
5. A.N. Jyothi, K.R. Rajeswari, Generation and implementation of barker and nested binary codes. J. Electric. Electron. Eng. **8**(2), 33–41 (2013)
6. R. Woods et al., *FPGA-Based Implementation of Signal Processing Systems*. Wiley (2008)
7. g897-vivado-sysgen-user guide Online
8. M.L. Skolnik, "Radar Handbook 3rd ed. McGraw-Hill", New York (2008)
9. M.C.A. Bhavani et al., "Design of less detectable RADAR waveforms using Barkar codes and polyphase codes."
10. Research and Implementation on Multi-DDS Technology in High-Performance Digital Up-conversion
11. L. Zhang, "System Generator Model-Based FPGA Design Optimization and Hardware Co-Simulation for Lorenz Chaotic Generator." 2017 2nd Asia-Pacific Conference on Intelligent Robot Systems (ACIRS). IEEE, 2017

Cloud Computing-Based Li-Ion Battery-BMS Design for Constant DC Load Applications

S. Madhankumar, S. Dharshini, N. Rohit Vignesh, P. Amrutha, and J. Dhanaselvam

Abstract Lithium-ion batteries find a wide range of applications in UPS, mobiles, electric vehicle, and in commonly used electronic items. Lithium-ion batteries are preferred because of their advantages of having low self-discharge, low maintenance, and also high energy density. Even though lithium-ion batteries have many advantages, it gets heated up when it is continuously used. So, monitoring of various factors such as charging/discharging cycle, temperature, input/output voltage and current using a battery management system is the main objective of this proposed system. Measurement of these parameters helps to identify the health of the battery and predict the life of the battery. When the temperature is increasing beyond normal, then the cooling system is activated so that the battery remains at normal temperature and works efficiently. The measured values of the parameters will be displayed in a liquid crystal display, and it will also be transferred to the cloud by ESP module through Arduino programming. The research work is experimentally verified for a 12 V, 2000mAh lithium-ion battery with 2A constant DC load conditions.

Keywords Battery management system · Lithium-ion battery · Life cycle prediction · Cloud computing · Constant DC load · State of charge · Temperature effect · State of cooling system

1 Introduction

The battery management system (BMS) is a simple electronic system that can be used for continuous parameter monitoring under normal and abnormal conditions. Previously, there was no battery management for lead-acid batteries. Because of this, various issues are encountered like performance degradation, maintenance problems, less accuracy, etc. [1, 2]. Later when lithium-ion batteries came into the picture, the development in battery management also emerged. A few managements such as measurement of voltage, current, internal resistance, charging, and discharging

---

S. Madhankumar · S. Dharshini (✉) · N. R. Vignesh · P. Amrutha · J. Dhanaselvam
Sri Krishna College of Technology, Coimbatore, Tamilnadu, India

G. Ranganathan et al. (eds.), *Soft Computing for Security Applications*,
Advances in Intelligent Systems and Computing 1397,
https://doi.org/10.1007/978-981-16-5301-8_22

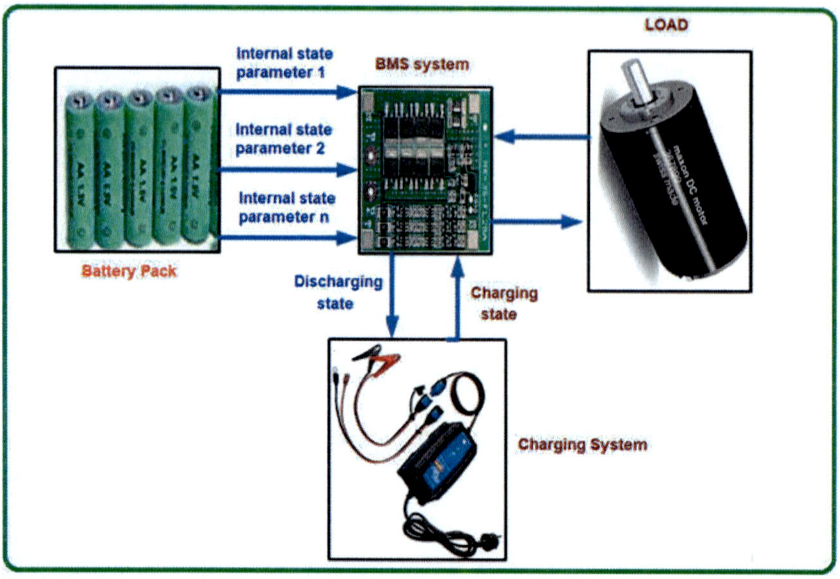

Fig. 1 Block diagram of battery management system

control were effectively done [2]. The continuous improvement and the recent technologies made BMS in a more effective way such as using advanced algorithms, data-driven methods, and implementation of battery technology to wider area applications [3]. So, in general, a BMS is used to monitor and regulate the battery's internal parameters such as charging/ discharging, voltage, temperature, internal resistance, etc. The generalized block diagrams of BMS are shown in Fig. 1. Most of the rechargeable batteries use different types of battery management systems in concern with various applications. Li-ion batteries, as one of the most advanced rechargeable batteries, have gotten a lot of coverage in recent decades [4].

Lithium-ion batteries provide high power, large capacity, quick charging, slow discharging, and long life. In spite of many advantages, safety is a very serious concern in lithium-ion batteries [5]. The production cost of a lithium-ion battery is high if it is compared to nickel-metal hydride, lead-acid batteries. When a lithium-ion battery is used for a specific application, it starts to degrade. The degradation starts from manufacturing to its secondary applications. It is very crucial to analyze battery performance in terms of battery degradation. The degradation is not only based on internal factors but also based on environmental factors, operating conditions, location, applications, etc. So, accurate measurement of various parameters is a crucial part of battery performance. In this proposed work, a BMS is designed for continuous monitoring of various factors so that the remaining lifetime of the battery can be estimated [6, 7].

This paper is organized as follows. In Sect. 2, various research and its findings have been given as a literature survey. The experimental setup and its components have been described in Sect. 3. Section 4 comprises results and interpretation for various parameters such as input and output state of charge, temperature changes, state of the cooling system when the temperature exceeds its limit. In Section 5, the findings of this research work and also the research gap of this work are summarized. The research work is experimentally verified for a 12 V, 2000mAh lithium-ion battery with 2A constant DC load conditions.

2 Literature Survey

A lot of research is going on to predict battery life, thermal behavior of lithium-ion battery, performance degradation, etc. In this section, some of the research findings have been summarized. Dongping Xu et al. described two different methods used to measure and monitor various cell parameters such as cell voltage, temperature, and current integration. This measurement is done by (i) software integration method and (ii) single-phase multifunction metering method [8]. The performance of the battery is also based on electrode materials so the choice of the materials plays an important role in pottery performance. Brandl M et al. describe the best choice for lithium-ion cell materials and also described various algorithms and battery utilization. A black-box model is based on voltage-current characteristics and used to extract the parameters which is used for the electro-chemical modeling [2]. Deng et al. discussed beyond Li-ion batteries and also clarified the need for new generation lithium-ion batteries with further advancements [9]. Nan et al. [10] explained the current trends in electric vehicle battery management systems and also some future trends. In addition to that, they discussed mostly on inventions in battery technology and its management system. The lifetime assessment of the lithium-ion battery is clearly stated in this research. Gaines et al. are discussing the life cycle performance by considering different battery materials. And also, this paper describes lithium-ion battery materials manufacturing and production. In EV industries, recycling of battery materials reduces the production energy by 50%, and it will be further increased if the recycling cycle is done for the active materials. Recently, cycle analysis is performed for the volt (a PHEV40), and the leaf (100-miall-electric), and it was known that recycling will have an impact on energy and environmental sectors as well as material supply [11]. MatsZackrisson et al. reviewed about impact of life cycle assessment in optimizing the designing of LiBs. For electric vehicles, the analysis is done for two different lithium-ion batteries (i.e., lithium-ion-phosphate) with two different solvents at the time of manufacturing. This study shows that water is preferred environmentally than N-methyl-2-pyrrolidone. Sensitivity calculation is very important to have better efficiency in terms of internal parameter efficiency and also battery weight [12].

A literature analysis is done by Sullivan et al. on cradle-to-gate life cycle inventory for various batteries including lithium-ion batteries. It is concluded that a large amount of material production data is required for recycling analysis and this influences directly on the environment [13]. Various research papers [14, 15], and [9]

Table 1 Properties of various battery chemistries [6]

Type of battery	PbA	NiCd	NiMH	LiB
Specific energy Wh/Kg	35–40	40–60	60–120	100–265
Energy density Wh/L	80–90	50–150	140–300	250–693
Power density	180	150	250–1000	1800
Open circuit voltage (V)	2.1	1.2	1.2	2.5–3.8
Efficiency (%)	50–90	70–90	66–92	80–90
Life cycle	500–1500	2000	180–2000	400–1200
Operating temperature range (°C)	35–45	−20–40	−20–45	5–45
Self-discharge rate/month	3–20%	10%	13.9–70%	0.35–2.5%
Cost	Low	Less	High	High
Thermal stability	Least stable	Least stable	Least stable	Most stable
Target applications	Automobile starting, lighting, and ignition, battery vehicles, electric scooters	Portable electronics and toys, aircraft starting batteries, electric vehicles, and standby power	Consumer electronics, electric vehicles	Electric vehicles, power tools, grid energy storage, cell phone, laptop

explained life cycle assessment for electric vehicle applications. Nowadays, a smart battery management system plays a vital role in analyzing battery performance. Data-driven methods along with advanced machine learning and deep learning algorithms make the predictions and estimation of various factors more efficient. The comparison of various battery chemistries on their properties is shown Table 1.

3 Experimental Setup

The source reduces the voltage from the 230 V alternating current electrical outlet to 12 V alternating current. Then the direct current-converter converts 12 V alternating current to 12 V direct current, and then direct current step down by regulator to 5 voltage direct current for MCU. A voltage divider measures the voltage value from the charging cycle, and a current sensor measures the amp from the charging cycle. The source that supplies power to the battery and the main control unit measures the following parameters during the charging cycle, and the data is sent to the MCU for

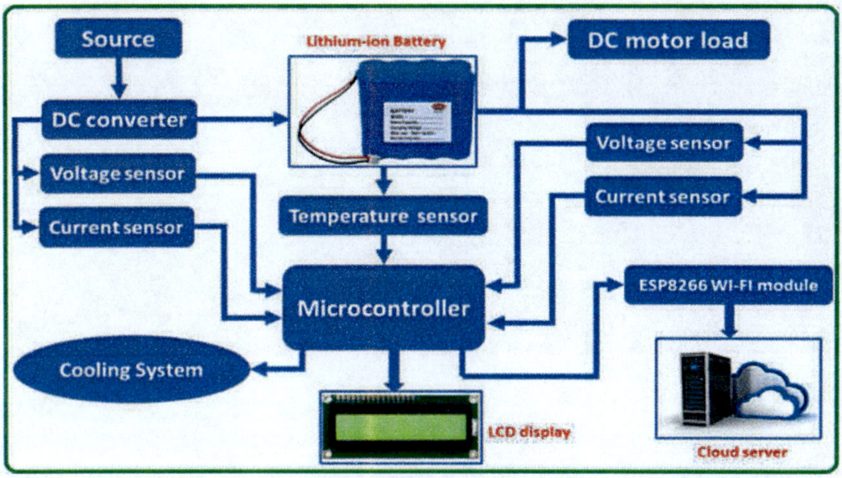

Fig. 2 Proposed BMS system

measuring and controlling, and the data is displayed on a liquid crystal display. At the same time, the temperature sensor measures the temperature of the battery and sensor data to operate the cooling system.

The proposed battery management system is performed under normal room temperature. And 36 °C temperature is fixed as the maximum limit. When the temperature exceeds the maximum limit, the cooling system will immediately be activated. So, the State of Cooling Fan (SCF) is 1 and 0 based on the temperature.

When the load operates from the battery during a discharge cycle, the current sensor measures the amp value of the load, and the voltage divider measures the voltage from the battery to the load and sends the data to the main control unit, and the calculated value is transferred to the cloud via the ESP module via Arduino programming [17, 18]. The application name is "thingSpeak.com" from which we can display the data of the battery management system. Figure 2 describes the block diagram of the proposed system, and Fig. 3 is the actual experimental setup of the proposed system.

4 Results and Discussion

The research work is experimentally verified for a 12 V, 2000mAh lithium-ion battery with 2A constant DC load conditions. The battery is charged and discharged under room temperature and constant load conditions. During charging and discharging, battery internal parameters are noted down, and also, the data is sent to the cloud server. Based on the data stored, it will be easy to predict the remaining useful life of the battery. And also, the data will be used for future references to analyze the battery

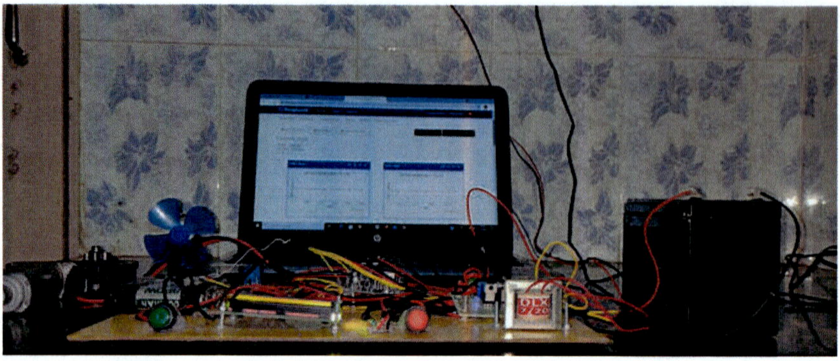

Fig. 3 Experimental setup

performance and degradation. The following section explains various observations from charging and discharging cycles.

4.1 Charging Cycle

In the charging cycle, the BMS monitors the value of voltage and current by using a voltage divider and current sensor. The measured data is transferred to the main control unit. The main control unit calculates the state of charge, state of health, and charge cycle by using input data. The output depends upon the input current, voltage, charging time, and transformer ampere rating [18] (Table 2)

Figure 4 is the analysis between state of charge and input voltage during charging cycle. The input voltage varies from 9.1 V to 12 V under different levels of state of charge from 0 to 100%. Since the experiment is done at a constant temperature, the effect of external parameters is less and the variation will be linear.

Figure 5 is the variation between the state of charge and input current during the charging cycle. It shows a nonlinear variation under this condition. The X-axis represents % SOC, and the Y-axis represents the input current. It is observed that during the charging cycle of the 2A lithium-ion battery, the input current varies from 1.7A to 1.9A with an accuracy of 95%.

Figure 6 is the graph between input state of charge and charging time for an entire charging duration. Since a 12 V, 2000mAh lithium-ion battery is considered for this proposed system, it takes 609 s to get charged completely and also at this time the SOC is at 100% (Fig. 7).

Temperature is an important factor for lithium-ion battery performance because it directly affects the performance. Both internal and environmental temperatures cause performance degradation. So, in this proposed work, temperature variation is observed for various input current, voltage for the fixed load. Even though the experiment is done under constant temperature conditions, the variation in environmental

Table 2 Battery charging parameter analysis

SOC%	IV	IC	T	SCF	Time
6	9.1	1.7	28.3	0	38
12	9.6	1.75	29.1	0	72
18	9.9	1.74	29.9	0	110
24	10.3	1.6	30.4	0	152
30	10.4	1.76	31.2	0	182
36	10.6	1.8	32.1	0	230
42	10.9	1.84	33.2	0	270
48	11.2	1.9	34.7	0	297
54	11.4	1.89	35.6	0	340
60	11.5	1.75	36.3	1	383
66	11.6	1.78	35.6	0	420
72	11.8	1.79	36.4	1	452
78	11.9	1.74	35.6	0	489
84	11.9	1.86	36.5	1	531
90	12	1.9	36.2	1	567
100	12	1.85	35.6	0	609

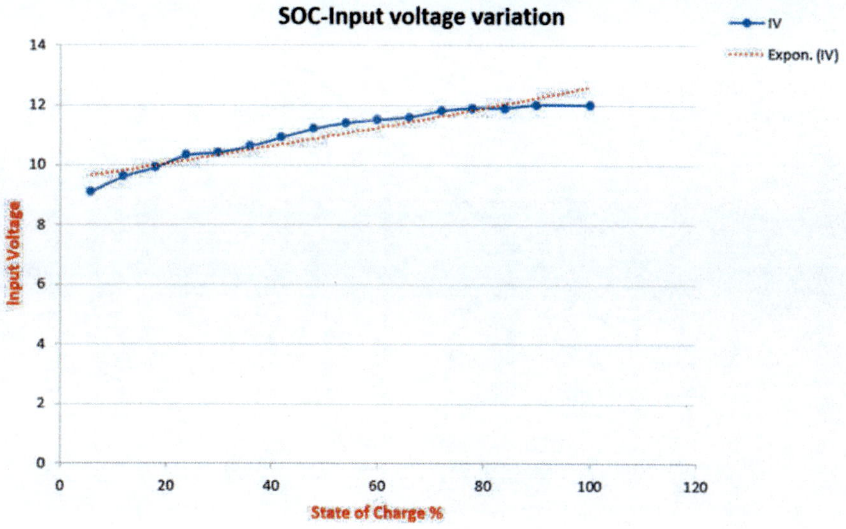

Fig. 4 Input SOC versus voltage

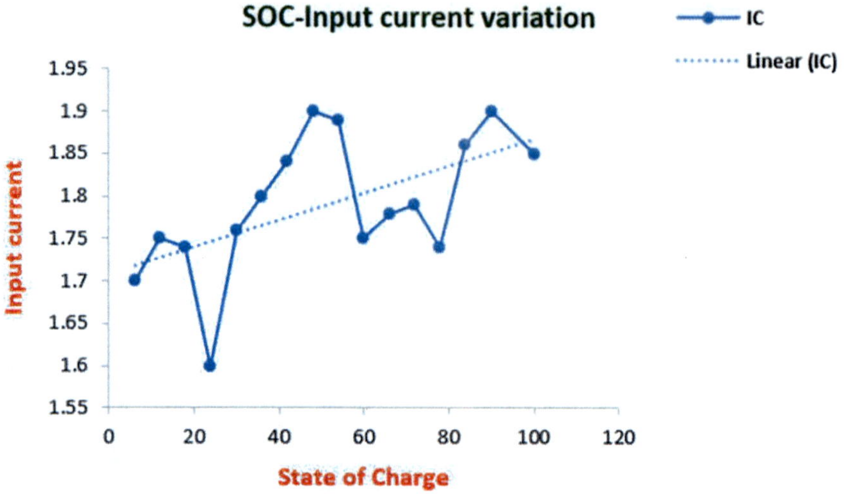

Fig. 5 Input SOC versus input current

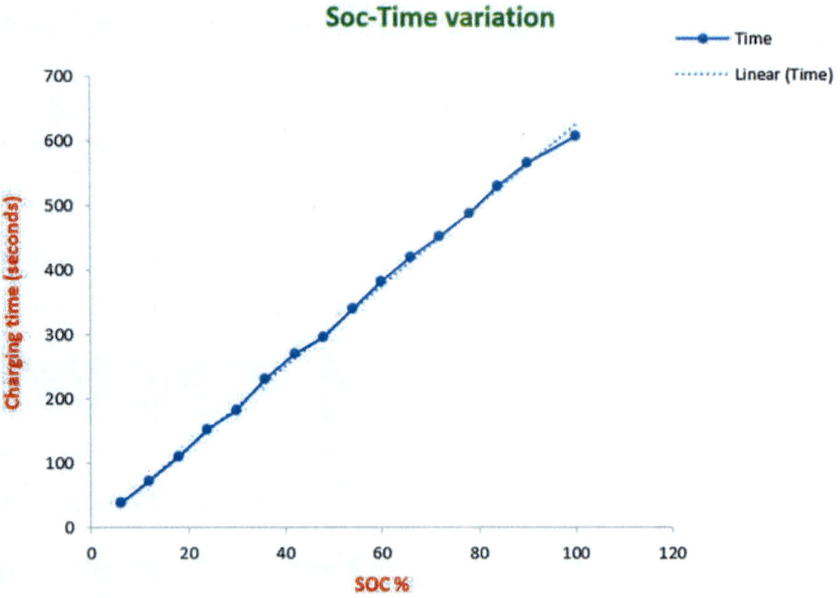

Fig. 6 Input state of charge versus time

and also internal parameter variations causes variation in the battery temperature. This change is expressed in the above graph.

In the proposed system, the temperature varies from 28.3 °C to 36.5 °C. The maximum temperature limit is set as 36 °C. So, whenever the temperature exceeds

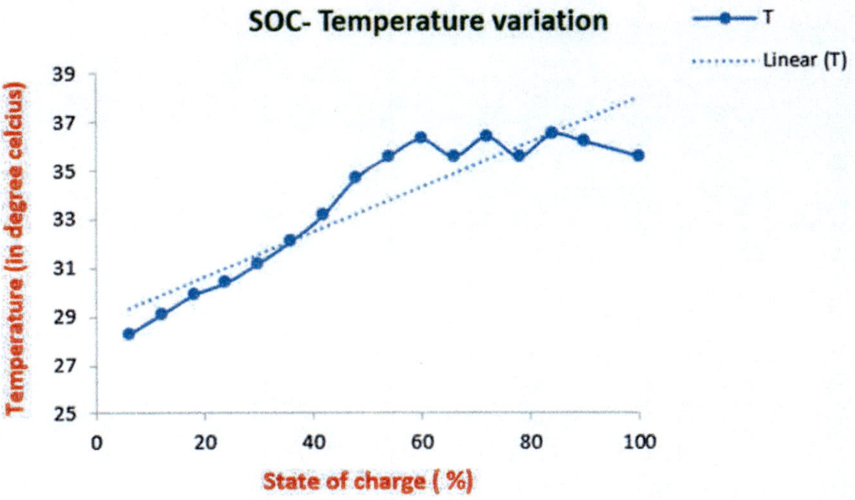

Fig. 7 State of charge versus temperature

36 °C, immediately the cooling system will be activated to reduce the temperature so that battery temperature will be maintained at a safe operating zone. Figure 8 is the graph between battery temperature and state of cooling system during charging cycle.

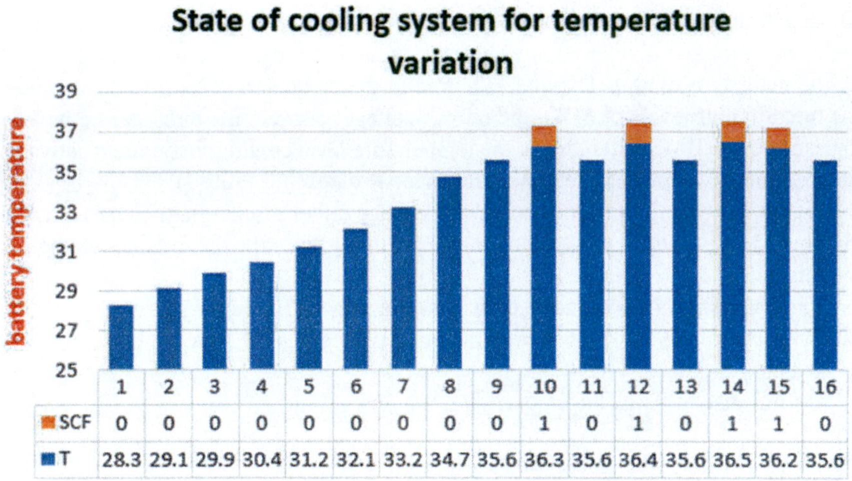

Fig. 8 Cooling system state under abnormal temperature

Table 3 Battery discharging parameter analysis

SOC %	OV	OC	T	CFS	Time
99	11.9	2.1	29	0	38
93	11.7	2.1	31	0	75
87	11.8	2	32.4	0	110
81	11.9	2.1	33.1	0	145
75	11.7	2	33.9	0	180
69	11.5	1.8	34.7	0	215
63	11.7	1.8	35.2	0	247
57	11.8	1.8	36.4	1	288
51	11.9	1.8	36.1	1	322
45	11.7	1.85	35.7	0	358
39	11.8	1.85	35.9	0	393
33	11.9	1.8	36.3	1	427
27	11.7	1.8	36.1	1	463
21	11.8	1.7	35.7	0	501
15	11.9	1.7	36.2	1	537
9	11.7	1.7	35.6	0	571

4.2 Discharging Cycle

During discharging cycle, the BMS monitors the value of output voltage and current by using a voltage divider and current sensor, and the measured data is transferred to the main control unit. The data is stored in the cloud server for future reference. At the different levels of SOC, the output voltage, current, the temperature will be measured. The BMS will monitor the temperature level continuously, and it activates the cooling system when the temperate exceeds the limit (Table 3).

Figure 9 shows the variation of output voltage under discharging. Due to internal parameter and load, a small voltage drop is there but still, the average voltage is maintained at 11.77 V (Fig. 10).

The graph shows the variation of discharging current for output state charge where SOC varies from 99 to 9%. The output current discharges 2.1 A to 1.7A non- linearly (Fig. 11).

During discharging, the output state of charge varies from 99 to 9%, and at different levels of SOC, there is variation in temperature. Minimum and maximum temperatures are 290C and 36.40C, respectively (Fig. 12).

From the above analysis, it is observed that when SOC varies from 99 to 9%, the time also varies linearly. The total time taken for full discharge is 571 s. The proposed system gives total time duration for one full cycle, i.e.,

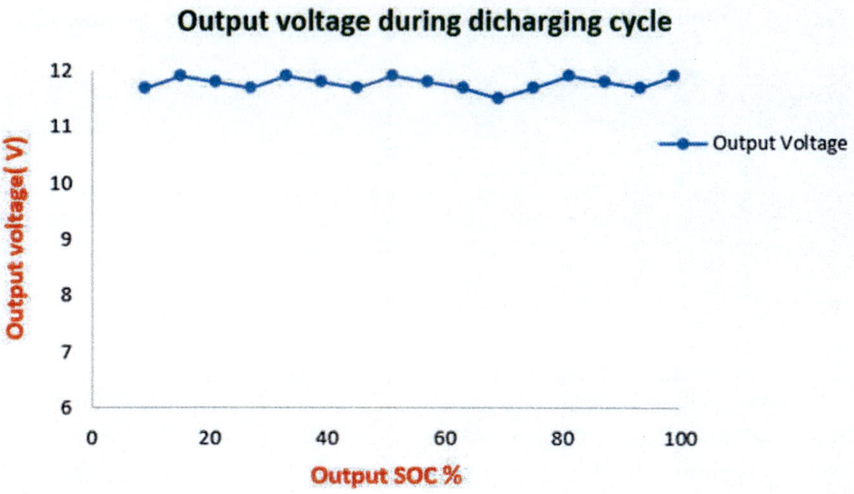

Fig. 9 Output state of charge versus voltage

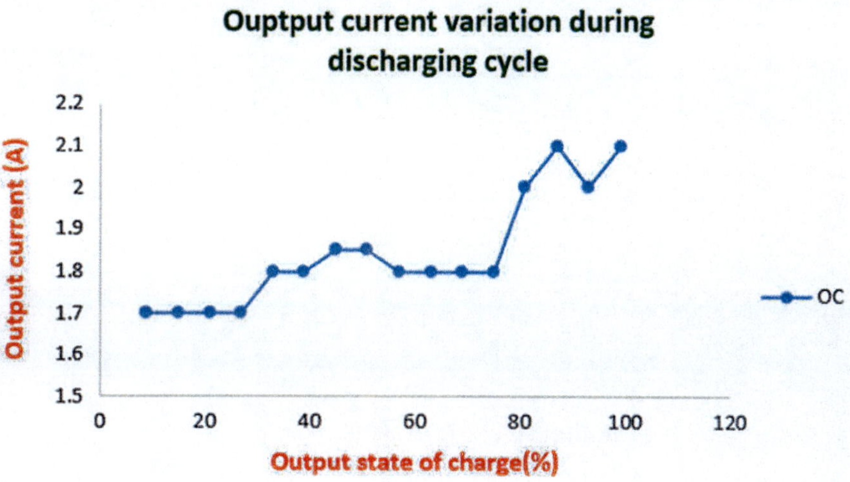

Fig. 10 Output SOC versus current

1full cycle = 1 charging cycle + 1 discharging cycle.

So, from the observations, the lithium-ion battery takes 609 s for charging and 571 s for discharging and the total time duration is 1180 s (19.8 min) (Fig. 13).

The graph shows the state of the cooling system under normal and abnormal temperature variations. During discharging of the battery, the internal parameter changes, and also, the external factors cause the temperature to increase. The maximum operating temperature is 360 °C. The graph gives a clear picture of the state of the cooling system when the battery temperature exceeds the limit.

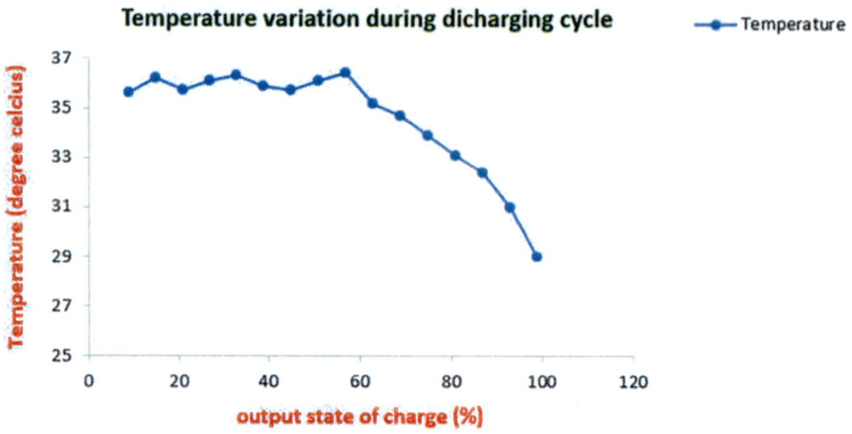

Fig. 11 Output SOC versus temp

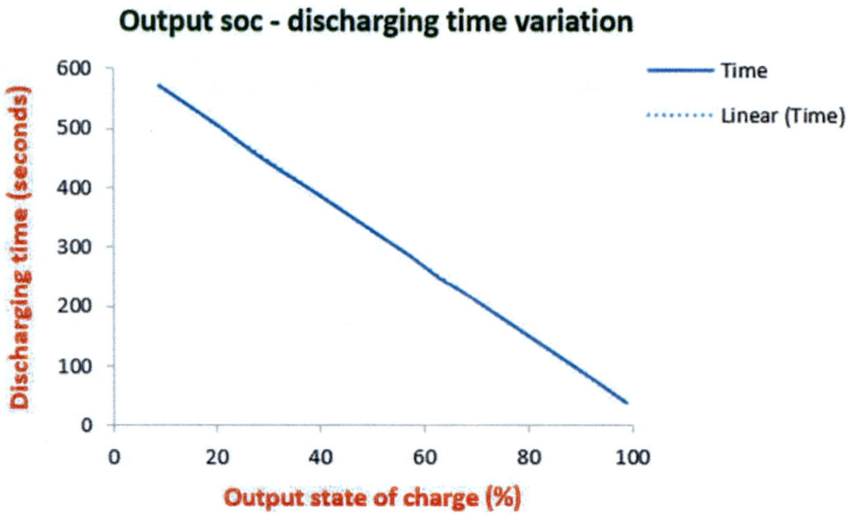

Fig. 12 Output SOC versus time

5 Conclusion

The BMS is a crucial part to give information about the past, present, and future state of the battery when LIBs are used for many of the primary and secondary applications. In the proposed system, it is experimentally verified for a 12 V, 2000mAh lithium-ion battery with 2A constant DC load conditions. During the charging cycle, it is observed that at a constant temperature, the effect of external parameters is less and the variation is linear. And, both internal and environmental temperatures

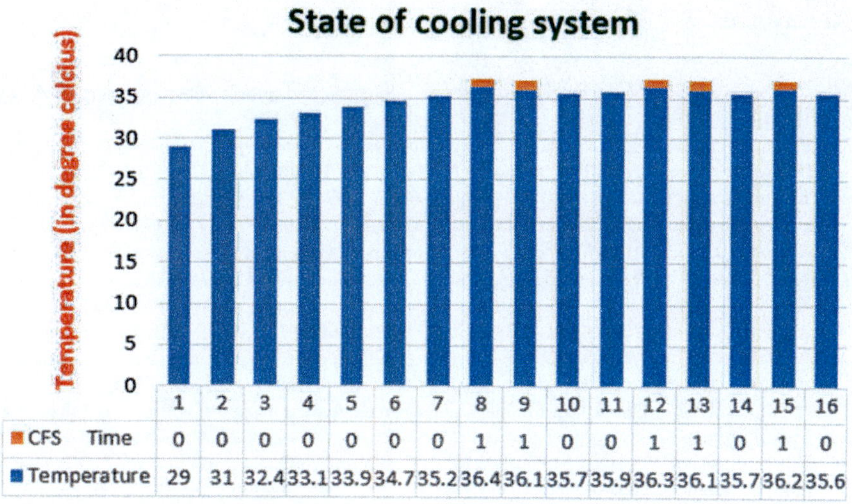

Fig. 13 Output temperature versus cooling fan state

cause performance degradation. So, in this proposed work, temperature variation is observed during charging and discharging for various current, voltage for the fixed load. Even though the experiment is done under constant temperature conditions, the variation in environmental and also internal parameter variations causes variation in the battery temperature. To analyze the battery life cycle, observation of charging and discharging time is very important. In this work, it is observed that when SOC varies from 99 to 9%, the time also varies linearly. The total time taken for full discharge is 571 s. The proposed system gives total time duration for one full cycle, i.e., 1 full cycle = 1 charging cycle + 1 discharging cycle. So, from the observations, the lithium-ion battery takes 609 s for charging and 571 s for discharging and the total time duration is 1180 s (19.8 min). With this analysis, the work can further be improved for large load, variable temperature conditions. According to the above analysis, the stored data in the cloud can be used for further prediction analysis. Cloud-based monitoring and analysis for battery management systems enable end-users to monitor and analyze the battery data as well as algorithms that offer information related to all the internal and external state of lithium-ion battery. Cloud-based battery management systems also allow the online identification of potential critical battery failures at various stages of its operation and the long-term tracking of cell aging.

References

1. Y. Xing et al., Battery management systems in electric and hybrid vehicles. Energies **4**(11), 1840–1857 (2011)
2. M. Brandl et al., "*Batteries and Battery Management Systems for Electric Vehicles*." 2012 Design, Automation and Test in Europe Conference & Exhibition (DATE). IEEE (2012)
3. T. Kim et al., "Cloud-based battery condition monitoring and fault diagnosis platform for large-scale lithium-ion battery energy storage systems." Energies **11**(1), 125 (2018)
4. R. Xiong, L. Li, J. Tian, Towards a smarter battery management system: a critical review on battery state of health monitoring methods. J. Power Sour. **405**, 18–29 (2018)
5. H. Rahimi-Eichi et al., "Battery management system: an overview of its application in the smart grid and electric vehicles." IEEE Ind. Electron. Mag. **7**(2), 4–16 (2013)
6. A.T. Elsayed, C.R. Lashway, O.A. Mohammed, Advanced battery management and diagnostic system for smart grid infrastructure. IEEE Trans. Smart Grid **7**(2), 897–905 (2015)
7. P. Liu, X. Zhang, The design of smart battery management systems. JCP **6**(11), 2484–2490 (2011)
8. T. F. Smith, M. S. Waterman, Identification of Common Molecular Subsequences. J. Xu, Dong-ping, L. Wang, J. Yang. "Research on Li-İon Battery Management System." 2010 International Conference on Electrical and Control Engineering. IEEE (2010)
9. Da. Deng, Li-ion batteries: basics, progress, and challenges. Energy Sci. Eng. **3**(5), 385–418 (2015)
10. J. R. Nan, F. C. Sun, J. Q. Wang., "Electric vehicle battery management system." J. Tsinghua Univ. (Sci. Technol.) **2**, 1831–1834 (2007)
11. L. Gaines et al., "Life-cycle analysis for lithium-ion battery production and recycling." Transportation Research Board 90th Annual Meeting, Washington, DC (2011)
12. M. Zackrisson, L. Avellán, J. Orlenius, Life cycle assessment of lithium-ion batteries for plug-in hybrid electric vehicles–critical issues. J. Clean. Prod. **18**(15), 1519–1529 (2010)
13. J. L. Sullivan, L. Gaines, "A review of battery life-cycle analysis: state of knowledge and critical needs" (2010)
14. R. Hawkins Troy et al., "Comparative environmental life cycle assessment of conventional and electric vehicles." J. İndus. Ecol. **17**(1), 53–64 (2013)
15. J. Gomez, R. Nelson, E.E. Kalu, M.H. Weatherspoon, J.P. Zheng, Equivalent circuit model parameters of a high-power Li-ion battery: thermal and state of charge effects. J. Power Sour. **196**(10), 4826–4831 (2011)
16. S. Saravanan, P. Pandiyan, T. Chinnadurai, T. Ramji, N. Prabaharan, R. Senthil Kumar, P. Lenin Pugalhanthi, Reconfigurable Battery Management System for Microgrid Application. Microgrid Technologies, (2021) pp. 145–176
17. B. Banuselvasaraswathy, R. Vimalathithan, T. Chinnadurai, *Analysis of BASNs Battery Performance at Different Temperature Conditions Using Artificial Neural Networks* (*ANN*). In Soft Computing for Problem Solving. (Springer, Singapore, 2020), pp. 329–340
18. P. Pandiyan, S. Saravanan, T. Chinnadurai, T. Ramji, N. Prabaharan, S. Umashankar, Mitigation techniques for removal of dust on solar photovoltaic system. Electric. Electron. Devices Circuits Mater.: Technol. Challenges Solutions pp 373–392 (2021)

Three-Level Hybrid Data Security

B. S. Spoorthi and S. Pushpa Mala

Abstract Due to the increase in information sharing over the Internet, there is a need to develop a hybrid technique that provides high-level security for the data. Data protection is a major concern in today's Internet era. The development of weaker algorithms could become hazardous for data security. Therefore, in this work, we proposed a hybrid method that provides three-level security for the data. The first level of security is provided by generating the ciphertext. The ciphertext is generated with the help of the Rivest, Shamir, Adleman (RSA) algorithm. In the second stage, we focus on the generation of quick response (QR) code, and the third level of security is provided by steganographic technique. The QR code is hidden inside the image using the least significant bit (LSB) method.

Keywords Cryptography · Steganography · RSA · Encryption · LSB · Message · QR code · Ciphertext · Stego image · Decryption

1 Introduction

To cope up with the usage of the Internet and sharing of information over the Internet, there is a need to develop a hybrid technique. People around the world are really worried about sharing their personal information like bank details, residential addresses, and some confidential data over the Internet. We often come across two security techniques that are cryptography and steganography which provide security for data. Cryptography helps to generate the ciphertext which will be in an unreadable form, whereas steganography focuses on embedding the information inside the cover. Many of the existing methods focused only on using either of the security techniques those are cryptography and steganography. By using alone cryptography or steganography, one cannot provide security for the data. It is recommended to use

B. S. Spoorthi (✉)
Malnad College of Engineering, Hassan, India

S. P. Mala
Dayananda Sagar University, Bangalore, India

G. Ranganathan et al. (eds.), *Soft Computing for Security Applications*,
Advances in Intelligent Systems and Computing 1397,
https://doi.org/10.1007/978-981-16-5301-8_23

both cryptography and steganography to protect the data from unauthorized access and secure them safely. The cryptographic algorithm helps to generate the ciphertext of the secret information, whereas the steganographic algorithm focuses on hiding the information inside the cover; cover can be an image, video, or text. If cryptography is only used to provide security for data, then it might be unsuccessful because of the brute-force attack. Hackers can easily get secret data by decrypting the ciphertext. Similarly, by using steganography alone it is harder to give high-level security for the data. Many existing methodologies used both the security techniques to provide security for the data and managed to keep the key used for scrambling secret to avoid unauthorized access. The accuracy of such methods did not meet the expectation. There is a need for a technique that provides the highest security for the data. The proposed method aims at giving high-level security for the data. The technique provides three-level security for the data by combining both cryptography and steganography techniques. On top of both the security techniques, we used QR code to provide extra security for the data.

2 Proposed Work

In the previous study, the security for the data is provided either by using only steganography or cryptography; the success rate of such methods is less when compared to the methods which were developed by combining both cryptography and steganography. To overcome the security issue of the data, in the proposed work three-level security is given for the data. The first level of security is provided by an encryption process. The ciphertext of the secret information is generated by using the RSA algorithm. The second level of security is provided for the data by converting the ciphertext into QR code. The third level of security is given by using a steganographic technique. The generated QR code is embedded inside the cover image with the help of the LSB algorithm. The encryption process is discussed in session 2.1, and the decryption process is discussed in Sect. 2.2

2.1 Encryption Process

The major concern in the proposed method is to encrypt the data with great care. The steps involved in the encryption process are as follows.

A. **Generation of ciphertext**:

The secret message is given as input to the encryption algorithm. RSA algorithm is used for encrypting the data. It generates the ciphertext which is in the unreadable form. The RSA algorithm is one of the popular asymmetric cryptographic algorithms. It uses both private key and public key [1]. The public is shared publicly, whereas private is unknown to everyone. The encryption key used in the algorithm is public, and the decryption key is kept private and it is distinct from the public key.

Operation of RSA algorithm.

The algorithm involves three steps: Key generation and distribution, Message Encryption, Decryption. Each step is discussed in detail in the below sessions.

i. Key generation and distribution

The keys required for the RSA algorithm are generated and distributed in the following manner.

- Prime numbers P and Q are chosen and kept secret. The length of P and Q should differ by few digits to make factorizing harder.
- Calculation of key length and totient function (TF).

$$N = P * Q \tag{1}$$

$$TF = (P - 1) * (Q - 1) \tag{2}$$

- Choose e such that $1 < e < TF$ and $\gcd(e, TF) = 1$. Calculate d as

$$d = e^{-1} \mathrm{mod}\ (TF) \tag{3}$$

- The sender should use the receiver's public key to encrypt the message, and the receiver should use his private key to decrypt the message.

ii. Message Encryption

Accept the secret message (m) and encrypt it to provide security, the output of the encryption process is ciphertext (c). The ciphertext is calculated as

$$c = m^e \mathrm{mod}\ (n) \tag{4}$$

iii. Decryption

The message is recovered back by using the below formula and the pseudocode for RSA algorithm is given in Table 1.

$$m = c^d \mathrm{mod}\ (n) \tag{5}$$

B. QR code generation

QR code is selected over the bar code because it is more flexible, reliable, and easy to use. The two-dimensional area of the QR code allows storing more data in a smaller area. QR code contains multiple rows and columns [4]. The combination of rows and columns makes a grid of modules which are also called squares. The

Table 1 Pseudocode for RSA algorithm

Showing the working of RSA algorithm by considering the example through pseudocode
1. Enter the prime number P = 27
2. Enter the prime number Q = 27
3. Calculate N = P * Q = 729
4. Find TF = 676
5. Choose e as 19 and calculate d = 427
6. The public key is (19, 729)
7. The private key is (427, 729)

maximum number of possible modules is 31,329 which consists of 177 rows and 177 columns. The proper arrangement of modules allows QR codes to encode data. QR code has many versions: the basic one consists of 21 rows and 21 columns, and the standard one consists of 177 rows and 177 columns. The lesser the modules the lesser the errors (Fig. 1).

C. **Generation of stego image** (Fig. 2)

Steganography is the security technique that helps to hide secret data inside the cover. The cover can be an image, video, or text. The main advantage of this method is that it helps to conceal any kind of digital content. The most popular algorithm used for the concealing purpose is least significant bit (LSB) algorithm. Each pixel

Fig. 1 QR code

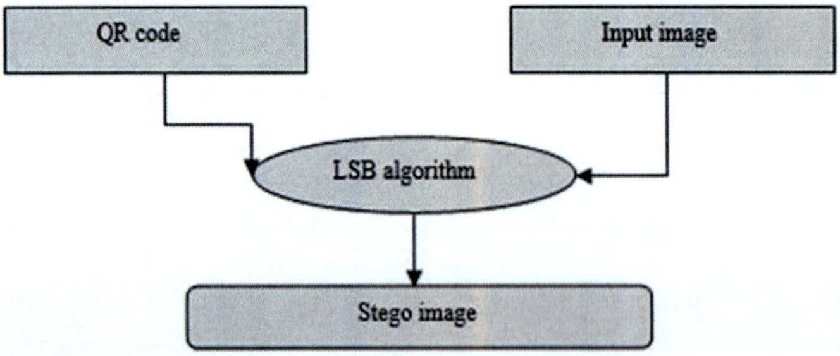

Fig. 2 Generation of stego image

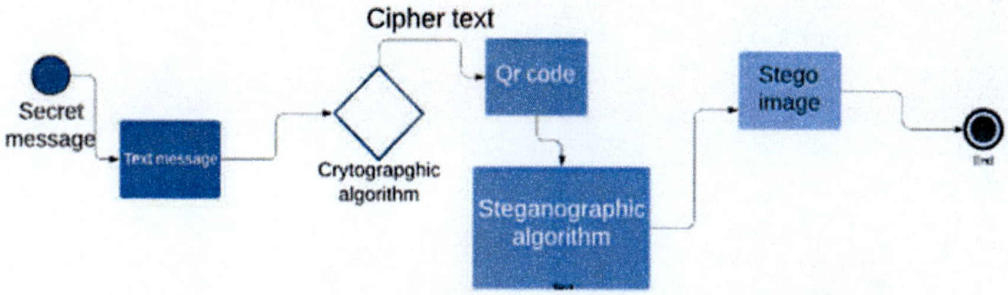

Fig. 3 Steps involved in the encryption process

value is represented using 8 bits in the image, and the last bit in the pixel value is called the least significant bit. The basic idea behind LSB embedding is that the last bit of pixel is altered, and in the place of the least significant bit, the new data is placed. Altering the last bit in the image pixel will not cause damage to image color, and the changes are invisible to naked eyes. İn the proposed method, the secret message is encrypted using the RSA algorithm. The generated QR code is embedded inside the image using the LSB algorithm. The advantages of the LSB algorithm are: it is easy to execute and runs very fast, the difference between the input image and output image is very less, and it allows to use last two bits in the pixel for embedding purposes. The generated stego image is the output from the transmitter model. The stego image is sent to the receiver. The storage capacity of the image increases when the image with a larger pixel value is chosen. Using steganography alone to conceal the secret message is hazardous for this purpose to protect the data from unauthorized access; it is first encrypted and generated as a QR code, and the generated QR code is embedded inside the cover image (Fig. 3).

2.2 Decryption Process

The stego image is received by the receiver at the receiving end. The stego image is given as input to the LSB algorithm for decryption purposes. LSB algorithm extracts the embedded QR code from the stego image. The ciphertext from the QR code is extracted . The extracted ciphertext is given as an input to the RSA algorithm. RSA decryption algorithm decrypts the ciphertext to get the original secret message (Fig. 4).

3 Experimental Results and Analysis

The proposed methodology is implemented in MATLAB. The three-level security of data is given by encrypting the data at the first level using the RSA algorithm. İn the second stage, the generated ciphertext is converted into QR code, and at the third level, the generated QR code is embedded inside the cover image using the LSB algorithm. Figure 5 shows the MATLAB window of the user prompt to enter

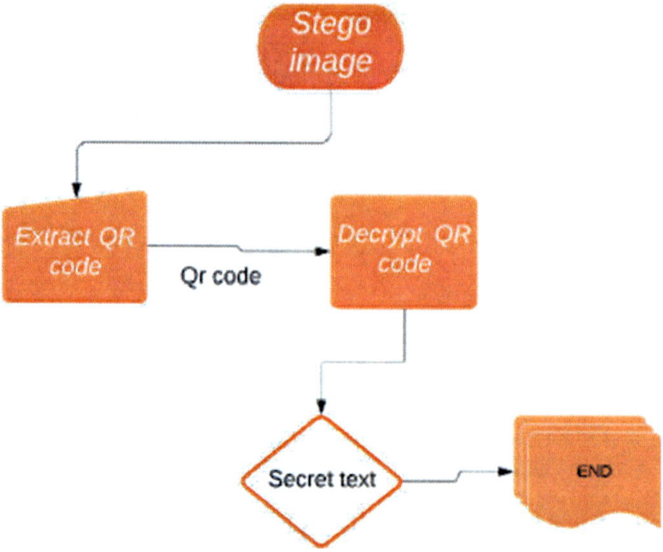

Fig. 4 Steps involved in the decryption process

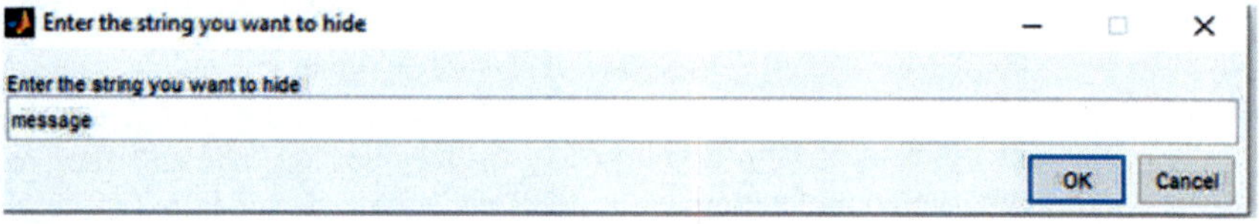

Fig. 5 User prompt to enter the secret message

the secret message. Using the RSA algorithm, ciphertext is generated for the entered secret message. The generated ciphertext is converted into a QR code. Figure 6 shows the generated QR code. The QR code is embedded inside the cover image using the LSB embedding steganographic technique. Figure 7 shows the stego image. At the receiver side, a QR code is extracted from the stego image. Figure 8 shows the decrypted QR code. From the QR code, the ciphertext is recovered. The complexity of the QR code depends upon the length of the secret message. The recovery of secret information is easier if the complexity of the QR code is less. The below figures show the results of the proposed method.

Fig. 6 Generated QR code

Fig. 7 Generated stego image

Fig. 8 Decrypted QR code

4 Conclusion

Due to the increase in message sharing over the Internet, data should be protected from unauthorized access. To provide high-level security to the secret information, there is a need to develop a hybrid method that helps to share the message securely over the Internet. The proposed hybrid method used both cryptography and steganography techniques to give three-level security for data. The generated QR code has been successfully embedded inside the input image. To increase the storage capacity, 512* 512 pixel image is chosen. The time taken to generate the ciphertext and stego image is 0.12 s and 0.066 s, respectively. This makes the proposed method superior compared to the existing methods. The extraction of the ciphertext from the QR code has been done successfully without any errors. The RSA algorithm encrypted the secret message with less time complexity and the LSB algorithm generated the robust stego image. To improve the execution time of the algorithm, the advanced encryption standard (AES) algorithm can be applied to generate the ciphertext of the secret information.

References

1. J. Bhadra, A.M. Bojamma, C.N. Prasad, M.N. Nachappa, An Insight to Steganography. Int. J. Innovative Sci. Eng. Technol. **1**(10), 29–42 (2014)
2. G. D. Rotel, Dr. A. M. Patil, "Steganography with cryptography technique for data hiding". Int. J. Sci. Res. **4**, 2312–2318 (2015)
3. P. Bharti, R. Soni, A new approach of data hiding in images using cryptography and steganography. Int. J. Comput. Appl. **58**, 1–5 (2015)
4. P. Sehgal, S. Rawat, S. Kaushik, S. Ali, R. Yadav, "Hiding encrypted text using text and image steganography: a dual steganographic technique". Int. J. Electric. Electron. Data Commun. 5, 54–57 (2017)
5. A. AL-Shaaby, T. AlKharobi, "Cryptography and steganography: new approach". Trans. Netw. Commun. **5**, 25–38 (2017)
6. K. N. Bangera, Y. Paddambali, Dr. N. V. Subba Reddy, G. Shivaprasas, "Multilayer security using RSA cryptography and Dual Audion Steganography". IEEE Int. Conf. RTEICT, pp. 492–495 (2017)
7. Y. Zhang, M. Zhang, X. Yang, D. Guo, L. Liu, Novel video steganography algorithm based on secret sharing and error-correcting code for H.264/AVC. Tsinghua Sci. Technol. **22**, 198–209 (2017)
8. J. Tao, S. Li, X. Zhang, Z. Wang, "Towards robust image steganography". IEEE Trans. Circuits Syst. Video Technol. pp. 1–7 (2018)
9. M. Okello, "A new timing steganography algorithm in real-time transmission devices", IEEE, pp. 880–884 (2018)
10. A. Shifa, M. S. Afgan, M. N. Asghar, M. Fleury, I. Memon3, S. Abdullah, N. Rasheed, "Joint crypto-stego scheme for enhanced image protection with nearest-centroid clustering", IEEE, **6**, 16189–16206 (2018)
11. Z. Wang, X. Zhang, Secure cover selection for steganography. IEEE **7**, 57857–57867 (2019)
12. K. Niu, J. Li, X. Yang, S. Zhang, B. Wang, Hybrid adaptive video steganography scheme under game model. IEEE **7**, 61523–61533 (2019)

Brain Tumour MRI Image Segmentation and Classification Using FCNN

Sanjay Kumar, Sanjeev Kumar Singh, Yash Gupta, Vasu Bajpai, and Piyush Tiwari

Abstract Brain tumours are one of the most prevalent and aggressive diseases across the globe, with a relatively short life expectancy in their most severe form. Thus, identification and treatment of brain tumour play an important role in improving the patients' quality of life. In general, the biomedical imaging methods such as computed tomography (CT), magnetic resonance imaging (MRI) and ultrasound images are used to assess tumours in different parts of human body such as brain, breast, lungs, liver, etc. Amongst the diagnostic technologies, magnetic resonance imaging is the most popular imaging approach offered by experts to analyze brain tumours. At the same time, image segmentation also plays a fundamental role in brain tumour analysis and classification. Recently, various image segmentation approaches are used to characterize the brain tumour in the early stages of development. However, the massive amount of data collected by an MRI scan makes manual categorization of tumour and non-tumour in a given period is impossible, and it also has several limitations (for example, reliable quantitative values are supplied for a restricted number of images). As a result, a reliable and automated image segmentation and classification technique are required to reduce human mortality. Automatic brain tumour detection using fully convolutional neural networks (FCNNs) classification is suggested in this paper. When compared to all other state-of-the-art technologies, experimental findings demonstrate that the proposed FCNN achieves a good accuracy with reduced complexity.

Keywords Intracranial · Convolution · Brain tumour · FCNN · Segmentation · Classification · Abstraction

1 Introduction

Brain is one of the major organs in the human body, and it is made up of billions of cells. The unusual group of cells generated by unregulated cell division is known

S. Kumar (✉) · S. K. Singh · Y. Gupta · V. Bajpai · P. Tiwari
Department of I.T, Galgotias College of Engineering and Technology, Greater Noida, India

G. Ranganathan et al. (eds.), *Soft Computing for Security Applications*, Advances in Intelligent Systems and Computing 1397,
https://doi.org/10.1007/978-981-16-5301-8_24

as a tumour, wherein low-grade tumour is referred as a benign brain tumour and a high-grade tumour is referred to as malignant. A benign tumour is not considered as malignant. On the other hand, the malignant tumour is considered as a cancerous tumour. As a result, it spreads quickly with unclear borders to other parts of the body and results in a sudden death. The brain MRI is primarily used to detect tumours and simulate the progression of brain tumours. The data obtained from MRI images is mostly utilized in tumour identification, classification and treatment. Also, MRI image contains more information about a specific medical imaging rather than other CT or ultrasound imaging techniques. Nevertheless, MRI scan leverages extensive information on brain anatomy as well as the identification of anomalies in brain tissue. In recent years, neural networks (NNs) and support vector machines (SVM) have become the most commonly utilized approaches for classifying the brain tumour images. On the other hand, deep learning (DL) models have recently established a significant research trend in image segmentation and classification to efficiently represent complex relationships without requiring more number of nodes as required in other architectures such as K-Nearest Neighbour (KNN) and support vector machine (SVM) [1].

In this research, we use a fully convolutional neural network (FCNN) technique, as well as image segmentation, to classify brain MRI images as malignant or non-malignant. The fundamental advantage of FCNN over its predecessors is that it automatically discovers the tumours and classifies it without any human intervention [2]. When compared to existing pre-trained models, the proposed model consumes far less processing resources and achieves higher accuracy outcomes.

2 Literature Review

The number of programmed tumour disseminations has increased rapidly during the recent decades. This agreement will not just need specialized equipment for tumour division; it also requires some in-depth research investigation [3]. The techniques for tumour segmentation, especially those that are occupied with RMN are typically separated into two segments: supported generative models and biased models. Generative models furiously swear spatially unequivocal past evidence concerning each tissue's vibrations. The tissue look is hard to explain, and some of the latest generative models are.

- In each comparison, the spatial details of the local image options are inserted and the enrolment operates along these lines to provide appropriate reparation for the impact of the turmoil [4].
- The neighbourhood of aeolotropism is aided by congruence segments and knows how to perform efficient segmentation without smoothing.

The results of the split with both falsified and legitimate images reveal that this expertise-based strategy protects the homogeneity of the spaces, and it has more strength than the connected FCM-based approaches. Maoguo Gong presented the

FCM algorithmic rule for tumour segmentation by exchanging a weighted feathery problem and component metrics. The commercially weighted fluffy problem is simultaneously based on the image segregation and the separation of each adjacent pixel [5]. Through the use of a speedy data measurement judgement law, the new algorithmic theory determined the part parameter accordingly, upholding the space distinction within the category of all the data focuses. Test outcomes of fake and real images demonstrate the success and efficiency of the recent algorithmic initiative, and it is equally autonomous of such disturbance.

2.1 *Methodology*

The proposed research work utilizes the convolutional neural network for MRI brain tumour segmentation by utilizing tensor stream. Typically, the segmentation is performed by utilizing different simulation platforms like MATLAB, LABVIEW and so forth, when the examination article utilizes the tensor stream based MRI brain tumour segmentation to improve the exactness, speed and affectability [6]. Segmentation can be performed on BRATS MRI brain images, and results are found to be more effective than any other existing methods.

3 Convolution Section

There are different layers in the convolution segment, which help to consecutively recognize the highlights based on the convolution activities. The obtained highlights are low-level highlights like edges and corner by establishing a connection. The component maps are the yield of the convolutional layer.

The complete convolution region contains three assorted ways. The way 1 and 2 use a similar nearby fix with different channel sizes to accommodate different area designs, and simultaneously sub way 3 uses down inspected fix to provide highlights [7]. The convolutional layer computes the yield of the neurons that are associated with one or the other locales present in the information. The convolution is the way towards performing an efficient operation between their data sources and open field to which they are associated with the data volume. For the proposed examination, L-be the profundity of the convolution channel stack a1 (1ε [1, L1]) that addresses a quantity of 1st layer maps, which are highlighted & Fi l (iε [1, α1 l]) addresses a ith iguide as a lst layers Fi l which is determined by g1 (.) in a PReLU work. The PRELU are neurons initiation work characterized by $g(x) = x$, $x > 0$ hatchet $x \leq 0$, which is considered as the organization boundary. Wj1 l is the channel associating the jth highlight maps on a l-1th layers & bi l area inclination of a fake neurons replica ReLU is the corrected straight units. It is quite possibly the most famously utilized enactment capacities. It is addressed as R1 (x) = maxi (0, X1), i.e. if X1 < 0 R1 (x) = 0 If X1 > = 0, R11 (x) = X1. Most of the AI strategies alludes to this capacity due to its straightforwardness, and furthermore, it tries not to disappear inclination issue. Yet, it ought to be utilized distinctly inside the secret layer of neural organization model. For instance, yield layers, SoftMax capacity will be utilized to ascertain the probabilities for the classes. During the preparation stage, a few angles can be delicate and pass on. Consequently, it needs a weight update, and further it never enacts on any information point again. It very well may be maintained a strategic distance from by another change called cracked ReLU. It gives little incline to keep refreshes alive [8]. Completely associated area the exceptional data is protected in the completely association segment by melding of the entirety of the component maps created by the convolution segment. The activity of completely associated area is like the convolution segment in which every co-effective hub performs like a convolution channel with the size (1, 1, 1) classification segment The order segment assists with giving likelihood to each voxel. A large portion of the convolutional neural organization utilizes SoftMax work to plan the element maps into absolute probabilities. Now x1v1 & Cs v is the 5th focus on voxels positions & grounds truths, where (F 1 L, F 2 L, F 3 L) are a lth layers include size map. s1 ϵ [1, b] now the b is a sizes of the bunch, PCs v are a yield on a SoftMax work. A boundary θ 1 is signified like channel coefficient predisposition. It tends to be found via preparing through the stochastic angle drop strategy [9]. SGD is one of the stochastic approximations of the inclination drop enhancement technique to limit a target work iteratively. For each cycle, the technique assesses the angle from a subset of named dataset, and furthermore, it assists with speeding up preparing in an immense preparing set. Preparation of organization the preparation model is made utilizing Keras. The 4 layer consecutive model was prepared on chosen 33×33 patches of MRI pictures to order the middle pixel. Each information has 4 channels, amongst four, one channel helps for each imaging arrangement and the net can become familiar with the relative pixel powers of each given class [9]. Approval of informational index the proposed strategy utilizes BRATS data set for assessing the mind tumour division techniques. Rascals furnishes every tolerant's T weights M.R.I. to gadolinium's contrasts (t1C) & t2 weights liquid lessened reversal recuperation (FLAIR) pictures [10]. This sort of informational index contrasts our programmed depiction technique and different calculations.

3.1 Convolution Neural Network Based on Python

Numerous scientists utilize MAT LAB to execute division measure. Our exploration work uses the python programming to execute the division of MRI brain tumour [10]. The highlights are recorded underneath to pick the python programming in order to execute the proposed research work:

1. Python code is smaller and coherent than MATLAB
2. The python information structure is better than MATLAB

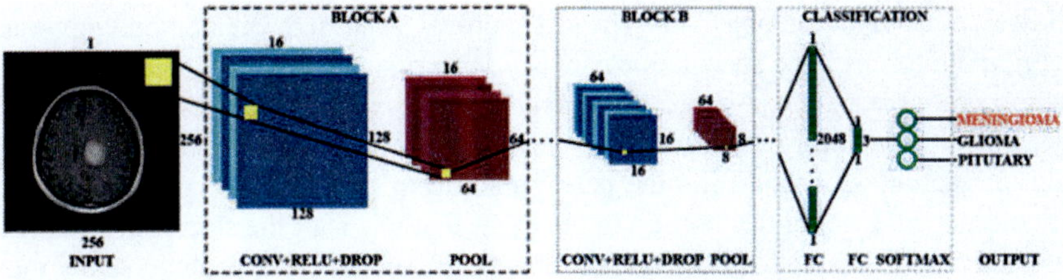

Fig. 1 FCNN architecture network [6]

3. It is an open source and at the same time it gives more realistic information collection.

Subsequently, the proposed research work uses python programming rather than MATLAB. Some extra python bundles are utilized during the execution the proposed research work through python.

4 Architecture Network

CNN was created in Matlab R2018a, and further it was utilized in the classification of tumour. Organization engineering is comprised of information, two principle boxes, and classification as demonstrated in Fig. 2. The box A includes a convolutional layer. Convolutional layers are trailed with direct units (RELU) initiation layer and dropout layer. At square, the additional maximum pooled layers yields less information [11]. A classification box is comprised of two completely associated (FC) layers of which the first one addresses the flattened yield of the last max pooling layer, whilst, in the second FC layer, the quantity of covered up units become equivalent to the quantity of the tumour classes. The entire organization design is comprised of the info layer, Block A and Block B, classification and yield layers as demonstrated in Fig. 1.

5 Network Training

The proposed research work has utilized a k-overlap cross-approval technique to test the organization execution [24]. Two distinctive 12 dropout half dropped out thirteen maximum pooled two by two maximum pooled like step and cushioning methodologies were carried out and both consist of 10-crease cross-approval. The first approach was to arbitrarily partition the information into 10 equivalent segments with the goal that every tumour classification at every segment recorded like crossing-approval. A subsequent methodology will haphazardly separate information to ten roughly equivalent parts here information like the solitary object must be contained at one set [12]. Along these lines it contains information, where several subjects consider tumour classes, alluded like an object like crossed approval. A subsequent methodology was carried out for testing a speculation capacity in an organization at clinical checking. A speculation capacity at clinic training addresses a capacity to foresee an analysis dependent on an information acquired with objects for null perceptions on preparing the interaction. Consequently, the perceptions from people in the preparation set should not show up in the test set. In the event that this isn't the situation, complex indicators can establish a connection amongst personality and analytic status that produce ridiculously high-expectation exactness [13]. To contrast the presentation of our organization and other best in class techniques, additionally the proposed organization which doesn't have K—overlay crossed approval (1 testing) has been utilized. Altogether from previously mention strategies, two information partitions were utilized for the test, two for approval and six for preparation. Both the datasets, which include ordinary and expanded were tried by utilizing every one of the strategies. The organization was prepared by utilizing an Adam analyzer, with a small group size equivalent to 16 and information movement in each cycle. Specifically, it has been designed to complete a preparation interaction later than 1 age, when the misfortune begins to increment [14]. A regularized factor was fixed at 00.00400. A load to a convolution layers was instated by utilizing the Glorot's initialization, otherwise called as Xavier's initialization [17]. An organization was prepared and tried in the solitary graph processing units (G.P.U.), C.U.D.A. gadget, Ge-Force G.T.X. 1050.00.

6 Conclusion

A partition method dependent on the K-shading is proposed in this article, which means grouping the MRI brain image for analyzing the tumour. The main investigation on the MRI frontal brain cortex image has shown empowering results by utilizing CIELab's features which provide an excellent efficiency in the division, and it can also help to consolidate the shading of the area of a tumour or injury. K—means is bundling around these lines, which makes it lucrative. The isolation of MR brain images is an important problem in medical imaging. The proposed research work has executed various computational frameworks to deal with the issue. A totally modified technique is used for gathering the cerebral tissue from anatomical images of MRI. The segmentation estimations can be extensively arranged and the request based, area based, or shape-based methodologies were examined. In this mission, three techniques are illustrated for segmenting the brain tissues from MRI models. The results and findings demonstrate a tumour that has provided the parameters to adequately choose the proposed technique.

Acknowledgements The creators might want to see the value in the exertion from editors and analysts. This examination got no particular award from government, business or non-benefit financing associations.

References

1. S. Kumar, A. Negi, J. N. Singh, "Semantic Segmentation Using Deep Learning for Brain Tumor MRI via Fully Convolution Neural Networks." *Information and Communication Technology for Intelligent Systems.* (Springer, Singapore, 2019), pp. 11–19
2. S. Kumar, A. Negi, J. N. Singh, A. Gaurav, "Brain Tumor Segmentation and Classification Using MRI Images via Fully Convolution Neural Networks." In *2018 International Conference on Advances in Computing, Communication Control and Networking (ICACCCN)*, pp. 1178–1181. IEEE (2018)
3. S. Kumar, A. Negi, J. N. Singh, H. Verma, "A deep learning for brain tumormri images semantic segmentation using fcn." In *2018 4th International Conference on Computing Communication and Automation (ICCCA)*, pp. 1–4. IEEE (2010)
4. K. Usman, K. Rajpoot, Brain tumor classification from multi-modality MRI using wavelets and machine learning. Pattern Anal. Appl. **20**(3), 871–881 (2017)
5. Z. Xiao, R. Huang, Y. Ding, T. Lan, R. F. Dong, Z. Qin, X. Zhang, W. Wang, "A deep learning-based segmentation method for brain tumor in MR images." In *2016 IEEE 6th International Conference on Computational Advances in Bio and Medical Sciences (ICCABS)*, pp. 1–6. IEEE (2016)
6. M. Havaei, A. Davy, D. Warde-Farley, A. Biard, A. Courville, Y. Bengio, C. Pal, P. M. Jodoin, H. Larochelle, "Brain tumor segmentation with deep neural networks." Med. Image Anal. **35**, 18-31 (2017)
7. H.-C. Shin, H.R. Roth, M. Gao, Lu. Le, Xu. Ziyue, I. Nogues, J. Yao, D. Mollura, R.M. Summers, Deep convolutional neural networks for computer-aided detection: CNN architectures, dataset characteristics and transfer learning. IEEE Trans. Med. Imaging **35**(5), 1285–1298 (2016)
8. J. Long, E. Shelhamer, T. Darrell, "Fully convolutional networks for semantic segmentation." In *Proceedings of the IEEE conference on computer vision and pattern recognition*, pp. 3431–3440 (2015)
9. H. Menze Bjoern, A. Jakab, S. Bauer, J. Kalpathy-Cramer, K. Farahani, J. Kirby, Y. Burren et al., "The multimodal brain tumor image segmentation benchmark (BRATS)." IEEE Trans. Med. Imaging **34**(10), 1993–2024 (2014)
10. O. Ronneberger, P. Fischer, TU-net Brox. "Convolutional networks for biomedical image segmentation." In *Paper presented at: International Conference on Medical Image Computing and Computer-Assisted Intervention* (2015)
11. L. Shen, T. Anderson, "Multimodal brain MRI tumor segmentation via convolution neural networks"
12. Bauer, Stefan, Lutz-P. Nolte, and Mauricio Reyes.: "Fully automatic segmentation of brain tumor images using support vector machine classification in combination with hierarchical conditional random field regularization." In International Conference on Medical Image Computing and Computer-Assisted Intervention, pp. 354–361. Springer, Berlin, Heidelberg, 2011.
13. K. Sharma, A. Kaur, S. Gujral, "Brain tumor detection based on machine learning algorithms." Int. J. Comput. Appl. **103**(1) (2014)
14. U. , Khalid, K. Rajpoot, "Brain tumor classification from multi-modality MRI using wavelets and machine learning." Patt. Anal. Appl. pp. 1–11 (2017)
15. X. Zhe, R. Huang, Y. Ding, T. Lan, R. F. Dong, Z. Qin, X. Zhang, W. Wang, "A deep learning-based segmentation method for brain tumor in MR images." In Computational Advances in Bio and Medical Sciences (ICCABS), 2016 IEEE 6th International Conference on, pp. 1–6. IEEE (2016)
16. H. Mohammad, A. Davy, D. Warde-Farley, A. Biard, A. Courville, Y. Bengio, C. Pal, P.M. Jodoin, H. Larochelle, Braintumor segmentation with deep neural networks. Med. Image Anal. **35**, 18–31 (2017)
17. S. Kumar, N. Kumar, I.K. Rishabh, V. Keshari, Automated brain tumour detection using deep learning via convolution neural networks (CNN). Int. J. Cur. Res. Rev. **13**(02), 148 (2021)

Decentralized Cloud and File System for Blockchain Environment

Kapil Aggarwal and Santosh Kumar Yadav

Abstract Blockchain is considered as an advanced technology that is always having issue related to data security and privacy and is widely used in cloud as well as fog-based environment for high-performance applications. Blockchain technology is now extensively employed in business and social sectors so that overall integrity can be maintained with high security standards. These segments include the Internet in general, social networking, e-commerce, transportation, logistics, professional communications, and e-governance. Blockchain is a high-performance security-aware technology used in this segment with the secured file distribution in decentralized environment with secure digital ledger. The digital ledger is very transparent, and there is no scope of manipulation by agent. The digital records of all the transactions are time-stamped in the blockchain. Ledger and transactions are irreversible once completed. This research manuscript is having the focus on the integration of InterPlanetary File System (IPFS) for the high-performance security and anonymity-based transactions with privacy. The IPFS protocol is the key technology and file system that is used in blockchain technology for the transmission of data via secured channels in which hacking or sniffing attempts can be avoided. The work integrates IPFS implementation with blockchain environment with demonstration toward the secured transmission.

Keywords Blockchain in cloud · Decentralized cloud · Decentralized cloud with blockchain · Digital ledger · InterPlanetary file system · Social networks

1 Introduction

1.1 Blockchain Technology

Blockchain allows history of digital properties unalterable and clear by usage of decentralization and cryptographic hashing. Blockchain functions like Google Docs.

K. Aggarwal (✉) · S. K. Yadav
Shri Jagdishprasad Jhabarmal Tibrewala University, Rajasthan, India

G. Ranganathan et al. (eds.), *Soft Computing for Security Applications*,
Advances in Intelligent Systems and Computing 1397,
https://doi.org/10.1007/978-981-16-5301-8_25

If a text is exchanged with a community of individuals, rather than being replicated or transferred, the document is transmitted instead of being transferred. This provides a distributed database for all at the same time. There are no contentious issues that remain unseen between departments or organizations.

Now, every corner of the globe is connected, thanks to the Internet. This is done using the state-of-the-art networks. In the present day, it is possible to utilize smartphones and access corporate, social and government services easily [1]. The high-performance operating systems include various flavors of Android, iOS, Windows, Blackberry, Linux OS, Kali OS, etc. "Blockchain" has emerged in recent years, with its technologies protecting the increased demand for services [2].

Blockchain may have a huge effect on enterprise usage cases including human resource management, e-governance, social networks, e-commerce and similar applications where privacy and integrity are quite important.

1.2 Mining in Blockchain

Mining is one of the key tasks in blockchain technology in which a block is used with security and block has a special ID and hash. Miners overcome extremely complicated math problems and discover difficult modules called nonces to integrate the security and privacy without fear of hacking attempts. The explanation this is taking a long time is that there are almost four billion variations of nonce and hash. It means that miners have discovered the golden nonce and their blocks were added to the chain.

A centralized web server structure is typically used for web applications. These web applications running on single-server cloud are typically used in variety of roles, including e-governance, medical records or other roles. Traditional web-based applications deployed on centralized web servers are at risk [3].

In above cases, all other clients or systems access the web services using their web clients. These web services are web browser applications. The main problem is the centralization which makes the server or server administrators prone to hacking.

A centralization of server storage is the major problem for backups. This way prevents issues with centralized servers by using decentralized web. Decentralized Web sites have no central data storage or applications. In blockchain technology, the web will be fully decentralized. Most data are stored at multiple points around the world using a blockchain to ensure the security and privacy of the information [4].

1.3 Use Cases

In the case of blockchain technology, the decentralization can lead to the emergence of decentralized apps or dApps. It is stored in each node of a decentralized web.

Blockchain takes the place of humans and cannot be tampered with (encrypted text). In the blockchain network, the chain network is constructed and programmed in such a way that it is secure and cannot be manipulated by anyone at all or by any sniffers [5].

A number of case scenarios exist where blockchain implementations are in practice including social networks of Mastodon, Matchpool, Minds, Steepshot, DTube, Sola. In addition, the entertainment industry is using blockchain technologies including Spotify, B2Expand, Guts, KickCity, Veredictum and many others. The retail sector is having the implementation patterns of EveryShop, Blockpoint, Warranteer, OPSkins, Loyyal, Portion, Shopin, Ecoinmerce.io, Buying.com. The cryptocurrencies are nowadays very prominent including Bitcoin, Ripple, Litecoin, Namecoin, Primecoin, Dogecoin, Ethereum, Stellar and many others which are widely used for corporate and social applications.

Blockchain allows history of digital properties unalterable and clear by usage of decentralization and cryptographic hashing. Blockchain functions like Google Docs. If a text is exchanged with a community of individuals, rather than being replicated or transferred, the document is transmitted instead of being transferred. This provides a distributed database for all at the same time. There are no contentious issues that remain unseen between departments or organizations.

2 Key Dimensions

There are blocks of different types of data in blockchain network. Each block is added to the blockchain and cannot be removed. The term immutable refers to reliable and indestructible. With blockchain, fraud, illegal copies or alteration would not be possible [6] (Fig. 1).

The first block in the blockchain is known as "the Genesis Block," which consists of all the future valid blocks. With the increase of encryption in every block, it becomes secure and impossible to crack previous states because of multiple encryptions in blocks [7, 8].

3 Decentralized Platform in Blockchain Implementations

There are number of blockchain platforms that can be used for the distributed network [9]. A number of decentralized cloud applications exist which are used for security and integrity including IPFS, Sia, Dfinity, PPIO, Opacity, Blockstack, BTFS, IExec, Safe Network and many others. Blockchain-based business networks require different applications for different applications. One of the main protocols and file systems for distributed, decentralized web or cloud is IPFS, that is, InterPlanetary File System. IPFS is a protocol designed specially for the decentralized applications [10] (Fig. 2).

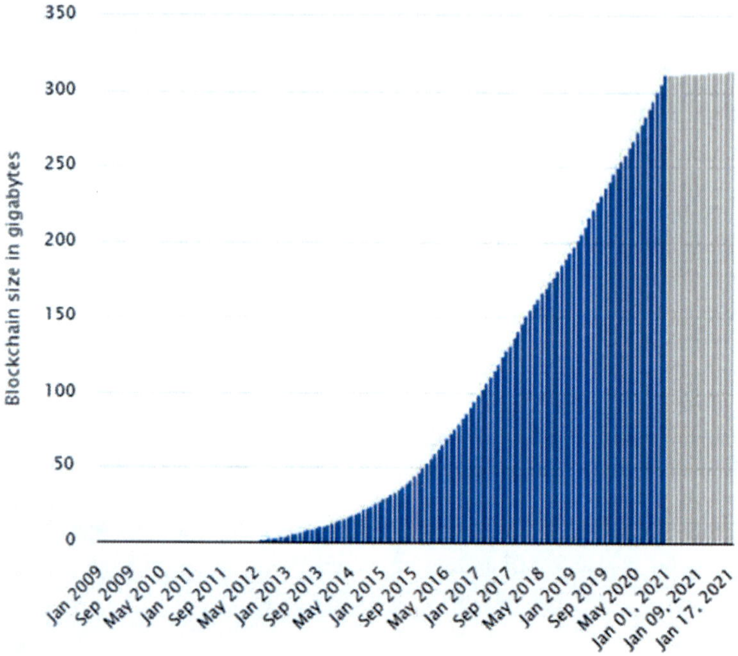

Fig. 1 Blockchain size from year 2009 to 2021 with Bitcoin-based platform

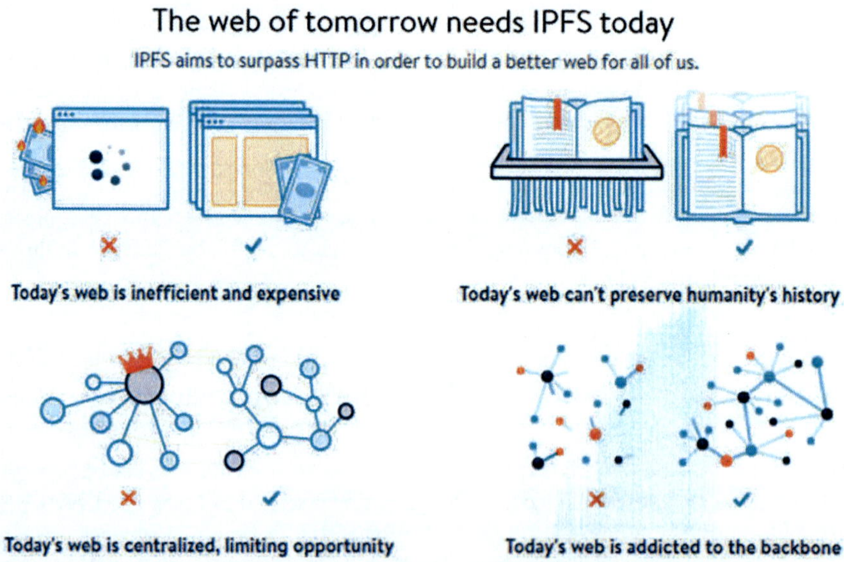

Fig. 2 IPFS-based blockchain patterns

4 IPFS Technology for Security in Blockchain

In traditional web-based environment, the web applications are accessed using Hypertext Transfer Protocol (HTTP) like https://www.google.com, https://www.nittrchd.ac.in, https://www.jnu.ac.in or any other web address in which specific address is there that is identifiable. Blockchain-based file system utilizes IPFS protocol with secured hashing for security and privacy. The hashing addresses make use of ipfs://QmHashFormat that is difficult to crack as having higher degree of encryption.

IPFS has language interfaces in Go and JavaScript implementations. A private blockchain environment is being used by companies to create and work on corporate blockchain-based and big data applications. Many blockchain-based projects are using Go as it possesses fast compilation speed, dynamic error-checking, scalability, simplicity, garbage collection and more [11].

4.1 IPFS Integration for Secured Transmission

The desktop version of IPFS can be deployed on multiple operating systems such as Linux and Windows. To install IPFS on Windows Operating System, you can download the IPFS application from https://github.com/ipfs-shipyard/ipfs-desktop.

IPFS makes use of advanced technologies in which the dynamic fragmentation of data is done and is transmitted over the network so that the security and privacy can be maintained effectively (Fig. 3).

Integrating cloud with IPFS can be possible with Ethereum blockchain. Once IPFS is integrated, a shared overlay dashboard which is used to share files on the blockchain network. By using this method, the entire IPFS-based blockchain network will be secured and all the addresses will be protected. In this way, file ss.txt has unique encrypted hash address. The final portion of filename denotes the file's varied properties [12]. This approach of data transfer will employ encrypted IPFS.

As an example, consider the Hash function, which is implemented on a standalone system. To keep blockchain distributed, it must be performed by all stakeholders (Fig. 4).

4.2 Python-Based Integration of Blockchain Security

The concept of proof-of-work (PoW) is one of the important algorithms in blockchain programming. This could be used to verify the blocks added to the block chain. The key consensus algorithm is used in determining transactions. In blockchain networks, different miners participate for the verification and transaction completion. Digital currencies like Bitcoin are given to miners when Bitcoin transactions are validated.

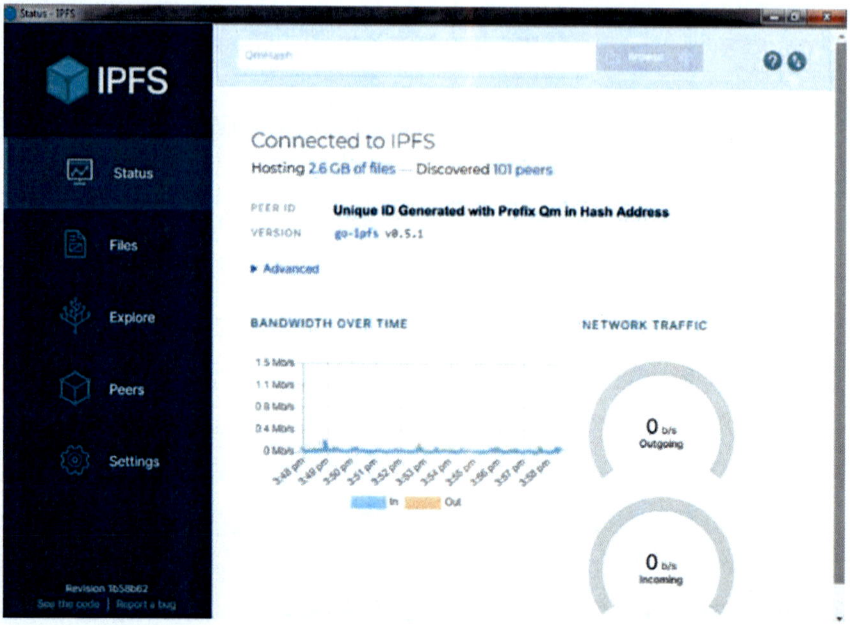

Fig. 3 IPFS dashboard

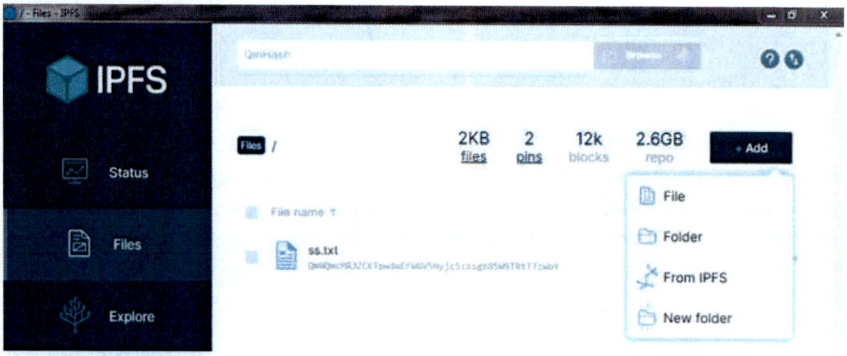

Fig. 4 Panel of IPFS

Digital currency avoids the double spending problem, thereby reducing the risk to the user. If A transferred a file to B, for example, the specific values for a record or currency of A should be removed and be reflected in the record of B. It is often done by banks as the intermediation method. In the case of blockchain network, intermediaries are out of the picture and it is checked and validated, automatically using specialized algorithms. One of the consequences for not deleting an accidental transaction from sender is that the currency could be rendered invalid.

```
miner_address = "***************************".
thisblchain = [].
thisblchain.append((create_genesis_bl()).
this_node_transactions = . [].
peer._node = [].
mining = True.
@node.route('/thisblchain', methods = [.'POST'.]).
def transaction.():
new_thisblchain = request.fetch__json().
this_node_transactions.append(new_thisblchain).
fetch and display "New transaction".
fetch and display "Sender: ".format.(new_thisblchain['from'].encode('ascii','replace')).
fetch and display "Receiver: ".format(new_thisblchain['t'].encode('ascii','replace')).
fetch and display "Amount: \n".format.(.new_thisblchain['.amount.'.]).
rreturn. "Transaction Successful\n".
@.node.route.('/bls', methods = .['FETCH']).
def fetch_bls():
chain_t_.send = thisblchain.
for i in range.(len(chain_t_send)):
bl = chain_t_send.[i].
bl_idx = str.(bl.idx).
bl_timestamp. = str.(bl.timestamp.)
bl_data = str(bl.data).
bl_thash = bl.thash.
chain_t_send.[i] =
"idx": bl_idx.,
"timestamp": bl_timestamp,.
"data": bl_data,
"thash": bl_thash.
```

The implementation patterns in Figs. 5 and 6 depict the generation of dynamic communication and security transmissions using blockchain-based environment. The impact on blockchain can be viewed through the application of curl command. The Windows tool cURL shall be downloaded from https://curl.haxx.se/windows/.

Fig. 5 Execution of custom blockchain script

Fig. 6 Blockchain transaction with cURL

Having a transparent picture of operations can be monitored and analyzed to prove there is no hacking in code. Using proof-of-work (PoW), the digital transactions on the blockchain are validated and successfully recorded.

5 Conclusion

Government organizations and corporations have been striving to make use of blockchain technology for securing the web application. In order to integrate, there is a need to link protected algorithms of IPFS-based network. Blockchain technology may be used by academics to predict identification of a homicide suspect and by forensic science to solve numerous identities the same case. Blockchain technology has a vast testing opportunity for academics, professors and professionals. It is one of the common goals that hackers, device crackers and sniffers have and as such must be protected through sophisticated cloud integrations. The implementation of high-performance security-aware applications like e-governance, social networks and similar applications can be integrated using IPFS protocol so that the overall communication and data transmission are secured and privacy aware.

References

1. H. Huang, X. Chen, Q. Wu, X. Huang, J. Shen, 'Bitcoin based fair payments for outsourcing computations of fog devices.' Future Gener. Comput. Syst. **78**, 850–858 (2018)
2. M. Green, S. Hohenberger, B. Waters, "Outsourcing the decryption of ABE ciphertext," In Proc. 20th USENIX Conf. Security (2011)
3. M. Asim, M. Petkovi, T. Ignatenko, "Attribute-based encryption with encryption and decryption outsourcing," In Proc. 20th USENIX Conf. Security, pp. 1–4 (2011)
4. C. Zuo, J. Shao, G. Wei, 'CCA-secure ABE with outsourced decryption for fog computing.' Future Gener. Comput. Syst. **78**, 720–738 (2018)
5. Y. Zhang, D. Zheng, X. Chen, J. Li, H. Li, 'Efficient attribute based data sharing in mobile clouds.' Pervasive Mobile Comput. **28**, 135–149 (2016)

6. Y. Zhang, X. Chen, J. Li, D.S. Wong, H. Li, I. You, 'Ensuring attribute privacy protection and fast decryption for outsourced data security in mobile cloud computing.' Inf. Sci. **379**, 42–61 (2017)
7. Q. Zheng, S. Xu, G. Ateniese, "VABKS: Verifiable attribute based keyword search over outsourced encrypted data," in Proc. IEEE Conf. Computing. Communication, pp. 522–530. https://doi.org/10.1109/INFOCOM.2014.6847976 (2014)
8. F. Zhao, T. Nishide, K. Sakurai, 'Realizing fine-grained and flexible access control to outsourced data with attribute-based cryptosystems.' Inf. Secur. **15**, 83–97 (2011). https://doi.org/10.1007/978-3-642-21031-07
9. X. Chen, J. Li, W. Susilo, 'Efficient fair conditional payments for outsourcing computations.' IEEE Trans. Inf. Forensics Secur. **7**(6), 1687–1694 (2012)
10. Y. Zhang, R.H. Deng, J. Shu, K. Yang, D. Zheng, 'TKSE: trustworthy keyword search over encrypted data with two-side verifiability via blockchain.' IEEE Access **6**, 31077–31087 (2018)
11. L. Zhang, "Attribute based encryption with privacy protection and accountability for cloud IoT," IEEE Trans. Cloud Computing., Early Access 2020. https://doi.org/10.1109/TCC.2020.2975184 (2020)
12. J. Li, Y. Wang, Y. Zhang, J. Han, 'Full verifiability for outsourced decryption in attribute based encryption.' IEEE Trans. Serv. Comput. **13**(3), 478–487 (2020). https://doi.org/10.1109/TSC.2017.2710190

A Network Investigation Framework Based on Deep Learning for Fraud Transaction Detection

Sonam Bhardwaj and Mayank Dave

Abstract The online banking system has made a way for the customers to make transactions from anywhere in the world with the availability of network. The banking sectors even communicate with the other branches by having an online banking management system. This management system communicates through network which seeks interest of many fraudsters and attackers. Therefore, to strengthen and implement the defense strategies, these suspicious activities need to be investigated. The paper discusses a case study, different risk factors and their investigation techniques. This work proposes a deep learning-based investigation framework for the online transaction system to detect fraud cases. The framework successfully detects the cases with an accuracy of 99.64%. The detailed process flow is discussed to explain the detailed work flow. In conclusion, future work is highlighted to give researchers direction in this area.

Keywords Network forensics · Investigation · Fraudulent transaction · Deep learning · Ensemble algorithm · Log analysis · Fraud operations

1 Introduction

In digital world, a number of industries are moving and growing online that were almost exclusively physical interactions previously, including online banking, gaming, cannabis distribution and prescription refills. The online banking is vulnerable to unauthorized credit and debits, data theft, privacy leakage and ransom wares. These attacks demand investigation to identify the criminal based on various evidence collected from network like network logs, trace logs, system logs, IP addresses, server logs and so on. After an attack on the network is executed, the attacker aims to destroy

S. Bhardwaj (✉) · M. Dave
National Institute of Technology, Kurukshetra, India
e-mail: sonam_6180082@nitkkr.ac.in

M. Dave
e-mail: mdave@nitkkr.ac.in

G. Ranganathan et al. (eds.), *Soft Computing for Security Applications*,
Advances in Intelligent Systems and Computing 1397,
https://doi.org/10.1007/978-981-16-5301-8_26

any digital traces or evidence left behind. If the manual digital investigation is opted, the procedure might lag in speed for collecting evidence. The consequence of this lag might result in successful modifications in evidence by the attacker and slow evidence retrieval [1].

Many techniques are adopted to secure the online transaction systems like installing a firewall, securing the database, installing the intrusion detection systems (IDS) and intrusion prevention systems as well [2]. These security techniques are limited to alerting the system, blocking some requests and supporting few file formats. Therefore, proper analysis of the collected evidence is required to trace the fraudsters. The required real-time analysis is simple with machine learning techniques. These techniques allow to build a model and train them on the basis of required dataset. It learns and grows in a more accurate sense by the cause a huge dataset and numerous iterations polishing the process flow of training, growing experience and utilizing that experience to give predictive results in identification of attacks [3].

The paper is organized as follows: Sect. 2 explains the operations of a banking fraud. In Sect. 3, related works to attack investigation have been discussed, Sect. 4 introduces the proposed model for investigation of fraud transactions and Sect. 5 depicts the results and their discussions. The last section concludes the work and discuss different research directions.

2 Operations of an Online Fraud

The study has been done on a cybercrime group to know their tactics and procedures they follow. The threat actors follow different techniques compromise payment card credentials. Once the credentials are acquired, the information is used to generate further card information. The primary technique used by the criminal groups includes online purchase and ATM withdrawals. The operations life cycle [4] followed by the fraudsters is depicted in Fig. 1.

Phase 1: Setting up the work place

The group of attackers follow the property of anonymity and to delete the evidence they use tools like CCleaner that allows them to clear the browsing history, temporary files, URLs, etc. The attackers also change their system's MAC address to avoid being uniquely identified. This hiding feature is also highlighted when they use virtual private networks (VPNs) for IPs based on different countries.

Phase 2: Data Acquisition

The payment card "dumps" are shared among the attacker communities through social media platforms. To obtain the stolen data from peers, freely available consolidations of email credentials, personal information, etc. are used. The attackers also collect payment card data from "dump shops," which is web-based platform that collects thousands of card data. These dump shops are responsible for stealing credentials that are stolen from other services. The available tools scan the websites for vulnerabilities to exploits them to access the database.

Fig. 1 Operation life cycle of a fraud case

Phase 3: Generating further card numbers

The community makes use of card generators to generate new cards based on the compromised ones, creating additional opportunities for monetization. The actors find the websites with vulnerable authentication systems to purchase data.

Phase 4: Validating new card numbers

After generating card data, the group validates them through tools. The vulnerable merchant websites accept payments by generated cards and criminals take advantage of these sites to purchase and bulk check card data for usability.

Phase 5: Laundering and monetization

The actors create cloned cards with stolen data. They exploit e-commerce sites to perform fraudulent transactions and validate cards and card holder data being legitimate applying different anti-fraud checks.

3 Related Works

PyFlag [5], an open-source network forensic framework, was to determine the files that have been encapsulated by malwares by operating at multiple levels as on disk, memory and network. It incorporates various forensic features in the same package to efficiently process very large capture files, extract high-level information and substantiate each deduction. The framework stands valuable for log analysis and disk forensics but only analyze the stored data and fails in the real-time environment. For responding to an incident, the documentation of all activities and collection must

Table 1 Risk factors of online transactions and their possible Investigation scenarios

Risk Factors	Investigation Technique
Cybercrime	Automatic ATM Fraud Detection as a sequence-based anomaly detection problem [8]
Money laundering	Transaction mining algorithm and frequent pattern mining techniques [9]
Black money	Building use cases for the fraud cases and making analysis [10]
Loan loss	Dominance-based Rough Set Balanced Rule Ensemble (DRSA-BRE) [11]
Credit growth	Transaction mining algorithm and machine learning-based technique [12]

be done properly without any negligence. The attack graph is useful in measuring the security of a network by calculating the critical nodes for an attack, expected path length that an attacker can take, and probabilistic path determining the likelihood to reach a particular state and impact analysis of the vulnerability. When the network is complex, building of attack graph is inefficient in terms of time and space complexity.

Alharbi et al. [6] study two types of processes for the digital investigation that are proactive and reactive investigation processes. The proactive digital component acts in real time to analyze evidence to identify an incident as it occurs. The reactive investigation approach for investigating a digital crime after an incident has occurred involves identifying, preserving, collecting, analyzing and generating the final report. The two processes are important but the visualization of the attacks is important to prepare the defense strategy. The various risk factors [7] and their investigation techniques are presented in Table 1.

4 Proposed Model

The cybercriminals adopt variety of techniques to breach the user data and perform fraudulent activities. The IP spoofing technique is used by the attackers to impersonate a system that can be used to commit the attack. These attacks mostly happen in the banking sectors where the customers are involved in online transactions. In Fig. 2, the model depicts the online transaction system with investigation and monitoring capabilities. The customers trying to avail banking facilities make use of different electronic systems like computers, cloud-based systems, tablets and mobile phones. The user inputs credentials to login to its portal. The credentials are verified based on two-way authentication system. Once the verification is done, the traffic passes through the internet firewalls.

The model involves third-party server to complete the request. The third-party server maintains the customer's data like customer ID, account number, account number and other details in the secure database. After the verification, that the request is coming from an authenticated source, it is forward to the bank for further processing of the request. The bank processes the request and response to the third party through a secure channel. The third-party server applies the privacy protection techniques to

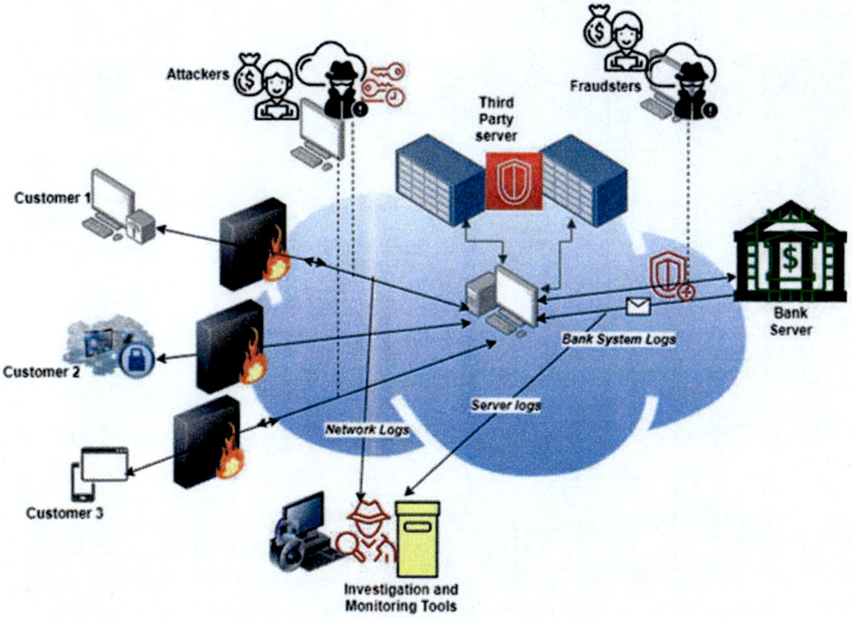

Fig. 2 Investigation framework for monitoring online banking system

secure the data and forwards the response to the customer. In this communication, the vulnerable points are:

(i) Customer to third-party server: When the customer makes a request to the bank, the credentials are verified by the third party. These credentials are continuously sniffed by the attacker and stolen from the network to commit theft or any other suspicious activity. The privacy breach by the attackers carried out allow them to access the user's account, transfer money, request money, etc.
(ii) Third-party to bank server: When the bank server responds to the customer, the attacker tries to sneak-peek into the response been sent to the third party. The advantage an attacker takes from this is it makes the bank transfer the funds into the attacker's account, as it is difficult to reverse.

To counter the suspicious activities and strengthen the defense system, it is important to investigate and analyze such activities. The continuous monitoring of such activities is done by the forensic investigator to observe any suspicious activity happening in the network. The process flow diagram in Fig. 3 explains the flow of the deep learning model for detection of fraud transactions.

In the banking system when a transaction takes place, the network investigator monitors it continuously and collects data from various sources. The data collected is treated as evidence for the investigation purpose, fed into the deep learning model for fraud detection. The model improves itself by training and testing on the recent and historical data and successfully detects the fraudulent activities in a short period

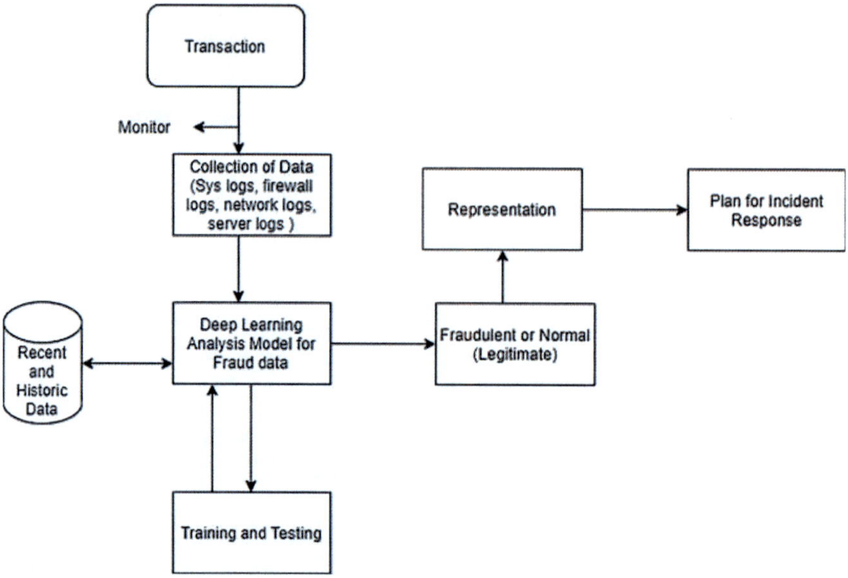

Fig. 3 Process flow of the proposed deep learning model for detection of fraud transactions

of time. After successful classification, the representation of the evidence is made based on the resultant output, making it liable to be presented in front of judiciary against the attacker. After the gaining an insight about the investigation outcome, the defense team makes preparation for responding to the incident.

In the analysis phase, there are different machine learning algorithms [13] that can be used to analyze the collected data. The algorithms that are used in the proposed framework to compare the performance of the model are:

i. Random forest: This algorithm is used for consists of a large number of individual decision trees that operate as an ensemble. Each individual tree in the random forest spits out a class prediction, the class with majority is chosen as the final outcome. This algorithm is used for classification and regression.
ii. Decision tree: The goal of using a decision tree is to create a training model that can use to predict the class or value of the target variable by learning simple decision rules inferred from prior data (training data). In decision, we compare the values of the root attribute with the record's attribute. On the basis of comparison, we follow the branch corresponding to that value and jump to the next node.
iii. XGB classifier: XGBoost is an implementation of gradient boosted decision trees. It is a perfect combination of software and hardware optimization techniques to yield superior results using fewer computing resources in the shortest amount of time.

5 Experimental Setup, Results and Discussions

The parent domain of machine learning and deep learning is artificial intelligence (AI). AI is any technology that supports virtual as well as real-time implementations in which it imitates human behavior and intelligence. Machine learning is a sub-domain of artificial intelligence in which machines learn and gain experience using statistical techniques over time. The proposed model was implemented on Lenovo core i5 (10th generation) machine and applied on the dataset [14] containing 32 features. The data was classified into fraud or valid transactions optimizing features based on the correlation value more than 0.8. The results are calculated with continuous training and testing of the historical and recent data with test size of 0.2. The deep learning [15] was applied on the dataset with its three main functions (dense, activation and dropout). In Table 2, the summary of the transaction amount and other details has been summarized.

In Table 3, the performance of the different models has been presented with different performance metrics and the performance of the same is graphically represented in Fig. 4.

Table 2 Details of the fraudulent transaction

Parameters	Value
Count value	594.000000
Mean value	126.211321
Standard value	258.67893
Minimum value	0.000000
25% of values	1.000000
50% of values	9.250000
75% of the values	105.890000
Maximum value	2125.870000
Amount	dtype: float64
The Matthews correlation coefficient is	0.8790748763736589
Outlier fraction	0.0017304750013189597
Fraud cases	492
Valid transactions	284,315

Table 3 Performance of different models

Model	Accuracy	Precision	Recall	F1 Score
Decision tree	0.979192	0.774648	0.748299	0.761246
Random Forest (n = 100)	0.988532	0.957265	0.761905	0.848485
XGB Classifier (n = 100)	0.98957	0.974358	0.775510	0.863643
Deep Learning model (epochs = 20)	0.99645	0.98345	0.96784	0.98342

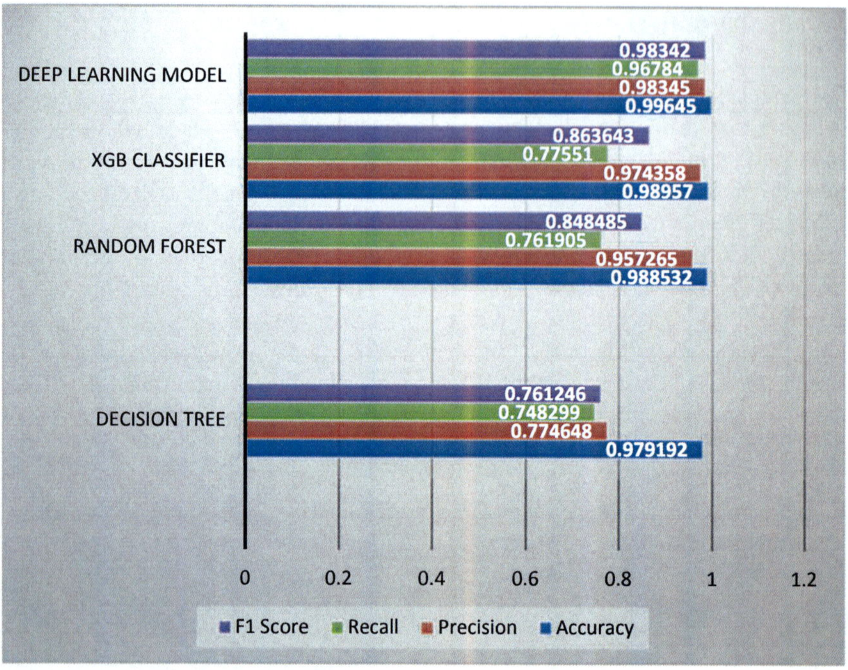

Fig. 4 Overall performance of different models and the proposed model

The performance shown by different model can be summarized in the decreasing order of the algorithms as follows:

$$\text{Deep learning} > \text{XGB} > \text{Random Forest} > \text{Decision tree}$$

6 Conclusion

The online transaction system in one way or the other vulnerable for attacks. The banking sectors keep updating their security measures but due to the new techniques of the attacker's, improper analysis leads to misguidance of the tools. The suspicious activities must be kept in check by the banking systems to avoid unnecessary malfunctioning of the system. The existing models fail to analyze and preserve the collected evidence. The tampered evidence may lead to broken chain of custody which leads to a false investigation. To strengthen the defense system, network investigation of the fraudulent activities is required. The work proposed a deep learning-based framework for analysis of the fraud data and training the data over collected logs. The model involves continuous monitoring of traffic and timely analysis of the collected evidence so as to maintain the credibility of the data. The deep learning model outperforms the various other models and has an accuracy rate of 99% which is a good score for detection of fraud activities. The work can be extended by obtaining real-time traffic and preserving the same with blockchain technology to maintain the chain of custody between the evidence.

References

1. A. S. Hussein, R. S. Khairy, S. M. M. Najeeb, H. T. AL Rikabi, Credit card fraud detection using fuzzy rough nearest neighbor and sequential minimal optimization with logistic regression. Int. J. Interactive Mobile Technol. **15**(5) (2021)
2. Talukdar, M. I., Hassan, R., Hossen, M. S., Ahmad, K., Qamar, F., Ahmed, A. S.: Performance improvements of AODV by black hole attack detection using IDS and digital signature. Wireless Communications and Mobile Computing, pp.1–13 (2021)
3. Li, Z., Huang, M., Liu, G., Jiang, C.: A hybrid method with dynamic weighted entropy for handling the problem of class imbalance with overlap in credit card fraud detection. Expert Systems with Applications, 175, 114750 (2021).
4. Syniavska, O., Dekhtyar, N., Deyneka, O., Zhukova, T., Syniavska, O.: Security of e-banking systems: modelling the process of counteracting e-banking fraud. In: SHS Web of Conferences, Vol. 65, p. 03004, EDP Sciences (2019)
5. M.I. Cohen, PyFlag–An advanced network forensic framework. Digit. Investig. **5**, S112–S120 (2018)
6. Alharbi, S., Weber-Jahnke, J.,Traore, I.: The proactive and reactive digital forensics investigation process: A systematic literature review. In International Conference on Information Security and Assurance, pp. 87–100, Springer, Berlin, Heidelberg (2011)
7. Fraudulent cases in UK: A Report. https://www.ukfinance.org.uk/system/files/Fraud%20The%20Facts%202021-%20FINAL.pdf
8. Anderka, M., Klerx, T., Priesterjahn, S., Büning, H. K.: Automatic ATM Fraud Detection as a Sequence-based Anomaly Detection Problem. In: ICPRAM, pp. 759–764 (2014)
9. Kim, H., Yun, U., Baek, Y., Kim, H., Nam, H., Lin, J. C. W., Fournier-Viger, P. Damped sliding-based utility-oriented pattern mining over stream data. Knowledge-Based Systems, 213, 106653 (2021).
10. M. Qiu, Y. Yang, Analysis of the current situation and characteristics of college student "Online Fraud Cases." Int. J. Mobile Comput. Multi. Commun. (IJMCMC) **12**(2), 56–73 (2021)
11. Błaszczyński, J., de Almeida Filho, A. T., Matuszyk, A., Szeląg, M., Słowiński, R. Auto loan fraud detection using dominance-based rough set approach versus machine learning methods. Expert Systems with Applications, 163, 113740 (2021)
12. Wu, J., Liu, J., Chen, W., Huang, H., Zheng, Z., Zhang, Y. Detecting mixing services via mining bitcoin transaction network with hybrid motifs. IEEE Transactions on Systems, Man, and Cybernetics: Systems, pp. 1–13 (2021)
13. Mao, C., Lin, R., Towey, D., Wang, W., Chen, J., He, Q. Trustworthiness prediction of cloud services based on selective neural network ensemble learning. Expert Systems with Applications, 168, 114390, (2021)
14. Dataset online available at: https://www.kaggle.com/mlg-ulb/creditcardfraud?select=creditcard.csv
15. J.V. Anand, A Methodology of Atmospheric Deterioration Forecasting and Evaluation through Data Mining and Business Intelligence. J. Ubiquitous Comput. Commun. Technol. (UCCT) **2**(02), 79–87 (2020)

Coverage Using Swarm Intelligence Family in Wireless Sensor Networks

S. Sumitha Pandit and B. Kalpana

Abstract Coverage is a fundamental challenge in wireless sensor networks (WSN's) since it has crucial characteristics like random deploy, restricted sensor nodes and a smaller number of sensors. Coverage is a term that describes how well the sensors are deployed in a monitoring area. It has a direct impact on the performance of network which is considered to be the major criteria to evaluate the Quality of Service (QoS) in WSN. Sensor nodes can be deployed either randomly or deterministically. Even though random deployment is preferred in large scale, it does not give guarantee to cover the whole ROI since it causes coverage holes, whereas deterministic is very costly and complex in nature. Thus, controlling coverage is the most significant initial task in WSN's. To reduce the unnecessary energy loss and to enhance the performance of network, energy efficiency and coverage velocity have to be considered. In this paper, an attempt has been made to review how swarm intelligence algorithms are used to produce an optimal coverage network. On study, it is understood that many researchers have chosen particle swarm optimization (PSO) technique in coverage control than other techniques. Finally, this paper describes about different properties, different methods and different problems in WSN's coverage along with a framework to solve the coverage problems in WSN.

Keywords Coverage · WSN · ROI · Particle swarm optimization (PSO) · QoS · Controlling

1 Wireless Sensor Network

Wireless sensor network has attracted the interest of many researchers and industries for the past few decades. The main reason is, it has potential applications like military operation, surveillance system, environment monitoring, health care and public care [1]. Each application requires number of sensors to cover its given region of interest in network field through deployment. These sensor nodes are used to determine various

S. Sumitha Pandit (✉) · B. Kalpana
Department of Computer Science, Avinashilingam Institute for Home Science and Higher Education for Women, Coimbatore, India

G. Ranganathan et al. (eds.), *Soft Computing for Security Applications*, Advances in Intelligent Systems and Computing 1397,
https://doi.org/10.1007/978-981-16-5301-8_27

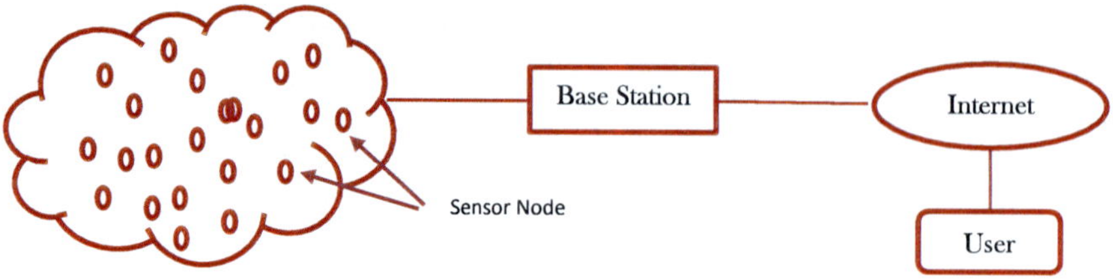

Fig. 1 Wireless sensor network

metrics like temperature, sound and pressure from different environment. Each sensor node is tiny in size, expensive, multifunctional and has low-power consumption. It consists of sensor unit, processor, A/D converter, memory and power supply [2] (Fig. 1).

In additional, sensor node has capability of tracking the location of a system. WSN face many issues, while implementing like scalability, data aggregation, and quality of service, power consumption, fault tolerance and node deployment [2]. Sensor nodes are categorized into static nodes and mobile nodes. In node deployment, either static node or mobile node is used based on applications. Compare to static nodes, mobile nodes are more flexible, and it is used by many advance and smart applications. Nowadays, many researchers focus on mobile nodes to collect information from their environment by using less energy level. So, in that case, optimized coverage becomes a primary problem in wireless sensor networks.

2 Overview of Coverage

Coverage reflects how well the sensor nodes are deployed and monitor the target area in the sensor network. The goal of coverage is to ensure whether each and every point in target is within sensing radius and with less coverage overlap and coverage holes. Constructing a fully connected coverage is a main challenge which has direct impact on connectivity and energy consumption in WSN.

2.1 Coverage Problem

The main cause of coverage problem is due to limited sensor nodes, limited sensing range and random deployment. When sensor nodes randomly deploy using limited power supply to cover full region of interest it dies after certain period of time.

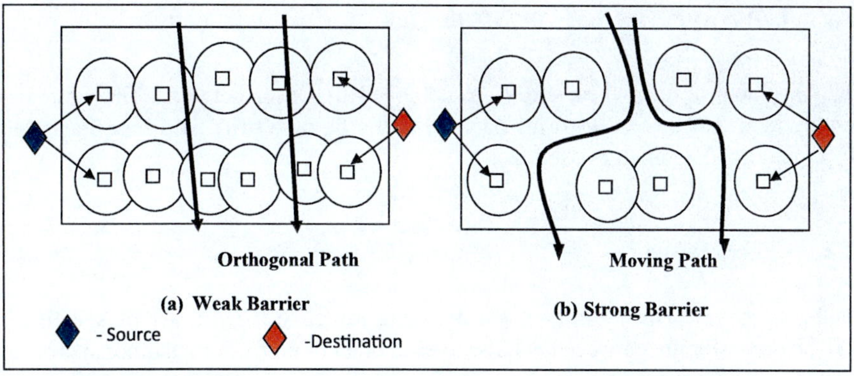

Fig. 2 Barrier coverage

Coverage is classified into 3 categories, i.e.:–

2.1.1 Area Coverage (Region)

This problem occurs when the whole sensing area has to be monitored (ROI-Region of Interest) or in other words, each point should be monitored at least by one sensor. Since each point has to be monitored the sensor nodes are densely deployed in this type which results into overlapped coverage: Forest fire accident (Fig. 2).

2.1.2 Point Coverage

In this type, only limited set of objects or target points (Point of Interest (POI)) are allowed to monitor. Here, nodes are divided into subsets under stochastic distribution. Each subset works in turn manner which helps to maximize the network lifetime.

2.1.3 Barrier Coverage

This type of coverage (otherwise called as Path) is not considered for monitoring the areas like ROI/POI. This type is completely used for detecting any intruder in the border of path of ROI. Barrier coverage further divides into weak barrier and strong barrier. Weak barrier is known for detecting any intruder crossing the path is detected by at least k-sensors using orthogonal crossing path, whereas strong barrier is guaranteed that intruders are detected by k-sensors using arbitrary moving path.

2.2 *Coverage Methods or Strategies*

This problem occurs when nodes are deployed in sensor network. The strategies are divided into three categories namely, (i) Force based (ii) Grid based and (iii) Computational geometry based [3].

2.2.1 Force Based

This strategy relies on mobility of sensors using attractive and virtual repulsive forces [3]. Here, sensor nodes are either close to each other or far from each other in order to increase the coverage area. It keeps moving until it attains equilibrium state; where attractive and virtual repulsive forces are equal [4].

2.2.2 Grid Based

Grid-based strategy is used for two purposes in WSN deployment, i.e., to calculate coverage area and to determine the sensor positions [4].

Coverage percentage is defined to be "ratio of area covered to the area of ROI." However, measuring coverage percentage becomes difficult when area covered with irregular shapes forming maximum overlap (Fig. 3).

To sort out such problems, grid based is divided into three categories namely,

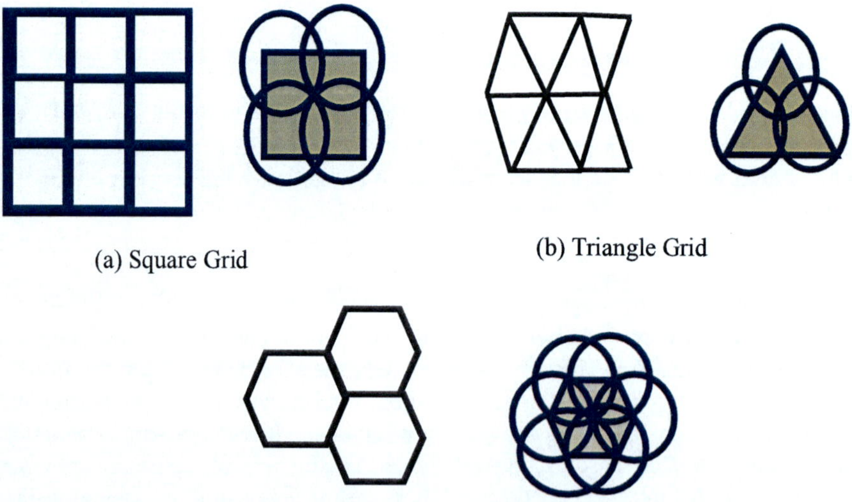

(a) Square Grid (b) Triangle Grid

(c) Hexagonal Lattice

Fig. 3 Types of grid

Table 1 Comparative study on different types of coverage

Type of coverage		Limitation	Benefits
Area coverage	Circular	Chances to occur holes	It covers only few areas around sensors
	Angled	It is better to avoid such coverage due to communication breakage	Best to cover only angle shape area
Grid-based coverage	Square	Less chance to occur holes	Easy to maintain and hole is small
	Triangle	Holes may distract the communication	Easy to maintain and hole is small
	Hexagon	It is not preferred mostly	Complex
Point coverage		Only few regions are covered	Easy to handle
Barrier coverage		Complex	Less bandwidth and more disturbance

(i) Square Grid—It provides better performance for any parameters.
(ii) Triangular lattice—It is the best among other types because it has the smallest overlap with least no of sensors.
(iii) Hexagonal Grid—It is the worst one where maximum area are overlapped (Table 1).

2.2.3 Computational Geometry Based

This type base is mostly used in wireless sensor network for optimizing coverage. The most frequently used approaches are Voronoi diagram and Delaunay triangulation.

- Voronoi diagram divides its region into partition of site which has polygonal shapes. If whole region is completely covered by polygonal vertices, then there won't exit holes in ROI.
- Delaunay triangulation is the partition of site in triangular form and circumcircle of one triangle which does not overlap any other triangle, i.e., circumcircle. To determine worst and best case of coverage, these two approaches are mostly used in WSN (Figs. 4 and 5).

Fig. 4 Voronoi diagram

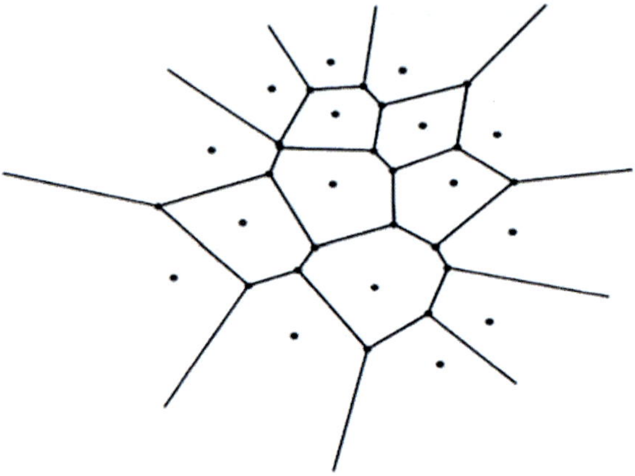

Fig. 5 Delaunay triangulation

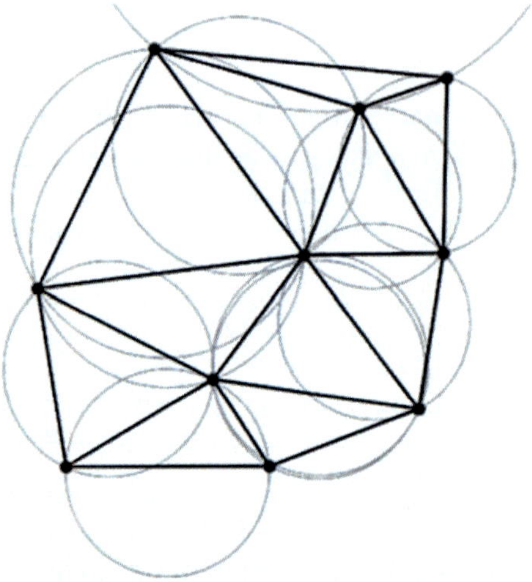

2.3 *Coverage Properties*

2.3.1 Full Coverage

It is the basic type of coverage excellence others are branches of this type. Here, detection model is used based on probabilistic method with predetermined threshold rate. A target is detected when detection model is within a sensing range simultaneously the detection probability on the target should be more than threshold value.

2.3.2 Partial Coverage

When it tries to extend the lifetime of network, this coverage weakens the coverage quality. The best and simple solution to solve this problem is coverage ratio. CR is the measure of target being covered for target coverage in percentage [5]. Among many problem-percentage, coverage is the most important problem in partial coverage. Partial coverage espouses quantitative ratio to control coverage quality because it produce ambiguous properties.

2.3.3 Probability Coverage

Probability coverage represents data in probabilistic manner to describe coverage quality. The probability coverage of an area is calculated as,

$$P(a) = 1 - \prod_{1}^{z}[1 - p(i)(a)] \tag{1}$$

where [5] a represents allocation point in the ROI, pi(a) represents the probabilistic of detection by the ith sensor node for a, z is the no. of sensors and on the basis of detection model $p(i)$ is calculated and finally $P(t)$ results the overall coverage probability.

The main aim of coverage probability is to extend the coverage. To measure maximum coverage probability,

$$\text{Max} \int \Omega R(a) P(a) \text{dt} \tag{2}$$

where $R(a)$ is a density function of an object, for some $t \in \Omega$ [5, 6].

Apart from the above three properties, few researchers have highlighted that connectivity and network overhead is an important element to be concentrated to have maximum coverage network.

3 Literature Review

3.1 Swarm Intelligence Family

Coverage is considered to be major issue, and it determines the Quality of Service (QOS) in wireless sensor network. In general, random deployment becomes fundamental issue for maximum coverage holes. On review, it is understood that many researchers have worked on coverage control using particle swarm intelligence algorithm compare to other techniques.

- *Related Work.*

Particle Swarm Intelligence (PSO)

Chuanyun Wang1 et al. [7] have designed on the basis of spatial neighborhood, the node coverage and area coverage model. Initially, spatial neighborhood is used to divide sub-local optimal values. Euclidean distance

$$d = \sqrt{(x_i - x)^2 + (y_i - y)^2} \tag{3}$$

is used for node coverage, whereas for area coverage ratio it is defined as combination of coverage node set X to the area targeted is expressed as,

$$R_{\text{area}} = \frac{\sum_x P_k}{m * n} \tag{4}$$

where $m*n$ is the monitoring target, P is the probability of target area k. The above two equation produce an output which is given as input for optimal function. Finally, particle swarm optimization with coherent velocity [7] is used to neglect solution falling into local optimal rather than better solution.

Manish Kumar et al. [2] proposed new approach, i.e., particle swarm optimization with Voronoi diagram to wrap the entire network area by optimal sensors nodes. Firstly, Voronoi diagram is used to divide the area into partition based on location of sensor nodes. Secondly, PSO is used where each node maintains a table containing its best location and its nearest best location. The velocity and location of sensor nodes are updated until it achieves best current location which in turn it triggers to find the globe best from the fitness function.

Dianna Song et al. [8] PSO algorithm is used along with particle flight velocity for WSN coverage. The flight speed velocity varies from 0 to vmax (100), where vmax is the maximum velocity. By setting appropriate vmax value, it is possible to attain highest percentage of WSN coverage.

Jin Wang et al. [9] have proposed new concept initially by dividing the network area into grid shape, with static nodes using random deployment. To optimize the coverage, coverage rate and energy cost of each grid are measured. In order to equalize the rate of coverage and energy consumption among all grids, sensing radius is adjusted for each grid. To obtain optimal coverage, all grids are adjusted using sensing radius.

Hongshan Kong et al. [10] have designed another way to enhance the PSO algorithm is, by adjusting inertia coefficient and mutation operators. On the other hand, these adjustments reflect on global convergence and particle diversity of PSO by preventing from local convergence.

Other Swarm Intelligence Algorithm.

Lei Wang et al. [11] have used very unique algorithm named Whale group algorithm which is one among swarm optimization algorithm. In this paper, author enhanced the model by using basic whale swarm optimization algorithm in the node deployment issue. Later using reverse learning algorithm, the network is protected from false movement of nodes in the adjacent distant node. This increases the speed of searching ability and improves the network coverage.

JingwenTian et al. [12] have introduced new technique, i.e., Genetic algorithm and Ant Colony optimization are improved and fused aiming at their disadvantages. This resulted in introducing an enhanced technique namely, improved genetic algorithm and binary ant colony algorithm. In this approach, binary ant colony solution is an initial population forwarded to the genetic algorithm. This method increases the accuracy, network coverage rate and effectively reduces the energy loss of wireless sensor network.

Osamah Ibrahem Khalaf [13] has implemented using BEE algorithm which has proved that compare to genetic algorithm, and it takes less amount of time to utilize the system resources with maximum percentage of coverage (Table 2).

YingyingFeng et al. [14] have proposed new technique by improving artificial fish swarm algorithm. This algorithm effectively improved network lifetime and reduces the energy consumption on the basis of Quality of Service (QoS).

HuynhThi ThanhBinh et al. [15] have used Cuckoo Search and Flower Pollination algorithm which is belong to the nature inspired algorithm to attain best coverage rate along with less computational time. This approach has proved that when network have different types of sensors these techniques have reduced the overall running time of a system.

Table 2 Comparative study of review article

Author	Techniques	Limitations	Inferences
Chuanyun Wang1 et al. [7]	PSO with coherent velocity	Avoided local optimal solution	Network Coverage improved
Dianna Song et al. [8]	PSO with particle flight velocity	Velocity ranges from 0 to 100	Coverage rate is increased
Lei Wang et al. [11]	Whale algorithm	Idea of reverse learning is included	Enhance the node search. Speedup global search
JingwenTian et al. [12]	Genetic algorithm and ant colony optimization	Less nodes	High coverage rate. Increased network lifetime
YingyingFeng et al., [14]	K-means and artificial fish swarm algorithm	Targeted clustering	Improves coverage rate. Running time is reduced
HuynhThi ThanhBinh et al. [15]	Cuckoo search and chaotic flower pollination	Complex	Reduced the energy level

4 Solution for Coverage Problems

In this section, we develop a framework to solve the coverage problems. This framework consists of three different categories for solving coverage problems, i.e., **Model, Objectives and Strategy**. In each category, various uncertain properties are focused.

4.1 Model Construction

Model construction consists of various models namely detection model, deployment model and network model. Each model is responsible for different task which range from small to sky-scraping level, i.e., i) Detection model detects sensor features, ii) Deployment model maintains relationship between sensors and iii) Network model plays the major role is maintain the component and formation of network. Figure 6 represents model construction in brief.

Detection model either can be deterministic or probabilistic which is responsible for sensing and detecting the quality of sensors based on geometric location of a sensor and target. Detection model divides into 2 types,' i.e., omni-directional and directional model. Omni-directional model further divides into 3 types; they are 0/1 model, attenuated model and truncated model. *Omni-directional 0/1 model* is the most traditional method where sensor detect the target else not [6, 16]. But attenuated model performance is superior to 0/1 model in sensing process because it receives the strongest signal.

Finally, truncate method is used only once the space is too far or too small to sense. *Probabilistic model* use probability method where detection level ranges from top to bottom bound to detect the target. In directional model, each sensor has only one working direction, i.e., θ $(-\pi \leq \theta \leq \pi)$ defined as angle value relative to the positive x-axis [16].

Deployment model is important for deploying sensor in the region of interest. As discussed above in section II, it consists of various categories namely triangle,

Fig. 6 Model constructions

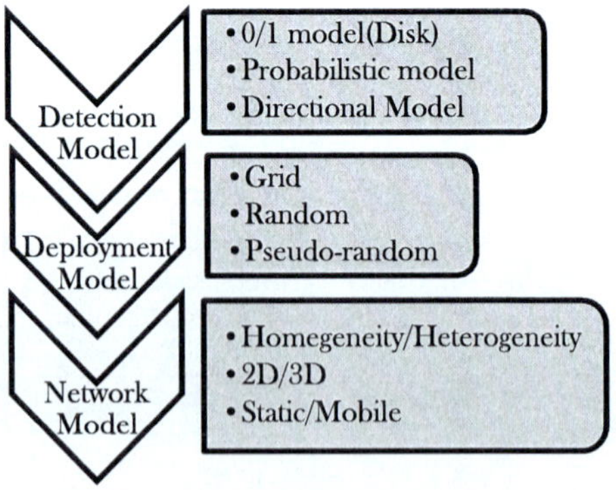

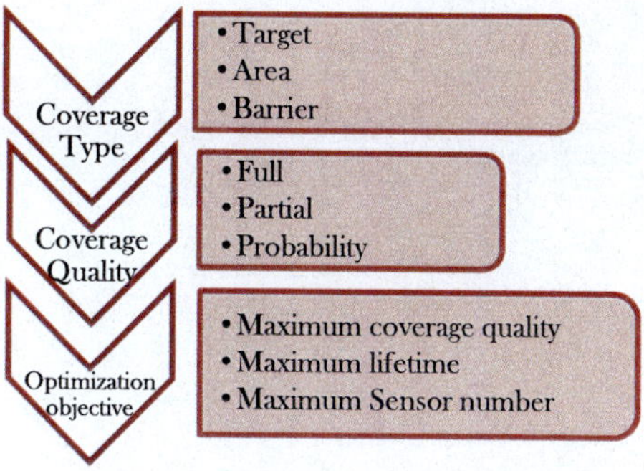

Fig. 7 Objective formulation

hexagon and square grid. Random deployment is the real time deployment which is used in rough and deep environment where continuous surveillance is maintained.

Homogenous models have equal range of detection, sensing choice and computation control, whereas in **Heterogeneous model—**It varies based on the position of region.

4.2 *Objective Formulation*

Coverage type, coverage quality and optimization objective are three different categories in objectives. Coverage quality is the base to solve the problems in coverage. It can be evaluated using coverage ratio and coverage probability. In partial coverage problem, coverage ratio declares how much ROI is covered in percentage, whereas in coverage probability it indicates in degree.

Optimization objectives—To achieve effective network, certain optimized goal has to be proposed. As shown in Fig. 7, to have optimized network, maximum coverage quality, maximum lifetime and maximum sensor number should be achieved. Many researchers introduced several techniques where homogenous and heterogeneous nodes are used in their WSN.

4.3 *Solution Design*

Deploy, schedule and movement are the different solutions to achieve the objective. Deployment is responsible for allocating sensor nodes to attain coverage quality. Based on original position, early location of sensor nodes must be deployed. But there are possibilities for sensor to deviate due to abnormal events. Scheduling strategy decides and changes the subset of nodes to be in active mode for certain duration and remaining nodes as sleep mode. This can be achieved once initial position of sensor nodes is known. Scheduling strategy aims to increase the lifetime of networks.

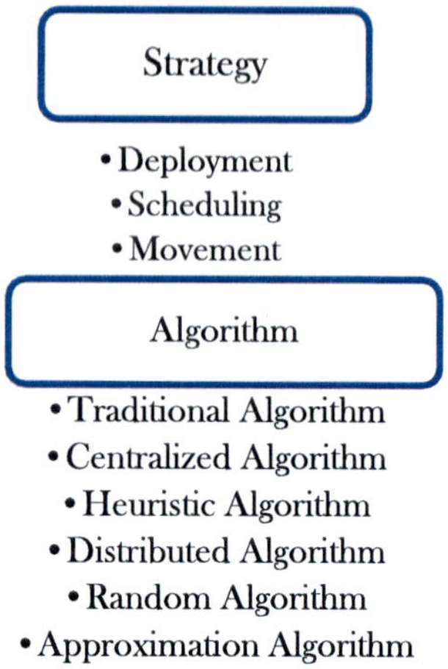

Fig. 8 Solution design

Selection strategy is one time task for subset of nodes for whole time. Nowadays, all researchers look for mobility components because mobile nodes can easily adjust and move the position of sensors. Thus, location of sensors is changed by dealing with movement model. But energy consumption is the major drawback faced in movement strategy (Fig. 8).

In location problem, rotating direction sensors is used. Two algorithms were proposed here: **Concurrent rotation and motion control** (CRMC) and **staged rotation and motion control** (SRMC).

5 Conclusion

Coverage is the most important investigation topic among researchers in wireless sensor networks (WSN's), and it is used to determine the Quality of Service of the network. The main objective of this paper is to deliver different category under network coverage components that has to be concentrated to increase the network lifetime using standardized techniques. So, in this paper, we contributed clear study on network coverage in different view, i.e., (i) **coverage problems** which is divided into area, target and barrier problem (ii) **coverage method** which is divided into grid, force and computational geometry method and (iii) **coverage properties** which includes full, partial and probability coverage. Further, we discussed how swarm intelligence algorithms are used for coverage control by many researches. Finally, a framework is described using three procedures to crack the coverage problems in WSN's.

References

1. R. Elhabyan, W. Shi, M. St-Hilaire, "Coverage protocol for wireless sensor networks: review and future directions". J. Commun. Netw. **17**(4) (Jan 2019)
2. M. Kumar, V. Gupta, "Benefit of particle swarm optimization and voronoi diagram for coverage in wireless sensor network". Int. Conf. Emerg. Trends Comput. Commun. Technol. (ICETCCT) (2017)
3. H. Kaur, Dr. S. Kautish, "An implementation of wireless sensor network using Voronoi_PSO (Particle Swarm Optimization)", Int. J. Res. Appl. Sci. Eng. Technol. (IJRASET), **4**(XI) (Nov 2016) ISSN: 2321–9653
4. N. A. A. Aziz, K. A. Aziz, W. Z. W. Ismail, "Coverage strategies for wireless sensor networks". World Acad. Sci. Eng. Technol. Int. J. Electron. Commun. Eng. **3**(2) (2009)
5. Y. Wang, S. Wu, Z. Chen, X. Gao, G. Chen, "Coverage problem with uncertain properties in wireless sensornetworks: a survey". Computer Networks(Elsevier) (2016)
6. A. More, V. Raisinghani, "A survey on energy efficient coverage protocols in wireless sensor networks" J. King Saud Univ.—Comput. Inf. Sci. (2016)
7. C. Wang, E. Sun, F. Tian, "Optimal coverage algorithm of wireless sensor networks based on particle swarm optimization with coherent velocity, "Int. J. Grid Distributed Comput. **9**(9) (2016), pp. 293–306. https://doi.org/10.14257/ijgdc.2016.9.9.25 ISSN: 2005–4262 IJGDC Copyright © 2016 SERSC
8. D. Song, J. Qu, "A fast efficient particle swarm optimization algorithm for coverage of wireless sensor network", Int. Conf. Comput. Syst. Electron. Control (ICCSEC) (2017)
9. J. Wang, C. Ju, " A PSO based energy efficient coverage control algorithm for wireless sensor networks", 2018 Tech Science Press, CMC, **56**(3), 433–446 (2018)
10. H. Kong, B. Yu, "An improved method of WSN coverage based on enhanced PSO algorithm". IEEE 8th Joint International Information Technology and Artificial Intelligence Conference (ITAIC 2019)
11. L. Wang, W. Wu, J. Qi, Z. Jia, "Wireless sensor network coverage optimization based on whale group algorithm", Comput. Sci. Inf. Syst. **15**(3), 569–583
12. J. Tian, M. Gao, G. Ge, "Wireless sensor network node optimal coverage based on improved genetic algorithm and binary ant colony algorithm". EURASIP J. Wireless Commun. Netw. **2016**:104 (Springer) (2016)
13. O. I. Khalaf, "Optimization of wireless sensor network coverage using the bee algorithm". J. Inf. Sci. Eng. (Research Gate) (May 2020)
14. Y. Feng, S. Zhao, H. Liu, "Analysis of network coverage optimization based on feedback k-means clustering and artificial fish swarm algorithm", Special Section On Green Communications On Wireless Networks IEEE Access 2020
15. H. T. T. Binh, N. T. Hanh, L. V. Quan, N. Dey, "Improved cuckoo search and chaotic flower pollination optimization algorithm for maximizing area coverage in wireless sensor networks". Neural Comput. Appl. **30**, 2305–2317 (Springer) (2018)
16. A. Maheshwari, N. Chand, "A Survey onWireless Sensor Networks Coverage Problems", *Proceedings Of 2nd International Conference On Communication, Computing And Networking*" ICCCN 2018, NITTTR Chandigarh, India
17. R. V. Kulkarni, Senior Member, IEEE, G. Kumar, V. Moorthy, Senior Member, IEEE," Particle Swarm Optimization inWireless-Sensor Networks: A Brief Survey", IEEE Trans. Syst. Man Cybernetics—Part C: Appl. Rev. **41**(2) (Mar 2017)

IOT-Based Remote Patient Monitoring System Using Microservices Architecture

B. A. Sujatha Kumari, K. S. Shreyas, M. S. Skanda, Manoj Kumar, and C. D. Prajwal

Abstract Medical and healthcare monitoring system has had rapid growth in not only hospitals but other health-related centers as well. Effect of a wrong prescription leads to death of a patient, also unnecessary medicine prescription is the burden for the guardians or caretakers. IoT (Internet of things) allows a variety of devices to interact with each other to enable monitoring, storage, analysis and display. This paper explores the developments taken in the field of patient monitoring systems to avoid wrong and unnecessary prescription, and through this research, we found that selecting IoT protocol to design the system is important and depends on the system requirements like number of data sending nodes (Data Producers), network traffic hitting the server, scalability, network latency and real-time requirement. This paper proposes the design of a real-time patient monitoring system with microservices architecture to send patient's data from ICU to the doctors and guardians. The proposed system also helps to reduce unnecessary prescription and billing.

Keywords Application programming interfaces (APIs) · Microservices · Web application · Websockets · Constrained application protocol (CoAP) · Message queue telemetry transfer (MQTT) · Real-time application · Representational state transfer (REST)

1 Introduction

Health can be defined as a state of physical, social and mental well-being and not necessarily a lack of illness. It is an essential component of people's need for a better life. Unfortunately, the global health problem has created a difficulty because of certain factors, such as poor health facilities, the remoteness of rural areas from urban areas, physicians, and nurse's unavailability during the hardest time.

B. A. Sujatha Kumari (✉) · K. S. Shreyas · M. S. Skanda · M. Kumar · C. D. Prajwal
Department of Electronics and Communication Engineering, Sri Jayachamarajendra College of Engineering, JSS Science and Technology University, Mysore, Karnataka, India
e-mail: sujathakumari@sjce.ac.in

G. Ranganathan et al. (eds.), *Soft Computing for Security Applications*,
Advances in Intelligent Systems and Computing 1397,
https://doi.org/10.1007/978-981-16-5301-8_28

IoT is enabling any objects internally connected in the recent past and it is being considered as the next technological revolution. Smart patient monitoring systems, smart parking, smart home, smart city, smart weather station, industrial automation, and smart irrigation system are some of the applications of IoT. The immense use of the Internet of Things is in healthcare management, which provides ICU sensors data tracking with low bandwidth utilization. IoT is simply the connection of computers to the internet utilizing sensors and networks. The components that are connected can be utilized for monitoring the vitals of a patient. It is a scalable, interoperable, energy efficient and robust way of tracking and optimizing care to any health problem. Now, present systems are equipped with a flexible interface, assistant devices, and mental health management to lead a superior life for the human being [3]. Medical sensors connected to the patient send all the information to the doctors directly to their assigned doctors who can monitor them from anywhere. This method for monitoring health greatly improves the responsiveness of treatments and keeps tabs on all the patients.

2 Related Work

Following papers were referenced to design the Remote patient monitoring system with microservices architecture.

Paper [1] presents an implementation of an automated IoT-based healthcare monitoring system for remotely located patients (or when the doctor is outside the hospital) which helps doctors to visualize the patient's body condition and guide them accordingly. Constrained Application Protocol (CoAP) is used to send data to the Central server. This system also sends Email alerts in abnormal cases only to the doctors. It provides an easy to use web interface that can be accessed by only doctors and hospital officials. Complex machine learning algorithms are used to analyze the real-time patient's data.

Paper [2] demonstrates an IoT system built with ZigBee mesh technology to send the body sensor data to the centralized gateway which broadcasts it to all the doctors and hospital faculties. This system periodically monitors the physiological parameters of the In-hospital patients.

A Smart Hospital System (SHS) architecture is proposed in [3] which uses RFID, WSN, and smart mobile, interoperating with each other through a 6LoWPANIoT protocol to send the patient's medical data. Sensed data are delivered to a central node (control center or server) where an advanced data visualization application makes them easily accessible by both local and remote (with the help of VPN) users via a REST web service. 6LoWPAN packets are sent to the control center through RFID communication.

Paper [4] demonstrates the patient centric agent based on healthcare architecture, where patient's sensor data are collected using smartphone (Bluetooth or WiFi). This data has been uploaded to the central server where processing is done. After processing, data will be sent to the doctor. All the communications are encrypted using lightweight Blockchain technology.

In [5] RFID tags are utilized to transmit information from the medical sensors to the Database server which adds a medical privacy protection scheme. The scheme ensures security privacy of the collected data via secure authentication. The security analysis and evaluation about the scheme indicate that the protocol can effectively prevent the risk of medical privacy data being easily leaked. RFID tag readability, Interoperability with other health information technology is a major flaw of the system.

Paper [6] proposes a system which can monitor a patient's vitas in real-time by using various sensors. A portal is created to relay the information to the medical staff who will analyze the information to take suitable steps to normalize the patient's condition if there is any anomaly.

The system that is tested here is bulky, but with the right manufacturing techniques, the volume can be decreased considerably. The system is very useful in the case of infectious disease like a novel coronavirus (COVID-19) treatment. The developed system will improve the current healthcare system that may protect lots of lives from death.

Wireless health monitoring of the patients has been proposed in [7]. The vital parameters are measured by the sensors such as pulse sensor, temperature sensor and SpO2 sensor. The data is stored and can be visualized on the web server. The drawback of this system is the results cannot be obtained for multiple patients at a time. In [8] the collection of data is made through wearable devices and sent through API endpoints. The data consisted of energy metabolism, heart rate and moments of sleep. The system was designed to be a mobile application and this system could help the users to monitor their health-related information and use the exhortation from the system for their health in daily life.

Summary of Literature Survey.

Sl. No	Paper title	Advantages	Drawbacks
[1]	Automated IoT-based healthcare system for monitoring of remotely located patients	Provides email alerts, data monitoring and generates reports of the patient and sends it to the doctor. Uses machine learning algorithm to detect the variation in patient health	Computationally intensive Only doctors can access the data, not the guardians
[2]	An implementation of IoT for health care	System is more energy efficient because of ZigBee	Not suitable for real-time monitoring applications Vulnerable to cyber attacks Scalability issues

(continued)

(continued)

Sl. No	Paper title	Advantages	Drawbacks
[3]	An IoT-Aware Architecture for Smart Healthcare Systems	Global access of data Data is secured because of VPN Easily usable web interface	Speed depends on uplink of the internet Not scalable, complex design Frequency band issues
[4]	Continuous patient monitoring with a patient centric agent: a block architecture	Provides real-time data on the dashboard. Security of the patient's health data is high because of Blockchain technology	Computational intensive system Bluetooth vulnerability
[5]	Lightweight RFID protocol for medical privacy protection in IoT	RFID is a simple, fast and reliable method of transmission. RFID tags and inexpensive	RFID tag readability, interoperability with other health information technology
[6]	Development of smart healthcare monitoring system in IoT	Monitor a patient's basic health signs and Authenticated users can see the dashboard	System is very bulky. Lack of encryption protocol
[7]	SemanMedical: a kind of schematic medical monitoring system model based on the IoT sensors	Takes all the readings from the sensors and stores it in two different database servers, Intercommunication algorithm reduces the time complexity	Not a real-time model that can be utilized for regular monitoring
[8]	An IoT-based CBL methodology to create real-world clinical cases for medical education	A medical expert can create real-world clinical cases for their students to rehearsal the CBL cases	Secure Transmission methods and protocols not defined
[9]	IoT-based real-time health monitoring system	The proposed system is a low-cost system	Proposed model doesn't include the blood pressure monitoring system
[10]	A Medical-IoT-based framework for eHealth care	This system is more energy efficient than the traditional approaches in IoMT	The proposed algorithm is not highly secured
[11]	Smart health monitoring system with IoT	Data recording and analysis using RapidMiner (Specialized data analysis program)	The accuracy from the results was 62.05%
[12]	IoT-based smart edge for global health: remote monitoring with severity detection and alerts transmission	Based on the alert levels, the low and normal patients can avoid unnecessary hospital visits and expensive medical procedures	The drawback for applications which have high requirements of latency

3 Software and Hardware Requirements

Raspberry Pi: In this proposed work Raspberry Pi 4 is used to host the Django and Node JS web application and also to run the databases (MongoDB and SQL) on Docker containers.

Python: Entire web application is written in Django, a web application framework written in Python.

JavaScript: It is used in both Backend and Frontend development. In the frontend Vanilla JS, JQuery and Plotly JS are used to enhance the interactivity, responsiveness and for Data visualization. In the backend multiple Node JS applications are used to receive MQTT data, handling real-time data storage with MongoDB.

ESP32: This microcontroller is connected with sensors that measure the data and send it to the MQTT broker.

Databases: MongoDB stores real-time patient's data, PostgreSQL connected with Django server stores users, patients and appointments-related information.

4 Proposed Work

The proposed system architecture is explained in this section. Raspberry Pi 4 Single Board Computer is used as the brain of the entire system. This hosts a Django Web Application, MongoDB database to store time series data, Django REST API to send time series data. It also runs MQTT Broker to receive data from MQTT nodes, a Web server to forward requests to the application. This project uses Django channels built-in ASGI web server to handle both HTTP and websockets requests.

4.1 Data Acquisition

Patient's health data is collected from the sensors attached to the ESP32 development board. Micropython is used to program the board. ESP32 connects to the network and the MQTT broker. Then periodically measures the data and sends it to the MQTT broker in JSON format.

4.2 Data Validation

NodeJS application on Raspberry Pi validates the incoming data then stores it in the database and sends it to the Web server through websocket connection. Sensed data is validated against its minimum and maximum value (e.g., negative values are not possible in health data). This application will not forward invalid data to the MongoDB database.

4.3 Data Storage

Validated data is stored in NoSQL database and the User and Authentication-related information is stored in the PostgreSQL database. Django directly stores data in the PostgreSQL database, to reduce the work on Django web server (One microservice) time series data is stored in MongoDB by NodeJS web server (Microservice container). Along with the body parameters medicines given to the patient are also recorded in the NoSQL database along with the prescription timestamp. PostgreSQL database schema is defined as models in Django application, and it is converted into SQL query when migrations are done. Patient's health data schema is defined in the ExpressJS web application. Mongoose library is used in the proposed system to interact MongoDB with the ExpressJS server.

4.4 Data Visualization

When the frontend receives time series patient's data through the websocket connection, it automatically refreshes the browser to update the graph in real-time as shown in Fig. 3. User gets a real-time plot of all the patients he subscribed to and are in the ICU. When the patient's appointment is closed (After his recovery), a report which contains the graph of body parameters (blood pressure, body temperature, etc.), medicines given, its composition, type of the medicine and the chat history. This data is important to analyze the effect of particular medicine on a patient's body.

5 Implementation

Implementation part of the proposed system is explained in this section.

The web application is designed with the Django web framework, Django project is divided into multiple applications which handle particular work independently without interfering with other applications. Django user URL type router to differentiate HTTP and websockets requests, forwards the requests to a particular application which defines view method implementation (Django uses Model-View-Template pattern). *Accounts* application handles log-in, log-out, register, adding patients to the database, adding patient subscriptions to clients and other user-related actions. *Data streamer* application initiates websocket connection with the logged-in user, adds him to the subscribed groups by fetching it from the database, initializes the dashboard, and starts streaming sensors value to the dashboard periodically. To achieve high throughput and to reduce load on the main server Redis is used as a message queue, which is a simple key-value database. *Chat* application initiates the websocket connection with the chat server and fetches previous sent messages. Doctors assigned to the appointment, Admin user and Guardians are group members of the Chat group. *REST API* listens for the incoming requests from the users for patient's data based on the input time and patient ID. It fetches data from MongoDB database and sends it back to the requested user to render it on the dashboard. This application also exposes data in the PostgreSQL database to other microservices. This flow is mentioned in Fig. 1.

'Mosquitto', MQTT Broker collects all the sensor data published by the MQTT nodes based on the topic (ICU number). Broker sends the same data to express server so that it saves real-time data in MongoDB and sends the same values to the Django server via a websocket connection (Established when server starts). In this case,

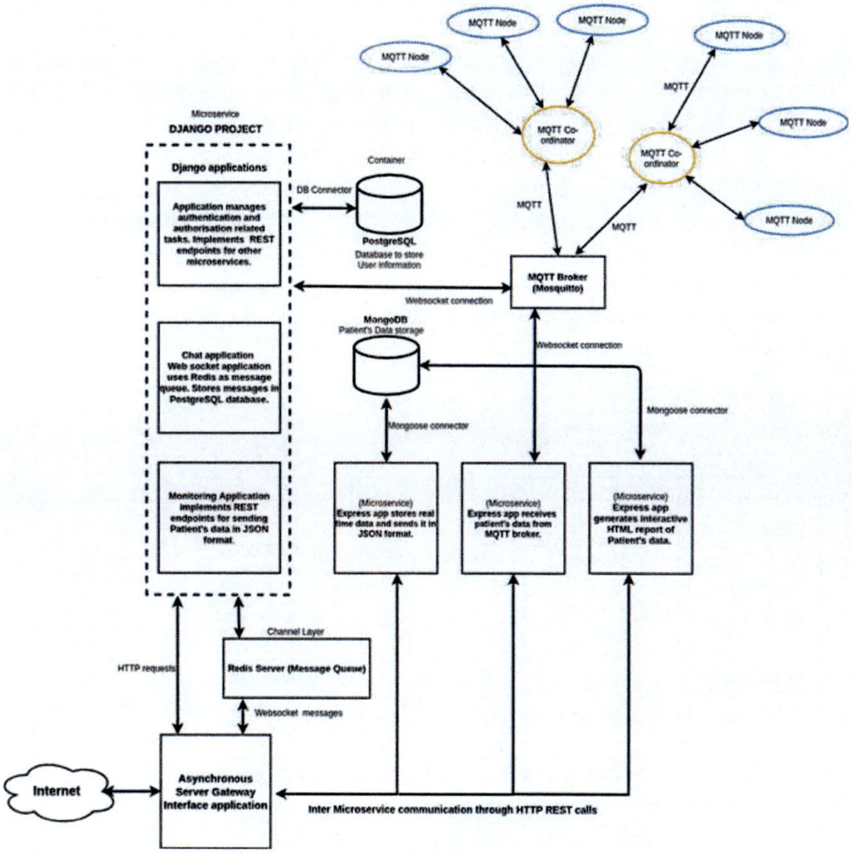

Fig. 1 Microservices architecture of patient monitoring system

publishers are sensor nodes and Subscriber is the Django Application and MongoDB database connection (Through NodeJS application using Mongoose library).

Requests come to the Django application through Asynchronous web Server Gateway Interface (ASGI) which separates the requests based on protocols (HTTP and websockets). HTTP requests are handled by Django's middlewares and websocket requests are handled by Django Channels. Django uses PostgreSQL database to store Session, Authentication and authorization-related information. Django has a built-in connector for PostgreSQL database and it is open source software. MongoDB is used to store the real-time patient's data which is a time series data. NoSQL database is used because data is time series with key-value pairs and data is stored using bucketing technique (Grouping multiple documents) which increases the database read efficiency. A medicine database has been created with all the information about each medicine (medicine type, dosage, drugs used, etc.). Further, Django application fetches the data about a medicine when the doctor prescribes it to the patient. Thereby guardians can check the details of the medicine and its price. Report and Bill Generator program generates the report of a patient which includes plot of the sensor values (Time series data), medicines given and the total bill amount, chat history. This generates the HTML file of the graph (Interactive) and PDF of the report using reportlab library.

Entire system is divided into 5 microservices as shown in Fig. 2. Each microservice is deployed on a Docker container. Except Django microservice remaining microservices do not allow Cross Origin Resource Sharing (CORS) hence increases system security. Databases can also be containerised to increase throughput and efficiency. Scalability of the system can be improved by using Container Orchestration and automation scripts can be written to increase or decrease the number of containers based on the incoming requests.

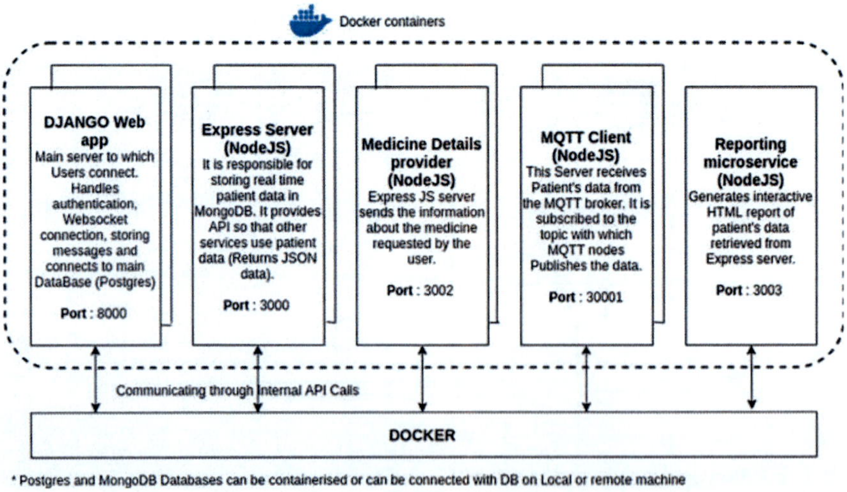

Fig. 2 Microservices containers

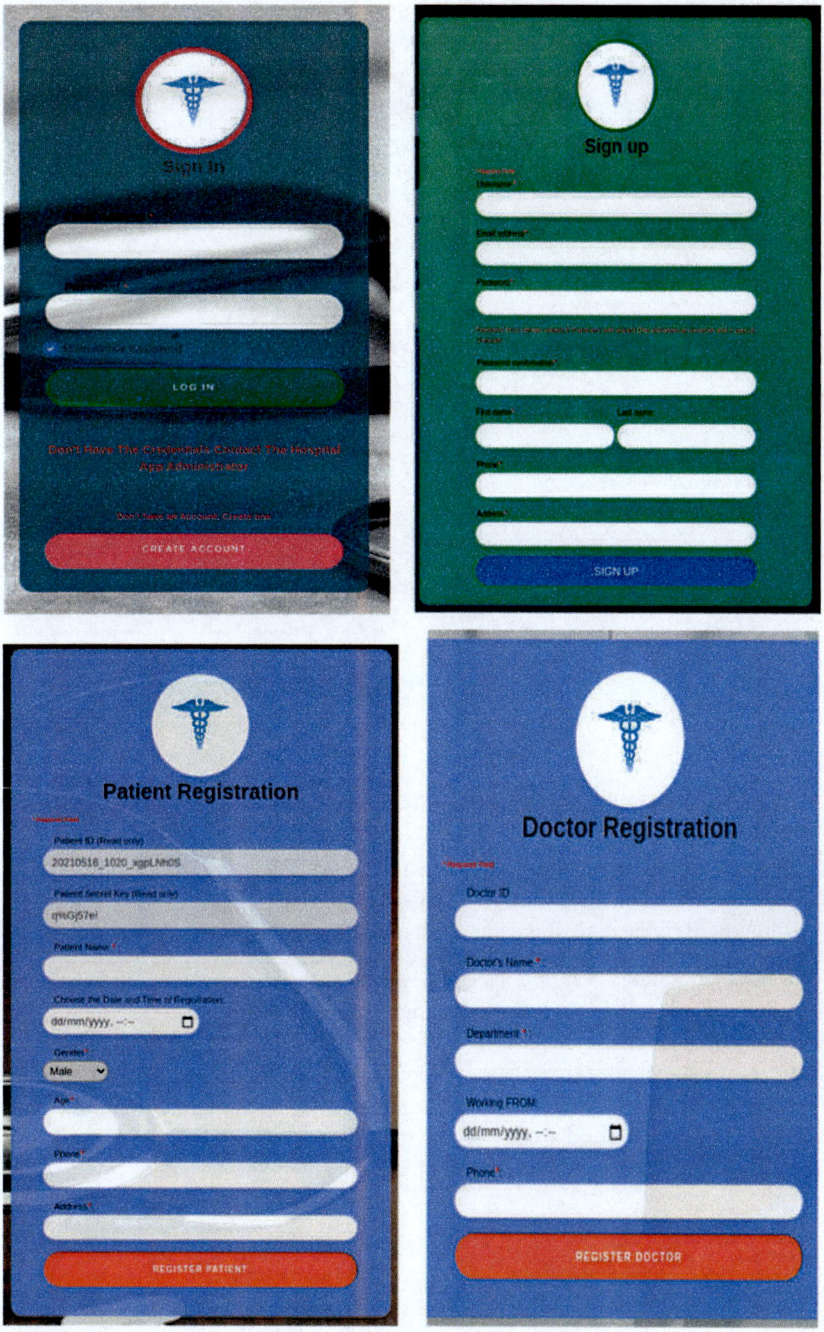

Fig. 3 User sign-in form, register form, patient registration form, doctor registration form, appointment booking form and patient subscription form

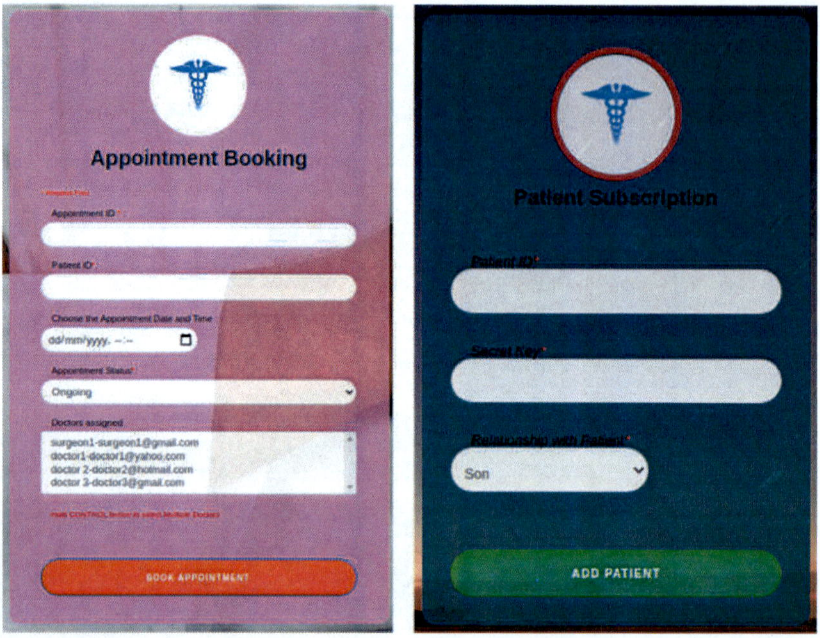

Fig. 3 (continued)

Docker Containers:

1. Django web application: This container runs Django web applications (Docker python ubuntu image).
2. Node JS application: This container runs Node JS Express application.
3. PostgreSQL: This container runs PostgreSQL database
4. MongoDB: This container runs a MongoDB database.

All these containers communicate with each other through APIs (HTTP methods with JSON data). APIs are designed using Django Rest Framework which returns the serialized data when a request is made to it. Since direct request to the API is not allowed from the User, authentication between internal microservices is not needed. Thus all the internal microservices communicate with each other without the use of Authentication Tokens and speeds up the request response cycle.

6 Results and Discussions

This section describes the web application output where the user gets the real-time patient data. In the designed web application, users can register through Sign-Up form. Admin users can register the patients and doctors as shown in Fig. 3. Each

Fig. 4 Dashboard of the authenticated user

patient is assigned with a secret key to avoid everyone to subscribe to him. User needs to enter the secret key and relationship with the patient to add a subscription.

Admin books the appointment by adding patient ID, date and time and doctors assigned to the appointment (Unique Appointment ID is generated by the system based on current date and time). Dashboard shows the maximum number of patients that can be subscribed, current subscriptions with patient details, subscribed appointments and the number of patients in the ICU as shown in Fig. 4. Dashboard page is dynamic and based on the User permissions and the group (Authentication group) sidebar links are generated.

Dashboard also contains 4 control buttons, (1) Start chat in the group (appointment group). (2) Subscribe to new patient. (3) Add appointment. (4) See real-time data of the patients in ICU.

In the real world, the system administrator of the hospital gives the secret key to the guardians during registration of the patient. Then registered users can access patient data by filling a patient subscription form, and the secret key provides a layer of security. To reduce the network traffic and load on the server a maximum of 5 Users can subscribe to a patient. This limit can be changed based on the capacity of the Django server. On the dashboard, admin user can add patient, add new ICU, update ICU when patient is discharged from the ICU. Patient's health data will be sent to authorized users in real time through the websocket connection, this data will

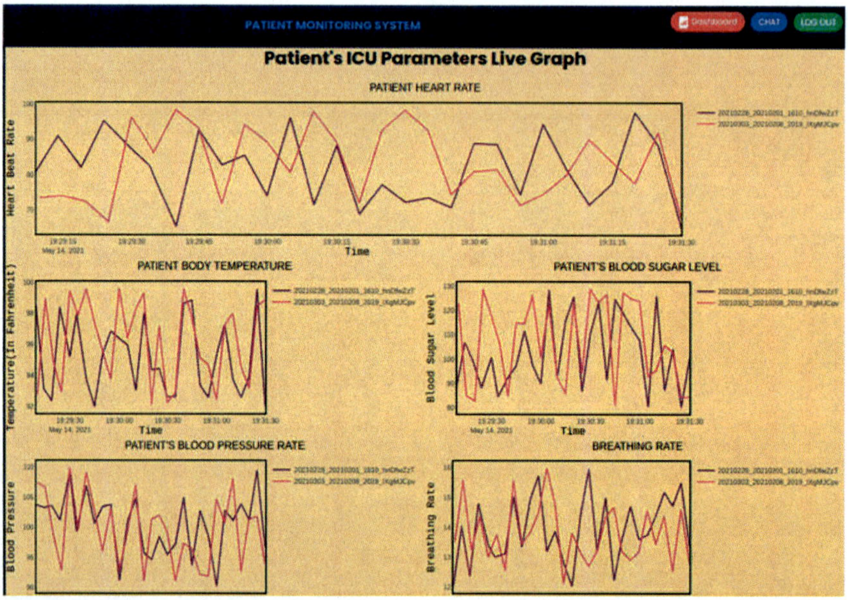

Fig. 5 Real-time graph view

not be persisted at the client end. The real-time graph is generated with the Plotly JS library as shown in Fig. 5 and the plot is interactive, responsive (Use of Bootstrap CSS). This plot refreshes when the data comes through the websocket connection. A simulator has been created with NodeJS which generates random numbers between maximum and minimum threshold value for every body parameter, which publishes to MQTT broker in JSON format. Django server has been tested with 10 MQTT Publishers and 20 registered users. Security of the data can be further improved by hosting this web application inside a hospital network with VPN server configured. When a new User gets registered a VPN profile can be created.

For the registered user there is at least one chat group present in the chat app, i.e., Admin. In this group system administrator is present, so the user can ask the admin to open any closed appointment for any queries with the doctors (All appointments will be closed after paying bills, so users cannot chat in the closed appointment group) or book the new appointment. It is possible to send text chat and images through this application as shown in Fig. 6. Chat group name is same as the subscribed Appointment name. Chat User Interface as shown in Fig. 5 fetches previous chat messages from the REST API. It only fetches 20 messages when the user hits the upper scroll bar of the message area. All the messages are stored in the PostgreSQL database connected to the Django server.

This entire monitoring system utilizes three websocket connections, to get health data, to get the push notifications and the other one is to get chat messages. An alerting system is implemented to send the push notifications when the health parameters go below or beyond the threshold values. These threshold values are set by the

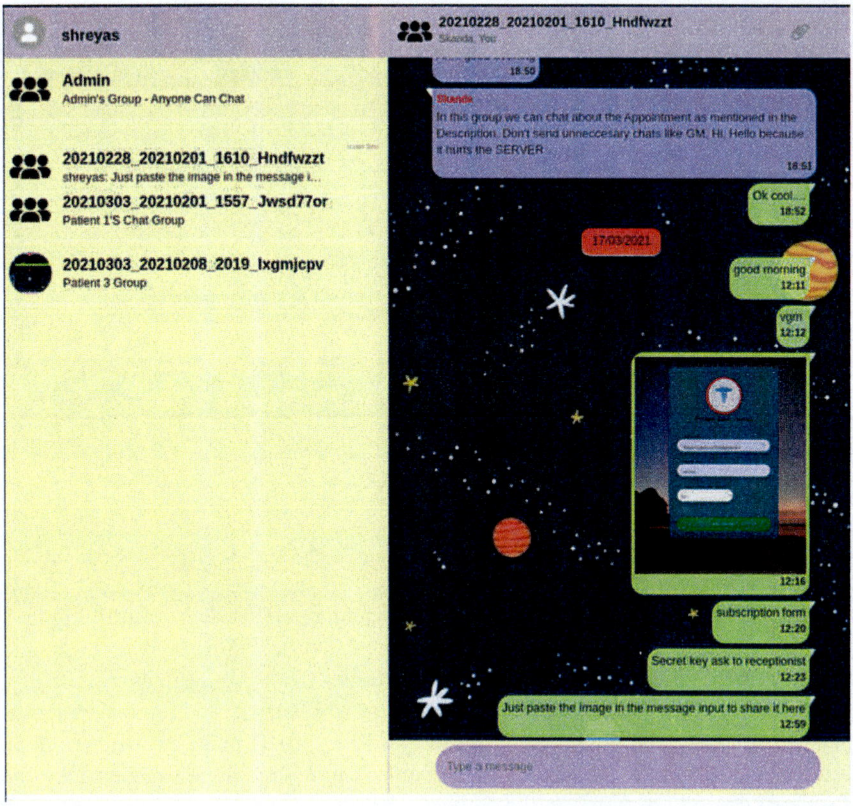

Fig. 6 Chat feature built into the application

doctors and different for different patients depending on their health conditions. Push notifications websocket channel is long running means it gets created when the user logged-in to the server and collapses when the user logged out or closes the browser tab, whereas the other two websocket connections open only when the user navigates to the particular tab (e.g., when the user goes to the chat tab it opens the websocket connection and fetches the previous messages). Chat application architecture can be further improved by creating a microservice which handles delivering messages, storing messages and creating groups and other chat application tasks. Django web application communicates with Chat microservice through API, this reduces work on the main server.

To analyze the patient's data, Fig. 7 shows the User Interface, where the user needs to enter the patient ID and appointment ID. It is also possible to plot the analysis graph from the downloaded JSON file. This graph contains multiple plots (body temperature, blood sugar level, etc.). Graph is interactive and the time scale can be varied with the slider. It also reports the medicines given to the patient along with the timestamp at which it was given.

Fig. 7 Patient's data analysis UI

7 Experimental Setup

In this section, we describe how the hardware is configured to test the results. Figure 8 is the Raspberry pi that acts as the server and the database. Figure 9 is the sensor and the transmission system that sits on the patient side.

The main server and the database are run on a raspberry pi and it receives sensor signals from BMP180 temperature sensor and MAX30102 pulse oximeter from an arduino nano through a ESP32 WiFi module. These signals are received by the Raspberry Pi and are stored in the database for it to be displayed on the dashboard for the doctors to see.

8 Security

8.1 System Security

All the microservices are deployed in a single network and the communication between each microservices is done through JWT authorization token. Requests to the microservices from other domains (Subnets) are prohibited hence the system is secure. Only HTTPS requests can be allowed by adding configuration in Django settings file, this further improves the system security by dropping HTTP only requests. Even websocket channels can be encrypted with SSL over TLS (websocket secured).

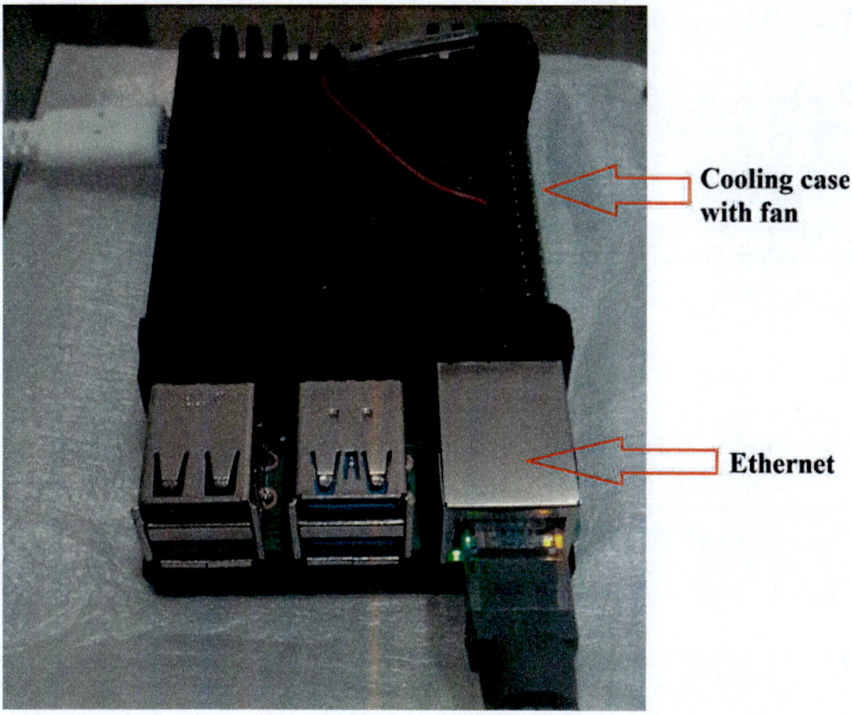

Fig. 8 Raspberry pi server

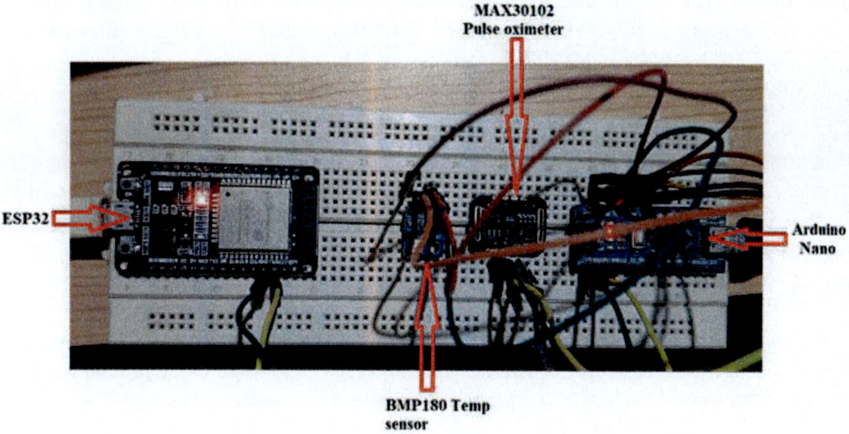

Fig. 9 Sensors and WiFi transmission system on the patient side

8.2 Application Security

Django uses Session token to handle authentication and authorisation, along with that it has built-in features to defend against Cross Site Scripting (using CSRF token in the forms), SQL injection (Using Django ORM). This makes the web application secure to some commonly known attacks. Maximum 5 users can connect to a single patient, to limit the network traffic, so the Django server does not overwhelm and thereby improves efficiency and performance.

8.3 Data Security

Each patient is given a security key (system generated) and the users with that key can only access the patient's data. NoSQL database is behind the Django server and its data can be queried only by the Django server and the data will be returned to the user in JSON format.

9 Conclusion and Future Work

Designed system sends ICU data to the subscribed Users and doctors in real time. This Microservices architecture reduces the load on a single server which improves the overall system performance and speed. But a major challenge is securing communication between each service. Proposed system uses REST APIs for inter service communication, security can be added using JWT authorisation token and cryptography (Another approach is to replace HTTP APIs with Message Queues). To scale the system as the network traffic increases Container Orchestration can be implemented as a future work.

This system provides message access to both doctors and guardians, this can be further improved by sharing medicine prescriptions with Druggists and Chemists present around the hospital so that the guardian can bring medicine from the medical shop which offers medicines with lower price.

Acknowledgements The Authors of the paper acknowledge and thank our Institution JSS Science and Technology University(Mysore), all the faculty and supporting staff of the Institution for the constant support.

References

1. R. G. Utekar, J. S. Umale, "*Automated IoT Based Healthcare System for Monitoring of Remotely Located Patients*", Publisher, IEEE 2018

2. R. K. Kodali, G. Swamy, B. Lakshmi, "*An Implementation of IoT for Healthcare*", Publisher, IEEE 2015
3. L. Catarinucci, D. De Donno, L. Mainetti, L. Palano, L. Patrono, M. L. Stefanizzi, L. Tarricone, "An IoT-aware architecture for smart healthcare systems", Publisher: IEEE Internet of Things Journal
4. M. A. Uddin, A. Stranieri, I. Gondal, V. Balasubramanian, "*Continuous Patient Monitoring With a Patient Centric Agent: A Block Architecture*", Publisher: IEEE (date of publication June 13, 2018)
5. K. Fan, Member, W. Jiang, H. Li, Y. Yang, "*Lightweight RFID Protocol for Medical Privacy Protection in IoT*", Publisher: IEEE (2017)
6. M. M. Islam, A. Rahaman, M. R. Islam, "Development of Smart Healthcare Monitoring System inIoT Environment", Publisher: Springer Nature Singapore Pte Ltd (2020)
7. V. Yeri, Dr. D. C. Shubhai, "IoT based Real Time Health Monitoring", Publisher: IEEE (2020.
8. S. Kajornkasirat, N. Chanapai, B. Hnusuwan, "Smart Health Monitoring System with IoT", Publisher: IEEE (2018)
9. S. Pirbhulal, W. Wu, S. C. Mukhopadhyay, "*A Medical-IoT based Framework for e-Health Care*", Publisher: IEEE (2018)
10. G. Xu," IoT-assisted ECG monitoring framework with secure data transmission for health care applications", Publisher : IEEE Access (2020)
11. G. Zhang, C. Li, Y. Zhang, C. Xing, J. Yang, "*SemanMedical: A Kind of Semantic Medical Monitoring System Model Based on the IoT Sensors*", Publisher: IEEE (2012)
12. B. Ramkumar, M. Sabarimalai, U. Satija, "Real-Time Signal Quality-Aware ECG Telemetry System for IoT-Based Health Care Monitoring", Publisher: IEEE (2012)

Detection and Prevention of Faulty Node in Heterogeneous Wireless Sensor Network

Santosh Anand and B. P. Adithi

Abstract Wireless Sensor Network [WSN] faces many issues in transmitting the data from source to sink node. The data or information must be securely transmitted. If the information is not transmitted securely, then there is a need to secure that node. In WSN, the communication takes place between one node and the other node with some energy sources. The node that communicates with the another node becomes faulty due to its low energy and less bandwidth. Hence, there is need to ensure the security of the node in order to avoid uninterrupted communication. The proposed research work detects the faulty node with the help of Honeypot nodes. The detected faulty node is provided with the suitable security to prevent it from attacks. The comparison of existing research works with the proposed work is done by using the Honeypot nodes in between the other nodes to find the best result in detecting the faulty node.

Keywords WSN · Cryptanalyst · Honeypot · Faulty node · Heterogeneous · Sensor node · DOS · False packet · UDGM · Mote

1 Introduction

Wireless Sensor Network (WSN) consists of several nodes, which have the capability of communicating with the other neighboring nodes. WSN has simplified the network monitoring and control operations. Attacks in the WSN can be divided into two types, where one is the attack on security mechanism and another is the attack on basic mechanism. During the transmission of packets, the cryptanalyst can listen to the transmitted packets and analyze the network operations, where the cryptanalyst uses alternate mechanism or technique, in which cryptanalyst can confuse the sensor network by introducing the false packets in the network. These false packets will contain undesirable data that is unrelated to the network; as a result, the sensor nodes will be confused; this is the situation in which the communicating node does not

S. Anand (✉) · B. P. Adithi
Department of Computer Science, Amrita School of Arts and Sciences, Amrita Vishwa Vidyapeetham, Mysuru, India

G. Ranganathan et al. (eds.), *Soft Computing for Security Applications*, Advances in Intelligent Systems and Computing 1397,
https://doi.org/10.1007/978-981-16-5301-8_29

know the destination node. With this mechanism, the cryptanalyst will obtain the node details and information of the node even if that node is not participating in the communication process. Cryptanalyst extracts all the secrets of the node and utilizes it. To prevent these types of attacks, the network must include a security mechanism. By providing security to the network, we can avoid the cryptanalyst from capturing the information. The network's security should be leveraged in such a manner that the data supplied by the node can only be understood by the destination node; giving security to the node will also entail stopping the cryptanalyst from modifying the node with undesirable data.

2 Objective

The nodes that are said to be in the normal state are considered to take part in the communication process and those nodes must interact with the neighboring node and it must utilize some amount of energy for the communication process. The node fails to interact with the neighboring node and it does not consume any energy during the process, which is considered as the faulty node. Faulty nodes can occur due to the external influences when the attacker may interrupt the process by getting into the network with the false identity. This identity will contain the packets, which is irrelevant to the process. These packets may lower the energy of the normal nodes and makes that nodes as the faulty node. These faulty nodes are detected by the Honeypot nodes present in the network. If the nodes find an external node by using the Honeypot, where it will stop the transfer of packets to the nodes.

3 Literature Survey

In [1], the author has identified fault detection methods that are found in WSN. They have used mobile fault detector in which the status of hardware and software is identified. Cluster-based fault detection algorithm is diagnosed and in this, the neighbor cluster head will take the work of the failed cluster node.

In [2], Logistical regression, machine learning algorithm is used to find any defect in the node. This work did not consider the large amount of training data because it reduces the calculation values of the sensor nodes. In the simulation part, they considered two parameters such as temperature and humidity. They have set certain range for the considered data. If the set range exceeds, then it is called as abnormal data. They have done test for the data that they have collected to know the accuracy of the collected data. They plotted the graph temperature and humidity in which the logistical regression and machine learning algorithm separates the selected data as normal and abnormal data.

In [3], author provides the implementation of prototype model by using zigbee wireless protocol, which is designed for locating and detecting a fault in the power.

This work clearly states the location, where the fault repeatedly occurs. This can be done since they have used current sensor node and zigbee wireless protocol as the combination to keep track of root mean square value even when there is a high voltage. They have used liquid crystal display screen for knowing the fault in the node.

In [4], the survey is based on different causes of faults in the sensor node, different types of faults that occur in the sensor node and various methods used for detecting failure nodes are given. They have used the method of faulty sensor node identification to detect nodes that does not work properly. The Cluster fault recovery technique is also employed. For the identification of faulty sensor node, path redundancy technique is used.

In [5], the authors have used an adaptive clustering algorithm. Here, the authors compare the static energy heterogeneity with the dynamic energy heterogeneity to propose a the model for dynamic heterogeneous WSN. The adaptive clustering algorithm is used in this work for increasing the lifetime of the network and data packets.

In [6] the authors focus on detecting the blackhole attack by using Honeypot. In this, the Honeypot by gathering the information of the neighboring node, and contacting with the attacker node will increase the rate of attack detection. Blackhole attack tree is designed for knowing the cause for the attack. The Honeypot interacts with the victim nodes and with the help of blackhole attack tree, it confirms the attack. This detection has two modules in which the nodes are kept for observation in the first module and another module performs the confirmation of attack. Authors are implemented the energy saving algorithm based on the link energy, in this link energy plays the major roles for the path selection and communication from the source and sink. This work also shows the reliability of the WSN is improved.

In [7], the comparison of Honeypots has been done in both Windows and Linux. In Windows, Kfsensor and in Linux, honeyd is installed. Various parameters such as TCP and UDP are considered as the results for the comparison. The results of honeyd are compared with the results of Windows. Honeyd and Kfsensor installation and simulation are explained clearly. To detect the intrusion detection in the network, Honeypot is set up in both the environments.

In [8], different techniques are applied on different transmission line in order to know the fault and their location of occurrence for short and long distance. Concurrent neuro-fuzzy techniques are used for the detection of fault in power transmission line. Two models are considered in which the power transmission line error and accuracy are calculated by using the amount of data utilized. First model is based on fault classification prediction and second model is based on fault prediction results.

In [9], authors have proposed a method to detect the sinkhole attack. This paper tells about how sinkhole attacks are detected. The idea behind this paper is to know the percentage of packets that are verified and the packets reaches the base station. In this work, the attacks are grouped into three types. The first type of attack is based on discovering the route in which the cryptanalyst causes a denial in the service. Second type is based on selection of route in which the cryptanalyst tries to get the legitimate node and use that node in routing the path. The third type of attack occurs once the routing path is available.

In [10], the authors mainly focus on two algorithms that are used for restoration of fault in the WSN nodes such as, "Full 2Connectivity Restoration Algorithm". This Full 2 Connectivity Restoration Algorithm can be extended further, when the available nodes are more in number. This algorithm is better and it can cover larger area of the network. This also focuses on quality of the network service.

In [11], TCSA algorithm is proposed with the use of fault tolerance method. The execution of this algorithm is based on cloud and in the environment of grid. When the work or job is completed, the scheduler split the tasks into various steps. Scheduler job is used to send the messages to the grid nodes in this task allocator to identify the fault in the jobs that is being executed.

In [12], the authors have proposed a hamming residue method for detecting the malicious attack and provide security to the WSN. The hamming code is used for correcting and detecting errors and hence one can choose the hamming code based on their network requirements.

In [13], this paper is based on providing security for the cloud by using Honeypot. Honeypot is being discussed, wherein the low interaction Honeypot and the interaction will be limited due to less attacker activity in the high interaction of Honeypot, where maximum number of information can be captured. In this, Honeypot uses less number of resources, where any attack that is encrypted is detected by the Honeypot. In, authors analysed the RREQ flooding (DOS) attack with the help of NS2. It shows that how sensor node's battery is consumed unnecessarily without doing any communications as well as increases the traffic in the WSN. Result of the work also shows the better performance compare to the existing works.

In this [14], authors are implemented for the identification of faulty nodes with respect to the QoS parameters such as energy. Once faulty nodes in the link is detected that nodes getting removed from the link and new link will be generated for the communications.

In this [15], authors implemented the innovative method for the detection and prevention of replication attacks in the wireless sensor networks. Centralized system is used for the identification, network will be fragmented into the several segments and each segment inspected separately with the help of cryptographic techniques.

In this work [16], authors did the observation on several security challenges in wireless sensor devices. Honeypot technique is implemented to attract the attacker and malicious nodes, once nodes will be trapped by the Honeypot, that nodes communication will be stopped from other nodes.

In this paper [17], authors did observation on several clustering techniques used in the wireless sensor network which enhanced the performance with several QoS parameters such as packet delivery ratio, packet drop ratio, delay and energy. In this authors also mentioned the limitation of existing works.

In this work [18], authors implemented the innovating approach of clustering which is energy efficient in the wireless sensor networks. Cluster selection is based on the energy of the node as well as cluster head will be analysed and change if its needed.

As the observations done in the survey, there is a lack of security in the node, which is detected as faulty node. Even if the node is detected as faulty node, it will remain in the communication process, which causes delay in the packet transfer. This proposed work will overcome this type of issues by providing suitable security measures for the node. To solve this type of problem, a Honeypot node is deployed into the network, where it collects the data from each node but it does not communicate with the other node. By this, the node information is collected and the node, which finds as faulty is removed from the network or that node is stopped from receiving packets.

4 Major Threats in Wireless Sensor Network

Denial of Service Attack (DoS): This type of attack can occur in any of the layer in the WSN. The nodes which contain limited energy are affected by this type of attack.

Attacks on the in-transit information: In this type of attack, the cryptanalyst is capable of altering some of the information or the cryptanalyst will repeat information, which is already present.

Sinkhole attack: In this, the cryptanalyst will introduce a new node, which looks attractive to all the other nodes present in the network..

Wormhole attack: In this simultaneously two cryptanalysts will perform attacking in the WSN. Cryptanalyst can get the position of its neighboring nodes and can listen to what is happening in the network.

Hello flood attack: In this attack, the attacker uses greeting packets to attract and convince the nodes, by doing this the node will be convinced that the node present will be its neighboring node. When the sender node sends the packet to the receiver node, the packets first reaches the attacker, and in this case, it can modify any changes to the packets and ends that node to the receiver nodes.

Sybil attack: In this attack, a single node in the network will exhibit more than one identity.

5 Proposed Work

The heterogeneous Wireless Sensor Network [WSN] node involving in the communication process will be of different sensing range and different computing power. The nodes remain as the part of communication will be of different range, where the range varies from one node to another. The range of the node has the ability to receive the packets sent from the other node with different range. The purpose of deploying heterogeneous sensor node into the network aims to improve the performance of sensing in the network. The use of heterogeneous wireless sensor network with the

Honeypot nodes will increase the sensing power of the nodes. When these nodes are inserted, the attacker will find it difficult to track the sensing range of the nodes present in the network. These nodes in between the Honeypot nodes will play a major role in collecting the information of the attacker node. In this work, Honeypot works as trap, where all the other internal node will be aware of the Honeypot nodes. When the external resources rushes to get into the network, these nodes are sent toward the Honeypot, where these nodes collect the external resources and get stored.

5.1 Proposed Algorithm

Step 1: Required number of heterogeneous nodes is created and is set for the simulation process.
Step 2: For every 60 s, the packets are sent to increase the performance of the network.
Step 3: Honeypot nodes are inserted in between the heterogeneous nodes.
Step 4: Honeypot nodes will collect the IP address of the node present in the network and it stores all the node information.
Step 5: Once the node is collected with the IP address, it verifies the information.
Step 6: Node with wrong information will be identified.
Step 7: Honeypot nodes to avoid denial of service (DOS), it just gets the other information and store it for further process.
Step 8: The attacker node is identified in the mote output window. This shows the node from which the attack take place and also it shows the node, which is being attacked and communicated with the other.
Step 9: When the faulty node is detected, it displays the output window.

Figure 1 represents the flow of proposed work in which the attacker tries to get into the network and tries to contact with the nodes present in the network. The attacker will not be aware of the Honeypot node and it tries to build communication with that node, where these Honeypot nodes just keep on responding the attacker node as the other node to avoid denial of service. The dotted lines present in the figure shows the coverage area of the Honeypot nodes.

6 Deployment of Honeypot in WSN

In WSN, all nodes working together in a network are called Honeypot. A Honeypot is a dummy node, which works as the other nodes present in the network, by this the attacker cannot understand the difference between the normal nodes and the Honeypot nodes. These Honeypot nodes are inserted to confuse the attacker. Honepot nodes will not be having the information of the other nodes in the network and also there will be no interaction with the other nodes, where it just behaves like a normal

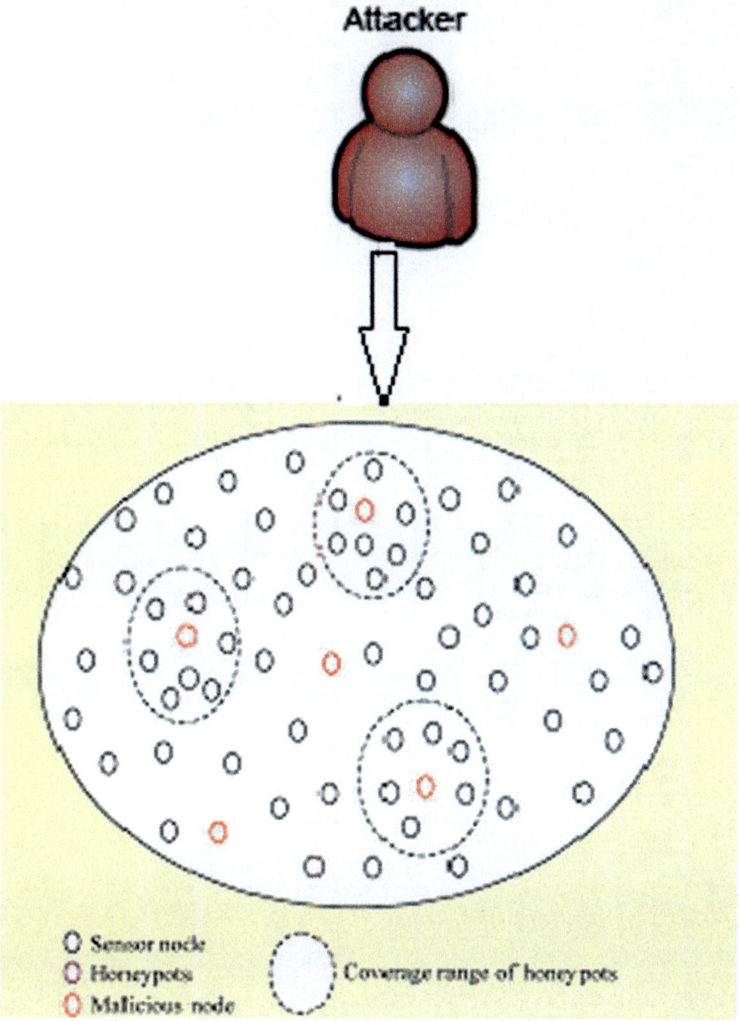

Fig. 1 Flow diagram

node even if it does not contain the information of the other node. These Honeypots are monitored by the base station continuously. Honeypot collects the attacker ID and pattern in which the attacker perform the attack and the amount of energy consumed to perform that attack to send the information to the base station without disturbing the other nodes. Base station will convey the presence of attacker to the other nodes present in the network.

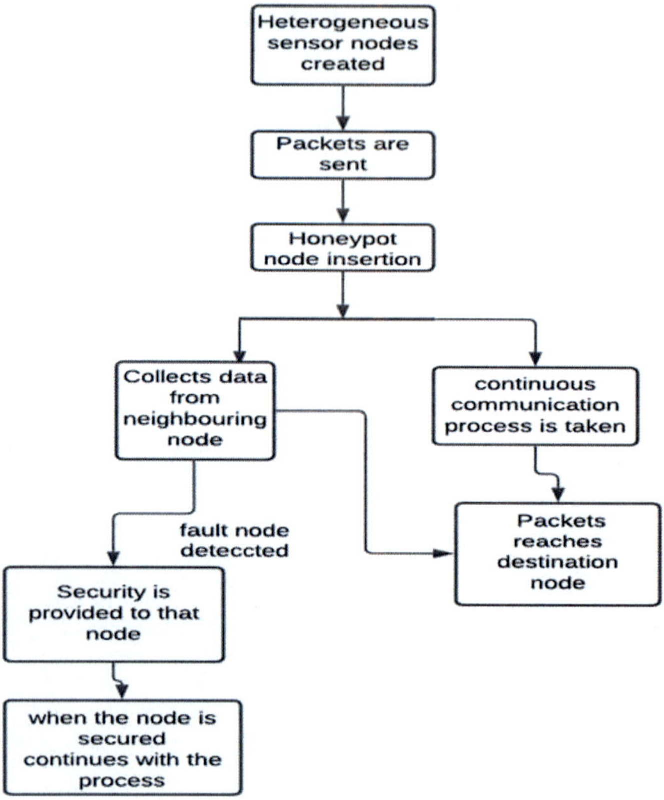

Fig. 2 Flow chart of proposed work

6.1 Node Parameters in Cooja

See Table 1.

Figure 3 shows creation of nodes, where the required number of nodes is created by using the mote type as sky mote. These created nodes can be dragged and placed in

Table 1 Node parameters

Number of nodes created	25
Honeypot nodes	5
Transmission range	(0–31)
Transmission medium	Unit Disk Graph Medium(UDGM)
Report interval	60 s
Randomness	60 s
Mote type	Sky mote
Node positioning	Random positioning

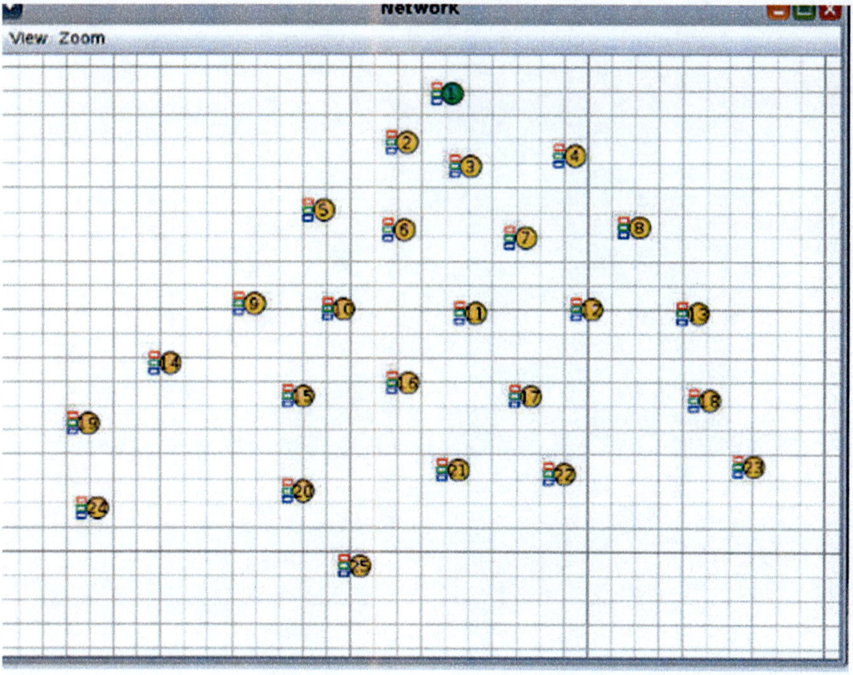

Fig. 3 Creation of N—nodes

any direction of the grid. When these nodes are created, the nodes start collecting the data from the other neighboring node. Once the nodes start the collection of packets, where it keeps on transferring the packets until the simulation is being stopped. The packets are transferred randomly in the network and the position of the node in the network remains in the random position. These nodes contain the 10 m background grid in which we can clearly know the transferring of packets and the range in which the packet is being transferred.

Figure 4 shows the process of transferring packets between the nodes present in the network. It also clearly represents the transfer of packets from one particular node to another. When the packet is received, the destination will be highlighted, every node present in the network will contain its own IP address, which are displayed below the node. Node present in the network will be communicating with the other node until the simulation is stopped. For every 60s, packets are sent to the node to maintain the ongoing communication process. In between these nodes, we insert the other node which is called as Honeypot nodes to get the information of the other attacker node and to avoid the denial of service. The process taking place in between the nodes are shown in the mote output window. In which it shows the from which node the process is being started, collected IP address of the node, for which node server connection is established, common process of the other node, IPV6 address of the node, etc. When these processes are written in the timeline window it also shows

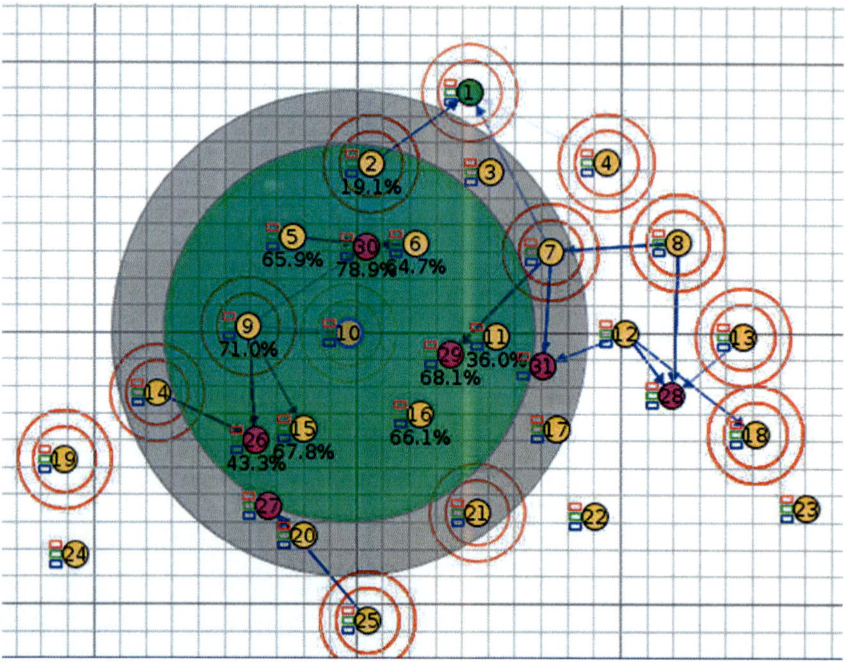

Fig. 4 Transferring of packets and deployment of honeypot

the node ID of the faulty node in which we can know the node which is reducing its range of communication.

Figure 5 clearly represents the transmitting of packets from one node to another with its range. In Fig. 4 the transmitting of packets stays for a second and disappears in this graph it clearly represents the packet transfer between one particular node and another. If the same node transfers the packet to the other this graph shows that node which is connected to the other node. These graphs are obtained when large number of packets is sent for every 60 s. By this graph, we can clearly track the path in which the node is receiving the packets (Table 2).

Figure 6 shows the packets delivery without Honeypots nodes and In the existing work the packet delivery was less, the packet sent from the source node was received by the attacker node present in the network all the packet sent was withheld by these attacker nodes and results in packet loss. Without Honeypot nodes, the amount of packets sent is more but the received packets are less.

Figure 7 shows the packets delivery of packets with Honeypot nodes. In the proposed work the attacker node will be identified by the Honeypot node and there is no chance of packet getting lost. Almost all the packets sent will be received by the destination node (Table 3).

Figure 8 shows the time taken by the packets to reach destination. In the existing work, the time consumed by the nodes to reach the destination node was more because the nodes taking part in the communication is not aware to which node the packets

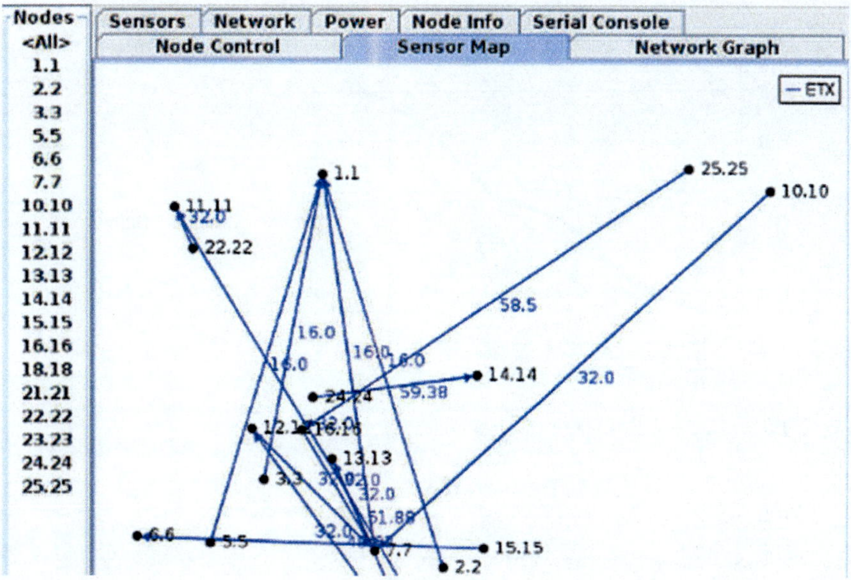

Fig. 5 Network graph

Table 2 Packets sent from source to destination

Number of packets sent	Packets sent from source to destination	
	Packets delivered without honeypot nodes	Packets delivered with honeypot nodes
100	55	90
200	120	195
300	245	290
400	300	390
500	440	480
600	523	590

has to be delivered instead it has to route to all the nodes and find its destination node due to this the packets took more time to reach destination.

Figure 9 proposed work shows the less time consumed by the packets to reach the destination. In the proposed work, the nodes transmitting the packets will be aware of the destination node in which the packets has to be delivered. The node which is found as faulty node will be recognized by the Honeypot nodes and stops that nodes taking part in the communication due to this the time taken is less in proposed work as compared to the existing work. This graph shows almost all the packets sent to are received with the less time as compared to Fig. 7. The packets took less time to reach destination due to the presence of Honeypot nodes. These Honeypot nodes will restrict the entry of the false packets that is being sent to the network. By this,

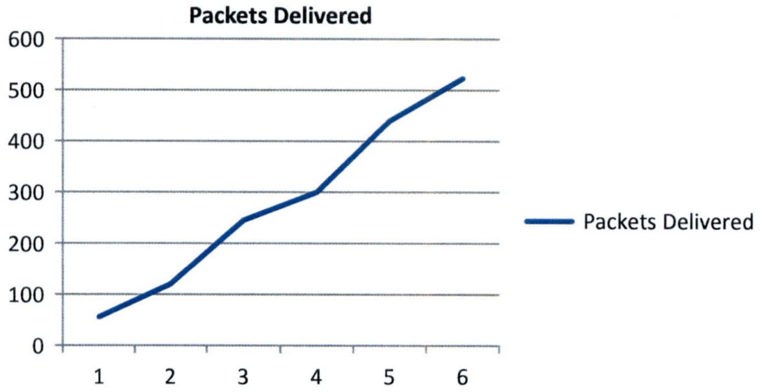

Fig. 6 Packets delivered without Honeypot nodes

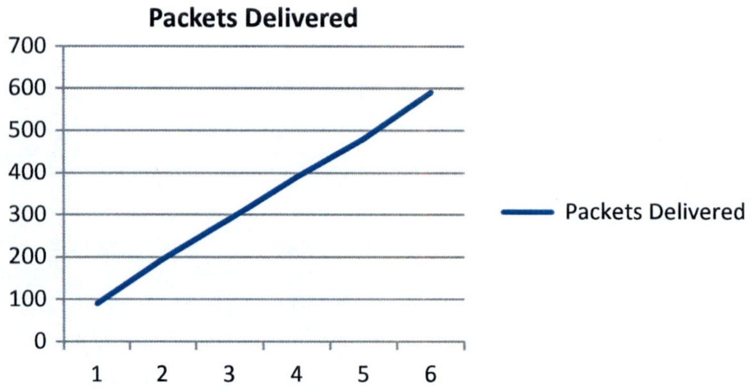

Fig. 7 Packets delivered when Honeypot node is inserted

Table 3 Time taken by packets to reach destination

Number of packets sent	Time taken by packets to reach destination (ms)	
	Time taken before inserting honeypot nodes	Time taken after inserting honeypot nodes
100	89	33
200	167	128
300	245	180
400	367	234
500	485	269
600	580	278

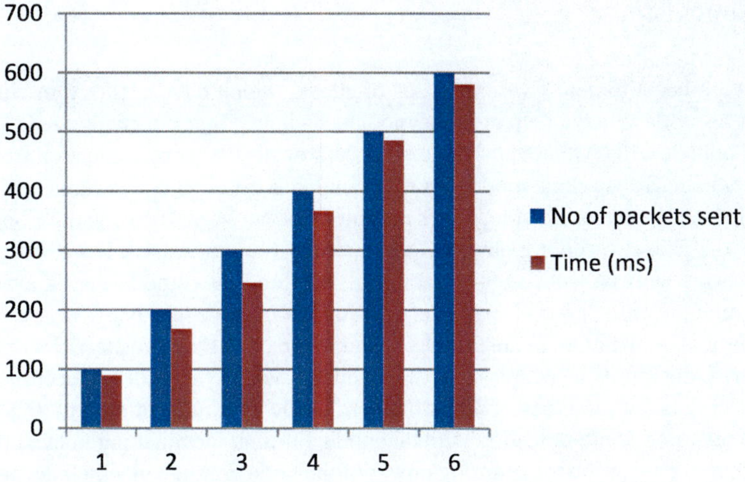

Fig. 8 Time taken before inserting Honeypot nodes

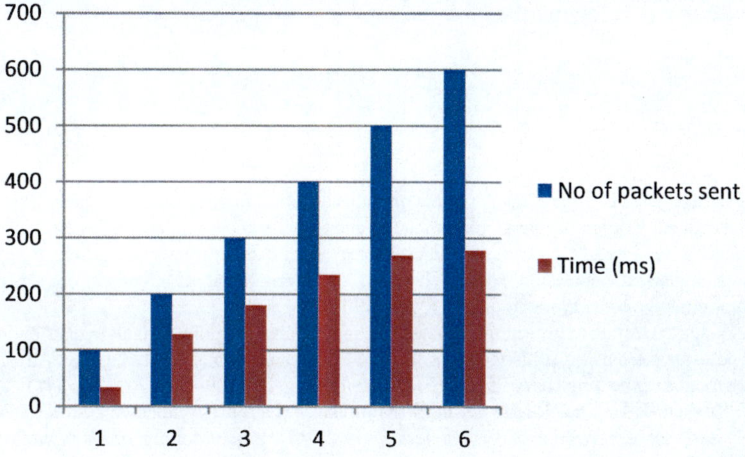

Fig. 9 Time taken after inserting Honeypot nodes

the false packets are withheld by the Honeypot nodes the other nodes present in the network will continue its process of transmitting the packets to the destination node.

7 Conclusion

This work helps to detect various types of attack, which have occurred in wireless sensor network with the help of Honeypot mechanism. Once attacker node is identified, that node will be removed from the network or stop the communication from the attacker node. These Honeypot nodes look similar to the other nodes and it cannot be distinguished between the other nodes present in the network. By using the Honeypot nodes, the packet receiving power of the nodes will be increased. Honeypot nodes in this work will collect the information of each and every node that is available for communication and collect the information from the nodes but do not transmit that message to the other nodes. Honeypot nodes analyze the characteristics of every node by colleting its IP address and the path from which the fault has occurred and then it checks for the other nodes, which are in the communication with that node and tries to stop communication with that node. Further, the other parameters can be considered, such as the transmitting power of the node or range in which the packets are transmitting by this node with low energy can be notified and that node can be stopped from taking part in the process. By maintaining the routing table, the path of the nodes traveled will be available, and by this, the security can be provided to the nodes that is receiving the false packets.

References

1. R. Sathiyavathi, B. Bharathi, "A review on fault detection in wireless sensor networks", 2017 International Conference on Communication and Signal Processing (ICCSP) (2017)
2. T. Zhang, Q. Zhao, Y. Nakamoto, "Faulty sensor data detection in wireless sensor networks using logistical regression", 2017 IEEE 37th International Conference on Distributed Computing Systems Workshops (ICDCSW) (2017)
3. B. Uddin, A. Imran, M.A. Rahman, Detection and locating the point of fault in distribution side of power system using WSN technology", 2017 4th International Conference on Advances in Electrical Engineering (ICAEE) (2017). https://doi.org/10.1109/icaee.2017.8255421
4. M. Royyan, J.-H. Cha, J.-M. Lee, D.-S. Kim, Data-driven faulty node detection scheme for wireless sensor networks. 2017 Wireless Days (2017). https://doi.org/10.1109/wd.2017.7918145
5. J. Zhang, J. Chen, An adaptive clustering algorithm for dynamic heterogeneous wireless sensor networks. Wirel Netw **25**(1), 455–470 (2019). https://doi.org/10.1007/s11276-017-1648-1
6. S. Anand, A. RR, A protocol for the effective utilization of energy in wireless sensor network. Int. J. Eng. Technol. **7**(3.3), 93 (2018). https://doi.org/10.14419/ijet.v7i3.3.14495
7. S. Manishankar, S. Sathayanarayana, "Enhanced big data platform for visualization of employee data", JOIV: Int. J. Inf. Visualization, **2**(3), 169–174 (2018)
8. D.S.K. Tiruvakadu, V. Pallapa, Honeypot based black-hole attack confirmation in a MANET. Int. J. Wireless Inf. Netw. **25**, 434–448 (2018). https://doi.org/10.1007/s10776-018-0415-2
9. Z. Zhang, A. Mehmood, L. Shu, Z. Huo, Y. Zhang, M. Mukherjee, A survey on fault diagnosis in wireless sensor networks. IEEE Access **6**, 11349–11364 (2018). https://doi.org/10.1109/access.2018.2794519
10. P.S.P. Eboule, J.H.C. Pretorius, N. Mbuli, C. Leke, Fault detection and location in power transmission line using concurrent neuro fuzzy technique. IEEE Electrical Power and Energy Conference (EPEC) **2018**, 1–6 (2018). https://doi.org/10.1109/EPEC.2018.8598311

11. A. Johnson, J. Molloy, J. Yunes, J. Puthuparampil, A. Elleithy, "Security in wireless sensors networks", 2019 IEEE Long Island Systems. Applications and Technology Conference (LISAT) (2019). https://doi.org/10.1109/lisat.2019.8817338
12. M. Alotaibi, Security to wireless sensor networks against malicious attacks using Hamming residue method. J. Wireless Com. Netw. **2019**, 8 (2019). https://doi.org/10.1186/s13638-018-1337-5
13. A. Prasad, S. Anand, S. Sinha "An efficient fault tolerance mechanism using TCSA to ensure reliabilty in cloud", 2019 Int. J. Eng. Adv. Technol. (IJEAT)
14. A.P. Prabhan, S. Anand, S. Sinha , Identifying Faulty Nodes in Wireless Sensor Network to Enhance Reliability International Journal of Recent Technology and Engineering (IJRTE) ISSN: 2277–3878, **8**(2) (Jul 2019)
15. M.N.I. Khan, M.S. Islam, A new scheme to detect and prevent node replication attacks for wireless sensor networks. International Conference on Computer Communication and Informatics (ICCCI) **2019**, 1–5 (2019). https://doi.org/10.1109/ICCCI.2019.8821839
16. P. S. Negi, A. Garg, R. Lal, "Intrusion detection and prevention using honeypot network for cloud security," 2020 10th International Conference on Cloud Computing, Data Science & Engineering (Confluence), 2020, pp. 129–132, doi: https://doi.org/10.1109/Confluence47617.2020.9057961.
17. S. Anand, K. C. Manoj, "A survey on clustering approaches to strengthen the performance of wireless sensor network," 2020 Second International Conference on Inventive Research in Computing Applications (ICIRCA), Coimbatore, India, 2020, pp. 814–820. https://doi.org/10.1109/ICIRCA48905.2020.9183174.
18. S. Anand, T. N. Gowthami, "Cluster-based energy efficient protocol for wireless sensor network," 2020 International Conference on Emerging Trends in Information Technology and Engineering (ic-ETITE), Vellore, India, (2020), pp. 1–7
19. S. Anand, P.A. Radhakrishna, A protocol for the effective utilization of energy in wireless sensor network. Int. J. Eng. Technol. **7**(3.3), 82 (2018). https://doi.org/10.14419/ijet.v7i3.3.14491
20. H.N. Lakshmi, S. Anand, S. Sinha "Flooding attack in wireless sensor network-analysis and prevention". 2019 Int. J. Eng. Adv. Technol. (IJEAT)

Classification of Fake News on Twitter by Using Naïve Bayes Classifier

R. Anand and Deepak S. Sakkari

Abstract The identification of fake news is generating new challenges for the global intelligence system. The algorithm uses web intelligence to recognize the false rumor categorization scheme and categorizes tweets as true or false. The proposed technology uses machine learning to classify the fake news by linking tweets with the proper source. Basically, false news means the news that includes unwanted or false information. False news is not just harmful to individuals and also the society but it necessitates the need to know what is false and factual. One of the significant drawbacks of fake news is that, it might undermine the status of individuals or things, which is not true. Henceforth, this paper will classify out of all the examples, which are fake news and real on the social media platform called Twitter. Based on this false news, analysis is performed on social media platforms.

Keywords Twitter · Social networks · Sentiment analysis · KNN · Latent semantic analysis

1 Introduction

Communal social networking is a challenge to the global information system. Specific person will use several communal broadcasting solicitations to linkage occupation prospects, which find persons across the world with like comforts, and stake their understanding and feelings. Communal broadcasting can be a valuable instrument for dealings and fetching benefits such as appealing with your spectators and improving social networking movement. Though nearby can also be drawbacks by including

R. Anand (✉)
Department of Computer Science and Engineering, BMS Institute of Technology and Management Bengaluru, Bengaluru, Karnataka 560064, India
e-mail: anandor@bmsit.in

D. S. Sakkari
School of Engineering, Presidency University, Rajanakunte, Yelahanka, Bangalore, Karnataka 560064, India
e-mail: deepakssakkari@presidencyuniversity.in

G. Ranganathan et al. (eds.), *Soft Computing for Security Applications*,
Advances in Intelligent Systems and Computing 1397,
https://doi.org/10.1007/978-981-16-5301-8_30

the incomes essential and bad reaction. Common broadcasting procedure is one of the largely accepted online actions. In 2020, over 3.6 billion people were using social media.

As of the third part of 2020, Twitter had 187 million monetizable day-to-day lively international operators. Worldwide social networks such as Twitter, Facebook, and LinkedIn have become crucial sources of social information and it has attracted increasing interest and curiosity in social research and commerce. As a social online micro-blogging service, Twitter allows users to interact with the posts called as "tweets". Since its inception in 2006, Twitter has grown significantly in popularity throughout the world. As some research says that, 29% adults use Twitter and over half of them are daily users. This fact makes Twitter a rich source for capturing information and due to a lot of information on Twitter, where they may contain fake and real news also. Whole dataset cannot be spread due to its confidentiality rules and news flash produces duplicate privileges. Communal activities and user evidence are not revealed due to its Twitter Rule.

Fake news is false or ambiguous information presented as news. The aim of the fake news is to damage the reputation of a person or organization or making money through advertising revenue. "False news" exists within a larger background of misinformation. Half-truths are false or incorrect information that is shared accidentally in order to promote the plan. False data that is comprehensive and reachable on purpose is referred to as untruth. Sentiment analysis has been performed on Twitter by identifying the material as false or not, and by utilizing non-fake news. The main objective is to understand about real data and fake data. The social media is the main source for collecting relevant information and categorization appropriately.

2 Data Extraction Methods

In general, a tweet by any user is public and readable by anyone. Registering an account on the Twitter developer platform provides access to all the tweeted data. Filter Stream is used to retrieve tweet data in the experiment. This function is provided by stream R, which connects to the Twitter Streaming API and opens a stream to capture the results based on specific conditions such as keywords and location. Three different types of data extraction techniques are applied to analyze Twitter plain text data that are efficient and effective.

A. **Text Mining**

Text mining is the data analysis process, which includes a collection of different text. Text mining is the procedure of altering free manuscript into organized documents for informal examination or derives. High-quality information is obtained from text. The amount of unstructured text data on Twitter is rapidly increasing day by day, accounting for 70–80% of all text data present in Twitter. Text excavating methods can examine the multifaceted and huge sets of records in a modest, firm and real way. Two methods are used in text mining.

- Analyzing the text that exists such as customer reviews.
- Structure the manuscript so that it can castoff in machine learning prototype.

The gathered facts remain appropriate for examination or to energize the machine learning (ML) procedures.

B. **Topic Modeling**

Topic modeling is an essential algorithm used in text mining. Subject Exhibiting is the diverse after rule-based manuscript excavating methods that use consistent terminologies or vocabulary constructed keyword penetrating practices. It is an unverified tactic used for judgment and spotting the group of arguments called subjects in great collection of manuscripts. It is a kind of arithmetic modeling for discovering the conceptual topics that happen in a gathering of credentials. Latent Semantic Analysis (LSA) computes how normally arguments happen in the forms. The typical way for calculating term incidences is what is known as tf-idf. It not only studies the recurrent arguments are in a specified article but also how recurrent arguments are in all the quantity of forms.

C. **Bag-of-Words Assumption**

The bag-of-words representation is usually used in systems of text documentation classifier, where the incidence of each appearance is used as a data feature for training a classifier. The bag-of-words model is a shortening illustration used in natural language processing and information retrieval (IR). The only thing that matters in the bag-of-words model is whether a word that appears in the text, and where does the text occur. It includes two types:

1. A vocabulary of known words.
2. A measure of the presence of known words.

3 Sentimental Analysis

1. **Data collection**: To extract data from Twitter, an automated web scraping tool called Octoparse can be used. Net Scrupling denotes the removal of data from a Web site in which the data is composed and then transferred into a API format. New scraping can also be done manually, in most cases, the automatic tackles are favorite, when grating web facts as they can be less expensive and can work at a quicker rate than manual mode [1].

By using Octoparse, web scraping simulates human interaction with a web page, which allows to pull all the information visible on any Web site like Twitter.

2. **Text Preparation**: The easiest way for text preparation is by using python or R language. We start with raw tweets, process them, and finally output the final tweet.

3. **Sentiment Identification**: Examine the opinions in the extracted text. The contents with opinions are retained, and the contents that actually occurring information are discarded.
4. **Sentiment Classification**: is to apply some approaches to classify the sentiments. Some approaches include the machine learning approach, the sentiment orientation approach.
5. **Results**: Results is to display the results in a direct way such as in graphs like pie charts, line graphs, etc.

4 Implementation

See Fig. 1.

A. **Data Collection**

First we retrieve the data from Twitter and create a dataset. Dataset is composed as of Twitter with false and real news. Twitter dataset contains 4048 information out of which 1865 are real article and 2118 are false news and we use some Twitter channels to do sentiment analysis.

B. **Preprocessing**

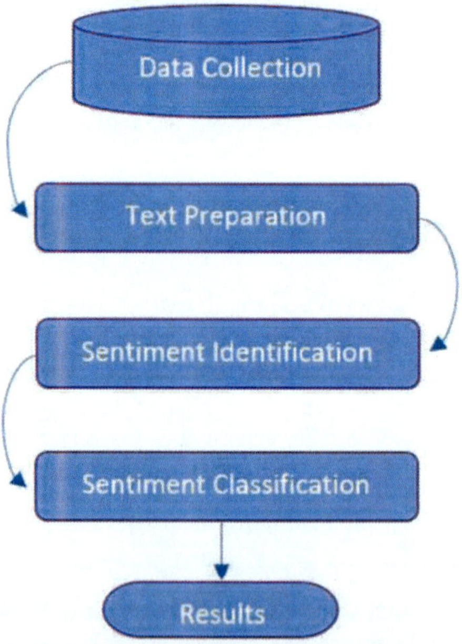

Fig. 1 Sentimental classification process of data

It is mainly transforming raw data into an understandable format, i.e., converting the uncooked data to a dirt free dataset. Data preprocessing prepares unprocessed information for further dealing out. Data preprocessing is used in record-motivated applications such as purchaser association supervision and decree-based applications.

1. **Stop Word Removal**—While computing the data, stop word removal is used to filter the data before or after the data is processed. It refers to the common words used in languages where natural language processing tool don't use same list of words.
2. **Tokenization**—Tokenization is the process of creating the tokens from the given data. The collected data is a very sensitive data, which needs to be divided into number of tokens to make the process of tokenization simple and effective. If required, the tokens are further used for data analysis.
3. **Stemming**—stemming is a procedure, where terms are summarized to a source by removing variation from end to end tumbling with a needless typeset, typically a suffix.

C. Feature Extraction/Reduction

See Fig. 2.

Feature extraction is one of the powerful techniques used to extract the data and understand the features of the extracted data. The extracted data should be non-repetitive; non-duplication of the data is strictly entertained. The reduction of the data

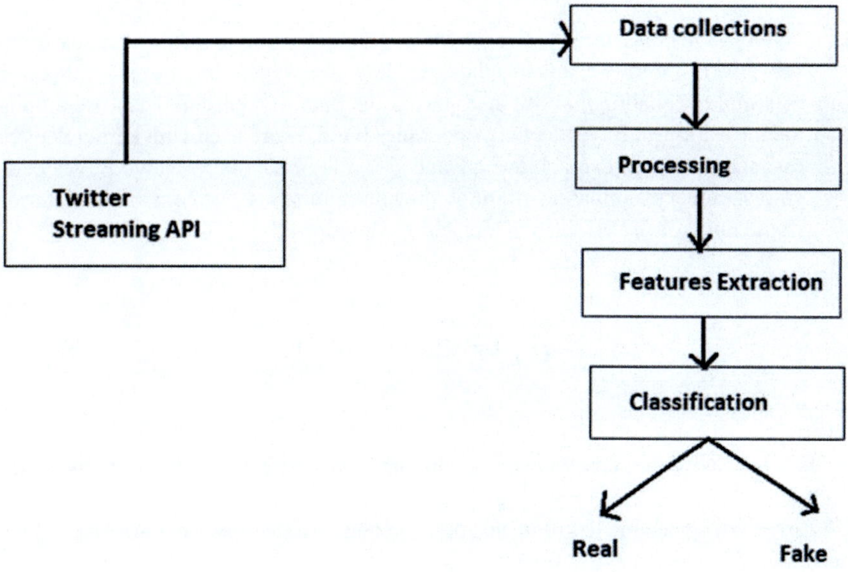

Fig. 2 Classification of real and fake data

is also very important so that the data is clear and error free by including the data, which is required for the experiment. The characteristics of the large set of extracted data depend on the variables associated with it. To compute this, much huge data with more number of computational resources and skills are necessary. Characteristic drawing out starts from an first set of calculated information by resulting standards proposed to be edifying. TF-IDF (Term Frequency—Inverse Document Frequency) has the main intention to reflect the importance of the word in the document. TF-IDF includes two metrics, named as

1. Term frequency (tf)
2. Inverse document frequency (idf).

TF-IDF is represented by
TF-IDF=tf(t,d)*idf(t,d)
tf(t,d) means that term t occurs in document d.

idf(t, d) means the **inverse document frequency** is a quantify of how far data the word provides. It is calculated using next method.

idf(t,d)=1+log T(1+df(t))

D. **Classification**

Classification is definite as the procedure of appreciation, sympathetic, and group of items and thoughts into pre-determined groups. Classification models attempt to illustrate some winding up from experimental values. Given one or supplementary inputs a cataloging representation will try to see the assessment of one or more outcomes. Outcomes are labels that can be applied to a dataset. Here, Naive Bayes classifier and sensitive aggressive classifier are used for the organization.

1. **Naive Bayes classifier**: Naive Bayes is a classification algorithm that assumes self-governing forecasters in a dataset. The naive Bayes classifier is a collection of simple probabilistic classifiers based on Bayes' theorem. They are among the more modest Bayesian network copies but they are used with kernel density estimation, which can achieve advanced levels of accuracy. Naive Bayes classifiers are very climbable, although they do require a set of constraints in terms of the number of variables in a teaching situation.

$$q\left(\frac{a}{b}\right) = q(a)q(\frac{b}{a})/q(b)$$

where

$q(a \mid b)$ = how frequently occurs given that b occurs; q(a) = how probable a will occur.

$q(b)$ = how probable b will occur; q(b |a) = how frequently b occurs given that a occurs.

2. **K-Nearest Neighbors**: It is used for the classification and regression, where the data is used in both inputs which consist of K closest training data. It is used

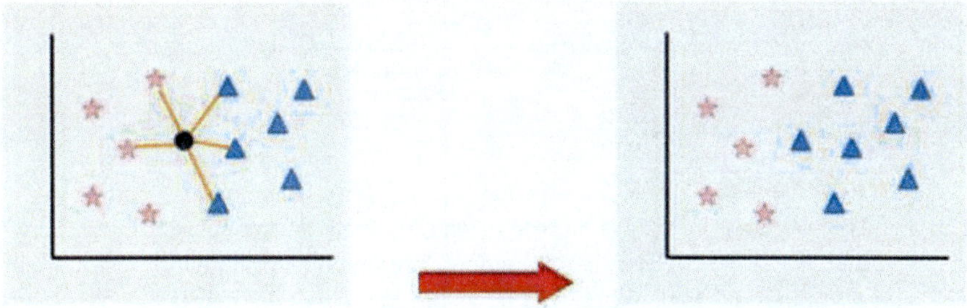

Fig. 3 KNN Algorithm

for pattern recognition and statistical Neighbor as a classification and prediction algorithm, which is used to divide data into classes based on the distance between the data points. K-Nearest Neighbor assumes that data points that are close to one another must be similar and hence the data point needs to be classified and it will be grouped with the closest cluster [2].

KNN algorithms use statistics and classify new data points based on resemblance events (e.g., distance function). Classification is done by a mass choose to its neighbors (Fig. 3).

Distance formula: Distance formula is defined as "the square root of the total square difference between a teaching point and testing position transversely at all the input.

$D = \sqrt{}$ [(X2-X1)^2 + (Y2-Y1)^2)

where D is the distance.

E .**Performance Evaluation Metrics**

1. **Naive Bayes Classifier**

Here to measure up to the performance of Naïve Bayes, Evaluation metrics are used for fake news detection by using confusion matrix.

Accuracy=True Positive True Negative

"True positive—counts the predicted news as 'Fake' when actually it is 'Fake'.

False Positive- counts the predicted news as 'Genuine' when actually it is 'Fake'.

False Negative—counts the predicted news as 'Fake' when actually it is 'Genuine'.

True Negative—counts the predicted news as 'Genuine' when actually it is 'Genuine'.

In Evaluation metrics there are three types of matrices: Accuracy, precision, recall, and F-score matrices" [3].

$$\text{Accuracy} = \frac{\text{TruePositive} + \text{TrueNegative}}{\text{TruePositive} + \text{TrueNegative} + \text{falsePositive} + \text{FalseNegative}}$$

$$\text{Precision} = \frac{\text{TruePositive}}{\text{TruePositive} + \text{FalsePositive}}$$

$$F1 - \text{Score} = \frac{2 * \text{Precision} * \text{recall}}{2a\text{Precision} + \text{recall}}$$

2. **K-Nearest Neighbors**

- True Positive (TP): When fake news predicted as fake news.
- True Negative (TN): When true news predicted as true news.
- False Negative (FN): When fake news predicted as true news.
- False Positive (FP): When true news predicted as fake news.

The misunderstanding environment gives you a portion of material but occasionally you may wish a brief metrics.

Precision

Precision = (TP)/(TP + FP)
A simple way to ensure flawless accuracy is to create a single positive forecast and ensure that it is accurate. (precision = 1/1 = 100%). This would not be actually valuable, since the classifier would ignore all except one positive event.

Recall: To measure compassion, recall is the total number of factual optimistics alienated by the total number of real optimistics.

Recall = (TP) / (TP + FN).

Accuracy is the entire no. of precise cataloging separated by entire no. of likely belongings.

Accuracy =(TP+TN)/ (TP+TN+FN+FP)

5 Results

This graph represents the number of tweets available in each day. Y axis means the number of tweets and X axis that signifies the sentiment analysis in which sentiment analysis includes negative, neutral, and positive. The recent February 2020 tweets are used for analysis. The data is analyzed to know whether it is fake, neutral, or genuine.

Confusion matrix: To summarize the performance of classification algorithm, the confusion matrix is used (Fig. 4).

A total of 96 cases (95 faulty cases, and 1 no-fault case) are used to notice and organize each of the three types of DC faults (Table 1).

Fig. 4 Number of Relevant Tweets/Day

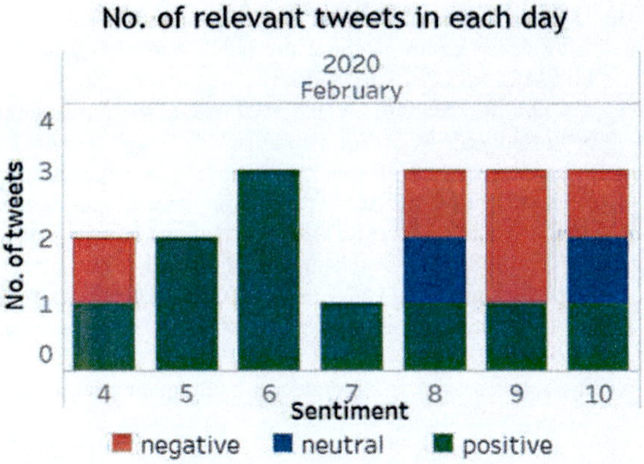

Table 1 Naïve Bayes classifier data

1	**96** **100%**	**0** **0.0%**	0 0.0%
2	**0** **0.0%**	**96** **100%**	**0** **0.0%**
3	**0** **0.0%**	**0** **0.0%**	**96** **100%**

The Bold Value indicates the Number of twetter cases under process. Out of which 95 fault cases are indentified

6 Conclusion

This paper has provided a systematic review on the data extraction techniques implemented in the social media. The Twitter data is considered as sample dataset for different data collection methods and classification techniques. Another goal of this work is to offer researchers a solid basis for making data extraction and classification procedures connected to machine learning easier and better understood by utilizing rapid machine learning models that can generate quick predictions.

References

1. A. Dey, R.Z. Rafi, S.H. Parash, S.K. Arko, A. Chakrabarty, "Fake news pattern recognition with linguistic analysis", 2018 Joint 7th International Conference on Informatics, Electronics & Vision (ICIEV)
2. K. Xu, F. Wang, H. Wang, B. Yang "Detecting fake news over online social media via domain reputations and content understanding" Tsinghua. Sci. Technol., ISSN 11007–02140 3/ 1 4, **25**(1), 2 0–2 7 (Feb 2020). https://doi.org/10.26599/TST.2018.9010139
3. T. Jiang, J.P. Li, A.U. Haq, A. Saboor, A. Ali, "A novel stacking approach for accurate detection of fake news", Dig. Object Identifier https://doi.org/10.1109/ACCESS.2021.3056079

4. M. Umer, Z. Imtiaz, S. Ullah, A. Mehmood, G.S. Choi, B.W. On "Fake news stance detection using deep learning architecture (CNN-LSTM)" Digital Object Identifier https://doi.org/10.1109/ACCESS.2020.3019735.
5. X. Zhou, R. Zafarani, "A survey of fake news: fundamental theories, detection methods, and opportunities". ACM Comput. Surv.1, 1, Article 1 (January 2020), p. 37
6. S.S. Nikam, Prof. R. Dalvi, "Machine learning algorithm based model for classification of fake news on twitter", 2020 Fourth International Conference on I-SMAC (IoT in Social, Mobile, Analytics and Cloud) (I-SMAC),**DOI:** https://doi.org/10.1109/I-SMAC49090.2020.9243385
7. A. Kesarwani, S.S. Chauhan, A.R. Nair, "Fake news detection on social media using k-nearest neighbor classifier", Conference: 2020 International Conference on Advances in Computing and Communication Engineering (ICACCE), https://doi.org/10.1109/ICACCE49060.2020.9154997
8. S. Kaur, P. Kumar, P. Kumaraguru, "Automating fake news detection system using multi-level voting model". ScI_J
9. A. Jain, A. Kasbe, "Fake news classification", 2018 IEEE International Students' Conference on Electrical, Electronics and Computer Sciences
10. S. Gilda, "Evaluating machine learning algorithms for fake news classification", 2017 IEEE 15th Student Conference on Research and Development (SCOReD)
11. H. Ahmed, I. Traore, S. Saad, "Classification of online fake news with n-gram analysis and machine learning method", 2017 Intelligent, Secure, and Dependable Systems in Distributed and Cloud Environments: First International Conference, ISDDC
12. N.J. Conroy, V.L. Rubin, Y. Chen, "Automatic deception classification: methods for finding fake news", 2015 ProcAssocInfSciTechnol **52**(1), 1
13. N.J. Conroy, V.L. Rubin, Y. Chen, "Towards news verification: deception classification methods for news discourse", 2015 In: Hawaii international conference on system sciences
14. Y. Chen, N.J. Conroy, V.L. Rubin, "Misleading online content: recognizing clickbait as false news". 2015 In: Proceedings of the 2015 ACM on workshop on multimodal deception classification. ACM, pp. 15–19
15. Mr. A.R. Priyanka, Dr. R.M. Patil, "Health monitoring in aerospace system". Int. J. Informative Futuristic Res. (IJIFR), (2017)
16. A.R. Pushpalatha, M. Dr. R.M. Patil, "A social networking for sharing infrastucture resources in the social cloud computing". Int. J. Informative Futuristic Res. (IJIFR) (2016)
17. Dr. M.V. Vijaykumar, P. Jagadish, K. Shryavani, R. Anand, "Authorized deduplication in hybrid cloud", IJCSN Int. J. Comput. Sci. Netw. (2016)
18. N. Chakrabarty, A regression approach to distribution and trend analysis of quarterly foreign tourist arrivals in India. J. Soft Comput. Paradigm (JSCP) **2**(01), 57–82 (2020)
19. V. Suma, S. M. Hills. "Data mining based prediction of demand in Indian market for refurbished electronics." J. Soft Comput. Paradigm (JSCP) **2**(02), 101–110 (2020)

Enhancing the Security in Wireless Sensor Network Using Hidden Markov Model

Santosh Anand and K. Muhammad Rafeeque

Abstract WSN is made up of nodes that communicate with one another with the help of a wireless channel. These days the WSN is used to communicate among two or more nodes that are linked in a network. So there should be a good security protocol to hold the confidentiality of the data which is transferred in between the source and the sink. In this proposed work, a method to detect and prevent blackhole attack using hidden Markov model (HMM) is used to identify malicious nodes. This work is implemented using simulation and result shows that malicious nodes are getting detected. After detection on malicious nodes, communication from malicious nodes are stopped and packet is getting rerouted with reliable and secure link.

Keywords WSN · Blackhole attack · Hidden Markov model · Dijkstra's algorithm · Deniel of service · Faulty node detection · Node failure prevention · Securitry enhancement

1 Introduction

WSN is a group of sensors that is used to monitor and record environment condition and forward the data to central location. WSN can be used to measure the environmental conditions such as temperature, humidity, sound, pollution levels and wind (Fig. 1).

The blackhole attack is prevalent in wireless sensor networks. Since the number of attackers grows every year, providing security is a research challenge. A packet drop attack, also known as a blackhole attack in computer networking, is a type of D-DOS attack in which packets are discarded instead of relayed by a router. This generally happens because a router is affected by a variety of different reasons. As compared to traditional networking technologies, WSN technology provides some advantages,

S. Anand (✉)
Department of Computer Science, Amrita School of Arts and Sciences, Mysuru, India

K. Muhammad Rafeeque
Amrita Vishwa Vidyapeetham, Coimbatore, India

G. Ranganathan et al. (eds.), *Soft Computing for Security Applications*,
Advances in Intelligent Systems and Computing 1397,
https://doi.org/10.1007/978-981-16-5301-8_31

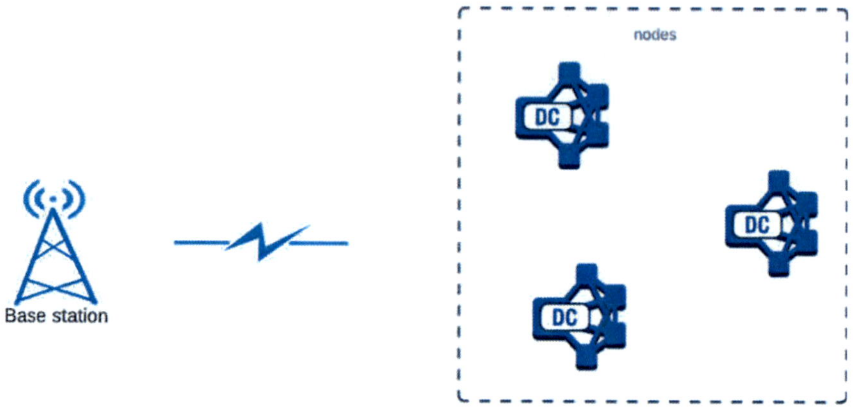

Fig. 1 Wireless sensor network

including cost savings, scalability, versatility, precision, and ease of deployment. Sensors are becoming smaller and cheaper as technology advances, and billions of them are being used in various applications. Military, climate, healthcare, and protection are some of the possible application domains. The various types of attacks are sinkhole attack, wormhole attack, blackhole attack, selective forwarding. When the blackhole attack occur will increase the number of drop packets and decrease the packet delivery ratio on mobile ad-hoc network performance. A blackhole assault is an attack carried out by an external adversary on a network's sensor nodes. A malicious node advertises an optimal route to the destination node without validating the routing database in the case of a Blackhole attack. Whenever the source node sends a packet to this malicious node it will accept that packet and ignore it or forward it to a different destination. That depends on the program done in that malicious node. This will lead to data loss or misusing of data. But the source node will not know this and they will send every packet to this malicious node all the time.

Here in our work, we use HMM (hidden Markov model) to find the probability of occurring malicious node by which we can improve the security of the WSN.

2 Literature Survey

Kalkha et al. [1] introduced a blackhole attack and provides AODV routing protocol avoidance measures for WSNs. Their preventative technique takes advantage of the shortest routes from point A to point B. We assume that neither the destination nor the source are harmed or malicious. Furthermore, their solution incorporates the process of locating a malicious path into the route discovery portion of AODV routing protocols.

Yasin et al. [2] proposed a solution by using timers and baiting messages to prevent intelligent blackhole attacks. The technique is divided into two stages: baiting and non-neighbor reply. During the Baiting process, various node has a timer for bait whose worth is randomly assigned B seconds, and when the timer reaches B, a bait request at random generated ID is generated and broadcasted the timer reaches B, it generates and broadcasts a bait request with a randomly generated fake id. When a blackhole node receives route requests, its natural behavior is to answer with a reply claiming that it has the correct route, despite the fact that it cannot exist when the blackhole gets the baited query, it submits a response pretending to get a path to just a fake node; once the cluster head gets the response, automatically takes into account the node that replied to be a blackhole and keeps adding it to the list since it assumed to have a path to a fake node and this may take some time.

Wazid et al. [3] proposed a new procedure in order to detect and prevent blackhole attacks in WSN. The simulation outcomes achieved by the proposed method display that throughout the event of the blackhole attack, the node which is malicious is correctly reported, but the use of the suggested methodology helps to prevent the WSN performance from degrading.

Kaur et al. [4] presented a new strong cluster strategy to detect and prevent numerous blackhole attacker nodes in WSNs. In their approach, the whole WSN is partitioned into too many cluster centers, and then each cluster has a great high cluster head that is in charge of sensing blackhole intruder nodes whether they are visible within this cluster. The system achieves a recognition rate of at least 90% and false positive rate of 3.75%, which really is significantly superior to the current similar measures. Besides that, our method is effective, making it suitable to be used in WSNs.

Sikarwar et al. [5] proposed the sensors are used in hostile environments, so strong authentication mechanisms are important to ensure data protection and safety. But for low power sensors, these powerful authentication mechanisms are computationally heavy. A small and stable WSN authentication protocol is therefore essential. Sensors range from medical treatment to petroleum and exploration applications. The data collected and distributed are of paramount importance because in some cases, they can be linked to lives and deaths.

AL-Shaihk et al. [6] proposed a method to DOS attacks, which can be detected but in many cases are still unpreventable, are the most powerful threat to WSN. To prevent the proposal of an alternative safe algorithm to bypass the attacked route, regulated by the cluster head, so that the False Node except DOS is mitigated because of a jamming attack. Using this approach they improved to 36 and 52% when the attack on a collection of nodes was performed.

Rani et al. [7] have examined many aspects of the security of wireless sensors in the form of cryptography, key management, and secured routing protocols. WSNs are vulnerable to a variety of assaults such as sinkhole, Sybil, selective routing, and so on. Confidentiality, honesty, authentication, and data freshness are all important considerations. any of the primary security issues in WSNs. Various aspects of encryption have been examined in this article to improve the safety method in wireless networks. According to their survey, most of protocols specified are there for static access points and static nodes. new protocols for mobile node protection should be developed; this high tech strategy would take to numerous advancements WSN security.

Grover et al. [8] investigated the numerous security problems and threats in WSNs. Also provides a brief discussion of some of the protocols used to achieve network protection. In addition, this paper compares the proposed methodologies analytically and displays the results in a table. These results may be used by other researchers or network implementers to make the WSN more reliable by selecting the best security mechanisms. These results may be used by other researchers or network implementers to make the WSN more reliable by selecting the best security mechanism.

Rajasekaran et al. [9] detected the failure of wireless connections between sensor nodes in the WSN topology. The author used a neural networks classifier with prearranged feature sets to test his hypothesis.

Kaur et al. [10] proposed wormhole and blackhole threats are hard to trace and safeguard against. Under this document, a procedure for defending against blackhole and wormhole attacks with much less operating costs is suggested. As a result, it conserves node battery power and contributes to network longevity. In the future, we enforce all this by taking real-world parameters into account. Also method can use for denial of service attack networks using AODV protocol.

Olakanmi et al. [11] discussed WSNs and their privacy and protection structure. This chapter provides readers with a thorough understanding of WSN-related security and privacy concerns.

Arulkumaran et al. [12] suggested executing value management within two nodes using fuzzy logic policy forecasting model to select the far more fast and effective MANET route. Each node in the proposed scheme retains the trust value of both its peer node. Until data forwarding in an AODV-enabled MANET, quantify the trust value, add up the path trust, and rather modify the trust value in each node's forwarding table If the reasonable pathway is by far the most trusted node way, then send signals; otherwise, compute the trust path for the specific data sending.

Hari et al. [13] proposed a new location approach has been developed to boost the central location algorithm. With relatively low communications and computer complexity, this approach will increase the position accuracy.

Arora et al. [14] focused on AODV protocol conglomeration by focusing blackhole attack. Blackhole and AODV attacks are examined and the implications of the attacks and effects on the output of MANET are defined. Our further analysis includes the review of mechanisms for recognition and avoidance of AODV blackhole attacks.

Motamedi et al. [15] have developed a scheme for locating blackhole attacks, that also mean the loss of core network data. Some practices for launching this attack on WSNs have been formulated, most of which make use of multipath routing and numerous base stations. Even so, these modalities waste a lot of time. In out method detects blackhole attacks using UAVs and the SPRT method, and it takes less time than other methods while providing reliable energy consumption. Furthermore, when compared to other techniques, the suggested model senses blackhole nodes with a better chance.

Yadav et al. [16] suggested an authenticated enrollment system for nodes using Blockchain, as well as identification and avoidance of blackhole and mutual black-hole attacks in WSN in this article. Blockchain provides a viable scheme for node enrollment as a decentralized, permanent, public ledger with anonymity and auditability in asymmetric encryption and authorization technologies. As a result, tracing the action of each node in a network is easy. Nodes can obtain complex and trustworthy knowledge on the blockchain network by making each node and route perceptible and temper-proof. The thorough AODV and Secure-AODV using blockchain select the best route and discard the routes with malicious nodes.

Shinde et al. [17] used the idea of active trust routing scheme protects against different varities of attacks during packet directing. Blackhole attacks, denial of service attacks, and selective forwarding attacks are the most common types of such attacks. The method also secures the data by covering it during routing with the ECC algorithm. The experiment results reveal that the proposed approach enhances network protection while also extending network lifetime, using less resources, and being more efficient during the network lifetime. In this work [18], authors implemented the aggregation techniques of the sensor node's data to enhance the availability of the data using centralized techniques of cloud. In this paper [19], authors are using the heterogeneous network components to improve the reliability using Deadline Hit Count algorithm. It is implemented with the GridSim tools to check the performance of the work. In this work [20], authors implemented the encryption and decryption algorithm to provide the secure data transfer from source to destination. It is based on the matrix of ASCII key. In [21], authors did the survey on various challenges of wireless sensor networks which affected the performance such as energy of the node and network, reliability and security related issues.

3 Blackhole Attack

In this above example A, B, C, D, and E are real nodes and M is the Malicious node. When A wants to send a packet to E, it still finds the routing path by sending dummy packet to adjacent nodes. And the adjacent nodes reply with their routing table details. But when the Malicious node gets this dummy packet it will reply the destination node is there in its routing table. A will take that data and store it in the routing table. And whenever A wants to send data to E, that will be redirected to the Malicious node. According to the program written for Malicious node, it will ignore or forward it. This is attack is called a Blackhole attack (Fig. 2).

4 Proposed System Architecture

This section provides an outline of our projected system attack prevention manner, which explains the mechanisms of detection, system structure, and algorithms is

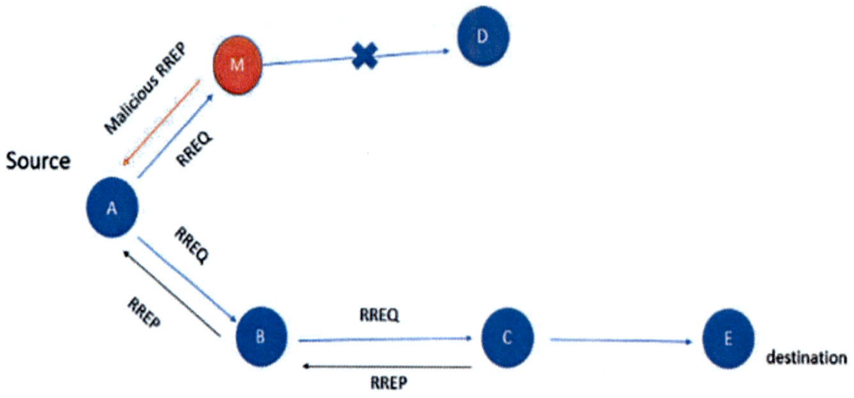

Fig. 2 Blackhole attack

used. The benefits and significance of employing a Hidden Markov Model to detect attacks in particular (Fig. 3).

Here we initialize nodes and send RREQ (Route request) to destination nodes from the source node. And then the source node will try to get the shortest path between the source and the destination using the Dijkstra algorithm. Later, system will detect an attack using the secret Markov model, which means the most likely blackhole attack is occurring. If there is a blackhole, altering the neighbor nodes and adding the nodes to the blacklist will prevent the method. And afterward, the safe source to destination routing path will be established shown in Fig. 4.

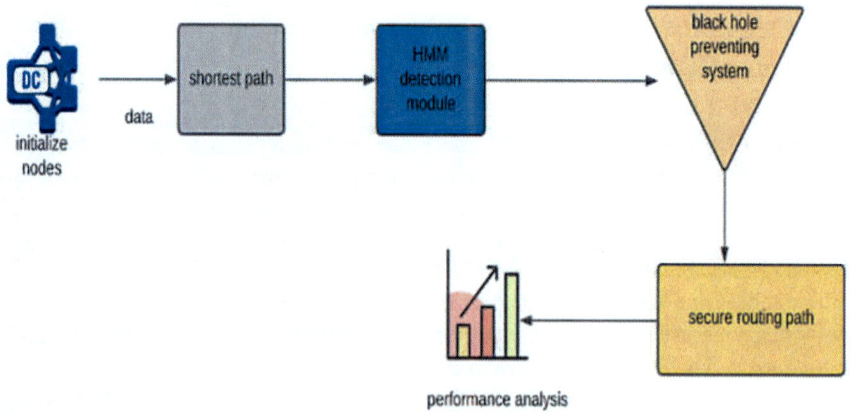

Fig. 3 Working of model

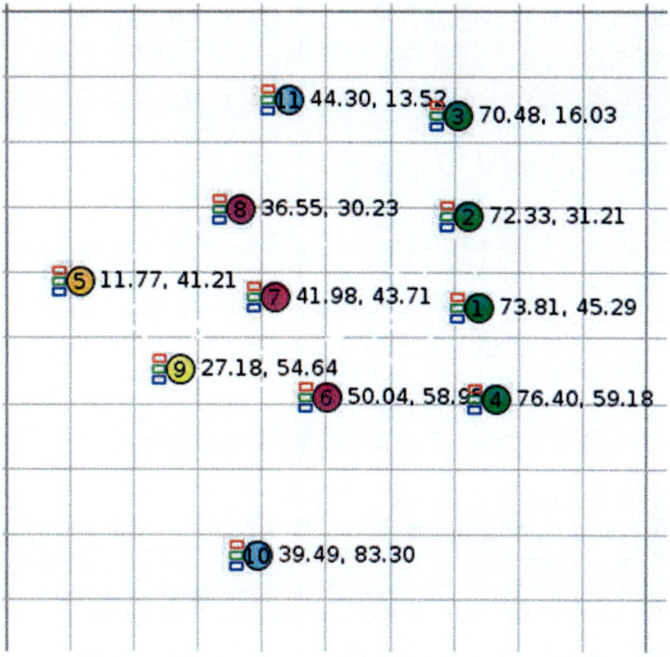

Fig. 4 WSN with 11 nodes

5 Proposed Methodology

In this work, the security of wireless sensors is enhanced with the help of HMM and Dijkstra algorithm. It will prevent from blackhole attack, as shown in the below steps.

Step1: Nodes are deployed in WSN.

Step2: Communication will start from the source node. Before starting the communication, The source node will route the RREQ to the sink via its adjacent nodes.

Step3: Source node will receive the RREP (Route reply) from the sink through all of its neighbor nodes.

Step4: When the RREP response is received, source node will have the multiple shortest path between source and sink by using the Dijkstra algorithm.

Step5: HMM model is deployed to detect the greatest probability of co-occurrence of probability.

Step6: If a Blackhole attack is detected then mitigation will start in the network to prevent the attack from the network.

Step7: Now Re-routing will be done in the network, which creates the secure shortest path from source to the sink.

From the above figure, sender nodes are node 1, node 2, node 3, node 4, node 6, node 7, node 8 and node 9. HMM nodes: node 10 and node 1 and Sink: node 5.

5.1 Proposed EHMM Algorithm

Dijkstra's algorithm is used to find the shortest path between two points A and B. It selects the unvisited vertex with the shortest distance, calculates the distance through it to each unvisited neighbor, and updates the neighbor's distance if the distance is tinier. HMM finds the probability of occurring blackhole attack from the paths given by Dijkstra algorithm.

$$\lambda = \{\Pi, A, B\}$$

A state transition probabilities
B observation probability matrix
π initial state distribution.

Step 1: Create a network with Z nodes, where Z = 1, 2, 3, 4, 5,…, and n is node count.
Step 2: Source Node *Sn* is used to start the route discovery process.
Step 3: RREQ Packets are sent from S_n to D_n (destination node).
Step 4: Using Dijkstra's algorithm, find the 5 shortest routes between the origin and the destination.
Step 5: The shortest routes and their associated prices (HMM model $\lambda = \{\Pi, A, B\}$).
Step 6: Use the HMM method.
Step 7: stop.
Step 8: which have the greatest probability of blackhole attack occurs.

In this work, HMM nodes work as the controller nodes, which can initially establish the communication from each and every nodes which is connected with the sensor network, shown in Fig. 4. After that it will check for the malicious nodes, once malicious nodes are getting detected it will inform the sink and stop the communication from the malicious nodes, which is shown in Fig. 5. Node information shows that what time malicious nodes are getting detected and how these nodes are getting g detached from the network.

Now malicious nodes have replaced with any other neighbor nodes, so controller will apply the Dijkstra's algorithm to calculate the shortest path from the source and sink.

Figure 5 shows that at the time of data transmission the HMM algorithm will help us to identify the malicious node and controller will use the Dijkstra's algorithm to get the shortest way from its neighbor node to destination node.

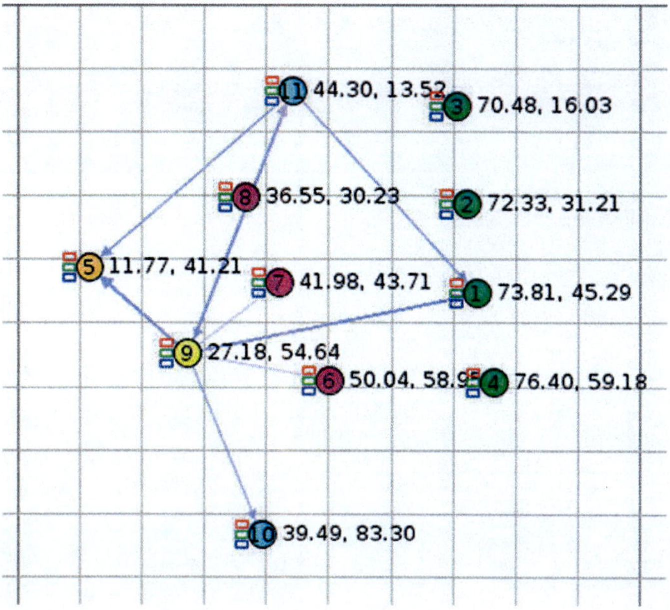

Fig. 5 Malicious node detection by the HMM

6 Result and Discussion

From Fig. 6, it shows that node 9 is detected as the malicious node. Once detection is completed, it will inform to the source and sink that malicious node is detected. So source can reroute the next shortest path for the communication (Fig. 7).

When a malicious node is detected, all of the network's nodes stop communicating. As a result, its received packet is zero. And begin transmission via a different path, attempting to complete the transmission (Fig. 8).

In the above figure, we can see that all the packets have reached the destination. Even though we had malicious node in the path. By using this technique, we were able to transfer all the packets with zero packet loss.

In Fig. 9, it shows the node information from cooja simulator, the highest packet deliver ratio and lowest packet drop ratio is achieved by EHMM model.

Packet Delivery Ratio

From Fig. 10 shows that how the packet drop is getting reduced by using EHMM model. Similarly packet deliver is getting increased, because this EHMM helps to detect the malicious nodes, and stop the communication with that node. So packet delivery ratio increases and packet drop ratio decreases (Table 1).

From Fig. 11 shows the malicious node detection by the EHMM algorithm.

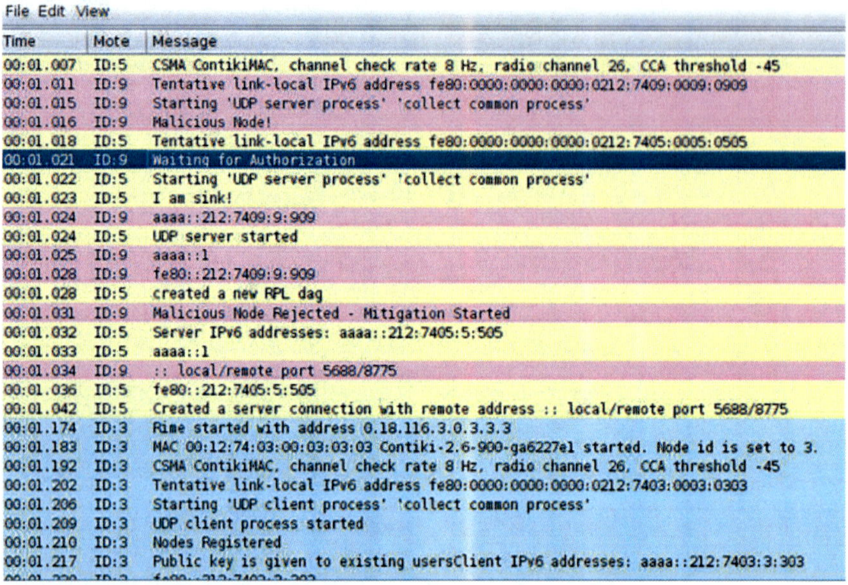

Fig. 6 Malicious node detection and mitigation process

7 Conclusion

In this work, WSN is implemented with the sensor nodes as source, sink node and malicious nodes. Malicious nodes are getting detected by the EHMM and path reroute will be done with the help of Dijkstra algorithm. By using these algorithms, we can getting a reliable, secure and shortest path from source and sink. This work can be used in any real-time scenario for monitoring and controlling using WSN with better secure and reliable communication such as border monitoring and weather monitoring.

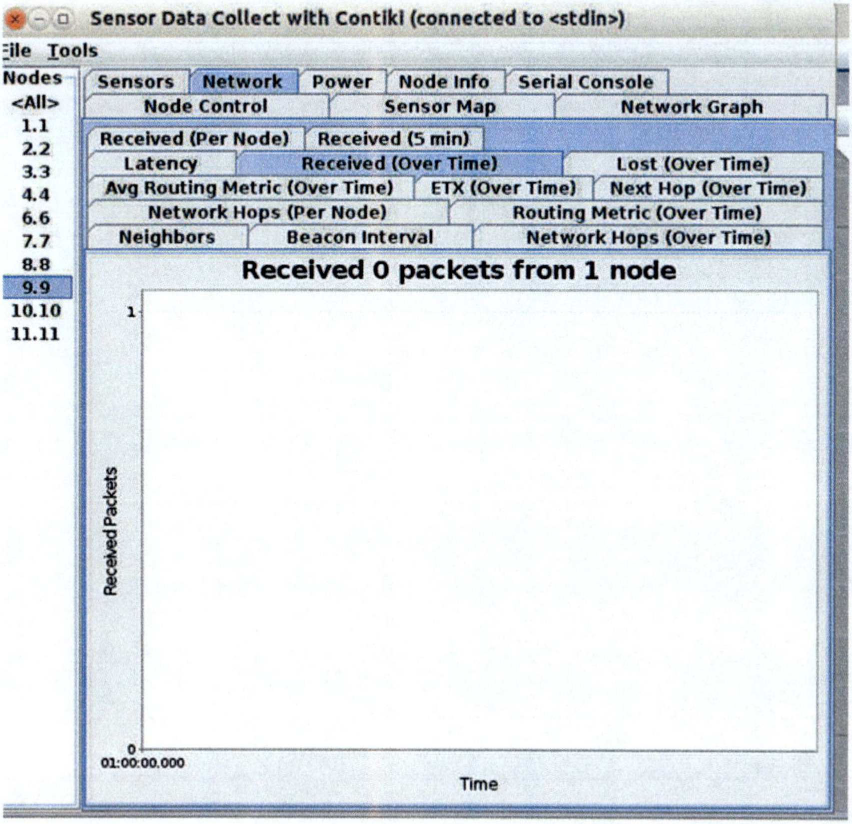

Fig. 7 Malicious node (Node 9) received packet

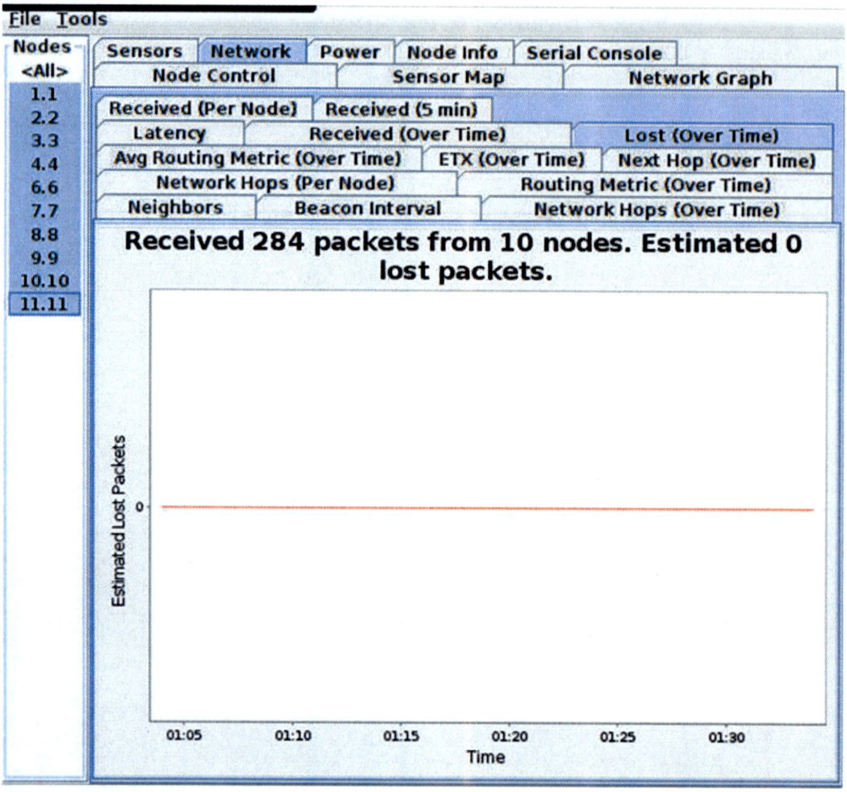

Fig. 8 Received and lost over time graph

Node	Received	...	Lost	Hops	Rtmetric	ETX	...	Beacon Interval
1.1	54	0	0	1.000	414.519	16.000	0	29 min, 36 sec
2.2	54	0	0	2.000	542.907	24.037	0	29 min, 36 sec
3.3	54	0	0	2.000	541.500	24.037	0	29 min, 36 sec
4.4	54	0	0	2.000	544.222	24.037	0	29 min, 51 sec
6.6	54	0	0	1.000	384.000	16.000	0	29 min, 55 sec
7.7	54	0	0	1.000	409.926	16.000	0	29 min, 46 sec
8.8	54	0	0	1.000	409.907	16.000	0	29 min, 51 sec
9.9	0	0	0	0.000	0.000	0.000	0	
10.10	54	0	0	1.000	409.907	16.000	0	29 min, 36 sec
11.11	54	0	0	1.000	412.833	16.000	0	29 min, 55 sec
Avg	**54.000**	**...**	**0.000**	**1.333**	**452.191**	**18.679**	**...**	**29 min, 45 sec**

Fig. 9 Node information

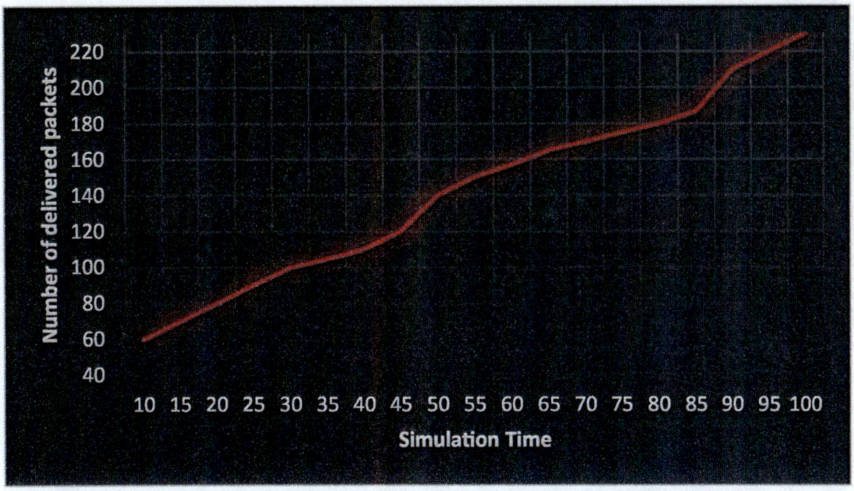

Fig. 10 Packet delivery by nodes

Table 1 Packet delivery by nodes

No. of delivered packets	Simulation time
15	70
20	80
25	90
30	100
35	105
40	115
45	120
50	140
55	155
60	157
65	165
70	170
75	175
80	180
85	187
90	210
95	220

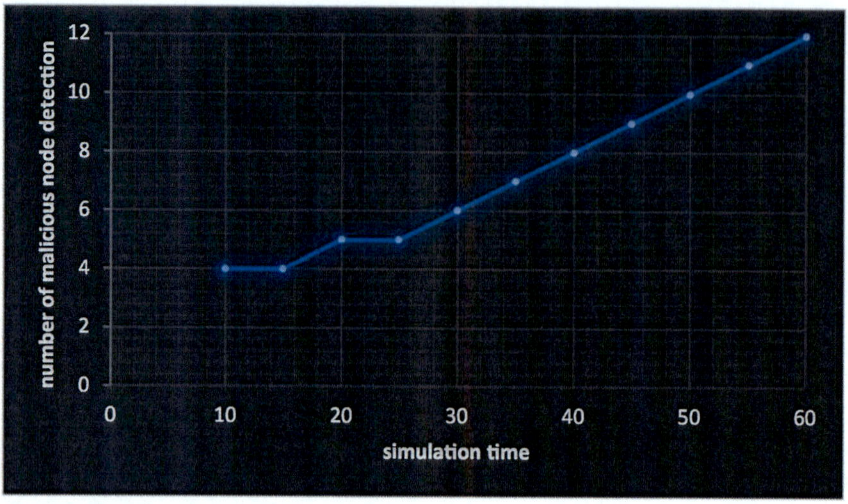

Fig. 11 Malicious nodes detection

References

1. H. Kalkha, H. Satori, K. Satori, Preventing black hole attack in wireless sensor network using HMM. Procedia Comput. Sci. **148**, 552–561 (2019)
2. A. Yasin, M. Abu Zant, Detecting and isolating black-hole attacks in MANET using timer based baited technique. Wirel. Commun. Mob. Comput. (2018)
3. M. Wazid, A.K. Das, A secure group-based blackhole node detection scheme for hierarchical wireless sensor networks. Wirel. Pers. Commun. **94**(3), 1165–1191 (2017)
4. G. Kaur, V.K. Jain, Y. Chaba, Detection and prevention of blackhole attacks in wireless sensor networks, in *International Conference on Intelligent, Secure, and Dependable Systems in Distributed and Cloud Environments* (Springer, Cham, 2017), pp. 118–126
5. H. Sikarwar, D. Das, A lightweight and secure authentication protocol for WSN, in *2020 International Wireless Communications and Mobile Computing (IWCMC)* (IEEE, 2020), pp. 475–480
6. N.F.A. AL-Shaihk, R. Hassanpour, Active defense strategy against jamming attack in wireless sensor networks. Int. J. Comput. Netw. Inf. Secur. **11**(11) (2019)
7. A. Rani, S. Kumar, A survey of security in wireless sensor networks, in *2017 3rd International Conference on Computational Intelligence & Communication Technology (CICT)* (IEEE, 2017), pp. 1–5
8. J. Grover, S. Sharma, Security issues in wireless sensor network—A review, in *2016 5th International Conference on Reliability, Infocom Technologies and Optimization (Trends and Future Directions) (ICRITO)* (IEEE, 2016), pp. 397–404
9. B. Rajasekaran, C. Arun, Link failure detection and classification in wireless sensor networks using classification method. Int. J. Innov. Technol. Exploring Eng. (IJITEE) **8**(12), 1–4 (2019)
10. T. Kaur, R. Kumar, Mitigation of blackhole attacks and wormhole attacks in wireless sensor networks using aodv protocol, in *2018 IEEE International Conference on Smart Energy Grid Engineering (SEGE)* (IEEE, 2018), pp. 288–292
11. O.O. Olakanmi, A. Dada, Wireless sensor networks (WSNs): Security and privacy issues and solutions, in *Wireless Mesh Networks-Security, Architectures and Protocols* (IntechOpen, 2020)
12. G. Arulkumaran, R.K. Gnanamurthy, Fuzzy trust approach for detecting black hole attack in mobile adhoc network. Mob. Netw. Appl. **24**(2), 386–393 (2019)

13. P.B. Hari, S.N. Singh, Security issues in wireless sensor networks: current research and challenges, in *2016 International Conference on Advances in Computing, Communication, & Automation (ICACCA)(Spring)* (IEEE, 2016), pp. 1–6
14. K. Arora, D. Kavita, V. Jain, Impacts of black hole attack on mobile ad-hoc networks. Int. J. Future Gener. Commun. Netw. **13**(4), 644–653 (2020)
15. M. Motamedi, N. Yazdani, Detection of black hole attack in wireless sensor network using UAV, in *2015 7th Conference on Information and Knowledge Technology (IKT)* (IEEE, 2015), pp. 1–5
16. R.K. Yadav, R. Mishra, An authenticated enrolment scheme of nodes using blockchain and prevention of collaborative blackhole attack in WSN (2020)
17. M. Shinde, D.C. Mehetre, Black hole and selective forwarding attack detection and prevention in WSN, in *2017 International Conference on Computing, Communication, Control and Automation (ICCUBEA)* (IEEE, 2017), pp. 1–6
18. S. Manishankar, P.R. Ranjitha, T. Manoj Kumar, Energy efficient data aggregation in sensor network using multiple sink data node, in *IEEE Xplore, International Conference on Communication and Signal Processing* (2017) [2018]
19. S. Gokuldev, C. Sowntharya, S. Manishankar, A pro-active fault tolerant deadline hit count based scheduling ın computational grid. ARPN J. Eng. Appl. Sci. © 2006–2017 Asian Research Publishing Network (ARPN), vol. 12, no. 23, December 2017. ISSN 1819-6608
20. B.J. Santhosh Kumar, K. Vijay, Symmetric key based encryption and decryption using Lissajous curve equations. Int. J. Electr. Comput. Eng. (IJECE) **7**(1), 285–288 (2017). ISSN: 2088-8708. https://doi.org/10.11591/ijece.v7i1.pp285-288. Journal homepage http://iaesjournal.com/online/index.php/IJECE
21. S. Anand, K.C. Manoj, A survey on clustering approaches to strengthen the performance of wireless sensor network, in *2020 Second International Conference on Inventive Research in Computing Applications (ICIRCA)*, Coimbatore, India (2020), pp. 814–820. https://doi.org/10.1109/ICIRCA48905.2020.9183174

Trust-Based Security Scheme Using Fuzzy Clustering for Vehicular Ad Hoc Networks

N. Kanthimathi, J. Roshini Roy, N. Saranya, and P. Sandhya

Abstract A vehicular ad hoc network (VANET) is a self-organizing network made up of interconnected vehicles that enables immediate sharing of valuable traffic data. It has the ability to transmit video files between nearby nodes. The video standard obtained is by using an advanced tool environment that allows an MPEG-4 compression codec, a system modeling, and video quality management software to transmit MPEG-4 video across UMTS networks. One of the benefits of such an advanced tool kit is the ability to analyze actual video sources compressed with an MPEG-4 encoder. UMTS radio link control (RLC) outperforms two modes named as acknowledged mode and unacknowledged mode. This unacknowledged mode guarantees on-time delivery but excludes errors retrieval. The acknowledged mode will provide strong perceived video quality for RLC blocking error rate of up to 30% by utilizing a replay buffer at the streaming client. Based on the performance data review, a self-adaptive RLC approved mode is proposed. In VANET, since there are several nodes, fuzzy clustering is used to minimize delay on each node. In proposed method, an effective fuzzy clustering technique is needed to improve network productivity and network stability.

Keywords VANET · Fuzzy clustering · Cluster head · Video files · RLC · UMTS

N. Kanthimathi (✉) · J. Roshini Roy · N. Saranya · P. Sandhya
Bannari Amman Institute of Technology, Erode, Tamil Nadu, India
e-mail: kanthimathi@bitsathy.ac.in

J. Roshini Roy
e-mail: roshiniroy.co19@bitsathy.ac.in

N. Saranya
e-mail: saranyan@bitsathy.ac.in

P. Sandhya
e-mail: sandhya.co19@bitsathy.ac.in

G. Ranganathan et al. (eds.), *Soft Computing for Security Applications*,
Advances in Intelligent Systems and Computing 1397,
https://doi.org/10.1007/978-981-16-5301-8_32

1 Introduction

The vehicular ad hoc network (VANET) is being touted as a promising technology for improving transportation system efficiency and safety. The vehicular ad hoc network has been taken by many researchers from various fields as a fundamental element of intelligent transport system, and massive research efforts have been made.

There are different mechanisms of communication in VANETs. Vehicle-to-vehicle (V2V) and vehicle-to-infrastructure (V2I) interactions are also possible with VANETs. Vehicles and transportation systems, like road side units and device servers, share data for navigation's secure driving and other purposes in VANETs.

In general, VANET communications can be divided into two categories based on the radio interface. Dedicated short-range communication (DSRC) is one type of approach [1]. The other class is focused on wireless infrastructure that already exists. Many techniques for managing VANETs have been suggested to allow the usage of existing wireless communication systems for data transfers. Low efficiency would be a major problem if VANETs are entirely operated by infrastructures, while completely decentralized VANETs would generate a lot of overhead. As a result, VANETs often merge centralized and decentralized components. VANETs are centralized via cellular-based connections and scheduling to reduce overhead through DSRC channels and the possibility of LTE channel congestions. Similarly, vehicles may use DSRC to communicate with their neighbors. In VANET operations, grouping vehicle into cluster is a standard and rational solution. Vehicles are assigned to clusters in a cluster-based system. A cluster's range is shorter or equivalent to that of 802.11p, allowing vehicles in the specific cluster to communicate through DSRC. Numerous clusters are managed by a single eNodeB. At least one vehicle acts as a cluster head (CH) within a cluster and collecting information from all cluster member nodes (CM) through DSRC and exchanging information with the eNodeB through TLE. While using both LTE and DSRC, this configuration reduces administration overhead.

Nodes in VANETs are more versatile and faster than other MANETs. Cluster reformation and CH shifting must happen even more often than in other MANETs. The clustering algorithm for VANET must be possible to shape stable clusters to reduce management overhead and improve communication efficiency. In order to accomplish this aim, we suggest a secure clustering algorithm for VANETs. To stop constant cluster reforming, we suggest a novel method for creating and maintaining balanced clusters for VANETs on roads. A fuzzy clustering algorithm is used to find the center of the initial clusters. The position, speed, and maximum acceleration of the vehicles in each cluster are used to choose an appropriate CH. A cluster management algorithm is proposed to maintain CMs within the propagation range of their cluster head (CH).

2 Literature Survey

Xin and Feng [4] suggested the industrial world is continually seeing technological advancements. Every day, a new technical innovation, challenge, or concept arises. Many Internet-enabled computers, also known as Internet of Things (IoT) devices, have emerged as a result of new computer applications and hardware advances. Users can also build networks using any computer to collect real-time data. In addition, new automobile manufacturers are being pushed by consumer interest toward fully autonomous vehicles. Vehicular ad hoc networks (VANETs) will be formed for a community of car and its surroundings in order to assure that accurate and appropriate information is conveyed in order to achieve autonomy efficiently and properly. Machine learning algorithms can be used to forecast situations and detect deviations using data obtained in a VANET. This research investigates various methods for clustering sets of vehicle, as well as machine learning algorithm for predicting where vehicles will be detecting false or unlikely facts.

In this article, Sari and Onursal [3] suggested that VANETs are a modern type of data transmission mechanism in which road vehicle provides assistance to end user. Because of the safety criteria, vehicular ad hoc networks (VANETs) are very probably to be implemented in the coming years, making them the most appropriate type of mobile ad hoc networks. Since the major need for VANETs is for safety-related applications, security is a major concern. In this case, the feasibility of security could endanger human lives. Here it examines the security problems that these networks present, as well as the implications of their overhead in VANET.

Nithya Darisini and Kumari [2] suggested a vehicular ad hoc network is a form of ad hoc network that in addition to find missing vehicle and allows communication organizations to travel at different speeds. As a result, creating secure end-to-end connectivity routes and having effective data transfer are hampered. As a result, VANETs face a variety of network issues and security problems in obtaining universal access, safe communication, and identity management schemes, both of which have an impact on confidence in collaboration and negotiation among mobile networking organizations, to identify security functions, threats, and vulnerabilities that VANETs face, as well as how to distinguish VANET security threats based on network layer.

Malik and Pandey [5] suggested the vehicular ad hoc network is a wireless network that connects several vehicles. The key goal of VANET safety is to have not only protection, reliable connectivity, and smart transportation but also entertainment, advertising, and special offers depend on locations. Since all communication-related facilities are more critical and vulnerable to threats, they require protection. Vehicles serve as nodes in a VANET, with contact occurring either between vehicles (V2V) or (V2I). It is a huge challenge to keep connections between vehicle and the RSU safe. Numerous scholars have put in a lot of effort to improve the security of VANET interaction, but it still has flaws. This examines several first-time threats, like hidden car, cave, and black hole attack, as well as potential cryptographic solutions. Here it would go through different security problems, routing protocols, VANET threats, various attacks, and their cryptographic solutions.

Luan [6] suggested the dedicated short-range networking and wireless cellular services are currently used to create communication in VANET. Many techniques for managing VANETs have been suggested to allow the use of current long-term evolution (LTE) networks in transmitting data. One of its most popular and efficient methods is to group the vehicle into cluster and organize the network through cluster. Since VANETs differ from these other ad hoc networks (MANETs) due to their high flexibility, past cluster-based mobile ad hoc methods fail in vehicular networks. It presents several metrics and a center-based clustering algorithms to assist self-organized VANETs in creating secure cluster and reducing the rate of system changes on highway. To decrease the effect of vehicle's speed differences, a unique cluster head (CH) filtering algorithm is also suggested. Researchers also insert dual metrics to help VANET be more safe. The findings show that the system has a high level of reliability and a low rate of packet loss.

3 Proposed Method

To control proper vehicular contact, VANETs have different safety requirements. In order to increase the level of confidence of the consumer, it guarantees that the message is created by legitimate users. Only data coming from an authenticated source would be trusted by the recipient. The availability of the node either node-supplied service is the object of numerous attackers. Attacker attempts to block or deny legitimate user services for a specified limited duration of time or indefinitely. The neighbor node collected data will be in the same type as that submitted by the sender. Try sharing with this intruder the data sent by the sender. For data security administration, digital signatures are used. After the transmission or receipt of transmitted messages, the transmitter or receiver cannot deny them. It is proposed that the vehicle's true identification should be secured from other vehicles, but that it can also be traced from the operational area. The minimum of one entity should be capable of identifying the original identity of vehicle nodes (Fig. 1).

VANET: VANET is an upgrade that is specifically intended for inter-vehicle connectivity over wireless ad hoc systems. Self-organizing systems built between vehicles are VANET. There are actually several common vehicular communication systems of which the most important is how to have the best protection for VANET without compromising efficiency and reliability. For VANET, two key forms of contact are possible: V2V and V2I. There is an OBU in each vehicle and each RSU-enabled cross-road unit. Attackers disrupt communications by obtaining partial or full access to the network communications.

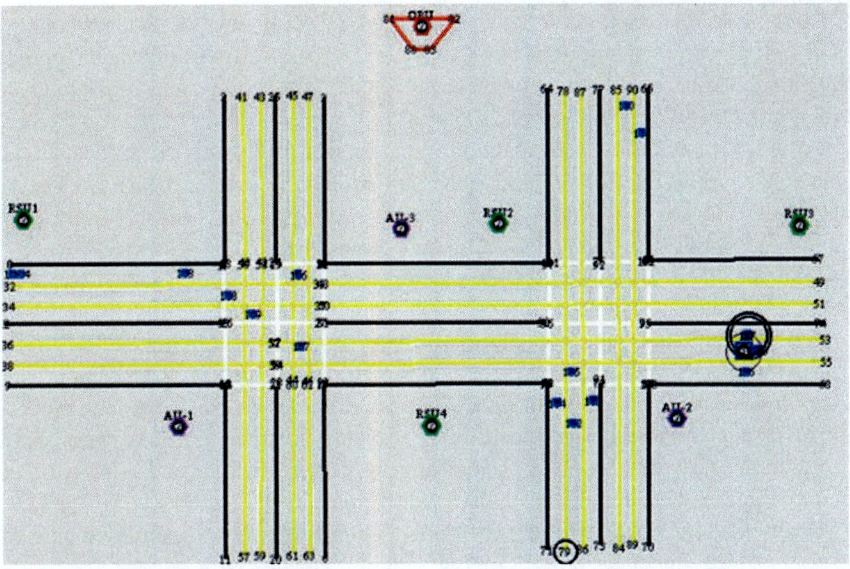

Fig. 1 Data transmitting in VANET

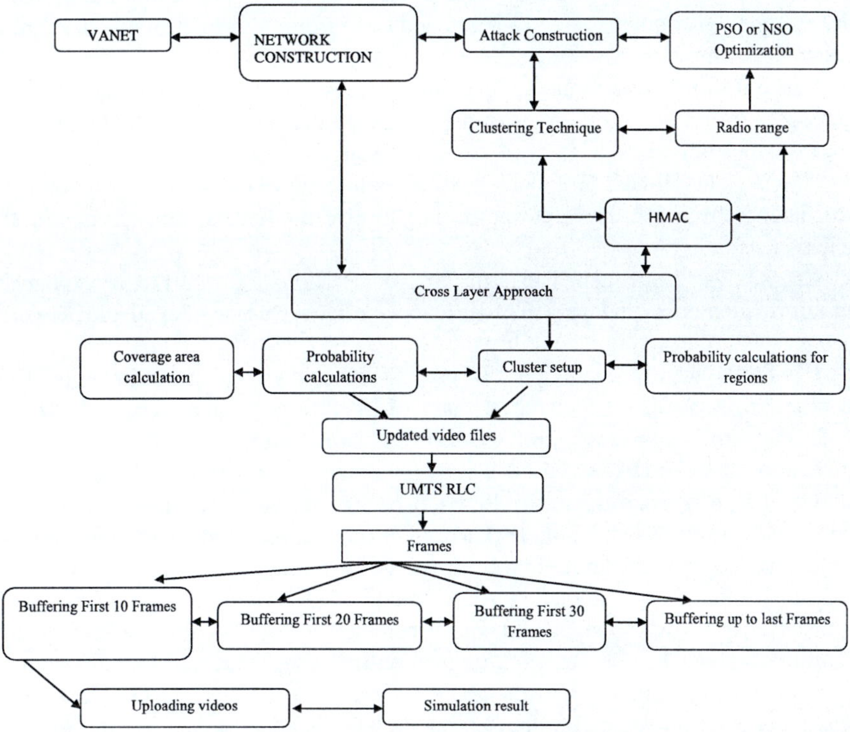

Network Construction: A network is composed of nodes (vertex) and connections (edges) connecting groups of node. Based on the database, node in wide range functional brain typically identifies regions of the brain, while connections indicate biological, logical, or productive associations.

Attack Construction: Information forging threats are really a threat to vehicular ad hoc networks. An attacker either generates a fresh message or recaps an existing one. Each vehicles in a VANET communicates shorter packet (beacons) in its code, period and actual geographical location on a regular basis.

NSO or PSO Optimization: In a wireless sensor network, node navigation attempts to calculate the locations of sensor node with the help of identified nodes. To fine-tune and gain precise position estimates, node swarm optimization is used. The algorithm's first step employs an enhanced model of the DV distance approach to generate coarse location estimates over all node. PSO is used in the second process to perfect and achieve correct position of estimated location. Particle swarm optimization of every particles attempts to change its present location and velocity based on the distances between the best possible location.

Fuzzy Clustering Technique: Every data point may belong to multiple clusters in fuzzy clustering. Clustering, otherwise known as clustering algorithms, is the process of grouping data points into clusters so that objects in the similar cluster are as identical as possible, while products from various clusters are as different as possible. Similarity tests are used to identify clusters. Width, connection, and strength are examples of similarity tests. Depending on the information or the method, various resemblance metrics may be selected.

Radio Range: Aerial mapping of a radio range is a network of radio transmission stations, including one that transforms a message that serves as both identity and intrinsic meaning to a navigator in determining his location.

HMAC: The HMAC protocol's key idea is to achieve energy consumption and low latency by using interrupt approaches to give the device priority once in an emergency.

Cross Layer Approach: The term "cross-layer approach" refers to the exchange of information between layers in order to make effective usage of available network services and achieve high flexibility.

UMTS: Users in the moving will use UMTS to communicate wirelessly. The device allows roaming, which means that UMTS customers can be accessed and can make calls from anywhere within the UMTS network range.

Radio Link Controller (RLC): The major aims of the RLC protocol is to transmit the upper-layer protocol data units (PDUs) in each of two modes: acknowledged mode (AM) and unacknowledged mode (UM). In acknowledged mode, radio link control (RLC) is a well-known service that ensures stability through retransmission. Retransmission is through the automatic repeat request (ARQ) protocol in the occurrence of a mistake for the acknowledged mode. When a received call completes safely, the session acknowledges receipt of a notification if messages are sent simultaneously to the applications. In unacknowledged mode, data is not given a header. RLC may not apply certain overhead to the service data unit's upper layer. Transfer of user information, differentiation and reassembly, error checking, duplication detecting, protocol error correction and retrieval, and ciphering are only a few examples.

Source: The video input can be in one of two formats: YUV QCIF (176 × 144) or YUV CIF (352 × 288).

Video Encoder and Video Decoder: Evaluate Vid actually provides two MPEG4 codecs: the NCTU codec and ffmpeg. For the purposes of this investigation, I chose the NCTU codec at random for video coding.

VS (Video Sender): The video encoder's output is read by the VS component, which compresses the video document; each wide video frame is broken down into small frames and then sends certain segments something over a real or virtual network in UDP packets. The framework keeps track of the timestamp, packet ID, and other information for each UDP packet sent out, and if the network is an actual source, the packets workload scale in the transmitter trace file can be determined using third-party software like tcp-dump or win-dump. The sender traces report is given by the simulation's sender agent if the system is simulated. A video traces report is also generated by the VS component, which provides data about each frame in the actual video files. Following that, the video tracing files and the transmitter tracing files are used to evaluate video accuracy.

ET (Evaluate Trace): After the completion of the video delivery, the task of assessment begins. The test happens on the sender's end. As a result, the timestamp, packet ID, and packets payload length relevant information at the recipient must be transferred back to the receiver. The ET portion produces a frame/packet loss and frame/packet jitter report and generates a replicated video folder following the original encrypted video files, the video trace file, the transmitter trace file, and the receiver trace file. This refers to the potentially distorted video located at the received signal as it can be repeated to a final user. In theory, creating the potentially manipulated video may be thought of as a method of transferring the actual video tracing file frames by frames, removing frames marked as missing or damaged at the receiving side. However, creating the actually manipulated video is more difficult, and the procedure is discussed in greater detail below. The combined inter-frame jitter algorithms for play-out buffers are also implemented in the new versions of the ET module. A frame is considered as a missing frame if it arrives after its specified playtime. This is a completely optional feature. The play-out buffer's size should be specified; otherwise, it would be considered to be infinite.

- **FV (Fix Video)**: Frame by frame, the accuracy of digital video is assessed. As a result, the maximum volume of video frame on the receiving side, including the inaccurate ones, should be same as of the transmitter end. When the codec is unable to accommodate missing frames, the FV portion can be used to solve the issue by adding the most recent accurately decoded frames in position of every lost frames as an errors concealment method.
- **Peak Signal–Noise Ratio (PSNR)**: PSNR is a widely used descriptive criterion for evaluating the implementation quality of service of video transmission. The PSNR between the luminance variable Y of source image S and destination is defined.

- **Traffic Trace**: The frame style and frame length of the video track file created by the VS portion of Eval Vid are extracted using the My Traffic Tracing device. Thirdly, according to the information systems defined in the simulations scripting code, this agent separates video frame into small frames and transfers such segments to the bottom UDP layers at the required period.

UDP

My UDP is responsible for a significant portion of the UDP agent. The duration of each forwarded packets, the packet ID, and the message payload scale are all recorded by this new server which enables users to determine the output data file of the sender traces file. In an actual network context, the role of the My UDP entity compares to the role of a tool like tcp-dump.

UDP Sink

The scattered video sample frames sent by My UDP are received by My UDP Sink. Every received packets timestamp, packet ID, and packet size are also recorded in the user-specified folder by this agent.

4 Simulation and Result

The simulated results and programmed analysis of fuzzy clustering algorithm are discussed in this section. The whole simulation is done in NS2-2.34 to implement the fuzzy clustering algorithm and through the assistance of a simulator will measure the energy consumption, accuracy rate, routing overhead, and delivery rate of the network.

4.1 Energy Consumption

The graph illustrates the calculation of the network's energy consumption. The x-axis and y-axis are the period given in ms and the energy usage given in the unit joules. Our recommended green color is fuzzy clustering algorithm as shown, and the existing process is shown in red color. The output shows that the energy consumption of the fuzzy clustering technique is greater than the existing multi-hop clustering algorithm (Fig. 2).

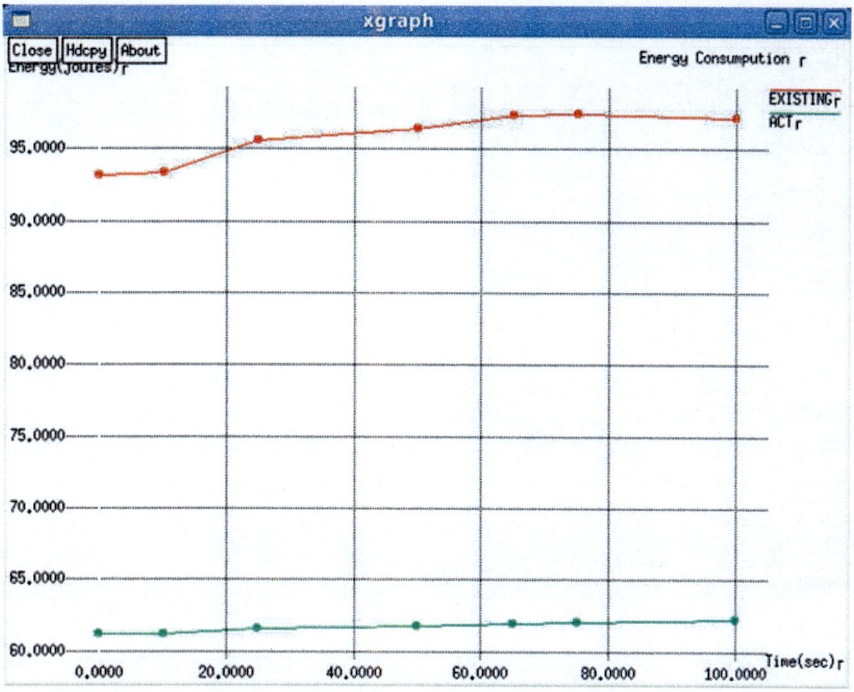

Fig. 2 Energy consumption

4.2 *Delivery Rate*

The graph shows the network's delivery rate. Here the *x*-axis is given in millisec for the unit packet and the *y*-axis in period. Our recommended green color is fuzzy clustering algorithm as shown, and the existing process is seen in red color. The outcomes show that the delivery rate measurement of the multi-hop clustering method is lower than the fuzzy clustering technique (Fig. 3).

4.3 *Routing Overhead*

The graph shows the network's overhead routing estimate. Here the *x*-axis is the overlap routing given in millisec for the unit packet and the *y*-axis in period. Our recommended green color fuzzy clustering strategy is shown, and the existing process is seen in red color. The outcomes show that the efficiency of routing overheads measurement of the multi-hop clustering technique method is lower than the fuzzy clustering technique (Fig. 4).

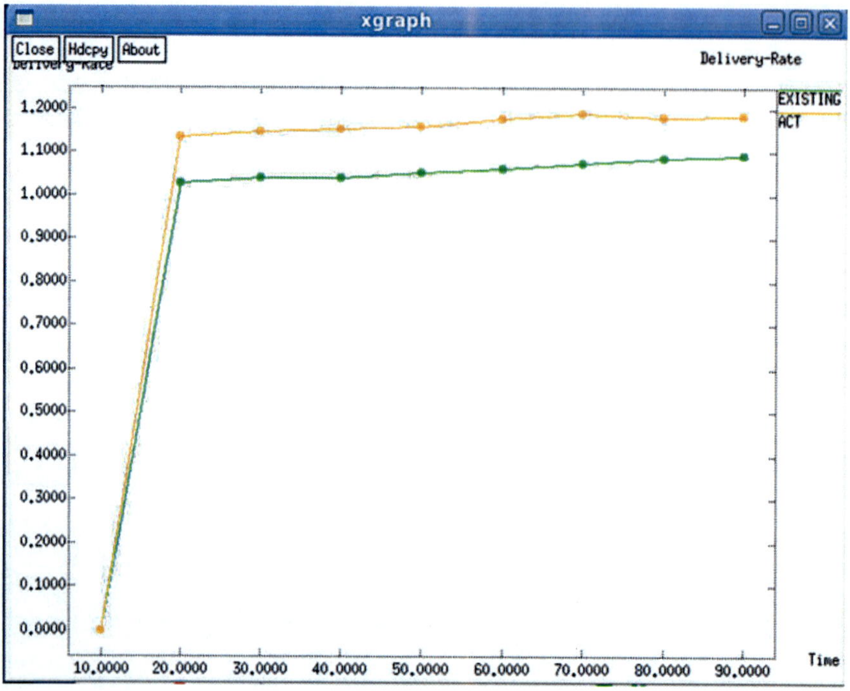

Fig. 3 Delivery rate

5 Conclusion

A fuzzy clustering has been proposed for creating a secure VANET cluster and removing fraudulent nodes from the networks using an integrated tool environment, the output of streaming MPEG-4 video over UMTS network. MOS, accumulated jitter, and video frame error rate are used to assess observed video quality. Two videos were considered, each with its own set of characteristics. The perceived video output obtained using UMTS RLC acknowledged mode is superior to that achieved using unacknowledged mode, according to simulation performance. Unacknowledged mode performs poorly than acknowledged mode because it does not have error recovery, and the MPEG-4 decoder is very vulnerable to incorrect video frames.

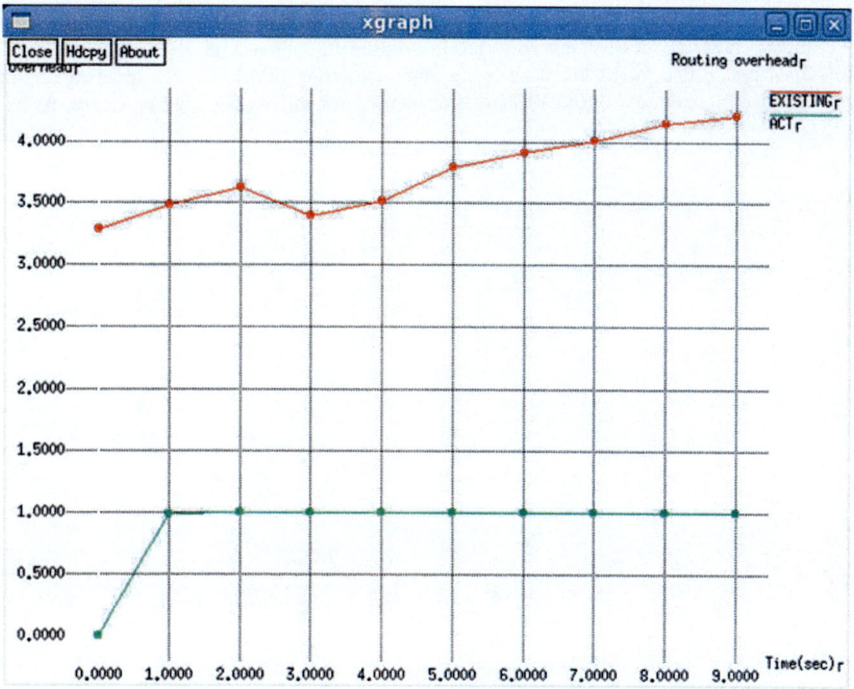

Fig. 4 Routing overhead

References

1. A.A. Khan, M. Abolhasan, W. Ni, An evolutionary game theoretic approach for stable and optimized clustering in VANETs. IEEE Trans. Veh. Technol. **67**(5), 4501–4513 (2018)
2. P. Nithya Darisini, N.S. Kumari, A survey of routing protocols for VANET in urban scenarios, in *2013 International Conference on Pattern Recognition, Informatics and Mobile Engineering (PRIME)* (2013), pp. 464–467
3. A. Sari, O. Onursal, M. Akkaya, Review of the security issues in vehicular ad hoc networks (VANET). Department of Management Information Systems, vol. 8, pp. 552–566 (2015)
4. Y. Xin, X. Feng, Replica attack detection method for vehicular ad hoc networks with sequential trajectory segment. Int. J. Distrib. Sens. Netw. **15**(02) (2019). https://doi.org/10.1177/1550147719827500
5. A. Malik, B. Pandey, An intelligent authentication based vehicle initiated broadcast-dynamic path data collection scheme in VANET. Indian J. Sci. Technol. **9**(16), 1–7 (2016)
6. Q. Luo, C. Li, Q. Ye, T.H. Luan, L. Zhu, X. Han, CFT: a cluster-based file transfer scheme for highway VANETs, in *Proceedings of the ICC 2017—IEEE International Conference on Communications* (Paris, France, 2017), pp. 1–6
7. E. Ben Hamida, M.A. Javed, Channel-aware ECDSA signature verifcation of basic safety messages with K-means clustering in VANETs, in *Proceedings of the 30th IEEE International Conference on Advanced Information Networking and Applications, AINA 2016* (2016), pp. 603–610
8. N. Ding, G. Tan, W. Zhang, Ia2p: Intrusion-tolerant malicious data injection attack analysis and processing in traffic flow data collection based on vanets. Int. J. Distrib. Sens. Netw. **12**(5), 5159739 (2016)

9. K. Sheridan, T. Puranik, E. Mangortey, O. Pinon, M. Kirby, D. Mavris, An application of dbscan clustering for flight anomaly detection during the approach phase, 01 (2020)
10. G.V. Rossi, Z. Fan, W.H. Chin, K.K. Leung, Stable clustering for Ad-Hoc vehicle networking, in *Proceedings of the 2017 IEEE Wireless Communications and Networking Conference, WCNC* (2017), pp. 1–6

AquaVision: Real-Time Identification of Microbes in Freshwater Using YOLOv3

Prajakta Vinay Yadav, Siddhant Prakash Dhokrat, Ajit Sheshrao Kadam, and Prashant Janardan Gadakh

Abstract According to World Health Organization (WHO), at least 2 billion people use drinking water contaminated due to different microorganisms and feces. Water pollution can be caused by industrial waste, oil pollution, various bacteria and microbes that can affect aquatic and human life. Considering water as one of the major aspects of life directly or indirectly, it must be preserved to avoid water-borne diseases. In this paper, the AquaVision application is built to detect the microbes present in water based on the water sample image uploaded by the user. Users can upload grayscale images, real-time Foldscope images, or microscopic images of the water. It uses deep learning methods like convolutional neural network (CNN) to detect the microbe called diatom using You Only Look Once version-3 (YOLOv3) algorithm. Diatom produces harmful toxins in the water that affects animals and human health leading to several acute and chronic diseases. The system uses the features of the YOLOv3 algorithm to detect the microorganisms more accurately, efficiently and generate more reliable results than the existing traditional laboratory methods. The reports generated by the system allow the user to monitor the type and size of diatom present in the water. It also helps to estimate the contamination of water due to the presence of microbes in less time.

Keywords AquaVision · Convolutional neural network · Deep learning · Diatom · YOLOv3 · Foldscope

1 Introduction

Water is an important part of our life, and every individual should have access to clean water. However, pollution and microbial contaminants present in water have adversely affected the quality of water, making it unhealthy and unsafe for drinking. Microbiological water pollution is a type of water pollution caused by microorganisms such as protozoa, bacteria, and viruses [1]. These microorganisms can cause

P. V. Yadav · S. P. Dhokrat (✉) · A. S. Kadam · P. J. Gadakh
Department of Computer Engineering, International Institute of Information Technology, Pune, India

G. Ranganathan et al. (eds.), *Soft Computing for Security Applications*,
Advances in Intelligent Systems and Computing 1397,
https://doi.org/10.1007/978-981-16-5301-8_33

severe diseases such as cholera and diarrhea. So, microbial contamination of freshwater is an intimidating challenge that has major impacts on aquatic as well as human life. In 2017, around 785 million people did not have access to drinking water facilities.

In India, every year nearly 37.7 million people are affected due to water-borne diseases. Such pollution makes water unfit for human consumption and possesses significant health hazards.

Diatom is a photosynthetic single-celled microorganism found across the water column of oceans, rivers, lakes, and ponds, etc. Diatoms produce neurotoxin domoic acid (DA) which is capable of causing acute and chronic diseases in human bodies. Domoic acid causes loss of short-term memory called amnesic shellfish poisoning. It also causes several diseases with symptoms like vomiting, diarrhea, abdominal cramps, short-term memory loss and can ultimately lead to coma or death. Hence, it is important to identify diatoms in water to save mammals including humans [2].

The proposed system aims to identify the diatoms present in water using YOLOv3. YOLO algorithm is a fast single-shot neural network algorithm used for object detection [3]. The input image to the system can be a Foldscope image or any other grayscale image. Foldscope can be used to capture real-time microscopic images of water. The predominant and widely practiced method of detecting water quality is performed in laboratories, which is time-consuming and requires skilled staff to conduct the test using the appropriate chemical reagents [4]. To overcome the drawback of the conventional methods used for water quality detection, we propose to use a Foldscope that captures real-time microscopic images of water.

The paper is divided into several sections as follows: Section 2 contains a Literature Survey. Section 3 explains the Proposed System. Section 4 and Sect. 5 describe the YOLOv3 algorithm and the result of the system, respectively, followed by the Conclusion and future work are presented in Sect. 6.

2 Literature Survey

This paper implements a smart system for the detection and identification of microbes in freshwater samples. The smart system uses a special microscopic device (Labomed Lx 300i) that overcomes the disadvantage of general microscopic devices. It contains a special Bacterial Image Analysis Software (BactImAS) open-source software for cell tracking, segmentation, and visualization of data. It also uses ANN for training several custom images. The results obtained using this technique are faster and easily accessible than traditional methods [5]. The paper identifies and classifies four different types of underwater targets using the YOLOv3 and YOLOv3-UW algorithm. The entire dataset of this paper is divided into four different classes. They have implemented these classes against the YOLOv3 and YOLOv3-UW algorithms. The primary goal of this paper is to compare the precision, accuracy, and speed of these two algorithms. The final result states that YOLOv3-UW has better speed and accuracy for underwater target detection [6].

This work discusses the turbidity units found using a turbidity meter. It also talks about the levels of impurity present in water and detects it using remote sensing methods. It reviews the existing turbidity detection methods like Sechi-3000. The system uses computer vision methods for detecting and identifying the level of turbidity present in water [7]. Several image processing techniques have been discussed in this paper that are applicable to wastewater quality detection. It talks about the flow of image processing using steps like image acquisition, image preprocessing, image segmentation, and image analysis. It explains different approaches that can be used for the implementation of the steps involved in image processing [8].

Digital image processing techniques have been discussed in this paper for water pollution control. The detection of pathogens like Escherichia, Bacillus, and Bordetella is done using enhanced image processing techniques. The system uses morphology that is a tool of MATLAB for the implementation of various image processing steps. It discusses how the image processing steps can be easily implemented for the detection of pathogens [9]. This paper assesses the water quality using MATLAB image processing techniques. This paper focuses on two major microorganisms Daphnia magna and Lemna minor. It focuses on the evolution and constant monitoring of these two microbes. The result of these papers finds which part of microbes is starting to become unhealthy (red) and which is still healthy (green) [10].

3 YOLOv3 Algorithm

YOLO is an intelligent and fast single-shot network for object detection [11]. It splits the image into a Size * Size grid and predicts the class probability using the bounding box and Intersection Over Union (IOU) metric. Since its original publication in 2015, five new versions of YOLO have been published. For this project, we have selected the YOLOv3 algorithm. It provides better accuracy and precision in detecting smaller objects [12]. It predicts bounding boxes at three different scales and uses a network called Darknet-53. It has 53 deep convolutional layers which are built using 3×3 and 1×1 convolutional layers. The images are trained in batches of three different detection layers. These detection layers are the 82nd, 94th, and 106th layers.

The images are down-sampled into 32,168 dimensions at each of these three layers, respectively. Up to the 82nd layer, each grid of the image is split 32 times. The same process is applied to the next 94th and 106th layers [13] (Fig. 1).

YOLOv3 uses metrics like anchor box and non-max suppression. Bounding boxes are predicted by the system using dimension clusters as anchor boxes for the given input. Four coordinates are predicted for each bounding box as t_x, t_y, t_w, t_h. Let the bounding box has width and height as *pw*, *ph,* respectively, and consider (*cx*, *cy*) coordinates of the image as cell offset from the top-left corner of the image, then the predictions correspond to (Fig. 2):

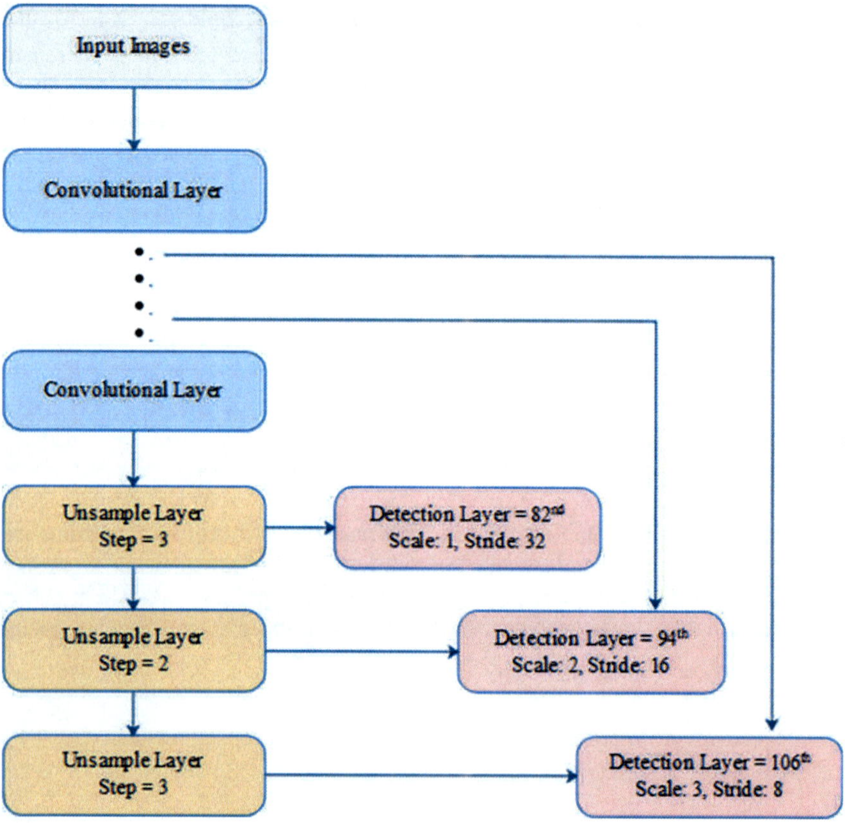

Fig. 1 YOLOv3 network architecture

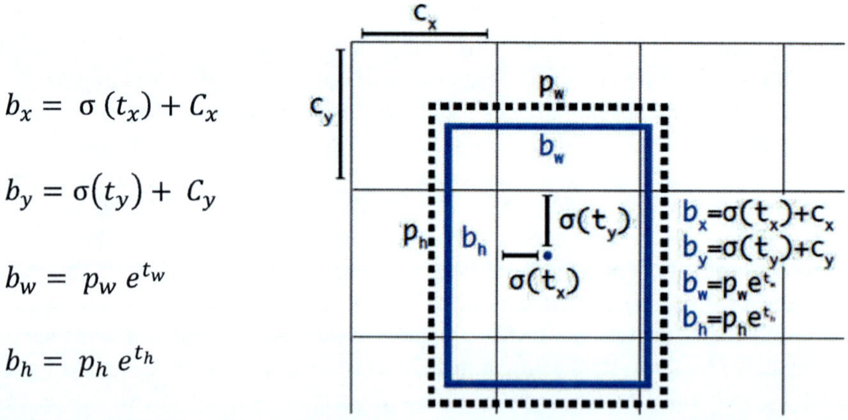

Fig. 2 Bounding boxes—location prediction with prior dimensions

$$b_x = \sigma(t_x) + C_x$$

$$b_y = \sigma(t_y) + C_y$$

$$b_w = p_w e^{t_w}$$

$$b_h = p_h e^{t_h}$$

Once the bounding box values are calculated, the IOU metric value is computed. The IOU gives the similarity between the predicted bounding box and the ground truth bounding box. This helps in identifying the class of images accurately.

4 Proposed System

The proposed system detects the presence of diatoms in freshwater based on the input image of the water sample. The system uses the YOLOv3 algorithm for training and model building of custom images. The image dataset contains several types and sizes of diatoms. Diatom contains domoic acid (DA) that is harmful to human life. Therefore, it is necessary to detect diatoms and reduce the number of affected people by several water-borne diseases. The input image for the proposed system is a grayscale image captured through a microscope or a Foldscope [14]. Foldscope is an inexpensive microscope, which is attached to a mobile phone to get a microscopic image of water. It is a lightweight microscope that can be used anytime and anywhere to get real-time images of water. The system is trained using custom images of diatoms using YOLOv3 algorithm, which is a clever CNN-based object detection algorithm (Figs. 3 and 4).

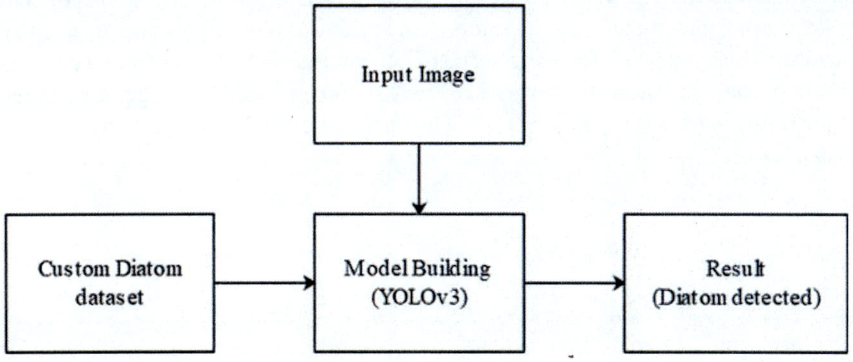

Fig. 3 Overview of the AquaVision system

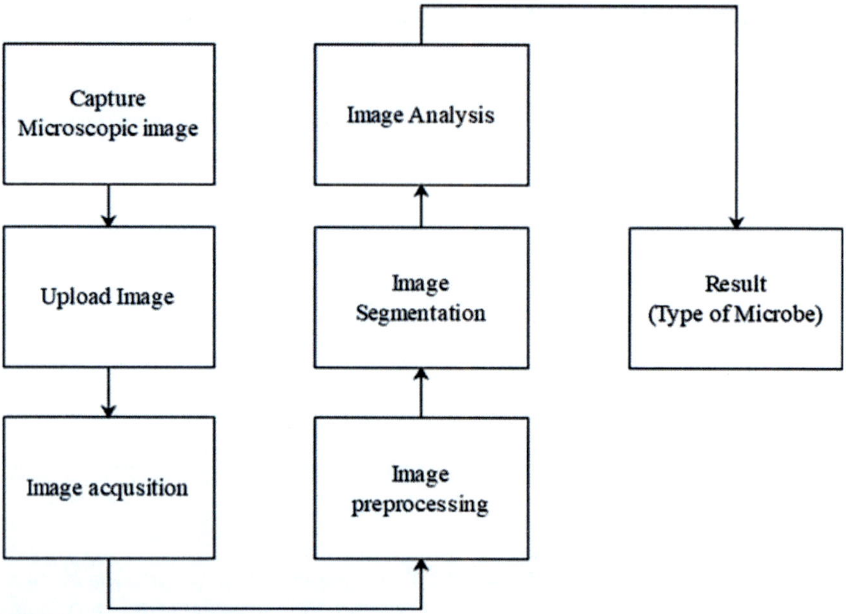

Fig. 4 Flowchart of the AquaVision system

Below are the detail steps that explain the process to detect the microbes present in water sample using YOLOv3 algorithm.

4.1 Data Preparation

The custom images of diatoms are taken from the Automatic Diatom Identification and Classification (ADIAC) project public data. ADIAC project has an efficient collection of image databases that are used for developing algorithms to identify diatoms. This system is built using a total of 640 images. The training dataset consists of 440 images, and the testing dataset consists of 200 images. The images are labeled using Labellmg software. Refer Fig. 5. The images are labeled based on their shapes and sizes.

4.2 Training Phase

The image classes and their name files are saved in different configuration folders for building the model using YOLOv3. The yolov3.cfg file is configured for the number of classes used and the number of filters to be applied for different batches. The

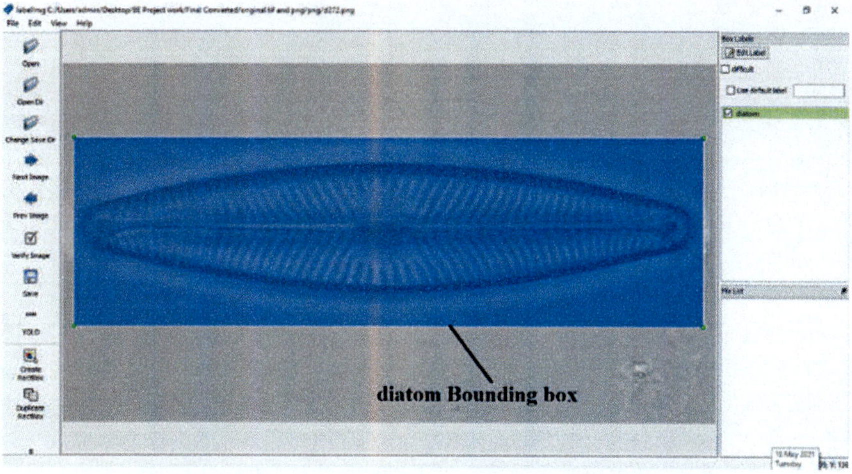

Fig. 5 Labeling of the diatoms

model is trained using the Darknet53, and post-training, the weights are stored in the yolov3.weights file. Darknet53 is a CNN with 53 fully connected deep layers. It is loaded using the YOLOv3 model. Once the weights are trained and checkpoints are saved, the model is ready to be tested on a test image. Refer Fig. 6. That explains the graph of loss versus number of iterations required for training the model.

Table 1 refers to the system configurations used to train and test the model.

4.3 *Testing Phase*

The system accepts input images in .jpeg and .png formats. The input image can be a grayscale image or an image captured through Foldscope or any other microscopic device. A Web application—"AquaVision"—is developed wherein users can upload the water sample image using graphical user interface (GUI). Once the image is uploaded successfully, the system will process and clean the image for detection. The user needs to click on the Detect button and wait for the results as shown in Fig. 7.

4.4 *Detection and Analysis Phase*

The proposed system trains the YOLOv3 model using the graphics card of the Nvidia series GeForce GTX 1650. Post the image is uploaded on the system, Darknet53 loads the weights from the YOLOv3 weights file for image classification and analysis. The weights, checkpoints, and IOU values are used to identify the input image.

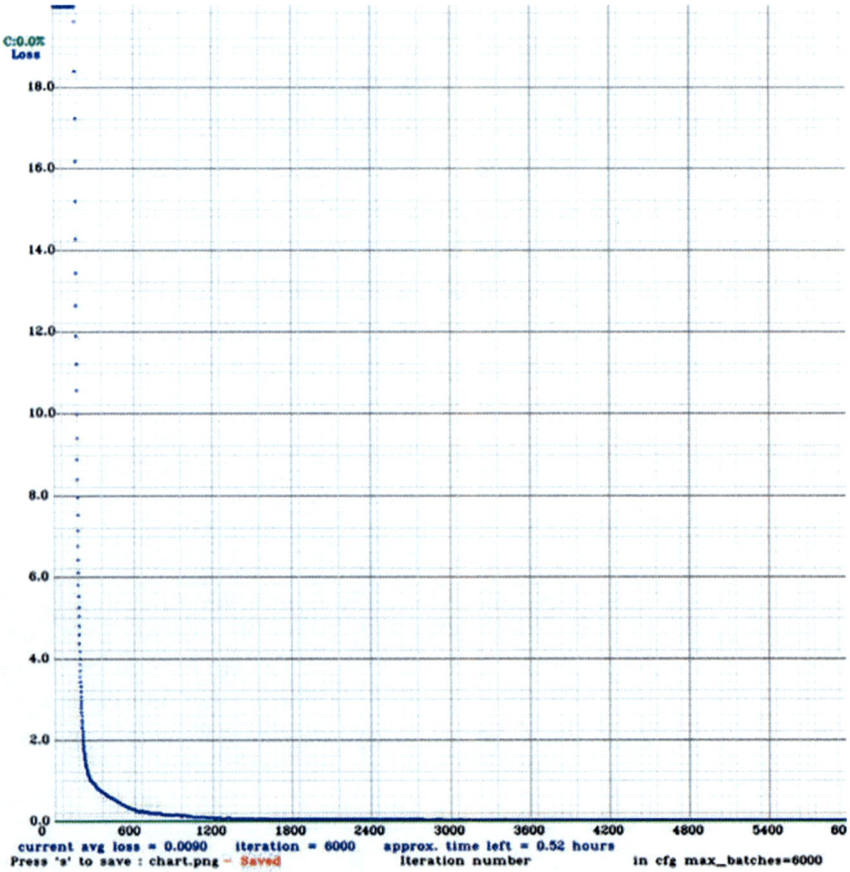

Fig. 6 Training result—graph of loss versus number of iterations

Table 1 Configuration details

Parameter	Configuration
Operating system	Windows 10
CPU	Intel(R) Core(TM) i5 103000H
GPU	NVIDIA GeForce GTX 1650
GPU library	CUDA 10.0, CUDANN 7.4
RAM	8 GB
Training framework	YOLOv3—Darknet-53

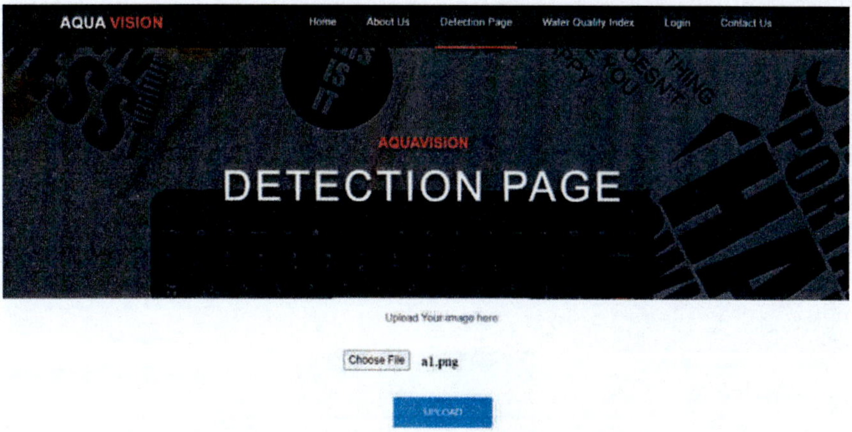

Fig. 7 Input test image

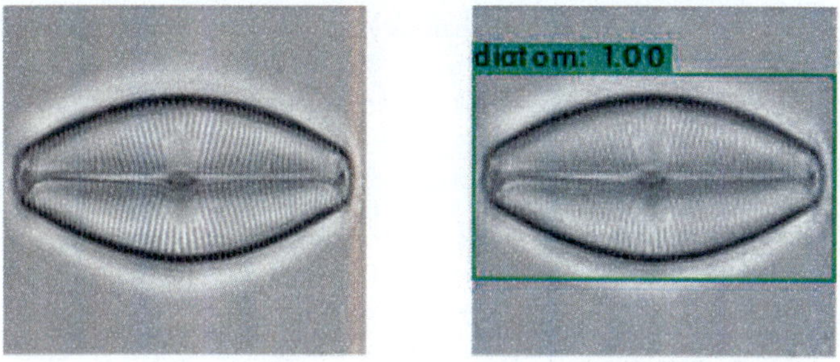

Before image   After image

Fig. 8 Detection of diatom

The system then generates an analysis report regarding the input image. This report contains details like type of microorganism detected, level of concentration of the microbe in the water, details of how this microbe can be harmful to human life [15]. Figure 8 shows the details of the Detection and Analysis phase [16].

5 Result and Discussion

As discussed in the earlier section, we collected water sample image from the user and uploaded on the "AquaVision" application. The system identifies the input image and returns the analysis regarding the image using the YOLOv3 algorithm which is efficient for microbe detection. The input image is converted into an S $\times$ S grid, and

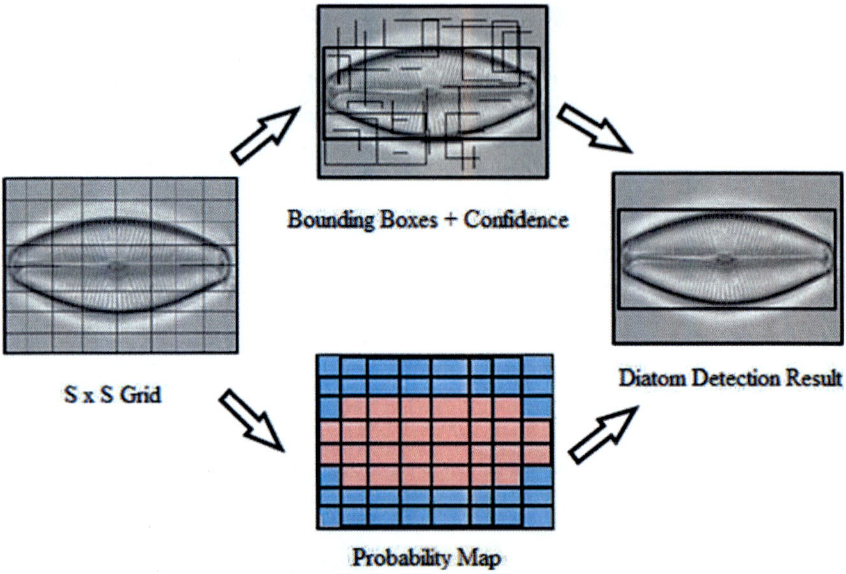

Fig. 9 YOLOv3 model detection

Table 2 Test result of model

	Precision	Recall	F1-score
YOLOv3	0.88	0.93	0.90

classes are identified for each bounding box along with the confidence score. Finally, diatoms are detected as shown in Fig. 9.

The performance of the model can be evaluated using the following metrics [17].

From Table 2, it can be seen that the precision value of the model is high which indicates a low false-positive rate. The recall value being 0.93 shows that the model is well trained. The F1-score for diatoms detection is 90% which depicts the overall accuracy of the model. The generated report also contains information about the type of microorganism identified, the level of concentration of the microbe in the water, how this microbe can be harmful to human life, etc. The system detects diatoms, but later it can be extended to detect other illness-causing pathogens present in water.

6 Conclusion

The purpose of this system is to develop a real-time approach for detecting diatom microbes in freshwater. This helps to identify the real-time water quality by determining the percentage of microbes present in water before using it for various purposes. It accepts microscopic images from the Foldscope or any microscope and determines the type and size of the microbe using the YOLOv3 algorithm. It uses the YOLOv3 algorithm for microbe detection and identification. Subsequently, this has reduced the time and cost required to identify the water quality. Digital image processing improves the process of identification and classification of microbes as compared to traditional laboratory methods with several hours of wait time.

This system can be further enhanced by using preprocessed images to get more accurate results. Also, the time taken to train a model can be reduced by using high-performance computing graphical processing units (GPUs). The scope of the system can further be extended to identify various illness-causing pathogens present in freshwater like protozoa, plankton, etc. Furthermore, the system can be used by different government and private organizations to monitor and determine the quality of water from different regions. This will help to check if water is free from microbes and suitable for human consumption, thereby reducing the threat of water-borne diseases.

References

1. N.J. Ashbolt, Microbial contamination of drinking water and human health from community water systems. Curr. Environ. Health Rep. **2**, 95–106 (2015)
2. B.C. Dhar, L. Cimarelli, K.S. Singh, L. Brandi, A. Brandi, C. Puccinelli, S. M.R. Spurio, Molecular detection of a potentially toxic diatom species. Int. J. Environ. Res. Public Health (2015)
3. J. Redmon, A. Farhadi, *YOLOv3: An Incremental Improvement*. arxiv (2018)
4. P. Vecchio, A. De Giovanni, C. Distante, M. Mirto, L. Conte, G. Aloisio, Measuring cells in phytoplankton images, in *IEEE Conference* (2015)
5. S. Levin, R. Patil, S. Rajkumar, T. Ajmal, *Design of a Smart System for Rapid Bacterial Test* (MDPI, Basel, Switzerland, 2019)
6. C. Wang, Y. Niu, M. Liu, Y. Huang, Y. Yang, T. Shi, Underwater dense targets detection and classification based on YOLOv3, in *IEEE International Conference on Robotics and Biomimetics Dali, China* (2019)
7. Y. Liu, Y. Chen, X. Fang, A review of turbidity detection based on computer vision. IEEE Access (2018)
8. N.H.Z. Suleiman, M.R.M. Sapiee, K.A. Aziz, A. Gomes, I.H. Shanono, C. Gomes, Image processing techniques applicable to wastewater quality detection: towards a hygienic environment. J. Mater. Environ. Sci. (2018)
9. R. Singh, T. Srivastava, A. Sharma, S. Sharma, S. Kumar, *Digital Image Processing-Based Water Pollution Control System* (Springer Science + Business Media, Singapore, 2017)
10. K. Horak, J. Klečka, *Water Quality Assessment by Image Processing* (2015)
11. J. Redmon, A. Farhadi, YOLO9000: Better, faster, stronger. CoRR, vol. abs/1612.08242 (2016)
12. S.C. Ponz, Detection of Mathematical Expressions in Scientific Papers. riunet (2019)
13. G. Liu, J.C. Nouaze, P.L.T. Mbouembe, J.H. Kim, YOLO-tomato: A robust algorithm for tomato detection based on YOLOv3. MPDI J. Sens. (2020)
14. T. Toivanen, S. Koponen, V. Kotovirta, M. Molinier, P. Chengyuan, Water quality analysis using an inexpensive device and a mobile phone. Environ. Syst. Res. Springer (2013)
15. A. Wong, J.L. Deglint, A. Chao, C. Jin, The feasibility of automated identification of six algae types using feed-forward neural networks and fluorescence-based spectral-morphological features. IEEE Access (2018)

16. J. Gong, J. Zhao, F. Li, H. Zhang, Vehicle detection in thermal images with an improved YOLOv3-tiny, in *IEEE International Conference on Power, Intelligent Computing and Systems (ICPICS)* (2019)
17. G. Fang, Y. Suhua, J. Shaofeng, Detection of white blood cells using YOLOv3 network, in *14th IEEE International Conference on Electronic Measurement & Instruments* (2019)

Gender and Age Recognition Using Audio Data—Artificial Neural Networks

Yashi Gupta, Kshitiz Gangwar, Muskan Singhal, and D. Hemavathi

Abstract Human voice is a result of movement of air in our vocal cords, and as we grow, different voice features tend to vary according to age and gender. In many criminal cases, the evidence is sometimes collected as voice samples which we can use to identify the speaker's features like age and gender. Hence, in our model, we are focusing on recognizing age and gender from audio samples using different machine learning algorithms and deep learning models. In the existing models, mostly work has been done using machine learning algorithms, but here we have tried to implement the same along with different dimensionality reduction techniques and deep learning models to figure the best one out of all. The main objective of this project is to identify age and gender by producing a robust feature and to design a good classifier that will improve the accuracy and performance of our system.

Keywords Speech recognition · Deep learning · Feature extraction · Machine learning classifiers · Neural networks

1 Motivation for Problem

In this modern age of data-driven solutions, the customer demographic attributes and characteristics like age, language, and gender play a crucial role that may not just enable multiple companies to improvise and enhance their services. Recognizing an

Y. Gupta · K. Gangwar · M. Singhal · D. Hemavathi (✉)
Department of Information Technology, School of Computing, SRM Institute of Science and Technology, Kattankulathur, India
e-mail: hemavathi.d@ktr.srmuniv.ac.in

Y. Gupta
e-mail: ym9743@srmist.edu.in

K. Gangwar
e-mail: ku4972@srmist.edu.in

M. Singhal
e-mail: mm4198@srmist.edu.in

G. Ranganathan et al. (eds.), *Soft Computing for Security Applications*, Advances in Intelligent Systems and Computing 1397,
https://doi.org/10.1007/978-981-16-5301-8_34

individual's gender as male and female on the basis of audio inputs sounds like an easy task. We, humans, can easily differentiate between a male voice and a female voice; however, designing a computer code to do the same is a bit trickier.

The problem that we are trying to address here is gender prediction using audio samples. Gender recognition has been proved to be useful in multiple ways. Making purchases online can be made easier by filtering out choices on clothing and accessories. Such predictions have helped emergency dispatching centers to easily identify the gender of the victim. Criminal acts and legal procedures will grow more transparent and personalized. With correct inputs and techniques, computer systems can easily use voice recognition for gender identification.

The purpose of this project is to examine the gender and age of individuals on the basis of the voice. This will help in making online purchasing easier by filtering out choices on clothing and accessories based on the gender and the age of the individual. Such predictions will help emergency dispatching centers to easily identify the gender of the victim. It can also help with criminal acts and legal procedures which will grow more transparent and personalized.

2 Literature Review

In the study of Feld et al. [1], we have ascertained that the author conferred a mathematician mixture model, a kind of supervector system usually helpful in estimating age and detecting gender of the speaker from progressive speaker recognition examination. They need an experimental study to study and assess the performance of their model likewise as exploring the choice of parameters.

Within the universal context model, the paper [2] uses the hypothesis that there is a chance of finding distinct Gaussians for each age bracket model. These Gaussians' weights will be set to a special mode to distinguish the age groups. The planned method was evaluated on two corpora with classification into four age groups, with weighted average recalls of 53.75% and 56.18%, respectively. As ascertained during a study of Bocklet [3], GMM supervector and support vector machine models were created to spot the gender and age of speakers in phone applications. It has been noted that the authors tried to perform two separate and different approaches to perform the required tasks of recognizing age and gender. One of the two includes the GMM with some touches of UBM, which is employed for speaker identification/verification. The secondary approach suggests coaching mathematician models to train and check sets of every speaker. Solely spectral options square measure being elaborated in Bocklet [3].

It has been noted in Bahari and Hamme [4] that the user uses models like Gaussian mixture model along with general regression neural networks and nonnegative matrix factorization. After testing their model on a database, the authors noted that the mean average error of calculating estimated age was found to be around 7.5 years. The authors seem to conclude that the use of specified models has increased the accuracy of the system, which is found to be 96%, in comparison with usual methods.

After performing a study on Faek [5], we found that the main aim of the authors was to recognize age and gender using acoustic data. For the gender prediction, the authors have used the support vector machine classifier where features and characteristic values of the acoustic signals are fed into the classifier to perform relevant modeling. It was noted that the accuracy levels were pretty impressive. For the gender prediction, the accuracy for classifying data into male and female classes was obtained as 96%. The model uses MFCC characteristics for the age estimation module, and it has been found that an accuracy of 94% has been obtained in identifying ages using data in the groups of teens, adults, and oldies. One of the major points to be noted is that the data used has a mixture of different clearance and tone signals that lead the authors to use different dimensionality reduction techniques to further improve the accuracy and performance of their system.

In this study of Mavaddati [6], the authors have used sparse nonnegative matrix factorization and atom correction. The Mel frequency cepstral coefficient is considered a good selection to present the voice signals in a compact form. As per the conclusion and results obtained by the proposed approach, the authors have stated that they have acquired the best possible accuracy after comparing results from other method applications. It was observed that some authors tried to study the results and performance to further test the proposed approach on a different set of conditions such as using a dataset with noisy and clean signals, and it was observed that the model outperformed other methods in all test conditions although it was also observed that this study works only for pre-specified voice recording attributes.

The authors of the paper [7] assume a speech signal within the time domain of male and feminine speakers and provide one worth within the type of an area underneath the modified voice contour (MVC). To tell apart between males and females, the prompt methodology calculates the region underneath the adjusted voice contour to work out voice pitch. The projected practicality has been proven to be independent of text, voice, vocabulary, nonstandard speech space, audio device, and atmosphere by utilizing two separate databases. For clean and creaking signals, the figures were 98.27% and 96.55%, respectively.

Instead of KNN, the paper [8] proposed using a smart classifier like support vector machines or artificial neural networks. The findings of this study show that MFCC is ineffective when dealing with data collected from various locations using various computers and microphones. During this project, a unique set of solid options is used to strengthen the accuracy of the gender and age classifiers, which are reinforced by the frequency variations depicted by the options.

It has been noticed that in paper [9], the authors have combined different classification techniques along with some matrix factorization like general regression neural networks and nonnegative matrix factorization including supervised/weighted principal component analysis and anchor modeling.

The authors of Ng et al. [10] published a systematic survey on human gender identification using computer vision-based methods, with an emphasis solely on 2-D approaches. An analysis of methods that use data from the face and whole body is also provided. They plan to concentrate on the issues raised and conduct a review of the various approaches. It is possible that smart success has been achieved for datasets collected in managed settings, but there is still work to be done to strengthen the robust nature of gender identification in real-life and corporeal things, according to the findings.

As we all are already aware of the advantages of using CNN model in various fields such as picture implantation and audio processing, we can note that the authors in Liao et al. [11] have also tried to extract and reduce some multivariate statistical models and powerful facial features into more efficient ones by using various deep learning approaches, including CNN. It was observed that the researchers have tried to implement a divide-and-rule approach, which is said to outperform many ML algorithms, for their model of age and gender recognition. It is observed that this approach is more useful in age estimation as it follows a series analysis of rank-based age modeling approaches. Several experiments done on animation and morphing show that the proposed method is more suitable and robust for the feature extraction process.

The authors in the study of Tatkare et al. [12] have described a way to recognize age and gender by using Haar cascade. The dataset consists of multiple images of different aged male and female participants. Facial attributes are calculated using feature extraction techniques. The authors have trained the model with the help of convolutional neural network in a way such that the classifier works effectively with limited data. The deep learning model is applied for estimating age groups. The proposed model has provided impressive accuracy and performance even with limited data.

It has been discovered that the authors in Ertham [13] used an extended remembering network model to acknowledge gender from audio information. The projected approach consists of three main steps. Initially, only ten information options and attributes were elite. Then, using a double-layer LSTM framework, a deep learning-based network was created. Accuracy performance metrics were measured concurrently with the performance comparison. Finally, the performance metric of the proposed approach was studied and presented in comparison with the results acquired from several standard machine learning models' classifiers.

In this study of Büyükyılmaz and Çıbıkdiken [14], the authors have introduced the multilayer perceptron (MLP), a deep learning model, to spot voice gender. The information set consists of three, 168 recorded acoustic samples of each male and feminine speaker. The required deep learning algorithmic program has been applied to acknowledge gender-specific characteristics. It was discovered that the model achieves 96.74 percent accuracy on the testing data set. An interactive Web site has additionally been designed to observe the gender of voice by mistreatment of the planned model.

This study [15] uses a brand new gender and age recognition approach supporting the HMM for frame-based dynamic spectral options like MFCC. Audio info is trained victimization HMM, and a few feature extraction from taking a look at audio are decoded. Initially, an acoustic model is trained for all speakers in an exceedingly coaching info as well as male and feminine speakers of various ages. Lastly, supervised HMM is applied to acknowledge the gender and cohort from the acoustic voice samples. Victimization HMM could end in an oversized range of unstructured parameters.

It has been observed that most papers solely determine speaker's gender or age and not both and have solely been valid on a smaller dataset. Models focusing on age estimation have mostly been fed with image dataset or facial characteristics. In many studies, no mention of features regarding noise issues while estimating the age and gender prediction has been noticed. Some authors have introduced MLP to implement their work; however, it ends up in superfluous parameters as this learning model is deeply connected. We also came to a conclusive point that use of high-dimensional data features and characteristics affects accuracy for age and gender recognition. We can finally say that there is room for improvement in the future studies, and we have tried to outperform most of the challenges.

3 Dataset Description

In our proposed model, the dataset (available as .wav files) is taken from kaggle.com, "Voice Recognition and Gender Classification in the Context of Native Languages and Lingua Franca," Johannesburg, South Africa, November 19–20, 2019, pp. 175–179 [23]. This dataset consists of voice utterances of 526 individuals out of which 336 are male voices and 190 are female voices. About 3964 recordings can be found that consist of single sentences as well as multiple sentences from a speaker.

Another dataset has been used and is taken from kaggle.com (uploaded by Kory Becker) [21]. 3168 audio files are collected which are registered voice notes of both genders' speakers. There are a total of 21 variables present in the form of columns in CSV format. The target variable in the dataset is label (this specifies the gender of the person).

Another dataset used, Common Voice Corpus of Speech Data, includes the audio (.wav) read by various users on a common voice Web site [22]. The main use of this dataset is to evaluate ASR systems. This dataset is used by us in order to predict the age of different speakers.

4 Requirement Analysis

4.1 Functional Requirements

The model should be able to classify the gender of a person—male or female when it is given an input consisting of various voice features (provided in the csv format).

Along with gender, the model should also be trained for classifying age according to the different age groups (Fig. 1).

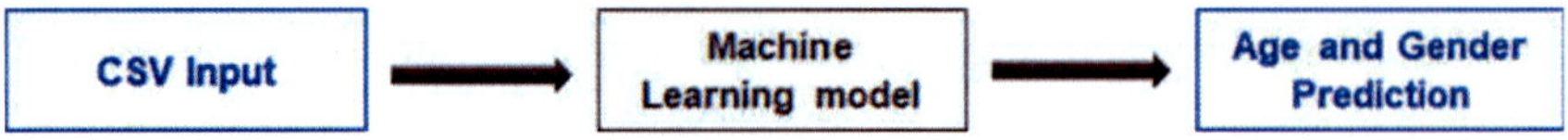

Fig. 1 Showcasing functional requirements

4.2 Non-functional Requirements

Usability (easy to use), Scalability (our model can effectively and efficiently be adapted in different environments and specifications), Operability (the process of recognizing age and gender from speech signals should be easy to operate and control), Availability (this represents how likely our model is accessible for a user at a given point in time) and Security & Privacy (this represents the degree to which our model protects information or data) are the non-functional requirements. Since a large amount of data is handled by the system, the data access should be appropriate to its type and the different authorization levels.

4.3 Hardware Requirements

An i3-i5 processor and 4–8 GB of RAM are bare minimum requirements to go ahead with the proposed model. Decent wireless/Ethernet to download and upload datasets for practice is always appreciated. Since a good amount of coding is involved in the project, a 2K screen resolution would be appropriate.

4.4 Software Requirements

- Spyder is an open-source cross-platform IDE for programming in the Python language.
- Python, an interpreted, interactive, object-oriented, dynamic type, easy to find out, and open-source programming language. Python combines remarkable power with very clear syntax.
- Keras is used for neural networks and usually runs along with TensorFlow.
- TensorFlow is a Python library for numerical manipulation using data flow graphs that is commonly used in neural networks. The graph nodes represent mathematical operations, and the graph edges represent the high-dimensional datasets that are communicated between them. Its versatile architecture helps us to perform machine learning and deep neural network research using either the Graphics or Central Processing Unit.
- librosa, a Python library used mainly for audio and music processing. It plays an important role in the initial stage of an audio processing model. It has been proved

effective in identifying and mathematically calculating the characteristics of an audio signal [20].

- pandas, an open-source, fast, and versatile Python library that gives high-performance computation of data and analysis tools using its powerful data storage structures. It allows data alignment and integrated handling of missing data. Also, one can reshape the data according to the requirements.
- NumPy is an open-source fundamental Python package commonly used for scientific computing. It allows us to define common types of data which efficiently integrates with different databases. NumPy is used in this project to extract the statistical features from the audio files.
- SciPy, an open-source Python library, is commonly used for solving mathematical, scientific, engineering, and technical problems. It allows data manipulation and visualization. SciPy works very well with the NumPy library, and this is one of the specific reasons that it is used. Functions like data integration and optimization go hand in hand with SciPy.
- sklearn, also known as scikit-learn, is one of the most robust and functional Python libraries for machine learning. It provides us with a selection of competent tools for machine learning and statistical modeling like clustering, classification, regression, and dimensionality reduction by a standardized interface in Python. This library is said to be built upon SciPy, NumPy, and Matplotlib.

5 Proposed Methodology

This model has been divided into three different continuous modules—data processing, gender prediction, and age estimation where the first module includes data collection, visualization, labeling of data. It further includes data cleaning—using imputation techniques, data transformation—scaling of data in order to normalize it. The next module consists of applying feature extraction techniques and performing a comparative analysis of various ML models, deep learning models, etc. The model that gives the highest accuracy on test data is chosen to predict the gender. Hence, the model is tested and evaluated. In the final module, different ages are first put into classes. A comparative analysis is performed on several feature extraction techniques, and the similarities and differences are observed in order to choose the best model for age prediction. The model is then tested and evaluated (Fig. 2).

5.1 Data Processing

The module has been further split into three submodules:

- Audio processing
- CSV creation

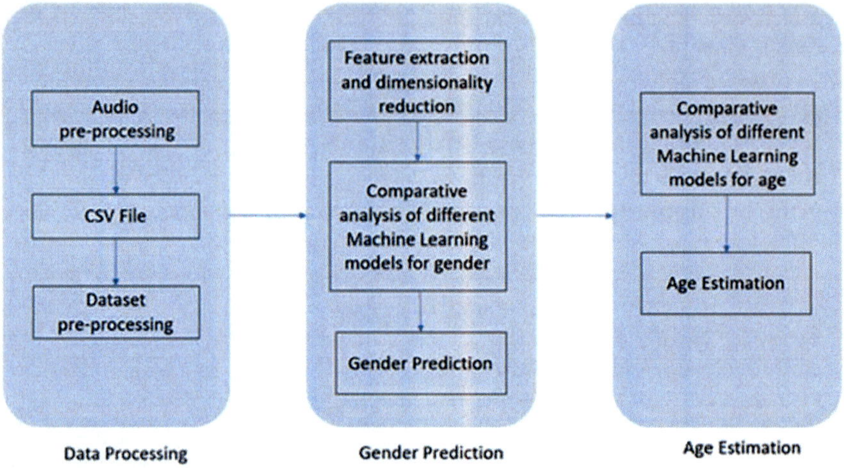

Fig. 2 Architecture for proposed model

- Data processing.

As specified in the previous section, data for our model is collected from three different sources, and most of them are available as .wav files; hence, to further proceed with our work we performed audio processing on the dataset to get the required acoustic characteristics for the next two models. During the first submodule, libraries like NumPy, librosa, and sklearn have been proved successful in obtaining the features shown in the table which are then stored in a csv file (Table 1).

In the csv file, the dependent variables are separated from the independent variables. Imputation techniques are used to fill up null values, if present. The next step includes encoding of categorical data where our gender variable is converted into 0s and 1s. Dataset is then split into testing dataset and training dataset, and the acoustic characteristics are then scaled using feature scaling techniques. The dataset is now ready to be used for the next modules (Fig. 3).

5.2 Data Visualization

The following are some of the kernel density estimate plots for the independent variables with respect to the dependent variable (gender—male and female). This visualization is done on BVC Gender & Age Estimation dataset. It helps us to analyze the relationship between the variables (Figs. 4, 5, 6, 7, 8, 9, and 10).

Table 1 Acoustic features obtained and stored in csv file

Mean	Mean of the frequencies
Std.	Standard deviation of frequency
Min. frequency	Minimum of the frequencies of audio signal
Max. frequency	Maximum of the frequencies of audio signal
Median	Median of the frequencies
Skew	Skewness of the signal
Kurt	Kurtosis of the signal
Q25	Quantile one value
Q75	Quantile third value
IQR	Interquartile bandwidth (range)
Sound energy	Energy of sound per unit of time
RMS energy	Root mean square energy
ZCR	Zero crossing rate
Tempo	Speed of acoustic signal
Mean of MFCCs	Mel frequency cepstral coefficients
Min. of MFCCs	Minimum of Mel frequency cepstral coefficients
Max. of MFCCs	Maximum of Mel frequency cepstral coefficients
Chroma frequency	Pitch class profile
Spectral centroid	Center of gravity of the spectrum
Spectral bandwidth	Bandwidth of spectral signal
Spectral roll-off	The signal's time-based roll-off point
Gender	Labeled as male or female
Age	Label for age of the speaker

5.3 Gender Prediction

In this module, different classification machine learning models are used, and a comparative study is performed. Along with this, feature extraction and dimensionality reduction are done to the dataset, and accuracies are calculated (Fig. 11).

Neural Networks—Deep Learning Every action we take is controlled by neurons, i.e., our nervous system. The actions depend on how the neurons are connected within the system and how strong is the connection between them. This is the main concept of deep learning. Several layers such as input, hidden, and output are contained. Some basic terminology regarding neural networks includes perceptron which can be defined as the simplest artificial neuron that only contains two layers—input and output. The first layer is processed by an activation function, and the output is attained.

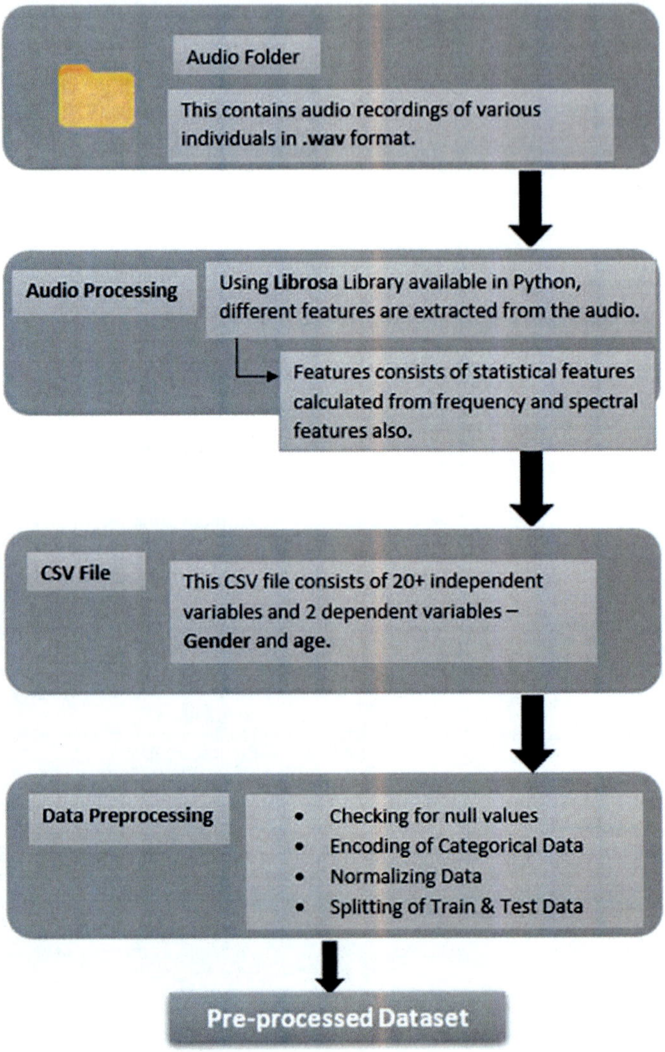

Fig. 3 Architecture for data collection and processing

Activation function is responsible for converting the sum of weighted inputs into outputs. The result of one layer is passed over to the next layer as the input. Some activation functions are sigmoid, Tanh, ReLU, Softmax.

A simple neural network consists of three layers. The first is the input layer, responsible for perceiving the input features of a problem. In certain cases, the number of nodes in this layer equals the number of input features. The data does not get changed when it passes through the input layer. The second one is the hidden layer, which is responsible for making the transformation in data. The processing in the

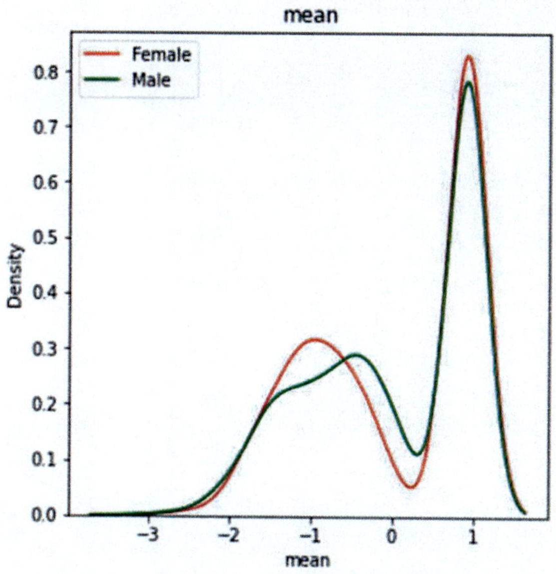

Fig. 4 Mean of frequencies

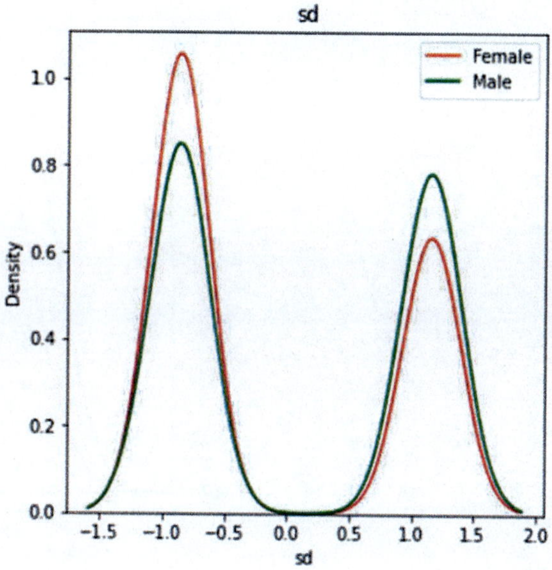

Fig. 5 Standard deviation

hidden layer is actually done by weighted connections. The hidden layers increase the predictive power of a neural network. The final layer is the output layer which finally receives the connections from the hidden layer or the input layer directly and is responsible for predicting the output. In classification problems, we give the number of nodes in an output layer as one (Fig. 12).

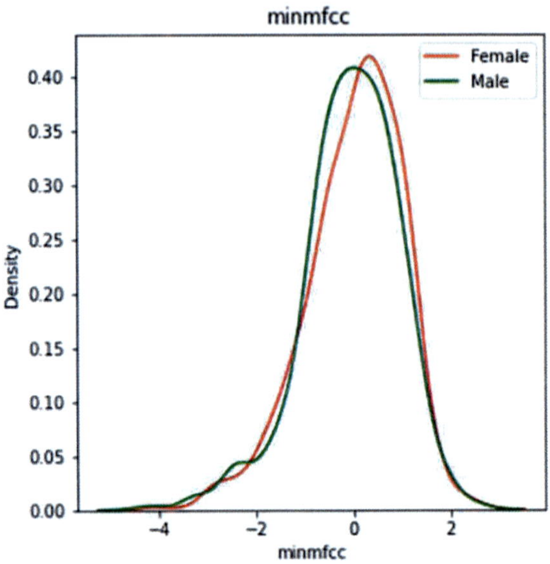

Fig. 6 Minimum MFCC

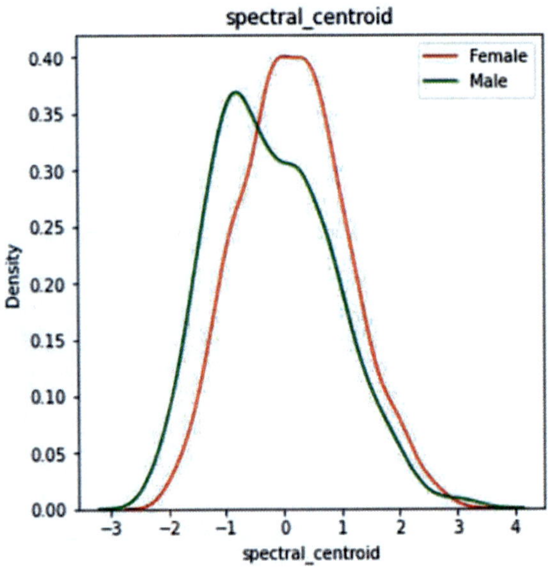

Fig. 7 Spectral centroid

The neural network learns to adjust the weights and also the threshold value (formula associated with the weights) in order to predict the correct outputs. Complex problems cannot be solved simply by using these three layers. Hence, more hidden layers are added to the neural network. Therefore, when we have more than three layers including the input and the output layer, it is referred to as deep neural network.

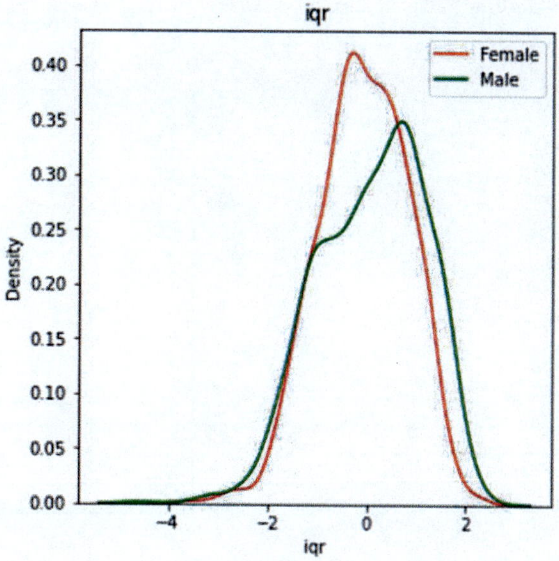

Fig. 8 Interquartile range

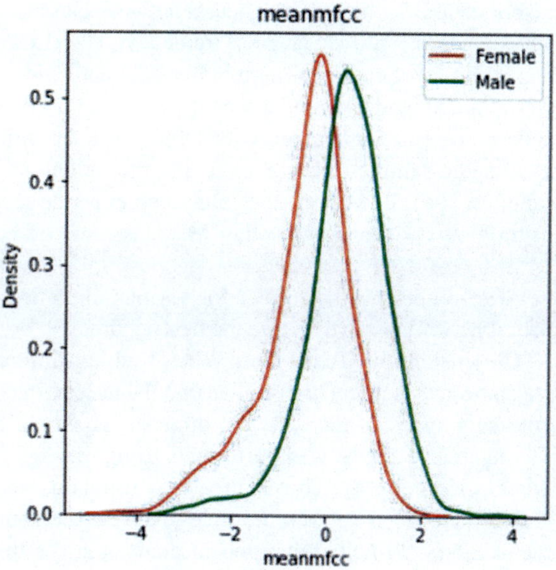

Fig. 9 Mean MFCC visualization

Also, training such complex networks is referred to as deep learning. The limitation which one comes across using neural networks is that neural networks need a large diversity of data in order to train themselves and give the correct predictions. Common applications of neural networks are financial forecasting (e.g., predicting the house prices, etc.), voice recognition, character recognition, signature verification, etc.

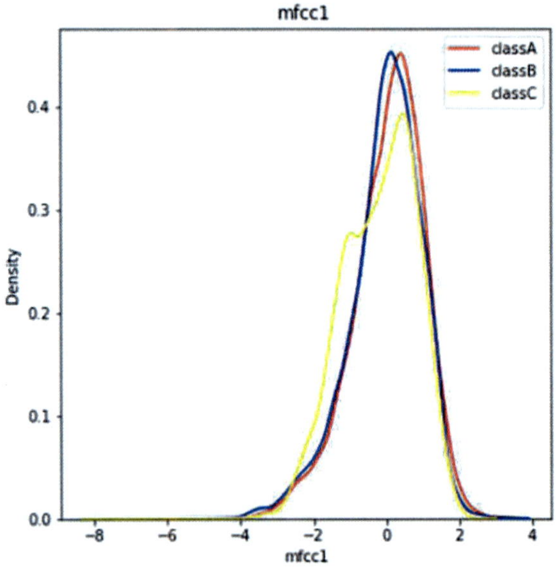

Fig. 10 MFCC1 (visualization for three age classes)

For dataset 1, the artificial neural network classifier is trained with an input layer with 22 nodes, 4 hidden layers is trained with "ReLU" as the activation function, and the output layer contains "sigmoid" as the activation function. 2534 audio samples were trained, and the model was tested on 634. For dataset 2, the artificial neural network classifier is trained with 1 hidden layer with "ReLU" activation function. 1959 audio samples were trained, and the model was tested and evaluated on 490 samples. The following are the graphs depicting the accuracy comparison for artificial neural network along with other ML classifiers on both the datasets (Figs. 13 and 14).

After observing the results for various classifiers, it is observed that artificial neural networks prove to give the best accuracy.

Dimensionality Reduction When training different machine learning models, we come across many high-dimensional datasets. In order to train models efficiently with such types of datasets, the dimensions should be reduced. For this purpose, a comparative study was performed using various feature extraction techniques. After performing the above three dimensionality reduction techniques, a considerable trade-off is seen in the time taken by the neural network to train itself and the accuracy. Principal component analysis gives the best result among the three when the number of components is chosen as 17. Hence, the gender prediction is done using artificial neural networks after applying principal component analysis. This prediction is done on the BVC Age and Gender Estimation dataset with an accuracy of 85.51%. Principal component analysis reduces the chances of model overfitting and also increases the algorithm performance, and data visualization is made easier since the dataset is low dimensional. Some disadvantages that come

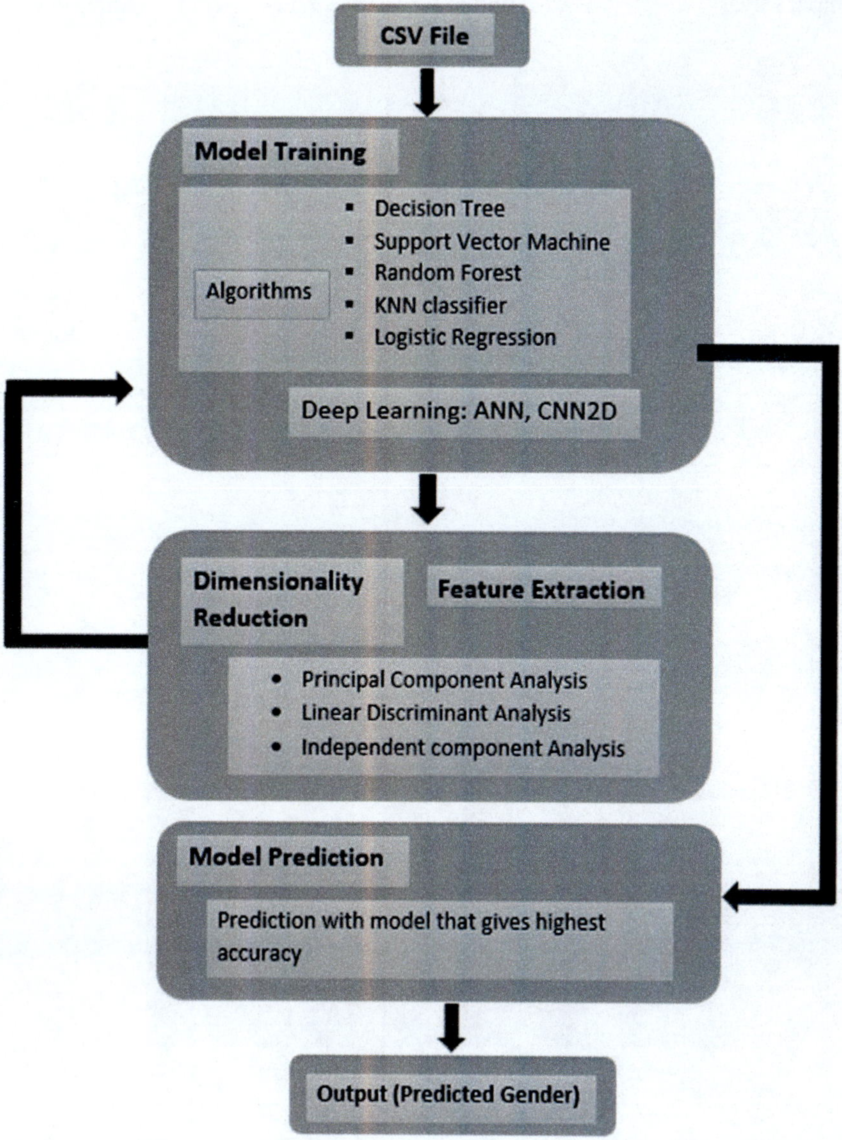

Fig. 11 Detailed architecture for gender prediction

along with PCA are: It is important to perform data standardization, and the data becomes less interpretable since the data is manipulated in PCA [24].

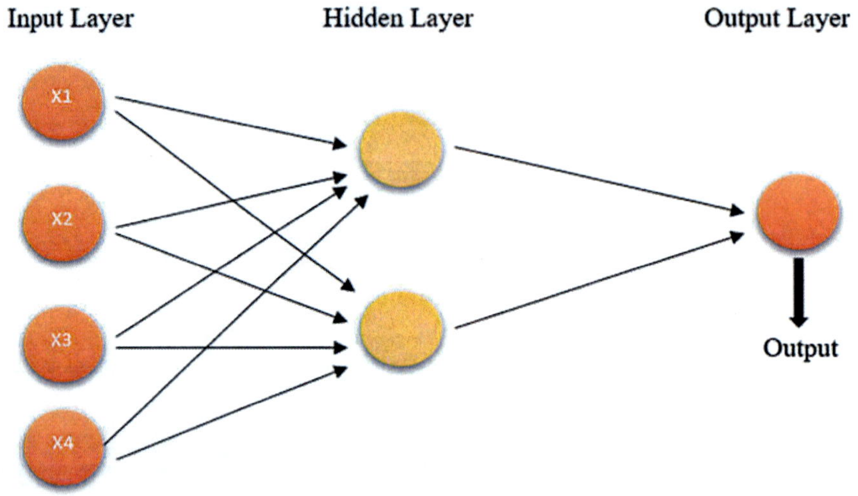

Fig. 12 ANN layers

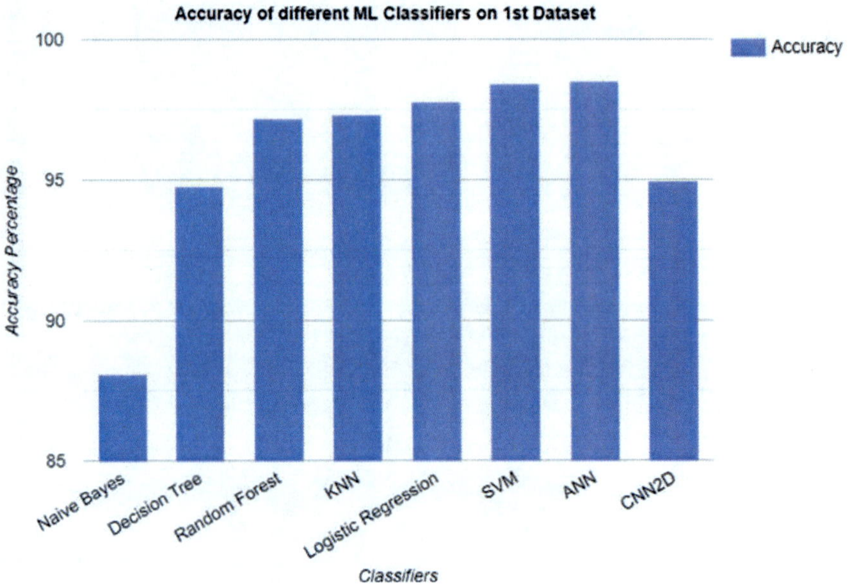

Fig. 13 Accuracy of different classifiers on first database

5.4 Age Prediction

The dataset used for age prediction is the "Common Voice" dataset which includes the speech data read by the users of common voice Web site. The Common Voice

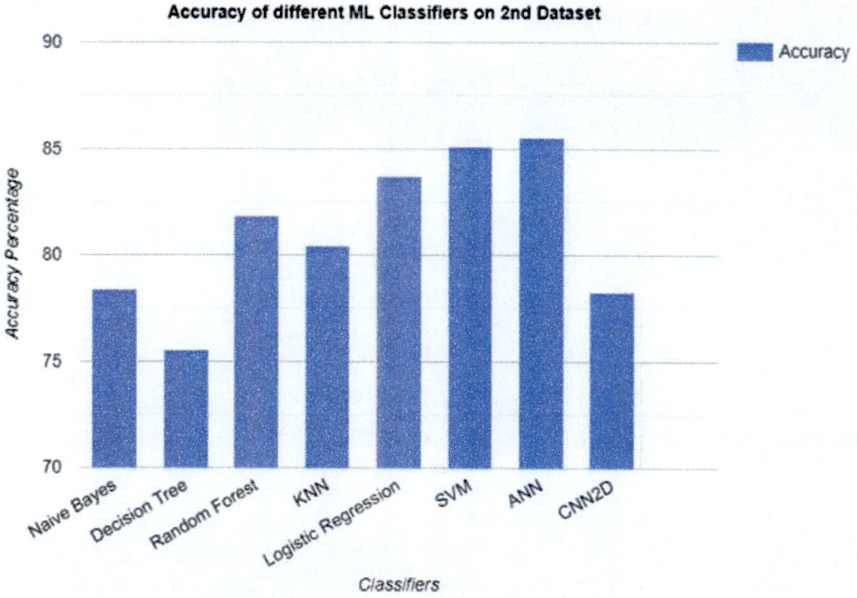

Fig. 14 Accuracy of different classifiers on second database

Team assembled this dataset. Approximately, 73 k audio files have been taken from it and processed in order to acquire audio features. 22 features, namely mean, median, mode of the frequencies, skew, kurtosis, tempo, Mel, quantiles 1 and 3, interquartile range, spectral features such as centroid, bandwidth, flatness, roll-off, and energy, have been calculated in the same way as for gender prediction. Before proceeding with other features, an understanding about human voice and the features that change as a result of growth is developed. Voice change is often accompanied by a significant drop in pitch to the vocal range. It has also been observed that as a human grows, spectral and temporal variability often tend to decrease (Fig. 15).

Considering the features that contribute to distinguishing one's age from another, Mel frequency cepstral coefficients (MFCCs) were calculated for each of the audio recordings.

In Fourier analysis, the cepstrum is calculated by calculating the in IFT of the log of the measured signal spectrum. The technique is used to look for periodic structures in frequency spectra. The power cepstrum has applications in human speech processing. By including this scale, our characteristics become more closely aligned with what humans hear. The MFCCs are the coefficients that make up an MFCC [19]. They are made up of a cepstral description of the audio file. The distinction between the cepstrum and the Mel frequency cepstrum is that the frequency bands in the MFC are evenly distributed on the Mel scale, which is more closely rather than the linearly spaced frequency bands used in the traditional cepstrum, and it approximates the human auditory system's response. This frequency warping can help with audio compression sound representation (Fig. 16).

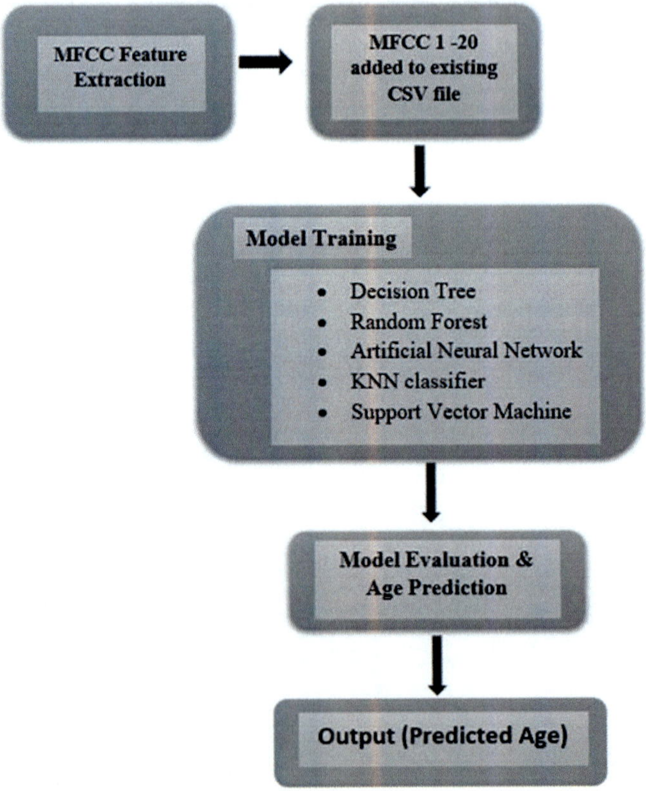

Fig. 15 Detailed architecture for age prediction

After calculating the MFFCs, we have a total of 45 features processed from the audio files. The ages of these speakers have been distributed into three groups as classA (ages from 11 to 30), classB (ages from 31 to 60), and finally classC (for ages above 60).

Multiclass Classification The term "classification" refers to the process of categorizing data and grouping it into categories dependent on similarities. The independent variables or features in a dataset are critical in classifying our data. When we speak of multiclass grouping, we mean that our dependent or target variable has more than two classes.

For example, a cybersecurity company may want to be able to monitor a user's mailbox and classify email messages as phishing attacks or not. To do so, it should teach classifiers on existing phishers scams' email texts and incoming email addresses in order to teach the model which URLs are likely to be assigned to abusive messages.

In order to predict age, we used the following classifiers:

- Decision tree
- Support vector machine (SVM)

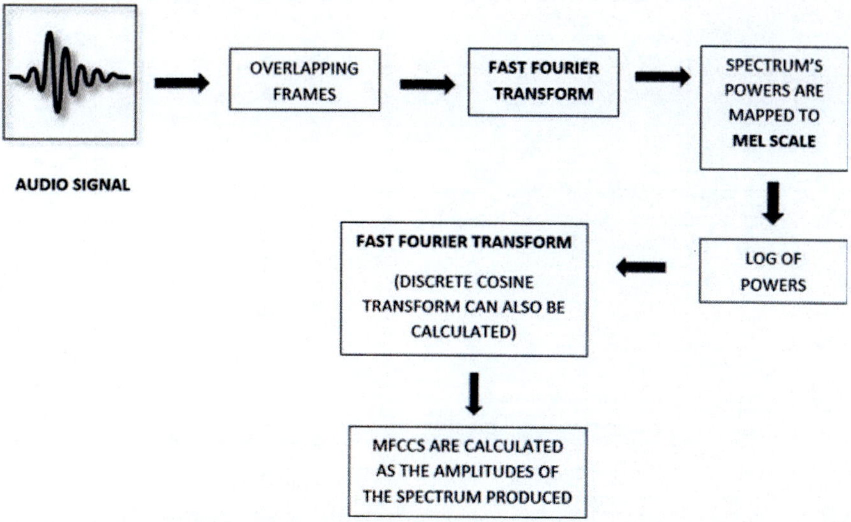

Fig. 16 Steps to calculate MFCCs from an audio signal

- K-nearest neighbors (KNNs)
- Random forest
- Artificial neural networks (ANNs).

58,774 audio samples were used for training the classifier, and 14,694 samples were used to test and evaluate the model. While training the decision tree classifier, the maximum depth of the tree is limited to ten. This is done in order to prevent overfitting of the classifier. The accuracy observed for this classifier is **64.87%**. For support vector machine, the accuracy for the classifier is **79.33%**. KNN proved to give considerably good accuracy of **87.08%**. The following is the accuracy plot for choosing the optimal number of neighbors (Fig. 17).

Hence, the number of neighbors chosen is nine. For artificial neural networks, the input layer is given 45 nodes as the number of independent features in the dataset is 45. Seven hidden layers are connected together with 200–300 nodes, and the activation function is "ReLU." Finally, the output layer has "Softmax" as its activation function and it consists of three nodes. The number of age classes is three. The accuracy of the ANN model is **89.20%**. Lastly, the accuracy for random forest classifier with a number of estimators as 10 is **73.39%** (Fig. 18).

After comparing the accuracies for above-mentioned machine learning models, artificial neural networks prove to give the best accuracy. Hence, ANN is used to predict the age groups.

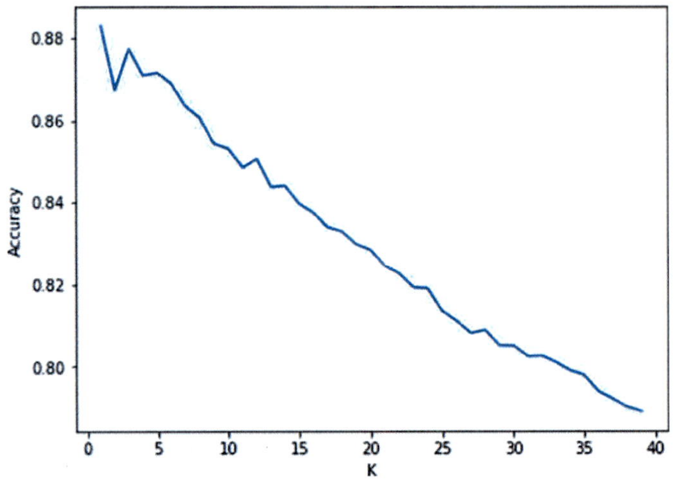

Fig. 17 Accuracy plot of KNN

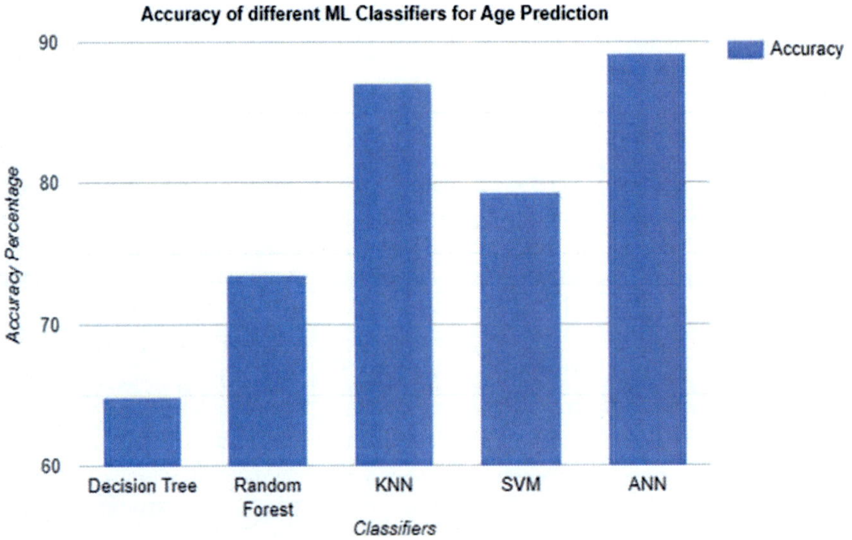

Fig. 18 Accuracy of different classifiers for age prediction

6 Conclusion

Human voice plays a vital role in determining the gender and age of the person. Hence, a model is proposed here to determine the gender and age group of a person using his/her voice. After training different machine learning classifiers for gender as well as age, artificial neural networks give the best results. Several feature extraction techniques were also comparatively studied. While feature extraction techniques resulted in an improvement in the training time of the model, the accuracy of the same was compromised a little bit. In the techniques applied, principal component analysis gave the best training time. The gender is thus predicted as either male or female, and the age is predicted as a group. Some common reasons which conclude that artificial neural networks prove to give the best accuracy are that even if some missing information is provided to it, the model can predict the output. The more training examples we provide to the network, the more it adapts itself to predict the result. Also, it works well with linear and nonlinear datasets.

References

1. M. Feld, C. Müller, F. Burkhardt, Automatic speaker age and gender recognition in the car for tailoring dialog and mobile services, in *INTERSPEECH* (2010)
2. R. Porat et al., Age recognition based on speech signals using weights supervector, in *INTERSPEECH* (2010)
3. T. Bocklet et al., Age and gender recognition for telephone applications based on GMM supervectors and support vector machines, in *2008 IEEE International Conference on Acoustics, Speech and Signal Processing* (2008), pp. 1605–1608
4. M.H. Bahari, H. Van hamme, ITN 2008, *Speaker Age Estimation and Gender Detection Based on Supervised Non-Negative Matrix Factorization.* Centre for Processing Speech and Images Belgium
5. F. Faek, Objective gender and age recognition from speech sentences. ARO Sci. J. Koya Univ. **3**, 24–29 (2015). https://doi.org/10.14500/aro.10072
6. S. Mavaddati, Voice-based age and gender recognition based on learning generative sparse models. Int. J. Eng. Trans. C: Aspects **31**, 1529–1535 (2018)
7. M. Alhussein, Z. Ali, M. Imran, W. Abdul, Automatic gender detection based on characteristics of vocal folds for mobile healthcare system. Mob. Inf. Syst. 1–12 (2016). https://doi.org/10.1155/2016/7805217
8. A.A. Assim, V. Davydov, V. Yushkova, A. Glinushkin, V. Rud, Age and gender recognition from speech signals. J. Phys. Conf. Ser. **1410**, 012073 (2019). https://doi.org/10.1088/1742-6596/1410/1/012073
9. S. Chaudhari, R. Kagalkar, A review of automatic speaker age classification, recognition and identifying speaker emotion using voice signal. Int. J. Sci. Res. (2014)
10. C.-B. Ng, Y.H. Tay, B.-M. Goi, Vision-based human gender recognition: a survey. Lect. Notes Comput. Sci. **7458** (2012)
11. H.-B. Liao, Y. Yan, W. Dai, P. Fan, Age estimation of face images based on CNN and divide-and-rule strategy. Math. Probl. Eng. **2018**, 1–8 (2018). https://doi.org/10.1155/2018/1712686
12. G. Tatkare, B. Patil, S. Patil, S. Salian (2020) Gender recognition and age approximation using deep learning techniques. Int. J. Eng. Res. **9**. https://doi.org/10.17577/IJERTV9IS040268
13. F. Ertham, *An Effective Gender Recognition Approach Using Voice Data Via Deeper LSTM Networks* (2019)
14. M. Büyükyılmaz, A. Çıbıkdiken (2016) *Voice Gender Recognition Using Deep Learning.* https://doi.org/10.2991/msota-16.2016.90
15. B.R. Jawale, S. Patil, M. Agrawal, Identification of age and gender using HMM. Int. J. Comput. Sci. Inf. Technol. (2015)
16. *Working with Audio Data for Machine Learning in Python | by Pragati Baheti | Heartbeat.* https://heartbeat.fritz.ai/working-with-audio-signals-in-python-6c2bd63b2daf
17. *Audio Data | Audio/Voice Data analysis Using Deep Learning| by Faizan Shaikh.* https://www.analyticsvidhya.com/blog/2017/08/audio-voice-processing-deep-learning/

18. *Audio Data Analysis Using Deep Learning with Python (Part 1) – Kdnuggets*. https://www.kdnuggets.com/2020/02/audio-data-analysis-deep-learning-python-part-1.html
19. *An introduction to audio processing and machine learning using Python*. https://opensource.com/article/19/9/audio-processing-machine-learning-python
20. *Librosa: A Python Audio Library. by: David Kaspar, Alexander Bailey | by Patrick Fuller | Medium*. https://medium.com/@patrickbfuller/librosa-a-python-audio-libary-60014eeaccfb
21. *Gender Recognition by Voice | Kaggle*. https://www.kaggle.com/primaryobjects/voicegender
22. *Common Voice | Kaggle*. https://www.kaggle.com/mozillaorg/common-voice
23. *Gender Age Classification from Voice Data | Kaggle*. https://www.kaggle.com/ogechukwu/voice/version/2
24. *In Depth: Principal Component Analysis | Python Data Science Handbook*. https://jakevdp.github.io/PythonDataScienceHandbook/05.09-principal-component-analysis.html

Merging Data Analytics and Machine Learning Algorithm for Home System Security-Based Internet of Things

Noor A. Ibraheem, Noor M. Abdulhadi, and Mokhtar M. Hasan

Abstract Due to the radical changes in the way of life and the enormous amount of obligations to which the individual is exposed, one of the most important obligations and responsibilities is to keep family members safe, especially children, after both parents have become at work and back at a late time. Children are alone at home without supervision or protection, and we found in this research that we could use technologies that have been invented to cope with the requirements of the times. Recently, the society is rapidly becoming highly integrated with the convergence of smart devices. This phenomenon is being spread more swiftly based on the wireless communication infrastructure provided to most homes, as terminal devices for communication have more various functions and are being supplied to homes through a multi-vendor system and as several terminal devices are installed in each home to provide services connected by Internet of Things (IoT). Digital information is the main component of signal control and flow in a smart home, while information security is another important aspect. Hence, in this research, we focused on the IoT technology by connecting multiple devices such as a control system that monitors the situation inside the house and in the case of any danger by determining one of the techniques of artificial intelligence (unsupervised machine learning). The proposed algorithm used to determine whether the situation in the home is safe or not, as it sends a warning signal to the sensor via the distributor device connects into the system, and through the sensor, it sends a wireless signal to the homeowner's mobile device to warn them in case the home is exposed to danger.

Keywords Internet of Things · Machine learning · Monitoring system · Data analysis · Distributor device · Sensor device

N. A. Ibraheem (✉) · N. M. Abdulhadi · M. M. Hasan
College of Science for Women/Computer Science Department, University of Baghdad, Baghdad, Iraq
e-mail: nooraibraheem@csw.uobaghdad.edu.iq

N. M. Abdulhadi
e-mail: noorma_comp@csw.uobaghdad.edu.iq

M. M. Hasan
e-mail: mokhtarmh@csw.uobaghdad.edu.iq

G. Ranganathan et al. (eds.), *Soft Computing for Security Applications*,
Advances in Intelligent Systems and Computing 1397,
https://doi.org/10.1007/978-981-16-5301-8_35

1 Introduction

The current conditions of life have imposed on many people to be outside the home constantly to work or to do various activities, which may cause many problems, especially in the presence of children in the home, in some cases circumstances dictate leaving children alone. Certainly, we do not need to mention how many problems that children alone at home may cause or the dangers that they may fall into. This problem is from the reality of our lives that we live in. This research is in line with the situation in which we live in terms of responsibilities and duties that have burdened people, and because technology has become an essential part of our lives and a magic solution to everything that stands before us, we decided to offer our smart solution to check on the children and keep your eyes on them wherever you are, to reassure when leaving your children alone at home with the design of a smart system that monitors children's movements while monitoring home systems to protect children from the problem of exposure to electrical devices at home. The research proposal is to design a monitoring and protection system for the child while home alone, using IoT technology by connecting the camera system and the home electrical system with a smart and sensitive device For any risk that children may be exposed to, as this system has a set of wonderful features that make you watch your children without staring at a screen throughout your absence from home in addition to solving all problems Because you may encounter your child at home and keep you away from continuous thinking, as it may happen to children during your absence. The applications for IoT spread out across an expansive assortment of utilization cases and verticals. In any case, all total IoT frameworks are something similar in that they address the combination of four particular segments: Sensors/Gadgets, Network, Data Processing, and User Interface. Figure 1 shows a block diagram of IoT system with input data.

This article is organized as follows: Sect. 2 introduces related works, Sect. 3 describes the idea of electricity for children. Section 4 demonstrates safety considerations of smart home and the tools required to achieve smart home. Section 5 explains monitoring Security system, while Sect. 6 defines machine learning for IoT. Section 7 System Requirements. Section 8 Development Environment. Section 9 proposes a

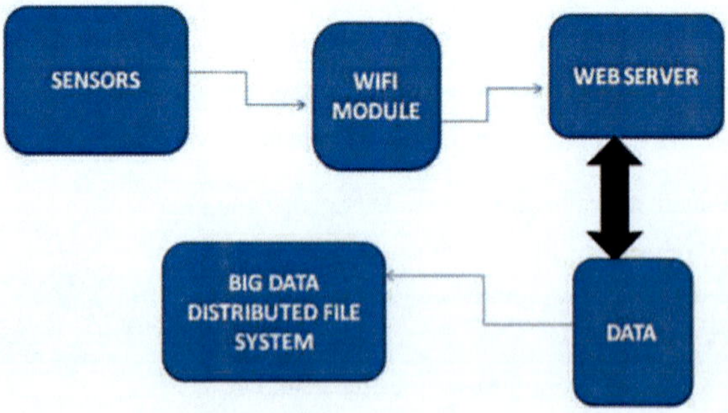

Fig. 1 Block diagram showing functioning of IoT with input data

framework for IoT monitoring system using machine learning, Sect. 10 concludes the conclusions.

2 Related Works

Many researches have been conducted in IoT and machine learning fields. In [2] researchers designed a system that can predict diabetes with high accuracy using the WEKA tool and feature reduction method, and they found 5 features (Glucose, BMI, Insulin, Pregnancy, and Age) and one output feature (outcome) in the PIMA dataset. Seven various machine learning algorithms have been used such as DT, KNN, RF, NB, AB, LR, SVM on the PIDD to predict diabetes. The performance is evaluated on various measures. Authors in [10] suggested a framework for a monitoring system to integrated home network and data collection management, which can control some existing home network systems by connecting API Adaptor connection technology to the universal terminal devices and IoT devices. Researchers in [19] inspected how IoT gadgets and the whole smart home will carry on in the event that they become survivors of a Distributed Denial of Service (DDoS) attacks focused on the smart home from the outside. The subject of safety was asked as far as whether a real client can proceed to control and get data from IoT sensors, which is accessible during the typical activity of the savvy home. In [20], four IoT smart home gadgets have been suggested (a Sense sleep screen, a Nest Cam Indoor surveillance camera, a WeMo switch, and, an Amazon Echo) and find that the traffic rates related to the network can uncover possibly delicate client cooperation in any event, when the traffic is scrambled. These outcomes demonstrate that a mechanical arrangement is expected to ensure IoT gadget proprietor protection. Reference [21] portrayed the activity of the proposed design and sums up the chance to accomplish network security in a more effective and adaptable with software-defined networking (SDN). An outline of existing SDN security applications was talked about and handled its issues, they considered the network access control and worldwide traffic checking for ad-hoc networks. At long last, they structured plan decisions for SDN utilizing Open Flow and examine their presentation suggestions.

In [9], an IoT-based payload cargo monitoring system (IoT-CMS) has been suggested to screen any natural changes of ESPs to guarantee their utilitarian quality all through the whole chain operational climate. Operational productivity, support procedure, natural change, and power utilization are considered, in actuality, cold chain activities. Through applying (i) a remote sensor organization to gather the ongoing item data, along with (ii) fuzzy logic and case-based reasoning strategies to propose fitting memory conditions for different ESPs, the viable capacity direction can be set up [9]. Authors in [22] introduced a comprehensive review of the empowering advances, conventions, and architecture for a metropolitan IoT. Besides, a specialized arrangement has been introduced and examined in details, best-practice rules embraced in the Padova Smart City project, a proof-of-idea arrangement of an IoT island in the city of Padova, Italy, acted in a joint effort with the city district

[22], while in [23, 24, 28] provided a survey on IoT privacy and security issues in smart home environments, and a security considerations for IoT and also presented an outline of the protection and security challenges coordinated toward the smart home space; likewise, distinguish limitations, evaluate arrangements, and examine various difficulties furthermore, research issues where further examination is required respectively. Shantanu et al. in [26] presented A Systematic study for IoT security. Valentina et al. [25] presented an observing apparatus for IoT frameworks dependent on the expansion of the Montimage network checking instruments. The support of the presented arrangement, approved inside the H2020 ANASTACIA project, end up being appropriate to screen IoT-level organizations on account of the misuse of convention explicit modules [25]. In contrast, authors in [27] provided an itemized outline of the diverse center units and the innovations utilized in smart city executions and explained the present status of their use/deployment, also authors discussed the crucial segments that created the IoT-based smart city landscape, the techniques that empower these fields to exist as far as models used. Authors in [29] proposed a topology framework that offers the use of IoT technology in Video Surveillance Systems, to work with the basic types of Video Surveillance technology with the aim to adopt the connection of video data.

3 Electricity for Children

We must be careful when we define the concept of electricity for children to convince them of the necessity and importance of electricity in relation to the presence of risks and damages to it when using the wrong way, so that children have the ability to deal with electricity without approaching electrical appliances in a worrying manner.

The electricity is very important to our daily life as it is indispensable for the uses of devices that use electricity, such as the heater, broom, fan, refrigerator, air conditioner and many others, and thus electricity has a very great importance in addition to it illuminating the streets at night, and thus it helps to light road is easy at night, but it must be dealt with very calmly, and that we do not use anything related to electricity and our hands are wet with water.

4 Smart Home

Smart homes are an effective part of developing smart cities [8]. It is the solution to the future trend, especially after the changes and development that took place in world, besides due to the huge number of tasks and requirements that led to the need to develop our homes where the need to use various electronic devices Internet of Things (IoT) sensors to collect data [6]. Where the insights obtained from that data are used at a time. To manage the system and services efficiently; and that this data is used in a smart system for processing and describing it in the process of monitoring and protection. Smart home is one of the most prominent applications in the IoT model. While it added a level of comfort and convenience in the users' daily life, it also brought a unique security challenge, which is to mitigate from inside the threats that house members are exposed to and how to protect them in the event that the house is exposed to one of the dangers such as electrical power problems in the home or a malfunction in the water system or Leakage of household gas valves. Such threats arise primarily due to the sharing of IoT devices and the existence of a relationship between smart systems in the home by linking them to the IoT, where the home IoT platforms manage control against internal threats based on a system managed by artificial intelligence using a machine learning and make them responsible for making decisions. Some of the leading IoT platforms that have emerged over the past few years are Smart things from Samsung, HomeKit from Apple and Android Things from Google, linking them with artificial intelligence technologies, IoT and monitoring devices and making them able to control access in order to protect home units.

(A) *Tools in Smart Home*

The idea of building an intelligent monitoring system will be explained in this paper. the system that controls the resources and sources of the home from electricity, water and gas systems, and connects them with a security control system using one of the artificial intelligence techniques through IoT and make them into a system in one. This system is built on the basis of monitoring children in the event that family members such as parents leave and leave their children at home, this system will monitor children's movements and in the event that they are exposed to any danger, the system will analyze their movements based on a database and link them with machine learning techniques to have the decision and mandate to control the situation and control devices. The house associated with the system to reduce the risk that they may be exposed to, with sending an alert signal to the parents that there is a danger inside the home to speed up action and take the necessary measures necessary to protect the children. Also, this system enables family members to watch the home and their condition when they are outside the house using the Web applications for mobile phone and mobile networks. This system is flexible in the sense that it allows a user to control multiple devices, or multiple users to control several devices at the same time.

5 Monitoring Security System

It is the control system in security systems and alarms through motion sensors. The security system triggers the alarm and sends a message to the owner of the house, and controls the home's systems in case the house is exposed to risks, especially when there are children during parents' absence or at any time [9].

The security system triggers the alarm through the sound system and calls the numbers set for the alarm in the event that children are exposed to risks or if there is

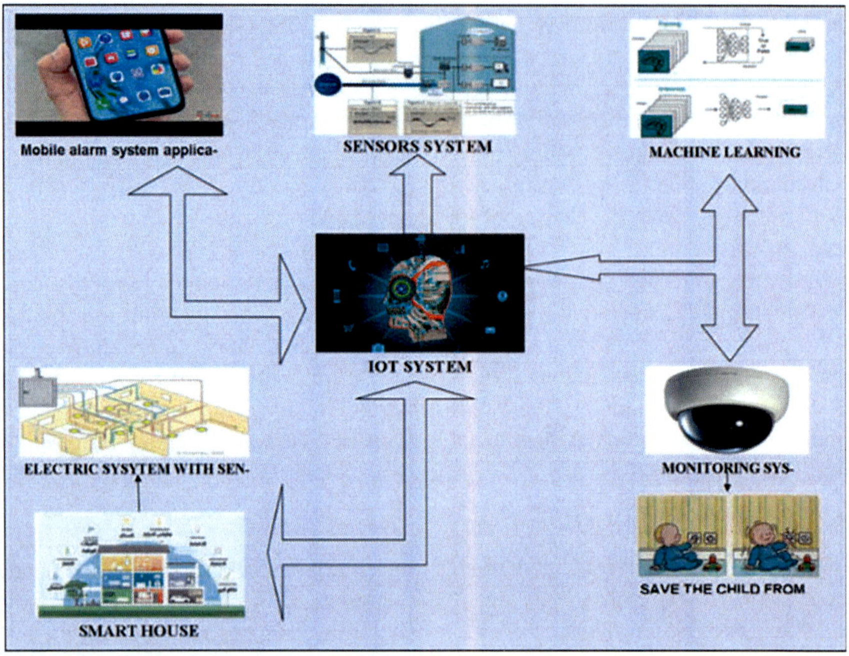

Fig. 2 Suggest monitoring system

a malfunction in the home's electrical systems [14]. There are also sensors for water leakage and play the same role in the event of any risk of water leakage. There are many types of sensors that can be included in the system, such as gas or fire sensors, and others. Our suggested monitoring system is explained in Fig. 2.

6 Machine Learning for IoT

Machine learning utilizes past conduct to recognize examples and fabricate models that help foresee future conduct and events [3]. The Internet of Things and Machine Learning offers covered up information bits of knowledge for quick, mechanized reactions and improved dynamic. AI of the Internet of Things can be utilized to highlight future trends, detect anomalies, and increase intelligence by absorbing images, video and audio. Various methods have been utilized for modeling systems by applying different ML classification techniques like Support Vector Machine (SVM), Decision Tree (DT) [4], neural network (NN), Gaussian Classifier Model [18] Genetic Algorithms (GAs) [5, 7], Random Forest, KNN, Logistic Regression, etc.

AI can help demystify covered up designs in IoT information by examining huge measures of information utilizing complex calculations. It can enhance or supplant

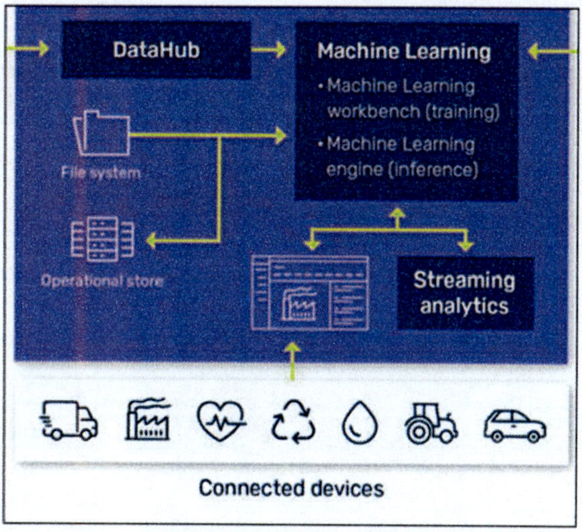

Fig. 3 Machine learning and IoT system [13]

manual cycles with computerized frameworks utilizing measurably inferred methodology in basic processes. Scientists applied machine learning and IoT to perform prescient capacities in a variety of utilization cases that empower organizations to acquire new experiences and progressed mechanization abilities, by developing a machine learning model that is associated with the cloud, edge and device as explained in Fig. 3.

7 System Requirements

When a traditional home environment develops into a smart home environment, new elements must be added (smart devices) such as communication devices, connecting the peripheral devices of the Internet of Things to complete the monitoring system, explaining the requirements of the monitoring system and we will define what each part of the system means and how they meet to form a complete IoT group. The detail description of these parts is explained in Table 1.

8 Development Environment

A monitoring system has been developed using a smart surveillance camera that sends data to the data processing system using an artificial intelligence algorithm (machine learning), which connects to a sensor via a device linkage device and then sends a warning message in the event that the home electrical system is exposed to danger, causing exposure to individuals. The family is to perish, and all of this is

Table 1 Requirements of the monitoring system

Device	Description	Type of device
Surveillance cameras	Definitely the most popular Products in the protection category	Surveillance camera with 24/7 Live Video and Night Vision
Distributer device	Control center for smart home devices that connects several different devices together	Samsung Smart Things Smart Home Hub 2nd Generation
Sensor device	sensors or devices collect data from their environment	Nest Secure (Nest Detect)
Connectivity	Connecting directly to the Internet	gateway/router/Ethernet
Data processing	Software performs some kind of processing on it	Machine learning algorithm (decision tree)
Mobile device	Used to inform if there is danger message	Hawaii mate nine

done by connecting to the home network based on the Internet of Things such as Sensors/Devices, Connectivity Data Processing User Interface, distributor device, processing device and the proposed environment has been referred to as shown in Fig. 4.

9 Proposed Framework for IoT Monitoring System Using Machine Learning

The proposed system is a home monitoring system using IoT technology, by making a group of devices that can communicate with each other to control the control of matters in everything related to monitoring the home from the inside by connecting it to the electrical system to protect the members of the house from any danger they may be exposed to. It has the latest electrical systems in the house. The smart home system mainly relies on a monitoring device, as it transfers information to a data processing device that applies one of the artificial intelligence algorithms.

In fact, this part is linked to another part, which is the sensor device, which is responsible for capturing any abnormal signal in the monitoring system, as it captures the physical changes from the monitoring and processing device and transfers them to the main part of the system to be properly interpreted and informs the party responsible for the house that there is some danger Happen in one way or the other; And all of these devices have been linked to a distributed device, where it connects all the devices and makes one interface that controls the system. In the following lines, we will talk about the group of devices that have been proposed with the technology used in the processing unit, and in the end, it will be an integrated picture of those systems for controlling the home.

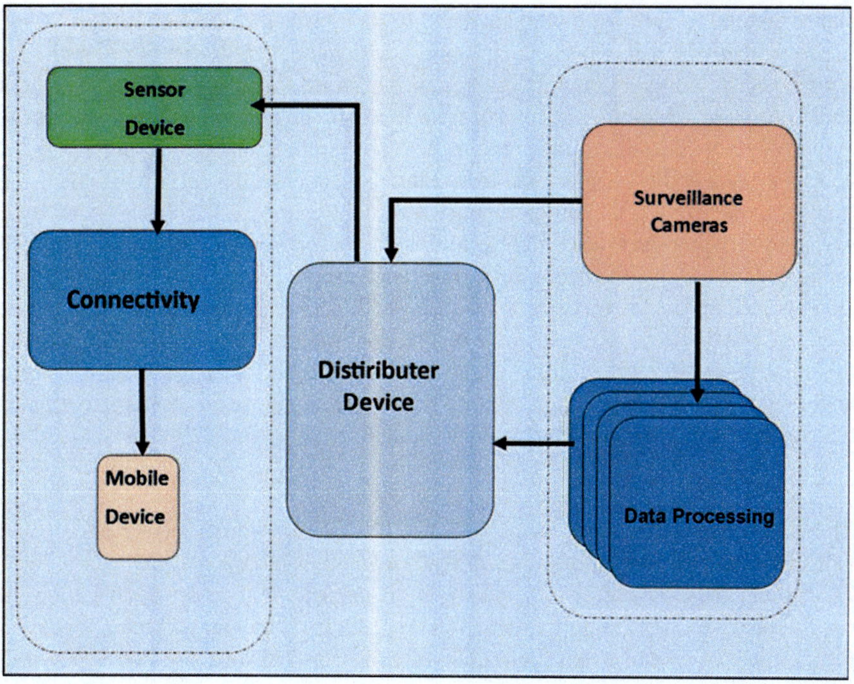

Fig. 4 Service structure

Surveillance cameras are definitely one of the most popular products in the protection category and the most important as well, as it will give you peace of mind and you will be able to know what is going on in your home.

Once the data reaches the data processing part through a database used to collect the monitoring information, where the program performs processing using one of the artificial intelligence techniques (tree decision algorithm) and the processing part is linked to the distributor device as it acts as the brain and the main control center for smart home devices, It links the proposed devices mentioned in Table 1 to control them through a single interface, and it also gives the system the ability to combine the functions of those different devices. To interact with each other under certain conditions the sensor, after receiving a signal from the processing device, through the distributed device, as it collects and sends data by linking to the cloud through a variety of methods including: cellular devices, satellites, Wi-Fi and Bluetooth.

As for the proposed method in the process of processing system data obtained from the monitoring device, the algorithm proposed in the system will be clarified.

The IoT creates huge measures of information from a huge number of gadgets. The network for the most part comprises of heterogeneous devices, with different protocols. Machine learning is supported by information and creates bits of knowledge from it. Machine learning utilizes past conduct to recognize examples and fabricate models that help foresee future conduct and occasions [15]. The Internet of Things and Machine Learning offers hidden data insights for fast, automated responses and improved decision making. Machine learning with IoT can be used to highlight future trends, detect anomalies, and increase intelligence by absorbing images, video and audio. To build an intelligent control framework for IoT with improved safety, the system developed in this study contains both a household electricity, water and gas control system and security monitoring features that define the decisions to be made by the system using artificial intelligence techniques.

This study proposes the design and implementation of the capabilities of having motion sensors inside the home for people, analyzing information and making decisions based on the existence of a database with a system that works on machine learning techniques, linking them with the IoT system, and sending signals via the smartphone to users holding authorization to enter the range system And to enable data control using a smartphone with a web application. An explanation of the proposed system is depicted in Figs. 5 and 6.

(A) *Supervised Learning*

This algorithm comprise of an objective/result variable which has be anticipated from a given arrangement of indicators (autonomous factors).

Utilizing these sets of factors, we produce a capacity that guides contributions to required outputs. The training interaction proceeds until the model accomplishes an ideal degree of exactness on the training set of data [12]. Instances of Supervised Learning: Regression, and Decision Tree. In this work, DT calculation has been utilized for classifying and recognition measure.

(B) *Decision Tree*

DT is a kind of supervised learning calculation that is generally utilized for classifying issues. It works for continuous dependent variables. In this algorithm, the population is divided into at least two homogeneous sets.

This is done dependent on most critical traits/autonomous factors to make as unmistakable gatherings as conceivable [16].

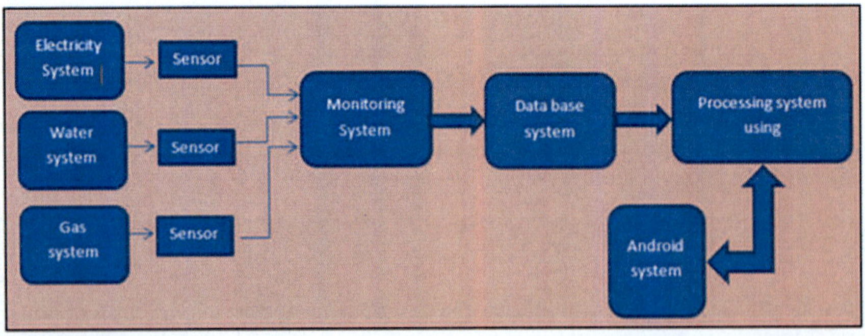

Fig. 5 System development stages

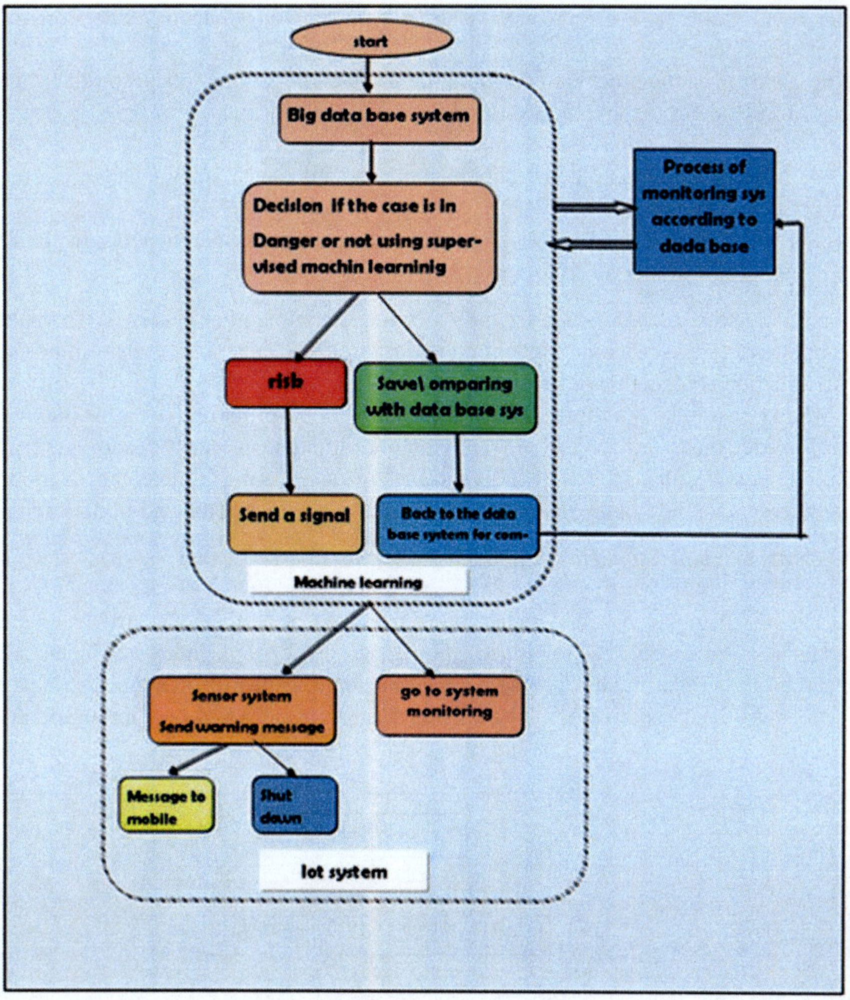

Fig. 6 Structure of proposed IoT monitoring system using decision tree

The significant segments of a decision tree model are; the nodes and their related branches with the fundamental vital strides for building a particular model, which are; parting, halting, and pruning [11].

There are several kinds of nodes can be defined briefly:

(a) The root node, it is a node that considered as a decision that came about because of the development into at least two elite subsets.
(b) The internal (chance) nodes, it is a set of nodes that address one of the potential decisions in the tree; the top node is connected to its parent node and is associated with its leaf nodes.

(c) Leaf (End) nodes, represent the final output result of a mix collections of decisions.
(d) Splitting is the process of dividing the decision node/root node into sub-nodes according to the given conditions.
(e) Branch/Sub Tree: A tree formed by splitting the tree.
(f) Pruning: Pruning is the process of removing the unwanted branches from the tree.
(g) Parent/Child node: The root node of the tree is called the parent node, and other nodes are called the child nodes.

A choice tree model is organized by a chain of command of branches. The path from the root node reaching the leaf node through the inner nodes to determine the classification choice guideline [17].

The DT assembles a choice tree by presenting the information given for training data as demonstrated in Fig. 7. The input training tests can be addressed as discrete qualities, the accessible individual credits address the arrangement of traits. Eventually, the yield of the decision tree is the end-product of a blend of decisions.

(C) *Assumptions Applied While Using Decision Tree Technique*

At the beginning, we consider the whole training set as the root. Attributes are assumed to be categorical for information gain and for Gini index, attributes are assumed to be continuous. On the basis of attribute values records are distributed recursively. We use statistical methods for ordering attributes as root or internal node.

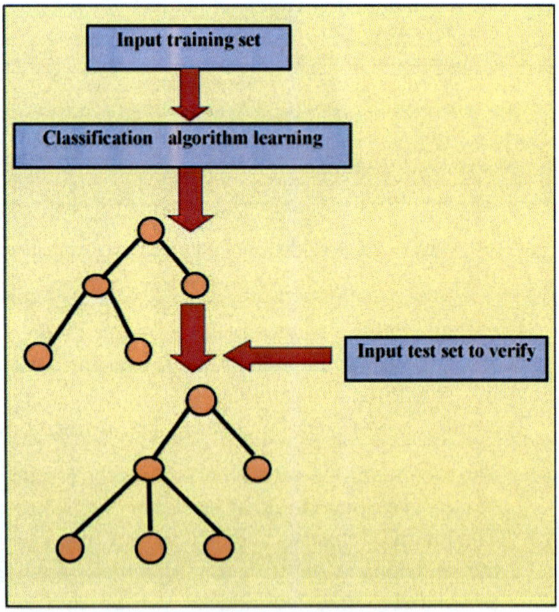

Fig. 7 General steps of the decision tree [12]

(D) *Algorithm Steps*

Step 1: Prepare the training set attributes.
Step 2: Collect the available data parameters from camera monitoring system.
Step 3: Select the best attribute fitting on it.
Step 4: Locate these attributes at the root node of the DT, or internal node according to statistical measures, as the testing dataset.
Step 5: Split the training set of the dataset into two subsets, and concentrate that each subset of the training dataset should have the same value for each attribute.
Step 6: Repeat step 4, and step 5, for each new subsets until we reach the leaf nodes in all DT branches.

10 Conclusions

Due to the urgent need at the present time to have smart systems to deal with vital sources in homes that represent sources of electricity and for security reasons, there is a need to use smart and sensitive elements in the management and control of homes, so this system is designed with the lack of resources and resources that allow us to use it in our homes. Through the research, several issues were revealed;

1. A smart monitoring system can be designed that works on the home's electrical system by connecting it with sensors to protect the home's members.
2. Several IoT technologies can be installed in a smart way to improve the quality of life of homeowners.
3. It is possible to connect more than one device with a device that creates a single control interface instead of making several control interfaces, which leads to a loss of effort and energy and the integration of benefit and safety.
4. The system consists of five parts, which are a monitoring device, a distribution device, a link device with the electronic cloud, a sensor, a device for making critical decisions based on one of the artificial intelligence techniques.
5. It is possible to provide safety for the members of the smart home in the event that the house is exposed to the danger of the electrical system associated with the home.
6. The system is designed for physical testing and for receiving IoT information from any device.
7. Ease of connecting this system with any external peripheral devices.
8. The system can be used anytime or anywhere on a smartphone or web browser via Wi-Fi connection link.
9. It is an experience of the modern system of IoT in real time, especially with the rapid development that we keep pace within this era.
10. The working steps and system installation showed clear potential benefits of a smart home, including strong security and low cost.
11. Finally, this research aims to show the great potential of all digital technologies that can be used in smart home technology.

Acknowledgements Declared none.

Conflict of Interest The authors declare that they have no conflicts of interest.

References

1. A. Pavithra, C. Anandhakumar, V. Nithin Meenashisundharam, Internet of Things with BIG DATA analytics—A survey. Int. J. Sci. Res. Comput. Sci. Appl. Manage. Stud. (2019)
2. J.J. Khanam, S.Y. Foo, A comparison of machine learning algorithms for diabetes prediction. ICT Express 1–4 (2019)
3. S.T. Siddiqui, S. Alam, R. Ahmad, M. Shuaib, Security threats, attacks, and possible countermeasures in Internet of Thing, in *Lecture Notes in Networks and Systems book series (LNNS)*. Advances in Data and Information Sciences, vol. 94, pp. 35–46
4. N. Ibraheem, Finger identification and gesture recognition using Gaussian classifier model. Int. J. Appl. Eng. Res. **11**(10), 6924–6931 (2016)
5. R.Z. Khan, N.A. Ibraheem, Genetic shape fitting for hand gesture modeling and feature extraction using variable length chromosome. Brit. J. Sci. **10**(1) (2013)
6. F. Masoodi, S. Alam, S.T. Siddiqui, Security & privacy threats, attacks and countermeasures in Internet of Things. Int. J. Netw. Secur. Appl. **11**(1), 67–77 (2019) https://doi.org/10.5121/ijnsa.2019.11205
7. N.A. Ibraheem, R.Z. Khan, Novel algorithm for hand gesture modeling using genetic algorithm with variable length chromosome. Int. J. Recent Innov. Trends in Comput. Commun. **2**(8), 2175–2183 (2014). https://doi.org/10.17762/ijritcc.v2i8.3678
8. T.A. Khoa, L.M.B. Nhu, H.H. Son, N.M. Trong, C.H. Phuc, N.T.H. Phuong, N. Van Dung, N.H. Nam, D.S.T. Chau, D.N.M. Duc, Designing efficient smart home management with IoT smart lighting: a case study. Wirel. Commun. Mob. Comput. (2020)
9. Y.P. Tsang, K.L. Choy, C.H. Wu, G.T.S. Ho, H.Y. Lam, P.S. Koo, An IoT-based cargo monitoring system for enhancing operational effectiveness under a cold chain environment. Int. J. Eng. Bus. Manage. **9**, 1–13 (2017). https://doi.org/10.1177/1847979017749063
10. C.-S. Ryu, C.-W. Hur, Monitoring system for integrated management of IoT-based home network. Int. J. Electr. Comput. Eng. (IJECE) **6**(1), 375–380. ISSN: 2088-8708 (2016). https://doi.org/10.11591/ijece.v6i1.9336
11. Y.-y. Song, Y. Lu, Decision tree methods: applications for classification and prediction. Shanghai Arch. Psychiat. **27**(2) (2015). https://doi.org/10.11919/j.issn.1002-0829.215044
12. Z. Lini, Application of research on decision tree algorithm for sports grade analysis. Open Autom. Control Syst. J. **7**, 2300–2305 (2015)
13. https://www.softwareag.com/en_corporate/resources/what-is/machine-learning.html#:~:text=The%20Internet%20of%20Things%20generates,predict%20future%20behavior%20and%20events
14. N. Ibraheem, M. Hasan, Combining several substitution cipher algorithms using circular queue data structure. Baghdad Sci. J. **17**(4), 1320 (2020)
15. G. Rajendran, R.S. Ragul Nivash, P.P. Parthy, S. Balamurugan, Modern security threats in the Internet of Things (IoT): attacks and countermeasures, in *International Carnahan Conference on Security Technology (ICCST)* (2019)
16. Z. Zhang, Z. Zhao, D.-S. Yeom, Decision tree algorithm-based model and computer simulation for evaluating the effectiveness of physical education in universities. Complexity **2020** (2020)
17. M.P.-L. Ooi, H.K. Sok, Y.C. Kuang, S. Demidenko, Alternating decision trees, in *Handbook of Neural Computation*, pp. 345–371 (2017)
18. J. Zhang, X. Hong, S.-U. Guan, X. Zhao, X. Huang, N. Xue, Maximum Gaussian mixture model for classification, in *8th International Conference on Information Technology in Medicine and Education* (2016)

19. L. Huraj, M. Šimon, T. Horák, Resistance of IoT sensors against DDoS attack in smart home environment. Sensors (2020)
20. N. Apthorpe, D. Reisman, N. Feamster, A smart home is no castle: privacy vulnerabilities of encrypted IoT traffic, in *Workshop on Data and Algorithmic Transparency (DAT '16)* (2017)
21. F. Olivier, G. Carlos, N. Florent, New security architecture for IoT network, in *International Workshop on Big Data and Data Mining Challenges on IoT and Pervasive Systems (BigD2M 2015)*, vol. 52, pp. 1028–1033 (2015)
22. A. Zanella, N. Bui, A. Castellani, L. Vangelista, M. Zorzi, Internet of Things for smart cities. IEEE Internet Things J. **1**(1) (2014)
23. H. Manoj, T. Gadiyar, G.S. Thyagaraju, T.P. Bhavya, M. Bhavana, R. Ahana, Privacy and security issues in IoT based smart home applications. Int. J. Eng. Res. Technol. (IJERT), ICRTT (2018)
24. A. Jurcut, T. Niculcea, P. Ranaweera, N.-A. Le-Khac, Security considerations for Internet of Things: a survey. SN Comput. Sci. **1**, 193 (2020)
25. V. Casola, A. De Benedictis, N. Riccio, D. Rivera, W. Mallouli, E.M. de Oca, A security monitoring system for Internet of Things. Internet of Things (7) (2019)
26. S. Pal, M. Hitchens, T. Rabehaja, S. Mukhopadhyay, Security requirements for the Internet of Things: a systematic approach. Sensors **20**, 5879 (2020)
27. A.S. Syed, D. Sierra-Sosa, A. Kumar, A. Elmaghraby, IoT in smart cities: a survey of technologies, practices and challenges. Smart Cities **4**, 429–475 (2021)
28. J. Bugeja, A. Jacobsson, P. Davidsson, On privacy and security challenges in smart connected homes, in *European Intelligence and Security Informatics Conference* (2016)
29. C. Stergiou, K.E. Psannis, A.P. Plageras, G. Kokkonis, Y. Ishibashi, Architecture for security monitoring in IoT environments, in *The 26th IEEE International Symposium on Industrial Electronics* (2017)

Multiple Approach of RFID-Based Attendance System Using IoT

Shruti S. Wakchoure, Priyanka S. Shewale, Janhavi G. Rajput, Sayali A. Gaupal, Mohan P. Thakre, and Minal R. Rade

Abstract The number of people present at a specific location or event is referred to as attendance. It is indeed a measure of the number of students/staff inside an institution. The traditional, manual method of marking attendance, which is a time-consuming task, necessitates more time to prepare, record, and calculate each under-study's normal attendance. The traditional method is more prone to human error and involves an excessive amount of paperwork. Emphasizing the most recent technological advancements can be beneficial in developing newer systems by eliminating the disadvantages of traditional methods and focusing on the enhancement of their advantages. The manual method is considered obsolete, indicating a need to increase the system's efficiency and productivity through automation. The main aim of this paper is to enhance the entire manual process of attendance marking using IoT. RFID tag detection, fingerprint recognition, and facial recognition technology are some of the approaches to the proposed system. The marked attendance would be saved on a Google Sheet through Wi-Fi, with parent notification sent via GSM. The ability to easily share data between multiple locations in real-time ensures that records are preserved appropriately with greater precision and error elimination. According to the current situation, the most common method of transmission of the COVID-19 virus appears to be through infected surfaces. In light of the coronavirus threat, the contactless and non-invasive process of face recognition using ESP-32 cam will be critical in almost every industry.

Keywords *Radio-frequency identification* (RFID) · Face recognition · Global system for mobile communications (GSM) · COVID-19 · Internet of Things (IoT) · ESP-32 cam · Wi-Fi

---

S. S. Wakchoure · P. S. Shewale · J. G. Rajput · S. A. Gaupal · M. P. Thakre (✉) · M. R. Rade
K.K. Wagh Institute of Engineering Education and Research, Nashik 422003, India

G. Ranganathan et al. (eds.), *Soft Computing for Security Applications*,
Advances in Intelligent Systems and Computing 1397,
https://doi.org/10.1007/978-981-16-5301-8_36

1 Introduction

Attendance is a critical component of evaluation in schooling institutions. The practice of maintaining attendance differs between institutes. The most common method is to use paper or a file-based approach, which has several disadvantages. The disadvantages include passing the attendance sheet to all students, which is inconvenient and disrupts the students' attention. In addition, whenever the strength is greater, a roll call becomes much more complex. In these cases, students are seen marking their classmates' participation while they are absent. If the teacher misplaces all such paperwork, all pertinent attendance records are demolished. Because the records exist on paper, they have quickly interfered with [1, 2]. The traditional method is incapable of resolving these issues. As a result, the authors require a computerized measure to eliminate the flaws related to the manual attendance monitoring system. RFID technology could be used to accomplish one such [3, 4]. RFID (radio-frequency identification) is the use of a radio-frequency wave to recognize as well as monitor a tag embedded inside an entity as well as a life form. It is just a wireless connectivity method that uses electromagnetic and electrostatic linkage within the radio-frequency portion of the spectrum to communicate between such a viewer and a tag, which together incorporates a coil that acts as an antenna for transmitting and receiving signals.

RFID readers produce radio waves that are transmitted to RFID tags. Backscatter modern technologies are used by these RFID tags to represent back to the reader the radio wave that has also been modulated with data [5, 6]. Using biometric data as well as RFID technology, it's indeed necessary to suggest a structure that does not require any human involvement [7]. Iris scanners are another example of biometric technology that can be used. Registered users must flash their RFID tag in front of the RFID reader, as well as scan their fingerprint in the Biometric fingerprint scanner, which validates their authentic identity [8, 9]. Furthermore, an SMS will be sent instantaneously to their parents' phone notifying the others of the same. RFID applications nowadays include health, transportation, network management, agricultural development, employment, and so many more [10–12]. Recently, hospitals have used active RFID to point out issues in their workflow, mainly in order to move patients in and out quicker and more efficiently. Further advancement could be the union of face recognition along with an RFID card reader. It involves the detection of authorized students and keeping the authentic record of every registered student [13, 14].

The attendance mark can be stored in a Google Sheet for the maintenance of record. An easy-to-use IoT-based attendance system reduces human intervention and inaccuracies. As a side benefit, improving the accuracy of attendance records [15]. Face database, face detection, face recognition, as well as marking participation are the stages of recognizing one's face. The evaluation of various face recognition systems takes into account various real-world scenarios [16, 17]. One such deployment eliminates the possibility of marking fake attendance [18]. With the development of new integrated system innovations as well as the enlargement of the web application 'Internet of Things,' it's indeed possible to develop a structure that can

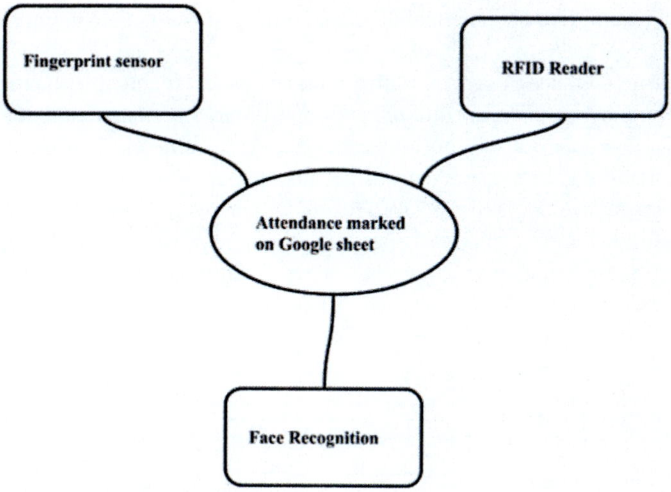

Fig. 1 Overview of the proposed system

address students' attendance structure concerns [19]. IoT is a network of physical objects embedded with sensors, software and other data connection technologies. It refers to a system of interrelated, Internet connected objects that are able to collect and transfer data over a wireless network without any human intervention. Whereas in the sensor business, there is no direct connection to the Internet. Hence, we can say that Sensor Business is a subset of IoT. Sensor business is a technology used within IoT. Sensors play an important role in creating solutions using IoT which detects information, replacing it with a signal that humans and machines can distinguish. Briefly, sensors used in IoT keep track of all data produced by thousands of connected devices, hence allowing scalability and connectivity for managing and analyzing data [20, 21]. The main aim is to build an automated system in the time post coronavirus using the microcontroller ESP 32 and to have a cost-effective system emphasizing the multiple steps for authentication [22] (Fig. 1).

2 Conventional Versus Smart Attendance

Tracking attendance has become a critical job in today's schools, offices, industries, and other sectors. People can choose between two types of attendance systems: manual attendance systems and automated attendance systems. The traditional system, on the other hand, necessitates constant human supervision. It entails punching clocks and signatures on attendance sheets, and a significant amount of time is spent on maintaining these attendance sheets, which is a time-consuming task. This scheme is also vulnerable to manipulation, and there is a high risk of human error. As a result, to reduce manual labor and eliminate human error, improved technology, such as an Automated Attendance System, is required. This system makes it much easier to manage and track the attendance of students in educational institutes and employees in organizations, particularly large organizations. It also keeps unauthorized people out. We no longer have to waste time on the time-consuming task of maintaining attendance records because the system automatically stores the attendance data on Google Sheets. As a result, we can retrieve attendance records and reports with a few mouse clicks whenever we need them. As a result, this system will improve accuracy significantly, and the time and effort saved will lead to increased productivity.

3 Proposed System

To improve system efficiency, the system is configured in the following manner: RFID tag detection, fingerprint recognition, and face recognition. The proposed system makes use of an ESP-32 microcontroller with built-in Wi-Fi technology (Fig. 2).

The R307 fingerprint sensor module is used and it is serially interfaced to ESP-32. After sensing the fingerprint of enrolled students, attendance will be marked. Here, two keypads are used having a 4 × 4 matrix to enroll new students in the system. The ESP-32 cam will detect the face and if matched with the database, attendance will be recorded (Fig. 3).

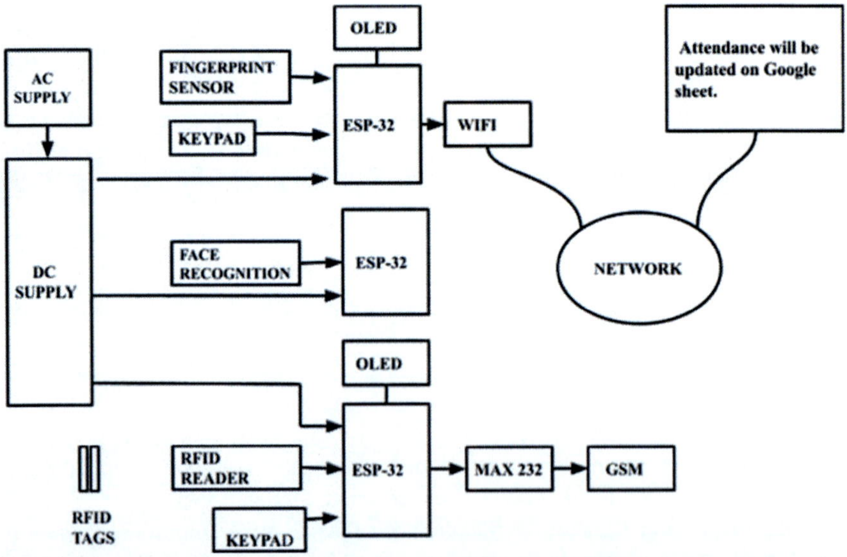

Fig. 2 Block diagram

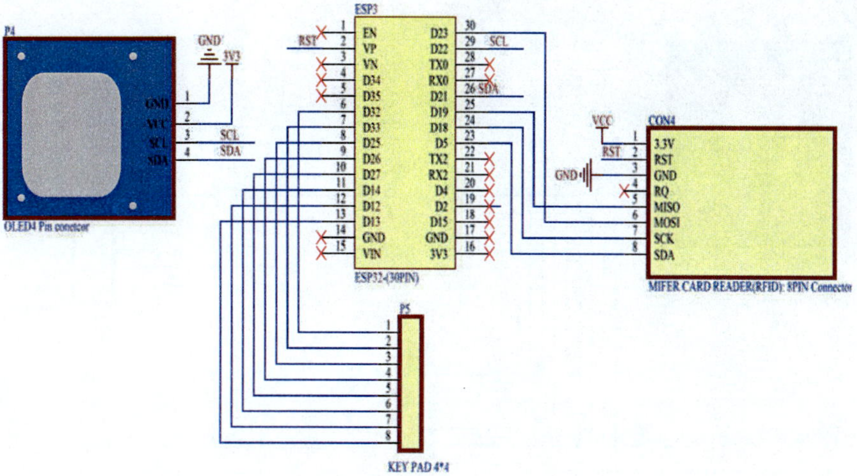

Fig. 3 Circuit diagram of RFID

The RC522 RFID reader module is serially interfaced to the microcontroller. The MAX-232 module is used to connect with the GSM module. OLEDs are interfaced between serial data pin, serial clock pin and ground pin and they are used to display the student's ID name and number. After retrieving the same information from students, the data will be transferred to Google Sheets via the web server cloud by using the ESP2866 module which is inbuilt into the ESP-32 microcontroller and at the same time, the SIM800L GSM module will send SMS notification to parents. This way, the attendance is tracked by the system (Figs. 4 and 5).

Here, the attendance system is designed using RFID technology, fingerprint sensor, and face recognition. The proposed system uses the following components:

- **ESP-32 microcontroller**: A moderate, limited system-on-a-chip microcontroller with interconnected Wi-Fi and dual-mode Bluetooth which has an onboard CP2102 USB to TTL converter, 3.3 V DC power, and 520 KB SRAM memory which utilizes less than 5A current. And through SPI/SDIO or I2C/UART interfaces, the ESP32 can communicate with other platforms to provide Wi-Fi and Bluetooth features.
- **R307 Fingerprint Reader Sensor Module**: The R307 fingerprint reader detector is a fingerprint sensor with a TTL UART interface for direct connections to a microcontroller UART or a PC via a MAX232/USB-Serial adapter. The module can communicate directly with a 3.3 or 5v microcontroller. The module has an operating voltage of DC 4.2 6.0 V, a current consumption of 75 mA, and the character size of the file of 256 bytes.
- **RC522 RFID Card Reader Module**: RC522-AN module uses Philips MFRC522 original chip design circuit card reader. The module uses 3.3 V voltage, 0.96 inch in size, 13-26 mA working current, and 13.56 MHz frequency with 128 × 64 resolutions.

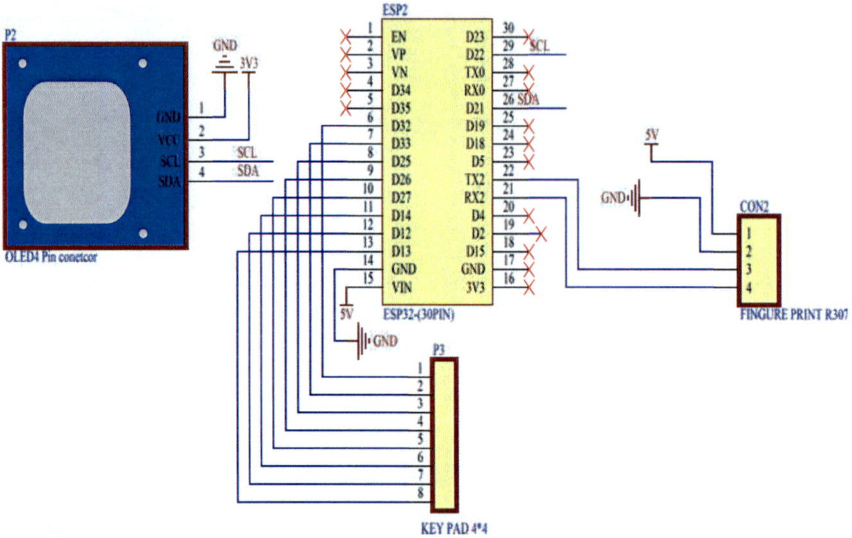

Fig. 4 Circuit diagram of fingerprint

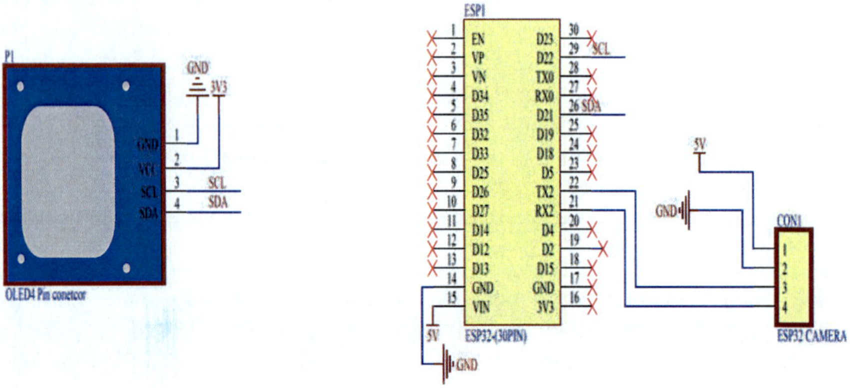

Fig. 5 Circuit diagram of face recognition

- **RFID tags**: This is a keychain type MIFARE tag provided with an RC522 Mifare reader-writer working at 13.56 MHz frequency. All the information is transferred into the chip and from the data memory (EEPROM). The compactness of the tag makes it highly portable and easy to attach.
- **Organic Light-Emitting Diode (OLED)**: An OLED is a self-light-emitting infrastructure made up of a thin, multi-layered organic film sandwiched between an anode and a cathode. The transfer material, emission layer material, and electrode selection are the most important factors in determining the quality of OLED components.

- **SIM800L GSM Module**: The SIM800L GSM Module is indeed a low-cost, small-form-factor GSM module based on Sitcom's SIM800L chipset. One such component is compatible with the quad-band GSM and GPRS networks. The whole module to send and receive text messages (SMS), contribute to making and receiving calls, connect to the Network via GPRS, and so on.
- **ESP32-CAM**: The camera on the ESP32 is used for face recognition. The ESP32-CAM is a low-cost microcontroller that includes a Wi-Fi camera. It includes a PCB antenna. This camera has a Micro SD card slot and a reset button, as well as Wi-Fi and Bluetooth enabled.

After implementation of hardware, various testing methodologies can be implemented which are as follows:

- **Security testing**: This testing is used to eliminate vulnerabilities and maintain the integrity of the data. From a testing standpoint, it checks if the data is protected when getting transferred to the other device.
- **Performance testing**: This test covers real-time aspects such as time bound outputs, load testing and timing analysis.
- **Compatibility testing**: There are lots of devices which can be connected through IoT system having a combination of hardware, software, protocols and operating systems. So, checking the compatibility of the system is important.
- **Functional testing**: This test examines the qualitative and quantitative functional deliverability of deployed IoT applications in the actual conditions.
- **Reliability and scalability testing**: This test involves simulation of sensors by utilizing virtualization tools and technologies.

4 Process Flowchart

There are 18 ADC channels and 2 DAC channels on the ESP-32 controller. An Arduino IDE can be used for integrations. Whenever the system is switched on, the RFID reader begins to monitor and detect any scanned card inside its electromagnetic area (Fig. 6).

The fingerprint sensor, as well as the webcam, are both activated at the same moment. Whenever a card has been scanned near the reader, a reader decodes the data and passes it to the microcontroller for encoding. The microcontroller processes and decodes the card number to determine whether that is justifiable. In the event that the card has been invalid, permission has been rejected. The user must also validate this same fingerprint and the data stored in the datasheet. Throughout the scenario of a fingerprint scanner, a new instance could be saved, as well as the user then must scan his face in the ESP-32 CAM. After successful verification, the microcontroller sends a signal to the GSM module, which further sends a short message (SMS) notification to the student's registered mobile number that also contains the names of the students, registration number, department, card details, and so on. This information is also saved on Google Sheets. The microcontroller will also send card information and the

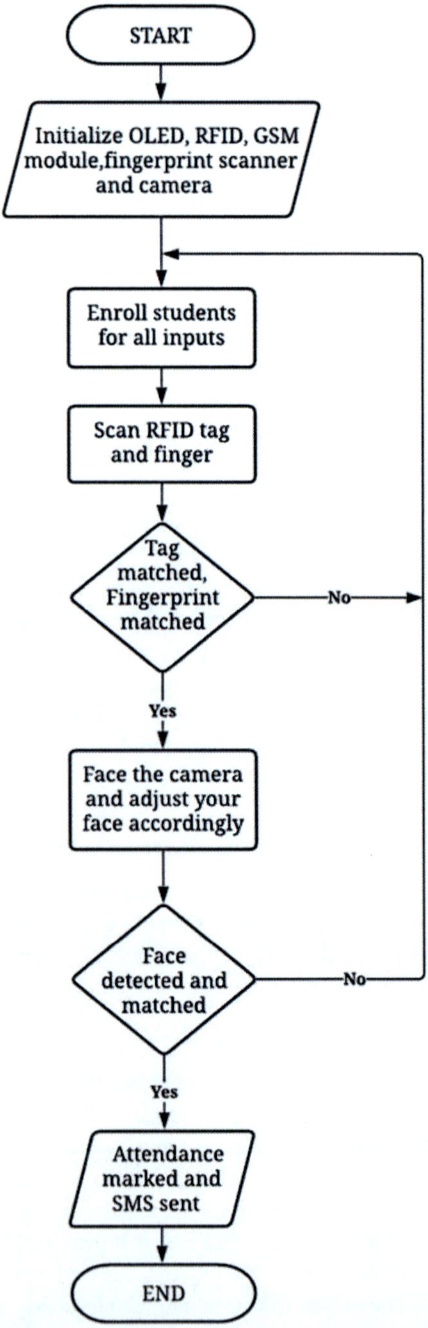

Fig. 6 Flowchart of the system

Fig. 7 Hardware module

number of times the card has indeed been scanned to a PC database. The information can be accessed at any time.

When the system receives power, all the components get initialized. Each student is then expected to register his or her RFID card, fingerprint, and face with the appropriate modules. After enrollment, the fingerprint and RFID card should be scanned. If matched, attendance will be marked. The student's face should be facing the camera appropriately for its recognition. If the face matches with that of the database, attendance will be marked; else the face should be enrolled again. The entire attendance will be marked on Google Sheet and an SMS will be sent to the respective parents (Fig. 7).

5 Hardware Implementation of Prototype Model

The proposed attendance system consists of multiple inputs. The multiple available inputs are RFID tag detection, Fingerprint Scanner, and Face Recognition. If any of the devices fail, the other available inputs will be marked the respective attendance. The ESP-32 microcontroller is used to interface these system components. The initial stage includes the enrollment of each student on the respective input devices. Every time a new enrollment is to be done, the keypad provided acts as an interface with the OLED. The OLED display is used to display the ID name and number of each student after enrollment. The next time one scans the RFID tag over the RFID reader or uses his/her fingerprint, the attendance is marked on the Google Sheet (Fig. 8, 9 and 10).

The ESP-32 camera module is being used to capture students' faces, recognize them, as well as make comparisons of individuals to those stored in the database to determine attendance. All of this data will be saved on Google Sheets, with the time and date noted. The student's attendance will be sent via SMS to the registered phone number via the GSM module. In the event of a system outage, an SMS would serve as a backup to store attendance (Fig. 11).

Fig. 8 A controller connected to Wi-Fi

Fig. 9 Indication of OLED to place the finger

Fig. 10 OLED displaying finger matched

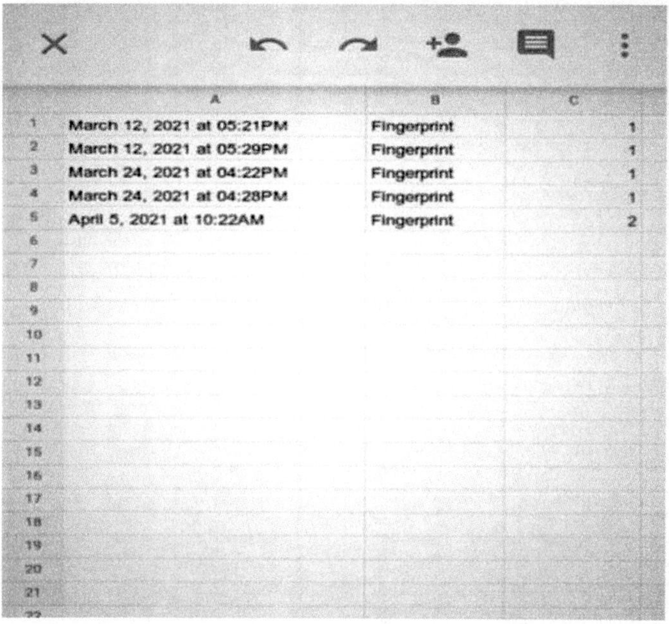

Fig. 11 Attendance on Google Sheet

6 Conclusion

A well-designed scheme ensures a quick, simple, and digitally dependable solution for twenty-first-century classrooms. A more refined system highlighting a highly autonomous process has substituted the manual technique of labeling attendance. The benefit includes a decrease in the number of false attendance entries. Face detection is simply a contactless method of making a system much more secure. This attendance system, which includes a short message service (SMS) notification backup, saves both students and teachers time. In an incident where data loss occurs suddenly, the SMS feature serves as a backup plan. As we know, the entire system works on power. In the case of interrupted power supply, the system will not function. Also, the absence of the Internet makes it difficult to update the attendance on Google Sheets. Further, with some more modifications, we can implement the proposed system in educational organizations, offices, etc.

Appendix

Components	Voltage	Current
ESP-32 microcontroller	3.3 V	<5 A
R307 fingerprint sensor	4.2–6 V	50 mA
OLED	3.3–5.5 V	20 mA
RC522 RFID reader	3.5 V	13–26 mA
SIM800L GSM module	3.8–4 V	500 mA–1 A

References

1. M.B. Shrinidhi, R. Roy, A web-enabled secured system for attendance monitoring and real-time location tracking using biometric and radio frequency identification (RFID) technology, in *2015 International Conference on Computer Communication and Informatics (ICCCI-2015)*. IEEE 978-1-4799-6805-3/15, January.
2. S. Bhattacharya, G.S. Nainala, P. Das, A. Routray, Smart attendance monitoring system (SAMS): a face recognition based attendance system for classroom environment, in *2018 IEEE 18th International Conference on Advanced Learning Technologies*
3. U. K. Nkalo, E.O. Agwu, E.C. Stanley, Radio frequency identification (RFID) based attendance system with short message service (SMS) backup. IOSR-JCE **21**(2, Ser. I) (Mar–April 2019), pp. 01–08
4. A. Kassem, M. Hamad, Z. Chalhoub, S. El Dahdaah, An RFID attendance and monitoring system for university applications (IEEE, 2010)
5. T.S. Lim, S.C. Sim, M.M. Mansor, RFID based attendance system, in*2009 IEEE Symposium on Industrial Electronics and Applications (ISEA 2009)*. IEEE 978-1-4244-4683-4/09, Oct 2009
6. M. Patil, V.N. Bhonge, Wireless sensor network and RFID for smart parking system. Int. J. Emerg. Technol. Adv. Eng. **3**(4), 188–192 (2013)
7. Y. Mittal, A. Varshney, P. Aggarwal, K. Matani, V.K. Mittal, Fingerprint biometric-based access control and classroom attendance management system. IEEE INDICON (2015)
8. S. Chennattu, A. Kelkar, A. Anthony, S. Nagdeote, Portable biometric attendance system using IoT, in *2019 4th International Conference on Information Systems and Computer Networks (ISCON)* (IEEE, 2019)
9. V.K. Shukla, N. Bhandari, Conceptual framework for enhancing payroll management and attendance monitoring system through RFID and biometric (IEEE, 2019)
10. D. Mijic, O. Bjelica, J. Durutovic, M. Ljubojevic, An improved version of student attendance management system and based on RFID, in *18th International Symposium INFOTECH-JAHORINA* (IEEE, 2019). 978-1-5386-7073-6/19
11. M.A. Meor Said, M.H. Misran, M.A. Othman, M.M. Ismail, H.A. Sulaiman, A. Salleh, N. Yusop, Biometric attendance, in *2014 International Symposium on Technology Management and Emerging Technologies (ISTMET 2014)* (IEEE, 2014)
12. K.P. Mohamed Basheer, C.V. Raghu, Fingerprint attendance system for classroom needs (2012 IEEE)
13. M.D. Sajid Akbar, P. Sarker, A.T. Mansoor, A.M. Al Ashray, J. Uddin, Face recognition and RFID verified attendance system (IEEE, 2018) 978-1-5386-4904-618

14. V. Soniya, R. Swetha Sri, K. Swetha Titty, R. Ramakrishnan, S. Sivakumar, Attendance automation using face recognition biometric authentication, in *2017 International Conference on Power and Embedded Drive Control (ICPEDC)*
15. U. Koppikar, S. Hiremath, A. Shiralka, A. Rajoor, V.P. Baligar, IoT based smart attendance monitoring system using RFID, in *2019 1st International Conference on Advances in Information Technology*. IEEE 978-1-7281-3241-9/19
16. G. Guptal, S. Jadhav, V. Gurav, A. Juwale, D. Bhosale, RFID based attendance system using face recognition. IOSR J. Eng. **4**, 23–27 (2018). ISSN 2278-8719
17. G.R. Pooja, M. Poornima, S. Palakshi, M. Bhanu Prakash Varma, A.N. Krishna, Automated attendance system using image processing, in *1st International Conference on Innovations in Computing & Networking*. ISSN 0975-0282
18. U.A. Patel, R. Swaminarayan Priya, Development of a student attendance management system using RFID and face recognition: a review. Int. J. Adv. Res. Comput. Sci. Manage. Stud. **2**(8) (2014). ISSN 2327782
19. S. Shinde, A. Patial, S. Chavan, S. Deshmukh, S. Ingleshwar, IOT based parking system using Google, in *International Conference on I-SMAC* (2017)
20. M.P. Thakre, P.S. Borse, N.P. Matale, P. Sharma,IOT based smart vehicle parking system using RFID, in *2021 International Conference on Computer Communication and Informatics (ICCCI)* (2021), pp. 1–5https://doi.org/10.1109/ICCCI50826.2021.9402699
21. H. Kumar, Involvement of cloud computing and IoT in the field of health care. Int. J. Res. Eng. Sci. Manage. **4**(3), 22–25 (2021)
22. D.K. Sarker, N.I. Hossain, I.A. Jamil (2016) Design and implementation of smart attendance management system using multiple step authentication, in *2016 International Workshop on Computational Intelligence (IWCI), 12–13 December 2016* (2016, IEEE)
23. S. Sujan, S. Shakya, Technical analysis of ZigBee wireless communication. J. Trends Comput. Sci. Smart Technol. (TCSST) **2**(04), 197–203 (2020)
24. A.R. Gidd, A.D. Gore, S.B. Jondhale, O.V. Kadekar, Modelling, analysis and performance of a DSTATCOM for voltage sag mitigation in distribution network, in *2019 3rd International Conference on Trends in Electronics and Informatics (ICOEI)* (2019), pp. 366–371. https://doi.org/10.1109/ICOEI.2019.8862554
25. A.K. Yadav, V. Kulshreshtha, The analysis of project auditing system. Int. J. Res. Eng. Sci. Manage. **4**(1), 88–92 (2021)
26. N.P. Matale, P.S. Borse,A survey based on PLL and its synchronization techniques for interconnected system, in *2021 International Conference on Artificial Intelligence and Smart Systems (ICAIS)* (2021), pp. 1011–1016. https://doi.org/10.1109/ICAIS50930.2021.9395770
27. J. Mane, V. Hadke, Performance analysis of SRM based on asymmetrical bridge converter for plug-in hybrid electric vehicle, *2020 International Conference on Power, Energy, Control and Transmission Systems (ICPECTS)*, Chennai, India (2020), pp. 1–6. https://doi.org/10.1109/ICPECTS49113.2020.9337059
28. T.S. Gaidhani, M.P. Thakre, Analysis of MMC coordination with hybrid DC breakers for HVDC grid protection, in *2020 4th International Conference on Electronics, Communication and Aerospace Technology (ICECA)*, Coimbatore, India (2020), pp. 333–338. https://doi.org/10.1109/ICECA49313.2020.9297443

Review of Deep Learning and Interpretability

Hrushikesh Joshi, Kannan Rajeswari, and Sneha Joshi

Abstract Deep learning has revolutionized machine learning. Deep convolution networks play a pivotal role in image classification. A conventional image classification technique requires extensive feature engineering; this approach is exhaustive as rigorous manual efforts are needed. Deep learning automatically detects the feature set required for classification. It uses backpropagation to achieve this. Deep learning is being used in a variety of applications including healthcare. The use of deep learning specifically in cancer biology is significant. For Pathologists, classifying tumors as benign or malignant is a complicated task. Classification of malignant lymphoma is challenging as features are not uniform across whole slides, and there are various patterns for classification of cancer. Deep learning is ideal for such a scenario where automated feature learning is required further deep learning detects features that are invariant to the location in an image. Tumor region-wise classification of a slide can also be achieved which can be further subjected to recurrent neural networks. Textual explanations of cancer regions can be generated, providing further insights into diagnosis. Here, three different cancer datasets are reviewed and subjected to deep learning methods. Explainable AI approaches can be applied to decision trees as well as on gradient boosted trees to obtain explanations.

Keywords Deep learning · Deep CNN · Cancer classification · Pathology · Explainable AI

H. Joshi (✉)
SY-ME Computer, Computer Department, Pimpri Chinchwad College of Engineering, Nigdi, India

K. Rajeswari
Computer Department, Pimpri Chinchwad College of Engineering, Nigdi, India

S. Joshi
Pathology Department, MIMER Medical College, Talegaon Dabhade, India

G. Ranganathan et al. (eds.), *Soft Computing for Security Applications*,
Advances in Intelligent Systems and Computing 1397,
https://doi.org/10.1007/978-981-16-5301-8_37

1 Introduction

Machine learning is used in many applications. There are mainly two types of machine learning approaches supervised and unsupervised learning. In supervised learning, basically model tries to find input (data) to output (class label) relationship. Various mathematical functions could be used to represent the relationship between data and labels. Various ML techniques use different approaches for detecting functions [1].

Deep learning is a supervised learning method. It is used to detect input to output relationship which is complex [2]. Deep learning is especially powerful in learning unstructured data (i.e., images, speech, audio samples, etc.). Learning features from data are key to build a classification model. For example, classifying a given image of a ball as a Tennis ball or Football requires certain features to be considered. Those features are not immediately present in an image as the image is in pixel values. The learning algorithm needs to extract hidden features such as ball shape and size, number of edges for correct classification. Deep learning excels in this task of detecting those features which are important for classification [2].

For many years, handcrafted features were used to classify data. Classification with handcrafted features set requires extensive domain knowledge and time. Deep CNN automates this process and able to detect subtle features for classification. Even the most subtle features can be learned from deep learning. Deep CNN is extensively used in many image processing applications.

Deep learning is supervised machine learning. Data is represented in multiple layers. Each layer represents data in a more abstract manner. The first layer which tries to learn edges in a given image subsequent layers tries to learn even more complex shapes, motifs, etc. [2]. Deep CNN layers are organized as convolution, RELU, and pooling layers to learn nonlinear data. Convolution is simply filtering operations in which specific features are learned which are not dependent on position in the input image. Kernels are learned by the backpropagation algorithm. Nonlinearity is introduced, and pooling operation is carried out. The purpose of pooling operation is to abstract learned features. Abstraction is a key process in learning. The model can learn features that are irrespective of noise present in those features.

Backpropagation is the heart of deep learning. Kernels that are used in feature extraction are learned by the backpropagation algorithm. An untrained neural network contains random sets of kernels and weights. Backpropagation algorithms find a proper set of kernels and can derive meaningful features. Gradient descent consists of many iterations. Error is calculated for every misclassification using a certain loss-function. The derivative of this loss-function for concerning weight is calculated. The algorithm converges when the derivative is zero, i.e., when a global minimum is achieved.

Deep CNN's are widely used in cancer classification. The role of CAD (Computer-aided diagnostic system) with deep CNN in cancer classification is significant [1] Classification of cancer is not a straightforward task. Pathologists need to undergo a complex process in decision making, and classification is a subject to change from person to person. Classification of malignant lymphoma is difficult for even experienced pathologists. There are many different types of lymphomas; here, only three types of CLL, ML, and FL are discussed. In addition to this breast cancer and B-ALL and multiple myeloma, datasets are studied here.

It is also possible to generate interpretable classification on cancer diagnosis. Whole slide cancer images can be made interpretable so that pathologists can understand why the model arrived at a particular conclusion [3]. This sort of interpretability is achieved by translating learned CNN feature maps to textual paragraphs. This can be done with the use of LSTM (Long short term memory). Interpretability is important in certain classification tasks [4].

With the use of interpretability, users can gain trust in the model. This is especially important in the medical domain. In complex decision making in medical diagnosis, there is no exact way or set of rules for classification such as cancer diagnosis. Generating interpretation of the underlying decision process in words or summary plots, graphs bridges the gap in AI and medical professionals. In deep learning, many intrinsic and subtle features are extracted. If such features are interpreted, pathologists or researchers could get further insight into the problem domain.

Shapely, value-based prediction can be used to understand feature interaction, feature impact on the global model. Shapely, value-based approach can be applied to decision trees, neural networks [5].

Interpretation techniques are also used in many other ML models such as decision tree, random forest, logistic regression, etc. Decision trees are inherently interpretable. This makes them an important model for interpretability. Game theory-based techniques can be implemented on DT's to explore feature interaction and feature importance on the global model.

2 Natural Inspiration for Deep Learning

The human brain is modular. The cerebral cortex has specific regions for specific functions, i.e., for audio-related task auditory cortex, etc. The brain has a layer-wise structure. Neuron in the brain does parallel processing, i.e., distributive computing. Information in the brain is distributed over many neurons. These characteristics are the foundation for artificial neural networks. The brain also contains layer-wise structures as the input layer, hidden layer, and output layer. Series of computations is involved in the visual perception process. Visual tasks are identified by the LGN-V1-V2-V4-IT ventral pathway. Conceptually, convolution neural networks are similar to the visual cortex [6].

3 Neural Network

The architecture of the neural network is shown in Fig. 1. Neural networks typically consist of an input layer then a hidden layer and an output layer. The hidden layer is a key layer where most of the calculation occurs. In the above example, input x consists of $\times 1$, $\times 2$, $\times 3$, and the output is given as a probability of class label (Fig. 1).

Input layer:

This layer is how we get input data (vectors) fed into network. The number of neurons in an input layer is typically the same as the number of input features to the network. Input layers are followed by one or more hidden layers. Input layers in classical feed-forward neural networks are connected to the next hidden layer.

Hidden layer:

There are one or more hidden layers in a feed-forward neural network. The weight values are by which importance to a particular node is given. Neuron in the hidden layer acts as a nonlinear neuron discussed earlier. It takes a linear combination of weight and inputs and applies some nonlinear activation functions over it. The activation function could be sigmoid, RELU, etc.

Output layer: Depending on the setup of the neural network the final output may be real-valued output or set of probabilities.

The difference between artificial neural network (ANN) and convolution neural network (CNN) are that last layer of CNN are fully connected, whereas in ANN, each neuron is connected to every other neurons.

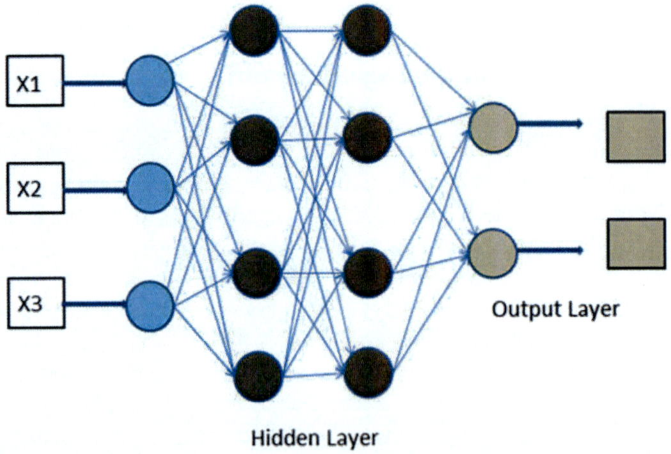

Fig. 1 Neural network architecture consists of an input layer, a hidden layer, and an output layer

4 Convolution Neural Networks

Layers on CNN are shown in figure. The output of the convolution network is forwarded to a fully connected layer. Loss is calculated for actual output, and label and is fed back to the network in backpropagation.

Convolution neural networks (CNN) are used to extract features from images and to classify them. It involves convolution, nonlinear function, and pooling operation. The convolution operation is carried out by taking the element-wise product of kernel and input image and then summing it using handcrafted features to develop classifiers requires domain knowledge, and deep CNN has a key aspect for finding those sorts of features automatically using backpropagation.

Layers on CNN are shown in Fig. 2. CNN layer consists of convolution, pooling layers. Deep CNN works in two phases, i.e., forward propagation and backward propagation.

The network is configured with hyperparameters such as the number of hidden layers, number of neurons in hidden layers, kernel size, kernel width, etc.

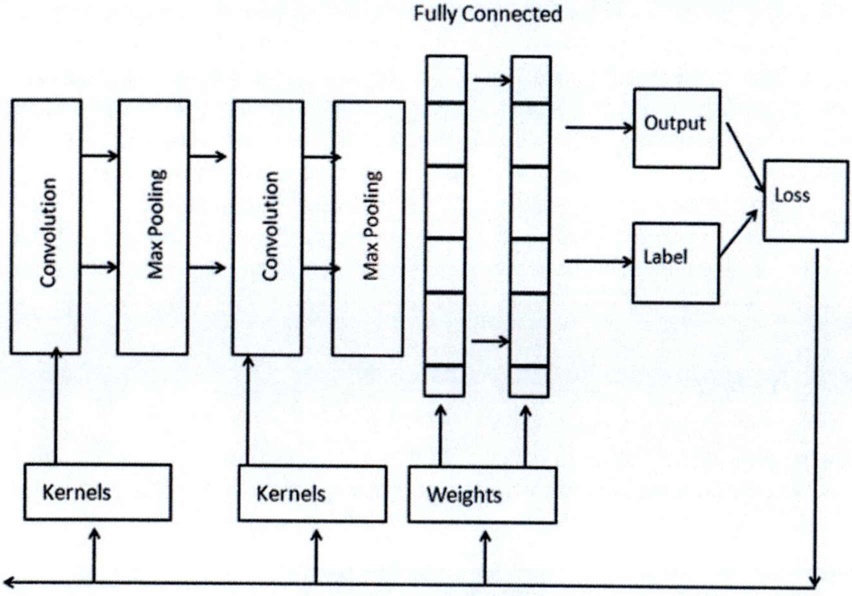

Fig. 2 CNN architecture

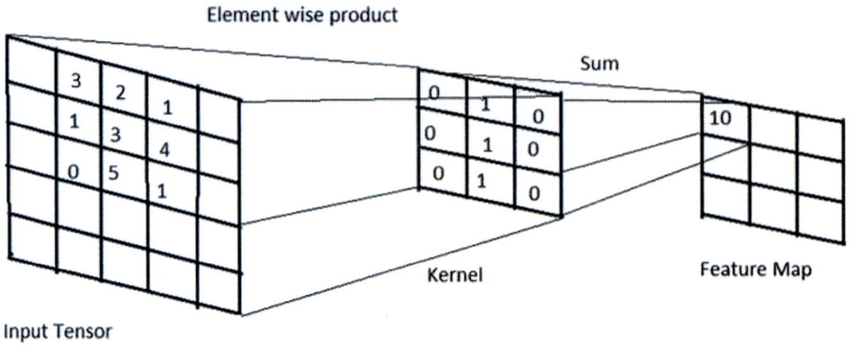

Fig. 3 Convolution operation consists of an input image, kernel which is used to detect certain features

4.1 Convolution

Convolution is a specialized linear combination used for feature extraction. As shown in Fig. 3, convolution consists of input tensor, kernel. A kernel is a feature extractor that is applied to the input tensor where the element-wise product is calculated, and then it is summed to get a feature map. The key feature of convolution is the same weight sharing is kernel is applied all over to the input tensor. So, it can extract local features. Different types of kernels are used for different purposes, e.g., extraction of horizontal edge, extraction of vertical edge, extraction of a border so on.

Different kernels are used to extract different features from input data.

Hyperparameters involved in the convolution procedure are stride, kernel size. Stride is the movement of the kernel window to do convolution.

4.2 Pooling

The purpose of pooling is to concise extracted features. Pooling is carried out to reduce noise in learned features and to abstract it. It is useful for generalization.

Fig. 4 Pooling operation with the filter size of two and stride two. The maximum element in a group is selected in max pooling

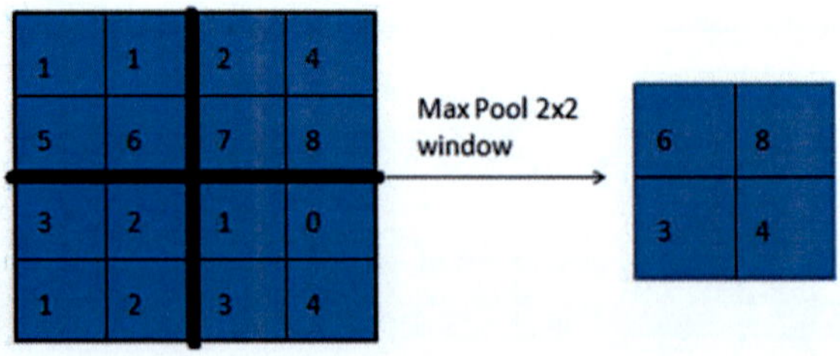

Hyperparameters related to pooling are filter size and stride. Figure 4 shows a max pooling with filter size 2 and stride 2.

4.3 Backpropagation

Backpropagation is the essence of gradient descent. With help, a backpropagation neural network can learn the proper set of weights and kernels. Backpropagation uses gradient descent. Error is calculated by a certain function known as loss-function. The partial derivative of the loss function for every weight is calculated to know, how an error is changing when weights are changed. In backpropagation, negative derivative is considered as the aim is to minimize error. Figure 5a depicts the idea of backpropagation. Green arrows in Fig. 5a show the flow of backpropagation.

Gradient descent is an iterative process. The network starts with a random set of weights. Error in classification is calculated by the loss function. The derivative is calculated at each iteration. The negation of derivative is taken as the aim is to arrive at global minima. Each step toward global minima is controlled learning rate. As shown in Fig. 5b, the model starts at point p, with the help of the gradient descent algorithm model finally converges at point q and able to learn the proper set of weights.

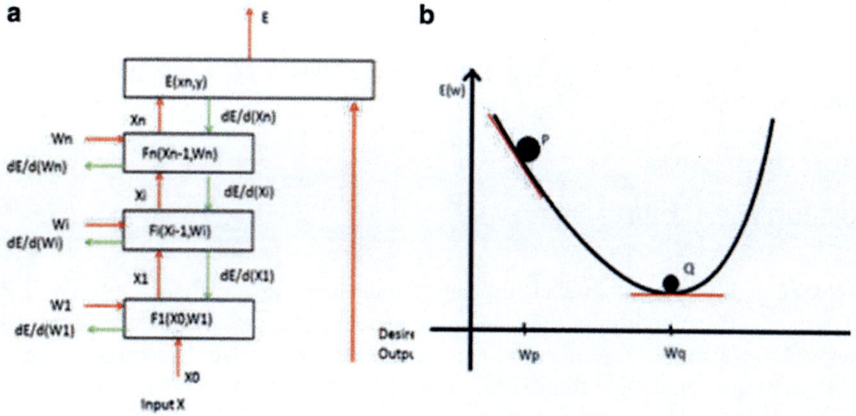

Fig. 5 **a** Backpropagation. Green arrows in figure show flow of backpropagation. Partial derivatives for weights are carried out at each layer with the chain rule. **b** Simplified error function is shown in figure. Gradient descent aims to arrive at a global minimum which is done by calculating the gradient at each step

5 Deep Learning for Cancer Classification

In this paper, mainly three different cancer classification problems are considered.

Malignant Lymphoma Classification

Lymphoma is a clonal malignancy from lymphocytes. There is a total of 38 different types of lymphoma. A pathologist studies tumor on hematoxylin (H)—and eosin (E) slides and classifies tumors, based on cytomorphological features. Dataset used in this article contains three lymphoma types, chronic lymphocytic leukemia (CLL), follicular lymphoma (FL), and mantle cell lymphoma (MCL).

Even for most experienced pathologists', classification of malignant lymphoma is a challenging task [7] as features may not be uniform across the whole area of slides. Individual tumors may contain a range of cell types. Follicular lymphoma contains two different types of cells also MCL has several patterns. Deep learning can be used to overcome these challenges in classification. Deep learning excels in finding complicated features and patterns.

Multiple myeloma and B-cell acute lymphatic leukemia classification.

Multiple myeloma

It is a malignant disorder of plasma cells. Plasma cells that show deep blue cytoplasm, eccentric nuclei, and a perinuclear halo are generally classified as multiple myeloma.

B-cell acute lymphatic leukemia

It is the neoplastic proliferation of bone marrow cells. Lymphoblast shows the following features (1) they are large cells with large nuclei, (2) scanty cytoplasm, (3) less than two nucleoli.

6 Interpretability

Interpretability is useful in many domains. Understanding why the model arrived at a particular decision is important in certain scenarios. Any machine learning model depends on features. Features contribute in certain ways to arrive at a certain outcome. Job of interpretable algorithms is to decode the black box of the machine learning model and to shed some light on feature contribution. Predicting hypoxia in surgery is a challenging task. Lundberg et al. [8] proposed and developed a system based to detect the risk of hypoxia during surgery and also explaining the risk factors associated with it, by using such method, it is possible to pinpoint which factors are contributing to hypoxia and to control them. Features produced by CNN can be made explainable with the use of a recurrent neural network (RNN). The image captioning system consists of an encoder and a decoder. The encoder usually relies on a convolution neural network, while the decoder relies on RNN. CNN operates on an input image, which generates feature vectors. These feature vectors act as input for the RNN. RNN predicts the next word associated with the feature vector. With the use of backpropagation, RNN can be able to generate textual descriptions. Even a set of paragraphs can be generated using RNN and LSTM [9].

6.1 Shapely Value-Based Interpretation

For any model, certain numbers of features are involved in model prediction. The basic intuition behind shape values is to consider each feature and to explain the model further by feature interaction and impact and explaining global model behavior.

SHAP, (Shapely Additive Explanation) by Lundberg and Lee [5], is a method to explain individual prediction. It is based on shape values from game theory [10]. This method can be used on tabular data and also on unstructured data.

7 Results and Discussions

Table 1 summarizes dataset that are used and their respective accuracy, precision and recall and F1 score. Accuracy for B-all and multiple myeloma classification is the highest among all datasets. Details about datasets are discussed further.

7.1 Malignant Lymphoma Classification

Dataset consists of three types of malignant lymphoma. Chronic lymphatic leukemia (CLL), follicular lymphoma (FL), and mantle cell lymphoma (MCL) [11]. A total of 374 images were used for classification. 250 images were used for training, while 124 were used for testing.

CNN architecture is shown in Fig. 6. It consists of four convolution layers. Dropout is introduced after each convolution layer.

Preprocessing

Table 1 Dataset used

Dataset	Accuracy (%)	Precision	Recall	F1-Score
Breast cancer	78	1	0.72	0.83
Malignant lymphoma	80	0.79	0.79	0.76
B-all and multiple myeloma	93	0.87	1	0.93

```
Model: "functional_9"
_________________________________________________________________
Layer (type)                 Output Shape              Param #
=================================================================
input_10 (InputLayer)        [(None, 64, 64, 3)]       0
_________________________________________________________________
conv2d_16 (Conv2D)           (None, 64, 64, 50)        1400
_________________________________________________________________
max_pooling2d_16 (MaxPooling (None, 32, 32, 50)        0
_________________________________________________________________
dropout_20 (Dropout)         (None, 32, 32, 50)        0
_________________________________________________________________
conv2d_17 (Conv2D)           (None, 32, 32, 100)       45100
_________________________________________________________________
max_pooling2d_17 (MaxPooling (None, 16, 16, 100)       0
_________________________________________________________________
conv2d_18 (Conv2D)           (None, 16, 16, 200)       180200
_________________________________________________________________
max_pooling2d_18 (MaxPooling (None, 8, 8, 200)         0
_________________________________________________________________
dropout_21 (Dropout)         (None, 8, 8, 200)         0
_________________________________________________________________
conv2d_19 (Conv2D)           (None, 8, 8, 300)         540300
_________________________________________________________________
max_pooling2d_19 (MaxPooling (None, 4, 4, 300)         0
_________________________________________________________________
dropout_22 (Dropout)         (None, 4, 4, 300)         0
_________________________________________________________________
flatten_4 (Flatten)          (None, 4800)              0
_________________________________________________________________
dense_12 (Dense)             (None, 300)               1440300
_________________________________________________________________
dropout_23 (Dropout)         (None, 300)               0
_________________________________________________________________
dense_13 (Dense)             (None, 3)                 903
=================================================================
Total params: 2,208,203
Trainable params: 2,208,203
Non-trainable params: 0
```

Fig. 6 The model used to classify malignant lymphoma is shown. Dropout is introduced to avoid overfitting.

Normalization of images is done by dividing pixels in images by 255, and then Z-score normalization is performed on data by using Eq. 1. This normalization scheme is used for all the datasets.

```
x = (x - np. average(x,0))/np.std(x,0)
```

Equation 1-Z-Score normalization

The model was trained on 150 epochs. CNN achieved training accuracy of 95% and achieved a testing accuracy of 80% for 150 epochs shown in Fig. 7.

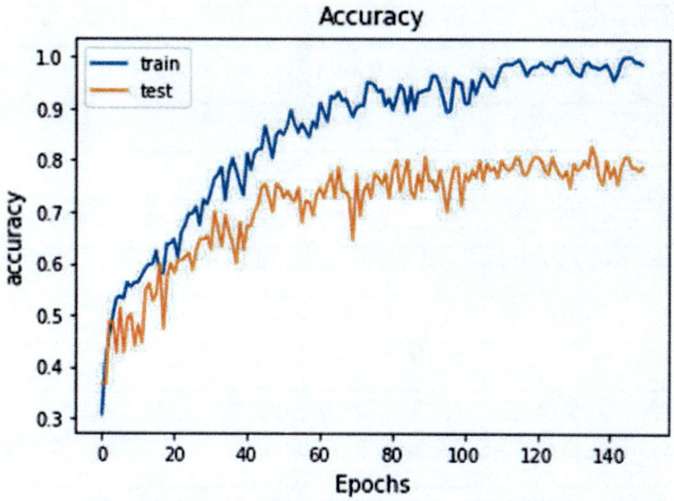

Fig. 7 CNN achieved training accuracy of 98% and testing accuracy of 80%

7.2 B-ALL and Multiple Myeloma Classification

The architecture of CNN used for B-ALL and myeloma classification is shown in Fig. 8. It consists of four convolution layer and one fully connected layer and one output layer. The first convolution layer uses the RELU activation function. Dropout is introduced after a fully connected layer to avoid overfitting. Max pooling is introduced after each convolution layer.

Dataset

CNN was used to classify between B-lineage acute lymphoid leukemia (B-ALL) and multiple myeloma (MM) [12]. CNN network is shown in Fig. 12. Out of a total of 58 images, 29 were used for training, and 29 were used for testing. CNN was trained on ten epochs. CNN achieved training accuracy of 96.55% as shown in Fig. 9. CNN achieved 93% testing accuracy.

The confusion matrix is shown in Fig. 10.

Sensitivity is given by the formula TP/TP + FN. The sensitivity of the model output is 0.875. Specificity is given by formula TN/TN + FP. The specificity of the model is 1.

CNN was able to predict all of B-All type cancer cases without any error.

```
Layer (type)                 Output Shape              Param #
=================================================================
input_1 (InputLayer)         [(None, 64, 64, 3)]       0
_________________________________________________________________
conv2d (Conv2D)              (None, 64, 64, 24)        672
_________________________________________________________________
max_pooling2d (MaxPooling2D) (None, 32, 32, 24)        0
_________________________________________________________________
conv2d_1 (Conv2D)            (None, 32, 32, 48)        10416
_________________________________________________________________
max_pooling2d_1 (MaxPooling2 (None, 16, 16, 48)        0
_________________________________________________________________
conv2d_2 (Conv2D)            (None, 16, 16, 64)        27712
_________________________________________________________________
max_pooling2d_2 (MaxPooling2 (None, 8, 8, 64)          0
_________________________________________________________________
conv2d_3 (Conv2D)            (None, 8, 8, 90)          51930
_________________________________________________________________
max_pooling2d_3 (MaxPooling2 (None, 4, 4, 90)          0
_________________________________________________________________
flatten (Flatten)            (None, 1440)              0
_________________________________________________________________
dense (Dense)                (None, 128)               184448
_________________________________________________________________
dropout (Dropout)            (None, 128)               0
_________________________________________________________________
dense_1 (Dense)              (None, 2)                 258
=================================================================
Total params: 275,436
Trainable params: 275,436
Non-trainable params: 0
```

Fig. 8 The architecture of CNN is used for classification. It consists of four convolution layer and one fully connected layer and output layer

7.3 Breast Cancer Dataset

Breast cancer dataset (BreakHis) [13] dataset was used for the classification of benign and malignant classes. 40X magnification images are used for classification. Out of a total of 350 images, 234 used for training CNN and 116 images used for testing. CNN model is shown in Fig. 11 used for classification. The model was trained with 10 epochs. CNN model was able to achieve a training accuracy of 91 and 78% testing accuracy.

8 Shapely Value-Based Interpretation

A shapely value-based approach is implemented in breast cancer Wisconsin (Dua and Graff 2019) (UCI Machine Learning Repository) dataset [14]. The gradient boosted

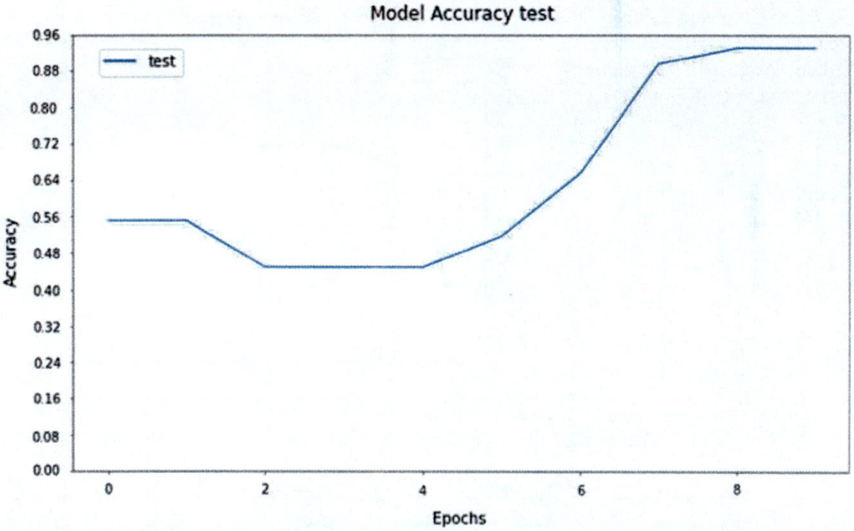

Fig. 9 After ten epochs, CNN achieved testing accuracy of 93%

Fig. 10 Confusion matrix

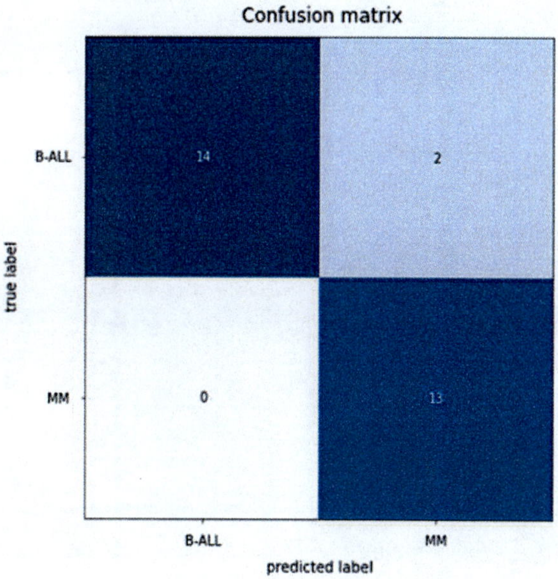

tree approach was used to classify tumors as benign or malignant (97% accuracy). Shapely valued-based interpretation methods are applied on xgb classifier (Fig. 12).

The breast cancer dataset includes certain features related to tumor dimension and shape, texture, etc.

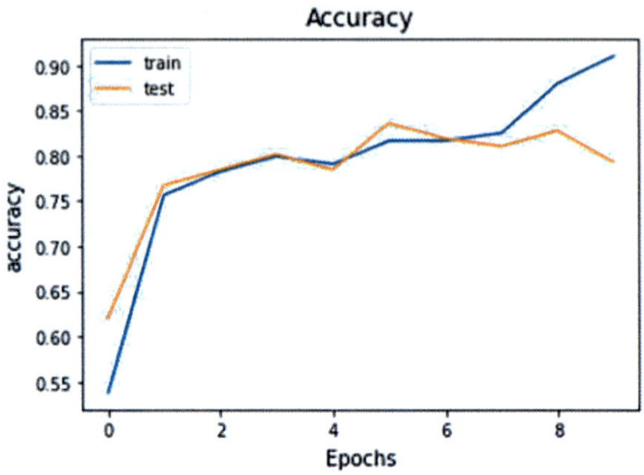

Fig. 11 Testing and training accuracy. CNN achieved training accuracy of 91% and testing accuracy of 78%

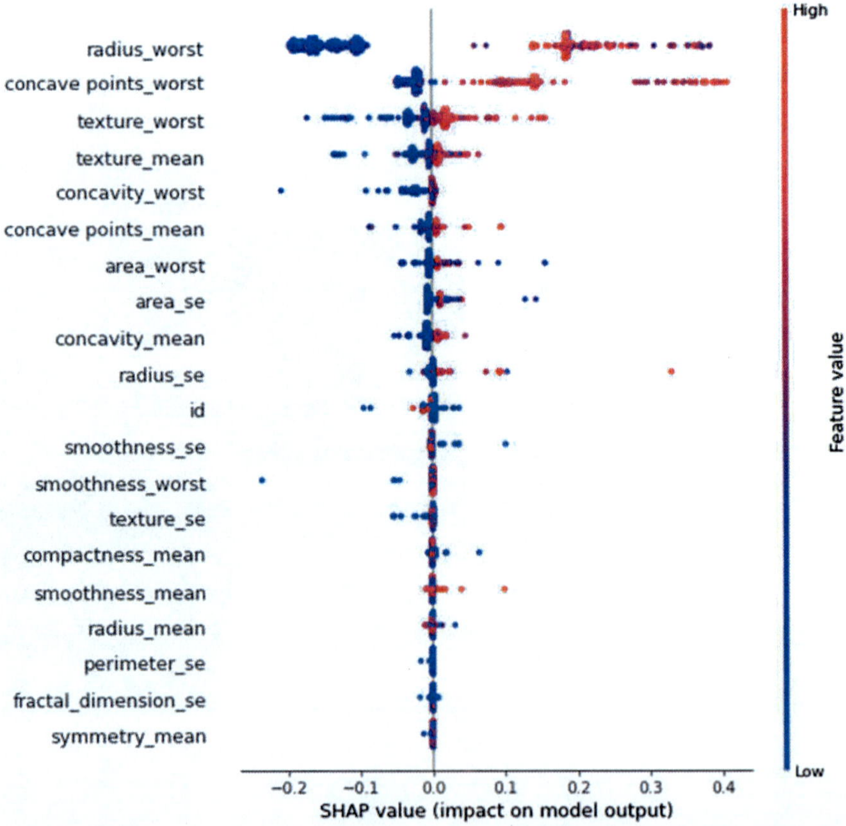

Fig. 12 Shap impact model for breast cancer dataset. The impact of each feature is shown. Feature value is color-coded from low (blue) to high (red). The X-axis represents Shap impact on model output (i.e., Malignant or Benign). Blue values indicate a negative impact on having breast cancer. while red values show a positive impact

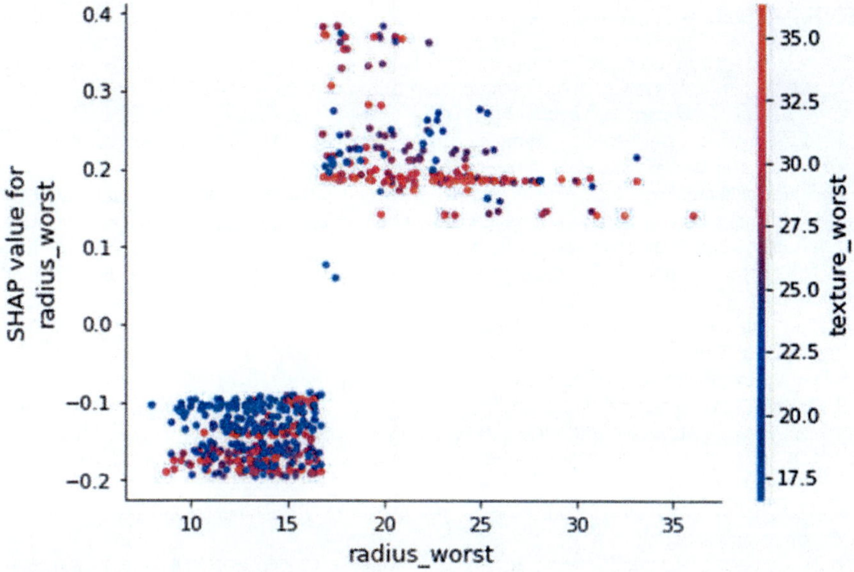

Fig. 13 Feature interaction of radius worst and texture worst. When the value of the feature radius worst is greater than 17 impacts on the global model from radius is higher. Feature interaction between radius worst and texture worst is shown.

8.1 Feature Interaction

Feature interactions can also be plotted with the help of Shap Explainable AI. As shown in Fig. 13, interactions between features radius and texture are plotted.

9 Conclusion

Deep learning is a powerful technique; especially, it is very useful in cancer classification. Deep learning excels in finding intricate features and complicated patterns in cancer classification. Classification of malignant lymphoma is a complicated task even for an experienced pathologist. Deep learning achieved 78% for breast cancer classification, 80% for malignant lymphoma classification, and 93% B-ALL and MM classification. Interpretability approaches are crucial in cancer diagnosis. A shapely value-based approach can be used to summarize the impact of feature and to know feature interaction. These concepts can be displayed through interactive graphs, which could be important in cancer diagnosis. Shapely value-based explanation methods excel in structured data; however, expansibility for CNN has great future scope.

References

1. B.E. Bejnordi, *Diagnostic Assessment of Deep Learning Algorithms* (2017)
2. Y. LeCun, Deep learning. Nature (2015)
3. Z. Zhang, Pathologist-level interpretable whole-slide cancer. Nat. Mach. Intell. (2019)
4. Lipton, *The Mythos of Model Interpretability* (ACM, 2018)
5. S.M. Lundberg, From local explanations to global understanding. Nat. Mach. Intell. (2020)
6. D.C. Van Essen, J.L. Gallant, Neural mechanisms of form and motion processing in the primate visual system. Neuron (1994)
7. N.V. Orlov, Automatic classification of lymphoma images with, IEEE Trans. Inf. Technol. Biomed (2010)
8. S.M. Lundberg, B. Nair, M.S. Vavilala, M. Horibe, M.J. Eisses, T. Adams, D.E. Liston, D.K. Low, S.F. Newman, J. Kim, S.I.L. Lee, Explainable machine-learning predictions for the prevention of hypoxaemia during surgery. Nat. Biomed. Eng. (2018)
9. J. Krause, A hierarchical approach for generating descriptive image paragraph, in *2017 IEEE Conference on Computer Vision and Pattern Recognition (CVPR)* (2017)
10. L.S. Shapley, A value for n-person games, in *Contributions to the Theory of Games* (1953), 307–317
11. *Malignant lymphomaclassification*, https://www.kaggle.com/andrewmvd/malignant-lymphoma-classification.
12. A. Gupta, R. Duggal, S. Gehlot, R. Gupta, A. Mangal, L. Kumar, N. Thakkar, D. Satpathy, SN-AM dataset: white blood cancer dataset of B-ALL and MM for stain normalization. *Cancer Imaging Arch.* https://doi.org/10.7937/tcia.2019.of2w8lxr.
13. F.A. Spanhol, L.S. Oliveira, C. Petitjean, L. Heutte, *A dataset for breast cancer histopathological image classification.* IEEE Trans. Biomed. Eng. (TBME) **63**(7), 1455–1462 (2016)
14. A. Asuncion, D. Newman, *UCI Machine Learning Repository.* http://archive.ics.uci.edu/ml. *Irvine, CA.*
15. M.a. Pitts, A logical calculus of the ideas immanent in nervous activity (1943)
16. O. Vinyals, Show and tell: A neural image caption generator, in *IEEE Conference on Computer Vision and Pattern Recognition (CVPR)* (2015)
17. F.A. Spanhol, Breast cancer histopathological image. IJCNN (2016)
18. G. Huang, Densely connected convolutional networks. *Computer Vision and Pattern Recognition (CVPR)* (2017)
19. O. Simeone, A very brief introduction to machine learning. IEEE Trans. Cogn. Commun. Netw. (2018)
20. S. Manoharan, Early diagnosis of lung cancer with probability of malignancy calculation and automatic segmentation of lung CT scan images. J. Innov. Image Process. (JIIP) (2020)
21. V. Bindhu, P.V. Saravanampatti, Semi-automated segmentation scheme for computerized axial tomography images of esophageal tumors. J. Innov. Image Process. (JIIP) (2020)

Enhancement of Network Security Through Intrusion Detection

Sameena Naaz, Khushboo Mir, and Iffat Rehman Ansari

Abstract Security has become the greatest issue, both within and outside the organizations which are on cloud. The use of user IDs, passwords, and firewalls are the common steps that are taken in order to secure the computers. However, these are not proven to be so effective solutions in current unsecured era. The need for efficient security systems led to development of Intrusion Detection Systems. Intrusion Detection System (IDS) aims at providing secure data transmission by detecting intrusions on the network that an organization is connected to or its database or the devices which are being used. Wireshark is used in this work for detection of intrusion in the network so that the data can flow in a safe manner. Main reason for using Wireshark is its ease of use, free availability, and better visualization due to color coding in graphical representation. First of all, the information about the packet is stored in a log file, and then it is examined to classify the data as normal or abnormal. The I/O graph is used to depict the total traffic flowing in terms of packet per second or byte per second.

Keywords Packet sniffer · Tcpdump · Intrusion detection system · Host based intrusions · Network based intrusions · DoS attack · Wireshark · TCP packet

1 Introduction

Cloud Computing is considered to be the next generation paradigm in Information Technology. Cloud computing environment provides both resources and applications through the Internet as service on demand. Cloud Computing is defined by NSIT (National Institute of Standards and Technology) as a service that is used to provide data storage capacity and computing power over the internet [1]. Five basic features

S. Naaz (✉) · K. Mir
Department of Computer Science and Engineering, School of Engineering Sciences and Technology, Jamia Hamdard, New Delhi, India
e-mail: snaaz@jamiahamdard.ac.in

I. R. Ansari
University Women's Polytechnic, Aligarh Muslim University, Aligarh, India

G. Ranganathan et al. (eds.), *Soft Computing for Security Applications*, Advances in Intelligent Systems and Computing 1397,
https://doi.org/10.1007/978-981-16-5301-8_38

of cloud as specified by The National Institute of Standards and Technology (NIST) are universal network access, on-demand service, measured service, flexibility, and resource pooling. The major services provided by cloud computing are Platform as a Service (PaaS), Software as a Service (SaaS), and Infrastructure as a Service (IaaS). Cloud Computing is an amalgamation of many computing concepts such as Web 2.0, Service Oriented Architecture (SOA), and virtualization on Internet. It helps the business houses by providing a common computing platform although their data and software are present on the remote servers [2].

Cloud Computing relies on virtualization technology where virtual copies of the actual hardware resources such as processor, memory, etc. are provided to the users in a dynamic manner. Due to this, the single physical resource can be accessed simultaneously by multiple users and hence the resource utilization increases. Due to remote access nature of the cloud computing services, it is very vulnerable to attacks. These attacks can range from being network based to host based to attack the data which is stored at various sites. The major reason for this vulnerability is the distributed nature of the network itself [3]. Therefore, there is a need for detection of the attacks.

One method of handling these attacks is by extending the capabilities of the firewalls. But there are some inherent problems with the capabilities of firewalls the most significant being that it can detect attacks that are coming from outside the network. It cannot detect a security violation that is internal to the organization. The other method is intrusion detection which complements the facilities provided by the firewall. It extends the security management aspect by empowering the system administrators to monitor the network so that they can recognize the attack and take corrective measures for the same.

An intrusion detection system analyzes a set of discrete events at regular intervals to find patterns of misuse. These systems are classified as host based and network based. The activities being analyzed in both types of system are sequential records that give us an idea about what activities are taking place and hence reflects the behavior. This in turn helps in detecting any sort of attack. Host-based approach looks at events like files that were accessed and applications that were executed. Network-based approach on the other hand looks at events as packets being exchanged between computers which means that it is monitoring the network traffic.

One of the most important aspects of cloud management is analysis of traffic over the network. This is necessary to make the system secure and to provide the guaranteed quality of services. The flow of network shows the behavior of users in cloud computing systems in service operation or their use. The required information can be obtained by creating a model for service usage or obtaining patterns that can further be analyzed for detecting normal and abnormal behavior of the system. Upon analysis, flow of network traffic gives an idea of how the applications are performing in the cloud environment. Development of these network traffic analysis techniques helps in improving the availability, security, and performance in the cloud environment [4].

One latest example of use of Wireshark is a project on software reverse-engineering by the US Army Research Laboratory with an aim to learn how a malware communicates over the network and then to develop software for detection and representation of the communication in plaintext [5].

Basic elements of Intrusion Detection System are:

Network Traffic Monitoring: The activities of the user and the system are monitored and data is collected for analysis.

Data Analysis and Detection of Possible Intrusion: The collected data is analyzed to build a prediction model which can help in detecting whether any intrusion has occurred or not.

Generation of Alert: In the event of any malicious activity being detected, an alert is generated to inform the administrator about the intrusion.

The rest of the paper is organized as: Sect. 2 gives the review of related work in the area of intrusion detection. Section 3 discusses in detail the various types of intrusion detection systems. The implementation details using Wireshark and the analysis of the results are given in Sect. 4. Section 5 gives the conclusions derived from the study and the future scope.

2 Literature Survey

Web applications contain many vulnerabilities which are used to cause many exploits and these can result in alarming impacts to the privacy of the organizations, as obtained by OSWAP communities [6]. Data traffic over the internet can be monitored by collecting the information of network packet flow as most of the web application traffic is HTTP traffic. It is possible to analyze this traffic because it can be intercepted and monitored. An effective forensic model for network was proposed by authors in [6] which was capable of acquiring and preserving digital evidence. According to the authors in [7] multitenant nature of the cloud environment is the major reason for security issues. They talk about the security issues in Infrastructure as a service (IaaS) clouds. The main concepts used in Infrastructure as a service model are that of machine virtualization and network virtualization. Most of these systems use access control and encryption techniques to secure the different components of a virtualized (multitenant) datacentre.

A security product of Alert Logic called Threat Manager which is a real-world cloud IDS is used by Amazon Web Services. According to the reports of Cloud Security 2014 given by Alert Logic, the threats in the cloud are growing. The attacks which previously targeted the host environment are now attacking the cloud and hence the total number of attacks is increasing at a very fast pace [8]. In the cloud environment, most of the resources are hosted at the service provider's remote server and hence the users of these services over the cloud have limited control over the resources [9]. Therefore, it is the responsibility of the cloud service provider to provide intrusion detection capability in the cloud environment.

Authors in [10] use SNORT is used to capture live packets from the internet with the help of predefined rules. These captured packets are stored in a log file and are then exported to Wireshark for detection of intrusion. A traceback mechanism has been proposed by authors in [11] to improve the capabilities of Wireshark so that it informs the victim about the source of DoS attack so that corrective measures can be taken appropriately.

Attacks in the data link layer such as DHCP Starvation attack, Rogue DHCP attack, DNS Cache Poisoning Attack, and DNS Hijacking Attack are studied by the authors in [12]. In their work, they have captured the DHCP and DNS packets and analyzed them using Wireshark. The focus of their analysis was the mechanism used to assign IP addresses from a DHCP Server and the mechanism of packet exchange and URL to IP address resolution.

A novel Ontology for analysis of packets captured through Wireshark has been proposed in [13]. Their Ontology based on OWL which includes packet analysis terminologies, their properties as well as restrictions is used for representation of knowledge and automated reasoning.

Integration of Behavior-based and knowledge-based analysis are used in Grid and Cloud Computing Intrusion Detection System (GCCIDS) in order to detect some specific types of intrusions on the cloud. But these systems cannot store the information about the attacks in a database and hence they cannot analyze the available data to create a model for intrusion detection. Therefore, they cannot detect new types of attacks which is a required feature of intrusion detection system (IDS) [14]. Vieira et al. proposed the Host Based IDS in 2010. In this model, an IDS is present in each node of the cloud which provides the IDS services and interaction between these services [9]. Another model was proposed by Arshad et al. which is comprised of six different components: system call handler, security analysis module, detection module, global components, profile engine, and intrusion response system [14]. This model is used for severity analysis along with intrusion detection.

According to Donadio [15] reliability and security has to be viewed as a process and not as a product. We cannot use the same processes in a cloud environment which were being used in traditional IDS. The methods employed here have to be very crisp which can cater to the various issues specific to the cloud environment. As suggested by Ambikavathi and Srivatsa [16] rule based IDS identifies attacks by analyzing elements such as variations in protocol behavior, signature, changes in files or directories, unusual system logs, etc. Dhage and Meshram [17] proposed an architecture which has a different intrusion detection system for each virtual machine and each of these IDS instances is managed by a separate controller. Intrusion detection is carried out by combining the signature and learning based methods. Distributed denial-of-service (DDoS) attack in the virtualized environment has been discussed in [18]. Only known attacks can be detected by using signature-based intrusion detection techniques. The external intrusions can be detected if this method is employed at the front-end of the cloud and external as well as internal intrusions can be detected if it is employed at the back-end. But as shown by authors in [19] it cannot detect unknown attacks in cloud as it was possible in traditional networks. In [20] authors have proposed a novel signature (misuse) based Intrusion Detection System and found it to perform better than the existing approaches with UNSW-NB15 (benchmark data set) as well as with real time data set (RTNITP18) generated in their own lab. Security is very important in wireless sensor networks and machine learning techniques have been used in [21] to secure the network data from attackers. The approach of deleting instructions in an event of hacking has been proposed in this paper.

3 Intrusion Detection System

Cybersecurity has become a need since society has become increasingly dependent on computerized environments in its application areas. Intrusion Detection becomes one of the most important aspects of cybersecurity. To prevent attacks, there is a dire need for awareness of an attack first hand. Awareness of attacks helps to react and defend against attackers. Cybersecurity involves security analytics of data to look for attack patterns which give insights into the attacks. Intrusion Detection is also important for forensic purposes as it enables identification of successful breaches even after they have occurred [22].

The role of intrusion detection is to keep a check on the use of network assets and detect any unusual behavior or use of network in an inappropriate manner [23]. Hence, there is an alarming need for security of these systems from intruders. The security in a computer network is enhanced by the use of these intrusion detection systems.

IDSs can be classified based on the method used for data collection. The Wireshark is placed before or after the firewall for Internet connectivity monitoring, it is connected to the LAN switch with port mirrored to the monitored router for WAN monitoring and connected to the LAN switch with port mirrored to the monitored server for server monitoring. The architecture of the Intrusion Detection System using Wireshark with all these three placements is depicted in Fig. 1.

- **NIDS (Network Intrusion Detection System)**: is the detection system based on analysis of network data. NIDS captures the traffic over the network for analysis and looks for abnormalities in the captured traffic which contains activities of monitored hosts on the network.
- **HIDS (Host-based Intrusion Detection System)**: is the intrusion detection system which keeps a check on the activities of a local host like personal computer. It can use databases or log files for analysis. The goal of HIDS is to track any attempts of unauthorized access to the system or any possible attack that has been done on the system. Here, machine monitor has the IDS on itself.

Integration of both IDS types (HIDS and NIDS) can also be done for implementation of cloud security as per the need. Both IDS types have the properties of anomaly-based and signature-based detection. Therefore, this approach will give better accuracy of detection in unknown as well as known attacks.

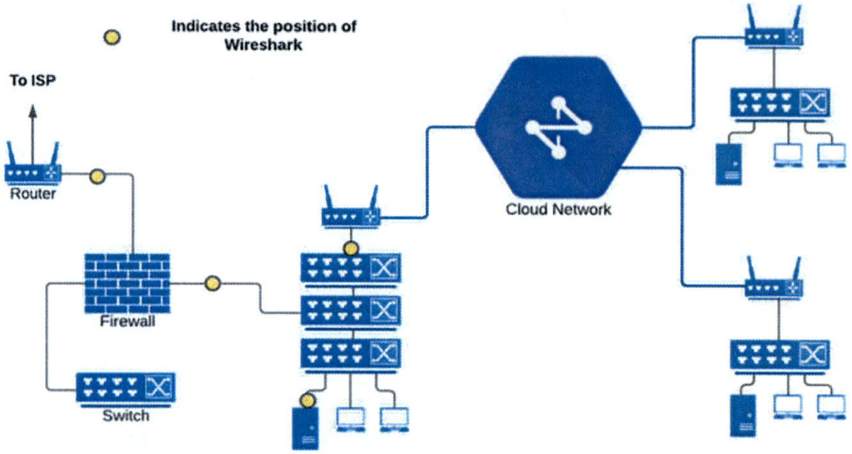

Fig. 1 Architecture of intrusion detection using Wireshark

4 Analysis of Intrusions on Wireshark

Wireshark is a network protocol analyzer which can be considered as an application having several tools. This tool helps us in analyzing the structure of network traffic in order to track errors in configuration and security attacks. It is compatible with most computing platforms and has rich features. Using Wireshark, one can troubleshoot the network, analyze the software and develop communication protocol. The various tools available in Wireshark help us capture, view, and analyze the data packets. There are packet filters in Wireshark which are used to take out some packets from huge volume of packets captured based upon specified search criteria. These filters act as micro-programming language for which the code is compiled and executed at runtime for the packets which are captured using tools such as tcpdump and Wireshark (OpenLogic, 2008).

The major advantage of the packet sniffing tool, Wireshark, is that it is an open source software with a very user-friendly GUI with color coding based on the protocol being used. Another good thing is that it is available for most modern operating systems. On the down side as with any open source software, it also has the disadvantage of not having any official support and hence the users have to rely on the open source community for any type of help.

Packets traveling across the entire network at any time can be captured using Wireshark. Log maintenance is the strongest feature of Wireshark over other IDS devices [20]. Figure 2 shows the network traffic captured by Wireshark interface. We are able to see the details of any packet by selecting that packet. From the header part, we can obtain source and destination IP address, protocol information, header length, and various other types of services. The data of the header field occurs in the decimal form whereas data of payload is displayed in hexadecimal form [24].

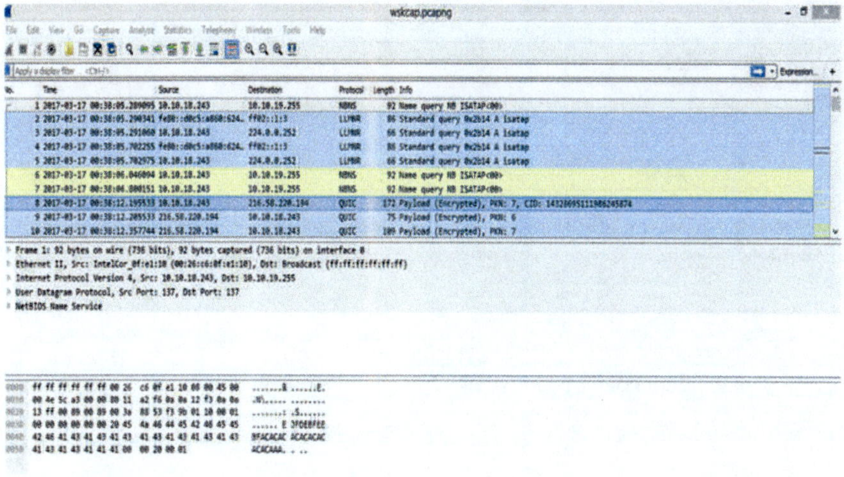

Fig. 2 Simple Wireshark capture

Authors in [25] have also studied various LAN attacks using Wireshark. Similarity measure has been used by authors in [26] for outlier detection.

Automatic generation of alarm in real time does not take place in Wireshark as in other IDS devices. Instead, the captured data is analyzed later on either manually or by using some applications. Intrusion information can be obtained in the following ways:

Expert information in Wireshark

The expert information gives us the log of anomalies. Expert information obtained by Wireshark is present in a capture file.

Figure 3 shows the expert information of our packet capture.

The contents of this expert information are:

Chat (grey in color): has the information about the normal flow of packets such as TCP packets with the SYN flag set.

Note (cyan in color): which contains any notable things, e.g., an error code.

Warning (yellow in color): which contains any warning, e.g., "unusual" error code as in the case of any connection problem.

Error (red in color): which shows any serious problem, e.g., Malformed Packet.

Intrusion detection using expert information

Chats are simply the steps or parts involved in any normal TCP connection. For usual TCP connection, the chat messages have the sequence of SYN, SYN + ACK, and ACK messages.

Figure 4 shows the sequence of TCP connection with all the three required steps, i.e., SYN, SYN + ACK, ACK messages which indicate normal connection. However, if there is any deviation in these steps like if there are only SYN messages observed,

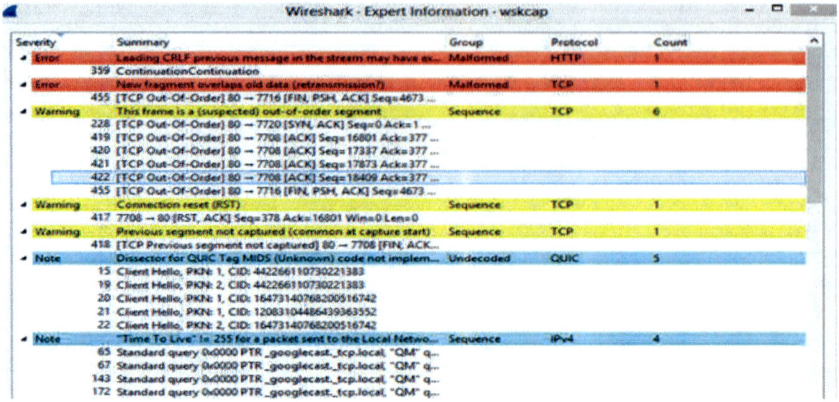

Fig. 3 Expert information

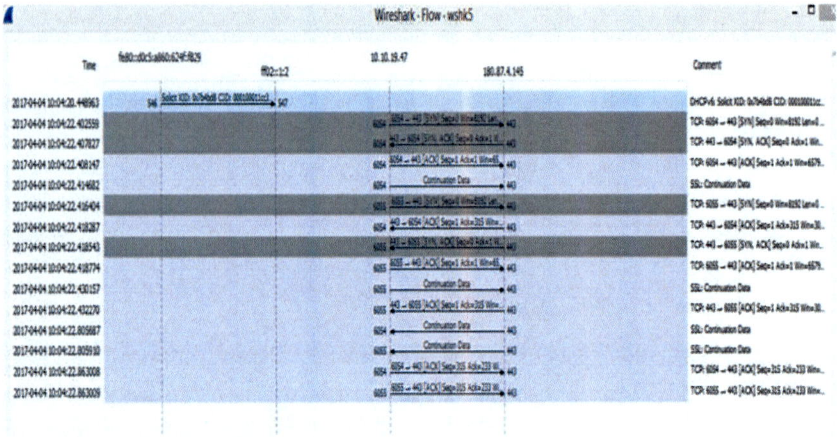

Fig. 4 Observing chat messages

there is a clear indication of DoS attack. In the case of DoS attack performed in our experiment, there is a continuous ping request shown as an echo message from the sender to the server.

The server due to DoS attack is not able to forward the request to the destination. In case of DoS attack, a flow graph as shown in Fig. 5 is obtained:

IO graphs on Wireshark

The IO graphs are the most important facility provided by Wireshark. IO graphs can be used to observe and compare the protocol performances and the normal packet flow can be observed by observing the peaks of the IO graph. Figure 6 shows the IO graph of the captured network traffic.

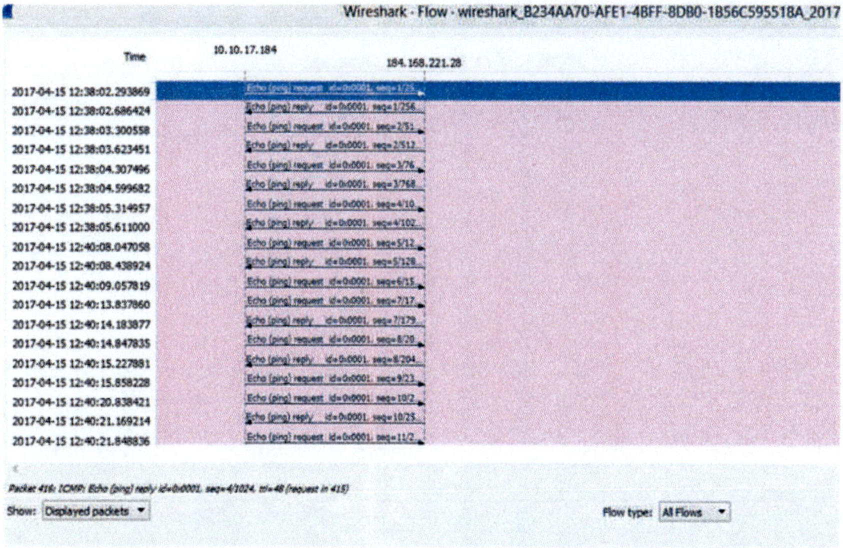

Fig. 5 Observing DoS attack

5 Conclusion and Future Scope

The aim of an administrator is not only to monitor if there is any network failure and fix the network problem on time but also to avoid failure of network due to outside threats. The information of network traffic is used to meet this requirement. Wireshark is used in this work for detection of intrusion in the network so that the data can flow in a safe manner. First of all, the information about the packet is stored in a log file, and then it is examined to classify the data as normal or abnormal. The I/O graph is used to depict the total traffic flowing in terms of packet per second or byte per second.

Wireshark has by now proved to be a very effective tool as a Network Protocol Analyzer. However, there are still many chances and capabilities of development in it. The developments can include alert generation and heuristic developments. Required work is being done to introduce these changes in Wireshark so that it becomes capable of alert generation for effective detection of many attacks well on time and thus making organizations and their data more secure.

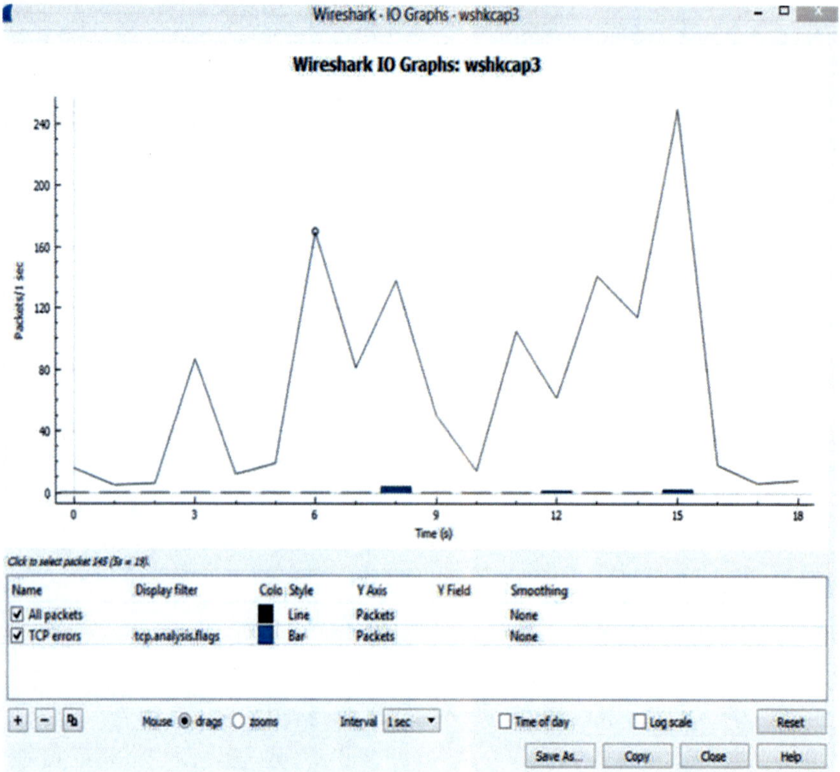

Fig. 6 Wireshark IO graph

References

1. E. Brown, NIST issues cloud computing guidelines for managing security and privacy. *National Institute of Standards and Technology, Special Publication* (2012), pp. 144–800
2. T. Halabi, M. Bellaiche, Towards quantification and evaluation of security of cloud service providers. J. Inf. Secur. Appl. **33**, 55–65 (2017)
3. S. Paquette, P.T. Jaeger, S.C. Wilson, Identifying the security risks associated with governmental use of cloud computing. Gov. Inf. Q. **27**(3), 245–253 (2010)
4. K. Barr et al., The VMware mobile virtualization platform: is that a hypervisor in your pocket? ACM SIGOPS Oper. Syst. Rev. **44**(4), 124–135 (2010)
5. J.C. Acosta, D.E. Krych, Hands-on cybersecurity studies: uncovering and decoding malware communications-initial analysis with Wireshark and volatility. *US Army Combat Capabilities Development Command Army Research Laboratory Adelphi* (2021)
6. M. Alqahtani, *Assessing Network Intrusion Detection System Performance: Forensic Implications* (Auckland University of Technology, 2013)
7. L.M. Vaquero, L. Rodero-Merino, D. Morán, Locking the sky: a survey on IaaS cloud security. Computing **91**(1), 93–118 (2011)
8. W. Huang, A. Ganjali, B.H. Kim, S. Oh, D. Lie, The state of public infrastructure-as-a-service cloud security. ACM Comput. Surv. **47**(4), 1–31 (2015)
9. S. Roschke, F. Cheng, C. Meinel, Intrusion detection in the cloud, in *2009 Eighth IEEE International Conference on Dependable, Autonomic and Secure Computing* (2009), pp. 729–734

10. G. Jain, Application of SNORT and Wireshark in network traffic analysis. IOP Conf. Ser. Mater. Sci. Eng. **1119**(1), 12007 (2021)
11. S. Pavithirakini, D. Bandara, C.N. Gunawardhana, K.K.S. Perera, B. Abeyrathne, D. Dhammearatchi, Improve the capabilities of Wireshark as a tool for Intrusion detection in DOS attacks. Int. J. Sci. Res. Publ. **6**(4), 378–384 (2016)
12. S. Naaz, F.A. Badroo, Investigating DHCP and DNS protocols using Wireshark. IOSR J. Comput. Eng. **18**(3), 1–8 (2016)
13. L.F. Sikos, Knowledge representation to support partially automated honeypot analysis based on Wireshark packet capture files, in *Intelligent Decision Technologies 2019* (Springer, 2020), pp. 345–351
14. K. Vieira, A. Schuler, C. Westphall, Intrusion detection techniques in grid and cloud computing environment, Proc. IEEE IT Prof. Mag. (2010)
15. P. Donadio, Virtual intrusion detection systems in the cloud. Bell Labs Tech. J. **17**(3), 113–128 (2012)
16. C. Ambikavathi, S.K. Srivatsa, Integrated intrusion detection approach for cloud computing, Indian J. Sci. Technol. **9**(22) (2016)
17. S.N. Dhage, B.B. Meshram, Intrusion detection system in cloud computing environment. Int. J. Cloud Comput. **1**(2–3), 261–282 (2012)
18. H.B. Baraka, H. Tianfield, Intrusion detection system for cloud environment, in *Proceedings of the 7th International Conference on Security of Information and Networks* (2014), pp. 399–404
19. C. Modi, D. Patel, B. Borisaniya, H. Patel, A. Patel, M. Rajarajan, A survey of intrusion detection techniques in cloud. J. Netw. Comput. Appl. **36**(1), 42–57 (2013)
20. V. Kumar, D. Sinha, A.K. Das, S.C. Pandey, R.T. Goswami, An integrated rule based intrusion detection system: analysis on UNSW-NB15 data set and the real time online dataset. Cluster Comput. **23**(2), 1397–1418 (2020)
21. E. Baraneetharan, Role of machine learning algorithms intrusion detection in WSNs: a survey. J. Inf. Technol. **2**(03), 161–173 (2020)
22. R. Zuech, T.M. Khoshgoftaar, R. Wald, Intrusion detection and big heterogeneous data: a survey. J. Big Data **2**(1), 1–41 (2015)
23. A.S. Ashoor, S. Gore, Importance of intrusion detection system (IDS). Int. J. Sci. Eng. Res. **2**(1), 1–4 (2011)
24. D.A. Bhosale, V.M. Mane, Comparative study and analysis of network intrusion detection tools, in *2015 International Conference on Applied and Theoretical Computing and Communication Technology (iCATccT)* (2015), pp. 312–315
25. H. Iqball, S. Naaz, Wireshark as a tool for detection of various LAN attacks. Int. J. Comput. Sci. Eng. **7**(05), 833–837 (2019)
26. H. Bin Fang, Outlier detection algorithm basing on similarity measurement relation, in *Advanced Engineering Forum*, vol. 6 (2012), pp. 621–624

Optimization of Cost of Smart Grid: A Case Study of Kutch, Gujarat Using HOMER Software

Neeraj Verma and Shelly Vadhera

Abstract This paper proposes economical method to reduce the cost of an area by using green energy technology, i.e., renewable energy. Here, this research work is more concerned with the optimization and simulation of a smart grid by the implementation of grid connected photovoltaic (PV) arrays, biomass and wind system at a same platform. The system also includes AC/DC converters and battery energy storage devices with suitable values. The major goal of this article is to evaluate various configurations of renewable resources with the grid in order to save costs. This study comprises a HOMER software-based simulation that produces multi-objective outcomes, with the primary goal of cost minimization and carbon dioxide emission reduction in the context of carbon economy. Hybrid optimization model for electric Renewables (HOMER) software is used here to analyze and identify the available information and data on a hybrid system in order to tackle the economic and environmental challenges present in an electrical network.

1 Introduction

The exploitation of renewable energy resources is increasing rapidly as the cost of electricity from the imported conventional fossil fuel energy sources becomes expensive and unsustainable. As a result, current researchers are seeking for renewable energy resources to provide energy in a cost-effective manner. The Indian government's ambitious Clean Energy Initiative aim of having renewable energy by 2022 has accelerated the development of renewable energy in India. Due to the current environmental conditions and energy crisis, there is a strong urge to investigate the renewable energy industry, since there is a large quantity of renewable energy accessible in Kutch, Gujarat, India. Renewable Energy Sources (RES) are on the verge of

N. Verma (✉)
School of Renewable Energy and Efficiency, National Institute of Technology, Kurukshetra, India
e-mail: neeraj_31910124@nitkkr.ac.in

S. Vadhera
Department of Electrical Engineering, National Institute of Technology, Kurukshetra, India

G. Ranganathan et al. (eds.), *Soft Computing for Security Applications*,
Advances in Intelligent Systems and Computing 1397,
https://doi.org/10.1007/978-981-16-5301-8_39

becoming the answer that meets the need of all stakeholders. In reality, major countries are enacting new legislation requiring the usage of renewable energy sources. By 2020, twenty-nine states in the US have set goals to produce between 20 and 33% of their power from renewable sources (Luis Valverde Felipe Rosa C. Bordons José Guerra) Similarly, India plans to increase renewable energy generation from 100 to 175 GW by 2022. It will be more useful to the country's economy if it is able to generate energy at a lower cost. The implementation of such sophisticated system necessitates the development of new control techniques that make the system "smart", resulting in the development of Smart Grid concept. This is followed by a summary of the approaches that are utilized to build dynamic models in order to study and refine the different controllers implemented within this work. Then, the implementation and validation of the developed Energy management techniques using the new lab scale micro grid are discussed. Experimental results are shown by comparing the response of simple strategies against complex online optimization techniques such as the Model Predictive Control [MPC]. The difference among different approaches is also extensively discussed. The results and findings demonstrate how alternative control strategies may have a significant impact on plant performance and also we conclude with a set of principles for developing and managing Smart Grids. This paper is dedicated to establish a location study in Kutch, Gujarat, India, by using each sort of power load and data acquired from GUVNL as average load.

2 System Description

In the HOMER software firstly, we need to collect the data of location for which we have to simulate the system. As per the requirement of Kutch Gujarat, India location (23°44.0′ N, 69°51.6′ E), which is a region of the Indian desert where abundant solar radiation, wind, and bio waste make it possible to build a renewable smart grid.

2.1 Load Profile

This paper considers all the domestic industrial load and municipal load as a combined average load. As per requirements of area, the average load requirement is 2424.24 kW/d and peak load demand is 348.08 kW.

2.2 System Components

This system retains the load according to local requirements and is connected by an AC/DC convertor string that converts the produced energy into useful form. On the DC bus bar, photovoltaic PV system is connected in order to convert thermal energy

Table 1 System components description

Components	Name	Size	Unit
Generator	Generic 500 kW biogas genet	500	kW
PV	Generic flat plate PV	String	kW
Storage	Generic 1 kW lead acid	String	kW
System convertor	System convertor	String	kW
Grid	Grid	999,999	kW
Wind turbine	Generic 10 kW	String	
Dispatch strategy	HOMER cycle charging		

from solar radiations into electrical energy, which is stored in the string of generic 1kWh Lead Acid batteries with which the stored energy can be used during nighttime. On the other hand, at the AC bus bar generic 500 kW biogas genet generator is using the available biomass to the supplying system and a windmill string of 10 kW is also placed by using the wind speed of producing energy and supplying to load. Also, an infinite bus grid supply is also connected to system as a backup resource. The prepared system consists of the following tools shown in Table 1 (Fig. 1).

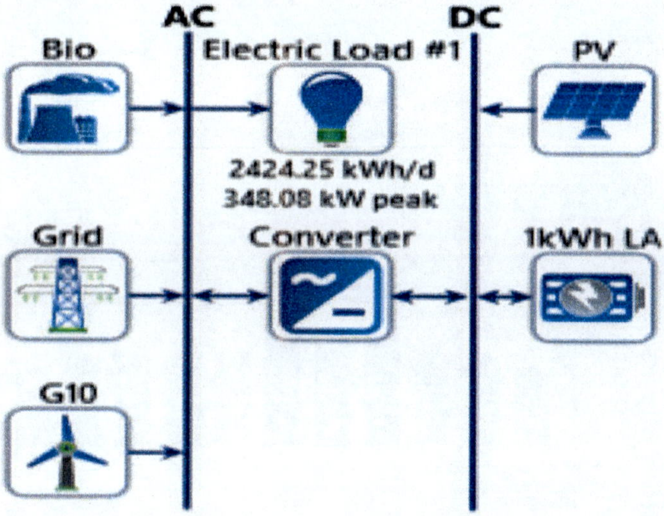

Fig. 1 Schematic diagram

2.3 Solar Radiation, Wind Speed and Biomass

According to NASA research, Gujarat is the state in India with the highest intensity of solar radiation, henceforth the installation of a PV array string in this location is considered as the most important instrument for producing green energy. In Kutch, Gujarat, the solar radiation intensity is 5.21 (kWh/m^2/day), which is the best optimal intensity for producing energy. These data are collected from NASA weather satellite resources and fed into the homer software.

The requirement of windmill to generate electricity with a minimum wind speed is 3.7 m/s^2 but as of desert area in that area, average wind speed is 4.12 m/s^2, which provides the efficient output from the system. All this wind speed data is also collected from the NASA weather satellite resources and fed into HOMER software. The biomass availability is 4.5 tons/day in this system (Figs. 2, 3 and 4).

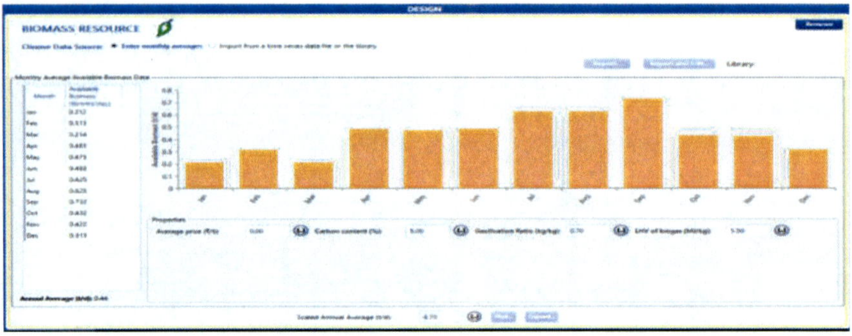

Fig. 2 Graph between available biomass (t/d) and month

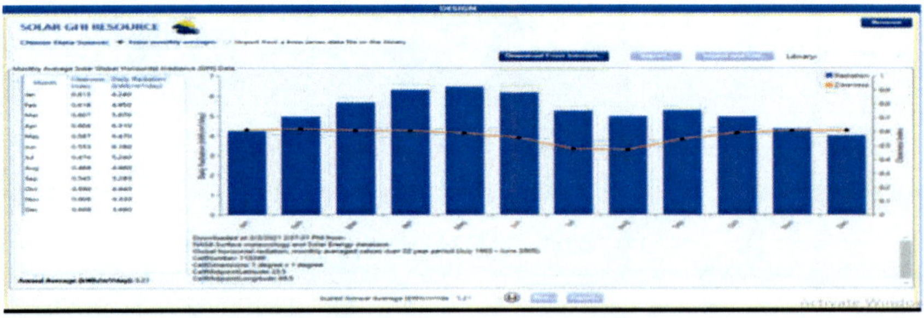

Fig. 3 Graph between daily solar radiation intensity (kWh/m^2/day) and month

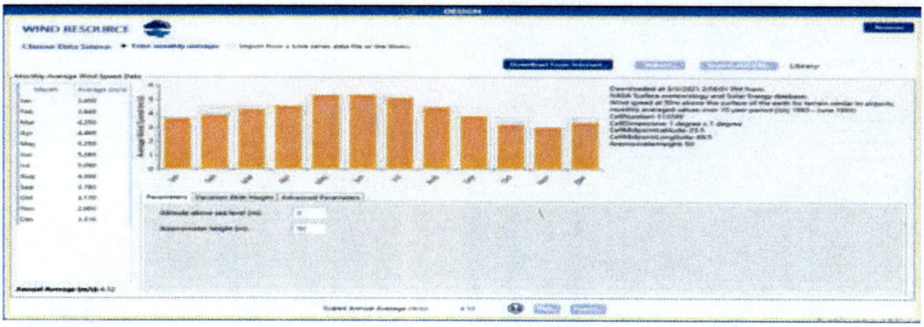

Fig. 4 Graph between average wind speed (m/s) and month

3 Methodology

In this HOMER software, suitable values of renewable resources are inserted to fulfill the load requirements in an economical and effective manner. In the simulation, software works on the possible ways to reduce the cost of electricity and thereby it results in developing a good carbon economy. During simulation, the proposed software combines all the available data on the wind speed, solar radiation, and biomass quantity at one part and on the other hand, it holds the values peak loads and average loads (yearly/monthly/weekly/daily/hourly).

In HOMER software, the inserted strings of PV array, wind, and other resources' string software pick the suitable string of resources, which obtains the best economical results. During simulation, each and every resource is picked in a suitable combination for providing effective data concern with the advanced economic and environmental aspects.

Simulation gives the information about resources with their prescribed configuration, which may provide the required economic and carbon dioxide emission results. Software presents the results based on all the configurations in a tabular form as shown below (Figs. 5 and 6).

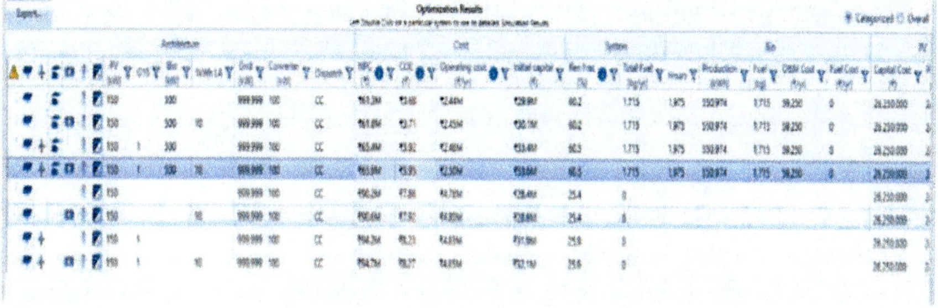

Fig. 5 Optimization results in tabular form

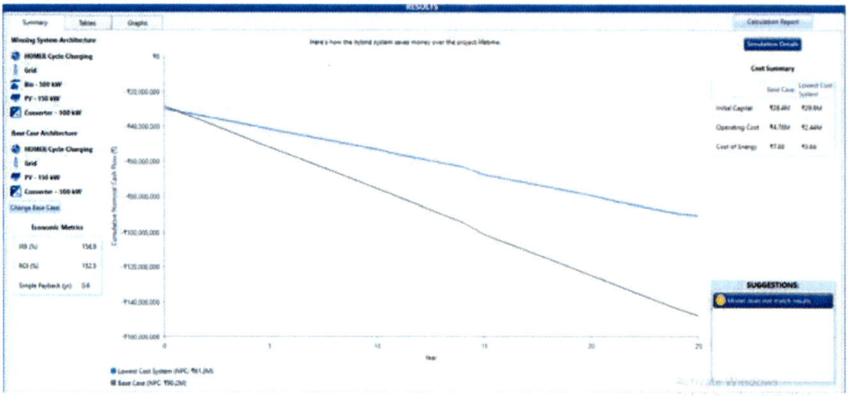

Fig. 6 Graph between cumulative nominal cash flow (rupees) and year

As the system gives the different configuration results for different combinations of resources, it makes the comparison easier. For these paper requirements, three cases are beneficial in order to study which is able to reduce COE (cost of electricity) and harmful gas (carbon dioxide) emission. As per the environmental consideration, carbon dioxide emission plays an important role in selecting suitable configurations. This paper compares the COE and carbon economy of three case configurations.

Case 0.1

This case configuration includes the Biomass generator, PV and grid purchased electricity. Most of the energy production part is taken from generic 500 kW biomass generator is 42.1% of total produced power. In this case, the COE is reduced to 3.68/kW and carbon dioxide emission is 324877 kg/year, which is the most economical configuration from the simulation results. Here, the generic 500 kW biogas genet is producing 550,974 kW/year. As biogas plant is the higher producer at this level, it consumes a total feedstock of 1715 tons, where the average feedstock consumption per day is 4.7 tons with 0.196 tons/h.

In this configuration, the production contribution of PV string of 150 kW is 18.7% and units obtained from grid are 39.2%. The operating hours for PV system are 4378 h/year, where the average solar radiation is 5.21 (kWh/m^2/day) and it produces 2, 45,013 kW/year. In order to fulfill the storage purpose, as per simulation results, the system requires 10 parallel strings of lead acid batteries (Fig. 7).

Case 0.2

This case is focusing on environmental aspects and attempts to reduce carbon emission by considering the windmill generating station or vice versa with an increase in the COE (cost of electricity). Case 1 is not considering the windmill because it increases the cost of electricity from RS 3.68 to RS 3.95/kWh but it has the ability to reduce the carbon emission from 324,877 to 322,868 g/year. The windmill reduced the burden of grid purchasing units and biomass generator loads. The contribution

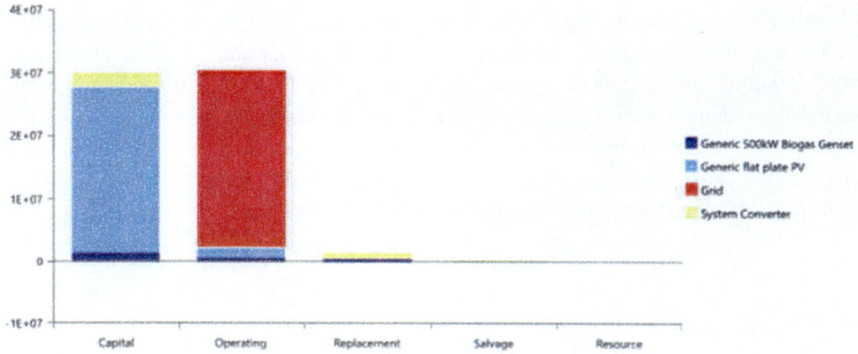

Fig. 7 Graph between cost components and energy contribution for Case 1

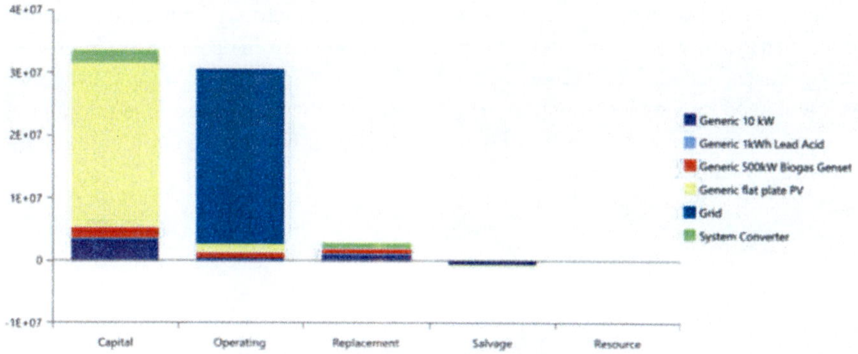

Fig. 8 Graph between cost components and energy contribution for Case 2

of windmill is 0.324%, where the biomass contribution is reduced to 42% and grid purchasing contribution to 38.9%. The PV is the most consistent and eco-friendly technology since its contribution remains the same at 18.7%. In this configuration, the windmill contribution is small but it shows the ability of windmill to overcome the cost of electricity, which is incremental by carbon economy (Fig. 8).

4 Results

This paper has presented all the possible efforts in simulation to get the most economical, eco-friendly, and optimized model. As per state electricity reports, the COE in Kutch, Gujarat is RS 4.35/kWh. But using this system configuration, we are able to reduce the COE to RS 3.68/kWh. From the tabular form of simulation results, case 1 is providing the most economical approach with COE is RS 3.68 as per the economical concern and environmental concern. As observed from the second scenario, current technology in the field of renewable energy predicts an inverse relationship between COE and carbon emissions. Implementation of windmills and other resources like hydrogen fuel cells in the smart grid will be able to reduce carbon dioxide and other harmful gases emission with an advantage of improved carbon economy. But on the other hand, the cost of electricity is also increased. The future goal is to reduce the production cost of these resources in order to obtain the reduced values in both the COE (cost of electricity) and carbon emission.

5 Future Scope

According to current energy needs, the future will necessitate massive amounts of energy production and quick response energy, such as hydrogen energy. Hydrogen energy is a type of energy with a high calorific value that may be used to generate green energy. However, current issues with safety and cost in dealing with hydrogen energy pose a barrier for smart grid. Authors are motivated towards hydrogen energy as it will become the most popular form of energy in the near future. The future goal is to reduce the production cost of Hydrogen energy and windmill in order to obtain the reduced values of both the COE (cost of electricity) and carbon emission.

References

1. M. Nurunnabi, N.K. Roy, *Grid Connected Hybrid Power System Design Using HOMER*
2. P. Kumara, R. Pukaleb, N. Kumabharc, U. Patild, *Optimal Design Configuration Using HOMER*
3. A. Singh, P. Baredar, B. Gupta, Computational simulation and optimization of solar cell, fuel cell and biomass hybrid energy using homer pro-software
4. L. Valverde, F. Rosa, J. Guerra, Energy management strategies in smart grid: a laboratory experience
5. www.nrel.gov/homer
6. Getting Started Guide for HOMER Version 3.3.1. https://www.homerenergy.com
7. NASA. Surface Meteorology and Solar Energy. Available online: https://eosweb.larc.nasa.gov/sse/
8. NASA. Surface Meteorology and wind Energy. Available online: https://eosweb.larc.nasa.gov/sse/
9. www.gercin.org>2020/03
10. www.irena.org>publication
11. www.En.m.wikipedia.org>wiki>smallwindturbine
12. L. Valverde, F. Rosa, C. Bordons, J. Guerra, Energy management strategies in hydrogen smart grids: a laboratory experience. Renewable Energy Sources (RES) are poised to be the solution that satisfies all parties. In fact, major countries are implementing new regulations that make the use of renewable sources mandatory. The implementation of such sophisticated systems drive to find out new control techniques that make the system "smart", bringing the Smart Grid concept. This is followed by a summary of the approach used towards obtaining dynamic models to study and refine the different controllers implemented within this work

13. https://energy.gov/sites/prod/files/oeprod/DocumentsandMedia/DOE_SG_Book_Single_Pages%281%29.pdf
14. www.economicstimes.com>power. Likewise, India aiming to attain 175 GW energy productions from renewable energy field which is 100 GW at present by the year 2022

Agriculture Protection from Animals Using Smart Scarecrow System

Rahul Mapari, Kishor Bhangale, Laukik Deshmukh, Prashant Gode, and Ankit Gaikwad

Abstract As our country's subsistence farmers continue to encroach on natural ecosystems, the interaction between farmers and animals grows, resulting in conflicts. We investigated how raiding patterns fluctuate with different seasons, farm land, and crop varieties. A smart scarecrow system has been proposed here to minimize crop raiding and man animal conflict from wild animals and birds. The scaring system works in three parts, such as object detection, message sending, and automatic height adjustment. Object detection is done by the PIR sensor and preventive action has been taken by initiating a servo, speaker, and blinking light. After object detection, alert will be sent by IoT cloud to the farm owner. Depending upon seasonal cropping, the pattern height of the system gets automatically adjusted by using motor, string pulley, and sliding drawer mechanism after the activation of ultrasonic sensor. The proposed study obtains the samples from different farms that are combined and observed how the object detection vary day and night in three seasons in order to provide the overall system efficiency. It is more convenient and cost effective than traditional scaring strategies like trapping, hunting, and wood fencing. The present system is made up of metal body, which does not require any manpower and at the same time, it can work under any worst climate conditions.

Keywords Precision agriculture · Smart scarecrow · Object detection · Smart farming · IoT cloud · Servo motor · Ultrasonic sensor · Wireless sensor network · Agriculture field monitoring

---

R. Mapari (✉) · K. Bhangale · L. Deshmukh · P. Gode · A. Gaikwad
Department of E&TC, Pimpri Chinchwad College of Engineering & Research, Ravet, Pune, India
e-mail: rahul.mapari@pccoer.in

K. Bhangale
e-mail: kishor.bhangale@pccoer.in

G. Ranganathan et al. (eds.), *Soft Computing for Security Applications*,
Advances in Intelligent Systems and Computing 1397,
https://doi.org/10.1007/978-981-16-5301-8_40

1 Introduction

The growing confrontation between animals and farmers, as well as crop raiding, is one of our fundamental issues. It has existed since the beginning of agriculture. Recently, Tamil Nadu has reported a total of 7562 incidents of crop raiding by wild animals in the last three years. There are many different situations and reasons, where wildlife-farmer come into conflict. Some of the reasons for that are in search of water, livestock predation, increase in human populations, increase in agriculture and reduction in forest land, the availability of palatable and nutritious foods near farm edges. As per 2018, agriculture employed more than 50% of the Indian work force and contributed 17–18% to country's GDP 70% of country's income is from agriculture and rank 2nd world wide in farm output so technological advancement is needed in this sector too [1–3]. For that farmers utilize strategies and traditional methods that are often cruel and ineffective. While arbitrary killing, trapping, or poisoning of suspect may provide short-term satisfaction but it fails to address long-term needs. So we must seek to understand sociological, economic, and cultural aspects to find solution. Researchers suggest managing pastures to reduce competition for forage between wildlife and domestic livestock [4–7].

In addition, shifting from farming system with perennial crops may reduce losses. But this cannot be the proper solution for the problem. In general, crop damage by monkey and wild-boar is more serious in the northern part epically in lower altitude areas. So farmers usually adopt some crop protection strategies guarded their crop by spending night out in field, wood fencing whereas wealthier farmers used imported barbed wire, trapping, hunting are popular. Use of plastic flags, brightly colored objects, scents, and fireworks work sometimes but again that is marginally successful. Some make system consists of sound clips for different animals and red lamp connected to stick which is not durable [8–10]. There are smartphone controlled systems but there was no provision of movement. For birds, unmanned aerial system (UAS) is used to deter birds. To detect birds, background subtraction algorithms have been used [11]. For security and detection, some use sensor cloud based architecture implemented using IoT. IoT provides extensive use of technology over the remote places and bringing automation [12, 13]. Many systems developed which detects the intruders, monitors any malicious activity, and then reports it to the owner of the system but these are unable to fear the wild animals and primates [14, 15]. There is by no means an easy solution to this problem. Wild animals nowadays are smart enough as they learn to overcome obstacles such as fences and scarecrows. So there is need to develop automation system in agriculture field in order to avoid man animal conflict and crop raiding [16–18].

So one of the technological solutions for this problem is smart scarecrow system, which uses high intensity light, sound sensor, arm movement to restrict the entry of the animal as well as birds. The common form of scarecrow is a humanoid structure dressed and placed in open field to discourage the birds and animals from disturbing and feeding on recently cast seeds and growing crops. This system consists of PIR sensor for object detection and then the servo motors perform motion operation in scarecrow arm and firecracker sound buzzer is used to demotivate the animal as well as birds and for the night time, high intensity lights placed on head is used to keep animals away from farm field. Also, the system consists of ultrasonic sensor for detecting the height of crop depending upon the different seasonal cropping patterns based on that the scarecrow height gets automatically set using motor and after object detection, the message will be conveyed to the farmer. Scarecrow will basically work in two different phases, where first is object detection and message transfer and second is automatic height adjustment using pulley arrangement. In the first phase after switching on the system, the PIR sensor gets activated and it is connected to the microcontroller once the object is detected by the sensor, the message is sent to the controller depending upon this controller, it activates other three peripherals, where two servo motors are connected on two sides of system, which works as the arms of scarecrow and normally it rotates with angle 120°, second is the activation of firecracker sound sensor and last is the activation of high intensity light at night time, furthermore the day and night is configured by LDR while doing this the alert is given by using IoT, which uses cloud platform for connecting sensors data with farmer mobile application. This is a very cost effective and secure way to send the message and also it stores monthly data on firebase for further manipulation [12]. In the second phase, ultrasonic sensor is connected to the microcontroller when any crop is detected in front of it, followed by that the message is given to the controller to activate the motor, which is placed at bottom of the system and further the motor is connected by string using pulley mechanism. When motor rotates clockwise and anticlockwise direction, it wound and relax the string so that the system will be able to adjust its height.

The rest of the paper is structured as follows. Section 2 gives the information about proposed scaring methodology for animal detection and prevention of farms by discussing the height adjustment in detail. Section 3 provides the specifications of the system components. Section 4 provides the flow chart and algorithms followed by experimental results and discussion. Finally, the last section presents the conclusion of the proposed work.

2 Proposed Methodology

In this paper, smart scarecrow system uses irritating sound, high intensity lights, and arm movement to fear the animal, as well as birds and automatic height adjustment mechanism, is also incorporated to increase the accuracy in object detection. The flow diagram of the proposed system is shown in Fig. 1. This system consists of PIR sensor for object detection. After detection, other three peripherals are activated and the servo motors present here performs the motion operation of scarecrow arm and also a firecracker sound buzzer is used to demotivate the animal as well as birds and for the night time, high intensity lights are placed on head to keep animals far away from the farm field. Simultaneously, the detection message is instantly sent by using IoT cloud platform to the farm owner's smart phone. Ultrasonic sensor is present here

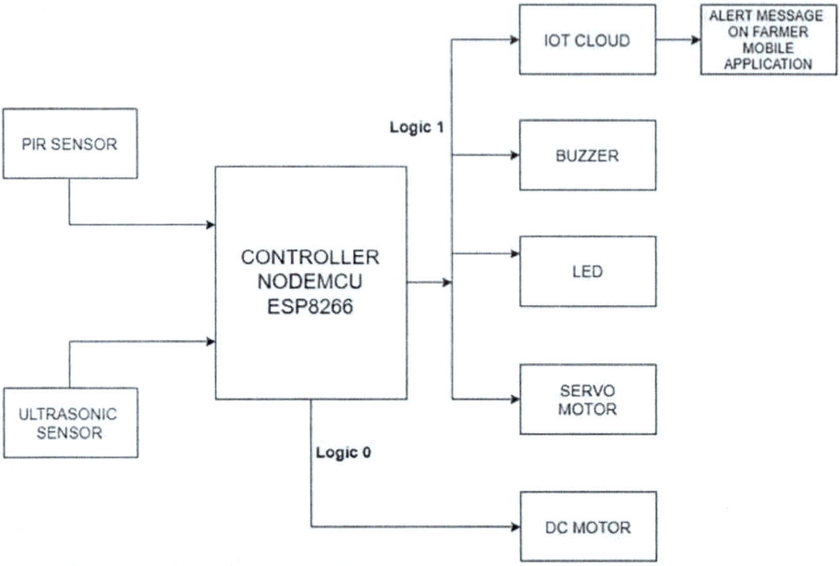

Fig. 1 Flow diagram of proposed system

for the detection of the height of crop depending upon different seasonal cropping patterns and based on that the scarecrow height gets automatically set by using pulley and motor mechanism. Nowadays, wild animals are smart enough to overcome the obstacles such as fences and scarecrows. This system utilizes the available technology to reduce labor. The automatic height adjustment concept is used here to detect the object clearly by varying the height of the crops. And this is achieved smoothly by using pulley arrangements. Main body of the system is made up of hollow metallic box in which the controller circuit is placed so that the system can work in different climatic conditions. Different scarecrow systems are available but these are very specific to particular animals or birds but the present system attempts to develop a scarecrow by considering most of the animals and birds. The whole operation is controlled by peripheral interface controller (NodeMCU ESP8266), which is fed by input power supply. Controller always monitors PIR and ultrasonic sensors for surveillance. When PIR sensor detects any type of infrared energy then it gives instructions to the controller in order to active the successors such as buzzer, high intensity light, and servo motor.

IoT model informs the owner that the object is detected and followed by that the buzzer will make a loud noise to scare the animals, high intensity light will help to visualize the field even in a fog condition, and finally servo motors are attached to the two sides of model so that it will rotate at a certain angle, which will look like the arms of the scarecrow (movement purpose only). When ultrasonic sensor detects any increase in crop height it gets activated and it will turn on the DC motor to rotate at a certain angle to align the height of bot higher than the object (crops).

Table 1 Specifications of the components

Component	Parameter	Specifications
DC motor	Rated voltage shaft diameter motor length rated	12 V DC 6 mm, 63 mm. 10
	RPM weight	180 gm 7.5A (Max.)
	Load current	800 mA (Max.)
	No load current torque	Up to 120 kg cm
	Motor diameter shaft length	28.5 mm 30 mm
Servo motor	Model weight	MG99555 gm
	Operating voltage	4.8–7.2 V
	Stall torque @4.8 V	13 kg-cm
	Stall torque @6.6 V	15 kg-cm
Ultrasonic sensor	Operating voltage	5 V
	Sonar sensing range	2–400 cm
	Max. sensing range	450 cm
	Frequency	40 kHz

3 System Components and Specification

The system components required for the implementation of the system and its specifications are given in Table 1.

4 Flow Chart

Here, the first step shows initialization of the system by switching on the supply. In the next step, the controller takes decision depending upon the sensor's detection. The next step defines the two operations, depending upon the sensor activations different peripherals are connected to controller. In this step, after PIR, four operations are performed, they are intensive blinking light, buzzer, servo motor, and firebase data changes and after completing these processes, the message is conveyed to the farmer's mobile application. Further, the DC motor is activated for automatic height adjustment of scarecrow after ultrasonic sensor detection, which shows the completion of the operation, whereas the last step shows the termination of the process. The flowchart of operation is as shown in Fig. 2.

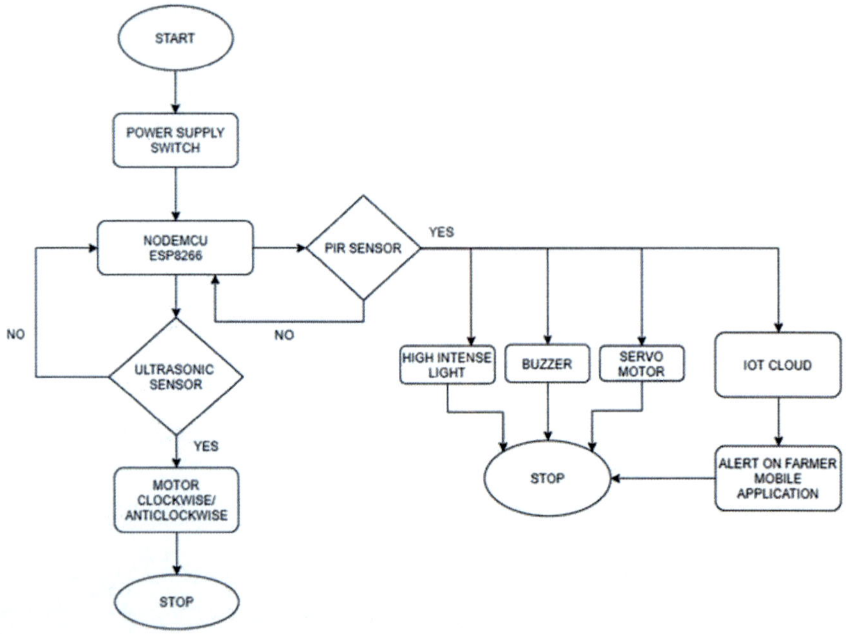

Fig. 2 Work flow chart of smart scarecrow system

5 Results and Discussion

This section provides the extensive experimentation of the scarecrow system in different environmental conditions.

5.1 Object Detection by PIR Sensor

As the object came in the detection range of PIR sensor, the PIR sensor will pass this activated signal to the controller in order to perform further operation. Table 2 shows the performance of the system by keeping the animal prototype in the detection range and out of the range.

5.2 Auto Height Adjustment

As from ultrasonic sensor, the controller receives the activated signal with auto height adjustment function, which will start performing by using motor, pulley, and belt. Table 3 shows the height adjustment scenario of the scarecrow robot. Table 3 shows

Table 2 Object detection by sensor at different distances

S. no.	Animal in detection range (m)	PIR sensor output
1	1	Object detected
2	2.58	Object detected
3	3.5	Object detected
4	5	Object detected
5	7.3	Object detected
6	8.2	Object detected
7	12	Object not detected
8	14	Object not detected

Table 3 Different height adjustment using motor depending upon sensor output

S. no.	Ultrasonic sensor detection (cm)	Height adjustment (cm)
1	Crop height increases by 15	System will increase its height by 20
2	Crop height increases by 20	System will increase its height by 25
3	Crop height increases by 25	System will increase its height by 30
4	Crop height increases by 30	System will increase its height by 35
5	Crop height increases by 35	System will increase its height by 40
6	Crop height increases by 40	System will increase its height by 45
7	Crop height increases by 45	System will increase its height by 50

the performance of the system for increasing the height of the crop. The proposed system is implemented in the farm consisting of cotton crops.

5.3 Graphical Representation of Season Wise Object Detection

In our country, different seasons and cropping patterns are responsible for attracting various animals as well as birds towards farm fields. So depending upon three seasons, the attempts of crop raiding varies. And also the differences on day and night time are also noted. To study this, the proposed system is implemented on three dissimilar farm lands with different types of crops and after averaging the sensor outputs, some results are obtained as shown in the form of responses below. Following graphs show the object detection at different time intervals in a particular season. Table 4 describes the variation in number of raiding attempts from different farm lands and seasons, which can be clearly observed. Here, 24 different readings are taken per day from three fields, and later the detection message is conveyed by IoT cloud. Monthly data of detection is stored on cloud and after manipulating these samples, i.e., for winter we take four months data from November to February by considering 30 samples from each month similarly for summer and monsoon and it will be combined for graphical representation.

Figure 3 implies the variation in animal detection at day and night in winter season. Here, animal detection is low after 12 am because of heavy winter but it starts increasing from sunrise, and around 12–5 pm it increases up to 82% as shown in graph, which means in day time the animal detection is maximum especially in winter season. And again it starts falling for night time. It is observed that in summer season major crop raiding is done during night time and early in the morning, i.e., performing object detection by using sensor is best-suited from 12 to 9 am duration and afterward it will decrease drastically up to 20% shown in response because of high temperature but from sunset again it will increase. The proposed system shows variation in animal detection in day and night in monsoon season. Object detection increases after sunrise and at day time between 1 and 7 pm and also at the time of clean and dry climatic conditions but if there is cyclonic wind and heavy rain, the sensor accuracy gets decreased and proper animal detection is not easily possible.

Hence, from the above three responses, different patterns of object detection at day and night are observed in Fig. 3. It is observed that maximum raiding attempts are detected by using our proposed system efficiently in winter. In summer, most attempts are detected to a maximum of up to 80%. In monsoon season, the attempts are effectively detected in good weather conditions with a maximum of up to 55%, henceforth its rate is less as compared to winter and summer seasons.

Figure 4 represents the monthly efficiency of the system. After studying this, it is observed that the smart scarecrow system performs its operations very efficiently in winter season as compared to summer and monsoon. In summer due to high temperature, sensors range reduces due to the difference between background temperature and body temperature. Similarly, in monsoon due to the bad weather conditions like high winds, thunderstorms, snowstorms affect the accuracy of sensor and overall system.

Figure 5 Shows the photograph of actual proposed system, here the height adjustment using mechanical pulley arrangement and sliding drawers are shown particularly Fig. 5a represent low level position of scarecrow for better detection of object when there is a low crop height and Fig. 5b represents the middle level position of scarecrow for the better detection of animals when crop height is moderate

Table 4 Readings of raiding attempts detected by system are taken from three farm lands

S. no.	Object detection in day and night by system	Season wise crop raiding attempts in three different farm lands in India with cultivation of various types of crops								
		Plain farm land			Irregular farm land			Terrace (hilly) farm land		
		W	S	M	W	S	M	W	S	M
1	1:00 AM	4	9	5	3	7	2	2	6	0
2	2:00 AM	3	8	4	4	8	1	3	7	1
3	3:00 AM	5	8	2	3	8	0	2	6	1
4	4:00 AM	2	10	1	2	9	1	3	6	4
5	5:00 AM	0	9	2	4	11	0	4	9	3
6	6:00 AM	4	7	5	4	6	4	3	8	2
7	7:00 AM	3	7	3	3	7	5	5	6	3
8	8:00 AM	6	5	6	5	6	4	5	5	5
9	9:00 AM	8	6	7	7	7	5	7	5	6
10	10:00 AM	7	5	7	8	5	5	8	5	6
11	11:00 AM	10	4	6	9	4	6	9	5	7
12	12:00 PM	12	4	5	9	5	7	8	4	6
13	1:00 PM	6	3	5	8	3	6	8	3	8
14	2:00 PM	6	5	7	5	4	6	6	2	6
15	3:00 PM	8	4	9	4	4	5	5	2	3
16	4:00 PM	9	6	8	9	7	9	7	5	4
17	5:00 PM	10	7	6	11	8	7	6	6	3
18	6:00 PM	5	8	6	5	8	4	4	10	2
19	7:00 PM	4	7	7	3	6	3	3	5	2

(continued)

Table 4 (continued)

S. no.	Object detection in day and night by system	Season wise crop raiding attempts in three different farm lands in India with cultivation of various types of crops								
		Plain farm land			Irregular farm land			Terrace (hilly) farm land		
		W	S	M	W	S	M	W	S	M
20	8:00 PM	4	6	5	3	8	0	4	6	3
21	9:00 PM	3	9	4	1	9	1	4	5	2
22	10:00 PM	1	11	2	1	7	0	2	6	1
23	11:00 PM	3	8	1	2	7	2	3	4	0

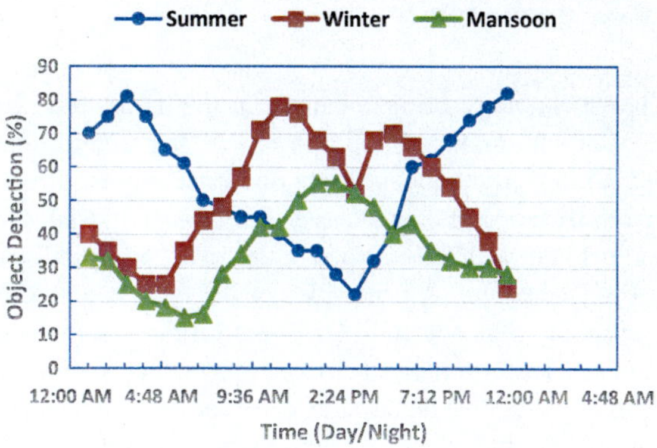

Fig. 3 Rate of object detection by PIR sensor on various time intervals in various season

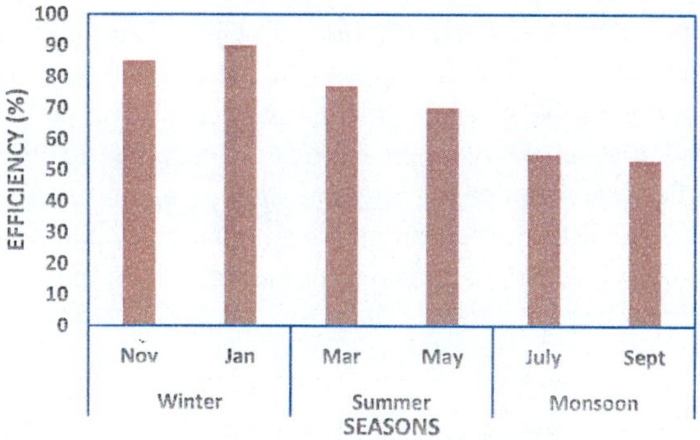

Fig. 4 Efficiency of the proposed system in three seasons

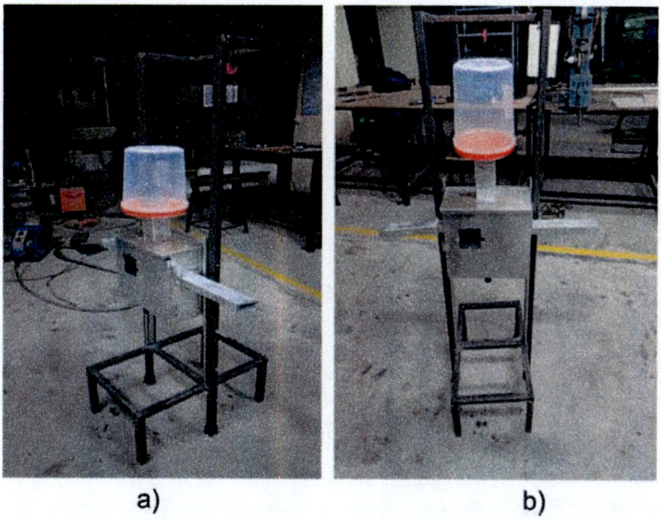

Fig. 5 Scarecrow model design (**a**) and **b** Height adjustment of the scarecrow robot

6 Conclusion

The technology brings changes in the agricultural sector in India. The scarecrow system can minimize the loss in the farm and ultimately provides a safer growing condition. Night out spending on farm is not required for crop protection. Detection from distance and preventive detection is possible by using this system. Height adjustment with the help of a pulley in the proposed system helps to leverage accurate object detection. All the sides of the farm can be covered by connecting two PIR sensors to the controller. Outer part of the system is made up of metallic boxes and rods; henceforth it is not easily damageable by wild animals. It will be more efficient and cost effective than traditional guarding methods. In general, the damage of crops by wild animals is more serious in the northern part, especially in lower altitude areas. Traditional methods of fencing are ineffective so in such a situation, Smart scarecrow system can resolve the problem. But, in summer and monsoon seasons, the efficiency slightly lowers summer due to the reduction in the range of high temperature sensors and intensive climatic conditions affect system performance in monsoon. This is a conflict-resolution approach that is amicable to both humans and animals. Further, the system can be improved by employing human intrusion detection techniques to distinguish between animals and humans [19, 20]. Also, further efficient routing and node placement is required for proper monitoring of agricultural fields.

References

1. C. Sillero-Zubiri, D. Switzer, *Crop Raiding Primates: Searching for Alternative, Humane Ways to Resolve Conflict with Farmers in Africa* (Wildlife Conservation Research Unit, Oxford University, Oxford, 2001)
2. S.W. Wang, P.D. Curtis, J.P. Lassoie, Farmers perceptions of crop damage by wildlife in Jigme Singye Wangchuck National Park, Bhutan. Wildl. Soc. Bull. **34**(2) (2006)
3. J. Beringer, K.C. Ver Cauteran, J.J. Millspaugh, *Evaluation of Animal-Activated Scarecrow and Monofilament Fence for Reducing Deer Use of Soybean Fields* (USDA National Wildlife Research Centre, 2003)
4. K. Sabic, *Human-Wildlife Conflicta in the Nanda Devi Biosphere Reserve* (Uttarakhand, India, 2011)
5. K.M. Lee, *Study on a Smart Phone Application Controlling Harmful Animals Scaring System Applicable to Farm Protection* (WCECS, 2016)
6. S. Goel, S. Bhusal, M. Taylor, M. Karkee, *Detection and Localization of Birds for Bird Deterrence Using UAS* (ASABE Washington, 2017)
7. C. Pornpanomchai, M. Homnan, N. Pramuksan, W. Rakyindee, Smart scarecrow, in *2011 Third International Conference on Measuring Technology and Mechatronics Automation*, vol. 3 (2011), pp. 294–297
8. S. Yadahalli, A. Parmar, A. Deshpande, Smart ıntrusion detection system for crop protection by using arduino, in *2020 Second International Conference on Inventive Research in Computing Applications (ICIRCA)* (IEEE, 2020), pp. 405–408
9. R.N. Brown, D.H. Brown, Robotic laser scarecrows: a tool for controlling bird damage in sweet corn. Crop Prot. **146**, 105652 (2021)

10. U. Manral, S. Sengupata, S.A. Hussain, S. Rana, R. Badola, Human wildlife conflict in India: a review of economic ımplication of loss and preventive measures. Indian For. **142**(10), 928–940 (2016)
11. A. Bashar, Sensor cloud based architecture with efficient data computation and security implantation for Internet of Things application. J. ISMAC **2**(02), 96–105 (2020)
12. P. Biradar, P. Kolsure, S. Khodaskar, K.B. Bhangale, IoT based smart bracelet for women security. Int. J. Res. Appl. Sci. Eng. Technol. (IJRASET) **8**(11), 688–691 (2020)
13. R. Sarraf, S. Ojha, D. Biraris, K.B. Bhangale, IoT based smart quality water management system. Int. J. Adv. Sci. Res. Eng. Trends **5**(3), 12–16 (2020)
14. V. Bindhu, Green cloud computing solution for operational cost efficiency and environmental impact reduction. J. ISMAC **1**(02), 120–128 (2019)
15. S.R. Chourey, P.A. Amale, N.B. Bhawarkar, IoT based wireless sensor network for prevention of crops from wild animals. Int. J. Electron. Commun. Soft Comput. Sci. Eng. (IJECSCSE), 57–60 (2017)
16. P. Navaneetha, R. Ramiya Devi, S. Vennila, P. Manikandan, IoT based crop protection system against birds and wild animals attacks. IJIRT **6**(11) (2020)
17. S. Giordano, I. Seitanidis, M. Ojo, D. Adami, F. Vignoli, IoT solutions for crop protection against wild animal attacks, in *2018 IEEE International Conference on Environmental Engineering (EE)* (2018), pp. 1–5
18. H.N. Shashank, S. Subash, R. Suhas, M.M. Savitha, Crop protection and animal detection system. IRJET **7**(7) (2020)
19. G.C. Jabshetti, M.S. Biradar, K. Bhangale, Object detection using Regionlet transform, in *2016 International Conference on Computing, Analytics and Security Trends (CAST)* (2016), pp. 600–604
20. K. Bhangale, R.U. Shekokar, Human body detection in static images using hog & piecewise linear SVM. Int. J. Innov. Res. Dev. **3**(6) (2014)

Literature Review on Redistribution of Routing Protocols in Wireless Networks Using SDN Along with NFV

Sanjay S. Kadam and D. R. Ingle

Abstract The Internet, metropolitan area networks (MAN), wide area networks (WAN), local area networks (LAN), wireless networks, and mobile networks have all seen technical breakthroughs in recent years. There are obstacles in developing and maintaining computer network infrastructures, as well as reducing operational and capital expenses, while achieving flexible convergence with these technologies. This paper focuses on reviewing the traditional computer networks and protocols to highlight the respective advantages and disadvantages. The advanced technology with OpenFlow networks includes the software defined networks (SDN). Such networks make use of modern environment due to which the competitiveness, scalability, applications and network services are achievable with the robust architectures. The underlying layer of network function virtualization (NFV) facilitates to use virtual machines such as switches, routers, firewall, and other network services. Some network administrators will only use NFV or SDN. The motivation of this paper is that to discuss various features of SDN and NFV that complement each other in terms of load balancing, scalability, virtual machines, network and application services, network security and reduced operational and capital expenditure cost in designing and maintaining the computer network infrastructures by using the redistribution of network routing protocol. Firstly, this paper discusses about the traditional wireless network technologies with Enhanced Interior Gateway Routing Protocol (EIGRP), Open Shortest Path First (OSPF), and redistribution of routing protocols on heterogeneous autonomous system. Followed by that, the OpenFlow protocol with SDN has been analyzed. In the third category, NFV is considered. As a continuation, the fourth category shows the trade-off to optimize the performance of wireless networks by improving the security, load balancing, and QoS along with the reduction in power consumption, capex and opex cost of computer networks.

Keywords WAN · LAN · Redistribution · Routing protocols · Heterogeneous · OSPF · EIGRP · VM · OpenFlow · SDN · NFV

S. S. Kadam (✉) · D. R. Ingle
Bharati Vidyapeeth College of Engineering, Navi Mumbai, Maharashtra 400615, India

G. Ranganathan et al. (eds.), *Soft Computing for Security Applications*,
Advances in Intelligent Systems and Computing 1397,
https://doi.org/10.1007/978-981-16-5301-8_41

1 Introduction

Traditional computer networks have a number of issues, including cost-effectiveness, operational expenditures, joining or merging different autonomous systems, load balancing in computer systems, and annual maintenance etc. Cable connections, such as cat 5, cat 6, and RJ45 connectors, are required in a wired LAN system. Infrastructures like switches, hubs, switches (8 port, 16 port, 24 port, and 32 port), routers, gateways, and other devices are required to achieve maximum scalability and network devices. This requires huge quantity of additional devices to establish the total infrastructure and the respective huge expenditure costs. This system is called as capex, which includes the designing, wiring, configuration, crimping of the connectors, cable fitting etc., in the infrastructure establishment. Experts are required to configure such networks, and these individuals create and install such systems on the basis of contracts and consultancy. The cost under this category is considered as opex.

To overcome the cabling configuration expenses and challenges, wireless local area network (WLAN) gives more flexibility. On other hand, the configuration of WLAN also requires opex. When these opex of LAN and WLAN are compared, the WLAN opex shows more relaxation. The mobility and flexibility is an added benefit in WLAN, as clients can move from place to place along with their devices. This kind of smoothness in operations also requires system administrators with good level of knowledge base.

In WLAN, there are two possible modes, namely ad hoc mode and infrastructure mode. In infrastructure mode, there is a need for administrators to manage the networks, whereas ad hoc networks are managed by individuals. In infrastructure mode, client devices have provided services via access points, which are usually connected via gateways for service-oriented interfaces. In ad hoc, the packet travels from device to device in multihop manner, whereas in infrastructure it always flows through the path via gateways. The packet delivery is mainly concerned with the routing in network infrastructure, and in most of the networks, it remains as the challenging task (Fig. 1).

The modern type of wireless mesh network ("multihop") has the capability to extend the network with hot spot into hot spaces, and hence, it provides better coverage region compared to traditional WLANs. The modern wireless networks are having more flexible interface architecture in their physical layers, which combine multiple types of wireless and wired networks [4, 5].

In existing WMN, the poor connectivity zones are remaining the significant challenge during installation. Also, peak time failures, latency issues, traffic inconsistency, high tolerance in bandwidth allocation, improper load balancing network congestion, and security threats are also considered as the major drawbacks in the existing WMNs. Existing EWMN can be used to solve the issue. This protocol is based on TCP/IP protocol architecture [4, 5]. In the existing EWMN, the architecture becomes complex with the use of QoS-oriented devices along with their security requirements. The recurring costs are associated with time to time up-graduation

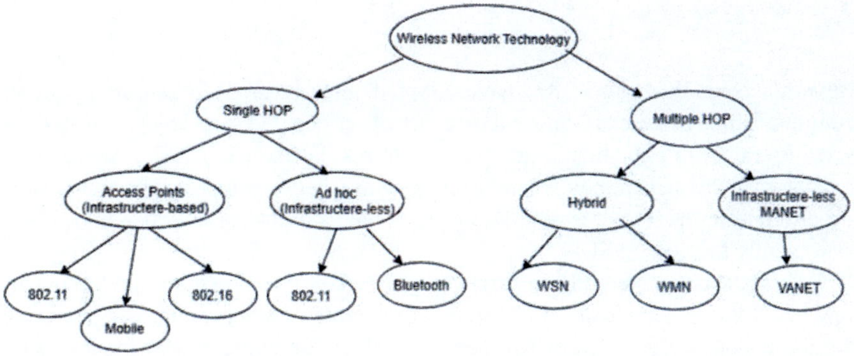

Fig. 1 Wireless network taxonomy [3]

requirements and also have less flexibility for changing the topologies and routing approaches. There is a very complex procedure, while merging the existing and new network infrastructures [4, 5].

During the process of multinetwork routing requirements, there are variety of challenges, which are also considered for discussion. In such redistribution strategy, various challenges are associated. This may include OSPF and EIGRP merging. In these protocols, use cost and K-values are considered, and hence, there is a compatibility issue among them. In the RIP protocol, hop count is considered. This shows that merging for redistribution leads to different challenges and issues [7, 8]. Hence, routing circle is the major solution considered while performing import and export of routing data from protocol to protocol.

The use of SDN and NFV in next generation wireless mesh network [WMN] is focused [9]. The network data traffic is handled independently with the use of SDN. The use of devices, as well as virtualization, is taken into account in this case [10]. This shows many advantages in terms of adaptability and control in designing the network assets, setting up policies, building new approaches, which further improves the performance. Also there is more flexibility in administrating the network with the use of SDN. The wide coverage requirements and enterprise administration with the use of firewalls and devices are the main point of interest, while choosing the SDN [10, 11].

2 Literature Review

In the consideration of SDN- and NFV-based network formations, the literature survey forms the platform of understanding the relevant details in the area and requirement of further researches in the field. To address various challenges and advantages related to works, articles, books, and different journals are discussed here in this section [11, 12].

2.1 Related Works

Researchers in this paper [37] have adopted three-layer SDN-powered satellite communication system to achieve the quality of service [QoS] and by proposing new satellite framework and in addition they use terrestrial service, which is powered by NFV. But in this paper, they are not proposing any network topologies, and also for experiment, researchers use simulators. They have not discussed much about NFV as it mainly focuses on SDN.

In this research paper [38], researcher discussed the use of Internet for the controlling and driving the vehicles, i.e., Internet of vehicles. Latency, reliability, and QoS are the major issues in Internet of vehicles, which are addressed by exploring the various features of SDN and NFV in networking domain. They have not discussed much on redistribution of routing protocols from the routing protocols perspective.

Researchers [39] explore the use of software-defined network/network function virtualization technologies in smart TV services such as miracast and cloud server. Here, researchers surveyed more on software-defined network than network function virtualization. They done all this on simulations also designed paradigm to use intelligent home IP gateway of the Internet of radio light (IoRL).

In this paper [6], researchers explore wireless local area working in the current networking technologies, and they have suggested the improved branch of enterprise WLAN to showcase the improved performance of Wi-Fi in terms of coverage area and robustness in mobility. The applicability of wireless mesh network is much more advanced when compared with Wi-Fi, and it also provides more bandwidth from which higher data rates are achievable even at larger area and with flexible mobility.

In this paper [10], there is an in-depth discussion on software-defined networks, OpenFlow protocol version 1.0 to 1.4. Also, researchers designed architecture based on various papers and articles. They also discussed about the issue of vendor-specific traditional network design problems, expenses, etc., but in this paper, they have not discussed about the redistribution and routing protocols. Researchers wants to prove that software-defined network and centralized routing are better than legacy routing protocols such as OSPF and BGP. SDN routing is better in delayed packet forwarding, and convergence time is better when there is a link or node failure.

2.2 Wireless Mesh Network

Traditional far-off associations have the different territories, subnetting and related with wired association for authentic working of enormous business, home, and neighborhood. Far-off clients center points partner directly with a far-off way to the wired association. Furthermore, the distant association decreases the dependence on costly wired association structure and sponsorship low compensation to get the access of web and PC network resources [1]. A wireless mesh network (WMN) is arranged by a social event of mesh switches, hubs, managed switches, entrances, network switches, distant clients, and cross-area clients. Switches and grid switches are used to plan the establishment of an association in which clients or far-off contraptions may be moved efficiently and effectively [1, 2]. Cross-segment clients can be CDMA, 2G, 3G, 4G mobiles, and in near future, 5G technologies will be adopted. Directed switch or layer 3 network routers are interconnected with the outside PC association (for instance, ISP) by sending each other's data (checking far-off data bundles) toward the PC network entrance or switch center, which are clearly facilitated with the wired infrastructure (LAN) [13]. In a WMN, each far-off layer 3 switches progresses data packs for other PC network centers (that may be outside of direct far-off transmission extent of their protests). Additionally, the switch functionalities license the mix of wireless mesh network with various existing distant associations, for instance, 802.11 Wi-Fi networks, mobile associations, IEEE 802.16 Worldwide Interoperability for Microwave Access (for instance WiMax), among other far-off devices. In such computer associations, the PC center points obviously develop and keep network accessibility among far-off devices [1, 13]. Exactly when associations crash and burn in distant associations, WMN offers remarkable contrast with other game plan for instance various abundance exchange ways which thus finds substitute ways by routing message across network. In the event of a hostile condition in the WMN, the devices present in the WMN truly send divides other without interfering with correspondence between each other. It works in a similar way to extend the idea of association by choosing the shortest distance between hosts (nodes), so that the recipient can get strong sign differentiating traditional far-off association on different occasions without increasing the transmitter's power and reducing the amount of electricity used. Likewise, a wireless mesh network (WMN) is in unconstrained solicitation, free, self-masterminded, self-healing, and monotonous condition (if presence network crashes and burns, duplicate organization is proficient and interconnects each other) [1, 13]. This gives most critical advantages, for instance, basic PC network fixing and backing, low cost, force, extreme, and strong assistance coverage [14].

The advantage of distant area network over ad hoc network is the use of various association devices like switch, section, multilayer switches, and their resources for performing more association limits. By using latest PC advancement, wireless mesh network is improved by including wireless personal area network (WPAN), wireless metropolitan area network (WMAN), and wireless local area network (WLAN). Today, various organizations in the field of a distant area are passing on wireless mesh network in inside and out with various colossal extension in large business grounds association, virtual private network (VPN), versatile association, neighborhood as well as in gigantic metropolitan far-off area [1, 13, 15].

Advanced wireless grid network improves the execution of multiplayer gaming and accessibility of different grounds, which are showing up at the inaccessible region with suitable information move limit, which is useful to execute distinctive government and online arrangement, in army (military) with the usage of WMN accepts the huge part to improve the trades between various sections [14].

As we understand that wireless mesh network is multibounce, far-off cross-sectional switches expect essential part to propel data to all centers in the scope

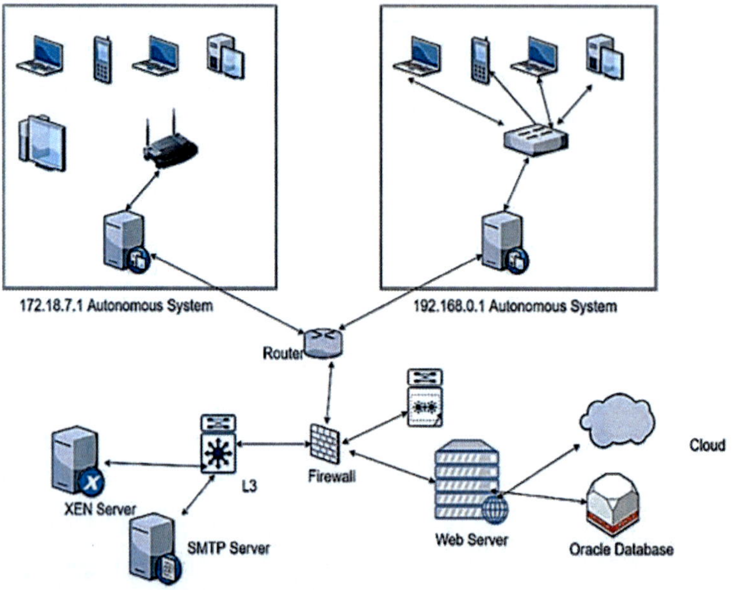

Fig. 2 Wireless mesh network

by making the coordinating decisions, for instance, picking the most ideal interface. Data is obtained from WMN switches and clients. For some circumstance, switch capacities work as far-off entry for various customers (for instance, compact PC, wise TV, cells or insightful devices) [1, 13]. Normally exceptional contraptions will obtain and transmit data via the backend of a remote grid connection. To connect to the Internet (Internet, VPN, Intranet) by utilizing at least one switch, a combination of remote and wired connections is sometimes required. At certain points, switches serve as entryways. As we can see, the IEEE 802.11 alliance is a far-reaching cross-border network. The IEEE 802.11 association makes extensive use of discretionary access techniques. In IEEE 802.11, there is a mechanism to avoid nonoverlapping channels that are utilized by multiple devices to deliver data with a smaller square in WMN [1, 2, 13]. In direct word, we can call it as the utilization of various switches, which are setting up the diverse radio interfaces that are tuned to different channels. Figure 2 shows the wireless mesh network [WMN].

2.3 Routing Protocols

As referred to routing shown in WMN, it can be isolated into two classes: proactive and responsive. Proactive routing incorporates a condition where the network center points reliably care for one, or more directing tables that store courses to all the center points of an association, and at the same time, it will be sent along the association to exchange and invigorate information in the connecting nodes [4, 13, 15]. Responsive shows along these lines get information on the way to the goal (center) of a package, when data transmission has to be influenced (on demand). These shows do not make additional traffic in the association, anyway the time required for data to be sent is attracted out when it is imperative to affect the information exchange by concerning the available route [13, 17]. The presentation of PC network totally depends upon the guiding show.

2.3.1 Metrics Used for Selection of Routing Protocol

An estimation is a cost (like way length) that find out approaches to an end point. Assurance of coordinating show relies upon various execution estimations like number of ricochets (hop count), nature of an affiliation (link quality metrics), network load rate (load-dependent metrics), expected transmission count (ETX), expected transmission time (ETT), energy use, and path openness/reliability. So the above estimations are differentiated and also for various traits, for instance, quality awareness, data rate, size of packet, intra-stream hindrances, inter-stream interferences, and medium instability. Following are estimations used by various coordinating protocols [1, 3].

Distinctive coordinating shows assesses their directing estimations from grouped limits and with different procedures. Some use standard estimations (RIP v1 and RIP v2), and others use complex estimations (EIGRP).

Tear v1 and RIP v2 use bounce check to choose the best available way. Since they never consider BW (information move limit), RIP v1 and RIP v2 are not important for PC networks that have distinctive transmission speeds on overabundance ways. PC networks use assorted transmission media on monotonous ways by guiding shows that ought to address speed and possibly the delay of the links [1, 3].

EIGRP uses the negligible exchange speed and accumulated concede of the path toward the end point in its metric computation by default (goal). Various limits (trouble and unflinching quality) can moreover be used, yet they ought to be organized with least troublesome if the effects are completely seen. If misconfigured, they could influence the blend and reason the coordinating circles.

In its metric assessment, EIGRP uses the base information transmission and assembled the deferment of the path toward target association. Various limits (trustworthiness and weight) can similarly be used and anyway they should be planned just if the results are totally seen. In case of misconfiguration, they may impact gathering and cause controlling in the circles [1, 3].

2.3.2 Criteria's Used for Classification of Routing Protocols

As we understand that, the coordinating cycle is finding the most ideal way (beginning to end path) between two centers (source node and target node). The following models and estimations are used for guiding shows.

1. Criteria used for the game plan of routing protocols: Routing shows can be perceived on various models like types of protocols (Philosophy of Routing). There are three essential sorts in the wireless mesh network, and they are generally proactive, reactive, and hybrid. For instance, position-based coordinating shows. Proactive controlling develops courses in topography with no consent [4, 13]. The name suggests that the connection of coordinating will begin when requesting comes. Additionally, creamer is just the blend of picking best beneficial piece of these proactive and responsive guiding in any given circumstance. WMN affiliation is a base for the centers' limits. When there is level affiliation all centers will have the same work. When there are some specific limit center points in affiliation then it goes under the reformist affiliation. Awareness of territory is also one of critical parts in coordinating shows. So the mobility load up, in light of everything, ought to be proper as adaptable centers are moving from one spot to other and recognize that time at the center is moving from strong sign direct association toward new point. So, the organization of flexibility is huge, when there is a discussion on coordinating protocols [4, 13, 15].
2. In a WMN network relationship, there are two standard level affiliations where all center points are likely to be the same and in a reformist affiliation where a couple of center points will have specific limits. This controlling show rules are considered and it accepts huge parts to overhaul the execution of association.

You can portray and investigate directing shows multiply.

- Static routing versus dynamic routing
- Interior routing versus exterior routing
- Distance vector versus link-state and hybrid routing
- Discussing diverse coordinating show estimations
- Importance of routing
- Comparing flat coordinating shows.

2.3.3 Static Routing Versus Dynamic Routing

As a static coordinating address, that courses which is actually conceived or inserted static courses for the purposes of sending packages. At various stages static coordinating is useful, for instance, when moderate bandwidth interface is there it is difficult to invigorate dynamic controlling. Right when organization head needs full scale authority over the association and when there is only a solitary way available that time static guiding is useful [18, 41]. In any case, static guiding is monotonous considering the way that it requires planning and keeping up task by doing it genuinely. So it requires part data on all endeavor association. In dynamic controlling, the association allows for upgrades in the geology of the association which has to be carried out without the involvement of a network administrator. If association misses the mark, static switch cannot do it intensely and that static course not, now fruitful at whatever point bombarded interface used. Sometime it is critical to add new switch in network that information ought to be revived capably and it is crazy to hope to add statically in quick time [18]. As a result of this development, the burden of work on network leaders has increased, and it takes some effort for each and every change to obtain information about new routers [15].

2.3.4 Interior Routing Versus Exterior Routing Protocols

The computer networking world is based on the domains or autonomous system (AS) idea. Two types of routing protocols are used: one is interior gateway protocols (IGP) and other one is exterior gateway protocols (EGP) [18].

Interior gateway protocols (IGP): IGP are intra-AS (inside an AS) routing protocols. Examples of IGPs include Routing Information Protocol (RIP) version 1 (RIPv1), RIP version 2 (RIPv2), Open Shortest Path First (OSPF), Integrated Intermediate System-to-Intermediate System (IS-IS), and Enhanced Interior Gateway Routing Protocol (EIGRP) [19].

Exterior gateway protocols (EGP): EGP are inter-domain (between various domain systems) routing protocols. Border Gateway Protocol (BGP) is the dominating exterior gateway protocols (EGP) on the computer networks such as WAN and Internet. BGP-4, which is the fourth type of BGP, is most sufficient transformation of Border Gateway Protocol on the computer network (Internet).

As all of you understand that in any domain network, there is a combination of switches which are obliged by single association manager, for instance, ISP and internal association. Henceforth, there is the assorted show required in affiliation which depends upon various conditions. When it comes to single space or inter self-administering systems, the selection of coordinating features requires the most limited decisions. By doing so, it will be able to complete a unique approach known as intermixing, as well as the scattering of direction in reward, and it will be able to choose the best open path to reach the goal [18, 19]. It is incredibly fundamental for intra-space guiding to find the best open way, so EGP (outside gateway show) controlling show has different estimations, which are used to pick the most ideal way, where these estimations join different limits to allow the association head to help in finding the best open coordinating ways. EGPs response to blend is moderate, and it is moreover amazingly tangled at the hour of arrangement. IGPs (interior section shows) use least estimations and less-obfuscated limits because of this advantage, and it is not hard for the network regulator to find the best coordinating ways for quick convergence [19].

2.3.5 Distance Vector Versus Link-State and Hybrid Protocols

Distance vector protocol (DVP): Jump by bounce premise is utilized in DVP (distance vector convention) steering. Here, the connection between neighbor switches is significant in light of the fact that any switch will rely upon neighbor switches data. The neighbor switch gives choice about the selected data on its directing table. Distance vector protocols (DVP) are normally delayed in merge, and furthermore, they do not scale great; be that as it may, they are easy to keep up and carry out. RIPv1, RIPv2, and Interior Gateway Routing Protocol (IGRP) are the instances of DVP [18, 19].

Link-state protocol (LSP): Each router sends full information about its states of link to other routers in the computer network or to a part of the domain. In LSP, every router makes its routing decision based on all the available information, which is sent by other routers and by using the Dijkstra algorithm [also called the shortest path first (SPF)], which estimates the shortest path to any end point (destination). Because of this, link-state protocols (LSP) converge very fast and also traffic overhead is also less so it also scales well. But LSP is more complex and so it is very difficult to maintain and implement. The examples of LSP are OSPF and Integrated IS-IS.

Hybrid interior gateway protocol (HIGP): In CISCO, the proprietary protocol has the characteristics of above mentioned with both the protocols called distance vector protocol (DVP) and link-state protocols (LSP), and its clubs distance vector conducts with useful link-state features. The name of this protocol is EIGRP, which is scalable and fast in converging the routing protocol [19].

2.4 *Open Shortest Path First (OSPF)*

Is an open standard convention for steering IPv4, and it will likewise connect state directing convention. OSPF was planned and produced for perplexing, adaptable, and enormous organizations, where RIP is ineffective to satisfy all prerequisites as a result of its basic constraints; it additionally utilizes the IP convention 89. In light of the different connections between AS (self-sufficient frameworks) and the fundamental for way the executives, static directing is discarded. OSPF is ideal and better than RIP in numerous angles. OSPF has a lot quicker union. OSPF is an awkward convention, and as a result of this, it helps Variable Length Subnet Mask (VLSM), progressive design and manual rundown. Quick union and best versatility will additionally settle on OSPF, which is considered as a most ideal decision for undertaking the steering convention and biggest organization. Open Shortest Path First [OSPF] practices the Dijkstra calculation to process the Sender Policy Framework (SPF) tree [20, 21]. The idea of a wide range of regions inside one self-governing frameworks (space) was carried out in OSPF to decrease and control the measure of directing traffic and make networks more solid and adaptable.

In OSPF, least interfaces work in that space switch. Also, if the switch is at the limit of non-backbone and spine territory is called as Area Border Router (ABR) with at least one interface. The limit between the spaces is made in the ABR itself. Between AS steering, ISPs have a number of important directing strategy options ranging from conforming to peering and other types of agreements. In the event that, outside courses are engendered into the OSPF AS, the switch that rearranges those courses is known as the Autonomous System Boundary Router (ASBR). Mindful arrangement and amend planning of zones to the organization geography are basic, since manual outline of courses can in a manner of speaking be performed on ABRs and ASBRs.

RFC 5340 depicts OSPF V3 as a steering convention for IPv6 organizations. Note that OSPFv3 is for IPv6 networks figuratively speaking which is not backward predictable with OSPFv2 (used in IPv4). OSPF parts and calculations, like flooding, kinds of switches, switches choice, districts, stub and NSSA, computations of SPF, stay something similar. Changes are made for OSPF to back IPv6 addresses, address hierarchy of leadership, and transport utilizing IPv6. OSPFv3 utilizes multicast pack FF02::5 for all OSPF switches and FF02::6 for all allocated routers [20, 21].

Following are the main advantages of OSPF:

- OSPF is an open source which means it is available to all public.
- As mentioned above, OSPF uses the link-state algorithm and link-state packet distribution.
- In OSPF network topology, mapping is done at each node.
- In OSPF route, calculation is done by Dijkastra algorithm.
- OSPF link-state advertisement in same domain (AS) is done router floods.
- OSPF without using TCP or UDP sends messages directly over IP.
- OSPF and IS-IS are almost identical.

Advanced features of OSPF are:

Security: All OSPF messages authenticated (to prevent malicious intrusion) ♣ multiple same-cost paths allowed (only one path in RIP) ♣ for each link, multiple cost metrics for different TOS (e.g., satellite link cost set low for best effort ToS; high for real-time ToS) ♣ integrated uni- and multicast support: • Multicast OSPF (MOSPF) uses same topology database as OSPF ♣ hierarchical OSPF in large domains.

2.5 Redistribution of Protocols in Networks

Why is it required in network?

Steering construction of colossal organizations, associations, and adventure network regularly involves various diverse sorts of areas or directing events because of different reasons such as company acquisitions, offices oversaw by various organization overseer groups, presence of new multimerchant kinds of hardware. The deliberate formation of disconnected directing occurrences to channel courses limit reachability and authorize new or presence approaches. Wary and careful development is starting with one steering convention and then onto the next convention. The cutting-edge directing convention will in all probability have particular necessities and capacities from the bygone one. It is basic for framework or organization directors to get it what should be changed and to shape a point by point mastermind some time as of late rolling out any improvements to move to new arrange from presence network [19, 22].

The utilization of a managing show to pitch courses that are learned by a couple other recommends, for instance, by another coordinating show, idle courses, or obviously related courses, is called modification. While running a lone planning show all through your whole IP internetwork is engaging, multishow controlling is fundamental for different reasons, for instance, association combinations, different work environments regulated by different association heads, and multitrader conditions. Running unquestionable coordinating shows is frequently somewhat of an association plan. Notwithstanding, having a different show environment makes modification a need. Differentiations in guiding show ascribes, similar to assessments, authoritative confined, classful and uncivilized capacities can influence redistribution [21]. Thought ought to be given to these separations for redistribution to succeed.

The division of an endeavor network in coordinating spaces is not inert. Different occasions can drive an undertaking network manager to change the constraints of its guiding regions. Separating and joining structures, e.g., to oblige solidifications and acquisitions, are two fundamental diagrams. Basically overriding a switch by another change from an obvious brand may oblige a cutoff change. At last, the events can be driven by transient focuses, for instance, an alter to the traffic planning requirements or switch upkeep, similarly as long stretch updates, for instance, changing piece of the association to the high-level programming described getting sorted out (SDN) and network function virtualization.

Figure 3 shows redistribution between coordinating shows, and we have different options, for instance, between guiding shows (OSPF, RIP, BGP, EIGRP). Static courses can be redistributed into a controlling show. Clearly related courses can be modified into a guiding show. Customarily you can utilize the association request to promote clearly related courses into your coordinating show. You will have the choice to utilize the revise-related request which can redistribute it into the controlling protocol [19, 22].

Tear and OSPF are two coordinating shows renowned in PC network area, and they are constantly used in frameworks organization. In Fig. 4, switches inside one guiding case regularly run the same guiding show to thoroughly share reachability information and do not exchange controlling information or data with switches or

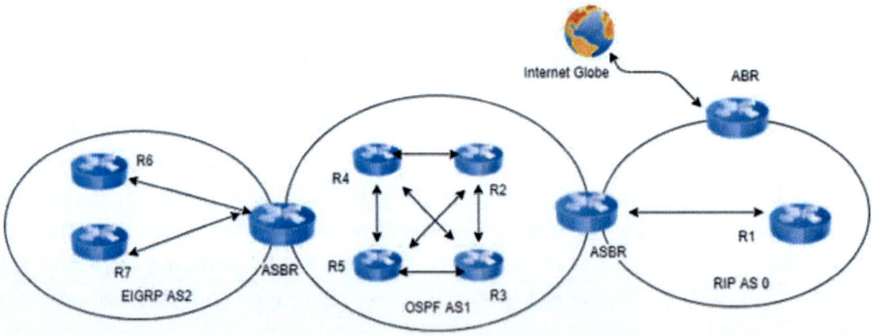

Fig. 3 Redistribution of routing protocol

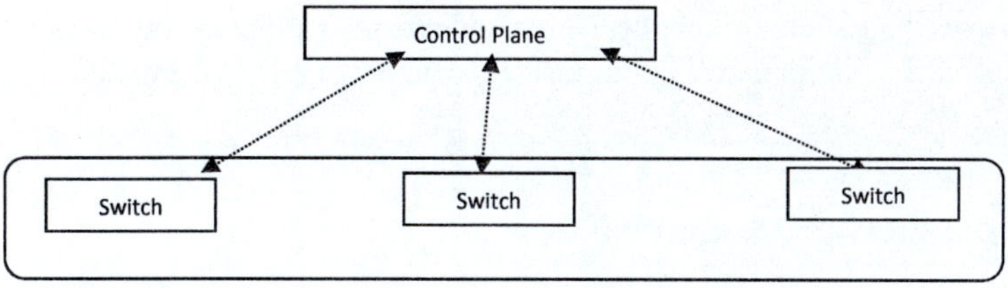

Fig. 4 Structural view of OpenFlow protocol

switches in other coordinating events. In Fig. 4, it incorporates two coordinating models. Switches inside the RIP event do not have unquestionable nature of the addresses and subnet prefixes inside the OSPF case and a contrary route around.

In course redistribution, each coordinating show depicts an estimation for each course. For instance, i.e., RIP—120, EIGRP—90, and OSPF—110. When a switch or switch improves courses from one controlling territory to the next, this data cannot be interpreted from one coordinating show to the next. A RIP hop cannot be proficiently recalculated to an Open Shortest Path First (OSPF) cost by the switch or switch doing redistribution. A seed metric is utilized to erroneously settle the distance, cost, etc., to each modified organization from the redistribution point [19, 22].

An estimation of limitlessness (infinity) tells the switch that the course is inaccessible, and thusly, it ought not to be advanced. When revamping courses into IGRP, RIP, and EIGRP, you ought to show a default metric. If you do not plan an estimate, the default will be tenacity, which implies that all courses improved by these three shows will be far off. For OSPF, the adjusted courses have a default type 2 estimation of 20, yet for redistributed BGP courses, which have a default type 2 estimation of 1. For IS-IS, the adjusted courses have a default metric of 0. Be that as it may, not regardless like IGRP, RIP, or EIGRP, a seed metric will not be treated as hard to reach by IS-IS. Coordinating a seed metric for revamp into is proposed. For BGP, the redistributed courses keep up the IGP coordinating metrics [19].

Find the cutoff switch between two directing measures. Sort out which directing measure is the center or spine measure. Sort out which coordinating connection is the edge or development measure. Select a system for permeating the predefined edge show courses into the middle. The nonexclusive advances recorded here apply to all controlling show or communication mixes. The articulations "focus" and "edge" are obscure terms that are utilized to change the talk almost improvement. Find the cutoff switch that requires an approach of redistribution [19, 22]. Picking a singular switch for redistribution restricts the probability of making managing circles that are achieved by input. Sort out which organizing show is the center or spine show. Expectedly, this show is OSPF, Intermediate System-to-Intermediate System Protocol (IS-IS), or EIGRP. Sort out which coordinating show is the edge or present second (inside the occasion of relocation) show. Choose if all courses from the edge show had the chance to be multiplied into the middle. Consider frameworks that decrease the amount of courses. Select a procedure for embedding the ideal edge show courses into the middle. Direct improvement utilizing outlines at network restricts the amount of new sections inside the guiding table of the middle switches.

3 OpenFlow, SDN, and NFV

3.1 OpenFlow

In above segments, we noticed that a PC networks switch's parcel forward choice has been founded mostly on a bundle's objective location. As we realize that switches different capacities perform on layer 3, for example, header IP locations and ports numbers modified by NAT boxes, Deep Packet Inspection (DPI) is sent by firewall for extra preparing in the wake of hindering or diverting the traffic which is based on IP address header field esteems. Burden balancers subsequent to getting http demand from set of workers it advances packets [13].

Joining layer 2 switches and layer 3 switches by utilizing its own different equipment, particular programming and different interfaces which is utilized for the board make parcel of torments to many organization overseers. SDN (software characterized network) makes this undertaking simpler as SDN has brought together methodology while giving this layer 2 and layer 3 capacities at different stages with cutting edge, complex, coordinated and strong way.

In match in addition to activity model of objective-based sending "match" is only the objective IP address and "activity" is yield port. In some time at least one yield ports ("actions") incorporates forward parcels which comes from the objective-based sending, NAT (rewriting header esteems), numerous active bundles of burden adjusting, firewall which squares or drops bundles and a lot more [13]. Objective-based sending tables are having the thoughts which are summed up from match in addition to activity table sending speculation. Source and objective location of layer 2 (connect layer) and layer 3 (Network layer) are utilized to settle on forward choice. So layer 2 (link layer) gadgets are designated "switches" and gadgets and layer 3 (network layer) called "switches".

From the start, this idea of OpenFlow was released by the fresh start program at Stanford and presently the subtleties are kept up by open systems administration discussion. A Ph.D. understudy of college of Stanford Martin Casado and his gathering in year 2006 proposes a fresh start security engineering (SANE) which describes a concentrated control of security [24]. The year 2008 OpenFlow paper in ACM SIGCOMM CCR is distributed. In year 2006, Martin Casado, a Ph.D. understudy of college of Stanford and gathering propose a fresh start security plan (rational) which describes a concentrated control of safety (instead at the edge as commonly done). Thus, the inquisitively truth is here that as of now, entire systems administration plan, they are having the chance to be open, they are moving from a shut local area or from a vender explicit local area to an open getting sorted out local area, where every merchants interface altogether [13]. Thus, the traders are building their fittings and the local area is building the open-source working system and the interface to relate with the comparing equipment. Thus, this has an effect in two different ways the headway speedy since by and by this entire PC program is available to the local area, you can plan yourself network convention and test on a gear for that, you simply do not need to look for an equipment of where you will be capable to purchase any SDN upheld switch, open switch. Also, by then you will be capable to do your convention execution on best of this open-source working framework, and the second benefit comes from the organization perspective, where the organization manager, they do not trouble roughly examining a lot more pages manuals for from different sellers.

OpenFlow is a show for controlling the sending conduct of Ethernet switches in software-defined network. In stream table of OpenFlow, each entry is the match notwithstanding action sending table [24]. A lot of header field regards to which a pushing toward group will be composed. As inside the occurrence of target-based sending, hardware-based getting sorted out is most promptly acted in TCAM memory, with more than 100 K target area passages being conceivable [25]. A group that organizes with no stream table passage can be dropped or dispatched off the removed controller for genuinely taking care of. In hone, a stream table may be finished by different stream tables for execution or cost reasons yet we will focus here on the thought of a singular stream table. A lot of counters that are overhauled as bundles are facilitated to stream table segments [23, 24]. These counters may join the amount of groups that have been facilitated by that table segment, and the time since the table passage was last revived. A lot of moves are to be made when a bundle arranges with a stream table area. These exercises can be to propel the package to a given o/p port, to drop the group, make duplicates of the bundle, and send them to different o/p ports, or conceivably to improve picked header fields [24].

Thusly, in the event that we see into the OpenFlow stream table, the OpenFlow table has four specific fragments; you have the norm, the taking a gander at movement, certain encounters around groups, the execution of a specific rule and a need regard, which is associated with a norm. Thusly, you have a standard 1 [24]. Along these lines, rule is just a lot of fields which field by and large says that where explicit field of an IP bundle or here really, in software-portrayed network you will see into MAC, IP, TCP, all of the headers. Along these lines, in your group header which unequivocal field to explore [9, 26]. Speculatively, you will have the choice to see into any field inside the package header. All things considered, you will be proficient to see into the package header and our standard essentially shows what ought to be or what is you are interested a motivating force for a particular field inside the group header? Like the switch port, VLAN ID, MAC source, MAC objective, Ethernet sort, IP source, IP objective, IP sort of benefit for nature of benefit, TCP source port, TCP objective port [13]. Finally, if there is a match with this norm; which implies that explicit fields fundamentally are deciding by then, there can be a lot of exercises, and the exercises can be orchestrated by you subject to your choice. Thusly, the action can forward the packages to nothing or more ports inside the switch epitomize the group and after that forward the package changes certain fields in the bundle and in forward the pack, drop the bundles, if you need to understand the firewall rule else, you can incorporate you are have augmentation, anything you will have the alternative to think of [24]. The information fields, it has a lot of explicit encounters like the package counter, the byte counter, the number of groups that have been composed with a specific norm, and so on. Subsequently, that it gets less hard for you to start the data from the association and a short time later there may be a need regard related, which is the need of a looking at rule. Thusly, because of an OpenFlow, at whatever point you have a lot of rules, on the off chance that there is a match with various rules, and the significant need rule is executed in general [24].

In Fig. 5, it shows the eleven group header fields and the inbound port ID that can be facilitated in an OpenFlow 1.0 match notwithstanding action rule. The figure depicts OpenFlow working in three layers: the associate layer, the network layer, and the transports layer. In the associate layer, the MAC tends to source and target these MAC addresses related to the packaging's sending and enduring the interface by sending the start of MAC address rather than IP addresses. In Fig. 5, we can see that an OpenFlow-enabled item or device can function as an L3 switch, which basically propels data as datagrams, and an L2 switch, which advances data in the form of edges [24]. The two octet fields in an Ethernet looks at the IP (for instance upper layer show) to which the data packaging's payload will be demultiplexed, and the VLAN fields are stressed over indicated virtual LAN [25]. The get-together of 12 characteristics that can be composed in the OpenFlow 1.0 assurance has grown up to 41 characteristics in latest OpenFlow specifications [13].

Data port of package switch is called entrance port were groups are gotten. The passageway port implies the data port at the pack switch on which a package is gotten. The source and objective of transport-layers port numbers fields can similarly be facilitated. In stream tables areas can use secret weapons 1 [24]. Finally, we perceive that not entire fields in an IP header can be composed. Rather than IPv4 and IPv6, OpenFlow does not allow organizing dependent on TTL field or datagram length

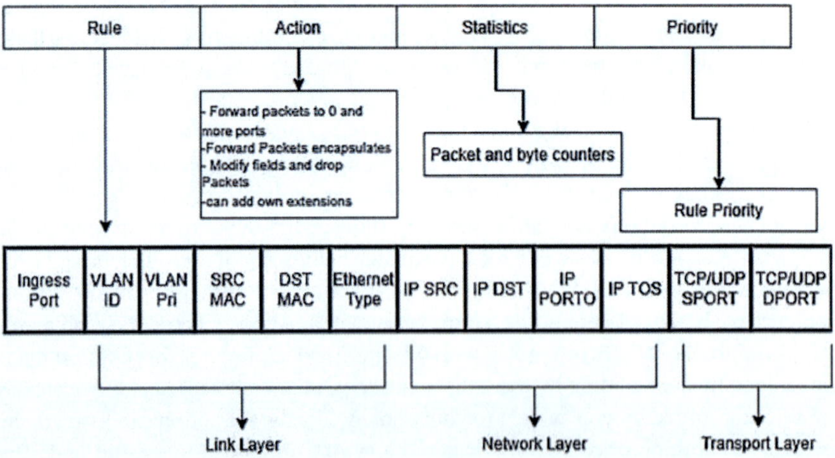

Fig. 5 Packet matching fields, OpenFlow 1.0 flow table

field. Each stream table area consolidates a once-over of nothing or more exercises that pick the taking care of that will be related with a package that facilitates with a stream table section. If there are different exercises, they are performed inside the coordinate appeared inside the list [13, 24].

3.2 *Software Defined Networking (SDN)*

Software-defined networking (SDN) is another advancement network model that executes the control occupation of the frameworks organization contraptions. So control plane needs to make close by data plane, showed up in Fig. 5. The control plane is in control for picking about each stream in the association, based on that decisions control plane is constrained in the frameworks organization devices of the data plane [28]. This parting of occupations licenses making the data plane programmable, appropriately discarding the need to actually organize (configure) every frameworks organization device [9, 27]. With a controller in the control plane, SDN works with the creation of new adaptable arrangement according to various applications and customer requirements, similar to improvement (handoff), security, and quality of service (QoS) [41] (Fig. 6).

OpenFlow is toward the south show used between the controller and the association contraptions (switches, switches, and entries) [27]. The target of OpenFlow was to offer the resources for test starter shows in make PC networks without hurting its movement. OpenFlow works by adding nuances of stream in stream tables put in every association equipment. These stream table data describe the activities to be taken by separate stream, for instance, package dropping and shipping off a network, sending groups to an exact port, execution of transmission, among various network [9].

Exactly when a SDN network starts working, void stream data is available. Right when the association nuances of new contraptions of association equipment, for instance, switch gets a package, without playing out any action since it does not know the first thing how to deal with this group and it progresses the group to the center controller, which examines the bundle header [30]. By finding the indistinct information about the package header with predefined rules, the controller takes decision about the group and updates this movement, through OpenFlow, to the switch. Another part is then fixed in the stream table. After that, whenever a bundle containing similar information (ascribes) occurs, the switch understands what action it must perform [31].

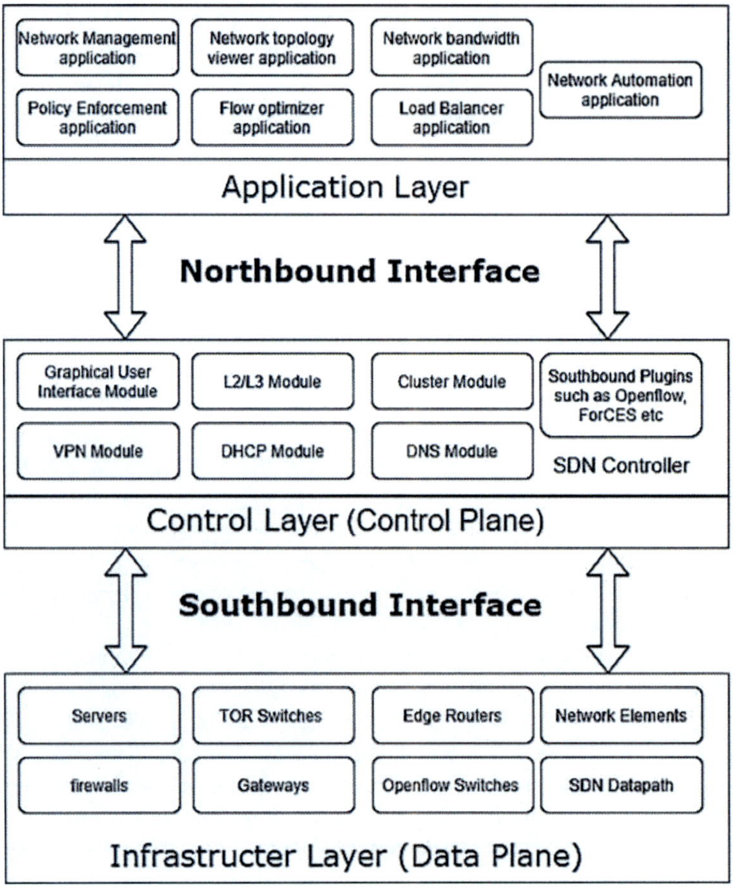

Fig. 6 SDN architecture

3.3 Network Function Virtualization (NFV)

They are software-based virtual devices, which are different form hardware devices. With NFV, network service providers can arrange various network roles, such as encryption or firewall, on virtual machines (VMs). Whenever a user or client wants new network roles, service providers give a virtual machine, which has automatic function. Applying this technology, networking companies will not have an intent to buy expensive and proprietary hardware devices in wireless network [32] (Fig. 7).

Network function virtualization offers another way to project, deploy, design, and manage networking services. NFV's main task is to optimize the network services in current network. NFV began when service providers attempted to speed up deployment of new network services in order to increase their revenue and development plans, and they found that hardware-based appliances limited their ability to achieve

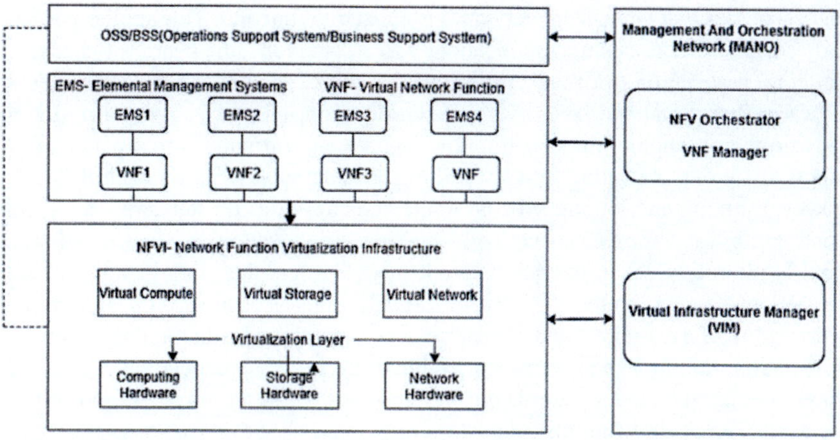

Fig. 7 NFV architecture

these goals [33]. They see to standard IT virtualization technologies and found NFV helped quicken service innovation and provisioning. Following are the important characteristics of NFV such as virtualization, orchestration, programmable, scaling, automation, visibility, network optimization, multitenancy, service integration, and openness [34].

4 Combining NFV and SDN with Traditional Network to Achieve the Optimization of Wireless Network

In any undertaking network plan, wired and far-off standard frameworks organization procedures are used to climb to latest development. To overhaul or do new way in the wired and far-off PC network someone uses just SDN. Besides, others favor the NFV alone. So this option is not sensible because SDN and NFV have their own advantages and a couple of injuries. But uniting SDN and NFV is better decision for suitable modification of coordinating shows, for instance, vector distance, interface state, proactive, open and new open stream protocol [30, 33]. If SDN and NFV work all the while together, they give the proper result in network. New model shows that it is plausible to continuously revamp of existing, new guiding shows with open stream show to achieve the objective of improving the show, low idleness, genuine exchange speed dissemination, secure, optimized, and nature of organization, power use the board, stream control in different sorts and various spaces distant associations with decreasing capital utilization and operational cost [30, 35]. Together, in truth, they address a way toward more nonexclusive association stuff and contraptions and more open PC programs, where the fused control and association communicated in software-defined network can in parcel be recognized through the virtualized limits and capacities that come from network virtualization function. This applies particularly for network association applications and associations, the board for checking, noticing, association, traffic examination, the chiefs, load changing consequently forward. Both SDN and NFV guarantee and depend intensely on virtualization to draw in their particular limits and to give upon their affirmations to separate affiliations and group managing from everything considered association control (SDN), however joining and joining specific limits and capacities on standard equipment, contraptions, and fragments (network function virtualization). Assurance of what Tom Nolle calls a "item overlay network model" infers that, in a blend of cloud-worked with virtual limits, NFV utilizes organizations as occupants on NFV frameworks [36]. Here, tunnels and virtual switches withdraw virtual abilities to avoid association, ruin malignant sneaking around or assault, and association with cloud-based virtual association interfacing, for instance, those portrayed in OpenStack Neutron [30, 31, 33] (Fig. 8).

The assessments show that OSPF presents higher association time in virtualized geologies. In any case, the association presents less deferral and RTT than the genuine association with a standard OSPF-based controlling plan. In addition, as for the video ongoing transmission, the SDN a few fundamental benefits over the common association. The SDN game plan has less setback rate and jitter, and better streaming. In like manner, the tall disaster rate appeared in the customary association diminishes the information move limit use, particularly on the host joins. Accordingly, the idea gives a steady guiding in a SDN, so later on, this plan can be utilized in more critical researches where searing coordinating in a SDN is required. As future work, we will finish more imperative geologies to check other controlling shows and their lead with respect to scalability [30, 31, 33]. Since the utilization of SDN is rapidly making, we should measure their power use when secure methods are completed in virtual topographies.

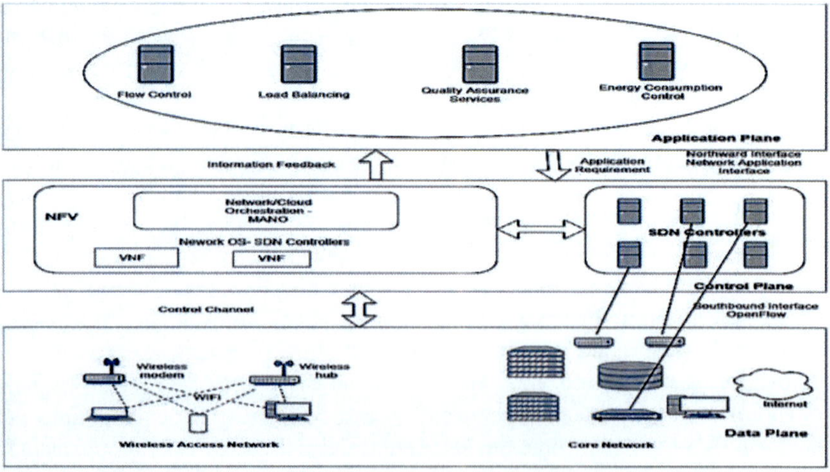

Fig. 8 SDN/NFV used model for optimization

5 Conclusion

Topics covered in this review article include wireless network technologies, wireless mesh network (WMN), various routing protocols, OpenFlow protocols, SDN, and NFV. As the digital technologies are becoming more ubiquitous, there is also demand for reliability, consistency, and load balancing in the computer network. So if good features of software-defined network [SDN] and network function virtualization [NFV] are combined together then the enterprise network (such as company, university, colleges, government departments) can work smoothly with proper load balancing, reliability, security, and consistency, which reduces the power consumption, reduces the cost expenditure (capex) and operational cost (opex). In this study, OSPF protocol for redistribution is discussed by exhibiting good results at various phases, and numerous researchers worked on it. The OpenFlow protocol remains as a key component of software-defined networking technology. So redistribution of routing protocols in wireless networks using SDN along with NFV gives optimization and improves the performance of the current advanced network technologies such as 5G, fast optical fiber Internet, and wireless technologies.

References

1. S. Waharte, R. Boutaba, Y. Iraqi, B. Ishibashi, Routing protocols in wireless mesh networks: challenges and design considerations. Multimed. Tools Appl. **29**(3), 285–303 (2006). https://doi.org/10.1007/s11042-006-0012-8
2. M. Piechowiak, P. Zwierzykowski, P. Owczarek, M. Waslowicz, Comparative analysis of routing protocols for wireless mesh networks, in *2016 10th International Symposium on Communication Systems Networks Digital Signal Processing CSNDSP 2016* (2016). https://doi.org/10.1109/CSNDSP.2016.7573902
3. Henning Schulzrinne FCC & Columbia University, *Mesh Networks (for Public Safety)—a taxonomy Henning Schulzrinne FCC & Columbia University*. ppt download. https://slideplayer.com/slide/5909032/. Accessed 24 Mar 2021
4. D. Benyamina, A. Hafid, M. Gendreau, Wireless mesh networks design—a survey. IEEE Commun. Surv. Tutorials **14**(2), 299–310 (2012). https://doi.org/10.1109/SURV.2011.042711.00007
5. P. Garnepudi, T. Damarla, J. Gaddipati, D. Veeraiah, Proactive, reactive and hybrid multicast routing protocols for Wireless Mesh Networks, in *2013 IEEE International Conference on Computational Intelligence and Computing Research IEEE ICCIC 2013* (2013). https://doi.org/10.1109/ICCIC.2013.6724255
6. S. Okada, C. Lee, H. Ueno, T. Ishihara, Detecting wireless LAN bottlenecks using TCP connection measurement at traffic aggregation point, in *2019 20th Asia-Pacific Network Operations and Management Symposium. Management in a Cyber-Physical World, APNOMS 2019* (2019), pp. 1–6. https://doi.org/10.23919/APNOMS.2019.8892945
7. D. Gopi, S. Cheng, R. Huck, Comparative analysis of SDN and conventional networks using routing protocols, in *IEEE CITS 2017—2017 International Conference on Computer, Information and Telecommunication Systems* (2017), pp. 108–112. https://doi.org/10.1109/CITS.2017.8035305

8. A. Rego, S. Sendra, J.M. Jimenez, J. Lloret, OSPF routing protocol performance in Software Defined Networks, in *2017 4th International Conference on Software Defined Systems*, SDS 2017, no. May (2017), pp. 131–136. https://doi.org/10.1109/SDS.2017.7939153
9. M. Gharbaoui et al., Demonstration of latency-aware and self-adaptive service chaining in 5G/SDN/NFV infrastructures, in *2018 IEEE Conference on Network Function Virtualization and Software Defined Networks, NFV-SDN 2018*, no. i (2018), pp. 1–2. https://doi.org/10.1109/NFV-SDN.2018.8725645
10. D. Kreutz, F.M.V. Ramos, P.E. Verissimo, C.E. Rothenberg, S. Azodolmolky, S. Uhlig, Software-defined networking: a comprehensive survey. Proc. IEEE **103**(1), 14–76 (2015). https://doi.org/10.1109/JPROC.2014.2371999
11. S. Rao, SDN and its use-cases-NV and NFV. Network **2**, H6 (2014)
12. H.S. Yang, L. Zheng, Y. Huang, Critical success factors for MES implementation in China, in *IEEE International Conference on Industrial Engineering and Engineering Management*, vol. 9 (2012), pp. 558–562. https://doi.org/10.1109/IEEM.2012.6837801
13. J.F. Kurose et al., *Computer Networking. A Top-Down Approach*, 7th edn (2017)
14. S. Garg, D. Arora, *Analysis of Routing Protocols in Wireless Mesh Networks*, vol. 1, no. 3 (2014), pp. 19–26
15. S.Q.A. Shah, F.Z. Khan, A. Baig, M.M. Iqbal, A QoS model for real-time application in wireless network using software defined network. Wirel. Pers. Commun. **112**(2), 1025–1044 (2020). https://doi.org/10.1007/s11277-020-07089-5
16. M.N. Siraj, Z. Ahmed, M.K. Hanif, M.H. Chaudary, S.A. Khan, N. Javaid, A hybrid routing protocol for wireless distributed networks. IEEE Access **6**(c), 67244–67260 (2018). https://doi.org/10.1109/ACCESS.2018.2875952
17. A. Zakrzewska, L. Koszałka, I. Poźniak-Koszałka, Performance study of routing protocols for wireless mesh networks, in *Proceedings of 19th International Conference on Systems Engineering. ICSEng 2008* (2008), pp. 331–336. https://doi.org/10.1109/ICSEng.2008.49
18. E. Akin, T. Korkmaz, Comparison of routing algorithms with static and dynamic link cost in Software Defined Networking (SDN). IEEE Access **7**, 148629–148644 (2019). https://doi.org/10.1109/ACCESS.2019.2946707
19. D. Teare, *Network Management Protocols and Features* (2008)
20. O.R. Protocol, G.K. Dey, M. Ahmed, Performance Analysis and Redistribution Among, pp. 26–27 (2015)
21. S. Sendra, P.A. Fernández, M.A. Quilez, J. Lloret, Study and performance of interior gateway IP routing protocols, *Network Protocols and Algorithms*, vol. 2, no. 4 (2011). https://doi.org/10.5296/npa.v2i4.547
22. E.W.M. Wong, A.K.M. Chan, T.S.P. Yum, Analysis of rerouting in circuit-switched networks. IEEE/ACM Trans. Netw. **8**(3), 419–427 (2000). https://doi.org/10.1109/90.851987
23. S. Vissicchio, L. Vanbever, L. Cittadini, G.G. Xie, O. Bonaventure, Safe routing reconfigurations with route redistribution, in *Proceedings—IEEE INFOCOM* (2014), pp. 199–207. https://doi.org/10.1109/INFOCOM.2014.6847940
24. S. Chakraboarty, Lecture 44: Software Defined Networking—II (Open Flow)—YouTube, *Department of CS, IIT Khargapur*. https://www.youtube.com/watch?v=l3E-C1j-SJg. Accessed 24 Mar 2021
25. P. Bosshart et al., Forwarding metamorphosis: fast programmable match-action processing in hardware for SDN. Comput. Commun. Rev. **43**(4), 99–110 (2013). https://doi.org/10.1145/2534169.2486011
26. A.K. Rangisetti, B.R. Tamma, Software defined wireless networks: a survey of issues and solutions. Wirel. Pers. Commun. **97**(4), 6019–6053 (2017). https://doi.org/10.1007/s11277-017-4825-8
27. R. Maaloul, R. Taktak, L. Chaari, B. Cousin, Energy-aware routing in carrier-grade Ethernet using SDN approach. IEEE Trans. Green Commun. Netw. **2**(3), 844–858 (2018). https://doi.org/10.1109/TGCN.2018.2832658
28. ONF, Software-Defined Networking (SDN) Definition—Open Networking Foundation. https://opennetworking.org/sdn-definition/. Accessed 24 Mar 2021

29. B. Dong, W. Wu, Z. Yang, J. Li, Software defined networking based on-demand routing protocol in vehicle ad hoc networks, in *Proceedings—12th International Conference on Mobility Ad-Hoc, Sensing and Networking, MSN 2016* (2017), pp. 207–213. https://doi.org/10.1109/MSN.2016.041
30. L.L. Zulu, K.A. Ogudo, P.O. Umenne, Simulating software defined networking using Mininet to optimize host communication in a realistic programmable network, in *2018 International Conference on Advances in Big Data, Computing and Data Communication Systems, icABCD 2018*, no. August (2018). https://doi.org/10.1109/ICABCD.2018.8465433
31. W. Braun, M. Menth, Software-defined networking using OpenFlow: protocols, applications and architectural design choices. Futur. Internet **6**(2), 302–336 (2014). https://doi.org/10.3390/fi6020302
32. R. Jain, OpenFlow, Software Defined Networking (SDN) and Network Function Virtualization (NFV), in *IEEE International Conference on Communication* (2014), pp. 1–102 [Online]. Available: http://icc2014.ieee-icc.org/2014/private/Tutorial11.pdf
33. R. Bruschi et al., An SDN/NFV platform for personal cloud services. IEEE Trans. Netw. Serv. Manag. **14**(4), 1143–1156 (2017). https://doi.org/10.1109/TNSM.2017.2761860
34. Q. Wang, G. Shou, Y. Liu, Y. Hu, Z. Guo, W. Chang, Implementation of multipath network virtualization with SDN and NFV. IEEE Access **6**, 32460–32470 (2018). https://doi.org/10.1109/ACCESS.2018.2842058
35. S.S. Kadam, Y.C. Kulkarni, Improving the performance of Squid proxy server by using SCSI HDD and blocking the media streaming. Int. J. Comput. Appl. **47**(25), 38–41 (2012). https://doi.org/10.5120/7540-0547
36. S. Ejaz, Z. Iqbal, P. Azmat Shah, B.H. Bukhari, A. Ali, F. Aadil, Traffic load balancing using Software Defined Networking (SDN) controller as virtualized network function. IEEE Access **7**(c), 46646–46658 (2019). https://doi.org/10.1109/ACCESS.2019.2909356
37. T. Li, H. Zhou, H. Luo, Q. Xu, Y. Ye, Using SDN and NFV to implement satellite communication networks, in *2016 International Conference on Networking and Network Applications (NaNA)* (2016), pp. 131–134. https://doi.org/10.1109/NaNA.2016.22
38. W. Zhuang, Q. Ye, F. Lyu, N. Cheng, J. Ren, SDN/NFV-empowered future IoV with enhanced communication, computing, and caching. Proc. IEEE **108**(2), 274–291 (2020). https://doi.org/10.1109/JPROC.2019.2951169
39. N. Jawad et al., Smart television services using NFV/SDN network management. IEEE Trans. Broadcast. **65**(2), 404–413 (2019). https://doi.org/10.1109/TBC.2019.2898159
40. H. Zhang, J. Yan, Performance of SDN routing in comparison with legacy routing protocols, in *2015 International Conference on Cyber-Enabled Distributed Computing and Knowledge Discovery* (2015), pp. 491–494. https://doi.org/10.1109/CyberC.2015.30
41. J.I.Z. Chen, S. Smys. Social multimedia security and suspicious activity detection in SDN using hybrid deep learning technique. J. Inf. Technol. **2**(02), 108–115 (2020)
42. J.S. Raj, A novel hybrid secure routing for flying ad-hoc networks. J. Trends Comput. Sci. Smart Technol. (TCSST) **2**(03), 155–164 (2020)

NewsCheck: A Fake News Detection and Analysis System

Ankush Shetty, Puneet Thawani, Aditya Rao, Aditya Uphade, and R. L. Priya

Abstract The growth of social media coupled with increased internet access has increased the reach of information and misinformation alike. The impact of fake news can be seen in various fields not limited to politics, entertainment finance, etc. To curtail the spread of misinformation, there is a need for a system to discern real and fake news. This paper demonstrates a complete web application to analyze and detect fake news. The proposed solution uses LSTM, CNN, and Hybrid models to classify the news. A web mining system to inculcate auxiliary facts for the user and an OCR module for text extraction for pieces of news transmitted via images.

Keywords Artificial ıntelligence · Fake news classification · Optical character recognition · Natural language processing · Deep learning · Convolutional neural network · Long short-term memory

1 Introduction

The internet and the world wide web have brought about a significant revolution in the propagation of news and information. Previously, people particularly relied

A. Shetty (✉) · P. Thawani · A. Rao · A. Uphade · R. L. Priya
Department of Computer Engineering, Vivekanand Education Society's Institute of Technology, Hashu Advani Memorial Complex, RC Marg, Collectors Colony, Chembur, Mumbai, Maharashtra 400074, India
e-mail: 2017.ankush.shetty@ves.ac.in

P. Thawani
e-mail: 2017.puneet.thawani@ves.ac.in

A. Rao
e-mail: 2017.aditya.rao@ves.ac.in

A. Uphade
e-mail: 2017.aditya.uphade@ves.ac.in

R. L. Priya
e-mail: priya.rl@ves.ac.in

G. Ranganathan et al. (eds.), *Soft Computing for Security Applications*, Advances in Intelligent Systems and Computing 1397,
https://doi.org/10.1007/978-981-16-5301-8_42

on traditional news channel broadcasts on television and radios, involving highly credible but relatively fewer established news sources. However, the modern technological advancements and feasibility of digital devices like mobile phones and computers have made social media platforms available at their fingertips. With this booming industry, people are not audiences anymore, rather they have the liberty to post and share news content on various social media platforms such as twitter, facebook, whatsapp, etc. without any forms of verification. Questions have always been raised about the authenticity of such posts, nevertheless, majority of the people believe what they see on the internet and share it further without reconsidering its credibility.

Fake news on social media is not just a profusely liked or shared post, rather it can be defined as a powerful technique of manipulating and multiplying cyber propaganda with its formidable effect on an extensive social media user base. According to research conducted about fake news proliferation in [1], people were determined to be the prime culprits when it comes to the propagation of misinformation through social networks. The study also revealed that tweets containing falsehoods reached about 1500 people on Twitter six times faster than truthful tweets.

Easily accessible and exponentially increasing information available on social media platforms has made it perplexing to distinguish between legitimate and false information. An example of it can be seen in [2] where multiple fake rumors were spread about nano GPS chips being embedded in 2000rs notes after the demonetization of the year 2016 in India. These astounding rumors and debates also made their way to mainstream media like ZeeNews. Later, RBI had to announce an official statement that such technology doesn't exist anywhere in the world to dismiss the rumors. Such incidents put limelight upon how major a concern, the credibility of news has become in the twenty-first century.

2 Literature Review/Related Work

We have performed a detailed literature review and published our findings in [3]. We will summarize the literature review required to implement our approach in the following section.

2.1 Models for Classification

Kaliyar [4] specifies the purpose of CNN and applied a neural network with convolutional, pooling, and dropout layers, which process vectors created with the help of the pre prepared GloVE embeddings. They have proved to be the gold standard by achieving an accuracy of 98% on their dataset, however, the scope of their data was only limited to the 2016 US Presidential Elections.

Another paper that inspired us to design our deep learning algorithms was [5], where they designed a Bi-directional Long Short-Term Memory (LSTM)—Recurrent Neural Network. For vector representation, they also have used GloVe pre-trained word embeddings. Here, they have implemented various other algorithms and compared them with the Bi-directional LSTM. They used two datasets to train, validate and test their models and for comparison among different models. The size of the combined dataset considered by this paper is lower compared to the dataset that we have used for training our models. They didn't consider news which was transmitted in the form of images. This system achieved an accuracy of 89.74% for the real_or_fake dataset [6] and 98% for Fake News detection [7] dataset.

The Multi Domain Visual Neural Network model in [8] utilizes the visual information in frequency and pixel domains to differentiate whether the given image is a fake or real news image. Weibo dataset consisting of around 4749 fake news and 4779 real news posts with corresponding images extracted using web scraping from trusted news sources was used. To clinch that each post corresponds to an image, posts containing only text were removed and a single image was saved for posts with multiple illustrations. Initially, a frequency domain sub-network with two fully connected layers incorporating 16 and 64 neurons respectively, along with a dropout layer of rate 0.4 was implemented, succeeding the last fully connected layer to avoid overfitting. The pixel domain sub-network consists of a number of hidden units of the GRU and the fully connected layer, which are 32 and 64 respectively, and every fully connected layer is accompanied by a dropout layer of rate 0.5. These sub-networks were pre-trained, furthermore fine-tuned during the joint training procedure. Additionally, data augmentation strategy for batch normalization was incorporated to enhance its generalization performance, accelerating the convergence, and batch size was set to 64. This CNN-RNN based MVNN has yielded an accuracy of 84.6%.

The implementation of an OCR model would provide us the flexibility of taking the news or facts to be verified in both, Textual as well as in the form of an image. As concluded in a comparative study in [9], EasyOCR outperformed Tesseract when an image containing a mixture of alphabets and numbers was given. Hence, it was selected as the substructure of our Enhanced OCR Module. With our implementation of numerous parameters, suggested in [10], along with additional features like a Dictionary or Smart Preprocessing block, The accuracy was enhanced further.

Along with the OCR Module, [11] intrigued us to implement a Web Mining stage to fetch auxiliary information when limited text is provided to the system. Addition of this block provided information like Source domain of the Article, Authors, Title, Timestamp, etc. which would provide the model with more context to the initial data. A comparative study was also provided, In which an LSTM Model and a GloVe Feed-Forward Neural Model witnessed an increase in accuracy, from 83.7% to 91.3% and from 83.3% to 84.3% respectively. These results concluded that Traditional Fake News Detection Systems when implemented with a Web Mining stage would yield a higher accuracy outperforming the prior system.

3 Proposed Solution

There is no need to include page numbers or running heads; this will be done at our end. If your paper title is too long to serve as a running head, it will be shortened. Your suggestion as to how to shorten it would be most welcome.

3.1 Data Preparation

In order to prepare a realistic dataset to emulate the real-life scenarios, authentic news was scraped from neutral sources, whereas conspiracy theory and satirical websites were scraped for fake news. The following websites were considered:

Authentic News: Reuters, AltNews, CNBC, Time.

Fake News: The Civilian, The Spoof, Waterford Whispers News, InfoWars.

The web pages were scraped with the help of.

BeautifulSoup, Requests, and Newspaper3k libraries of python. The scraped corpus contains a total of 16,496 articles, with 8248 articles each in the 'Real' and 'Fake' labels. The dataset consists of the article, heading, date of publication, and author's name. An additional dataset of 1606 articles was prepared using the web scraping technique, which was used for testing the accuracy of our system.

3.2 Data Processing

After data collection, the data was prepared before feeding it into our model for classification. The articles and titles contain URLs, punctuation, and special characters, which needed removal. This process is completed in the preprocessing stage with the help of Regular Expressions and the NLTK (Natural Language Toolkit) library of python. Subsequently, these pre-processed words from the article and title are tokenized and stemmed. These tokenized and stemmed words are used to prepare the tokenizer for our system, which is used for vector formation.

The same procedure is followed for the news articles which need to be classified. The texts are first pre-processed and vectors are created using our trained tokenizer. This vector is then fed into our deep learning model to get classified.

3.3 Deep Learning Models

3.3.1 CNN Model

We have trained a CNN model for the classification of news articles using the article text. The detailed diagram of our CNN model is shown in Fig. 1. A 32 neuron CNN is used to learn patterns from the news article.

The following layers were employed in our model:

Embedding Layer: When a user enters the text input, the input is in the form of alphabets and words. So in the first stage, we need to convert it into a matrix of numbers. When the text is converted into a matrix, there will be a high number of zeros in it. This matrix is called a sparse matrix. The embedding layer is used when the matrix is sparse and is filled with zeros. This layer is generally used to reduce the size of the matrix and convert them into a matrix of fixed size. This is a memory-intensive approach.

Convolutional Layer: The convolutional layer comes from the mathematical operation, convolution, where two functions produce a third function, which is a combination of the first two functions. In the convolutional layer, we have a filter which is a 3X3 matrix. The final result is a dot product of the input data and the filter data. The filters are always smaller compared to the input data because by doing so, we can use the same filter multiple times.

Max Pooling Layer: The Max Pooling layer selects the maximum element among the feature map, which is the result generated from the convolutional layer. In our model, we are using a 2X2 Max Pooling layer that selects the highest element from the 2X2 matrix. By doing this, we were able to select the most prominent feature of the input data.

Flatten layer: The Flatten layer is used to produce a 1-dimensional result. Here the 2-dimensional matrix is converted into 1 dimensional by traversing through row by row. The output of the Convolutional layers are flattened and are prepared for inputting it to the next layer for the process of classification.

Dense Layer: The Dense layer is used to change the dimensionality of the result generated from the past layers. The dense layer receives all the inputs from past neurons and then uses the activation function to produce a result with a different dimension. Generally, this layer applies scaling, rotation, and other vector transformations to the input data.

3.3.2 LSTM Model

We used a long short term recurrent neural network with 100 neurons to classify the data. The self connected neurons allow a constant flow of error as opposed to a naive approach[12].

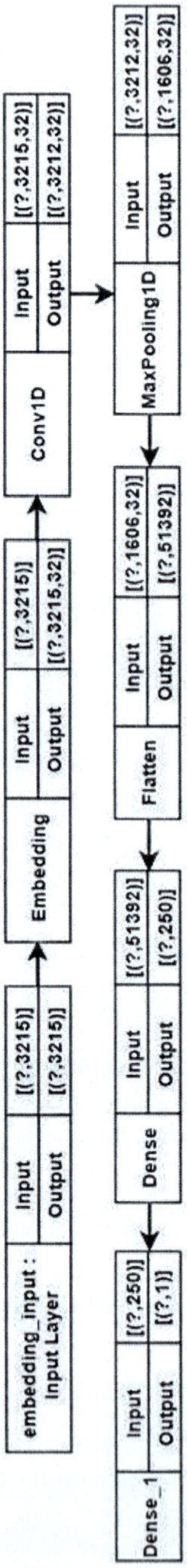

Fig. 1 Architecture of CNN

3.3.3 Hybrid Model

Having studied and performed traditional approaches to classification ie: a Convolutional Neural Network and a Long Short Term Memory Recurrent Neural Network and looking at the data for our case, we decided to develop a model that separately parses the text and title and merges the data at a later stage. The architecture of this model is shown in Fig. 2. a 100 neuron LSTM is used to learn patterns in the text of the news whereas a 24 neuron LSTM learns from the title. The results from both these LSTMs are concatenated by means of a dense layer followed by the output neuron.

3.4 Input Formats

Our proposed system uses different deep learning models to classify news. Our system takes input in three formats as follows:

a. **Title and text**:
   Here, the user needs to enter text and title, which is then used by our hybrid model to classify news. Consequently, this data is augmented with supplementary information through web mining. The architecture of the hybrid model was mentioned before.
b. **URL**:
   When the user enters a URL, the first step is to scrape text from the website using the BeautifulSoup, Requests, and Newspaper3k libraries.
c. **Image as an input**:
   In this type of user input, the user is allowed to upload images that need to be classified. The proposed system is integrated with an Optical Character Recognition (OCR) to extract text from the image.

For input in the form of images, we have integrated an OCR system to extract text from images. The following section explains the OCR implementation.

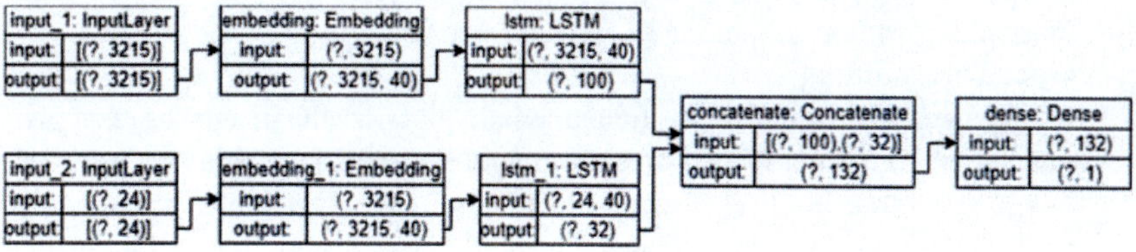

Fig. 2 Architecture of hybrid model

3.5 Optical Character Recognition (OCR) Module

3.5.1 Image Acquisition

Acquisition of Image by the OCR Model. This Image can be of various File Formats like JPG, PNG.

3.5.2 Smart Preprocessing

Smart Preprocessing allows the model to select the Most Accurate preprocessing approach out of Otsu Thresholding, Grayscaling & Image Sharpening for the given image input. The primary objective of preprocessing is to improve image data that suppresses undesired distortions and enhance some image features relevant for further processing and analysis tasks. For extracting the text from images efficiently, grayscaling and image sharpening combined with otsu thresholding were researched to be the best techniques. Furthermore, selecting the best among these techniques based on confidence level for individual images has also been implemented.

a. **Grayscaling**
Grayscale is a collection or range of monochromic gray shades, ranging from pure white on the lightest end to pure black on the opposite end. These achromatic images only contain luminance, i.e., brightness information and no color data. Anything between maximum luminance and zero luminance is a shade of gray. When a grayscaling algorithm is applied to an image consisting of a broad spectrum of colors, the text content gets highlighted and thus making it more convenient for the OCR engine to detect it.

b. **Sharpening**
The original image is first filtered by a high-pass filter that extracts the high-frequency components, and then a scaled version of the high-pass filter output is added to the original image, thus producing a sharpened image of the original. The homogeneous regions of the signal, where the signal is constant, remain unchanged. Image sharpening basically makes the edges of the text content sharper consequently enhancing visibility and detectability. A 5×5 sharpening mask along with a sharpening kernel has been used.

c. **Otsu Thresholding**
Image thresholding algorithms fundamentally binarize the image based on pixel intensities. The input to such thresholding algorithms is generally a grayscale image and a threshold. The output is a binary image. If the intensity of a pixel in the input image is greater than a threshold, the corresponding output pixel is marked as white representing foreground, whereas if the input pixel intensity is less than or equal to the threshold, the output pixel location is marked as black constituting the background. Otsu's thresholding method involves iterating through all the possible threshold values and calculating a measure of

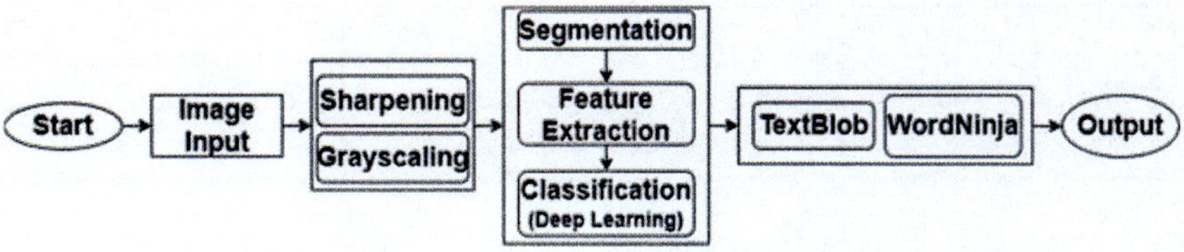

Fig. 3 Architecture of OCR module

spread for the pixel levels on each side of the threshold, i.e., the pixels that either fall in foreground or background.

d. **Segmentation**
The Input Image is firstly separated into its constituent components. This is done by performing Implicit & Explicit Segmentation.

e. **Feature Extraction**
This stage is used to extract geometrical & Statistical features, like Loops, corner points, moments, from the input image to identify the characters in the text.

f. **Classification**
Algorithms like Neural Networks, Bayesian, Nearest Neighborhood are used to categorize the identified character into its respective class.

g. **Post Processing**
Post Processing was implemented mainly to detect & rectify linguistic errors incurred in the output of our OCR model. This was done by implementing a text processing model named TextBlob, and a dictionary-based approach, named Wordninja. Further enhancement of the OCR result would help in the accurate detection of fraudulent text extracted from the image.

TextBlob: TextBlob is used for processing textual data provided from the OCR Model. It provides a simple API for diving into Natural Language Processing (NLP) tasks such as spelling correction, tokenization, parsing, word inflection, and lemmatization. It uses both NLTK and pattern identification for performing the tasks mentioned above.

Wordninja: Wordninja is also used for error detection of text but does this with the help of a pre-existing dictionary. It concatenates the string & then separates it into words on the basis of the pre-existing Wordninja Dictionary. Words can be added into or removed from this dictionary as per our need (Fig. 3).

3.6 Web Mining

Our system is integrated with the web mining stage. The whole process of adding auxiliary information is subdivided into 3 stages which are represented in Fig. 4.

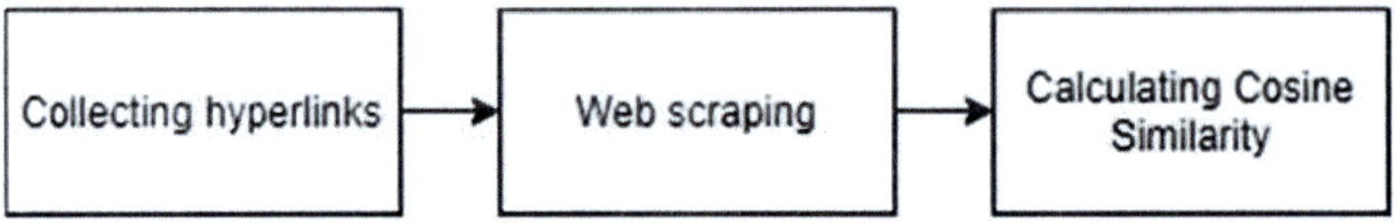

Fig. 4 Architecture of web mining module

3.6.1 Collect Hyperlinks for Specific Data

The hyperlinks for a specific search are collected in this stage. It is crucial to utilize texts which are present on the world wide web. To implement this stage we are using urllib, BeautifulSoup, and request libraries of python for returning various search results.

3.6.2 Scrape Text from Each Link

This stage is used to scrape text from various websites that were returned from stage 1 of web mining. For web scraping, we are using BeautifulSoup, Request, and Newspaper3k libraries.

3.6.3 Calculate Cosine Similarity

Now before adding auxiliary information to a user's input, we need to determine that the user's input and the scraped texts are similar. Now to perform that we are calculating cosine similarity between two texts. To calculate the cosine similarity, we need to create vectors of the user's text and the scraped text from various websites. Here we are using GlOve pre-trained word vectors to create vectors out of scraped texts and user's input. The reason for using a pre-trained word vector is that there were 840 billion tokens. After creating vectors we are calculating the cosine similarity between the tokens that were created.

4 Implementation Setup

For the successful implementation of our system, we need to install various libraries: starting from web scraping to the final web application. For the preparation of our dataset, we have used BeautifulSoup, Newspaper3k, and Requests libraries. Moreover, for successfully extracting text from an image, we have implemented an OCR module using Jupyter Notebook along with EasyOCR being the substructure of this module. With the intent to rectify errors, which might be detected in the extracted text, python libraries like TextBlob and Wordninja were implemented, enhancing the

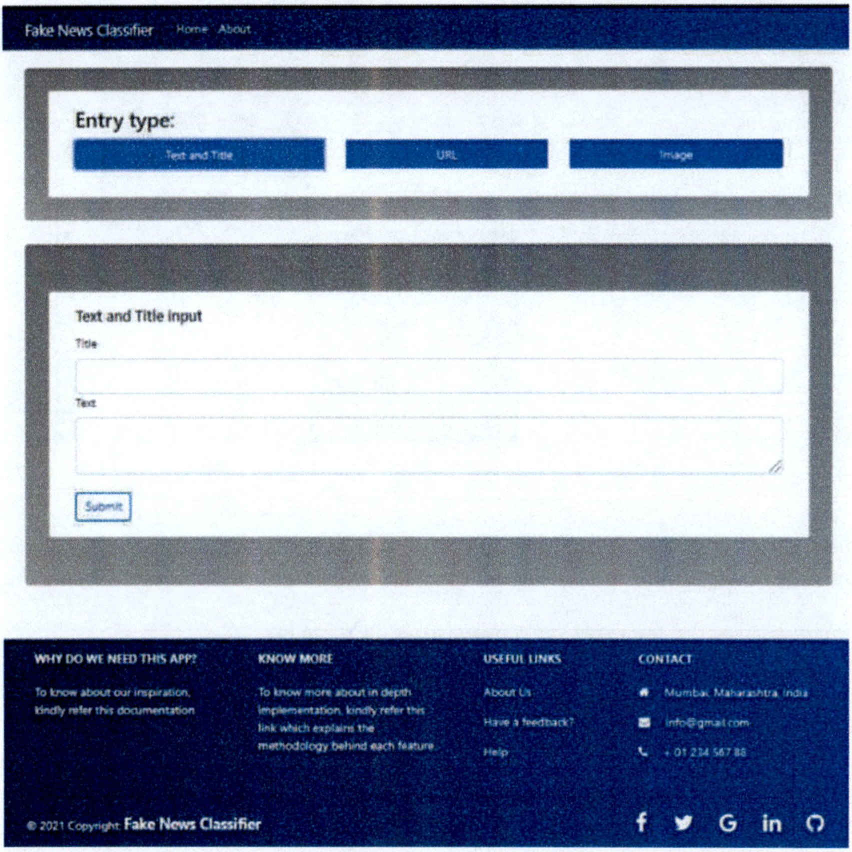

Fig. 5 User interface

accuracy even further. The chief module responsible for the classification of news articles was implemented using TensorFlow, NLTK, Pandas, Sklearn, and Matplotlib libraries. Furthermore, we have implemented a web mining module, which uses URLlib and Scipy libraries. In the end, we integrated all the different modules to produce a web application using Python's Flask framework. The UI of the same is shown in Fig. 5.

5 Results and Analysis

Having trained our models on our dataset of 16,496 articles, we performed the testing on these models using articles from similar sources and new sources for fake news like breitbart.com. A dataset of 1606 articles was prepared with 803 articles in each of the real and fake classes. We plotted the confusion matrices for all our models where

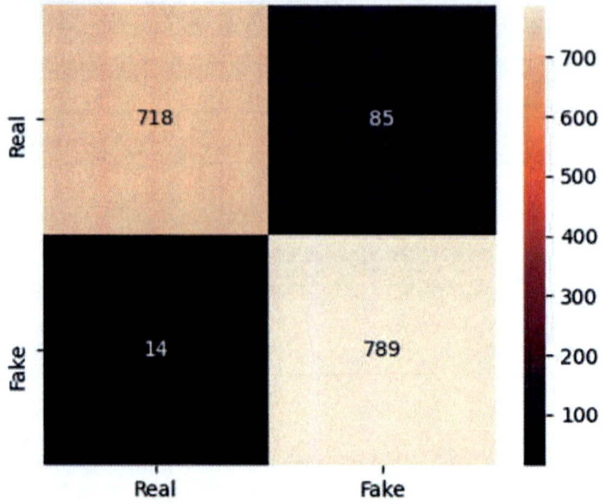

Fig. 6 Confusion matrix for CNN model

the X axis represents the labels predicted by our model and the Y axis represents the true labels. We will be measuring the accuracy of the models as follows:

$$\text{Accuracy} = \frac{\text{True Positives } + \text{True Negatives}}{\text{Total Samples}}$$

5.1 Results of CNN Model

The Fig. 6 represents the confusion matrix of our CNN model. According to this matrix, we achieved an accuracy of 93.83%. This model uses just the article text to classify considered news. This model yields the last number of false negatives (Fake news which is classified as real) and hence is the most favorable.

5.2 Results of Hybrid Model

The hybrid model which parses the text and title separately yielded an accuracy of 92.65% (Fig. 7).

5.3 Results of LSTM Model

The LSTM Model yielded an accuracy of 92.46% on the test data (Fig. 8).

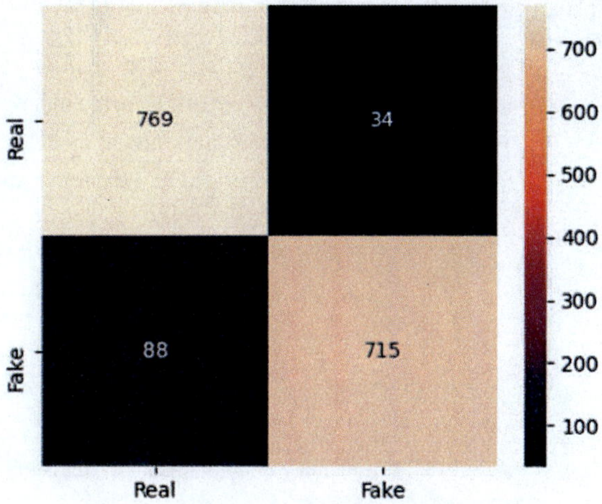

Fig. 7 Confusion matrix for hybrid model

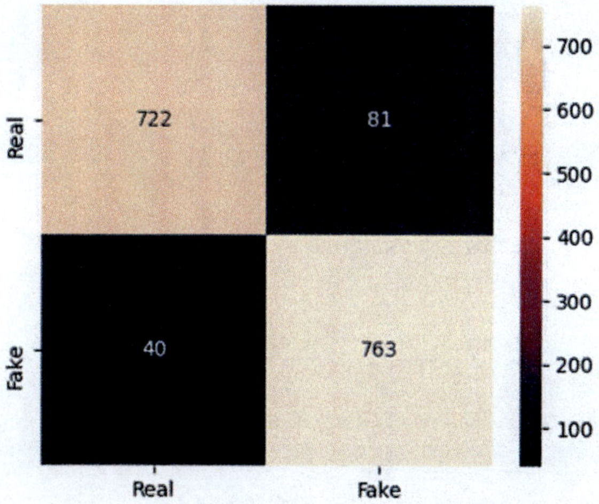

Fig. 8 Confusion matrix for LSTM model

6 Comparative Analysis with the Existing System

6.1 Comparison of Other System

Table 1 compares the implemented model with other systems used as reference.

Table 1 Comparison of models

System	Accuracy (%)
NewsCheck CNN model	**93.8**
NewsCheck Hybrid model	**92.65**
NewsCheck LSTM model	**92.46**
Bilateral Multi Perspective Matching (BiMPM) [13]	78
Deep neural approach to fake news identification [11]	91
Identification of fake news using machine learning [14]	85
FNDNet [4]	98.36

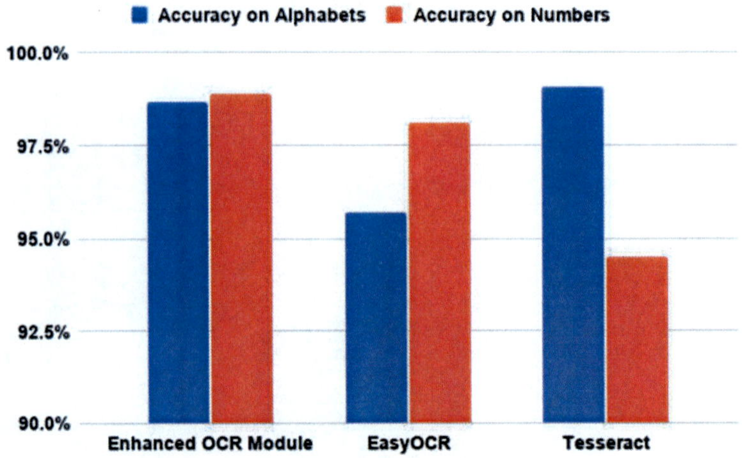

Fig. 9 Comparison of OCR models

6.2 *Comparison of OCR Models*

The Fig. 9 compares the accuracy on numbers and alphabets of our OCR module with the Easy OCR and Tesseract Models.

7 Conclusion and Future Work

In the contemporary world, the internet has become an integral part of our life. With the increase in internet access, it has now a huge impact on our day-to-day life. The users now have access to news through social media platforms, which they believe hastily. The containment of the spread of fake news is possible only when we can classify a piece of news as fake or real. So, there is a need for a system that could classify a piece of news. Therefore, we have proposed a web application to analyze and classify news. The classification of news was done with the deep learning model. The Convolutional Neural Network with one dimensional convolution and max pooling layers to extract features from the word embeddings gave us the best performance on our self prepared dataset from known authentic and fake pieces of news. Our system is integrated with an Optical Character Recognition (OCR) module, which would extract text from the image, owing to the fact that images with texts serve as a considerably large source of fake news. Further, due to the fact that it is important to ascertain additional facts and enhance the input to our neural network, we prepared a web mining stage which adds supplementary data to the system for the user to analyze their input and that adds to the overall applicability of the system. The system covers various aspects, which are crucial for fake news classification. Overall, the neural network provides satisfactory classification over a generalized news dataset and the web application provides useful features in addition to providing a classifier. Therefore, this system will surely assist the users.

References

1. K. Langin, Fake news spreads faster than true news on Twitter—thanks to people, not bots. https://www.sciencemag.org/news/2018/03/fake-news-spreads-faster-true-news-twitter-thanks-people-not-bots
2. R. Venkataramakrishnan, No, the RBI hasn't announced a Rs 2,000 note with a 'nano GPS chip' that it can track. Scroll, https://scroll.in/article/821078/no-the-new-rs-2000-note-will-not-have-a-nano-gps-chip-that-can-track-it-anywhere
3. A. Rao, A. Shetty, A. Uphade, P. Thawani, R.L. Priya, A Proposal for a novel approach to analyze and detect the fake news using AI techniques. 2020. 3rd International Conference on Intelligent Sustainable Systems (ICISS) (2020)
4. R.K. Kaliyar, A. Goswami, P. Narang, S. Sinha, 'FNDNet—A Deep Convolutional Neural Network for Fake News Detection. Cognitive Systems Research (2019)
5. P. Bahad, P. Saxena, R. Kamal, Fake news detection using bi-directional LSTM-recurrent neural network. Procedia Computer Science **165**, pp. 74–82 (2019). ISSN 1877–0509
6. real_or_fake dataset. https://www.kaggle.com/rchitic17/real-or-fake
7. Fake News detection dataset. https://www.kaggle.com/jruvika/fake-news-detection
8. P. Qi, J. Cao, T. Yang, J. Guo, J. Li, Exploiting multi-domain visual ınformation for fake news detection. IEEE International Conference on Data Mining (ICDM) (2019)
9. C. Liao, OCR engine comparison—Tesseract vs. EasyOCR. https://medium.com/swlh/ocr-engine-comparison-tesseract-vs-easyocr-729be893d3ae
10. API documentation. Jaided AI. https://www.jaided.ai/easyocr/documentation/
11. S. Deepak, C. Bhadrachalam, Deep neural approach to Fake news identification. International Conference on Computational Intelligence and Data Science (ICCIDS 2019)
12. S. Hochreiter et. al., Long short-term memory
13. K. Kim, C. Jeong, Fake news detection system using article abstraction. 2019 16th International Joint Conference on Computer Science and Software Engineering (JCSSE) (2019), pp. 209–212, https://doi.org/10.1109/JCSSE.2019.8864154.
14. R.R. Mandical, N. Mamatha, N. Shivakumar, R. Monica, A.N. Krishna, Identification of fake news using machine learning. 2020 IEEE International Conference on Electronics, Computing and Communication Technologies (CONECCT) (2020)

A Predictive Model of Handwritten Digits Recognition Using Expert Systems

P. Praveen, N. Abhishek, K. Jayanth Reddy, C. Tanishq Raj, and A. Yashwanth Reddy

Abstract Converting real written data and information into digital form for storage in a digital format necessitates the use of expensive manual labor. The paper aims to debate a response to some of the issues since the spectrum has been limited to just handwritten digits (0–9). The 'MNIST DATASET,' which includes training and test datasets for handwritten digits (0–9) with a resolution of (28 × 28), or 784 pixels, was used. The MNIST data set contains 60,000 training data and 10,000 test data.

Keywords Machine learning · Artificial neural networks (ANN) · Deep learning · SVM · MNIST · Preprocessing · Handwritten · Resolution · Pixel

1 Introduction

Computer Vision and Deep Learning technologies are constantly improving, and it all comes down to one algorithm: the CNN. It's a deep learning algorithm that takes an image as input and assigns important (learnable weights and biases) to different aspects/objects within the image, allowing it to distinguish between them. Deep CNN (Convolutional Neural Networks) have achieved human-level precision in fields like handwritten digit recognition [3–6].

Everyone now uses images to reflect and convey knowledge, including students, children, and others. It's also vital to extract data from the picture, which may involve numbers or digits written in a variety of ways by various people. In a handwritten digit recognition method, it's important to consider how the data is interpreted; the data is expressed in pixels here [1, 2].

Character images are provided as input to the model in handwritten digit recognition. The machine is capable of recognizing the model. A few hidden layers between the input and output layers along with input and output layer make up a basic artificial neural network (ANN). CNN and ANN have somewhat similar architectures. Each layer of the ANN contains some neurons. The input to the next layer of a neuron

P. Praveen (✉) · N. Abhishek · K. J. Reddy · C. T. Raj · A. Y. Reddy
Department of Computer Science and Artificial Intelligence, SR University, Warangal, Telangana, India

G. Ranganathan et al. (eds.), *Soft Computing for Security Applications*, Advances in Intelligent Systems and Computing 1397,
https://doi.org/10.1007/978-981-16-5301-8_43

is the total/weighted sum of all the neurons in that layer. The layer would be three-dimensionally visible in CNN [9–11]. The neurons in this region aren't all linked in the same way. Instead, every neuron in the layer is connected to the receptive field in its immediate surroundings. We applied a special type of CNN algorithm to the MNIST (Modified National Institute of Standards and Technology) dataset using TensorFlow, a Python-based neural network library [15–18].

Handwritten digit recognition is now extensively used for automated processing in bank, such as processing payment checks, postal addresses, and other types of data. If you look at a few of the current systems, you'll find that they use ANN which is a computational intelligence technique, as well as massive lookup tables similar to those used in a handwritten digit recognition system [12–14].

Artificial neural networks, also referred to as Neural Networks (NNs), are developed since the 1940s, but they need been utilized in several broad disciplines over the past 15 years. Although different forms of neural networks have emerged in recent years, these are the most basic mathematical models of biological neurons. While some are implemented in hardware, the large percentage are software-based. This project aims to manipulate images that contain data such as numbers or digits using a variety of techniques. This will help us figure out how much we can increase our productivity by using very simple and straightforward techniques [7, 8].

2 Literacy Survey

CNN is known for detecting and categorizing flaws. Many papers and articles have been written on this topic. Machine learning algorithms such as SVM, KNN, and RFC are less accurate when compared to Deep Learning algorithms such as multilayer CNN using Keras with Theano and TensorFlow have the higher precision, according to the study. Because of its high precision, the Convolutional Neural Network (CNN) is commonly used in image recognition, video processing, and other applications. CNN plays an important role in a variety of areas, including image processing. It has a big impact on a number of areas. CNN is also a forerunner of nanotechnology, such as semiconductor manufacturing. By varying different parameters, CNN is also used in nanotechnologies, such as semiconductor manufacturing, for tongue processing and emotion recognition. Since the large-scale neural network has more parameters, it's more difficult to achieve good results. All of us are attempting to improve CNN's accuracy while lowering error rates.

Handwritten digit recognition has gotten a lot of attention because of its various applications in pattern recognition and machine learning. Various methods are proposed for identifying digits in handwritten systems. Despite the fact that many studies and publications describe how to convert textual material from a document paper to a machine-readable language. The most important aim of this recognition system is to create a system with improved preprocessing and recognition accuracy, they can be achieved using ANN algorithms.

In another analysis, they discovered that when deep nets are trained using simple back-propagation. Unlike NORB and CIFAR10, their design results in a very low error rate on MNIST. It has around 99.8% accuracy rate and is capable of handling large volumes of data. CNN's applications are evolving with each day, and a wide range of research is being conducted. Various techniques are used to adjust the slope and slant of handwritten text within the sort of digits and reduce the size of the text using supervised learning methods.

3 Existing Systems

Existing algorithms are mentioned below:

a. KNN-K Nearest Neighbors
b. SVM-Support Vector Machine
c. NN-Neural Networks

3.1 K Nearest Neighbors (KNN)

K Nearest Neighbor algorithm is a basic Machine Learning algorithm. It is based on the Supervised Learning approach. The KNN algorithm believes that the new case/data and existing cases are identical, and it assigns the new case to the category that matches the existing categories the best. The primary purpose of KNN algorithm is for classification, but it is used for regression also. As it does not learn immediately from the training set, it is also called as lazy learner algorithm; instead, it retains the dataset and uses it to interpret it later. During the training process, the KNN algorithm stores the dataset, and when it receives new data, it classifies it into a gaggle of existing similar data.

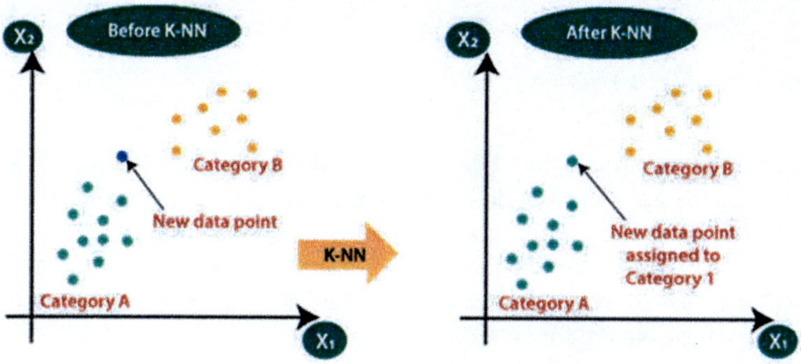

Fig. 1 K nearest neighbors

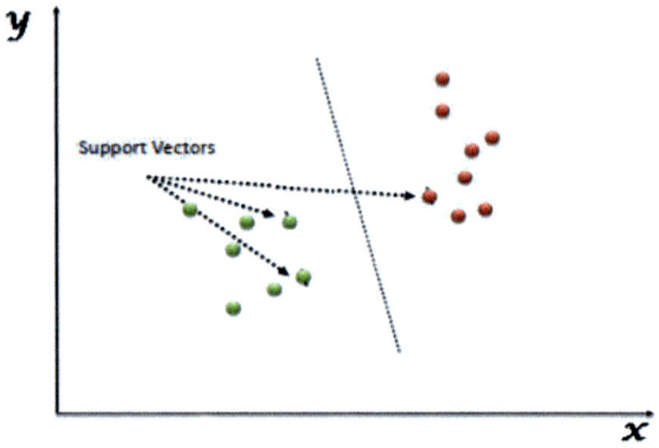

Fig. 2 Support vector machine

Figure 1 shows how the classification in KNN is completed, with two groups of information points, A and B. If we add a replacement datum anywhere in the aircraft, it will fall into the category that is closest to it, i.e., it will help the room.

3.2 *Support Vector Machine (SVM)*

SVM is a supervised machine learning algorithm that is used to solve regression and classification tasks. It is, however, more often used to address classification issues. The value of each function is the value of a chosen coordinate in the SVM algorithm, and each data object is defined as an extent in n-dimensional space (where n is the number of features). Then, as shown in Fig. 2, we classify the data by finding the hyper-plane that best separates the two classes.

The coordinates of each individual observation within the plane are called Support Vectors. The SVM classifier could function as a frontier, separating the two classes (hyper-plane and line) as much as possible.

3.3 *Neural Networks (NN)*

Our brains and neural networks both function in the same way. They have come a long way in the era of increased computing power. The abbreviation for Neural Networks, a multilayer network, is deep learning. Nodes are formed by connecting the layers. A node is a perception that refers to a process, evaluates it, and then sends it through a node's activation process to demonstrate how the background signal flows through the network for classification.

Layers of computational units called neurons, each with its own set of connections, make up neural networks. These networks also data transformation before classifying it as an output. Each neuron multiplies an initial value by weight, applies the results

Fig. 3 Neural networks

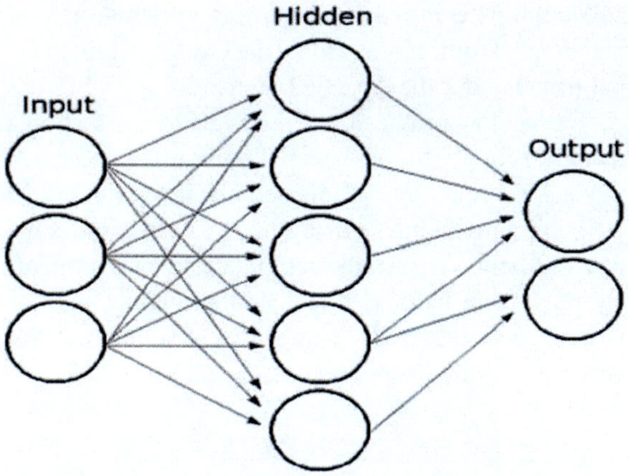

to additional values are being added an equivalent neuron, adjusts the obtained result with the neuron's bias, and then finally normalizes the performance with a function of activation as shown in Fig. 3.

4 Adopted Methodology

Throughout this project, we will design, practice, and test our models using the TensorFlow library, with data from the MNIST database. A Convolutional Layer would be used as the algorithm. A Convolutional Neural Network (ConvNet/CNN) is a Deep Learning algorithm which uses a pixel and assigns value (learnable weights and biases) to different aspects/objects within the image while also distinguishing between them. The amount of preprocessing required by a ConvNet is significantly less than that required by other classification algorithms. Despite the fact that ConvNets are hand-engineered using primitive methods, they can learn these filters/characteristics. Deep Learning has emerged as a key method for self-perception problems such as image comprehension, human voice recognition, and robot exploration of the world. We want to use the Convolutional Neural Network principle for digit recognition. The proposed model's goal is to understand CNN and apply it to a handwritten digit recognition system. The features maps are extracted from the 2D images using a Convolutional Neural Network.

4.1 Tensor Flow

TensorFlow is a free and open-source dataflow and differentiable programming library that can be used to solve a broad variety of issues. It is a symbolic math collection used in machine learning systems such as neural networks. TensorFlow was created for internal use by the Google Brain team. On November 9, 2015, it was published under the Apache License 2.0.

TensorFlow provides several levels of abstraction, enabling us to settle on the one that most closely fits our needs. The high-level Keras API often will not build and train models, making TensorFlow and machine learning simple to know. TensorFlow is a programming language that allows you to express machine learning algorithms and execute them. TensorFlow computations can be run on a wide range of heterogeneous systems, from handheld devices like phones and tablets to large-scale distributed systems with hundreds of machines and thousands of computing devices like GPU cards, with little to no changes.

4.2 MNIST Dataset

Without altering the model's description in some way. The MNIST is a vast database of handwritten digits that is widely used to train image processing systems. The database is commonly used for machine learning testing and research.

There are 60,000 instructional images and 10,000 research images in the MNIST database. The instruction data collection from NIST included half of the training set and half of the test set, while the testing dataset included the other half of the training set and half of the test set. Attempts to achieve a near-zero error rate are recorded in a series of research papers (Figs. 4, 5, 6, 7 and 8).

Fig. 4 Differing types of handwritten digits

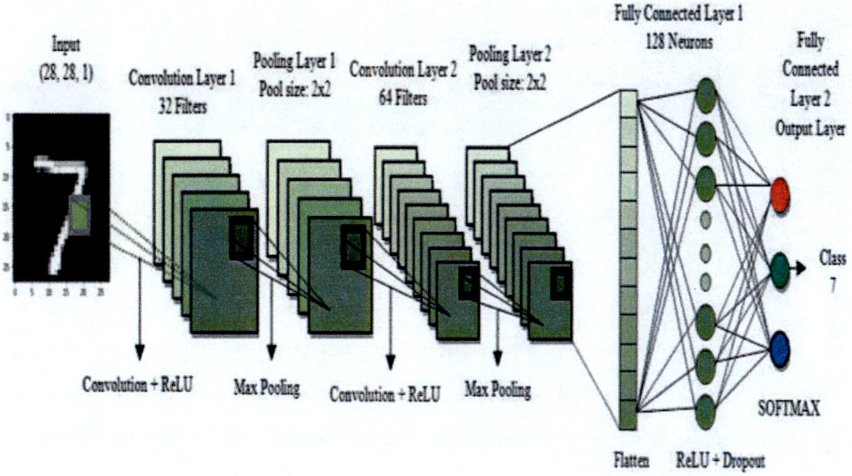

Fig. 5 Layers of CNN

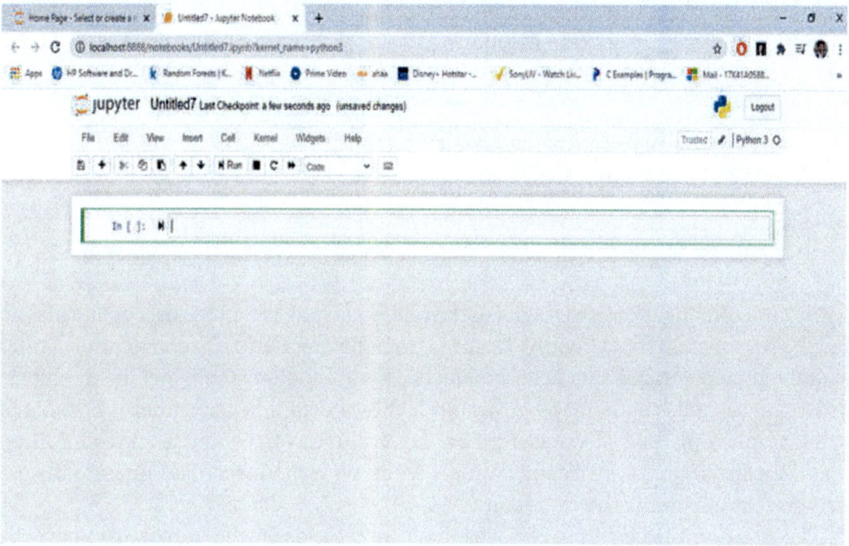

Fig. 6 Juypter notebook environment

4.3 The Algorithm CNN

With one input layer, five hidden layers, and one output layer, a seven-layered CNN is programmed to detect handwritten numbers. The network's input layer is made up of 28 by 28-pixel pixels, resulting in 784 neurons. The input pixels are greyscaled, with white pixels having a worth of 0 and black pixels having a worth of 1. Five secret

```
Train on 60000 samples, validate on 10000 samples
Epoch 1/10
60000/60000 [==============================] - 39s 656us/step - loss: 0.3793 - accuracy: 0.8828 - val_loss: 0.0660 - val_accura
cy: 0.9800
Epoch 2/10
60000/60000 [==============================] - 35s 586us/step - loss: 0.1078 - accuracy: 0.9718 - val_loss: 0.0380 - val_accura
cy: 0.9881
Epoch 3/10
60000/60000 [==============================] - 36s 601us/step - loss: 0.0773 - accuracy: 0.9802 - val_loss: 0.0321 - val_accura
cy: 0.9897
Epoch 4/10
60000/60000 [==============================] - 36s 604us/step - loss: 0.0617 - accuracy: 0.9839 - val_loss: 0.0334 - val_accura
cy: 0.9897
Epoch 5/10
60000/60000 [==============================] - 47s 786us/step - loss: 0.0536 - accuracy: 0.9865 - val_loss: 0.0346 - val_accura
cy: 0.9908
Epoch 6/10
60000/60000 [==============================] - 64s 1ms/step - loss: 0.0472 - accuracy: 0.9879 - val_loss: 0.0339 - val_accurac
y: 0.9902
Epoch 7/10
60000/60000 [==============================] - 44s 737us/step - loss: 0.0399 - accuracy: 0.9895 - val_loss: 0.0272 - val_accura
cy: 0.9927
Epoch 8/10
60000/60000 [==============================] - 47s 789us/step - loss: 0.0354 - accuracy: 0.9905 - val_loss: 0.0269 - val_accura
cy: 0.9930
Epoch 9/10
60000/60000 [==============================] - 46s 765us/step - loss: 0.0331 - accuracy: 0.9916 - val_loss: 0.0237 - val_accura
cy: 0.9934
Epoch 10/10
60000/60000 [==============================] - 49s 822us/step - loss: 0.0298 - accuracy: 0.9919 - val_loss: 0.0299 - val_accura
cy: 0.9917
The model has successfully trained
```

Fig. 7 Number of epocs of the model

```
score = model.evaluate(x_test, y_test, verbose=0)
print('Test loss:', score[0])
print('Test accuracy:', score[1])
```

```
Test loss: 0.029874360483574003
Test accuracy: 0.9916999936103821
```

Fig. 8 Model accuracy

layers structure the proposed model. Convolution layer 1 is that the first opaque layer, and it's responsible for extracting features from the input info. By convolution a filter with the previous layer, this layer conducts convolution on small, localized regions. It also has several feature maps of various levels of detail. When a filter is convolved with the previous one, it creates a new layer, this layer conducts convolution on small, localized areas. Learnable kernels along with these rectified linear units are also present in certain function maps.

The location of the filters is determined by the size of the kernel. To boost the model's efficiency, ReLU is used as both an activation feature and a totally linked layer at the top of a convolution layer. The next hidden layer is the pooling layer 1. By reducing the output information from the convolution layer, it reduces the parameter count and also the computational complexity of the model. To subsample the dimension of every function index, max pooling is employed. Level 2 of convolution and layer 2 of pooling have the same purpose as convolution layer 1 and pooling layer 1 respectively, with the exception that their feature maps and kernel sizes are different. After the PO is applied, A sheet of flattening is added. Following the layer of pooling, the flatten layer transforms the 2D feature to convert a matrix to a 1D function vector, allowing output management to be managed by fully linked layers. Another secret layer, also known as the dense layer, that may or may not be fully connected.

It's close to Artificial Neural Networks' hidden layer (ANNs), except it's fully wired and connects all of the network's neurons layer by layer. Overfitting can be avoided at totally connected layer of level 1, the dropout regularization technique is also used. It switches off those neurons at random during preparation to increase the network's efficiency by making it more stable. As a result, the network has a good chance of succeeding.

As a result, the network is more generalizable and less prone to overfitting the training data. The output layer of the network, which consists of ten neurons, defines the digits 0 to 9. The output layer classifies the output digits from 0 to 9, with 9 being the best worth of activation, since it uses an activation function like SoftMax to enhance the model's accuracy.

5 Experimental Setup

We used Jupyter Notebook for this project, which is a non-profit organization whose mission is to "develop open-source software, open standards, and services for interactive computing through dozens of programming languages." It's an interactive computing environment for creating Jupyter notebook documents that run on the internet (Web Application). This will also connect to a variety of kernels, allowing for programming in a variety of languages. IPython 0.12 (December 2011) introduced the Notebook gui, which was renamed to Jupyter notebook in 2015 (IPython 4.0—Jupyter 1.0). Jupyter Notebook is a computational interface style that began with Mathematica in the 1980s and is similar to the notebook interface of other programs such as Maple, Mathematica, and SageMath. Jupyter's success surpassed that of the Mathematica notebook gui in early 2018, according to The Atlantic.

6 Results and Discussion

In the preceding segment, CNN was applied to the MNIST dataset in order to monitor the difference in digit accuracies. TensorFlow in Python is used to calculate the accuracy. Using a batch size of 100, For ten different epochs, the preparation and validation precision was observed. We have taken 60,000 samples on training dataset and 10,000 on validation dataset, at 1st epoch the accuracy is about 88.28% and the test loss is about 6.6%, at second epoch the test accuracy is about 97.18 and test validation is about 3.8%.

At epoch 1, for 5 secret layers with a batch size of 128 and 10 epochs. The entire training accuracy was observed to be 98.00% at epoch 10, resulting in a maximum validation loss of 0.83%.

During testing, the model achieved a 99.16% accuracy rate over 10,000 test images, as shown in Fig. 7.

7 Conclusion

In this study, an implementation of Handwritten Digit Recognition based on Deep Learning is presented to identify the handwritten digits, we built a convolutional neural network with a complete of seven layers during this project. CNN outperforms other classifiers by a wide margin. It has the potential to eliminate the need for typing entirely. Digit identification is an excellent prototype dilemma for learning about neural networks, and it provides an excellent way to practice. To create more sophisticated deep learning techniques The MNIST dataset is used to train the algorithm. During testing, we discovered that our model outperformed the previously existing versions (K Nearest Neighbors, Support Vector Machine, and Neural Networks), with an accuracy of 99.16%.

References

1. Wu M, Zhang Z, Handwritten digit classification using the MNIST dataset (2010)
2. A. Dutta, Handwritten digit recognition using deep learning. Int. J. Adv. Res. Comput. Eng. & Technol. (IJARCET) **6**(7) (2017)
3. N. Dalal, B. Triggs, Histograms of oriented gradients for human detection. Proceedings of the IEEE Computer Society Conference on Computer Vision and Pattern Recognition (CVPR '05), vol. 1 (San Diego, CA, USA, 20–26 June 2005) pp. 886–893
4. D.G. Lowe, Distinctive image features from scale-invariant keypoints. Int. J. Comput. Vis. **60**, 2 (2004)
5. Y. LeCun et al., Backpropagation applied to handwritten zip code recognition. Neural Comput. **1**(4), 541–551 (1989)
6. R Ravi Kumar, M. Babu Reddy, P. Praveen, Text classification performance analysis on machine learning. Int. J. Adv. Sci. Technol. **28**(20), pp. 691–697, (2019)
7. https://derektanderson.com/FuzzyLibrary/customnetmnst.html
8. "MatConvNet Convolutional Neural Networks for MATLAB" Andrea Vedaldi, Karel Lenc, Ankush Gupta
9. K. Jarrett, K. Kavukcuoglu, M.A. Ranzato, Y. LeCun, What is the best multi-stage architecture for object recognition, CCV'09, IEEE (2009)
10. R. Hadsell, S. Chopra, Y. LeCun, Dimensionality reduction by learning an invariant mapping. CVPR **2**, 1735–1742 (2006)
11. H.W. Ian, E. Frank, *Data Mining: Practical Machine Learning Tools and Techniques*, 2nd edn. (Morgan Kaufmann, San Francisco, CA, USA, 2005)
12. N. Otsu, A threshold selection method from gray-level histograms. IEEE Trans. Syst. Man Cybern. **9**(1), 62–66 (1979)
13. F.C. Ribas, L.S. Oliveira, A.S. Britto, R. Sabourin, Handwritten digit segmentation: A comparative study. Int. J. Doc. Anal. Recognit. IJDAR **16**(2), 127–137 (2013)
14. Praveen, C. Jayanth Babu, Big data clustering: applying conventional data mining techniques in Big Data environment. Innovations in Computer Science and Engineering, Lecture Notes

in Networks and Systems 74, ISSN 2367-3370 (Springer, Singapore, 2019). https://doi.org/10.1007/978-981-13-7082-3_58
15. K. Bansal, R.G. Kumar, "Cleaning and recognition of numerals in a handwritten Devnagari document consisting of roll numbers and marks, PhD Thesis (2012)
16. J. Schmidhuber, Deep learning in neural networks: An overview. Neural Netw. **61**, 85–117 (2015)
17. P. Praveen, C.J. Babu, B. Rama, Big data environment for geospatial data analysis. 2016 International Conference on Communication and Electronics Systems (ICCES), Coimbatore, 2016, pp. 1–6. https://doi.org/10.1109/CESYS.2016.7889816
18. B. Rama, P. Praveen, H. Sinha, T. Choudhury, A study on causal rule discovery with PC algorithm. 2017 International Conference on Infocom Technologies and Unmanned Systems (Trends and Future Directions) (ICTUS), Dubai (2017), pp. 616–621. https://doi.org/10.1109/ICTUS.2017.8286083

A Review of Deepfake Technology: An Emerging AI Threat

Mridul Sharma and Mandeep Kaur

Abstract Fake digital content disseminated over the internet has been posing a great challenge for the various social media platforms and real-world applications. The technological advancement has facilitated the generation of such content, such that differentiating between fake and factual content has become ever more challenging. One such emerging technology, known as *deepfake*, can create some new hyper-realistic but fraudlent material that cannot be easily captured by human eyes and seems utmost realistic. The technology relies on AI and its subsets, namely machine learning, and deep learning. It is pertinent to note that the research in this direction is meager and there is an instant need for novel approaches that can reliably discriminate between genuine and fake multimedia content. The current paper aims to review the deepfake technology along with its associated threats, challenges, and future directions that can help cope with fake information circulated online. The accuracy and robustness of available methods for detecting such fake videos based on the available two generations of datasets are deliberated. Constructive criticism is presented that also highlights some of its useful applications for the digital society.

Keywords Deepfake · Deep learning · Artificial intelligence · Fake information · Face swap · Lip synching · Machine learning · Deepfake detection

1 Introduction

Deepfake [1] is a technology that is used to create fraudulent material that would seem real to human eyes. One of the most common examples is swapping the face of a target person with the original image. This technology has both constructive and harmful implications in real-life scenarios. However, ill-effects of this technology could pose a danger to the personal reflection of humans majorly seen in pornography,

M. Sharma (✉) · M. Kaur
Department of Information Technology, U.I.E.T. Panjab University, Chandigarh, India

M. Kaur
e-mail: mandeep@pu.ac.in

G. Ranganathan et al. (eds.), *Soft Computing for Security Applications*,
Advances in Intelligent Systems and Computing 1397,
https://doi.org/10.1007/978-981-16-5301-8_44

and many fraudulent materials that are circulated over the internet. Deepfakes were primarily seen in 2017 where famous celebrities were made controversial by generating deepfake versions of them in pornographic videos. Similarly, in 2020 during the US election conflicting trolls were circulated on social media referring to Trump and Biden. Different actors, companies hire people to create false content to outstand amongst their competitors. Besides many negative impacts some of the optimistic usages of this technology can be reflected in fields of art, education, industry where making digital voices for people who lost theirs or updating film footage instead of reshooting it if actors trip over their lines. In the case of businesses that are interested in the potential of brand applicable deepfake technology, it can transform e-commerce and advertising in significant ways. In 2021, deepfake showed constructive implication were deepfake videos of deceased ones brought an emotional connect with their family members. This technology primarily allows the user to swap the faces of two actors in such a manner as it appears realistic. With the advancement in technology, many variants of deepfake could be visualized making it merely impossible to detect at one look. Deepfake in still images, audios or videos, or any multimedia source represent forgeries and tampered visual information, and detection of such media inorder to maintain authenticity is of great concern. Deepfake has become an emerging center of fake news which has become a hot issue, posing a threat to public discourse. The visual quality of videos and efficiency would determine how good a deepfake video is being created. Many such deepfake generated videos do not produce much better results in terms of details directly linked with skin tones and facial hairs. These synthesized media can be turned into realistic sources as in [1] if real voices are accompanied by video making audio and video synthesis together. Besides the generation of powerful deepfake videos much more important aspect is how to detect what is real and what is fake as it is being increasingly tough to know what to actually trust upon.

The remainder of this paper is organized as follows: Sect. 2 gives a brief description of the Deepfake technology. Section 3 presents the state-of-the-art of Deepfake generation, followed by review of detection methods till date in Sect. 4. İn Sect. 5, future research directions and constructive criticism is presented that highlight some of its useful applications for the digital society Finally, Sect. 6 concludes the paper.

2 Deepfake Technology

Deepfakes [1] are the visual combination of AI (artificial intelligence) and its applications that can create such content that is merged, somewhat replaced, and then superimposed to images or videos to create fraudulent material that could hoax social media users. Deepfake videos actually depict the users saying or doing things that never happened in real scenario. Deepfake was first seen in 2017 which increasingly began to circulate over the internet and till date many such videos with better accuracy rates are being created and witnessed by millions which could not be easily detected by humans [1]. Deepfake comes in many variants where three major types

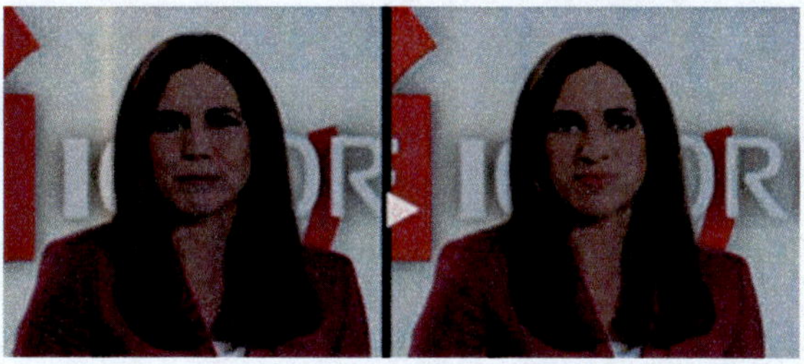

Fig. 1 Example of a deepfake image: the left-hand image is authentic; on the right is the fake (spliced). *Source* SRF [3]

are lip syncing [2], face-swapping [3] (depicted in Fig. 1), and working on the head and whole upper body including shoulders movement swap [4]. Most commonly and widely used face-swapping videos are seen on social media targeting millions of users with misinformation. In other terms, deepfake are falsified videos using AI and its applications for the swapping face of one person with another person in the video. As in [5] deepfakes surfaced to publicity in 2017 in which a Reddit user posted videos that were depicting celebrities in sexual content showing how the face of a celebrity was manipulated to highlight them in a pornographic video clip [3, 5]. Deepfakes are used to generate fake information which is very difficult to detect at a single go, as they use real videos, accompanying authentic-sounding audios, which spread on social platforms very quickly posing threat to privacy [5].

2.1 Creation of Deepfake Content

Creating a deepfake video involves swapping of the original image with a target image wherein a large number of images were required as a dataset to create realistic forgeries which are now replaced with such a technology that requires less information to generate a fake video. The facial expressions, movements of existing images/videos are replicated with target images/videos. Briefly describing, the technology of generating deepfakes, this technology makes it possible to swap one's face in an image/video with the face of another person [3] to generate so-called deep fakes videos. Deepfakes are generated by two artificial neural networks namely 'the generator' and 'the discriminator' Generative adversarial network (GAN) [6]. It works together in such a manner that realistic media are generated making them indistinguishable to humans. GAN is a type of AI technique which comprises two networks in which the first network namely the *generator* is given the input on randomized manner based on some datasets of images or videos which produces output that is fed to the second network namely *discriminator* whose focus is at flagging which output received by it have been artificially created. Creating a deepfake takes several successful steps to swap the faces of individuals as shown in Fig. 2.

The first two *autoencoders* are trained to generate images of the face of both individuals precisely. Autoencoder is simply a NN that learns to copy its input to

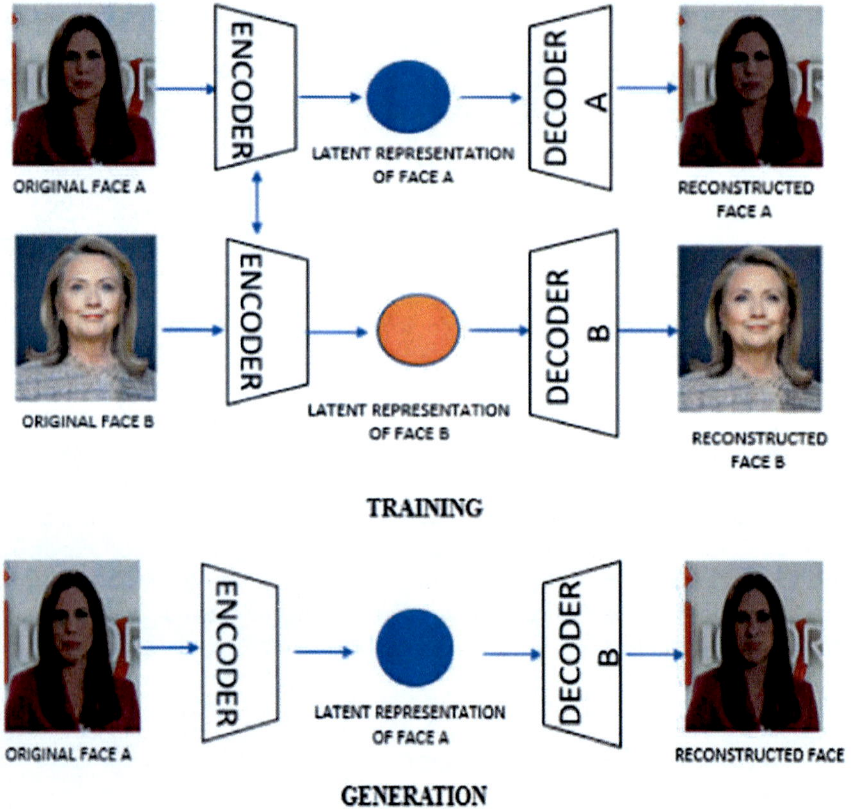

Fig. 2 Creation of deepfake by training the autoencoders (top), generation process where decoders are swapped to create fake image (bottom)

its output. An encoder takes the input and copies the same reducing it to a lower-dimensional latent space and a decoder, which restores the image from smaller representation fetched from encoder's end to the reconstructed image as its output. After the successful reproduction of faces of an individual by encoders; next step comes to the swapping of the output generated by encoders and in this step simply the decoders are swapped so that the facial expressions of the first individual replicates the second individual and vice versa [3] in terms of facial expressions such as eyes, mouth, nose.

2.2 *Constructive Applications of Deepfake*

Deepfake technology, besides its deleterious effects in terms of fake news, extortion, or pornography, also have some spectacular potential for constructive uses in the realm of film and art industries, educational sources and digital communications, games and entertainment, social media and healthcare, material science, and various business fields, such as fashion and e-commerce [3]. This technology can widely be used in the film industry to avoid reshoots or resuscitate a dead actor or could be used to dub the movies in different languages to allow the varied audience to enjoy the art. Similar case was seen in political speech in India where Indian politician Manoj Tiwari in his original video speaks in English, criticizing the opponents and trying to gather and encourage voters to vote him. Further same video was manipulated and made viral on social media that hit around 15 million just to target voters over 20 different languages used in India. This technology can also be used in video conferencing or teleservices where different languages translation is done with proper lip syncing in order to build the interest of the members. Similarly, deepfakes could be better used in business fields where advertising a special brand on various shopping apps allowing the customer to view product onto his own body which would draw the interest of the customers and thus enhancing the business. Similar applications are seen at healthcare, entertainment sources, and educational ends. Recently, a video surfaced this year over internet where people reacted to deepfakes created of their deceased family members and relatives. People could watch their deceased one move again for first time in years. Deep nostalgia is a deep learning AI technique that automatically animate faces in user's photographs. Bringing back emotional touch to loved ones can be creepy to some users.

2.3 *Threats of the Deepfake Technology*

Deepfakes are generally created without consent from the person in the fake video and these videos are fraudulent; but appear to be hyper-realistic which are not easily identified by millions of users surfing the internet. Scenarios where these realistic fake videos are made viral create political distress, blackmailing someone, posing harm to privacy, or any fake events can easily be visualized [7]. The potential threats range from revenge pornographic videos of a victim whose face is synthesized and spliced into realistically looking videos [1], to politically creating provocative synthesized videos of leaders causing election manipulation. Deepfakes in form of trolls are increasingly spread over social media. Bots were used by a popular messaging app named telegram to generate fake nude images of women [8]. Blackmailing individuals for monetary benefits or evading from some crime scenes are majority some threats of deepfake generation. The quality and accuracy of generating these fake videos and images are significantly improving with the advancement in technology and users are more inclined to fake information rather than real and thus exposing these fake videos become a major concern. Thus, Deepfakes have the potential to ruin privacy by creating fake news and malicious hoaxes.

2.4 Existing Deepfake Databases

Many deepfake datasets of 1st generation and 2nd generation were used in the detection and extraction of deepfake. Comparing the results of datasets on basis of accuracy of detection of deepfakes in terms of facial regions and performance, both generation databases in [9] showed better results which were more inclined towards a database of 2nd generation which significantly improved at image level in some facial regions such as mouth, nose, and edge of the face as compared to 1st generation databases. Table 1 summarizes existing deepfake datasets of first and second generation. Existing deepfake datasets of both generations only include visual forgeries and ignore the fact that audio forgeries are also becoming a major concern of detection as deepfake videos not only contain visual manipulations but also audio forgeries [8]. Another lack of potential attributes can be seen in [10] as only a few datasets exist for particular feature extraction such as eye blinking. No existing video datasets are specially designed for the same purpose.

3 Categorization of the State of Art Technique

Deepfake images/videos/audios are found in many variants. Consequently, detection algorithms are proposed by researchers from time to time to detect these variants, which are summarized and depicted in Fig. 3. Social media platforms mainly consist of fake videos or images made out of face-swapping techniques in which deepfakes

Table 1 Comparison of deepfake datasets

	UADFV	Deepfake TIMIT	DFDC	Celeb-DF	FaceForensics++
Year of release	2018	2018	2019	2019	2019
Visual quality	Low	Low	High	High	Low
Real data	48	NIL	1131	408	1000
Fake data	48	620	4119	795	3000
Generation	1ST	1ST	2ND	2ND	2ND

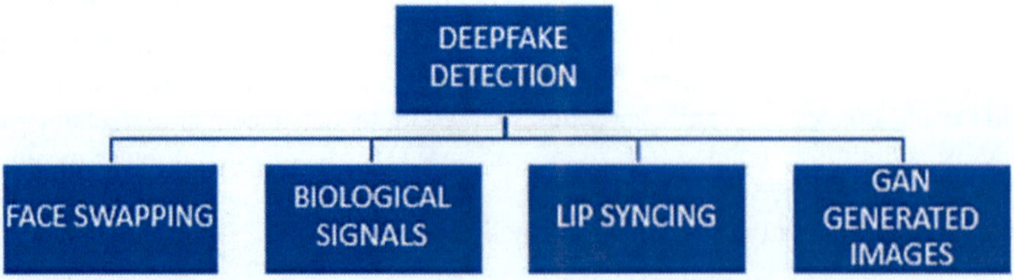

Fig. 3 Deepfake detection techniques

are created by splicing synthesized face region of target image into the original image discarding any changes in facial expression.

Such videos/images can be easily mainstreamed using freely available software. Detection of these manipulations mainly uses CNN (Convolutional Neural Network), which tries to learn discriminative features that are left from the GAN-synthesized images.

Detection of fake content needs various techniques which could expose deepfakes in terms of biological signals whether it is eye blinking, facial hair, facial expressions, double eyebrows, jawline, or chin movements. In case of images/videos basic machine learning models or GAN models can help tackle deepfakes. Basic approach of detecting biological signals in synthesized videos were to train the models by giving them large dataset for specific landmarks in order to detect whether video is subject to manipulation or not. Moreover, with the advancement in technology the GAN models as in recent works [6, 11] have demonstrated performance in recovering these minute details of the face such as skin tone, mismatched color profiles, facial expression, and facial hair. The majority of detection methods use StarGAN [12] which uses only one model for training multiple attributes instead of previous system that requires different generators for every domain.

Another variation is seen by taking a speech or audio and manipulating it onto the original unsuspected person's face especially the lip region. Adobe created a Voco tool which is an audio editing and creation tool which simply requires text string to be given as input and within few seconds text string is converted into audio of person who has never spoken such words out. This manipulation is used with videos specially done in the lip area only to create a falsified video of a person. Blockchain, watermarks, and smart contracts can be an efficient combating strategy against such content. Figure 4 describes some deepfake content in videos related to head movements, face swapping, and lip syncing.

4 Review of Deepfake Detection Methods

Exposing AI-generated fake face images/videos/audios has become a major concern with the advancement in technology. Deepfake detection methods as in Table 2, have significantly progressed in last few years, and detecting such fake information through variants of deep neural networks or based on deep learning could effectively be seen [1, 4, 7, 10, 13, 14] which was further detected by a first forensic tool developed by US Defense Advanced Research Project Agency (DARPA) which was able to distinguish between fake and real videos [3], and the Facebook DeepFake Detection Challenge [15] could also do the same [1, 3, 16]. The availability of several large-scale DeepFake video datasets from the 1st generation including UADFV introduced in 2018 and FaceForensics + + up to the latest databases of the 2nd generation including Celeb-DF and DFDC introduced in 2019 [9, 10] has achieved improvements at both visual, as well as performance level but still poor

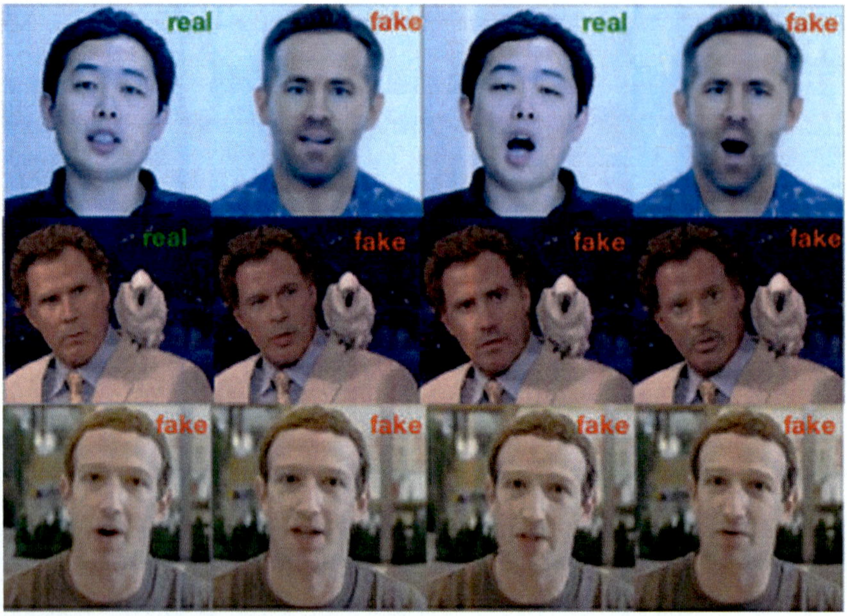

Fig. 4 Head movements (top), face swapping (middle), and lip syncing (bottom) [1]

results, are encountered which can further be improved in next deepfake generations to come.

Deepfake detection is generally regarded as a binary classification problem where basic classifiers like SVM are used to classify the authentic and spliced content. Such detection models need a large training dataset to train model. General processing pipeline of deepfake detection is depicted in Fig. 5. To address this issue, in [17] method is based on a classical frequency domain analysis at pre-processing phase followed by various classifiers SVM, logistic regression, and k-means at training phase. Accuracy results were quite promising as method is able to detect deepfakes of high- and medium resolution with 100% accuracy but low resolution images were tough to find due to smaller frequency spectrum [17]. Use of deep learning methods such as CNN, GAN can help detect these low- resolution images as well as high and medium resolution videos/multimedia content. Deepfake is posing a challenge to forensics analysis to detect fake videos as deep learning models preserve lightning, poses, and expressions very well. Majorly, methods used for detection of fake images can't be strong enough to detect the deepfake videos due to varied temporal characteristics of the frames.

In most of these scenarios of deepfake content generation, the videos are created which are primarily based on face-swapping technique and the spread of such misinformation is leading its way. So, it becomes highly significant to detect these deepfake either by biological/physiological signals, looking at videos on a frame-by-frame basis, introducing watermarking, blockchain technology [18] and smart contracts to impose privacy to content, or through variants of deep neural network.

4.1 Exposing Deepfake Using Biological Signals

Exposing of the deepfake videos which are generated with deep neural networks could be detected with physiological signals that are not well introduced in the

Table 2 Summary of prominent deepfake detection methods

Methods	Technique	Features	Dataset used	Accuracy results	Dealing with
Using face warping artifacts [2]	VGG16 and ResNet50, 101 or 152 [2, 21]	Artifacts are discovered using CNN model based on resolution inconsistency between the warped face area and the surrounding context [2, 21]	UADFV consists of 49 deep fake videos and their respective real videos and another dataset as Deepfake TIMIT	ResNet50 method showed 97.4% AUC for UADFV dataset and 99.9% for low quality and 93.2% for high quality Deepfake TIMIT	Videos
Head poses [4]	SVM classifier	Each image or video frame, are run with face detector and extracted 68 facial landmarks further used SVM to classify using the extracted features	UADFV and a subset of DARPA MediFor GAN Image/Video Challenge	For UADFV dataset, SVM gives AUROC Of 0.89 and for DARPA GAN it is 0.843	Videos and images
Using recurrent neural networks [7]	CNN RNN	Convolutional neural network (CNN) are used to extract frame level features which are then used to train a recurrent neural network (RNN) that learns to classify if a video is subject to manipulation	600 videos in total. 300 deepfake videos from multiple video-hosting websites and 300 randomly from HOHA dataset	Conv-LSTM model (80 frames) testing accuracy was 97%	Videos

(continued)

Table 2 (continued)

Methods	Technique	Features	Dataset used	Accuracy results	Dealing with
Eye blinking [10]	LRCN CNN model	Use of LRCN to learn the temporal patterns of eye blinking as blinking frequency of deepfakes is much smaller than normal. Evaluated LRCN method with comparison to other methods: the eye aspect ratio (EAR) based method and a CNN-based method	40 videos as training set for the overall LRCN model and 10 videos as the testing set are used. CEW dataset used together with training set to train the front-end	LRCN—0.99 CNN—0.98 EAR—0.79 İn terms of AUC	Videos
MesoNet [14]	CNN	Two deep networks, i.e., Meso-4 and MesoInception-4 are introduced to detect forged videos of faces examined by placing the method at a mesoscopic level of analysis [14]	Deepfake and the FaceForensics dataset. FaceForensics created through Face2Face approach and online videos used to create other deepfake datasets	Deepfake videos with 98% average detection rates for MesoInception-4 dataset followed by 95% for Face2Face	Videos
Using spatiotemporal features [22]	RCN	RCN is used in order to explore the video frames for temporal discrepancies that integrate convolutional network DenseNet and the gated recurrent unit cells [21, 22]	FaceForensics + + including 1,000 videos overall are used as datasets	For Deepfake manipulation 93.46% accuracy results seen	Videos

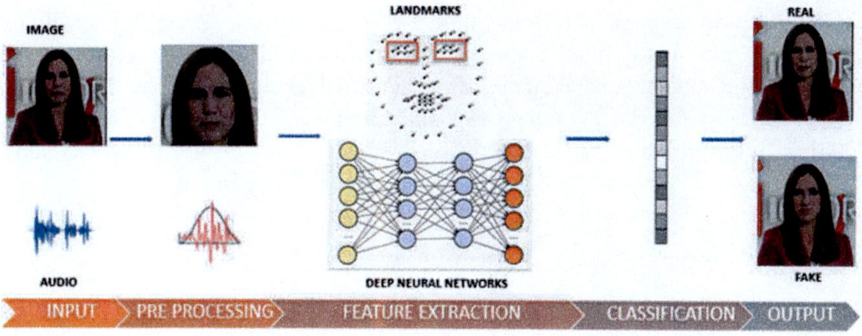

Fig. 5 General processing pipeline of deepfake detection

synthesized fake videos due to loss of information in the encoding step of generation. These signals could be the lack of realistic eye blinking [10], as blinking rates in deepfake videos are much lower than those in normal scenarios. Blinking which is not precisely presented in synthesized fake videos due to lack of closed eye dataset can be used to detect fake content, or it could be inconsistent head poses that are generated in deepfake videos which are used to expose fraudulent material [4]. It depicts deepfake videos are generated by splicing the face region produced by artificial means into the original face image and process of generating so would sometimes lead to introduction of minor errors that could be easily disclosed with 3D head poses [4] and mismatched color profiles [19] which could be used to spot deepfake videos. The visualization and study [14] show that the eyes and mouth play a pre-eminent role in exposing deepfakes. Moreover, with the advancement in technology the GAN models as in recent works [6, 11] could be used to combat these minute details of face such as skin tone and facial hair.

4.2 Blockchain and Smart Contracts

With the advances in AI techniques, there is a proliferation in fake videos, images, and audios generation. It is indeed becoming scarier and needs an efficient combating strategy in terms of the authenticity of the content delivered online or an efficient strategy for online source verification. As in [3] adding watermarks to original content delivered online or videos are recognized only when they have a publicly available signature, this would help in verifying the videos by comparing the signature which would add power to the authenticity of the content. If the signature mismatches, then we can question that content made available is altered in some manner. Blockchain technology [18] is not only highly resistant to forgeries and tampered content but is also proof of authenticity and can store data in a secure, safe, transparent, and traceable way. It can also track and certify the origins and history of the data as well as determine if a video or digital content is traceable to a trusted source [5,

18]. This would give a check to its authenticity and privacy. Great tool for digital provenance solutions is blockchain and watermarks as blockchain would create a chain of unique unchangeable blocks of metadata. Blockchain and other Distributed Ledger Technologies (DLTs) can provide authenticity and traceability of data online. Another future direction for detection of deepfake could be use of blockchains, watermarks, or DLTs to combat and tackle the digital fake content.

4.3 Variants of Deep Neural Networks

With various variants of neural networks like CNN, RNN (Recurrent Neural Network), deep learning, autoencoders have been practiced for extracting deepfakes that are invincible. Many neural networks and other technologies are launched to detect digital deception. Neural network having more than one hidden layer makes up a deep neural network. As discussed in study [7, 20] system uses a convolutional neural network (CNN) for extracting the frame-by-frame level features and those extracted features are then used to train the recurrent neural network (RNN) [7] which learns and trains the model to analyze if a video has been subject to manipulation or not to automatically detect the deepfakes. Deep neural networks would give more promising results as in [17] classical frequency domain analysis is used at preprocessing where feature extraction is done with DFT and new transformed features are then fed to classifier at training phase. Power spectrum for high and medium resolution images give promising results as compared to low resolution images due to frequency spectrum being smaller. As higher the frequency spectrum easier is the detection of fake content. Practicing such pipeline in near future can give promising results in detecting the deepfake videos and fine-tuning the model can bring the highest level of performance and effectiveness in detecting fake content.

5 Discussions and Future Research Directions

Deepfakes are eroding the trust of individuals who are much into social platforms as media content delivered online are full of distress and false information. Thus, the question arises that what should people believe to be real or fake, especially when digital media is presented as evidence in the court of law. Proving the authenticity and integrity of the digital content is highly significant. In view of this arising problem, many such detection variants are introduced which lack appropriate approaches that could give better results in terms of the detailed description of what is authentic. Limitations in terms of accuracy rate of algorithms provide ample opportunities for future direction to come up with anti-deepfake technology which includes deepfake extraction, its detection, content authentication, and prevention of fake news to be spread online. Deepfake videos with low visual quality are unlikely to have a real impact on users and the first generation deepfake videos were mainly characterized by low quality visual synthesized videos, mismatched color profiles of original and fake videos, rate of errors in poses, and movement of body parts. Many aspects have been successfully improved in second-generation databases but still poor results are achieved in [9] which could be improved with better accuracy rates for new datasets which could ultimately increase the overall running efficiency.

On the other hand, detecting deepfakes online is crucial, but understanding the real intent behind its creation is also equally important. Deepfakes, besides their negative impact, can hopefully give a beneficial contribution in our lives whether it is in fields of education, entertainment, or political insights. In 2021 deepfake videos went viral where emotional connect of families were seen with their deceased ones. People could watch their deceased one move again for the first time in years. People's judgment behind creation and publishing deepfakes is even more important than detecting deepfake content. Adobe previewed its Voco tool which is an audio editing and creation tool. Voco was able to generate the fake content within seconds after being given a text string. But if the content created is not harming any sentiments, then deepfake can be thought to be a powerful tool, but the question arises whether the content created is appreciable or not. A few months ago famous news reporter from South Korea was replaced by her deepfake version delivering news related to COVID-19 and pandemic updates to which mixed responses of people were seen. An AI-based approach known as deepfake can generate synthesized videos of any person saying anything which would seem real and it is highly anticipated that detection methods would lag behind these deepfake creation technologies. Generation methods of deepfake videos barely produce good details of facial regions such as facial hair and skin tones. Recovering of such details can be incorporated by GAN models [6, 11].

The detection methods need to be more vigorous in order to prevent forgery makers to attempt strong deepfake generations. Protecting people from this fake content and individuals becoming victims of cyber-attacks needs to be seriously taken into account. Privacy being a major concern of individual, cybersecurity can prove best to combat deepfakes. Watermarks can be introduced at the time of the creation of content and before its distribution. Watermarks introduced in videos act as a unique identifier that links it to blockchain technology. Content with watermarking technology is detected at nodes and backend of network portals and the user gets alert to inauthentic content. Digital watermarks also help to track and verify the content for forensics analysis. This would also ease the distress caused to the individual. Compared to previous systems [4, 7, 10, 14, 17] combating fake content through machine learning and deep learning models could be improved to more accurate results by building models and training them by giving few annotated data instead of large datasets at training phase. Principles of information fusion are also found to be highly significant in the forensic analysis of multimedia content. Hence, multiple cues can be examined to prove authenticity of digital media. As in the case of image forensics, a simple feature level fusion [23] or measurement level fusion techniques [24] can be exploited by audio/video deepfake detection methods. Further fine-tuning the model can yield more desirable accuracy results in near future.

6 Conclusion

In this paper, we reviewed deepfake technology along with its associated threats, challenges, benefits, and future directions that can help cope with fake information circulated online. Authentication of digital content has great social and legal implications in this digital era. The real challenge for detection method of deepfake development and future research needs to focus majorly on introducing more robust, scalable, and efficient methods. Detection methods as of now focus mainly on AI and ML models which classify fake and real content. Ease of access to technology such as pre trained models, transfer learning, Tensorflow or keras, and many deep learning and machine learning models especially the GAN's have made generation of deepfakes much easier. The paper provides comprehensive review of AI-generated synthetic media and ways to combat it through cybersecurity, models, and AI techniques. Not all manipulations in digital content are harmful and understanding the real intent behind its creation is much needed. Deepfake detection methods majorly focus on detecting visual forgeries such as face swapping, lip sync, head movements through various ML and DL models. There exist a need to develop such detection methods that focus both on visual as well as audio manipulations at the same time. Deepfakes will continue having disrepute but surely it has clear benefits which would provide excellent opportunity to have constructive impact in our lives. Deepfakes seem threatening but educating individuals and trusting content delivered by authentic and reputable sources would give good image to deepfakes outweighing its deleterious effects.

References

1. S. Lyu, Deepfake detection: Current challenges and next steps. IEEE International Conference on Multimedia & Expo Workshops (ICMEW), London, United Kingdom., pp. 1–6 (2020).https://doi.org/10.1109/ICMEW46912.2020.9105991
2. L. Yuezun, L. Siwei,Exposing deepfake videos by detecting face warping artifacts. IEEE Conference on Computer Vision and Pattern Recognition Workshops (CVPRW) (2019)
3. A. Hauser, M. Ruef (eds.), Deepfake—An introduction. https://www.scip.ch/en/?labs.20181004
4. X. Yang, Y. Li, S. Lyu, Exposing deep fakes using inconsistent head poses. IEEE International Conference on Acoustics, Speech, and Signal Processing (ICASSP) (2019)
5. W. Mika,The emergence of deepfake technology: A review. Technol. Innov. Manag. Rev. **9**, 39–52 https://doi.org/10.22215/timreview/1282. (2019)
6. I. Goodfellow, J. Pouget-Abadie, M. Mirza, B. Xu, D. W. Farley, S. Ozair, A. Courville, Y. Bengio, Generative adversarial nets. Advances in Neural Information Processing Systems (NIPS), pp. 2672–2680 (2014)
7. D. Guera, E.J. Delp, Deepfake video detection using recurrent neural networks. AVSS(2018)
8. M. Masood, M. Nawaz, K. Malik, A. Javed, A. Irtaza, Deepfakes Generation and Detection: State-of-the-art, open challenges, countermeasures, and way forward (2021) https://arxiv.org/abs/2103.00484

9. T. Ruben, S.R. Tapiador, J. Fierrez, R. Vera-Rodriguez, Deep-fakes evolution: Analysis of facial regions and fake detection performance. In Biometrics and data pattern analytics—BiDA Lab, Universidad Autonoma de Madrid (2020) arXiv:2004.07532
10. L. Yuezun, M.-C. Chang, S. Lyu, In ictu oculi: Exposing AI generated fake facevideos by detecting eye blinking. IEEE International Workshop on Information Forensics and Security (WIFS) (2018)
11. T. Karras, T. Aila, S. Laine, J. Lehtinen, Progressive growing of gans for improved quality, stability, and variation. The International Conference on Learning Representations (ICLR) (2017)
12. Y. Choi, M. Choi, M. Kim, J.W. Ha, S. Kim, J. Choo, Stargan: Unified generative adversarial networks for multi-domain image-to-image translation. Proceedings of the IEEE Conference on Computer Vision and Pattern Recognition, pp. 8789–8797 (2018)
13. F. Matern, C. Riess, M. Stamminger, Exploiting visual artifacts to expose deep-fakes and face manipulations. IEEE Winter Applications of Computer Vision Workshops (WACVW) (2019)
14. D. Afchar, V. Nozick, J. Yamagishi, I. Echizen, Mesonet: a compact facial video forgery detection network. IEEE International Workshop on Information Forensics and Security (WIFS) (2018)
15. B. Dolhansky, R. Howes, B. Pflaum, N. Baram, C. Ferrer, The deepfake detection challenge (DFDC) preview dataset. arXiv:1910.08854 (2019)
16. https://www.technologyreview.com/s/611726/the-defense-department-has-produced-the-first-tools-for-catching-deepfakes/
17. R. Durall, M. Keuper, F. Pfreundt, J. Keuper, Unmasking deepfakes with simple Features. Retrieved 17 April 2021, from (2021) https://arxiv.org/abs/1911.00686
18. H.R. Hasan, K. Salah, Combating deepfake videos using blockchain and smart contracts. IEEE Access(2019)
19. L. Haodong, L. Bin, T. Shunquan, H. Jiwu, Detection of deep network generated images using disparities in color components (2018). arXiv:1808.07276
20. I. Amerini,L. Galteri, R. Caldelli, A. Del Bimbo, Deepfake video detection through optical flow based CNN. IEEE/CVF International Conference on Computer Vision Workshop (ICCVW), Seoul, Korea (South) (2019), pp. 1205–1207. https://doi.org/10.1109/ICCVW.2019.00152
21. T.T. Nguyen, C.M. Nguyen, T. Nguyen, T. Nguyen, S. Nahavandi, Deep learning for deepfakes creation and detection (2019). arXiv:1909.11573
22. E. Sabir, J. Cheng, A. Jaiswal, W. AbdAlmageed, I. Masi, P. Natarajan, Recurrent convolutional strategies for face manipulation detection in videos. Proceedings of the IEEE Conference on Computer Vision and Pattern Recognition Workshops (2019), pp. 80–87
23. M. Kaur, S. Gupta, A passive blind approach for image splicing detection based on DWT and LBP histograms. International Symposium on Security in Computing and Communication (2016), pp. 318–327. https://doi.org/10.1007/978-981-10-2738-3_27
24. M. Kaur, S. Gupta, A fusion framework based on fuzzy integrals for passive-blind image tamper detection. Cluster Computing **22**.https://doi.org/10.1007/s10586-017-1393-3 (2019)

Robust Image Watermarking Using LWT Transform and Stacking Ensemble Classifier

Rakesh Kumar Verma, M. Sivakkumar, and Varsha Namdeo

Abstract To maintain the relationship between robustness and imperceptibility is very challenging for conventional digital watermarking methods. The proposed digital image watermarking algorithm efficiently manages the value of robustness and imperceptibility. The proposed algorithm has applied the lifting wavelet transform to extract the features from the source image and symbol image. The LWT transform overcomes the limitation of the wavelet transform. The proposed algorithm focusses on pattern-based embedding process involved in the watermark. For the generation and formation of pattern, stack-based ensemble classifier has been applied. The stack-based ensemble classifier is derived in two-stage, and it uses three classifiers, such as SVM, KNN and DT. The glowworm swarm optimization is applied to the symbol image in order to select an optimal coefficient for the embedding process. The proposed algorithm is then simulated in MATLAB software, and further the parameters such as PSNR, NC, SSIM and BER are measured. The proposed algorithm is compared with DWT-SVD and SVM. From the results, it is observed that the proposed algorithm is better than the existing digital image watermarking algorithms.

Keywords Watermarking · ımages · Feature extraction · Wavelet LWT · SVM · KNN · DT · SVD

1 Introduction

The explosive growth of multimedia data such as images, video and audio is compromised with easy distribution, copy paste and alteration process [1, 2]. The prevention and maintenance of digital multimedia copyright act face various types of tampering and loss their integrity and authentication. The process of integrity and authentication needs multimedia cryptography, image steganography and watermarking methods—the digital image watermarking plays a vital role in multimedia data authentication

R. K. Verma (✉) · M. Sivakkumar · V. Namdeo
Sarvepalli Radhakrishnan University, Bhopal, MP, India

G. Ranganathan et al. (eds.), *Soft Computing for Security Applications*,
Advances in Intelligent Systems and Computing 1397,
https://doi.org/10.1007/978-981-16-5301-8_45

and integrity [3–5]. The process of authentication and integrity provides the security to multimedia data over the other communication media. The processing of image watermarking techniques is applied in two ways, namely spatial domain and transform domain. The spatial domain is compromised with security risk and the methodology shifts the pixel from left to right, which means a reversible watermarking approach [6, 7]. The transform-based watermarking methods provide the characteristics of imperceptibility and robustness. The property of imperceptibility and robustness reduces the risk of geometrical and non-geometrical attacks with loss of information [8–10]. The family of transform function provides great protentional in the development of watermarking algorithms. The family of transform functions such as discrete cosine transform (DCT), discrete wavelet transforms (DWT), wavelet packet transforms (WPT), and many derive transform function such as lifting wavelet transform (LWT) are applied for the processing of image watermarking. The transform-based methods are dominated in terms of feature coefficients [11, 12]. The texture features are the majorly dominated features of multimedia data. Various authors focus on the improvements in the number of correlation (NC) parameters for the enhancing the robustness of watermarking images. The increased value of NC indicates the strength of watermarking methods [13–15]. However, this imperceptibility deals with the visual context of watermarking image in peak parameters to signal noise ratio (PSNR) and similarity index matrix (SSIM) of watermark image. The transform-based methods are promising factors in the process of watermarking. The bottleneck issue related to transforming-based methods is the minimization of the host image's impulsive noise and boundary edge value [16]. Various authors resolve this issue with SVD (singular value decomposition) methods. The singular value decomposition methods transform data processing methods and enhance the processing of watermarking methods [17, 18].

Despite SVD methods, various authors applied swarm intelligence-based optimization methods for the optimization of features coefficient [19]. The family of swarm intelligence algorithms such as particle swarm optimization, firefly, glowworm optimization and many more algorithms [20–22]. The nature of iteration improves the selection of optimal coefficient for the processing of embedding. The watermark image's randomness increases robustness factors, for the randomness factor of watermarking methods applied a pattern-based approach [21, 23]. The pattern-based approach applied the methods of machine learning and neural network methods. The machine learning-based algorithm influences the capacity of watermarking algorithms [24, 25]. The various authors applied support vector machine (SVM), ensemble-based classifiers, and many more machine learning algorithms and neural networks [24]. The support vector machine applied as a binary classifier for selecting the pattern for embedding in digital image watermarking [25]. This paper proposed feature optimization-based digital image watermarking methods. The proposed algorithm applied LWT (lifting wavelet transform), as feature extractor of host image and source images. The LWT transform decomposed images into different energy bands in terms of the high filter and low filter. The algorithm process also encapsulates the glowworm optimization algorithm for the optimization of features of host and source images [26]. The processing of embedding done by ensemble-based classifier algorithm. The design ensemble-based classifier on three classifiers, such as SVM, KNN and decision tree. The proposed ensemble-based classifier improves the robustness factor of the watermarking algorithm—the rest of the paper is organized accordingly to section 2.

2 Feature Extraction

This section describes two process, lifting wavelet transform and glowworm swarm optimization algorithm.

a. Lifting Wavelet Transform (LWT)

The extraction of features in transform-based digital watermarking methods is the primary function. The process of feature extraction is applied to lifting wavelet transform (LWT). The process of lifting wavelet transform proposed by Sweldens in 1995 [13, 16]. The lifting wavelet transform overcomes the limitation of wavelet transform in terms of reducing computational complexity and requirements of memory. The lifting wavelet transform mapped the transform value integer to integer, and it also maintains the forwarded and reverse transform. The processing of lifting transform describes in three sections, as splitting, prediction and update. The more information about the LWT transform [16, 18].

Consider $K(x, y)$ be an image matrix of source and host image. The mapping of two-dimensional transform into one-dimensional wavelet transforms [18, 25].

1. Process of split

The given matrix of data K(x, y) split into two subset as even and odd parts as Ke(x, y) and Ko(x, y)

$$K(x, y) = \begin{cases} k_e(x, y) = K(m, 2n) \\ k_o(x, y) = K(m, 2n + 1) \end{cases} \tag{1}$$

2. Process of predict

In the process of predict operation, the odd part of transform predicts from the group of even coefficients. The HF coefficient $E(x, y)$ is estimated the error in process of predicting odd part from even using prediction operator P as

$$E(x, y) = K_o(x, y) - P[K_e(x, y)] \tag{2}$$

From Eq. (2), the estimated odd part as

$$k_o(x, y) = E(x, y) + P[K_e(x, y)] \tag{3}$$

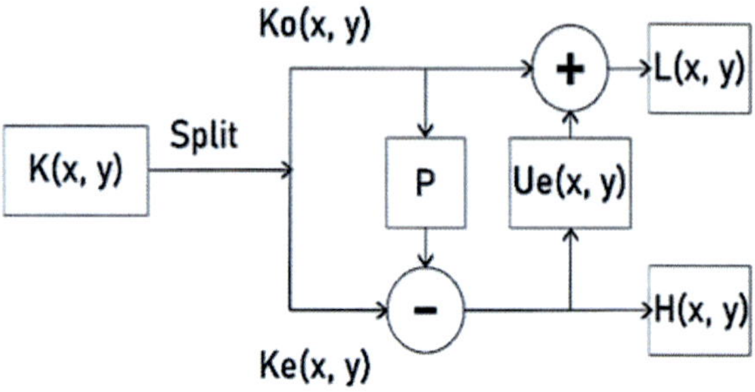

Fig. 1 Shows that the processing of lifting wavelet transforms in three different stages split, predict and update with the function of odd and even

3. Process of update

The process of update estimate the low-pass coefficient $l(x, y)$ and the even part is updated with value of $U_E(x, y)$ (Fig. 1).

$$l(x, y) = k_e(x, y) + U_E(x, y) \tag{4}$$

b. Glowworm Swarm Optimization

Glowworm swarm optimization is bioinspired objective-based optimization algorithm. The processing of algorithm based on the behaviors of glowworm [5, 6]. The position and sharing of information of glowworm based on flaring of attracters. The position of glowworm changes to their brightness factor [6, 17]. The distance between glowworms is the influence factor of glowworm optimization algorithm. The position of glowworm updates according to their value of luciferin of glowworm agent. The major steps of glowworm algorithm are luciferin update, neighborhood selection, movement of glowworm, velocity and decision of radius update [17, 22, 27].

2.1 Process of GSO Algorithm

1. Luciferin update: The value of luciferin update depends on fitness value and pervious value of luciferin [15, 16]

$$l_i(t+1) = (1-\rho)l_i(t) + \gamma fitness(xi(t+1) \tag{5}$$

here $l_i$denotes the luciferin value of glowworm i at time t
2. Neighborhood selection
The selection of neighborhood $N_i(t)$ as

$$N_i(t) = \left\{J : d_{ij}(t) < r_d^i(t); l_i(t) < l_j(t)\right\} \tag{6}$$

here $d_{ij}(t)$euclidean distance between glowworm i and j at time t

3. Compute probability
   Glowworm applied function of probability to measure the movements of glowworm

$$P_{ij}(t) = \frac{l_j(t) - l_i(t)}{\sum_{k \in Ni(t)} l_k(t) - l_i(t)} \tag{7}$$

4. Movement process

$$x_i(t+1) = x_i(t) + s\left(\frac{x_j(t) - x_i(t)}{\left|x_j(t) - x_i(t)\right|}\right) \tag{8}$$

5. Decision rule update

$$r_d^i(t+1) = \min\left\{r_{s,max}\{\ 0, r_d^i(t) + \beta(n_t - |N_i(t)|)\}\right\} \tag{9}$$

Here, β is the constant, $r_{s,}$ shows radius of glowworm.

3 Proposed Methodology

The proposed algorithm is based on the concept of stacking-based ensemble classifier. The stacking-based ensemble classifier is very efficient instead of bagging, boosting and other ensemble methods. The process of stacking applied three classifier support vector machines (SVM) [13], nearest neighbors (KNN) [14] and decision tree (DT) [14]. The process of stacking applied in two stages, first stage combines two classifiers namely KNN and SVM. In second stage, decision tree (DT) classifier is applied to predict the class based on the guidance of leaf node. The support vector machine is very efficient in performing digital image watermarking for increasing the robustness and imperceptibility. The KNN and DT also contribute to digital watermarking domain for the generation and formation of patterns mainly for the watermark embedding process.

The process of stacking is described here.

The feature matrix of source image is categorized into training dataset D_r and testing dataset D_s with proportion of 1:3. "0" and "1" represent the feature coefficient of high and low intensity, respectively, $y_i \in \{0, 1\}$ the process of learning define the cross fold into two base classifier. Measure the probability of base classifier (P_1, P_2) of each.

Input training data: D_r and testing data: D_s.

Two classifier in 1st stage: $C_1\{C_1 = \text{SVM}, C_2 = \text{KNN}\}$.

The estimation of feature entropy in 2nd stage: $C_{12}\{C_3 = \text{DT}\}$.

Begin:

For C_i in C_1;

$$D_s = \emptyset;\ D_r = \emptyset$$

For M = 1,…M // cross validation

$$\frac{(M-1)}{M} \text{of} D_s \text{training of} C_i;$$

$$\text{otherwise} \frac{1}{M} \text{of} D_r \left(D_{r(M)}\right) \text{predicting} C_i$$

$P_{\text{m}}\{P_1,\ldots P_{\text{n}}\}$ $(m = D_s/k)$.
Now, predict test data D_s

$$F_m\{f1, f2, \ldots, fn\}(n = D_s);$$

$$D_i = D_{\text{stage 1}} \cup P_m$$

$$D_i = D_{\text{stage 2}} \cup F_m;$$

End for
End for
Process of 2nd layer
Stack Di input data $D_{\text{stage 2}} = [D_1 \text{and} D_2]$.
Average of $D_s = [d_1, d_2, \ldots, d_n]$; using $D_{\text{stage 2}}$ in C_3.
Output: patterns of host image (Fig. 2).
Process of watermark embedding
Begin

1. Apply 3-level LWT on source image and symbol image
2. The selected sub-band converted into feature matrix
3. The host image feature matrix passes through stack ensemble classifier (SEC)
4. Return the class patterns c1, c2,…,cn
5. GSO apply on symbol image feature matrix
6. Update the position and luciferin value of new feature points
7. If feature value if high intensity mark as 1 and if low mark 0
8. Randomized the value of 0 and 1
9. Select SEC pattern and optimal coefficient
10. Embedding done
11. Apply watermark attack
12. Measure NC
13. End

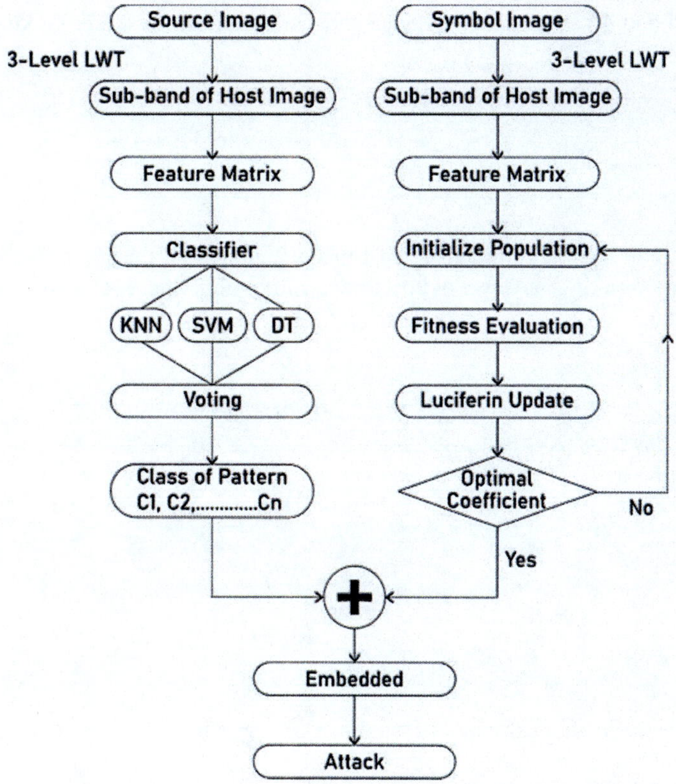

Fig. 2 Process block diagram of proposed model of ensemble-based watermarking based on GSO

4 Experimental Analysis

To assess the validity of the proposed digital image watermarking algorithm, which is based on a stack ensemble classifier and the LWT transform [13, 14, 16]. The major concern is regarding the robustness and imperceptibility of digital watermark. All simulation process runs over a personal computer I7 processor, 2.84GHZ, 16 GB RAM and using MATLAB software. The used host image is Leena, Barbara, cameraman, Peppers. The symbol image is shown in figure. The size of host image is 512×512 and symbol image 128×128. The following parameters are estimated

$$\text{PSNR (peak to signal noise ratio)} = 10 \log 10 \frac{M \times N}{\sum_{m=1}^{M} \sum_{n=1}^{N} [A(m,n) - A^*(m,n)]} \tag{10}$$

Here, A and A^* show the host image and symbol image for watermarking M and N is size of matrix.

Now

$$SSIM(A, A*) = \frac{(\text{UAUA}^* + c_1)(2\sigma AA * + c_2)}{\left(U_A^2 + U_{A*}^2 + c_1\right)\left(\sigma_A^2 + \sigma_{A*}^2 + c_2\right)} \tag{11}$$

where UA and UA* are average of host image and symbol image, respectively. σA and σA^* are variation and error of host image and symbol image. C_1 and C_2 are GSO update value.

$$NC(A, A*) = \frac{\sum_{m=1}^{M}\sum_{n=1}^{N} A(m,n)A*(m,n)^{Z}}{\sqrt{\sum_{m=1}^{M}\sum_{n=1}^{N}[A(m,n)]^2}\sqrt{\sum_{m=1}^{M}\sum_{n=1}^{N}[A*(m,n)]^2}^{Z}} \tag{12}$$

$$BER = \frac{R(Xi)}{\sum_{J=1}^{M} R|(Xj)} \tag{13}$$

where, R is the index coefficient of watermark image (Table 1, Figs. 3, 4, 5, 6, 7, 8).

5 Conclusion and Future Work

This paper has proposed a stack-based ensemble classifier for performing advanced digital image watermarking. The proposed algorithm also encapsulates the lifting wavelet transform methods to extract the host image and symbol image features. The embedding process is performed by using GSO algorithm. GSO proceeds with the updated feature coefficient of symbol image and correlates it with the ensemble class of pattern. The process of ensemble is derived from 2 stage learning models. The two-stage learning model has efficiently performed the digital image watermarking process. The proposed algorithm is very efficient by considering the imperceptibility and robustness factor. The increased value of NC has increased the robustness of digital image watermark. BER's minimized value shows that the processing of the transform function is fair in terms of scaling. The proposed algorithm compares with DWT-SVD and SVM algorithms. The analysis of results indicates that the proposed algorithm is efficient. The scaling factor controlled with GSO luciferin value. The controlled population of the watermark embedding point is referred to as robust embedding. The algorithm's strength is measured using geometrical and non-geometrical attacks. The applied attack decreases the value of NC. However,

Table 1 The comparative analysis of DWT-SVD, SVM and proposed techniques with Leena, cameraman image using NC and BER parameters

	LEENA Image			Cameraman Image			Parameter
Type of attack	DWT-SVD [7]	SVM [13]	Proposed	DWT-SVD [7]	SVM [13]	Proposed	
Median filter (MFA)	0.997	0.998	0.990	0.992	0.933	0.969	NC
Rescaling (RA)	0.990	0.991	0.993	0.980	0.984	0.990	
Salt and pepper (SPNA)	0.986	0.981	0.988	0.995	0.991	0.996	
Gaussian (GNA)	0.994	0.990	0.995	0.994	0.994	0.995	
JPEG compression (JCA)	0.998	0.996	0.998	0.994	0.998	0.999	
Counterclockwise rotation (CCRA)	0.392	0.395	0.436	0.541	0.519	0.667	
Speckle noise (SN)	0.995	0.946	0.996	0.995	0.996	0.998	
Gamma correction (GCA)	0.763	0.760	0.764	0.760	0.764	0.764	
Median filter (MFA)	0.009	0.020	0.0024	0.108	0.032	0.023	BER
Rescaling (RA)	0.006	0.040	0.0015	0.153	0.021	0.015	
Salt and pepper (SPNA)	0.009	0.016	0.0018	0.006	0.008	0.002	
Gaussian (GNA)	0.008	0.032	0.0058	0.010	0.015	0.008	
JPEG compression (JCA)	0.0007	0.007	0.0003	0.0001	0.0002	0.0005	
Counterclockwise rotation (CCRA)	0.778	0.745	0.509	0.702	0.719	0.682	
Speckle noise (SN)	0.013	0.081	0.008	0.010	0.002	0.001	
Gamma correction (GCA)	0.068	0.072	0.045	0.042	0.070	0.070	

in the case of the proposed algorithm, the value of NC is remained constant and shows the strength of digital image watermarking algorithm. The iteration process of transform and GSO increases the algorithm's time complexity, minimizing the number of iteration and increasing the efficiency of the algorithm.

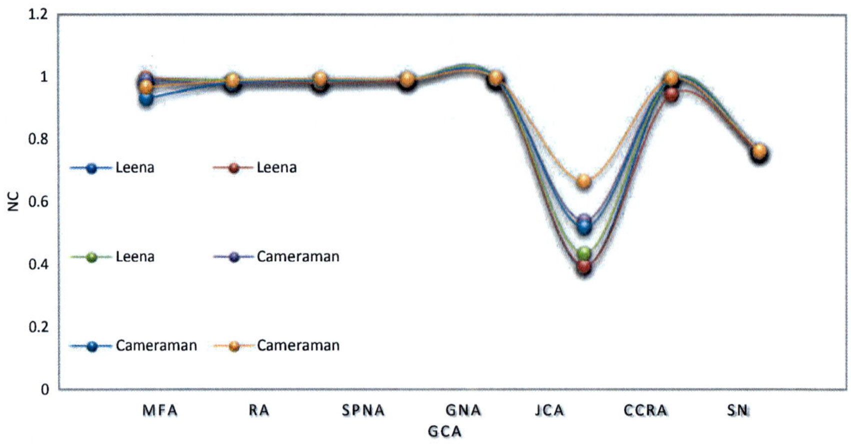

Fig. 3 Performance analysis of comparative analysis of DWT-SVD [7], SVM [13] and proposed techniques with Leena, cameraman image using NC with Median filter (MFA), Rescaling (RA), Salt and pepper (SPNA), Gaussian (GNA), JPEG compression (JCA), Counterclockwise rotation (CCRA), Speckle noise (SN), Gamma correction (GCA), Median filter (MFA), Rescaling (RA), Salt and pepper (SPNA), Gaussian (GNA), JPEG compression (JCA), Counterclockwise rotation (CCRA), Speckle noise (SN), Gamma correction (GCA) attacks

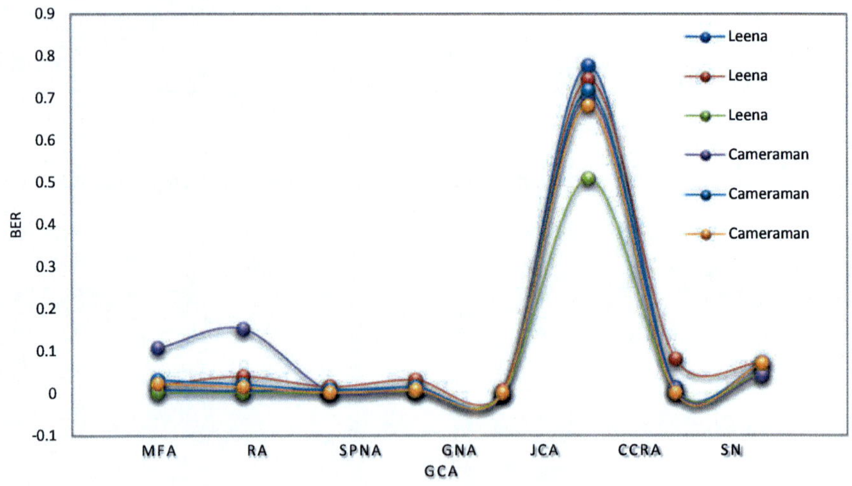

Fig. 4 Performance analysis of comparative analysis of DWT-SVD [7], SVM [13] and Proposed techniques with Leena, cameraman image using BER with Median filter (MFA), Rescaling (RA), Salt and pepper (SPNA), Gaussian (GNA), JPEG compression (JCA), Counterclockwise rotation (CCRA), Speckle noise (SN), Gamma correction (GCA), Median filter (MFA), Rescaling (RA), Salt and pepper (SPNA), Gaussian (GNA), JPEG compression (JCA), Counterclockwise rotation (CCRA), Speckle noise (SN), Gamma correction (GCA) attacks

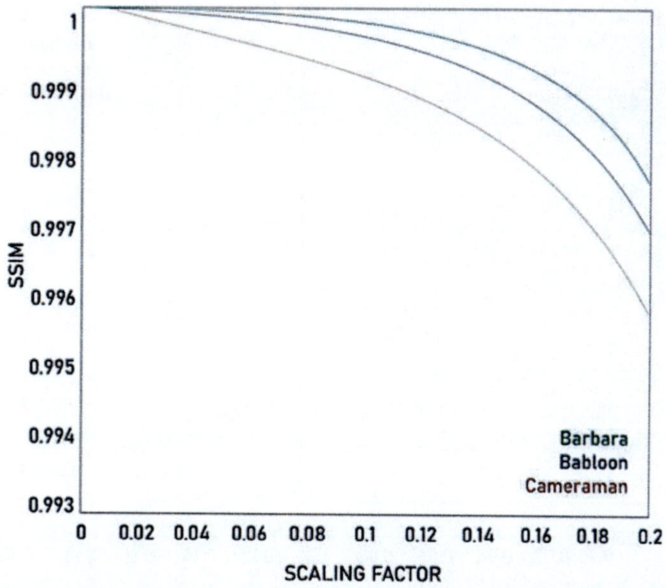

Fig. 5 Performance analysis of proposed techniques with Barbara, baboon, cameraman image using SSIM parameter with respect to scaling factor and range of scaling factor start from zero to 0.2

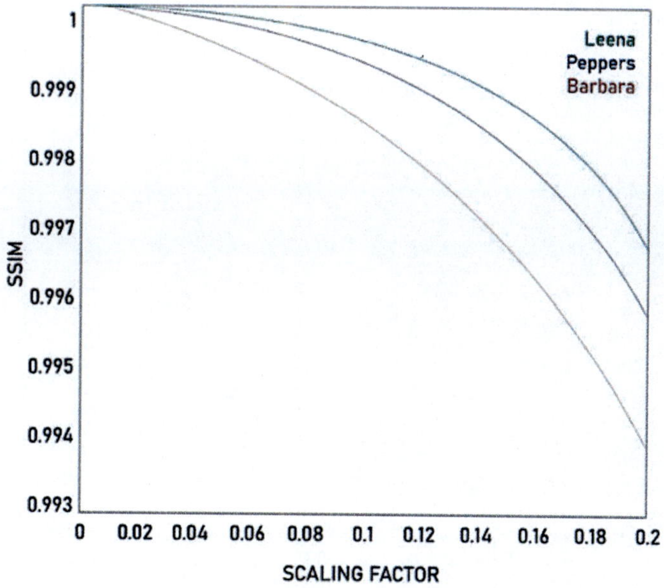

Fig. 6 Performance analysis of proposed techniques with Leena, peppers, Barbara image using SSIM parameter with respect to scaling factor and range of scaling factor start from zero to 0.2

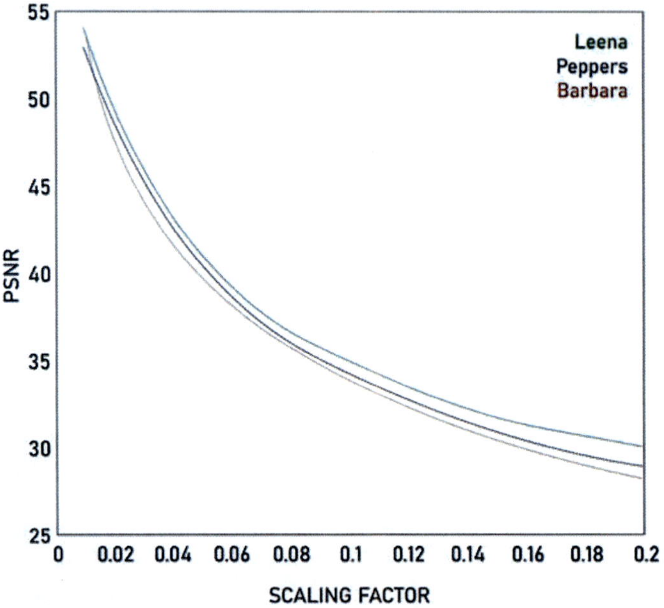

Fig. 7 Performance analysis of proposed techniques with Leena, peppers, Barbara image using PSNR parameter with respect to scaling factor and range of scaling factor start from zero to 0.2

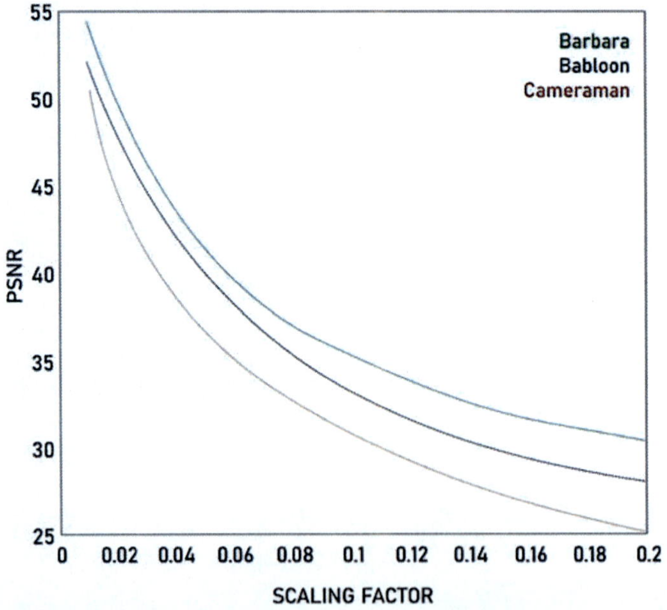

Fig. 8 Performance analysis of proposed techniques with Barbara, baboons, cameraman image using PSNR parameter with respect to scaling factor and range of scaling factor start from zero to 0.2

References

1. S.M. Pourhashemi, M. Mosleh, Y. Erfani,A novel audio watermarking scheme using ensemble-based watermark detector and discrete wavelet transform. Neural Comput. Appl., 1–21 (2020)
2. Y. Gao, X. Kang, Y. Chen,A robust video zero-watermarking based on deep convolutional neural network and self-organizing map in polar complex exponential transform domain. Multimedia Tools Appl, 1–21 (2020)
3. A. Ray, S. Roy,Recent trends in image watermarking techniques for copyright protection: a survey. Int. J. Multimedia Info. Retrieval, 1–22 (2020)
4. M. Begum, M.S. Uddin,Digital image watermarking techniques: a review. Information **11**(2), 110 (2020)
5. Z. Yu, X. Yang,Full glowworm swarm optimization algorithm for whole-set orders scheduling in single machine. The Scient. World J. (2013)
6. Z. Li, X. Huang, Glowworm swarm optimization and its application to blind signal separation. Mathematical Problems in Engineering (2016)
7. J. Liu, J. Huang, Y. Luo, Su. Lvchen Cao, D.W. Yang, R. Zhou, An optimized image watermarking method based on HD and SVD in DWT domain. IEEE Access **7**, 80849–80860 (2019)
8. P. Kumar, A. Sharma,A robust image watermarking technique using feature optimization and cascaded neural network. Int. J. Comp. Sci. Info. Security (IJCSIS) **18**(8), 36–45 (2019)
9. A. Fatahbeygi, F. Akhlaghian Tab, A highly robust and secure image watermarking based on classification and visual cryptography. J. Info. Security Appl. **45**(2019), 71–78
10. A. Soualmi, A. Alti, L. Laouamer, A new blind medical image watermarking based on weber descriptors and Arnold chaotic map. Arab. J. Sci. Eng. **43**(12), 7893–7905 (2018)
11. F. Deeba, K. She, F. Ali Dharejo, Y. Zhou, Lossless digital image watermarking in sparse domain by using K-singular value decomposition algorithm. IET Image Process. **14**(6), 1005–1014 (2020)
12. F. Deeba, K. She, F. Ali Dharejo, H. Memon, Digital image watermarking based on ANN and least significant bit. Info. Security J.: A Global Perspective **29**(1), 30–39 (2020)
13. R. Mehta, K. Gupta, A. Kumar Yadav,An adaptive framework to image watermarking based on the twin support vector regression and genetic algorithm in lifting wavelet transform domain. Multimedia Tools Appl., 1–22 (2020)
14. A.J. Nair, R. Rasheed, K.M. Maheeshma, L.S. Aiswarya, K.R. Kavitha, "An ensemble-based feature selection and classification of gene expression using support vector machine, k-nearest neighbor, decision tree.2019 International Conference on Communication and Electronics Systems (ICCES), 1618–1623 (2019)
15. K. Fares, K. Amine, E. Salah,A robust blind color image watermarking based on Fourier transform domain. Optik **208**, 164562 (2020)
16. S.P. Singh, G. Bhatnagar, A simplified watermarking algorithm based on lifting wavelet transform. Multimedia Tools and Applications **78**(15), 20765–20786 (2019)
17. M. Marinaki, Y. Marinakis, A glowworm swarm optimization algorithm for the vehicle routing problem with stochastic demands. Expert Syst. Appl. **46**, 145–163 (2016)
18. M. Islam, A. Roy, R. Hussain Laskar, SVM-based robust image watermarking technique in LWT domain using different sub-bands. Neural Comp. Appl. **32**(5), 1379–1403 (2020)
19. H.I. Shahadi, A. Toman Thahab, H.R. Farhan, A novel robust approach for image copyright protection based on concentric rectangles. J. King Saud University-Comp. Information Sci.(2020)
20. S. Kannadhasan, R. Suresh,EMD algorithm for robust ımage watermarking, in *Advanced Materials Research*, vol. 984 (Trans Tech Publications Ltd., 2014), pp. 1255–1260
21. B. Wang, P. Zhao, An adaptive image watermarking method combining SVD and Wang-Landau sampling in DWT domain. Mathematics **8**(5), 691 (2020)
22. P. Qiong, Y. Liao, P. Hao, X. He, C. Hui, A self-adaptive step glowworm swarm optimization approach. Int. J. Comput. Intell. Appl. **18**(01), 1950004 (2019)

23. H.-H. Tsai, H.-C. Tseng, Y.-S. Lai, Robust lossless image watermarking based on α-trimmed mean algorithm and support vector machine, J. Systems Software **83**(6), 1015–1028 (2010)
24. Z. Han, J. Gao, H. Sun, R. Liu, C. Huang, L. Kong, H. Qi,An ensemble learning-based Model for Classification of Insincere Questions. In FIRE (Working Notes), pp. 482–488 (2019)
25. X. Wang, Y. Shen, Z. Zhou, L. Fang,An image fusion algorithm based on lifting wavelet transform. J. Optics **17**(5), 055702 (2015)
26. Y. Zhou, G. Zhou, J. Zhang, A hybrid glowworm swarm optimization algorithm to solve constrained multimodal functions optimization. Optimization **64**(4), 1057–1080 (2015)

Deep Learning Approach for Stock Price Prediction and Trading

Nilesh Patil, Jeswin Thomas, Ethan Palani, and Chinmay Gawde

Abstract In today's world, a lot of businesses are using Machine Learning and Artificial Intelligence to increase their efficiency and profits. To cater to the needs of stock brokers, our research aims to provide predictions of any stock price in the Indian Stock Market. This research includes an informative user interface where the user is able to see the stock price predictions along with a suggested trading strategy which would hypothetically allow him to make the most profit on that particular stock. Also, we have included a section where we provide a sentiment score on the recent news articles about a particular stock which in turn helps the investor to foretell whether these current affairs will affect the stock price. Our research brings forth to the users a full-fledged system that would benefit not only an experienced trader but also a layman who has just begun his trading career.

Keywords Stock price prediction · Trading agent · Machine learning · Deep learning · Long short term memory

1 Introduction

Machine Learning has far surpassed the limits of speed and computations set by the human mind. It is a powerful concept that has recently come into the boom, there are countless applications of its technology in various domains. One such application of machine learning is stock price prediction. Algorithmic Trading was legalized in India in only 2008 whereas in foreign countries trading using algorithms control nearly one-third of the total trading. This concept of algorithmic trading was introduced nearly thirty years ago in the US (United States). The predictions are based on past and present data and most commonly by analysis of trends. It can be broken down into the following terms [1]:

N. Patil · J. Thomas (✉) · E. Palani · C. Gawde
Fr. Conceicao Rodrigues College of Engineering, Mumbai 400050, India

N. Patil
e-mail: nilesh.patil@fragnel.edu.in

G. Ranganathan et al. (eds.), *Soft Computing for Security Applications*,
Advances in Intelligent Systems and Computing 1397,
https://doi.org/10.1007/978-981-16-5301-8_46

1. Short (Up to 52 weeks)
2. Medium (52–104 weeks)
3. Long (More than 104 weeks)

These predictions help the brokers in the stock market in planning recuperative courses of action in cases of utmost trouble [2]. If algorithms are able to provide close to accurate results on price predictions, traders can maximize their return on investment [3]. Some people are of the opinion that one cannot predict the flow of the stock market, others might argue that it can be predicted with a fair amount of accuracy [4]. It is however difficult for investors to analyze and predict market trends according to all of this information [5]. This led us to develop a solution for the common man to be able to make the best use of such technology. Users can use the website which will fetch predicted data and other information as per user's needs. Users can search for a particular stock along with an amount they can invest. On the other hand, machine learning models will predict values for the searched data and give the necessary output to the user. Users can use a chatbot for basic information like stock price value, stock news, stock market information, etc.

2 Review of Literature

In our attempt to implement a model to predict the prices of various stocks, we dedicated some time towards reviewing the contributions and analysis of other researchers. As we learned from one of the very first papers we read, A new model for stock price prediction [6], though the approach was a simple one, the results found were definitely bound to improve the prediction accuracies. It is found that deep learning models have the ability to identify the underlying trends in different stock markets [7]. Moving forward, we learned that Deep Neural Network (DNN) models perform better as compared to other models and these models can be applied to even portfolio management and trading strategies [8]. Various studies also mention there being a relation between the news articles and trend of the price of the stock and analyzing this trend to predict the future trends in the market [9]. Some research also throws some light on how textual information can be vital as compared to numerical-data-only methods and Bag-of-Words based methods [5]. Jeong [10] proposed the Deep Q-Learning (DQL) model and other techniques to overcome challenges in the Deep Q-Network (DQN) algorithm used for portfolio trading strategy practically and the results show that DQN strategy is superior compared to other benchmarks. Whereas Yuming Li [11] for stock transaction strategy performed a unique deep reinforcement learning and proved practicality in bartering with financial plan issues. Three classical models of DRL were also compared and the results on developing these three algorithms were capable with the DQN model giving the most reliable performance in judgment. Yuyu Yuan [12] to show the deficiency of training data in financial deep reinforcement learning DARL framework was proposed. DARL is a Data Augmentation Reinforcement Learning. To choose financial assets for trading skewness based access mechanism, as well as kurtosis, is used. Zhong [13] proposed machine learning based hybrid algorithm to calculate the everyday return path of the SPDR (Standard & Poor's Depositary Receipt) S&P 500 ETF. In his research, he showed that DNN based classification has higher accuracy compared to Principal component analysis (PCA) representing data over the data set which is untransformed. He observed that patterns of classification accuracy emerge as the number of hidden layers in the DNN model increases. Naik [14] mainly focused on predicting stock prices for short term trading. Technical indicator feature selection and accurate prediction model for stocks. Results showed that Artificial Neural Network (ANN) outperformed compared to other existing works.

3 Methodology

In order to make the predictions, we can make use of technical indicators which helps study certain elements from the historical data of the stock, such as, trend, volatility, and oscillations. These technical indicators are majorly used to produce better insights into the movement of the stock in the future. And hence play a vital role in prediction, when passed with the stock prices. After we generate the indicator values using the TA-Lib library in python we create a final data frame which is then passed to our machine learning models. We use 6 models which independently predict the closing price for a given stock. The models we use are:

3.1 Long Short Term Memory (LSTM)

The issue of disappearing gradients is solved by LSTM and gives us efficient accuracy compared to recurrent neural networks (RNN). LSTM consists of one cell state and three gates. The given cell states are added interactions. The forget gate takes the previous state and the input and multiplies it with the respective weights and us just to through a sigmoid activation. And we just process with the input gate and output gate C that is the intermediate cell state and then we calculate Ct which is the cell state using the function the input gate and intermediate cell state are added with the old cell and forget gate. (See Fig. 1.) Then we pass the cell state through the tanh activation and then multiply with the output gate.

3.2 Support Vector Regressor (SVR)

We use SVR for constant values as an alternative to classification. Support vector machine (SVM) supports both linear and non-linear regression that can be stated as SVR. The thickness of the road is controlled by a hyperparameter Epsilon. SVR

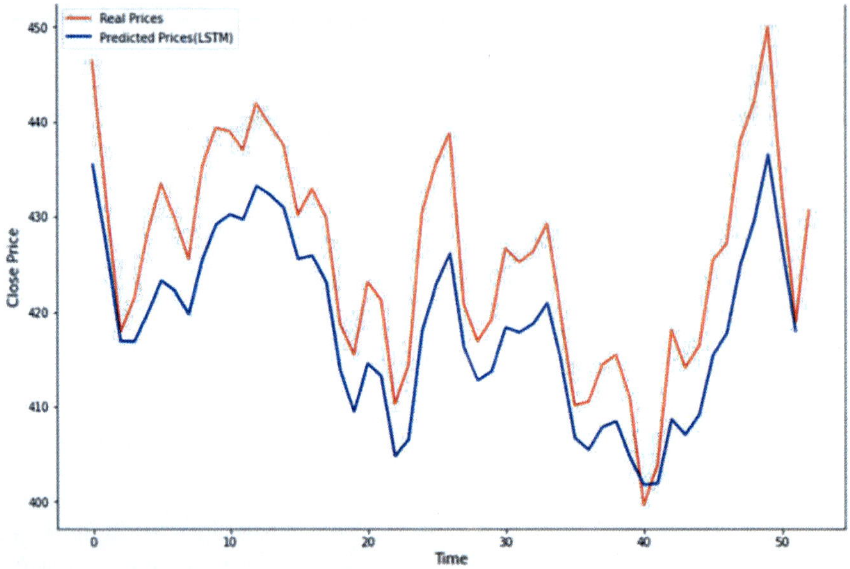

Fig. 1 LSTM stock predictions

accomplishes linear regression in higher (dimensional space) and each data point in the training signifies its own dimension. When we evaluate our kernel among a test point and a point in the resultant values gives you the coordinates of our test point in that dimension. In SVR we effort to fit a convinced threshold. Linear, Polynomial, and Radial basis function (RBF) are different in the case of making the hyperplane decision boundary the classes. RBF gives more correctness than Linear and Polynomial which take less time.

3.3 Random Forest Regressor (RFR)

Random Forest is a method to achieve both regression and classification use of multiple trees to get output results In RF x value to consider as a technical indicator (independent variable) and y value as strategy signal (dependent variable).

3.4 Gradient Boosting Regressor (GBR)

GBR works with the Boosting Technique that computes the base model and will give us one output (average of dependent). It works with many Decision Trees and will compute the residual that is errors. It relies on the best next model, and combined with the previous model, and minimized the prediction errors that are overall. The key reason for GBR is to set the goal output for other models.

3.5 Trading Agent

A stock trading bot has been implemented which uses Deep Q-learning. Deep Q-learning is a part of Deep Reinforcement Learning. Deep Reinforcement Learning belongs to machine learning techniques. It allows us to make brilliant agents that can learn through interactions from the surroundings. By trying and making mistakes it improves itself. When the supervised learning approach is not the top approach for actual activities where there are no proper tags for the data or missing information, this technique is very useful. The main perspective here is to understand that this technique can be used in any real-life activity. This can be said as a Markov process.

In this method of Deep Q-learning, the agent first goes through the initial state, chooses and performs an action (either buy, hold or sell), then watches the adjacent state, obtains signal, and accordingly finalizes its parameters to the gradient.

Some of the Q-learning algorithms which we have executed in this research are:

1. Vanilla Deep Q-Networks
2. Double Deep Q-Networks
3. Dueling Network Architectures
4. Deep Q-Networks with fixed target distribution
5. Prioritized Experience Replay

Drawbacks

1. To keep things simple, the broker can only do one thing at a time that is, either buy or sell one share. Deciding how many stocks to sell is another problem encountered with it.
2. N-day window function, we are trading the closing price, after which a sigmoid function is implemented to simplify values in the interval [0,1].
3. In a sequential manner, training of the agents is done on the central processing unit (CPU). The experiences are reproduced following each trading episode and then updating the parameters.

4 Results

After conducting our research, we conduced the following results.

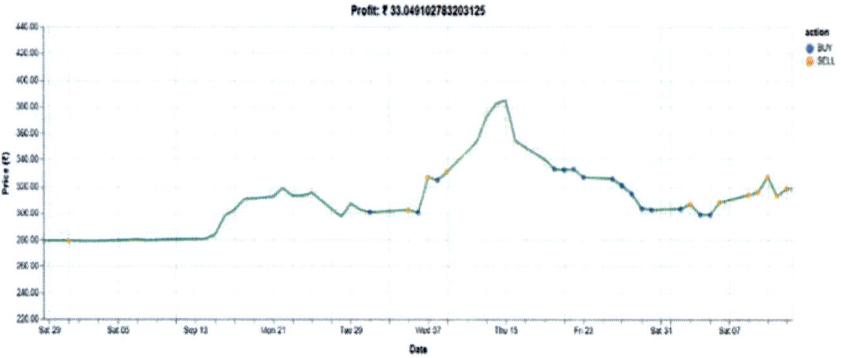

Fig. 2 LSTM buy/sell points

4.1 LSTM

As you can notice from Fig. 1, LSTM is able to make predictions with good directional symmetry but a slightly high mean absolute error (Fig. 2).

4.2 SVR

Our Support Vector Regressor is the least performing model among all the other models (Figs. 3 and 4).

4.3 Random Forest Regressor

The random forest regressor is able to make quite accurate predictions as you can notice in Fig. 5 below, with slight noise for a few data points (Fig. 6).

4.4 Gradient Boosting Regressor

This model is able to make the most precise predictions for most stocks as compared to any other model (Figs. 7 and 8).

Performance Metrics

We evaluate the performance of the models based on 3 metrics: Mean Absolute Error, Mean Squared Error, and Directional Symmetry (Tables 1, 2, and 3).

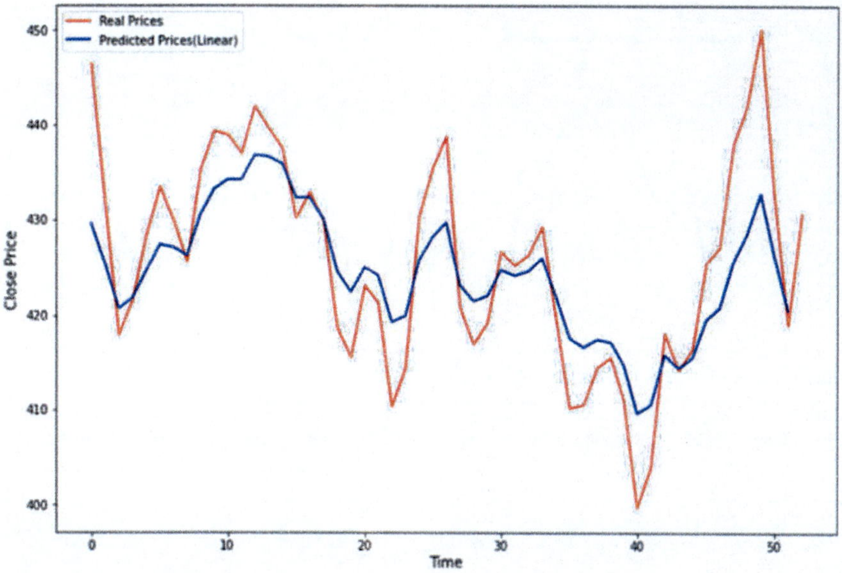

Fig. 3 SVR stock predictions

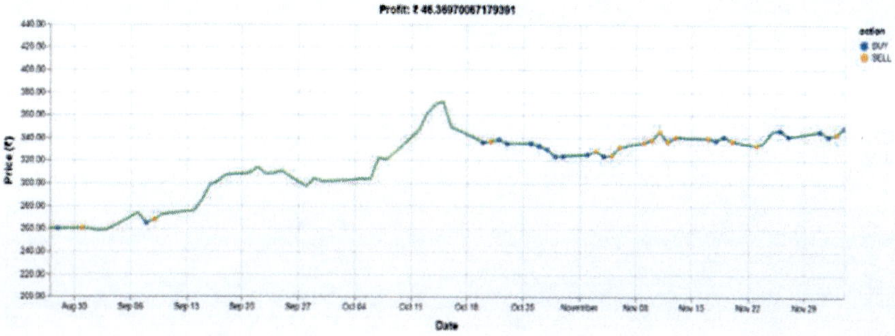

Fig. 4 SVR buy/sell points

5 Conclusion

The proposed system has scope in terms of ease of use, enhanced visualization, user-friendly interface, and integration with a wide range of services. In order to stay ahead in the stock market, it is crucial to be well aware of the various news updates, which are provided to the user conveniently. Also, providing links to the articles to get detailed information pertaining to the article.

In addition to the updates on the news articles, our system is capable of running sentiment analysis tests using Natural Language Toolkit (NLTK) to prompt the user

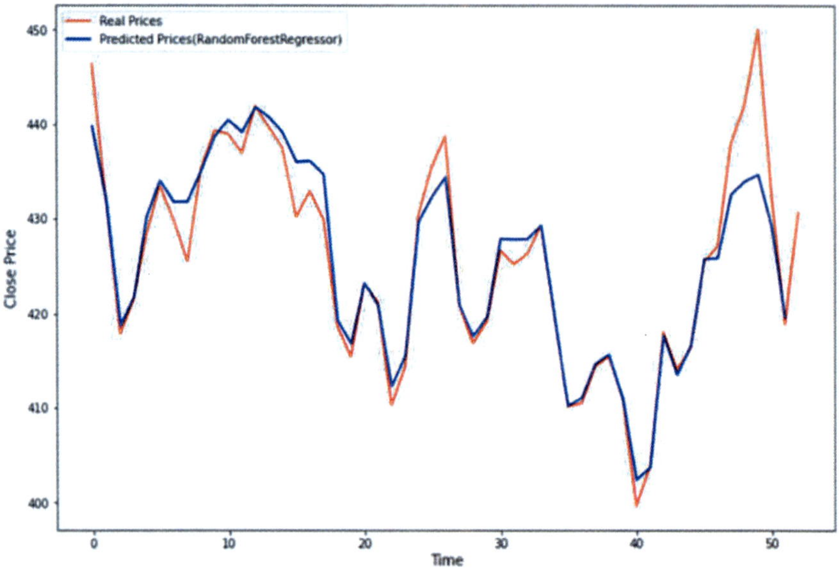

Fig. 5 Random forest regressor stock predictions

Fig. 6 Random forest regressor buy/sell points

regarding the effect of the news (positive or negative). This can provide the investor with a slight edge over the others.

In order to provide a user-friendly workflow, we have also integrated a chatbot in the application to guide the user with basic queries. This proves advantageous as it can save a significant amount of time and enhances the user experience. The chatbot is capable of providing quotes, indexes, and stock prices at an instance provided by the user.

Coming to the center of attraction of our application, the stock prediction and trading agent: Our system is capable of predicting prices for any stock listed in the National Stock Exchange (NSE). We run predictions on 6 different Artificially Intelligent models namely—LSTM, SVR (with Linear Kernel), SVR (with Polynomial

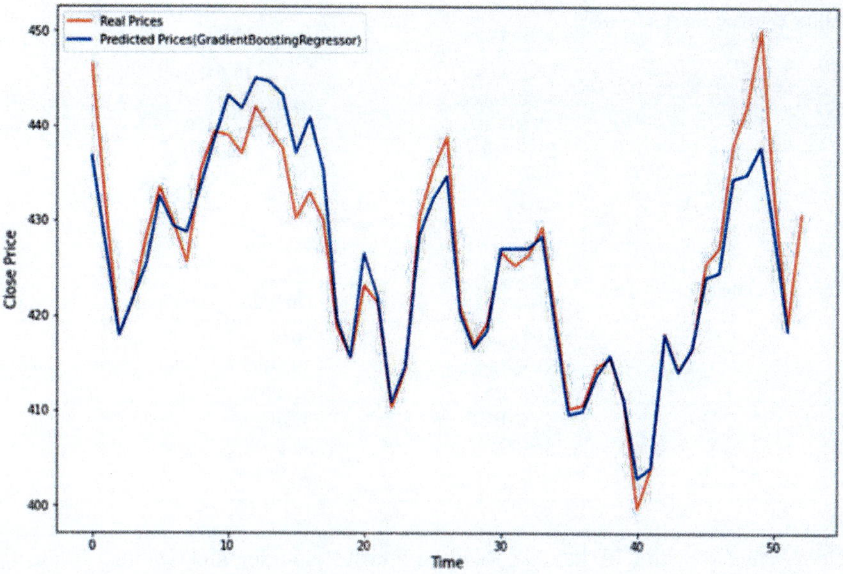

Fig. 7 Gradient boosting regressor stock predictions

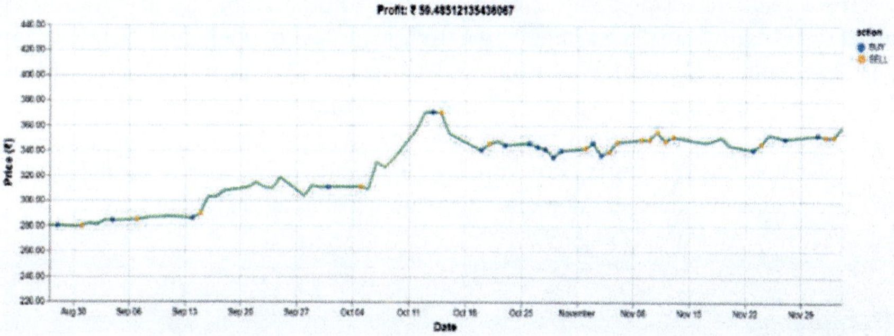

Fig. 8 Gradient boosting regressor buy/sell points

Table 1 Mean squared error

Model name	MSE
LSTM	527.0538
SVR-Linear	171.3700
SVR-Poly	481.5234
SVR-RBF	178.8041
RFR	58.9714
GBR	64.2883

Table 2 Mean absolute error

Model name	MAE
LSTM	18.6550
SVR-Linear	11.4427
SVR-Poly	17.1552
SVR-RBF	11.3080
RFR	5.6481
GBR	5.7667

Table 3 Directional symmetry

Model name	Directional symmetry
LSTM	39.7059
SVR-Linear	39.7059
SVR-Poly	51.4706
SVR-RBF	39.7059
RFR	52.9412
GBR	52.9412

Kernel), SVR (with RBF kernel), Random Forest Regressor, and Gradient Boosting Regressor. Once we predict the closing prices for the provided stock name, we pass these predictions to our trading agent to generate a strategy to gain maximum profit.

The additional features and enhancements would be involving the latest news articles and relevant updates related to the stock in the task of predicting the prices. We can also optimize the stock prediction model's accuracy by the use of reinforcement learning to learn from the faulty predictions and improve on them. The proposed system lacks maintaining the state of the user's work (user cannot view the previously analyzed stocks). Another feature worth adding would be the ability for the system to place trades in the live market after displaying a simulation to the user.

References

1. S. Selvin, R. Vinayakumar, E.A. Gopalakrishnan E.A. Vijay Krishna Menon, K.P. Soman K.P, Stock price prediction using LSTM, RNN and CNN-sliding window model
2. K. Khare, O. Darekar, P. Gupta, V.Z. Attar, Short term stock price prediction using deep learning. 2nd IEEE International Conference on Recent Trends in Electronics, Information & Communication Technology (RTEICT), Bangalore, India, pp. 482–486 (2017)
3. M. Nikou, G. Mansourfar, Bagherzadeh, Stock price prediction using DEEP learning algorithm and its comparison with machine learning algorithms. Intelligent Systems in Accounting, Finance and Management (2019)
4. S. Mehtab, J. Sen, A Robust predictive model for stock price prediction using deep learning and natural language processing (2019)
5. R. Akita, A. Yoshihara, T. Matsubara, K. Uehara, Deep learning for stock prediction using numerical and textual information, 1–6 (2016)
6. H. Huynh, L.M. Dang, D. Duong, A new model for stock price movements prediction using deep neural network (2017)
7. M. Hiransha, E.A. Gopalakrishnan, V. Menon, K.P. Soman, NSE Stock market prediction using deep-learning models. Procedia Computer Science (2018)

8. R. Singh, S. Srivastava, Stock prediction using deep learning. Multimedia Tools and Applications (2017)
9. A. Yoshihara, K. Fujikawa, K. Seki, K. Uehara, Predicting stock market trends by recurrent deep neural networks, 8862, 759–769 (2014)
10. G. Jeong, H. Kim, Improving financial trading decisions using deep Q-learning: Predicting the number of shares, action strategies, and transfer learning. Expert Systems with Applications (2018)
11. Y. Li, P. Ni, V. Chang, Application of deep reinforcement learning in stock trading strategies and stock forecasting. Computing **102** (2020)
12. Y. Yuan, W. Wen, J. Yang, Using Data augmentation based reinforcement learning for daily stock trading. Electronics **9**, 1384 (2020)
13. X. Zhong, D. Enke, David, Predicting the daily return direction of the stock market using hybrid machine learning algorithms. Financial Innovation (2019)
14. N. Naik, B. Mohan, Optimal feature selection of technical indicator and stock prediction using Machine learning technique (2019)

Robust Symmetric Encryption for Public Key Infrastructure

Anubhav Mattoo, Nikhil Samleti, Puja Aher, Shubham Patil, and Ashwini Jarali

Abstract In every encryption suite, symmetric ciphers are an essential tool. They provide safe data transmission with minimal processing effort for both encryption and decryption. The randomness supplied by a decent cipher makes it impractical to reverse by brute force attacks. This paper describes a method of using Double-Ratchet algorithm, error-correcting codes and existing symmetric encryption algorithms to form a robust symmetric cipher. This can be used to verify transferred data, detect modifications in the data and correct modifications done to the data to a certain tolerance level, whereby improving the robustness of a cipher suite in case of modification or mutilation (such as minor bit-flip attacks and insertion or deletion errors). Applications of these ciphers are in the field of public key infrastructure.

Keywords Symmetric cipher · Double-Ratchet algorithms · Hybrid key infrastructure · Cryptography · Public key cryptosystems · AES · Error correction codes

1 Introduction

Symmetric encryption has been proven to be effective in various media. However, shortcomings of the symmetric key system start to show when the channel of communication is not trusted [1]. In the modern world where people, isolated by various bounds, can have no knowledge of each other's existence, can communicate. This poses a challenge of authenticity which is a demanding challenge for symmetric encryption and hence asymmetric encryption was introduced.

A. Mattoo (✉) · N. Samleti · P. Aher · S. Patil · A. Jarali
International Institute of Information Technology, Pune 411057, India

G. Ranganathan et al. (eds.), *Soft Computing for Security Applications*,
Advances in Intelligent Systems and Computing 1397,
https://doi.org/10.1007/978-981-16-5301-8_47

1.1 Asymmetric Encryption

A trapdoor problem based on the *Euler's Totient Function* was considered by R. L. Rivest, A. Shamir, and L. Aldeman. They argued in their 1977 paper that this resulted in a function that was easier to calculate one way but took enormous computational effort to reverse, and the approach became known as the RSA algorithm. Despite the existence of numerous different asymmetric key systems and their utility in niche applications, the RSA algorithm [2] has become the de facto standard for all modern asymmetric encryption system.

2 Hybrid Key Infrastructure

The use of an asymmetric encryption in order to preserve the authenticity of a symmetric cipher is the basic principle of a hybrid key system. Systems made over this consideration make the hybrid key infrastructure. In a hybrid key infrastructure, roles and responsibilities can be assigned as per requirement and can also be revoked in case of failure of compliance. This property allows a system to be secure by removing roles or users that fail to comply to a standard and use only complaint roles or users who can be verified by *User Authentication* [3]

2.1 Advantages of Asymmetric Encryption

The major advantage of using an asymmetric encryption scheme is that the authenticity of a message and a conversation can be uniquely identified by *Message Authentication* and *User Authentication*. This ability provides an additional layer of authenticity to users involved in a communication and makes it harder for a person to forge an identity in digital keys. It makes a hybrid key infrastructure robust against forgery and reduces the reliance on Web of Trust system, instead using a validating proof [4].

2.2 Limitations

The size of the data that may be effectively encrypted is minimal as compared to symmetric encryption, which is a fundamental disadvantage of asymmetric encryption [5, 6]. This indicates that using a symmetric cipher with an asymmetric key is more efficient, and it also makes cryptological sense because trapdoor problems are solvable problems [7]. It lowers the risk of data leakage due to brute force attacks, but does not completely mitigate the risk.

3 Ratcheting Key Generation

Ratcheting key derivation is a system which allows for continuously and efficiently changing symmetric keys by using the previous key and a *Key Derivation Function (KDF)* to generate random keys in a linear, on-demand fashion. Similar to a mechanical ratchet, the KDF can't be reversed easily, it requires immense amount of computational work in order to reverse a generated key. This property ensures backward secrecy of messages, in essence, that if a packet is compromised in the chain encrypted by a particular key, the packets before this packet will still remain intact. However, this method does not ensure forward secrecy, and all packets after the compromised packet are compromised [8]. Ratcheting is essential in asynchronous communication. It allows the communicating user the access to a particular key only when the message is available or expected, or to keep a one way communication explicitly to the intended receiver even if the receiver is not actively a part of the conversation.

3.1 *Double-Ratcheting*

As explained by the Signal Protocol Suite, the Double-Ratchet System or Axolotl Algorithm is a combination of symmetric and asymmetric key derivation functions that use the transfer state messages in order to change the generated public and private keys. Despite the system not being capable of providing complete forward secrecy, it provides crucial self-healing in the form of changing asymmetric keys. This improves the overall security of the system however adds a computational overhead that may not be suitable for small and micro systems.

The generation of random primes has to be limited to a certain frequency and hence, the forward secrecy provided by this system is compromised for communications of longer duration [9] (Fig. 1).

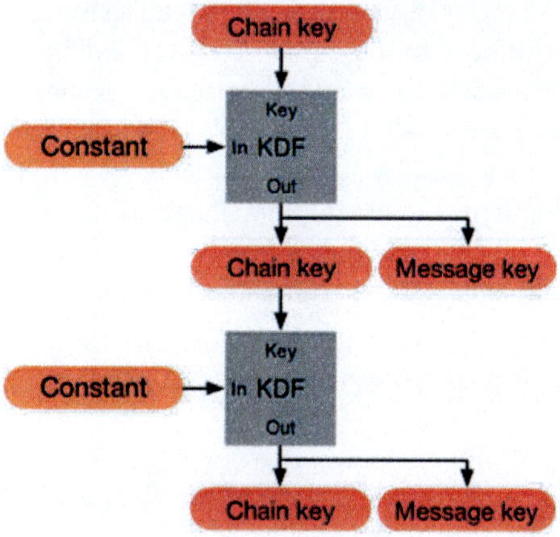

Fig. 1 KDF as illustrated in signal protocol [9]

4 Error-Correcting Codes

Not to be confused with elliptic curve cryptography, an error-correcting code (ECC) or error correction code is a type of block code or continuous code that has the property of correcting errors present in data. This is used to return the data to its original condition after it has been encoded, in case an error occurs during transmission or by manipulation. It is widely employed in physical storage medium, with varying degrees of tolerance according to the application. There are various ECC that are used in data encoding and each one has their own unique properties. Block codes such as Hamming codes and parity check codes are used to detect errors and Reed–Solomon Codes, Bose–Chaudari–Hocquenghem (BCH) codes [10], Turbo codes and Low-Density Parity Check (LDPC) codes are used correct multiple errors [11, 12].

4.1 ECC in Cryptography

Despite being commonly used with physical media, such as microwave radios, magnetic and optical discs, ECC has seen relatively little use in cryptography, specifically symmetric cryptography, despite some implementation in asymmetric key systems like the McEliece Key System which uses BCH codes and Golay codes [13, 14]. Such systems are good for usage on channels of communications where error can occur due to the nature of the channel(such as microwave radio), it does not provide any additional resilience to the cipher itself apart from correcting errors that occur in the lower communications channels, which can be modified in order to carry a mutilated packet correctly [15].

4.2 ECC with Cryptographic Hash

Adding a ECC to data along with a strong cryptographic hash can provide additional resilience to the cipher itself by adding a property of correcting any errors within the data to a certain tolerance and providing a method to identify between accidental and perpetrated mutilation or modification of data over a channel.

The disadvantage of using an ECC in a symmetric cipher is that it may become inefficient to use additional data in order to keep track of existing data and its validity. Additional data can make the package size larger than expected and, in some implementations, can undermine the efficiency of a symmetric cipher.

5 Methods and Tools

5.1 *Super Block Encryption*

Block ciphers use blocks of data which are encoded by substitutions and permutations using an encryption scheme. Some of these ciphers explicitly tie proof of existence of a specific block to succeeding blocks, which is troublesome for decryption of vast volumes of data formatted by relative proofs of existence; these ciphers are referred to as chaining ciphers [16].

If such encryption is required, an alternate technique is to utilise a super block, which consists of one or more small blocks connected only by the initial parameters used for encryption, such as the base key.

Advantages of Super Blocks The main advantage of employing a super block encryption over a normal chaining cipher is that the chains are contained within a super block. This allows for indexed ciphertexts with known or derivable keys and verifiable authenticity. Since all data before a given block can be bypassed in order to check that block, propagation of an error is confined to that block and does not affect any other block in the file, such a methodology allows faster and secure decryption of enormous volumes of data and enhances system performance for decryption. Explained in Algorithms 1 and 2, a Hash-based Key Derivation Function (HKDF) can be used to mitigate the problem of long key schedules. Since hashing is a faster operation than decryption and does not leave traces of data while decryption of a specific block while also authenticating the specific block correctly, unlike plain Advanced Encryption Standard (AES) in Galois Counter Mode (GCM) which requires complete decryption in order to authenticate data.

Limitations of Super Block Encryption Super blocks are self-contained blocks; if data that is longer than the block size is used, it will be fragmented. As a result, even if the isolated fragment is correctly decrypted and verified, it will be of little service in understanding the entire contents. As a result, using super blocks with data as fixed-size records that can be accessed individually is the ideal option. Since each block is separately encoded the size of the keys is long due to the fact that each block's key is to be preserved separately, or a method of key derivation is to be used without decryption of data.

5.2 *Double-Ratcheted Key Derivation using Parity*

Using the principles of Double Ratchets, error-correcting codes and super block encryption, we have developed a system that can efficiently increase the robustness of a hybrid key infrastructure by making symmetric ciphers robust against modification and mutilation of the ciphertext.

Algorithm 1: Super Block Encryption with AES-GCM

```
Input  : Data to be Encrypted, Super Block Size and Base Key used for Encryption
Output: Encrypted Super Blocks
def encrypt_AES_GCM_SB(data : bytes, blocksize : int, key : bytes):
    blocks ← [ ]_{n×blocksize}
    if length of data= 0 then
        raise Error
    end
    if length of data  mod blocksize ≠ 0 then
        data ← data ‖ padding
    end
    if blocksize ≥ length of data then
        blocks_0 ←AES_GCM(key, block)
    end
    else
        for x ← 0 to length of data step blocksize do
            tempblock ← data_{x x+blocksize}
            n ← ⌊x ÷ blocksize⌋
            blocks_n ←AES_GCM(key, tempblock)
            key ← HKDF(key)
        end
    end
return blocks
```

Algorithm 2: Super Block Decryption with AES-GCM

```
Input  : Encrypted Super Blocks, Super Block Size, Starting Block Number, Ending Block
         Number and Base Key used for Encryption
Output: Decrypted Super Blocks
def decrypt_AES_GCM_SB(blocks : bytes, blocksize : int, start : int, end : int,
  key : bytes):
    dblock ← NULL
    if start > end then
        raise Error
    end
    for x ← 0 to start do
        key ←HKDF(key)
    end
    for x ← start to end do
        dblock ← dblock ‖ AES_GCM_REV(key, blocks_x)
        key ←HKDF(key)
    end
return dblock
```

The system we propose relies on the use of *Double Ratcheting Key Derivation Over Error Correcting Code Exchange* or *Double Ratchet Over Parity Exchange (DOPE)*, since error-correcting codes are a form of parity, they are used in a way to make the additional information useful.

Advantages of DOPE The main advantage of employing DOPE over regular super block encryption is that if a block is modified, a hash-based system can only verify that the message's authenticity is compromised, but an ECC may repair faults in a cipher block to a specific tolerance and try to retrieve information that was modified or mutilated or determine if the information was altered on purpose. In order to do that, it is recommended to use a specific balance of error correcting strength and overhead measured by *Error Correcting Strength to Total Size Ratio (ECTR)*. Since both ends keep a record of each other's keys and both Ratchets can move freely with respect to each other, the asynchronous communication formed will be independent despite having a shared secret, i.e. even if one side of the conversation is compromised in some way, the other side will take at least the same amount of computational work to be compromised. The subsequent sections of this paper describe, in detail, the use of these methods to create a robust symmetric cipher for public key infrastructure.

6 DOPE Key

A DOPE key is a combination of encapsulated asymmetric keys which are transferred between users in order to begin a DOPE session. Every session has a unique key associated to it and can be reused after it has been discarded. Since the ratchets depend on the parity of data sent over a communication, the sequence can be duplicated only if bit exact same data is sent again, in case if a session key is reused by the same user. In DOPE, we refer to the outer key as *Identity Key* or *Strong Key* and inner key as the *Session Key* or *Weak Key*.

In order to generate a DOPE key, user should have the access to the receivers' public keys. The user then generates another key pair which is of a relatively small size as it is only used as a *Pre-Key* for the actual transfer. This means that no symmetric keys are sent over the communication channel, encrypted or otherwise.

$$\text{WPriK}, \text{WPubK} = \text{RSA}(80) \tag{1}$$

The (WPriK) is the weak private key generated for the session or session key and (WPubK) is the weak public key for the weak private key. These keys need to be stored for a particular session and can be negotiated again if needed.

$$\text{DOPEKey} = \text{PubK}_{\text{R}}.\text{Encrypt}(\text{WPubK}) \tag{2}$$

The (DOPE Key) is the weak public key encrypted with the receiver's public key (PubK_{R}) usually using a scheme like described in Public Key Cryptography Standards

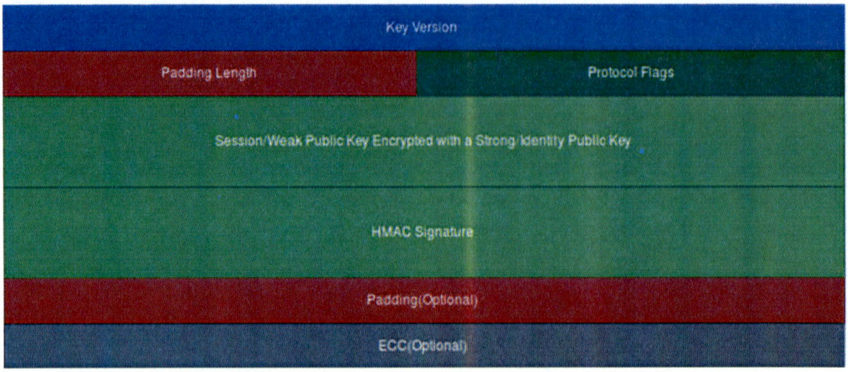

Fig. 2 Ideal DOPE key packet

(PKCS#1) [5, 6]. Furthermore, digital signatures should be utilised to sign the key as confirmation of authenticity. It is analogous to a Diffie–Hellman key exchange [17] (Fig. 2).

7 DOPE Key Exchange

7.1 Overview

A DOPE Key Exchange happens as 4 Phases:

1. Alice sends Bob her public key
2. Bob sends his public key to Alice in response.
3. Both generate and exchange DOPE Keys.
4. Both send a synchronisation message to confirm the pairing.

They following sections explain each of them in detail. Usually, a DOPE key exchange should be preceded by a Diffie–Hellmann or an X3DH Key Exchange [3].

7.2 Request Phase

Alice sends Bob her public key which can be used to identify Alice uniquely under the constraints that Alice and Bob are within the same Peer Group. The Peer Group is a subset of the entire network of communication where each person involved has to have a unique public key. These are used to relatively establish identity without necessarily giving out the users' information as the user can have a different public key in another peer groups. Once Bob receives the public key, he can choose to proceed with the request or deny the request at the instant, which will silently reject Alice's public key.

7.3 Acceptance Phase

Bob sends Alice his public key in response, this means that Bob is ready for a conversation and has acknowledged Alice's presence.

In case Alice has dropped the Request, Bob's public key will be silently discarded. If the key is accepted a Session Key is generated on Alice's end of the conversation. Bob also prepares a Session Key for the conversation. The Session Keys are designed to be small as they will only be used when required and are required for non-essential data.

7.4 Fixation Phase

Both create and exchange DOPE Keys, which are then used to fixate the ratchet at both ends. Each end of the conversation has a HKDF-based Double Ratchet, one for the home keys and one for the keys on the other end of the conversation. A total of four Ratchets are utilised to maintain track of a dialogue, two at each end.

7.5 Synchronisation Phase

Both send a synchronisation message to confirm the pairing. This causes the ratchets to align at either ends and confirms that either ends generate the same symmetric keys.

In case there was an error while transmission or authenticity was compromised during the exchange, the ratchets won't align, and the system will effectively cease the pairing.

Once keys synchronise, any symmetric block cipher or stream cipher can be used to encrypt data and is assured to decrypt at the other end faithfully. Additionally, protocols can be sent over DOPE Keys in order to configure the ratchets to work in case both ends use a different version or configuration of the DOPE system.

Algorithm 3: DOPE Key Generation

```
Input  : Sender's Private Key, Receiver's Public Key, Size of Session Key, Type of HMAC
         to use and Size of a Normal Packet
Output: DOPE Key
def DOPE_generate_key(public_key_r : bytes, private_key_s : bytes,
 weak_size : int, hmac_type : str, packet_size : int):
    key ← NULL
    wprikey, wpubkey ←RSA_Keygen(weak_size)
    protocols ← Protocol Flags
                                    /* Store Session Key for use later */
    key ←encrypt(public_key_r, wpubkey)
    hmac ←HMAC(hmac_type, key, private_key_s)
    key ← key || hmac
    if key mod packet_size ≠ 0 then
        key ← key || padding
    end
    key ← protocols || pad_len || key
    /* Additionally an ECC can be used here as a trailing
       Synchronisation Message                                  */
return key
```

8 DOPE Data Exchange

The exchange of data in DOPE is done as super block encryption and each super block is encoded with its own ECC. The data exchange can be started after the keys are fixated and synchronised as stated in Algorithm 3.

The exchange for DOPE is described in Algorithms 5 and 6. While decryption, if the underlying layers' error detection or correction fails or if the packet is modified or mutilated, then instead of discarding the packet DOPE tries to use the ECC to recover the packet's state. If that recovery fails, then the packet is discarded.

8.1 DOPE with Block Ciphers

In case of block ciphers like AES, DOPE can be used as an additional check major and an efficient method of encrypted indexing of data, especially for large volumes of fixed block storage. It also addresses problems with AES in cipher block chaining (CBC) mode, which is susceptible to mutilation, by using a balanced ECC which can correct small errors but isn't designed to correct all errors within the data, this allows for an ECC to have a hash like capability to discard data in case the encountered faults are more than the design limit of the ECC [8, 18].

DOPE also provides efficient indexing of long strings of data, due to the fact that ECC for the data can be accessed independently and does not require decryption of the block necessarily.

Algorithm 4: DOPE Key Fixation

```
Input  : Sender's Public Key, Receiver's Private Key and Received DOPE Key
Output: HKDF Ratchets for Conversations
def DOPE_fixate_key(DOPE_key : bytes, public_key_s : bytes,
  private_key_r : bytes):
    proto, pad_len, key, hmac ←Unpack_DOPE(DOPE_key)
    verification ←verify(public_key_s, key, hmac)
    if verification then
        wkey ←decrypt(key, private_key_r)
        wkey_end ← hash(wkey)
        wkey_home ←hash(wpub_key)
                                    /* Use the stored Session Public Key */
        skey_end ← hash(public_key_s)
        skey_home ← hash(public_key_r)
        home_key ← HKDF(skey_home, wkey_home)
        end_key ← HKDF(skey_end, wkey_end)
                                                /* Synchronise Ratchets */
    end
    else
        raise Error
    end
return home_key, end_key
```

8.2 *DOPE with Stream Ciphers*

Stream ciphers encode data in a bit-for-bit fashion and hence are faster than a block cipher. However, stream ciphers are prone to modifications to the ciphertext itself. Most stream ciphers use a Message Authentication Codes (MACs) such as Poly1305 which provides immense capabilities for error detection but fail to provide a robust solution to attacks that intent to corrupt information repeatedly as a Denial of Service (DoS) or a Distributed Denial of Service (DDoS); such attacks are called bit-flip attacks and are commonly seen in controller area networks.

Bit-flip attacks can also be used for DoS since consistently flipped bits will cause the communicating ends to reject messages consistently and will not allow the communication to continue faithfully. DOPE is fault tolerant to such attacks only to some extent [19].

Protection Against Bit-Flip Style DoS Attacks DOPE provides a system to handle such attacks till a certain threshold, and identifies if a channel has crossed the threshold to inform about the possible attack over the channel. It can be used in cases where communication is scarce and re-transmission of messages is expensive.

Algorithm 5: DOPE Data Encryption with AES-GCM

Input : Data to be Encrypted, Block Size, Base Key used for Encryption, ECC Polynomials and Correction Strength

Output: Encrypted DOPE Blocks

```
def DOPE_AES_GCM(data : bytes, blocksize : int, end_key : bytes, ecc_poly : int,
    error_s : int):
    blocks ← [ ]_{n×blocksize}
    if length of data = 0 then
        raise Error
    end
    if length of data mod blocksize ≠ 0 then
        data ← data || padding
    end
    if blocksize ≥ length of data then
        blocks_0 ← AES_GCM(key, block)
    end
    else
        for x ← 0 to length of data step blocksize do
            tempblock ← data_{x x+blocksize}
            n ← ⌊x ÷ blocksize⌋
            tempblock ← AES_GCM(key, tempblock)
            blocks_n, ecc ← ecc_encode(tempblock)
            blocks_n ← blocks_n||ecc
            key ← HKDF(key, ecc)
        end
    end
return blocks
```

Algorithm 6: DOPE Decryption with AES-GCM

Input : Encrypted Super Blocks, Super Block Size, Starting Block Number, Ending Block Number and Base Key used for Encryption

Output: Decrypted Super Blocks

```
def DOPE_AES_GCM_REV(blocks : bytes, blocksize : int, start : int, end : int,
    key : bytes):
    dblock ← NULL
    if start > end then
        raise Error
    end
    for x ← 0 to start do
        key ← HKDF(key, ecc_x)
    end
    for x ← start to end do
        blocks_x ← ecc_decode(blocks_x, ecc_x)
        if decoding fails then
            raise Error
        end
        dblock ← dblock|| AES_GCM_REV(key, blocks_x)
        key ← HKDF(key)
    end
return dblock
```

9 Security Considerations

9.1 Authentication

Users can identify a particular communication is legitimate by either sending synchronisation messages or by verifying ratchets via out-of-scope communication (such as QR Codes or OTPs). Without the out-of-scope authentication, the users cannot guarantee that communication is legitimate.

9.2 Reusing Session Keys

Due to limitations in the DOPE key generation methods, it is advised to not reuse session keys if the data to be transferred is exactly the same. This can expose previous messages unintentionally.

Keys, if needed, should be reused at a very low frequency if at all.

9.3 Compromised Peer in a Group

Since all communications are dependent on a peer group, a compromised peer can leak information about other users in a group. For such scenarios, it is better to use a central server for negotiating the keys via DH or X3DH key exchange and then starting the Request Phase.

9.4 Compromised Keys in Transit

If DOPE keys are compromised in transit, they can't be detected in case they are packed again with similar characteristics. In order to mitigate this, an HMAC is used with the key itself to sign it. This establishes a trust between the users for the communication. If a system uses centralised servers for key negotiation, it is appropriate to use a separate HMAC to verify the authenticity of the server that is communicating the requests and replies.

9.5 Revocation and Denial

User keys can be revoked if the user fails to comply to the peer group or server standard. In such cases, the user bears the burden of upgrading and re-verifying itself.

In case if a request is denied or discarded, the communication within the channel should not occur, if a central server is used, the DOPE keys negotiated should be discarded.

9.6 Hash Functions

Separate hash functions must be used for verification and key derivation. Using the same hash functions can compromise the system due to the current nature of HKDF Ratchets used.

10 Conclusion

1. A robust symmetric cipher was designed for public key infrastructure that involves the use of ECC.
2. Error correction capability of such system can be used for communication where re-transmission is expensive or scarce.
3. Despite its ability an ECC can't entirely correct data(and it should not be done either) that has been doctored for an attack, in such case a hash can verify the authenticity and data will have to be re-transmitted.
4. Super block encryption is an efficient method of indexing symmetric block ciphers, whereas it can compromise the efficiency of stream ciphers by pal-letisation of data which incurs unnecessary overheads. In such cases, an ECC can be appended to a message which will serve the same purpose as that of an ECC in a super block.
5. Super blocks can be designed specifically in order to transmit large volumes of data and allow the receiver to only request specific blocks in case of corruption of a block.
6. Double-Ratchet Over Parity Exchange (DOPE) was designed as an alternative to PGP system for transport of large volumes of data securely. It uses super block encryption, error correction codes for every block and hashes to verify data at the receivers' end.
7. DOPE system is fault resistant but not fault proof, in case if data is corrupted beyond repair, DOPE cannot recover it.
8. DOPE should be designed to only recover ciphertext and not plaintext as that can compromise the communication.

11 Appendix

11.1 Choice of BCH Codes for DOPE

Despite being an arbitrary choice, BCH codes have some interesting properties that made them a strong candidate for such applications. BCH codes are powerful, random error-correcting and variable length codes which have a relatively small size. Since errors may or may not have a random pattern, having a code that corrects random patterns is an advantage.

These codes are computationally inexpensive for encoding and decoding and can be used with small and microdevices. This allows for applications in Internet of Things (IoT), controller area networks (CANs), NAND storage devices, and micro-controller interfaces.

Some common uses of BCH codes are in multimedia transport such as ITU-T Rec. H.261 Standard for p x 64kbit/s streaming audiovisual transfer [20].

11.2 Probability of Error and Error Correction

Since ECC is relatively smaller than the packet, it is less probable to be affected by an attack.

$$\text{Packet} = \text{Data} \parallel \text{ECC}$$
$$\text{Size(ECC)} < \text{Size(Data)}$$
$$P(n - \text{Bit Error in ECC}) = \frac{\binom{\text{Size}_{\text{bits}}(\text{ECC})}{n}}{\binom{\text{Size}_{\text{bits}}(\text{Packet})}{n}}$$

As mentioned above, ECC can only correct fault till a certain design limit; usually, this limit should be higher than the maximum possible message length so as to accommodate for insertion faults caused by an attacker or through the medium.

$$\text{Maximum Codeword Size} = 2^m - 1$$
$$\text{Maximum Size}_{\text{bits}}(\text{Data}) = 2^m - 1 - \text{Size}_{\text{bits}}(\text{ECC})$$
$$\text{Design Size}_{\text{bits}}(\text{Data}) \leq 2^m - 1 - \text{Size}_{\text{bits}}(\text{ECC})$$

where m = Order of the Galois Field of the BCH Code
t = Maximum Error that the Code can correct (in bits)
Size_bits (X) = Size of X in bits.

Figure 3 gives the percent composition of ECC with respect to the block size of the Super Block and the Packet Size, which is the total size of all data required for decryption serialised.

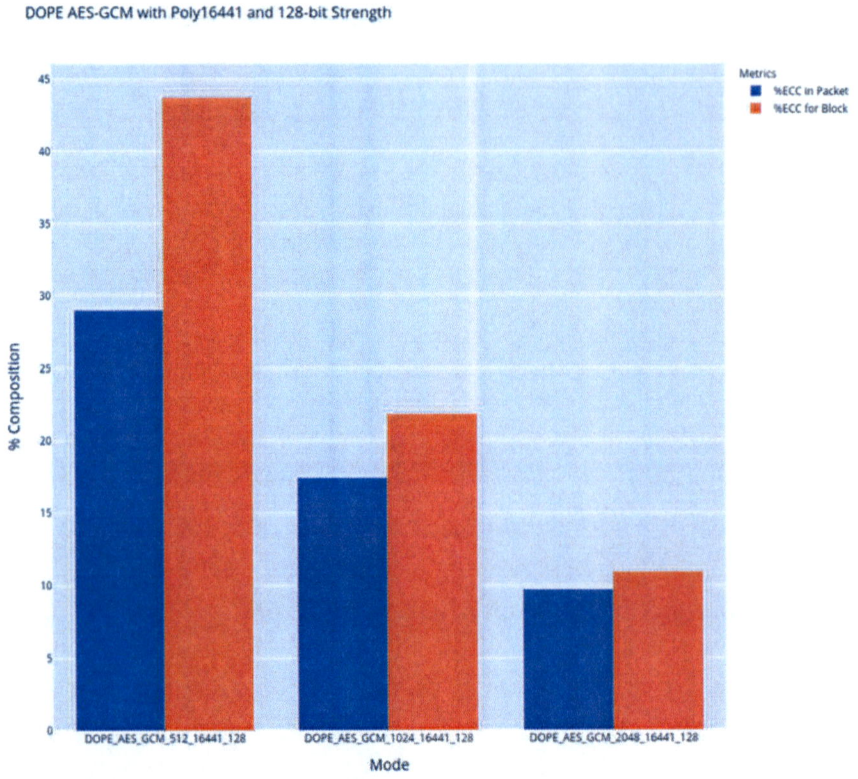

Fig. 3 DOPE AES-GCM ECC composition with different block sizes

11.3 Super Block AES-GCM Encoding Times Versus DOPE Encoding Times

Figure 4 gives a run-time comparison of Super Block AES-GCM Encoder with 512 byte super block size(SB_AES_GCM_512) with comparable DOPE AES-GCM Super Block Encoders with BCH generator Polynomial 8219(Poly8219) and Polynomial 16441(Poly16441) with 128-bit error correction strength. Poly8219 version being DOPE_AES_GCM_512_8219_128 and Poly16441 version being DOPE_AES_GCM_512_16441_128

The major part of the run time is consumed for generation of the generating polynomial in the case of DOPE Encoders.

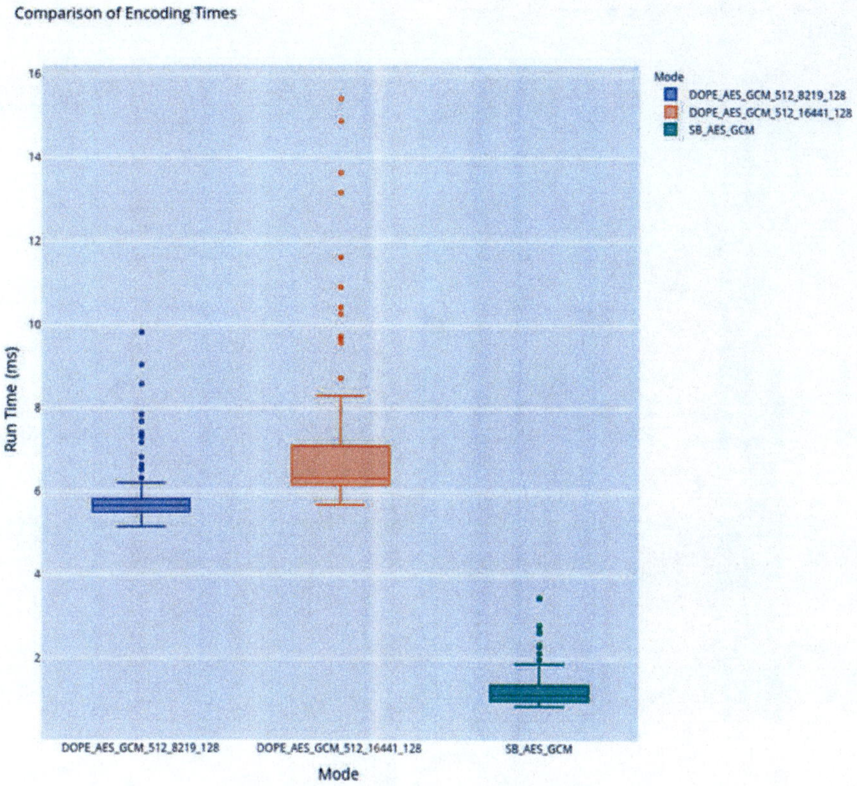

Fig. 4 Average running time for encoders

11.4 Encoding overheads with DOPE

Figure 5 shows a comparative view of the block size and the packet size of various DOPE AES-GCM configurations in Poly16441 with varying error-correcting strengths and packet overheads. As a rule of thumb, it should be considered that ECC with higher error threshold also add greater packet overhead. A balance between the encoding overheads and error-correcting strength can be achieved by using the ratio of bits corrected to the total size of a block with the ECC, this ratio is called the ECTR.

$$\text{ECTR} = \frac{t}{\text{Size}_{\text{bits}}(\text{Block}) + \text{Size}_{\text{bits}}(\text{ECC})} \tag{3}$$

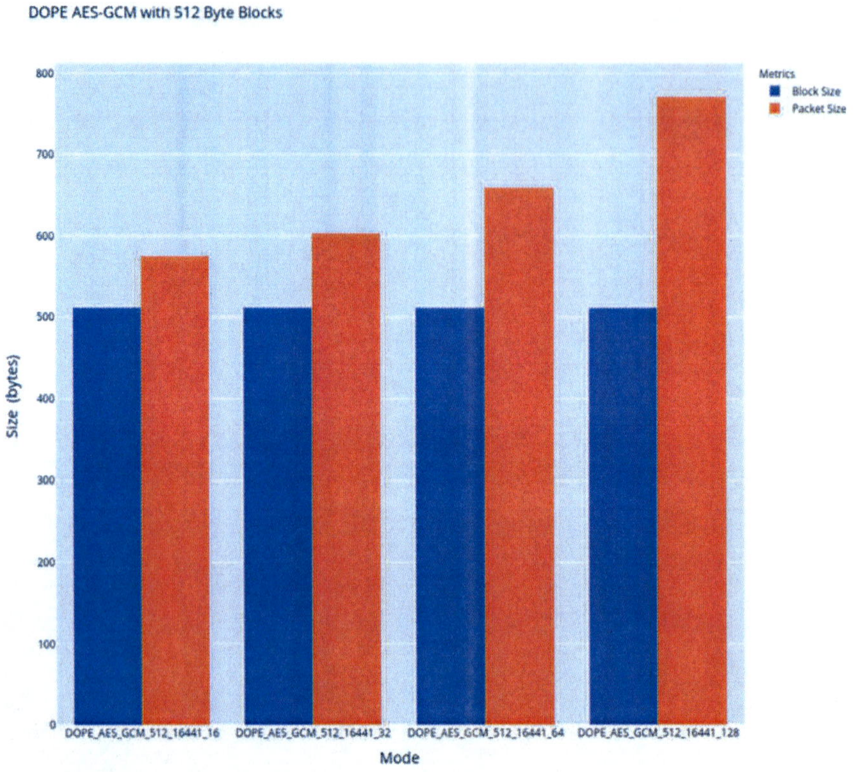

Fig. 5 Packet overheads with different modes of DOPE

ECTR is error-correcting strength to total size ratio and (t) is the total number of error-correcting strength in bits.

References

1. S. Chandra, S. Paira, S.S. Alam, G. Sanyal, A comparative survey of symmetric and asymmetric key cryptography. https://ieeexplore.ieee.org/document/7086640
2. R.L. Rivest, A. Shamir, L. Aldeman, A method for obtaining digital signatures and public-key cryptosystems (1977). https://people.csail.mit.edu/rivest/Rsapaper.pdf
3. M. Marlinspike, T. Perrin, *The X3DH Key Agreement Protocol* (2016). https://signal.org/docs/specifications/x3dh/x3dh.pdf
4. RFC 7748, Elliptic Curves for Security, IETF. https://tools.ietf.org/html/rfc7748
5. RFC 8017, PKCS#1, IETF. https://tools.ietf.org/html/rfc8017
6. RFC 3447, PKCS#1 v1.5, IETF. https://tools.ietf.org/html/rfc34473
7. Bhatia, V., Ramkumar, K.R.: An efficient quantum computing technique for cracking RSA using Shor's algorithm. https://ieeexplore.ieee.org/document/9250806

8. A. Barenghi, G.M. Bertoni, L. Breveglieri, M. Pellicioli, G. Pelosi, Fault attack on AES with single-bit induced faults (2010). https://www.researchgate.net/publication/224183202_Fault_attack_on_AES_with_single-bit_induced_faults
9. M. Marlinspike, T. Perrin, The Double Ratchet Algorithm Protocol (2016). https://signal.org/docs/specifications/doubleratchet/doubleratchet.pdf
10. O. Ghazi, BCH codes (2020). https://doi.org/10.1007/978-3-030-33380-5_6
11. B. Johnston, F. Richman, Error correcting codes, numbers and symmetry, pp. 209–240. https://www.researchgate.net/publication/342989653_Error-Correcting_Codes
12. N.J.A. Sloane , Error-correcting codes and cryptography, in *The Mathematical Gardner*. ed. by D.A. Klarner (Springer, Boston, MA, 1981). https://doi.org/10.1007/978-1-4684-6686-7_30
13. M. Repka, P. Zajac, Overview of the McEliece cryptosystem and its security (2014). https://www.researchgate.net/publication/273902880_Overview_of_the_Mceliece_Cryptosystem_and_its_Security
14. M. Repka, P.-L. Cayrel, Cryptography based on error correcting codes: a survey (2014). https://www.researchgate.net/publication/277019334_Cryptography_Based_on_Error_Correcting_Codes_A_Survey
15. U. Meyer, S. Wetzel, On the Impact of GSM encryption and man-in-the-middle attacks on the security of interoperating GSM/UMTS networks. https://www.cs.stevens.edu/~swetzel/publications/gsm.pdf
16. D. Blazhevski, A. Brozhinovski, B. Stojcevska, V. Pachovski, Modes of operation the AES algorithm. https://www.researchgate.net/publication/236656798_MODES_OF_OPERATION_OF_THE_AES_ALGORITHM
17. N. Li, Research on Diffie-Hellman key exchange protocol. https://www.researchgate.net/publication/224146008_Research_on_Diffie-Hellman_key_exchange_protocol
18. T. Fuhr, E. Jaulmes, V. Lomné, A. Thillard, Fault attacks on AES with faulty ciphertexts only (2013). https://www.ssi.gouv.fr/uploads/IMG/pdf/Fault_Attacks_on_AES_with_Faulty_Ciphertexts_Only.pdf
19. S. Duval, V. Lallemand, Y. Rotella, Cryptanalysis of the FLIP family of stream ciphers (2016). https://eprint.iacr.org/2016/271.pdf
20. ITU-T Rec. H.261, ITU. https://www.itu.int/rec/T-REC-H.261-199303-I/en

Big Social Data Analytics

Paresh Gupta, Shruti Khanduja, Suresh Kumar, Monika, and Kamlesh Sharma

Abstract In recent years, social media has become the best platform to advertise products or gain popularity in one way or another. The only way to figure out the public demand is by analyzing the advertisements or product launch, which is gaining increased popularity. On various social media sites, there are billions of users who regularly share their opinions. As a result, a vast amount of data is generated. To identify useful patterns from this data, it must be analyzed. Henceforth, this paper discusses the importance of social media data analysis and its benefits to the business. Also, this paper has utilized the customer support dataset from Twitter for further analysis. This research work considers a large, modern corpus of tweets from customers and their replies. The pandas-profiling tool is used to perform the analysis. The overall outcome of this analysis is about the improvements required in the services of different companies which in return if addressed timely, can increase the overall profit of the companies.

Keywords Social media · Big data · Text mining · Process model · Analysis · Dataset information · Platforms · Techniques

1 Introduction

The social media businesses are experts in data, and they require a lot of it. As a result, they are able to persuade consumers to disclose as much data as possible. This information might include any conversations that individuals consider to be particularly important. This shows that an immense amount of data is being produced

P. Gupta (✉) · S. Khanduja · S. Kumar · Monika · K. Sharma
Manav Rachna International Institute of Research and Studies, Faridabad, India

S. Kumar
e-mail: suresh.fet@mriu.edu.in

K. Sharma
e-mail: kamlesh.fet@mriu.edu.in

G. Ranganathan et al. (eds.), *Soft Computing for Security Applications*,
Advances in Intelligent Systems and Computing 1397,
https://doi.org/10.1007/978-981-16-5301-8_48

everyday. The volume of data produced and the flow of this immense data basically conveys the importance of big data [1, 2].

So, if a large amount of data, which is being produced every single second is considered for the analysis process with a deep analysis on the massive data, it can directly affect the performance of any industry. For example, If we analyze the reviews of the launch of any new product, then we can easily configure the changes required in the product and demands of the people leading to a more profitable other launchs [3].

Further with associate implementation of analytics, firms are providing their staff a factual understanding of customers' necessities, thereby boosting a cohesive read of their business's operations and therefore the direction of their mission in relevancy their exploration for dynamic knowledge [4].

Actually like the other exchange, web-based media companies discover enormous information helpful for dissecting markets and foreseeing customer conduct. In 2012, Jay Parikh, designing VP at Facebook, revealed that Facebook handles more than 500 terabytes of information daily, approx 30 billion photographs, 2.6 billion 'Preferences', and five billion substance transfers. All this data is prepared in precisely minutes giving Facebook understanding into client responses and in this way the capacity to carry out or change its giving.

In addition, Analysis can be done on the basis of gender, age, client status, financial background, locality, city. The company can target an audience which is called hyper targeting. Choosing a target audience can be very helpful for the sellers as they can target that specific group by delivering relevant messages or by advertising and thus leading to increased profit. For example, A product based company which deals mostly with young people carried out an investigation of web-based media data & stated that bereaved females under 25 may be suitable as models. Upholding this information, the company may mastermind to focus on these applicants with rebate offers through Twitter, Facebook, and other unique media. In the event that investigation shows the take-up and remarks region unit perilous, the inventory will be refined to improve execution.

There a lot of companies which uses the study and research created by these kinds of data analysis techniques which will be explained further in the paper in order to create strategies for digital marketing. For example; On Instagram, there are a lot of posts and ads regarding a particular product which contains URLs in order to reach that particular product page or a particular website of that company, so the number of hits on that particular URL shows the interested users and then further data is analyzed on the basis of gender, age, locality, etc.

Enormous data conjointly makes it to accomplish understanding into the jobs people play among web-based media groups. Clients with a larger than the usual scope of devotees as a model, will be considered to be influencers. By singling out such people, an association will screen drifts in conversation strings and even take an interest in such conversations.

2 Literature Review

Jenq-Haur Wang and Kuan-Ting Chen [5], Gigantic volume of social Web information in different structures are quickly produced all throughout the planet at a quicker speed than those of some other media. This can prompt the troubles in friendly information investigation: loud information and productivity. In the paper, the authors proposed a conveyed information examination stage for online media. To begin with, information from different sources were gathered and put away in a distributed record for proficient recovery. At that point, a conveyed examination system is worked from memory-put together bunch processing based with respect to the MapReduce worldview. At last, factual investigation is performed and incorporated for show.

Chengang Zhu and Guang Cheng [6], discussed the improved forecast approach dependent on pattern identification. Initially, a powerful time warping distance-based K-medoids calculation is applied to bunch projects' ubiquity advancement into four patterns. At that point, four pattern explicit expectation models are constructed independently utilizing arbitrary woods relapse. As per the highlights extricated from an electronic program control and early survey records, recently distributed projects are grouped into the four patterns by an angle boosting choice tree. At last, by joining estimating values from the pattern explicit models and the order likelihood, the proposed approach accomplishes better expectation results. The trial results on an enormous arrangement of genuine Video on demand information from the Jiangsu Broadcasting Corporation show that contrasted and the current expectation models, the expectation precision is expanded by over 20%, and the anticipating time frame is viably abbreviated.

Han Hu et al. [7], have presented a writing study and framework instructional exercise for enormous information investigation stages, intending to give a general picture to nonexpert perusers and impart a do it yourself soul for cutting edge crowds to alter their own huge information arrangements. In the first place, they present the meaning of large information and talk about enormous information challenges. Then, they present a precise structure to deteriorate enormous information frameworks into four successive modules, in particular information age, information procurement, information stockpiling, and information investigation. These four modules structure a major information esteem chain. Following that, they present a definite overview of various methodologies and components from exploration and industry networks. Also, they present the predominant Hadoop system for tending to enormous information challenges. At last, they tell about a few assessment benchmarks and potential exploration bearings for enormous information frameworks.

Jadon and Kumar [8], processed twitter data for identifying people's reactions, their opinions, and opinions. They have processed tweets related to Covid-19 pandemic. When the whole world is facing this epidemic and it has a great impact on everyone's life, business, and future, people are still positive and more focused on their safety and by spreading positive views on COVID-19 Taking care of others.

3 Significance of Social Media Analytics

Big Data social network analysis may be a part of an essential tool which may be used to apprehend the insights behind the data or information posted by a person on social media. It is a plan of action that syndicates varied data processing techniques, applied math strategies, sentiment analysis, and machine learning for statement predicting and analyzing the present trends & social information intelligence [9].

The application of precise big data tools is required to utilize info and drive analytics from social information to the fullest. This perceptive and authoritative information will drive brands to line up actions and enlighten business ways. Social data will assist firms in revealing innovative opportunities in addition to discovering incompetence in their business services. This will be drawn from conversations associated with specific merchandise, the brand, and also the management of their client services [10].

Apache Hadoop, the principal technology utilized is an associate degree open source and steadfast tool that consists of two constituents, namely, Hadoop Map scale back framework and Distributed filing system [11].

Apache Storm and Apache Spark square measure utilized for a period of time distributed computation. Facebook utilizes fast, that is once more open supply engine that assists in querying information victimization Cassandra and Hive. Twitter non inheritable Gnip, a social media analytics tool that assembles the tweets, filters, and resells this information that is extremely valuable for marketers and advertisers to execute digital campaigns. IBM assembles the information distributed on Facebook and Twitter regarding human thoughts, comments, and complaints a few definite subjects to drive a period of time information analytics. This assists in up their promotional campaigns, discount and valuation offers [8].

4 Key Technologies and Working

There is no single innovation that envelops monstrous data analytics. Obviously, there is cutting edge competition that will be applied to enormous data, notwithstanding basically numerous sorts of innovations working together to help you get the principal cost from your information. Some of the main players are:

- **Information management:** It refers to basically the management of data and must be top quality also, very much represented before it will be constantly broken down for analysis. With data continuously streaming all through an organization, it is critical to decide repeatable cycles to make and keep up norms for data quality. Once data is dependable, associations should set up an expert data board program that gets the total organization on an identical page.
- **Information mining:** Data preparing innovation helps you look at monstrous measures of data and information to get designs inside the information. This data will be utilized for any examination to help answer progressed business inquiries.

With information preparing programming framework, you'll filter through all the turbulent and dreary commotion in data, pinpoint what is significant, utilize that information to survey without a doubt results, and so speed up the decisions [12].
- **Predictive investigation:** Prophetical examination innovation utilizes information, applied science calculations, and artificial intelligence (AI) procedures to recognize the likelihood of future results upheld chronicled data. It's all with respect to giving a best appraisal on what will occur inside, what's to come, consequently, associations will feel extra guaranteed that they are making the best potential business call. Some of the prominent regular utilizations of prophetical investigation are a perfect example of extortion discovery, hazard, tasks, and advancing.
- **Text mining**: With text mining innovation, you'll examine text data from the web, remark fields, books, and distinctive content—based sources to reveal bits of knowledge you hadn't taken note of previously. Text mining utilizes AI or tongue measure innovation to go over archives messages, websites, twitter channels, studies, serious insight, and extra to help you break down monstrous measures of information and see new points and term connections [13].

5 Tools Used for Big Data Analytics

Algorithms, techniques, and programs are used to analyze data. Rather than considering online media investigation as a thing, consider the big picture as an action word. In particular, the gathered data from various social stages help to manage your advanced system. This technique starts by focusing on business objectives. For business model, your aim is to increase the amount of visitors on your site. The subsequent step would be key performance indicators (KPI). This comprises of:

- Likes and offers your post gets.
- Answers and remarks.
- Clicks your connections and substance acquire.

By gathering this information, you can sort out how web-based media factors into meeting your business objective. From that point, you can continue to go toward the path you're going or change your approach. The acclaimed tools for enormous information examination are:

- Keyhole
- AgoraPulse
- Brandwatch
- Cushion
- BuzzSumo
- Crowdbooster
- Edgar
- Google Analytics

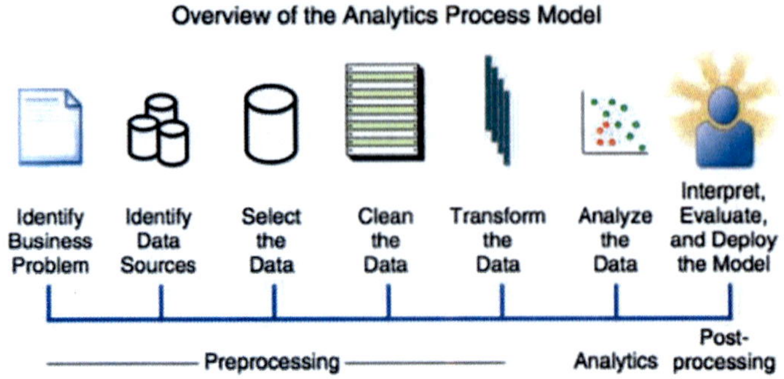

Fig. 1 The analytics process model

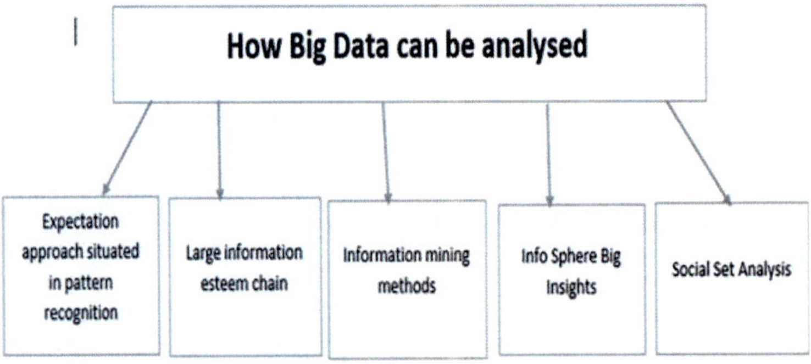

Fig. 2 The different methods for social data analysis

Large information is investigated by utilizing various procedures which give results in the required format. The general steps used for the data analytics are shown in Fig. 1. Further, different strategies which can be used for data analysis are shown in Fig. 2. Which strategy is to be applied, is subject to the client's requirement for dissecting the huge information.

Expectation approach based on pattern recognition: It presents the improved assumption approach with a maintained pattern detection. Starting, a dynamic time reshaping distance-based "K-medoids rule" is applied to bunch ventures' quality headway into some examples. By then, some examples express conjecture models are arranged independently using self-assertive forests backslide. Steady with the decisions eliminated from an electronic program immediately and early review records, recently printed programs are gathered into some examples by an inclination boosting choice tree. At last, by joining estimate values from the example unequivocal models and the portrayal likelihood the engineered approach accomplishes higher prediction comes about.

Large information esteem chain: It presents a logical system to disintegrate gigantic data frameworks into four progressive models, especially data age, data securing, data stockpiling, and data investigation. These four modules structure a monster data esteem chain.

Information mining methods: The extending reliance on social organizations needs data taking care of systems that are most likely going to empower changing the unstructured information and spot them among an intelligent model. The objective is to inspect the "information mining strategies" that were utilized by online social media.

InfoSphere BigInsights: It presents the use of Twitter app during scope of arranged subjects that will be the biggest person to person communication site any place Twitter data is in expanding at higher rates a day that considers it as monstrous data supply. At that point, depicting great the way inside which gigantic data innovation, such as InfoSphere BigInsights permits a cycle of this data, that are basically gathered from informal communities by Apache Flume and kept in Hadoop stockpiling [14].

Social set analysis: It gives a shiny new way to deal with enormous data investigation known as friendly set examination. Social set investigation comprises of a generative system for the methods of reasoning of methodology science, hypothesis of social data, unique and formal models of social data, and a scientific structure for consolidating monstrous social data sets with design and social data sets [15].

6 Challenges of Bıg Social Data

As the analysis of big social data is mostly based on the likes and dislikes of the user thus it can violate the privacy policy of any social media platform very easily. An investigation printed by analysts from Cambridge and Stanford Universities shows that Facebook will utilize its data to anticipate individuals' personalities with extra precision than shut loved ones as shown in Fig. 3.

Each data of the user for example:"like", "share", "follow "and "comment", is information that tells online media endeavors what the user like or dislike, what will work out for the user, that cause or the user will keep up and what user is,

Fig. 3 Security at the risk

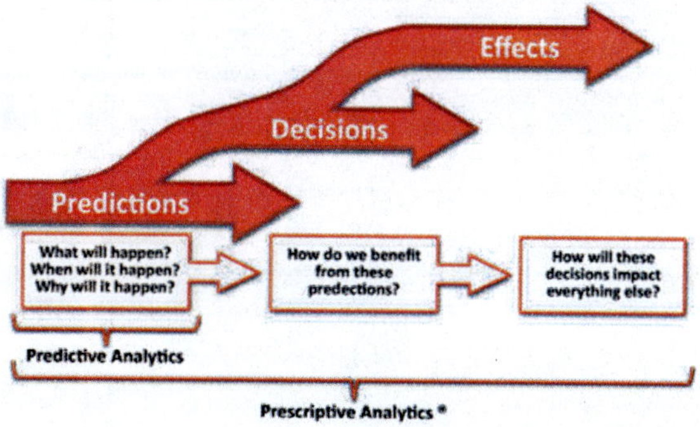

in reality, to explore for. Not to see, any move users are making on projects and browsers these days would be able to back to your social media profile, surrendering a comprehensive method of modernized impression which will be used for your best game-plans. This case can be exclusively increased as people become extra reliant via web-based media stages for getting to sharing information hence compromising the security of the user. The second challenge of big data can be maintaining the quality of data as there is a lot of data hence requiring a lot of time to remove errors and inconsistencies.

7 Data Preparation and Results Analysis

We have used customer support dataset from Twitter. This is a large, modern corpus of tweets from customers and their replies. This dataset can be used to study innovation in natural language understanding and conversational models. Further, this dataset can also be used for the study of modern customer support practices and its impact on the business. The sample of the dataset is shown in Fig. 4.

The dataset is in.csv format, where each row is a tweet. The different columns are described below. Every conversation in the dataset has at least one request from a consumer and at least one response from a company.

tweet_id: A unique, anonymized ID for the Tweet. Referenced by response_tweet_id and in_response_to_tweet_id.

author_id: A unique, anonymized user ID. @s in the dataset have been replaced with their associated anonymized user ID.

Inbound: Whether the tweet is "inbound" to a company doing customer support on Twitter. This feature is useful when re-organizing data for training conversational models.

created_at: Date and time when the tweet was sent.

First rows

	tweet_id	author_id	inbound	created_at	text
0	119237	105834	True	Wed Oct 11 06:55:44 +0000 2017	@AppleSupport causing the reply to be disregarded and the tapped notification unde
1	119238	ChaseSupport	False	Wed Oct 11 13:25:49 +0000 2017	@105835 Your business means a lot to us. Please DM your name, zip code and add
2	119239	105835	True	Wed Oct 11 13:00:09 +0000 2017	@76328 I really hope you all change but I'm sure you won't! Because you don't have
3	119240	VirginTrains	False	Tue Oct 10 15:16:08 +0000 2017	@105836 LiveChat is online at the moment - https://t.co/SY94VtU8Kq or contact 03
4	119241	105836	True	Tue Oct 10 15:17:21 +0000 2017	@VirginTrains see attached error message. I've tried leaving a voicemail several tim
5	119243	VirginTrains	False	Tue Oct 10 15:25:14 +0000 2017	@105836 Have you tried from another device, Miriam ^MM
6	119244	105836	True	Tue Oct 10 15:26:44 +0000 2017	@VirginTrains yep, I've tried laptop too several times over the past week and again
7	119245	VirginTrains	False	Tue Oct 10 15:33:22 +0000 2017	@105836 It's working OK from here, Miriam. Does this link help https://t.co/0m2mpH
8	119242	105836	True	Tue Oct 10 15:09:00 +0000 2017	@VirginTrains I still haven't heard & the number I'm directed to by phone is a d
9	119246	VirginTrains	False	Tue Oct 10 10:13:19 +0000 2017	@105836 That's what we're here for Miriam 🙂 The team should send you an email

Fig. 4 The sample dataset

Text: Tweet content. Sensitive information like phone numbers and email addresses are replaced with mask values like _email_.

response_tweet_id: IDs of tweets that are responses to this tweet, comma-separated.

in_response_to_tweet_id: ID of the tweet this tweet is in response to if any.

We have used Pandas-profiling software for the analysis of the dataset. The time stamp of the analysis, i.e., when the analysis started and when it is finished is shown in Fig. 5. The duration of the analysis is also given, i.e., 11.81 s.

In Fig. 6, the statistics of the tweet_id is presented. In the Quantile statistics, the value of the terms like Percentile, range, medium, etc. is measured. Further, in the Descriptive statistics, the values of the terms like mean, kurtosis, skewness, etc. are presented. Further, the tweet_ids having maximum frequencies are presented in a histogram as shown in Fig. 7. The results show that the frequency of all the tweets is two, which signifies, most of the tweets addressed.

The correlation between various parameters in the dataset is presented in the form of a heat map as shown in Fig. 8.

Overview | Warnings 19 | Reproduction

Reproduction

Analysis started	2021-05-07 07:27:03.360962
Analysis finished	2021-05-07 07:27:15.169325
Duration	11.81 seconds
Software version	pandas-profiling v2.12.0
Download configuration	config.yaml

Fig. 5 The experimental setup time stamp

Statistics | Histogram | Common values | Extreme values

Quantile statistics

Minimum	119237
5-th percentile	119241.6
Q1	119262
median	119285
Q3	119309
95-th percentile	119329.4
Maximum	119335
Range	98
Interquartile range (IQR)	47

Descriptive statistics

Standard deviation	28.31404481
Coefficient of variation (CV)	0.0002373637726
Kurtosis	-1.146282437
Mean	119285.4516
Median Absolute Deviation (MAD)	24
Skewness	0.02893190178
Sum	11093547
Variance	801.6851332
Monotonicity	Not monotonic

Fig. 6 The tweet_id statistics

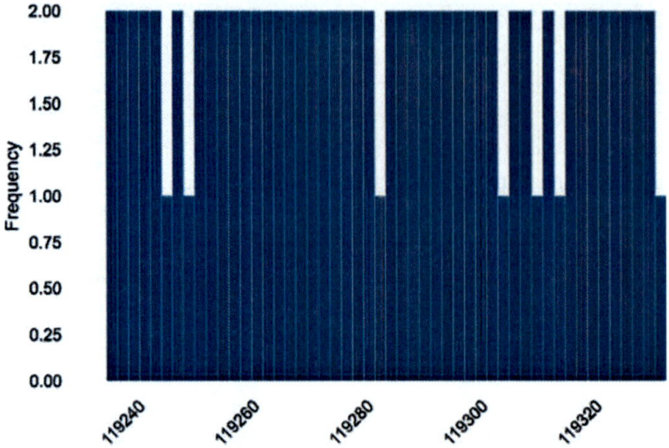

Fig. 7 Tweet_id histogram

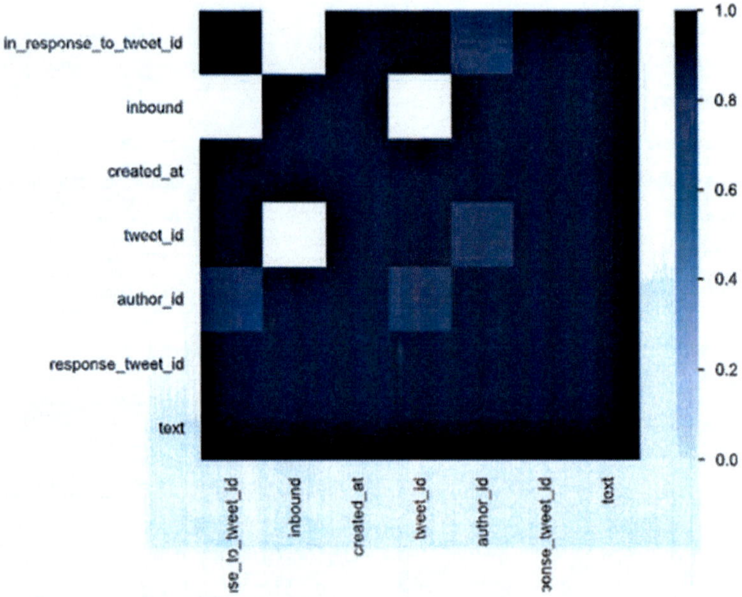

Fig. 8 Heat map

8 Conclusion

The social media platforms generate massive amounts of data every day. This data needs to be analyzed to extract valuable patterns. There are various investigation strategies, tools, and techniques which can be used for the social platforms data analysis. This paper discusses the importance of social media data analysis and its benefits to the business. In this paper, we have used the customer support dataset from Twitter for the analysis. This is a large, modern corpus of tweets from customers and their replies. This dataset includes the queries raised by the users of different companies. The pandas-profiling tool is used to perform the analysis. The overall outcome of this analysis is the improvements needed in the services of different companies which in return if addressed timely can increase the overall profit of the companies.

References

1. D.F. McCaffrey, G. Ridgeway, A.R. Morral, Big data and social media analytics. Psychol. Methods **9**(4), 403–425 (2014)
2. P. Sahatiya, Big data analytics on social media data: a literature review. Int. Res. Journal Eng. Technol. (IRJET), e-ISSN (2018): 2395-0056
3. T.S. Kumar, Data mining based marketing decision support system using hybrid machine learning algorithm. J. Artif. Intell. **2**, 185–193 (2020)
4. A. Mitra, Sentiment analysis using machine learning approaches (Lexicon based on movie review dataset). J. Ubiquitous Comput. Com-Munication Technol. (UCCT) **2**(3), 145–152 (2020)
5. J.-H. Wang, K.-T. Chen, Towards an efficient platform for social big data analytics. In: 2015 24th Wireless and Optical Communication Conference (WOCC), pp. 175–179 (2015)
6. C. Zhu, G. Cheng, K. Wang, Big data analytics for program popularity prediction in broadcast for TV industries. IEEE Access **5**, 24593–24601 (2017)
7. H. Hu, Y. Wen, T.-S. Chua, X. Li, Toward scalable systems for big data analytics: A technology tutorial. IEEE access **2**, 652–687 (2014)
8. A.K. Jadon, S. Kumar, Predicting the Transmission of Covid-19 outbreak on Social Media Platform. Int. J. Adv. Sci. Technol. **29**(4s), 3129–3144 (2020)
9. N.H. Egebjerg, N. Hedegaard, G. Kuum, R.R. Mukkamala, R. Vatrapu, Big social data analytics in football: Predicting spectators and TV ratings from facebook data. In: 2017 IEEE International Congress on Big Data (BigData Congress) (pp. 81–88). IEEE (2017)
10. M.K. Yahya-Imam, F.O. Aranuwa, An empirical study on big data analytics: Challenges and directions. In: Intelligent Computing and Innovation on Data Science, pp. 669–677 (2020)
11. M. Afzali, N. Singh, S. Kumar, Hadoop-MapReduce: A Platform for Mining Large Datasets. In: Third International Conference on Computing for Sustainable Global Development, pp: 5578–5582 (2016)
12. M. Injadat, F. Salo, A.B. Nassif, Data mining techniques in social media: A survey. Neurocomputing **214**, 654–670 (2016)
13. J.N. Ochin, S. Kumar, SRPF interest measure based classification to extract important patterns. Proceedings of 2nd International Conference on Communication, Computing and Networking (2019), pp 523–530
14. M. Birjali, A. Beni-Hssane, M. Erritali, Analyzing social media through big data using infosphere biginsights and apache flume. Procedia computer science **113**, 280–285 (2017)
15. R. Vatrapu, R.R. Mukkamala, A. Hussain, B. Flesch, Social set analysis: A set theoretical approach to big data analytics. IEEE Access **4**, 2542–2571 (2016)

A Study on Cloud Environment: Confidentiality Problems, Security Threats, and Challenges

Megha Gupta, Laxmi Ahuja, and Ashish Seth

Abstract In terms of computing services, cloud computing has expanded significantly. It is a set of computing tools that may be accessed through the Internet. The cloud computing security, a field rife with challenges and critical significance, is still in its early stages of growth, and many research questions remain unanswered. The IT industry has benefited immensely from cloud computing. It is like the next step in the development of the Internet, with a controlled and powerful data storage network with enormous capabilities and applications. It allows authorities to gain access to a user's file or data without having to install the information on the user's computer using Internet services. Cloud computing allows for on-demand community access or pooled access to computing resources that have been configured. Having multiple operating systems on a computer allows you to get the most out of that resource. It refers to the development of virtual version resources such as storage devices, a virtual operating system, or the deployment of an application or software on the cloud in this context. Cloud computing is no exception; it has both advantages and disadvantages. Cloud protection, manageability, honesty, fraud, data loss, and other issues are some of the implications. In the presented paperwork, we reviewed different challenges and issues faced in the data security over cloud with an objective to identify security challenges for adapting cloud computing.

Keywords Cloud computing · Security services and polices · Client management issues · Cloud configuration and exploitation of authentication

M. Gupta (✉) · L. Ahuja
AmityInstitute of Information Technology, Amity University, Noida, UP, India
e-mail: megha.gupta@s.amity.edu

L. Ahuja
e-mail: lahuja@amity.edu

A. Seth
School of Global Convergence Studies (SGCS), Inha University, Incheon, South Korea
e-mail: a.seth@inha.uz

G. Ranganathan et al. (eds.), *Soft Computing for Security Applications*,
Advances in Intelligent Systems and Computing 1397,
https://doi.org/10.1007/978-981-16-5301-8_49

1 Introduction

The demand for data has risen significantly in recent years, and the number of people who use the Internet has increased faster than anticipated. Furthermore, conventional computing infrastructure has become prohibitively costly and difficult to maintain, making data access anywhere and at any time impossible. As a result, the external storage solution has proven to be a requirement for data storage. The topologies capacities of processing devices can at this point do not adapt to the expanding number of web clients on systems administration locales, long range interpersonal communication destinations, media broadcasting destinations, and different destinations. Because of the large number of users and resources available, a new term known as cloud computing has emerged, as global Internet use has steadily increased [1].

In the same way as other specialized offices, distributed computing has a scope of benefits. It permitted the capacity of a lot of information and the arrangement of a scope of administrations, for instance. Besides, this stage tackled the issue of restricted assets and diminished the expense of administrations by dividing important assets between different clients. To give asset unwavering quality and productivity, the stage should be protected against security dangers [2]. As of late, distributed computing has been quite possibly the main subjects of safety research. These studies cover a variety of subjects, including digital storage security, network monitoring, and user authentication. "According to the National Institute of Standards and Technology, 'a model for enabling easy, resource sharing, widespread, and on-demand access that could be effectively supported with various types of service provider interaction' (NIST)." In cloud computing, customers just pay for the services they use, which are known as pay-as-you-go (PAYG). Customers can configure applications, storage, production platforms, and computing resources based on customer or end-user needs with the PAYG model. These advantages are the reason why the scientific community has put so much effort into developing this cutting-edge principle [4].

End-user service availability has increased thanks to the use of virtualization. Versatility, reasonability, and moderateness are three principal attributes of distributed computing. Practically speaking, these procedures join an organization's accessible assets by parting accessible transmission capacity into isolated and unmistakable diverts to allocate to a specific worker or gadget or remain totally unallocated [5]. It also provides a cost-effective demand service, as well as elasticity and flexibility. The three fundamental assistance conveyance models given by distributed computing are programming as a administration (SaaS), framework as a administration (IaaS), and stage as a administration (PaaS). There are broad capacities for utilizing applications, for example, strategy controls are focused on. An individual is just permitted to download each record in turn such data from applications, for example. In this way, a single instance of the service benefits multiple end users. Companies such as Google, Microsoft Office 365, Drop box, and others include SaaS. PaaS stands for "platform as a service," which refers to the use of a layer of the improvement climate as a stage on which other more significant level administrations can be constructed. In the PaaS model, clients create the applications themselves that work on the supplier's framework. Microsoft Azure, Google App Engine, and the LAMP structure (Apache, Linux, Php, and Msql), among others, offer a blend of working frameworks and application workers (Fig. 1).

It is important that one of the principle focal points of the PaaS model is information assurance. This is particularly obvious with regards to capacity as assistance. This model ought to have the option to scramble information when put away on an outsider stage, and it ought to know about administrative worries what influence information accessibility in various geologies. With IaaS, processing and capacity assets are conveyed as incorporated administrations all through the organization. Virtualization's intrinsic system is to make independent virtual machines (VMs). Firewalls, intrusion detection and prevention mechanisms, and virtual machine monitoring are all part of the virtualization process. However, there are a slew of other security concerns to address. Such a safe environment is preferred by businesses.

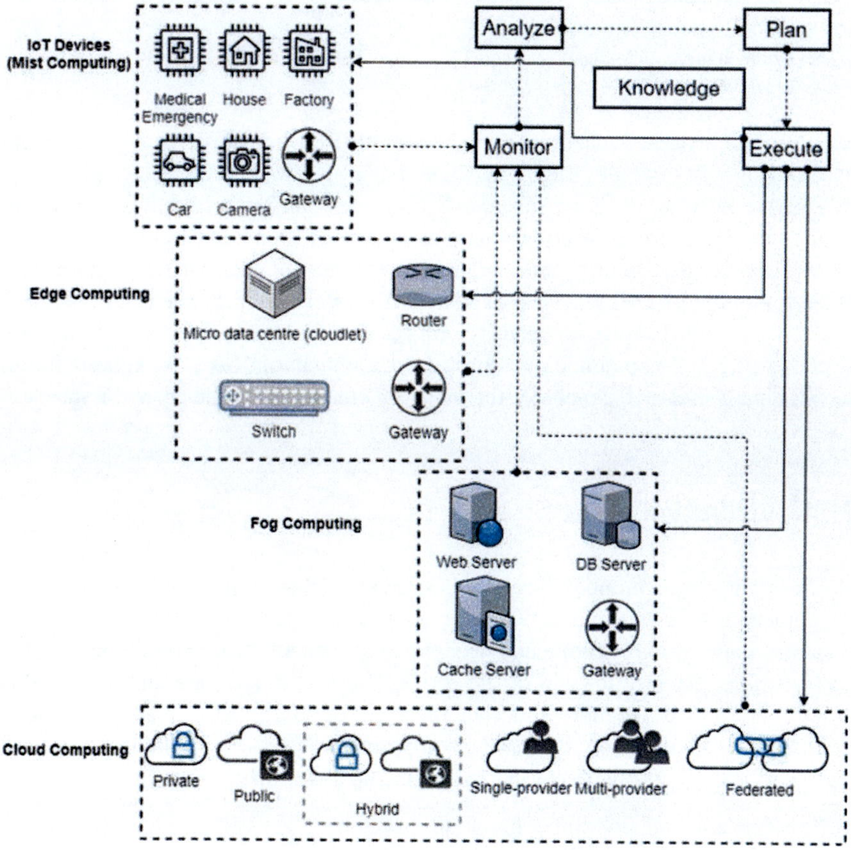

Fig. 1 Conceptual of overview of cloud computing [3]

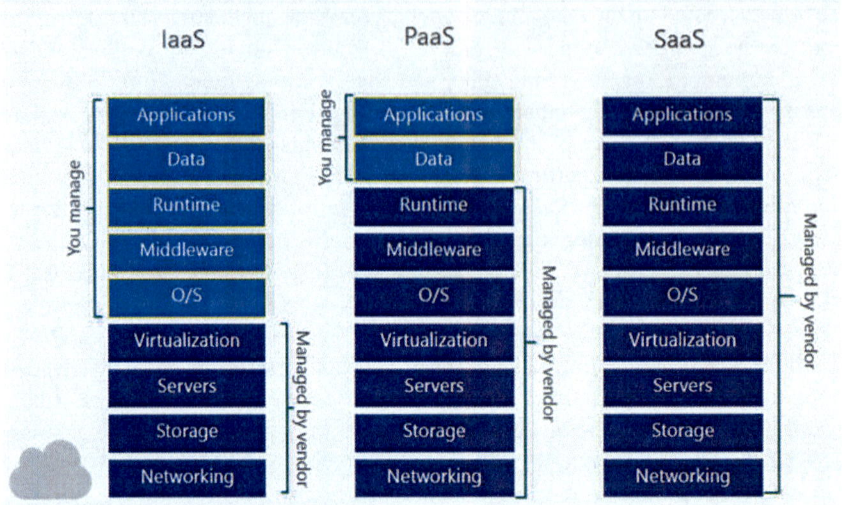

Fig. 2 Configuration of cloud deliverymodel [6]

IaaS services include Amazon Web Services (AWS), Microsoft Azure, Cisco Meta pod, and Google Compute Engine (GCE), to name a few (Fig. 2).

The rest of the article is organized as follows. Section 2 provides the contribution of this survey. Section 3 describes the related work on cloud and security problems. In this Sect. 4, cloud security and challenges are discussed. Section 5 depicts security policies on cloud-based services. Section 6 discusses different client management issue over cloud. In Sect. 7, security attacks and threats in cloud computing are discussed; Sect. 8 explains exploitation of authentication; Sect. 9 explains future scope. Finally, Sect. 10 concludes the paper on cloud-based security and issues.

2 Contribution of the Survey

However, there are a number of security issues and vulnerabilities with this common phenomenon. It is undeniable that the cloud environment faces numerous threats. As a result, using this phenomenon necessitates sufficient information as well as the best possible response to each of these risks. Several articles are analyzed in this article, and some of the more important ones will be addressed. In reality, this paper discusses how security concerns, cloud computing problems, and risks are expressed. This work's key contributions can be summed up as follows:

- The fundamental principles of cloud computing, as well as all organizations involved with cloud computing system architecture, are discussed.
- In the security status of cloud conditions, it has made another various-classification grouping of cloud security issues and difficulties.

3 Related Work

The context section addresses topics such as data protection preliminaries and provides details on cloud computing system architecture.

3.1 Security Service

All techniques for securing, restoring, and sustaining the stability of data in network devices among multiple threats are referred to as protection. In reality, security networks impose security services that implement security policies. Integrity, confidentiality, authentication, non-repudiation, and perhaps other resources are used to secure computer services and network systems [6].

- Informed consent ensures that the information is not shared or made accessible to individuals, organizations, or procedures that are not approved. Data must be sent (and received) without being exploited by unauthorized parties during transmission. User authentication is a good way to keep the information private. A centralized or distributed key approach can be used to access information [7].
- Relevant entity that data obtained by a trustworthy individual is identical to evidence sent. To put that in perspective, it ensures that a third party has not tampered with the results (accidentally or intentionally). The connection is broken upon an intrusion, and the transmitting of error messages is halted [8].
- Configuration guarantees that authorized users can access resources and that data can be accessed and used by an authorized individual on trend. A distributed cognitive dissonance threat disables the ability of a computer to transmit data [9].
- The sender and receiver of data must be who they claim they are to be considered reliable. In reality, integrity of the data and authentication is only required once the identities of the senders and receivers have been established [10].
- The inability of senders and receivers to contest actions taken is referred to as pseudo. Source repudiation and destination repudiation are the two types of repudiations available. The sender or receiver of a message cannot refuse its transmission in the former, and they cannot dispute its receipt in the latter [11].

3.2 Cloud Configuration

To better comprehend security problems, we need to understand the cloud setup described below. A cloud field contains demand-driven services. Cloud computing setups, according to NITS, have five key actors, as shown in Fig. 3. Cloud consumer risks and risk perceptions are the subject of the classification.

Cloud user: A cloud consumer is an individual or organization who makes use of a cloud provider's services. Indeed, by evaluating the services provided by cloud

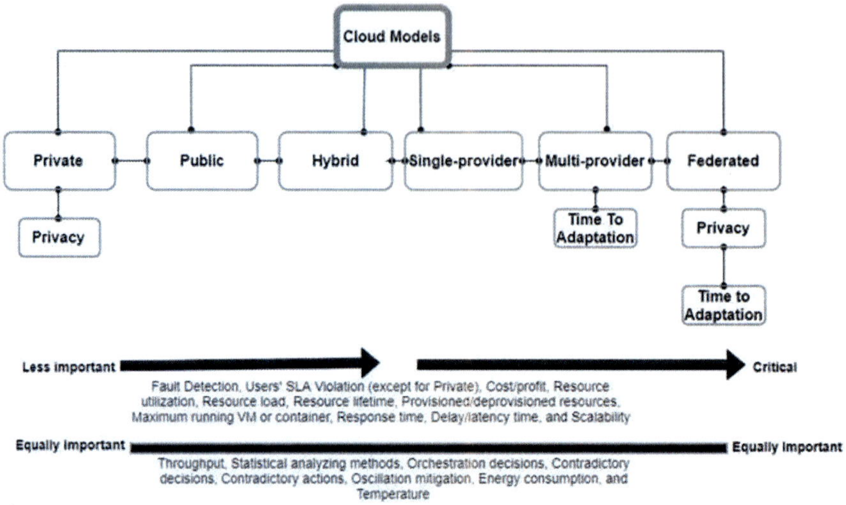

Fig. 3 Taxonomy of metrics based on cloud models [8]

providers and signing a contract, a cloud user may choose the most appropriate services. A cloud customer must assess technical efficiency and sign a service-level agreement (SLA) with a cloud provider to complete this contract. Service-level agreements (SLAs) deal with things like system development, reliability, prevention, and performance degradation. A cloud customer, on the other hand, has a wider range of providers to choose from many of which offer better offerings and lower revenue [9].

Cloud contributor: An individual or company that makes a service accessible to a data center is known as a cloud service provider. Through buying and handling cloud services, the cloud provider combines and arranges cloud applications. The cloud provider offers SaaS services by installing, configuring, managing, and upgrading software applications on a regular basis [10]. The majority of the controlling and regulating functions in the infrastructure and technology are in the services of SaaS providers, according to cloud constrained organizational applications. The platform's cloud provider is managed by the network operator, and the platform's components are supported by software. The cloud service provider retains physical computing tools such as networks, databases, server farms, and hosting platforms while achieving IaaS.

Cloud assessor: A cloud auditor's job is to perform an unbiased evaluation of online services. The auditor examines a variety of scientific data to ensure compliance with regulations. Cloud auditors will evaluate cloud service providers in terms of safety, security, pricing, and other conditions [11].

Cloud adviser: Customers order cloud services via a cloud broker rather than directly calling the cloud provider because integrating cloud services are too difficult for them to handle. The cloud broker, in reality, is in charge of cloud service use, performance, and distribution, as well as cloud consumer-cloud vendor relationships. Cloud brokers' services can be divided into categories of three [12].

- **Service intermediation** a community cloud could improve services by including value-added functionality and offering value-added services to cloud customers. Access regulation, performance monitoring, identity management, and improved protection are only a few of the enhancements.
- **Service Aggregation** is the mechanism by which a cloud service combines or merges several services under one or even more new ones. The service provider, in reality, collects data and acts as a link between the data owner and numerous data centers.
- **Service arbitrage although** it functions similarly to service restructuring, it would not necessitate the fixing of networks. In reality, service diversification allows the vendor pick services from either a variety of agencies. A network, for example, may identify and select the right agency, using a service that provides credit-scores.

Cloud carrier to deliver cloud services, a cloud carrier acts as a middleman between the cloud client and the cloud provider. Consumers are connected to cloud services through the network and perhaps other devices that allow. As previously stated, a cloud provider can provide SLA-compliant services to cloud customers by forming SLAs with a cloud carrier. Furthermore, the cloud carrier is in charge of maintaining a stable environment [13] (Table 1).

4 Cloud Security Issues and Challenges

Cloud computing—primarily based computing paradigm introduces a slew of latest issues in information security, access management, and different areas. The protection issues of cloud computing are the topic of many analysis papers over the last decade. Consequently, the majority of peer-reviewed papers are critical in cloud security, and so, these notable peer-reviewed articles reproduce a significant amount of useful and elaborated research in this field [14].

In [15] presents a taxonomy of virtualized PC attacks supported the extent of target, the source, and also the attackers' objectives. In fact, they need to demonstrate; however, threats, connected protection, and trust expectations have modified over time in virtualized systems at varied levels, together with hardware, software package, and application.

A study of cloud computing security problems was mentioned in [16]. This paper mentioned the problems of knowledge location, storage, confidentiality, accessibility, and honesty. Indeed, this research relies on one of the most critical privacy concerns; however, it is important to note that perhaps the authors only discuss security issues, not possible solutions.

In [17] show variety of knowledge security issues in cloud computing, likewise as a mechanism for addressing security problems in multi-tenant environments. This

Table 1 Survey of different techniques used in cloud computing

S. No.	Techniques/methods	References
1	Cloud computing is a model for enabling convenient, on-demand network access to a shared pool of configurable computing resources (e.g., networks, servers, storage, applications, and services) that can be rapidly provisioned and released with minimal management effort or service provider interaction	[14]
2	Cloud computing really comes into focus only when you think about what IT always needs: a way to increase capacity or add capabilities on the fly without investing in new infrastructure, training new personnel, or licensing new software. Cloud computing encompasses any subscription-based or pay-per-use service that, in real time over the Internet, extends its existing capabilities	[15]
3	I view cloud computing as a broad array of web-based services aimed at allowing users to obtain a wide range of functional capabilities on a 'pay-as-you-go' basis that previously required tremendous hardware/software investments and professional skills to acquire. Cloud computing is the realization of the earlier ideals of utility computing without the technical complexities or complicated deployment worries	[16]
4	When virtualizing applications to be used by people who care nothing about computers or technology - as is mostly the case with Clouds - the key thing we want to virtualize or hide from the user is complexity. Most people want to deal with an application or a service, not software. The more intelligent we want [computers and computer applications] to be - that is, intuitive, exhibiting common sense and not making us have to constantly take care of them - the more smart software it will take. But with cloud computing, our expectation is that all that software will be virtualized or hidden from us and taken care of by systems and/or professionals that are somewhere else -out there in The Cloud	[17]
5	Today's combination of high-speed networks, sophisticated PC graphics processors, and fast inexpensive servers and disk storage have tilted engineers toward housing more computing in data centers. In the earlier part of this decade, researchers espoused a similar, centralized approach called "grid computing." But cloud computing projects are more powerful and crash-proof than grid systems developed even in recent years	[18]
6	The web fanatics and blogosphere would have you believe that all applications will move to the web. Some will, most will not. Reliability, scalability, security, and a host of other issues will prevent most businesses from moving their mission critical applications to hosted services or cloud-based services. The risk of failure is too great. Amazon is the leader in cloud-based services, but even Amazon has experience down times for its own business. Cloud services will continue to improve. But my guess is the uptake will take longer than most people predict	[19]
7	Most computer savvy folks actually have a pretty good idea of what the term "cloud computing" means: outsourced, pay-as-you-go, on-demand, somewhere in the Internet, etc.	[20]

(continued)

Table 1 (continued)

S. No.	Techniques/methods	References
8	In order to discuss some of the issues surrounding the cloud concept, I think it is important to place it in historical context. Looking at the cloud's forerunners and the problems they encountered gives us the reference points to guide us through the challenges it needs to overcome before it is adopted	[21]
9	I view cloud computing as a broad array of web-based services aimed at allowing users to obtain a wide range of functional capabilities on a 'pay-as-you-go' basis that previously required tremendous hardware/software investments and professional skills to acquire. Cloud computing is the realization of the earlier ideals of utility computing without the technical complexities or complicated deployment worries	[22]
10	Cloud computing overlaps some of the concepts of distributed, grid, and utility computing; however, it does have its own meaning if contextually used correctly. Cloud computing really is accessing resources and services needed to perform functions with dynamically changing needs. An application or service developer requests access from the cloud rather than a specific endpoint or named resource. What goes on in the cloud manages multiple infrastructures across multiple organizations and consists of one or more frameworks overlaid on top of the infrastructures tying them together. The cloud is a virtualization of resources that maintains and manages itself	[23]

article concentrates entirely on information security problems and techniques for information and privacy protection.

An analysis of questions regarding cloud computing protection and privacy is presented in [18]. Throughout this work, various types of known cyber threats, as well as various types of cloud limitations, are defined. More precisely, this research looks at the issues in current solutions and discusses potential security implications.

In [19] provide a rundown of the foremost vulnerable security risks in cloud computing. Furthermore, by offering insight associated with the varied security models and tools, the key security issues associated with cloud computing are considered in this analysis work by both application developers and suppliers.

In [20] provide a rundown of the foremost vulnerable cloud computing security threats. Through offering insight into the various security models and tools, this research work also considers both end users' and suppliers' main security concerns relevant to cloud computing Since the majority of cloud-related (Fig. 4).

Technologies have advanced, and new challenges have arisen today, and the issues regarding data security in cloud-related technologies have risen and new issues have arisen. As a result, thorough analysis, identification of potential problems and the development [12] is needed. Additionally, the bulk of the above surveys focused on cloud security rather than detailing the concerns, challenges, and risks associated with cloud computing. We have covered every aspect of cloud security and its limitations, and the current solutions for them, as thoroughly as possible.

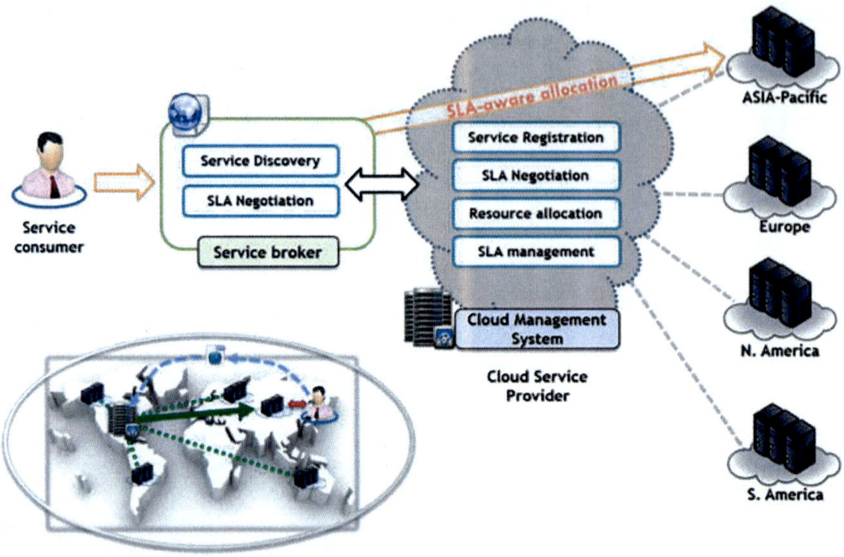

Fig. 4 SLA [23]

5 Security Policies

Security policies specify the conditions that must be fulfilled in order to prevent attacks from occurring. These guidelines should ensure that the cloud-based working environment is protected without compromising its efficiency or stability [23, 24]. Security policies, such as service-level agreements (SLAs), queries relating to managing clients, and antecedent trust, are governed by regulatory authorities [22] (Fig. 5).

Cloud security issues classification

Security policies	**User-oriented security**	**Data storage**	**Application**	**Network**
Service level agreement [47]	Authentication	Data warehouse	Operating systems	Intrusion prevention system
Client management issue [48]	Authorization	CIA Tired	Frontend/Back end	Intrusion detection system
Antecedent trust [49]	Identity and access management	Malware	Application vulnerabilities	Firewalls
		Metadata		

Fig. 5 Security policies

5.1 *Service-Level Agreement (SLA)*

A service-level agreement is linked to the customer-provider partnership which involves all parties. Providers of services are planned.

Customer standards are influenced by SLAs. In reality, service-level agreements (SLAs) specify the circumstances in which providers are not really responsible for service interruptions or poor performance. SLAs also represent the characteristics of facilities that can be compared in terms of efficiency. SLAs, on the other hand, cannot guarantee the delivery of a particular service [23]. To put it another way, SLA makes it possible to choose a subpar service, nor does it turn a subpar service into a profitable one. A service-level agreement (SLA) usually includes a declaration of goals as well as a collection of contract services. It also outlines the service provider's and client's responsibilities under SLA. The service-level agreement (SLA) defines the service provider's availability, consumption figures, and response time, as well as basic performance metrics.

6 Client Management Issue

Most businesses have attempted to concentrate on the consumer in recent years. User experience, client-centric anonymity, a scheme to provide authentication to clients, and a client service-level agreement are all part of client management. It is one of the most serious issues in cloud protection. In the following chapters, we will go through the different aspects of cloud security [24].

6.1 *Client Authentication*

By displaying a digital certificate, users can access a remote device or a server system. This digital certificate serves as a form of digital identification, cryptographically tying customer and employee identities together. A digital certificate can be used to control access to cloud services [25]. Unauthorized users are also prevented from accessing services provided by cloud. Users could manage their applications through any machine that includes televisions, laptops, tablets, smart phones, etc. Cloud service providers face a challenge in ensuring approval to limited users to access the services. Authentication risks and attacks in the cloud, however, that include cookies reply attacks, password discovery attacks, login holdup (for example, social engineering, or even phishing), etc. [27].

6.2 Client-Centric Privacy

The most serious vulnerabilities arise when privacy is violated and data is leaked in the cloud [28]. In reality, data collection by providers is the source of this problem and puts clients' privacy at risk. This issue is a significant impediment to cloud service adoption. A user has no control over how data is processed when it is sent to a cloud machine for processing. The locality of the data storage remains unknown to the users. For example, without the user's consent, the service provider might be able to exchange sensitive data with unapproved third parties, resulting in data leakage and privacy violations. As a result, addressing this issue requires a client-centered approach. Through this, users should have access to the information they have sent. As a result, addressing this problem requires a client-centric approach. This problem could lead to a slew of issues, including issues with security, safety, confidence, and customer relationship management (CRM).

7 Cloud Computing Security Attacks and Risks

The storage system in cloud, which is made up of several different hardware and software modules, is vulnerable to both old and new security threats. A hazard is a term used in information security to describe anything that could damage a computer system. Threats may, in other words, be used to target computer systems, networks, and other forms of communication infrastructure. Trojans, viruses, and outright hacker attacks are all threats. Due to its effective management and cost-economic model, data privacy and software protection are becoming increasingly important [29] (Fig. 6).

7.1 Threat Model and Compromise Data Attribute

Due to dynamic scaling, there are various benefits in cloud computing, including improved speed and effectiveness. Security breaches, user error, insider threats, and distributed denial-of-service (DDoS) attacks are all security issues. A threat risk model is a way of evaluating the security from any Internet system, finding areas of weakness, and measuring the costs involved with each one. The threat model took into account steps in the process such as defining security goals, risks, and vulnerabilities. Threat models are a method of detecting and mitigating security risks in networks that is systematic and organized [30]. Defines a threat model identifying attacks from one virtual machine to the other. Furthermore, a malicious tenant user can exhaust the server's resources by flooding the virtualization layer containing ICMP/UDP packets with a spoofed address for the source. It categorizes threats into six categories and provides a snapshot of threats in a given region (tampering,

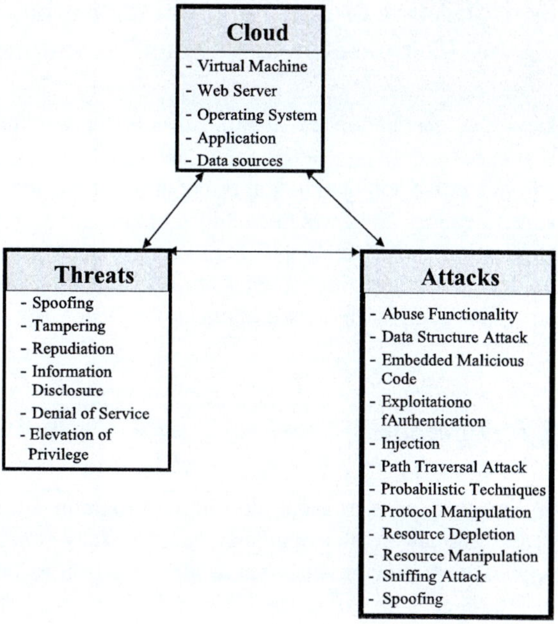

Fig. 6 Classification cloud security issues [26]

spoofing, information disclosure, repudiation, elevation of privilege, and DoS). The characteristics of attacks that pose a specific threat are defined in each STRIDE classification. Indeed, the STRIDE threat model divides threats into categories based on the outcome or impact of their realization. The threat model STRIDE has produced an elevated assessment of the issues that are being faced. In addition, this study expands the threat scope to include advanced cloud components [31].

7.2 *Spoofing*

Spoofing is when an unknown source sends a message that appears to come from a trusted and known source. Spoofing is actually often used to obtain access to a user's confidential information, malware infection by corrupted attachments or links, circumvented controls for accessing network, or reallocate traffic to execute a DoS threat [32]. Spoofing can be achieved using a variety of technical factors on emails and websites, such as a software that spoofs an IP address or a web address. Since a spoofing attack cannot be prevented, steps can be taken to keep spoofed packets out of a network. Ingress filtering, a kind of packet filtering, is a well-adapted spoofing defense. The source headers of incoming IP packets are verified by ingress filtering, which is typically implemented on something like a network edge system. The packets must be rejected if the source headers do not suit their origins or otherwise seem suspicious.

8 Utilization of Full-Potential for Authentication and Comparision of Various Techniques

Device authentication and identification systems, without a doubt, must fix vulnerabilities in order to prevent sensitive data from being exposed. These attacks disclosed administrative and operational frameworks, duplicate user accounts, and weak user authentication, all of which could lead to backdoor vulnerabilities [64]. Web-based applications, which are popular among developer firms, are used in most of contemporary cloud computing systems. Furthermore, brute-force authentication attacks are a popular aim for web-based applications [33].

8.1 Injection

By injecting code or a question into a programmed, or by installing one or more pieces of malware on a computer, attackers may alter a database. Depending on the application's execution context as well as position, the error in program that yields an attack, code injection attacks take different forms. This is a common type of attack in system cracking or hacking, which entails obtaining access to a device without authorization for collecting data. SQL and cross-site scripting are most well-known injection attacks (XSS) [34].

8.2 Path Traversal Attack

A path traversal attack's main objective is to gain access to directories those of which are not in the web-root folder. To gain access to on the file system, arbitrary files and folders, this attack manipulates variables with "dot-dot-slash" sequences as references. This attack also has been known to gain access to sensitive system data, such as source code or configuration. A path traversal attack, on the other hand, allows authorized access to cloud settings, such as VM escape, through file servers [35].

8.3 Protocol Manipulation

The protocols used for networking are vulnerable to DoS, exploiting various application flaws, and tampering with the details of XML data transferred between servers and users in order to learn about the target's protection. Adversaries can crash or hijack victims' computers by delivering malformed messages that exploit vulnerabilities in implementing protocols [36].

8.4 *Resource Depletion*

A species extinction attack is one in which a compromised node is used to produce excess traffic in networks, using the resources of the node. Threats due to resource depletion, for example, are attempted at the routing protocol layer with the aim of depleting node resources and thereby disabling networks. In the cloud, attackers would deplete all computational resources, including bandwidth, memory, and other computing capacities are all factors to consider. Despite the ability to scale to handle a broad variety of workload sizes, resource depletion attacks including volume-based flooding protocol exploiting or leveraging application communication vulnerabilities are still likely to occur [37].

8.5 *Resource Manipulation*

Resource manipulation is the abuse from one or even more resources to compromise the integrity of a system, such as a cloud system. This form of attack can compromise data integrity as well as the resource availability such as files, programmers, libraries, configuration data, and infrastructure. Parameter tampering may also be used by attackers to compromise the cloud's data and infrastructure. Parameter tampering is a type of Internet intrusion that seeks to alter some variables in the form field data of a web page without the user's consent. Resource manipulation attacks include manipulating an object relationship to gain unauthorized access data and XML manipulation [38].

8.6 *Sniffing Attacks*

A sniffer is used in sniffing attacks to detect network traffic. A sniffer a programmed that catches network packets as they move through networks. This type of attack can also sniff network traffic or allow user data kept remotely in cloud to collect sensitive data if security mechanisms are incorrectly configured. There are two forms of sniffing attacks: passive sniffing and active sniffing. Active sniffing uses methods and strategies for finding information about the device, while passive sniffing collects data that is communicated between the two parties. Internal system vulnerabilities and misconfigurations would be exposed as a result of this form of assault. In order to gain access to networked classified details, these attacks look for open ports in order to explore resources and vulnerabilities related to them. Additionally, attackers will sniff features, criteria, and profiles on a customer's device to circumvent standard authentication. Authentication, but in the other hand, is one of the more reasonable ways to keep a packet sniffer at bay. Security ensures holding computers secured on the network protects private data from malicious intruders [39–41] (Table 2).

Table2 Comparison study of various techniques

	Study									
Performance evaluation metrics	[4]	[18]	[14]	[37]	[1]	[36]	[9]	[13]	[38]	[12]
Resource utilization	✔	✔	✔		✔	✔	✔	✔	✔	✔
Resource load	✔									
Throughput			✔				✔	✔	✔	✔
Resource lifetime										
Max. running resource										
Response time	✔	✔	✔	✔			✔	✔	✔	✔
Delay time (or latency)						✔				
Network congestion/traffic control										
Damaged tasks		✔					✔			✔
SLA violations				✔			✔			✔
Fault tolerance							✔			✔
Performance evaluation metrics	[26]	[39]	[35]	[19]	[40]	[41]	[42]	[2]	[43]	
Resource utilization							✔	✔		
Resource load								✔		
Throughput										
Resource lifetime								✔		
Max. running resource								✔		
Response time				✔			✔	✔		
Delay time (or latency)				✔	✔			✔		
Network congestion/traffic control										
Damaged tasks							✔	✔		
SLA violations	✔	✔						✔	✔	
Fault tolerance										

9 Cloud Security Issues in the Future

The cloud computing model includes end-user servers, networking networks, access management systems, and cloud infrastructures. Furthermore, as new technologies emerge, cloud computing will become more relevant as new technologies emerge, such as 5G Internet, smart city, and Internet of Things than ever before for processing and storing more data. The heterogeneity of today's market climate has resulted in a new wave of vulnerabilities and security issues. Organizations were in the dark as to how much data and information they have stored in recent years, finding it challenging to detect and mitigate growing security threats [33–36]. Security organizations face issues ranging from data replication to the failure to detect threats in a complex environment, reasonable timeframe without a detailed understanding of the network infrastructure, as well as a loss of control over data transfer and the security required to meet regulatory enforcement. To create robust cloud security, information and cloud resources must be protected from known and unknown threats across all cloud solutions. In order to address concerns about protection in cloud environments, several studies have been performed. However, in order to provide a robust cloud infrastructure, there are still a range of open issues that must be tackled [37–41]. Cloud networking, network, data protection, computer, and web service security problems are all common concerns when it comes to cloud computing. Security issues arising from shared pool, virtualization, and multi-tenancy infrastructure are all emerging security concerns. When working in a cloud computing system, a variety of programs and resources available, but the degree of resource security is dictated by the resource's sensitivity and value. In cloud computing, computation protection is a prime issue [42, 43]. The bulk of the information in the storage is protected by encryption. The sensitive information in the storage, on the other hand, receives no operations. The majority of operations during computation needed plain text data. Hackers can gain access to the memory allocated to an outside or even inside of a processor, which is used to store temporary data. As a result, researchers are working to create a comprehensive solution that guarantees privacy during computation. A security solution for cloud storage is also needed to protect against insider threats. The detection of insider attacks in cloud computing is still a work in progress. This indicator might enhance the cloud gadget's safety capacity. Similarly, any other open issue in a cloud setting is figuring out who's a legitimate person and who's a malicious consumer [44]. AI and machine learning are increasingly being used by security systems to simplify responsibilities and provide a higher degree of intelligence for detecting intruder and outsider attacks. This covers user behavior analytics, which could be used to identify unusual cloud interest and reveal possible security risks by creating a usage baseline over time. Too many organizations are struggling to understand the cloud security concept gap and massively underestimating brand-new risks, making them susceptible to cloud account compromises and statistics breaches which could result in tremendous reputational and monetary harm [45]. An easy way to automate defenses and implement information governance requirements is to invest in cyber security structures that use AI and automation to supplement limited human capital. Using behavioral analytics to automate the collection and modeling of current community records now not simplest makes it less difficult for organizations to stumble on and classify capacity threats, but it also makes them greater effective [46–51].

10 Conclusion

In truth, cloud computing has numerous blessings for corporations, organizations, and even international locations. In spite of its many benefits, the cloud continues to be susceptible to a spread of protection threats. As a result, protection is the maximum widespread barrier to cloud adoption. Security dangers are properly understood by way of both the customer and the vendors. The current primary aim includes all capacity protection problems in the cloud computing world, as well as appropriate solutions to these issues. In reality, the purpose of this study was to demonstrate the numerous security troubles, vulnerabilities, assaults, and threats that stymie cloud computing adoption. The purpose of our paper became to behavior a survey of cloud safety troubles and challenges that emerge from the cloud's particular characteristics. To emphasize the importance of figuring out the safety vulnerabilities inside the cloud computing machine and framing appropriate countermeasures for them, a simplified view of those issues has been offered right here. It recommend examine of recent protection architectures in terms of lowering vulnerabilities in an effort to deter destiny assaults based on this line of research. We gift a fixed of recorded regulations, tactics, and methods that describe the secure control manner within the cloud environment to minimize threat and vulnerability even as growing agree with in an ever-linked international in the article. These worries consist of facts and carrier protection on cloud structures. We categorize protection threats and conduct a comparative examine of safety troubles and proposed countermeasures to address them.

Acknowledgements Author express their deep sense of gratitude to the Founder President of Amity University, Dr. Ashok K. Chauhan for his keen interest in promoting research in the Amity University and have always been an inspiration for achieving great heights.

References

1. A. Jeyaraj, N. Subramanian, Recent security challenges in cloud computing. Comput. Electr. Eng. **71**, 28–42 (2018)
2. P. Mell, T. Grance, SP800145,The NIST Definition of cloud computing CSRC (online) Csrc.nist.gov.https://csrc.nist.gov/publications/detail/sp/800145/final. Accessed 11 December 2018
3. Xu X (2012) From cloud computing to cloud manufacturing. Robot Comput. Integr. Manuf. **28**(1), 75–86
4. D.S. Kushwaha, S.K. Pippal, A simple, adaptable and efficient heterogeneous multi-tenantdata base architecture for adhoc cloud. J. Cloud Comput. Adv. Syst. Appl. **2**(1), 5 (2013)
5. B. Shi, L. Cui, B. Li, X. Liu, Z. Hao, H. Shen, Shadow monitor: an effective in-VM monitoring framework with hardware-enforced isolation, in *International Symposium on Research in Attacks, Intrusions, and Defenses* (Springer, Berlin, 2018), pp. 670–690
6. D. Bhamare, M. Samaka, A. Erbad, R. Jain, L. Gupta, H.A. Chan, Optimal virtual network func-tionplacement inmulti-cloud service functionchaining architecture. ComputCommun **102**, 1–16 (2017)
7. A. Alzahrani, N. Alalwan, M. Sarrab, Mobile cloud computing. In:Proceedings of the 7th Euro American Conferenceon Telematics and Information Systems (EATIS'14)
8. Deka GC, Das PK (2018) Application of virtualization technology in IaaS cloud deployment model.In:Design andUse ofVirtualization Technologyin CloudComputing: IGIGlobal, pp29–99
9. Oracle.com (2018) The Oracle and KPMG Cloud Threat Report 2018 | Oracle (online). https://www.oracle.com/cloud/cloud-threat-report.html.Accessed 11 Dec 2018
10. I.A.T. Hashem, I. Yaqoob, N.B. Anuar, S. Mokhtar, A. Gani, S.U. Khan, The rise of "big data" oncloudcomputing: reviewand open researchissues. Inf Syst **47**, 98–115 (2015)

11. R. Roman, J. Lopez, M. Mambo, Mobile edge computing, fog et al.: a survey and analysis ofsecuritythreats and challenges. Future GenerComput Syst **78**, 680–698 (2018)
12. G. Ramachandra, M. Iftikhar, F. A. Khan, A comprehensive survey on security in cloud computing.Proc Comput Sci **110** , 465–472 (2017)
13. Csrc.nist.gov, SP 500–299 (DRAFT), NIST Cloud Computing Security Reference Architecture, CSRC (online) (2018). https://csrc.nist.gov/publications/detail/sp/500299/draft. Accessed 11 September 2018
14. S. Yu, C. Wang, K. Ren, W. Lou, Achieving secure, scalable, and fine-grained data access control in cloud computing. In: Proceedings of the IEEE INFOCOM, March (2020)
15. D. Sgandurra, E. Lupu, Evolution of attacks, threat models, and solutions for virtualized systems. ACM Comput. Surv. **48**(3), 1–38 (2016)
16. M. Kaur, H. Singh, A review of cloud computing security issues. Int J Adv Eng Technol **8**(3), 397–403 (2015)
17. P.R. Kumar, P.H. Raj, P. Jelciana, (Exploring data security issues and solutions in cloud computing. Proc. Comput. Sci. **125**, 691–697 (2018)
18. K.A. KhalilI, M. Azeem, Cloud computing security:a survey.Computers **3**(1), 1–35 (2014)
19. S.F. Bashir, S. Haider, Security threats in cloud computing. In: Proceedings of the International Conference for Internet Technology and Secured Transactions, pp. 214–219 (2011)
20. M.D. Ryan, Cloud computing security: the scientific challenge, and a survey of solutions. J. Syst .Soft. **86**(9), 2263–2268 (2013)
21. Wang C, Wang Q, Ren K, Lou W (Mar 2010) Privacy-preserving public auditing for data storage security in cloud computing. In: Proceedings of the IEEE INFOCOM
22. S. Singh, Y.-S. Jeong, J.H. Park, A survey on cloud computing security: issues, threats, and solutions. J. Netw. Comput. Appl. **75**, 200–222 (2016)
23. M. KhalilI, A. Khreishah, M. Azeem, Cloud computing security:a survey. Computers **3**(1), 1–35 (2014)
24. M. Ahmed, A.T. Litchfield, Taxonomy for identification of security issues in cloud computing environments. J. Comput. Inf. Syst. **58**(1), 79–88 (2018)
25. N. Fotiou, A. Machas, G.C. Polyzos, G. Xylomenos, Access control as a service for the Cloud. J. Internet Serv. Appl. **6**(1), 11 (2015)
26. B. Sumitra, C. Pethuru, M. Misbahuddin, A survey of cloud authentication attacks and solution approaches. Int. J. Innov. Res. Comput. Commun. Eng. **2**(10), 6245–6253 (2014)
27. D.A. Fernandes, L.F. Soares, J.V. Gomes, M.M. Freire, P.R. Inácio, Security issues in cloud envi-ronments:a survey. Int. J. Inf. Secur. **13**(2), 113–170 (2014)
28. S. Subashini, V. Kavitha, A survey on security issues in service delivery models of cloud computing. J. Netw. Comput. Appl. **34**(1), 1–11 (2011)
29. X. Chen, Y. Zhang, D.S. Wong, J. Li, H. Li, Ensuring attribute privacy protection and fast decryption for outsourced data security in mobile cloud computing. Inf. Sci. **379**, 42–61 (2017)
30. H. Abbas, O. Maennel, S. Assar, *Security and Privacy Issues in Cloud Computing* (Springer, Berlin, 2017)
31. TechRepublic, Building Trust in a Cloudy Sky (online) (2018). https://www.techrepublic.com/resource-library/whitepapers/building-trust-in-a-cloudy-sky/. Accessed 11 September 2018
32. S. Basu et al., Cloud computing security challenges and solutions—a survey, Proceedings of the IEEE 8th Annual on Computing and Communication Workshop and Conference (CCWC), pp. 347–356 (2018)
33. S. Dzombeta, V. Stantchev, R. Colomo-Palacios, K. Brandis, K. Haufe, Governance of cloud computing services for the life sciences. IT Prof. **16**(4), 30–37 (2018)
34. M. ButunI, B. Kantarci, H. Song, Cloud-centric multi-level authentication as a service for secure public safety device networks.IEEE Commun. Mag. **54**(4), 47–53 (2016)
35. N. ClarkeN, H. Saevanee, V. Biscione, S. Furnell, Continuous user authentication using multi-modal biometrics. Comput. Secur. **53**, 234–246 (2015)
36. K.A. KhalilI, M. Azeem, Consolidated identity management system for secure mobile cloud computing. Comput. Netw. **65**, 99–110 (2014)

37. T. Faber, S. Schwab, J. Wroclawski, Authorization and access control:ABAC. R. McGeer, M. Berman, C. Elliott, R. Ricci (eds.), *The GENI Book* (Springer, Berlin, 2016), pp. 203–234
38. M.A.. Khan, A survey of security issues for cloud computing. J. Netw. Comput. Appl. **71**, 11–29 (2016)
39. F. Cai, N. Zhu, J. He, P. Mu, W. Li,, Y. Yu, Survey of access control models and technologies for cloud computing. Clust. Comput. **22**(S3), 6111–6122 (2018)
40. M.P. Joshi, K.P. Joshi, T. Finin, Attribute based encryption for secure access to cloud based EHR systems. Proceedings of the International Conference on Cloud Computing (2018)
41. P.R. Indul, V. Bhaskar, Identity and access management in cloud environment:mechanisms and challenges. Eng. Sci. Technol. Int. J. **21**(4), 574–588 (2018)
42. G. Biswas, P. Mohit, Confidentiality and storage of data incloud environment, Proceedingsof the 5th International Conference on Frontiers in Intelligent Computing: Theory and Applications. (Springer, Berlin, 2017), pp. 289–295
43. S.I. Khan, A.S.L. Hoque, Privacy and security problems of national health data warehouse: aconvenient solution for developing countries. Proceedings of the IEEE International ConferenceonNetworking Systems and Security (NSysS), pp. 1–6 (2016)
44. J. Tang, Y. Cui, Q. Li, K. Ren, J. Liu, R. Buyya, Ensuring security and privacy preservation forcloud data services. ACM Comput Surv (CSUR) **49**(1), 13 (2016)
45. M.A. Islam, S.V. Vrbsky, Transaction management with tree-based consistency in cloud databases. Int. J. Cloud. Comput. **6**(1), 58–78 (2017)
46. C.-Y. Ku, Y.-S. Chiu, A novel infrastructure for data sanitization in cloud computing. Diversity, Technology, and Innovation for Operational Competitiveness:Proceedings of the 2013 International Conference on Technology Innovation and Industrial Management, pp. 3–25 (2013)
47. H.J. Singh, S. Bawa, Scalable metadata management techniques for ultra-large distributed stor-agesystems—asystematicreview. ACM Comput. Surv. (CSUR) **51**(4), 82 (2018)
48. N.K. Sehgal, P.C.P. Bhatt, *Cloud Computing Concepts and Practices* (Springer,2018)
49. V. Prokhorenko, K.R. ChooK, H. Ashman, Web application protection techniques:a taxonomy. J. Netw. Comput. Appl. **60**, 95–112 (2016)
50. S. Setal, Rosemary:a robust, secure, and high-performance network operating system.Proceedings of the 2014 ACM SIGSAC Conference on Computer and Communications Security. ACM, New York, pp. 78–89 (2012)
51. N. Chauhan, L. Ahuja, S.K. Khatri, Secure data in cloud computing using face detection and fingerprint. International Conference on Inventive Research in Computing Applications (ICIRCA), pp. 231–234 (2018)

Security Issues of IoT in Healthcare Sector: A Systematic Review

Mueed Ahmed Mirza, Rao Muhammad Mahtab Mahboob, Hassan Jahangir, Majid Khawar, and Mubashir Imam

Abstract Internet of Things (IoT) has become a new trend in the healthcare system which provides a high diversity of efficacious services for healthcare professionals as well as patients' treatments for various diseases. In IoT healthcare system application design consists of devices, network, layered model, and Communication Application/Protocols. Mostly in studies, in fact, IoT provides several benefits, and it also facing different security challenges in their patient record hierarchy. Due to the fact of private and sympathetic data of patients, required complete security and privacy are a key issue in the formation of IoT-based systems in the healthcare sector. The aim of this study is to investigate the security-related issues of IoT in the health sector. In this study, we compiled searched articles that are published in well-reputed journals from 2015 to 2021 for constructing the summary of related challenges. We construct a table that analyzes the security-related issues of IoT in the healthcare sector. Moreover, an IoT healthcare architecture has been proposed that represents the effective patient discovery solution in the field of health care. Our study sketches the open and recent security issues of IoT in the healthcare sector.

Keywords Internet of Things · Health sensors · Healthcare devices · Security threats · Layer model

1 Introduction

Digital health care is a complex system that aims to diagnose, treat, and prevention of health-related issues in humans. Health professionals, health facilities, diagnosis centers, and financing institutions are a major part of the healthcare system. Each part

M. A. Mirza (✉) · H. Jahangir · M. Khawar
NFC-IET, Multan, Pakistan

R. M. M. Mahboob
UAF, Faisalabad, Pakistan

M. Imam
GCU, Faisalabad, Pakistan

G. Ranganathan et al. (eds.), *Soft Computing for Security Applications*,
Advances in Intelligent Systems and Computing 1397,
https://doi.org/10.1007/978-981-16-5301-8_50

has its importance and depending levels. In healthcare professionals, several kinds of information are captured such as prescriptions, diagnosis, laboratory, imaging, personal and private related medical data. Medical records of patients capture by using different types of devices, after processing the data it will be used for future prediction or treatment purpose [1].

mHealth provides different services for improving the healthcare system by using IP connectivity devices that provide mobile-based services for users in the healthcare sector. Data that is a capture from the sensor communicated electronically with the help of the Internet the increase Health Privacy and security issues rapidly, Mobile devices use the smart application that informs the patients after analysis according to captured data (Oxygen monitoring system, ECG, Body temperature, blood pressure, glucose level) of their diseases. However, the expected research concerned with security issues in the healthcare sector using the concept of IoT-based devices [2].

The idea of the connected devices "Internet of Things" was invent by Kevin Ashton in 1999 in the beginning it is used to boost RFID technology. The IoT carried the information from different sources like RFID, Smart devices, smart grids, cloud computing, sensors, vehicle network, and a number of different information carriers. Basically, IoT is a network in which all physical objects are connected to the internet through network devices or routers and exchange data. IoT allows objects to be controlled remotely across existing network infrastructure. IoT is a very good and intelligent technique which reduces human efforts as well as easy access to physical devices. This technique also has autonomous control feature by which any device can control without any human interaction [6].

Healthcare sector is the most common area where demand of IoT application increase day by day. In healthcare sector, Ambient Assisted Living (AAL) and Smart Health are few names of IoT application used for making a report of patients [7]. IoT makes sure the early stage detection of diseases and help out in the process of diagnosis and treatment. Instead of advantages, IoT also faces the security challenges and vulnerabilities in healthcare system [3].

Nowadays, various organizations generate a massive amount of data. In the health care department, various types of records fetch like patients' medical records, hospital records, medical examination records, and biomedical research records. All types of records based on management and analysis by using heterogeneous devices. This clinical data maintained and accessed by different healthcare staff for the treatment of the patients. Many challenges and issues are associated with capturing and cleaning data. Protecting patient data is a key issue while sharing information between health professionals and health insurance companies. One major challenge is in the analysis of the use of data, how much EHR data originally used, and compromises the validity and reliability of their results [9]. IoT network encompasses a variety of things smart devices and sensors. Security and privacy are the major concern about key issues in IoT security. It is a very difficult task to provide security for a large amount of data that exists in huge velocity and volume facing the different security threats like Denial of Service, Access control. For managing the high data storage devices that share the sensor data of a patient it is difficult to communicate and processing the voluminous data, for the solution and searching of health data information, IoT introduces the architecture components like smartphone, analytic module. Some major challenges facing IoT security systems are Spoofing, Injection of false signals, Replay attacks, Eavesdropping. A different solution to cryptographic security can be applied over IoT for maintaining authenticity and reliability. Many researchers provide a relevant solution for improving healthcare issues [4].

However, in the real-time environment, many things are interconnecting that might possibly become a cause of various attacks and threats. Therefore, an effective and efficient defense strategy is the most important for the security of IoT devices in the healthcare sector. After reading this proposed study we have a sterling understanding of IoT security issues in our healthcare sector. The main aim of this review paper is to spark new research attention in this field that is helpful for the researcher. The objective of this review paper was achieved through two phases. We investigate the security challenges of IoT, especially in the healthcare system. Beside it, the proposed patient discovery architecture will be present after it we summarize our research work that shows the comprehensive study of current challenges in the IoT-based healthcare system. Classification of relevant articles based on inclusion and exclusion criteria with the help of the Mendeley manager. The selection of studies articles is within the period from 2015 to 2020. In inclusion criteria of this research focus on those issues that are related to our topic. This paper will fulfill exceptional information for the interested student and researcher, in this abundant recent research area.

Collecting the direction toward the healthcare professionals, academics, designers, and also nontechnical non-medical audience for easy to understand the issues in this domain. Section 2, a literature review of IoT healthcare technology will be discussed. Section 3, Research Methodology with the research question, exclusion/inclusion criteria, and search keywords that are used to collect the effective and comprehensive study about IoT security domain that observed in various research journals are introduced. Section 4, Results of this research will be discussed in systematic tabular form. Aside from some latest security issues to IoT in the healthcare system to support the researchers, devices, and sensors are also discussed in this section. Section 5, a summary of our finding will be discussed with IoT-based health care proposed patients discovery architecture. Section 6, the conclusion of this article will be present.

2 Literature Review

The main aim of IoT is to focus on interconnection devices worldwide by using the Internet. IoT provides a vast range of different applications like smart cities, agriculture, logistics, and other demands to make life modern or comfortable. IoT expedites the diseases on an early base that supports the process of treatment and diagnosis. Besides the above advantages of IoT application, it is also facing security challenges and protection of healthcare systems. Due to the nature of Medical data faces security issues like integrity, availability, and confidentiality are increasing. Cyber security has become a big concern in the healthcare domain. HIPPA "Health

Insurance Portability and Accountability Act" provides the technical and physical safeguards but that is not sufficient and powerful to protect the updated security challenges. HIPPA uses a preventive or protective approach for cybersecurity in the health care of IoT systems but these are not responsive because they only protecting known threats. On the other hand, security resiliency has greater capabilities for prevention, fault tolerance, prediction and know or unknown attacks [9]. IoT systems typically used sensor devices, machine algorithms, cloud-based services, and Wireless Sensor Network (WSN). The sensor is used to collect the data from the human body like heartbeat rate, blood pressure, and temperature, then WSN makes a smooth communication between different devices, after collecting the data, different algorithms are applied for analysis, and collected data is stored on cloud storage and permits the patients and healthcare professional to get the data. WSN used sensor nodes that have a unique ID for maintaining smooth communication. Due to various device security challenges increased in IoT health systems, major issues are data exchange, data collection, and data storage [5]. Due to the remote patient monitoring system constant communication is needed between IoT devices. MEDiSn and MobiCare are the best application at present but they also facing common threats: eavesdropping, data leak, and impersonation, all these threats change the Medical data of patients. IoT health care uses a different application like Assisted living (Apple Healthkit, Google Fit, Fitcoracy), Clinical care (BSN-Care, Codeblue, HetNet, eHealth), Context-awareness (mHealth), and remote monitoring (iMedBox, k-healthcare, MiiM, RunKeeper) for the collection of patient data but ECG sensor found the most used device and application layer have the highest security threats [6]. According to the Massachusetts Institute of Technology (MIT) major problem in IoT systems is Malware attacks on end devices. Traffic aware scheme and traffic management are the major problem in IoT, but it can be managed by the operation research technique [7].

There are different protocols are used in IoT like Bluetooth low energy (BLE), Multicast DNS, etc. IoT has not a standardized security architecture it varies from different vendors, generally, it works on three levels (1) Device security level that deals on Perception Layer mostly face the physical threat in IoT security. (2) Communication Security is used for communication and network node protection deals on Transport and Network layer 3) Cloud Security ensures the protection of data mostly facing SQL injection that occurs on the Application Layer [8]. Recent research in IoT visualized that IoT controls the all world through its infrastructure and assumes that IoT devices make the gigantic network that encapsulates all IoT devices and sensors. Mostly IoT networks used components are RFID, smart card, smartphone, and computers [9].

Furthermore, the individual security of devices are most important, two actual components are mentioned, Privacy & Security. Security is accessible by two subsequential categories: (1) Infrastructure security (2) Information security. However prevention of user data in an IoT-based architecture is a challenging task, some major security threats are (1) Functionality of encryption (2) Access control [10].

3 Methodology

A Basic aimed of SLR is the evaluation and presentation of literature about the research topic by the use of the verifiable methodology. This review adopts the SLR, a procedure given by [11] (Fig. 1).

3.1 Stage 1: Research Planning

3.1.1 Need for Systematic Review

In the introduction section, we already discuss the importance of IoT in the healthcare system. Hence there are three objectives follows that describe the literature review process.

- The latest research work has been focused on the Security area of IoT health care.
- Characterize the current IoT healthcare security threats and issues.
- Proposed the typology that highlights the quick discovery of patients in IoT health care.

3.1.2 Defining the Research Questions

The preparation of the recent research status and the meaning of the research question are defined in this section.

RQ.1 What are the major security threat are targeted in IoT health care?
RQ.2 What are the most measurable threats solution in IoT health care?
RQ.3 Which IoT security protocol is used in health care for communication?
RQ.4 How many types of IoT sensors and devices are used in health care?
RQ.5 How many types of application are used in IoT healthcare sector ?

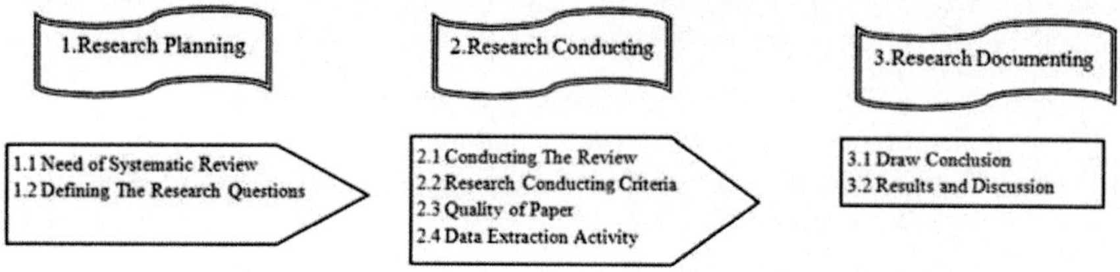

Fig. 1 SLR methodology

3.2 *Stage 2: Research Conducting*

3.2.1 Conducting the Review

For getting the current challenges of the IoT in the healthcare system, we examine with the help of keywords that are related to the topic. Journals names that will help researchers are Bioinformatics, Digital Health, IGI Global, ACM, IEEE, Springer, Elsevier, MDPI, Science Direct. In this research, we enlist the IoT device and sensors that are used in the healthcare sector. Findings of this paper were collected manually from the authentic source of information to defining the research question, for that purpose Mendeley software used for managing the article. Tableau generates the results of Mendeley's articles based on their published year/source of paper/app/sensor/network/reference in the form of Table or Graphs.

3.2.2 Research Conducting Criteria

The main aim of this review paper is to spark new research attention in this field that is helpful for the researcher. We achieved the objective of this paper by investigating the challenges of IoT security especially in the healthcare system for fulfilling our research question. Classification of relevant articles based on inclusion and exclusion criteria with the help of the Mendeley manager. The selection of studies articles is within the period from 2015 to 2021. In inclusion criteria of this research focus on those studies related to the research, the domain was taken from 2015 to 2021. Exclude those articles that are not related to the research question. This paper will fulfill exceptional information for the interested student and researcher, in this abundant recent research area.

3.2.3 Quality of Paper

In this phase, we analyze the quality of the included paper for ensuring the validity of the SLR. Here we apply the standards of [12] to increase the quality of a paper.

- A key research method is used in this study to get precision results.
- The significant and rational conclusion is finding.
- The findings of this paper are useful for healthcare professionals (Fig. 2).

3.2.4 Quality Assessment

To achieve the quality of the chosen paper, a set of questions has been sketched that measure the effectiveness of the selected paper. This QA is following the (Table 1) mapping study.

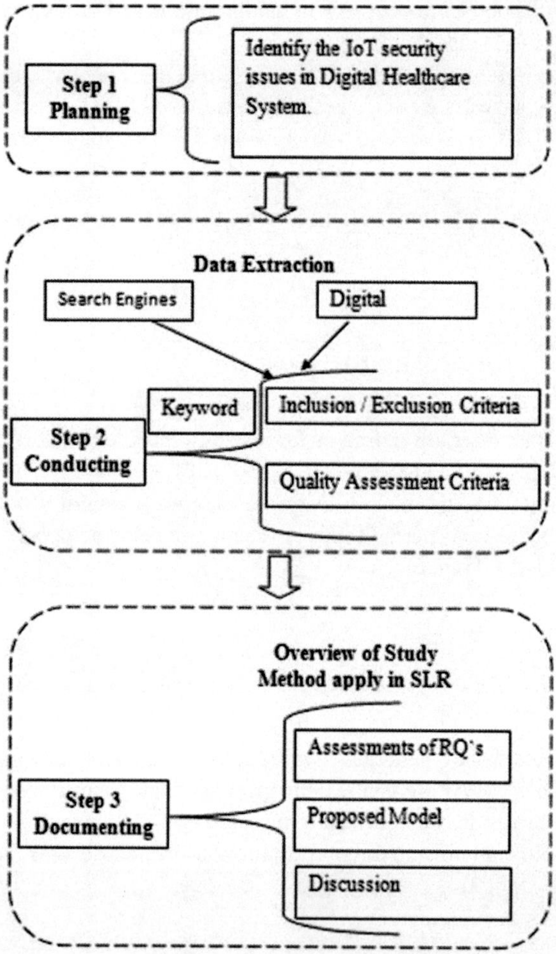

Fig. 2 Study method

Table 1 Data schema

Source	Bioinformatics, Digital Health, IGI Global, ACM, IEEE, Springer, Elsvier, MDPI, Science Direct
Searching keywords	Healthcare IoT apps, IoT security, IoT protocols, Security layer model, IoT healthcare architecture, healthcare challenges
Domain	Digital Medical Health System
Sub-domain	Internet of Things

- Contribution of IoT in health care based on two possible answers were if "yes" (+1) else "No"(0).
- Absolute solution of security in health care by using IoT based on three possible answers were if "yes" (+1) else "No" (0) and "Partially" (0.5)
- The published journal how much meet with a research question. Grading of these questions were: No (0), Yes (+1), and Partially (0.5).

Accepted answers according to our research conducting criteria were presented in Table 2

4 Results and Discussion

Outcomes are defines with research question in Table 2. A number of papers have been studied to show the outcome of each research questions. The search gives 50 citations in which 30 were considered potentially related to our topic and met the inclusion criteria (Table 1). The list of selected papers is shown in Table 2 with their QA and outcomes.

4.1 RQ #1 Security Threats in IoT Health Care

Health care generates a most sensitive data that is very useful for patients in various aspects. Although the rapid growth of data security becomes vital challenge, different threats are categorized in our study on the basis of previous research. According to our study inside threat from the staff is most serious threats to the CIA of data (Table 3, Fig. 3).

4.2 RQ #2 Threats Solution in IoT Health Care

Security is the key problem in IoT health care, our research found different standards or solution in Table 1 to overcome the security and privacy problem in healthcare field. In this research, we divide the threats into several measureable components (Table 4).

4.3 RQ #3 Security Protocols for Reliable Communication

IoT devices use the Internet protocol for Smart Objects (IPSO) that adopt on different layer and provide the smooth communication between smart devices. In this study,

Table 2 Summarized finding

Classification							Quality assessment grading			
Year	S. paper	Area of focus	Type of protocol	Healthcare application	Senor/network	References	a	b	c	Total
2019	Journal	Proposed an intrusion detection system to monitor the patients	CoAp	e-Medicine	Flexed pulse/WSN	[1]	1	0.5	1	2.5
2018	Journal	A wireless mobile robot system has been developed to perform multiple monitoring and controlling tasks	MQTT	MEDiSN	ECG/ ZigBee	[3]	1	1	0	2
2019	Journal	A survey of different monitoring applications and communication protocols	LPWAN	MySignals	FDA/M-Protocols	[2]	1	0.5	0.5	2
2017	Journal	Developed an application for disease detection at early stages	CoAp	Google health	PRO/WSN	[10]	1	1	1	3

(continued)

Table 2 (continued)

Classification							Quality assessment grading			
Year	S. paper	Area of focus	Type of protocol	Healthcare application	Senor/network	References	a	b	c	Total
2020	Journal	An agent-based system that is used to maintain the security of EHR data	SigFox	MiiM	CVD /WSN	[13]	1	0	0.5	1.5
2020	Conference	Proposed a method to measure the security cost in the health area by using IoT such as patients emergency	NB-IoT	MySignals	SPO2/WIFI	[14]	1	0	0.5	1.5
2019	Conference	Blockchain is used to solve the IoT security-related problems of networks, users, and devices	PoW,PoS	mHealth	SWS/WIFI	[15]	1	0.5	1	2.5

(continued)

Table 2 (continued)

Classification							Quality assessment grading			
Year	S. paper	Area of focus	Type of protocol	Healthcare application	Senor/network	References	a	b	c	Total
2018	Conference	Cloud-based technology has been compiled automatically for patients' data taking along with sensors of medical devices	IoT MS	SecureData	EKG/WSN	[16]	1	1	1	3
2016	Journal	Narrowband (NB-IoT) has a good combination with healthcare applications	MQTT	MedShare	PDCP /LTE	[17]	1	0	1	2
2018	Journal	Mostly cloud computing adopts for accessibility of IoT-based health system data	CoAp	UbiMon CodeBlue	LOS	[18]	1	1	0	2

(continued)

Table 2 (continued)

Classification							Quality assessment grading			
Year	S. paper	Area of focus	Type of protocol	Healthcare application	Senor/network	References	a	b	c	Total
2019	Journal	HIPPA privacy and security rule are followed mostly in the healthcare sector	CoAP-HTTP	MySignals	QSH	[6]	1	0	0.5	1.5
2017	Journal	Sensors are embedded into an elder body that transfers the biological data for measuring on the cloud	IPSO	kHealth	ECC	[19]	1	0.5	1	2.5
2020	Journal	Adoption of proper security policies can prevent the outsider to access the data	MQTT	UbiMon	SWS/WIFI	[20]	1	0.5	1	2

(continued)

Table 2 (continued)

Classification							Quality assessment grading			
Year	S. paper	Area of focus	Type of protocol	Healthcare application	Senor/network	References	a	b	c	Total
2016	Conference	Cloud services that focus on EHR data for analysis report using Application	CoAp	Secure Data	QSHM/WSN	[8]	1	0.5	0.5	1.5
2018	Conference	Proposed an analysis report model on Health record by using Blockchain technology	PoA,PoB,PoI,	iMedBox	SWS/WIFI	[21]	1	1	1	3
2020	Journal	Analysis of various Security architecture methods of blockchain	PoD,DPoS,PoW	MySignals	SWS/WIFI	[22]	1	0	0.5	1.5

(continued)

Table 2 (continued)

Classification							Quality assessment grading			
Year	S. paper	Area of focus	Type of protocol	Healthcare application	Senor/network	References	a	b	c	Total
2020	Journal	Auto adjustability and flexibility helpful in the IoT application operation of EHR	LPWAN	UbiMon	AMON/WSN	[23]	1	0	0.5	1.5
2019	Journal	IoT in health care improves through smart devices and wearable sensors	MQTT	S-Health	Melbourne Victoria	[24]	1	0.5	1	2.5
2016	Journal	Introduced the security model that minimized the overhead of real-time data	NB-IoT	MiiM	AliveCor's/WSN	[25]	1	1	1	3

(continued)

Table 2 (continued)

Classification							Quality assessment grading			
Year	S. paper	Area of focus	Type of protocol	Healthcare application	Senor/network	References	a	b	c	Total
2018	Journal	Healthcare data become more secure by using the blockchain framework because it does not involve a third party (TTP)	PoW,PoA,PoB	UbeHealth	SWS/WIFI	[26]	1	1	1	3
2020	Journal	IoT in healthcare security needs an inter-domain cryptographic algorithm	NB-IoT	Apple MS Kit	EEG/WSN	[4]	1	1	0	2
2020	Journal	In fact, Adopting unique policies that cover every feature of Security and privacy can minimize IoT Threats	CoAP-HTTP	Mio	BSC/WSN	[27]	1	0	0.5	1.5

(continued)

Table 2 (continued)

Classification							Quality assessment grading			
Year	S. paper	Area of focus	Type of protocol	Healthcare application	Senor/network	References	a	b	c	Total
2019	Journal	IoT devices are immature resources and incapable to deliver the security mechanism for global robustness	MQTT	Fitocarcey	MOD/Wifi	[9]	1	0.5	1	2.5
2017	Conference	Access control of private data becomes more secure with the usage of control decision components	CoAp	Google health	Mica2/WSN	[7]	1	1	0	2
2018	Journal	The novel framework proposed for authentication of health data by using DUKPT advanced encryption method	LPWAN	BSN-Care	P-Oximeter/LTE	[28]	1	1	1	3

(continued)

Table 2 (continued)

Classification							Quality assessment grading			
Year	S. paper	Area of focus	Type of protocol	Healthcare application	Senor/network	References	a	b	c	Total
2017	Conference	Develop a privacy base tracker for an elder person that cares on the daily basis	Desearch	kHealth	Desearch/GPS	[14]	1	1	1	3

Table 3 Security threats in IoT health care

Threats	Categorization
Internet of Things (IOT) devices	Hackers, security breaches, power breakdown
Inside threat from Staff	Stealing the data, password misuse, misuse of data, lack of staff training
Third party software's	Vendors (hardware and software), outdated antivirus, system bugs
Network	Lack of encryption, neglecting proper security configuration, security alert, software outdated, network breakdown

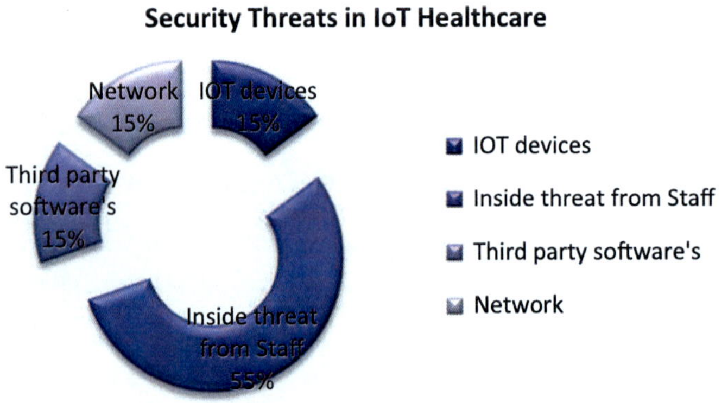

Fig. 3 Security threats in IoT health care

Table 4 Threats solutions in IoT health care

Measureable components	Descriptions
User access control	Strong authentication method for employee Weekly update password Don't forget the password
Internet security	Use a security firewall Restriction on other web links
Data storage security	Apply a authentication password on storage devices Encrypt all storage data
Update security policies	Security alert Used different authentication factors

we review that mostly researcher discusses blockchain technology for enhancing the clinical trials, medical billing, credentials, accuracy in EHR system and transferring patients data in the HCP field. Blockchain-based protocols (proof-of-work (PoW), proof-of-stake (PoS), delegated-proof-of-stake (DPoS), proof-of-importance (PoI),

Table 5 Uses of H-APP in IoT

Category of IoT healthcare applications	H-Application
Assistant application	Apple Healthkit, Fitcoracy, Google Fit, Google Health
Clinical application	BSN-Care, Codeblue, eHealth,e-Medicine,MEDiSN MeDShare, SecureData
Location-based application	mHealth
Remote-based application	iMedBox,k-healthcare, Microsoft health, MiiM, MobiCare, MySignals, S-Health

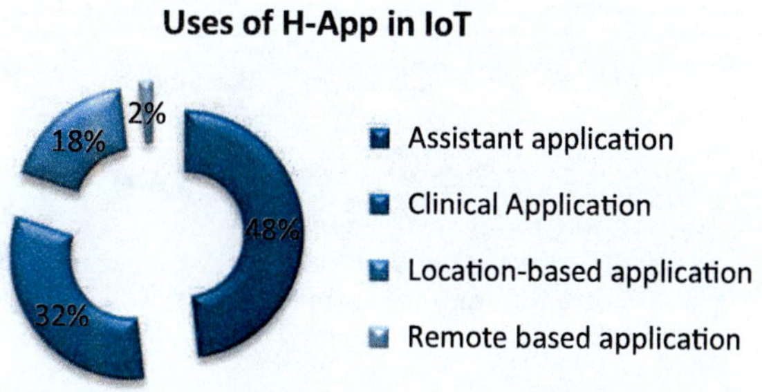

Fig. 4 Uses of healthcare application in IoT

proof-of-activity (PoA), proof-of-burn (PoB), proof-of-deposit (PoD)) were more secure and reliable as well as support the patients data sharing.

4.4 RQ #4 Uses of H-Application in IoT

After reviewing different research papers application of IoT health care Table 5 is categorized on its usage, Assistant application, Clinical Application, Location-based application, and remote-based application. Figure 4 displays the usage of IoT applications in health care, in which 48% use of remote-based application, clinical and Assistant based application shows 32% and 18% respectively. Location-based shows the lowest 2% usage.

4.5 RQ #5 Healthcare Sensors/Devices in IoT

Different types of medical sensors are used to capture the healthcare data. Figure 5 shows the ECG sensor has more active part of health sector, while other sensor is relative usage with aspect to different authors.

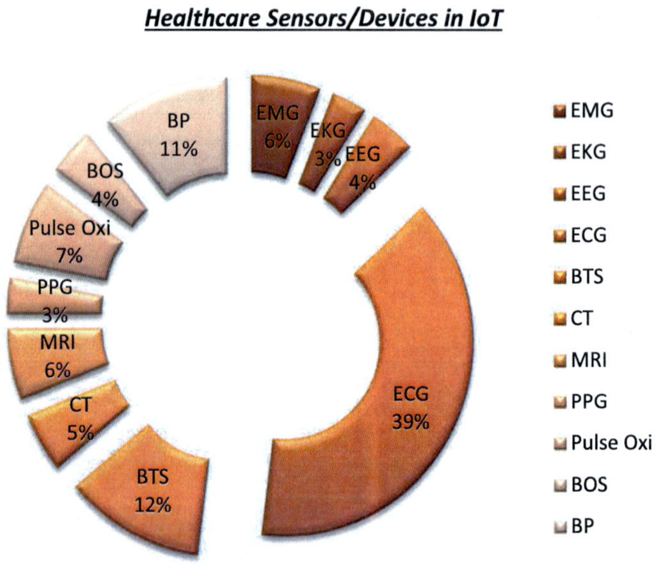

Fig. 5 Healthcare sensors/devices in IoT

5 Discussion and Conclusion

5.1 Security Issues of IoT in the Medical Sector

Nowadays smartphones are working with intelligent sensors and wearable gadgets that collect patients' data. Generally, biometric data is collected through connected devices with a different application for interpreting and processing the signal and data. Six main security prospect in the Medical sector, These are (1) Confidentiality (2) Correct uses of information (3) Legal issues (4) Sensor Devices (5) Data Transmission (6) Cloud storage.

It is observed that data extortion from the devices is less mostly data is theft during the transmission using different hacking techniques. In fact, efficient authentication methods and encryption techniques are required to avoid data extortion. Another vulnerability of data is different cloud vendors' service providers there is more chance of stolen data. For resolving these challenges we needed some complex passwords, versatile authenticity, and access control mechanisms that turn transliterate the system more complex to use.

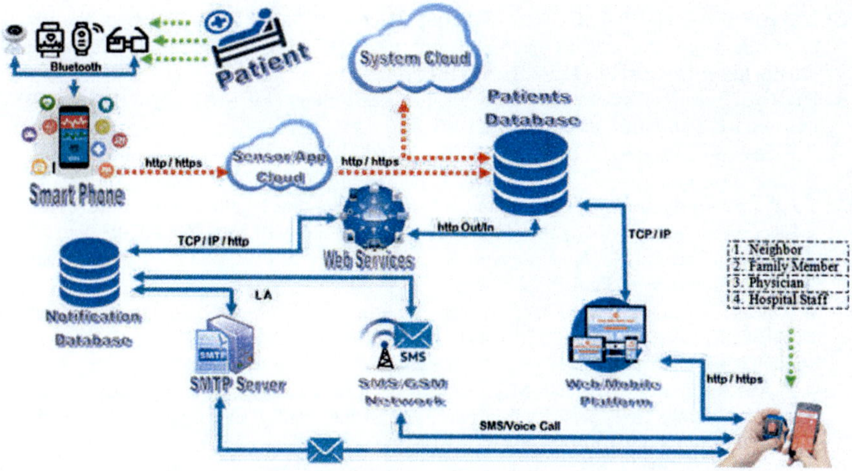

Fig. 6 Fast patient recovery model IoT health care

5.2 Proposed Model for Fast Patients Recovery Model in IoT Health care:

High-Performance Computing (HCP) devices are required for data storage, sharing, and processing of the data in health care architecture. Figure 6 Logical solution is provided based on a comprehensive study, in which registered patients and other people are search and retrieve their health information, it also allows nearest neighbor for rescue the patients with the help of security camera between this time sensor sends the message to their families and blood relation members to communicate the hospital's staff or concreted doctor.

Some key components of the proposed architecture are as followed:

1. Smartphone
2. Cloud-based infrastructure
3. Analytic module
4. Access element
5. Messaging platform
6. IP Camera

References

1. S. Dash, S. K. Shakyawar, M. Sharma, S. Kaushik, Big data in healthcare: management, analysis and future prospects. J. Big Data **6**(1), 2019. https://doi.org/10.1186/s40537-019-0217-0

2. S. Nazir, Y. Ali, N. Ullah, I. García-Magariño, Internet of Things for healthcare using effects of mobile computing: A systematic literature review. Wirel. Commun. Mob. Comput. **2019** (2019). https://doi.org/10.1155/2019/5931315
3. P.P. Ray, A survey on Internet of Things architectures, Comput. Inf. Sci. **30**(3), 291–319 (2018). https://doi.org/10.1016/j.jksuci.2016.10.003
4. P. Pandey, S.C. Pandey, U. Kumar, Security issues of Internet of Things in health-care sector: An analytical approach, 307–329 (2020). https://doi.org/10.1007/978-981-15-1100-4_15
5. J.S. Raj, A novel information processing in IoT based real time health care monitoring system. J. Electron. Informatics **2**(3), 188–196 (2020). https://doi.org/10.36548/jei.2020.3.006
6. S.P. Amaraweera, M.N. Halgamuge, in *Security, Privacy and Trust in the IoT Environment* (Springer International Publishing, 2019)
7. C. Cheng, R. Lu, A. Petzoldt, T. Takagi, Securing the Internet of Things in a quantum world. IEEE Commun. Mag. **55**(2), 116–120 (2017). https://doi.org/10.1109/MCOM.2017.1600522CM
8. A.F. Abdul Rahman, M. Daud, M.Z. Mohamad, Securing sensor to cloud ecosystem using Internet of Things (IoT) security framework. ACM Int. Conf. Proceeding Ser. 22–23 March (2016). https://doi.org/10.1145/2896387.2906198
9. A. Maria, I. Nazurl, J. NZ, A lightweight and secure authentication scheme for IoT based e-health application. IJCSNS Int. J. Comput. Sci. Netw. Secur. **19**(1), 107–120 (2019)
10. Y. Zhang, M. Qiu, C.W. Tsai, M.M. Hassan, A. Alamri, Health-CPS: Healthcare cyber-physical system assisted by cloud and big data. IEEE Syst. J. **11**(1), 88–95 (2017). https://doi.org/10.1109/JSYST.2015.2460747
11. S. Hamid, S. Bukhari, S.D. Ravana, A.A. Norman, M.T. Ijab, Role of social media in information-seeking behaviour of international students: A systematic literature review. Aslib J. Inf. Manag. **68**(5), 643–666 (2016). https://doi.org/10.1108/AJIM-03-2016-0031
12. T. Dybå, T. Dingsøyr, Empirical studies of agile software development: A systematic review. Inf. Softw. Technol. **50**(9–10), 833–859 (2008). https://doi.org/10.1016/j.infsof.2008.01.006
13. M.A. Alanezi, Z. Faizal Khan, in *A Framework for Privacy and Security Model Based on Agents in E-Health Care Systems*, vol. 1073 (Springer International Publishing, 2020)
14. Z. Ajazmoharkan, T. Choudhury, S.C. Gupta, G. Raj, Internet of Things and its applications in E-learning. 3rd IEEE Int. Conf., January (2017). https://doi.org/10.1109/CIACT.2017.7977333
15. H. Suryotrisongko, Y. Musashi, Review of cybersecurity research topics, taxonomy and challenges: Interdisciplinary perspective. Proc.—2019 IEEE 12th Conf. Serv. Comput. Appl. SOCA 2019, pp. 162–167 (2019) https://doi.org/10.1109/SOCA.2019.00031
16. O. Ali, A. Shrestha, J. Soar, S.F. Wamba, Cloud computing-enabled healthcare opportunities, issues, and applications: A systematic review. Int. J. Inf. Manage. **43**(July), 146–158 (2018). https://doi.org/10.1016/j.ijinfomgt.2018.07.009
17. A.K. Jha et al., Use of electronic health records in U.S. Hospitals. N. Engl. J. Med. **360**(16), 1628–1638 (2016). https://doi.org/10.1056/NEJMsa0900592
18. A. Fatima, R. Colomo-Palacios, Security aspects in healthcare information systems: A systematic mapping. Procedia Comput. Sci. **138**, 12–19 (2018).https://doi.org/10.1016/j.procs.2018.10.003
19. J. X. Hu, C.L. Chen, C.L. Fan, K.H. Wang, An intelligent and secure health monitoring scheme using IoT sensor based on cloud computing. J. Sensors (2017). https://doi.org/10.1155/2017/3734764
20. M. N. Birje, S.S. Hanji, Internet of things based distributed healthcare systems: a review. J. Data, Inf. Manag. **2**(3), 149–165 (2020). https://doi.org/10.1007/s42488-020-00027-x.
21. M.A. Khan, K. Salah, IoT security: Review, blockchain solutions, and open challenges. Futur. Gener. Comput. Syst. **82**, 395–411 (2018). https://doi.org/10.1016/j.future.2017.11.022
22. J.R. Gil-Garcia, T.A. Pardo, M. Gasco-Hernandez, *Internet of Things and the Public Sector BT—Beyond Smart and Connected Governments: Sensors and the Internet of Things in the Public Sector* (2020)

23. S. Gupta, L. Goel, A.K. Agarwal, Technologies in health care domain: A systematic review. Int. J. e-Collaboration **16**(1), 33–44 (2020). https://doi.org/10.4018/IJeC.2020010103
24. G.V. Angelov, D.P. Nikolakov, I.N. Ruskova, E.E. Gieva, M.L. Spasova, *Healthcare Sensing and Monitoring*, vol. 11369 (Springer International Publishing, LNCS, 2019)
25. K. Natarajan, B. Prasath, P. Kokila, Smart Health Care System Using Internet of Things. J. Netw. Commun. Emerg. Technol. **6**(3), 37–42 (2016)
26. M. Hölbl, M. Kompara, A. Kamišalić, L.N. Zlatolas, A systematic review of the use of blockchain in healthcare. Symmetry (Basel) **10**(10) (2018). https://doi.org/10.3390/sym10100470
27. Y. Koumpouros, A. Georgoulas, A systematic review of mHealth funded R&D activities in EU: Trends, technologies and obstacles. Informatics Heal. Soc. Care **45**(2), 168–187 (2020). https://doi.org/10.1080/17538157.2019.1656208
28. M. Pasha, S.M.W. Shah, Framework for e-health systems in IoT-based environments. Wirel. Commun. Mob. Comput. **2018**(2018). https://doi.org/10.1155/2018/6183732

Resume Screening Bot Using RPA

Nilam Khatal, Siddhesh Chaughule, and Nikhita Mangaonkar

Abstract With the rise in technology, most job seekers prefer to find jobs via some Web site. Many submit their Resume's/CV's online. The organizations receive countless of those every day on their Web site or portal. The process begins with identifying the needs within an organization. This need could be to fill a vacant position in an organization or strengthen the team or expand the reach of organizational tasks. The organization's recruiters should start by generating a job description that includes a prioritized list of job requirements, special qualifications which are needed for the selection of candidates suitable for them. The actual scale of the applicants submitting their resumes makes it an impossible task for any organization to sort it manually. Not only is it a time-consuming task but it's tedious too. So, to manage this task Robotic Process Automation was introduced in the job recruitment process. Using RPA we can reduce the manual work and also the time required to sort the candidate's resumes according to organization needs. Therefore, we thought of an idea to use this technology and automate this whole process of selection including accepting resumes from students and also forwarding them Emails of whether they are selected or not using Robotic process automation.

Keywords Robotic process automation (RPA) · Screening · Automate · PDF-extraction · Threshold · Automation anywhere · Java

N. Khatal (✉) · S. Chaughule · N. Mangaonkar
Sardar Patel Institute of Technology, Mumbai, India
e-mail: nilam.khatal@spit.ac.in

S. Chaughule
e-mail: siddhesh.chaughule@spit.ac.in

N. Mangaonkar
e-mail: nikhita.mangaonkar@spit.ac.in

G. Ranganathan et al. (eds.), *Soft Computing for Security Applications*,
Advances in Intelligent Systems and Computing 1397,
https://doi.org/10.1007/978-981-16-5301-8_51

1 Introduction

The rapid development of the modern world has introduced many technologies so that human resources are dependent on every task for hardware devices, software technologies as well as smart devices. One of these growing technologies is RPA. RPA makes it easy to build and manage software robots that can replace human workers doing redundant tasks. Just like people these robots understand what's on the screen, navigate systems, identify and extract information and perform many other such tasks. The robots designed are far more consistent and faster than an average individual.

RPA has so many reliable advantages, such as it has low error probability compared to manual processes. It utilizes high-level repetitive tasks, resulting in the early completion of tasks. More accuracy and quality check rather than errors. Consistency in task with reconciliation from multiple sources. RPA also reduces the chances of biased results. It allows users to follow the deadline and complete tasks in a given time.

The current system involves an organization needing to describe a job description based on the job profile, then collecting the resumes of all the people appearing for that particular job profile interview selection process. The organization then assigns a team that thoroughly checks those resumes and sorts them according to the requirements as per the job description of the job profile. This isn't only time-consuming but also involves a lot of manpower. Also, as there are humans involved in this activity there are further chances of human error. The current aspect of handling this work is surely enervated and can cause blunder that will not be favorable for the company. So, we decided to replace this whole system using a bot designed with the help of RPA as a technology with the help of Automation anywhere.

RPA reduces a lot of manual work to speed up business growth and also helps to identify the right applicant for the job profile. Organizations big or small can easily automate the resume screening with an RPA bot. In RPA we create an automated shortlisting applicants bot. In this work, our approach is different from conventional e-recruiting systems in that we don't accept Resume/CV directly from the applicants rather we provide an applicant predefined web form that models the applicant's Resume/CV into PDF format. The bot helps the recruiters to get the responses from the applicants automated on their folder with email, and also help to eliminate applicants that are not suitable for the job profile based on the job description and also sending automated communication for the selected applicants through the email and also store their information in the organization database for future use. This bot will not only reduce the requirement of human resources but also reduce the time and will be much more efficient.

2 Literature Survey

Due to the increase in technology in various fields, there is an increase in demand and the speed of execution for certain activities. Workers are finding it difficult to match with the increasing expectations which lead to an increase in the stress level of employees. Workers in all sectors are experiencing an enhanced time pressure. Not all are lucky enough to get their desired jobs that easily. It takes lots of effort and wastage of time by pursuing every possible opportunity to obtain the desired result. Looking at the growing competition in every industry that's very essential that one must find recruitment according to their capability.

Automation has evolved. Some difficult and time-consuming tasks have been replaced with heavy hardware robots (industrial machinery), whereas complex and mental tasks are executed through software robots (AI/Machine Learning). Automation based on RPA is generally made for fully focusing on specific tasks. It has been estimated that using robotics cuts operational costs by as much as 25–50%. Having robots take over some of the manually intensive work from humans could result in visible gains for the business. Automation helps increase the growth of business within less cost estimation. RPA helps in giving more accuracy to the task while the absence of human error. And optimizes capabilities that multiply the organizational capacity. As bots are handling the execution here, a larger amount of work can be done in a relatively much shorter period. More accuracy results in fast delivery.

In the current scenario, industries are looking for candidates with different skill sets to fulfill the needs of their organization, depending on the organization's recruitment needs. The recruiters in an organization for this purpose collect the resumes from students for recruiting immediately or for future use. Students submit their resumes either online or offline. Then some individual screens those resumes compares them and filters the students according to company/organization needs. Organizations generally receive a very large number of resumes, and the current manual process is bound to miss out on potential candidates who may have been the best fit for the job profile. When we develop this type of project on a Web site or a web app, we require a database and also network congestion can happen in some cases, another option is we can use AI/ML algorithms, but then they are expensive and require prerequisite data. Therefore, we use an alternative technique that is feasible financially and can be used in small-scale organizations as well as big ones. Robotic process automation involves a bot for this tedious and time-consuming work activity. Most importantly, Automation is not meant to replace recruiters, but rather to empower them to work more efficiently and make better decisions at a faster pace. Automation helps to improve the quality of results, by including more efficient candidates with more experience. One of the main aspects of automation in the recruitment process is bias-free hiring. It is completely based on the candidate's given aspect of abilities.

3 Methodology

Resume automation is carried out for three main reasons:

- Manually screening resumes is still the most time-consuming and tedious part of the recruiting process.
- Out of the total resumes obtained during a hiring process up to 88% are considered unqualified.
- An average of 23 h is spent by a recruiter while screening resumes while searching for candidates in a single hiring process (Figs. 1 and 2).

1. Applicants get to know about the new opening in the organization for a particular job profile.
2. Applicants then need to fill the web form with legal information within a specific period and these web forms would be in a predefined structure.
3. Then the filled web forms are converted into PDF format and applicants need to mail that PDF document with the subject as “CV/Resume” to the respective email id.
4. When the applicant’s documents are received by email id then the bot starts its work.
5. The bot searches the keyword CV/Resume and only under the subject CV/Resume checks the format of the document and it then further checks for the documents that are in PDF format, rest of the documents are ignored. The PDF documents are then stored in the respective folder of the employee’s machine.
6. If the subject is not mentioned as CV/Resume then that application is not considered and rejected without screening.
7. The bot then retrieves the documents one-by-one that are initially stored in a folder. The bot uses a PDF-extraction function to extract the content from the PDF. The content is then checked against the given criteria by an organization.
8. Based on criteria, the bot sorts the documents into eligible/non-eligible candidates and selects the applicants that are best suited for the job and stores their

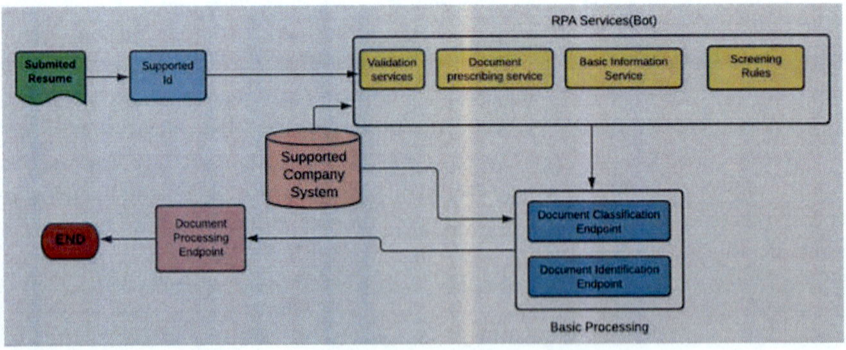

Fig. 1 System architecture diagram

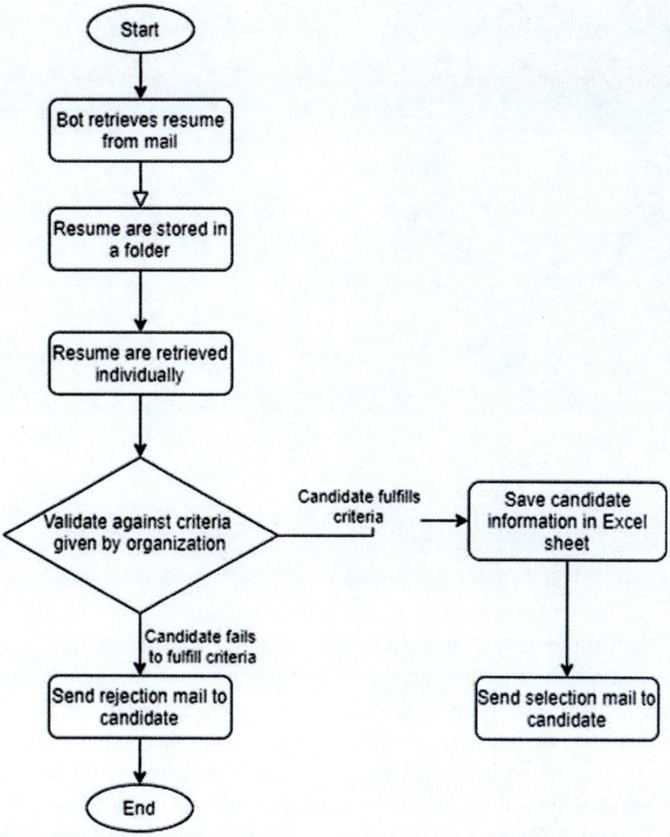

Fig. 2 Flowchart diagram

information in an excel file, and sends an email to these selected candidates, that they are selected for the interview. The excel file can also be used for further use by the organization.

3.1 Process Description in Diagrammatic Format

Figure 3 plays an important role in understanding the motive behind automating the hiring process using RPA. Whereas, Fig. 4 gives us the hint about how the process is carried out using a three-step process diagram which helps us understand the algorithm in a better way (Table 1).

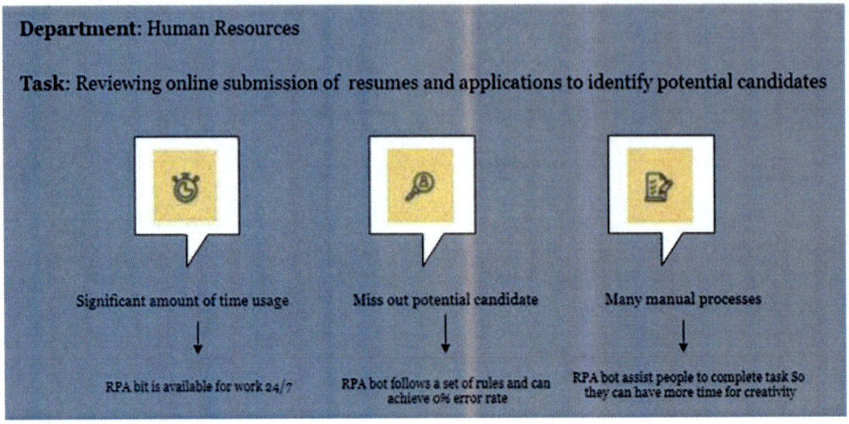

Fig. 3 Use of RPA in our project

4 Results

We performed an experiment in this and we took resumes from 30 real users and calculated the time required to process the same. We took a test of the number of documents against the time required and as we can see in the table takes an even amount of time to screen these resumes.

During this test, we have set the sample criteria that the organization would set while recruiting a java software developer fresher candidate. As the selection is for an entry-level job, the sample criteria involve the selection of a candidate based on his/her CGPA score, the projects that the candidate has designed/worked on which involves checking if the candidate has worked on any Java projects as in this case or mentioned Java in their technical skills list. So, the focus was on technical as well as non-technical skills which included his/her information about extra-curricular activities, and further the information about any educational gaps and also the candidates' date of birth for eligibility. For this, we took the help of the college Training and Placement Coordinator and set a certain threshold for every criterion. The bot was made to test the received documents in the form of PDF or docs against the sample criteria, using that we have got the result which is depicted in the below pie chart (Fig. 5) that out of the total candidates 60% of students are selected who match the criteria.

Fig. 6 displays an excel sheet which is created by the bot during the process of selection, it contains the data of selected students. After shortlisting candidates, the bot stores their email addresses and names in an excel sheet.

After storing the emails, the candidates are sent the details of the further selection process on the email by the bot. In this case, we have sent a mail containing a congratulations message on selection for round 2. The format of the email is displayed in Fig. 7.

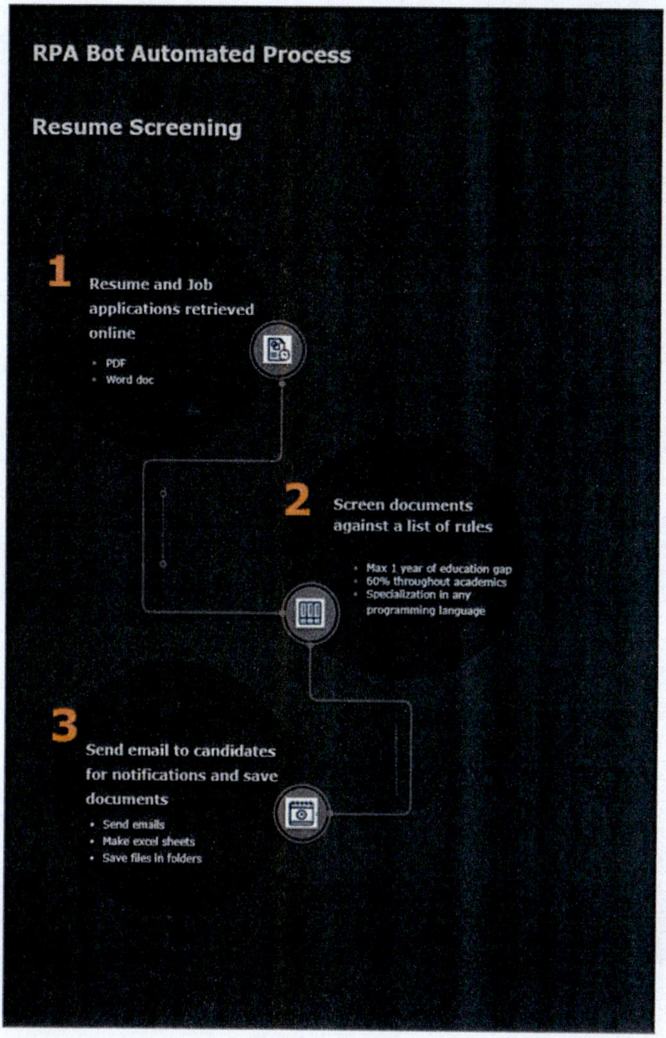

Fig. 4 Automated process flow

Table 1 Time required to process documents

No. of documents (resume/CV)	Time required (sec)
5	17.53
10	23.21
15	29.32
20	33.57
25	40.61
30	44.67

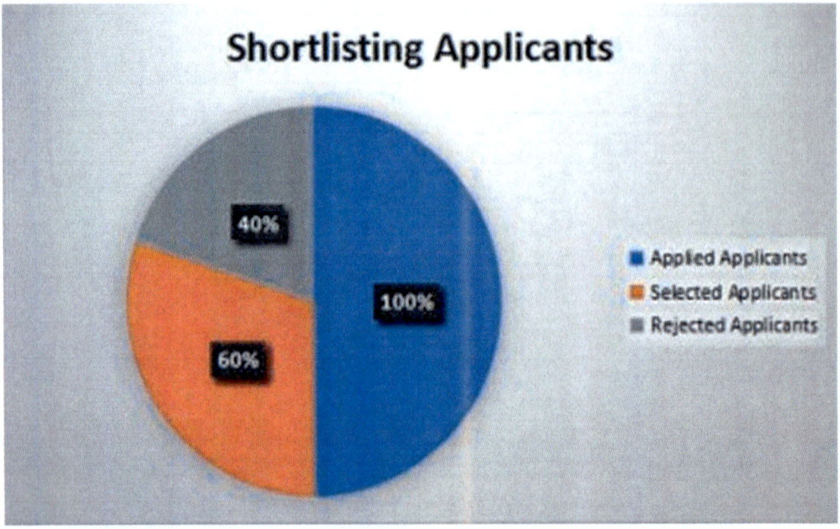

Fig. 5 Pie chart for selected/rejected candidates

FirstName	SurName	Email
Ajinkya	Sathe	ajinkyaanilsathe@gmail.com
Anushka	Tarphe	anushkatarphe16@gmail.com
Diksha	Bandagale	dikshab97@gmail.com
Divya	Hatkar	divya.hatkar@spit.ac.in
Jayesh	Pednekar	jpednekar7@gmail.com
Kirti	Pedamkar	kirtipedamkar17@gmail.com
Manish	Jha	Manish.jha@spit.ac.in
Maulika	Arekar	maulikaarekar321@gmail.com
Neha	Rane	neha.rane@spit.ac.in
Pradnyot	Morwekar	pmorwekar@gmail.com
Prajakta	Shinde	Prajakta2498shinde@gmail.com
Pranav	Soneji	pranav.soneji@spit.ac.in

Fig. 6 Excel sheet of selected students for sample criteria

5 Conclusion

The Shortlisting applicant using RPA is a very important and more sophisticated technology that would be an essential part of the upcoming days of the recruitment and hiring process. The process is rapid and does not require much manpower. So, employees can focus on their other activities. Also, applicants do not need to wait for too long to get to know the result. RPA allows companies to have a more efficient

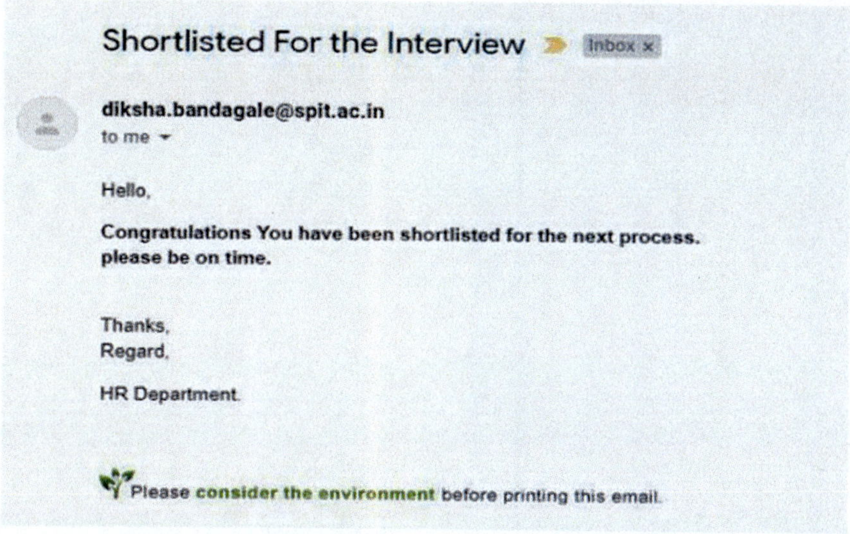

Fig. 7 An email obtained after selection

and convenient way. RPA bot is user-friendly from both the side of the recruiters and applicants. The system offers a great efficiency that will help boost applicants' engagement and experience, plus saves their time for indulgent creativity. RPA technology has a significant impact on an organization to find the right balance between talent and human interaction.

References

1. M. Deckard, Modern companies automate hr processes with RPA, uipath. Published May 2018
2. fylim, RPA use case: Using a bot for resume screening, Medium, 2021
3. L. Ivančić, D. Suša Vugec, V. Bosilj Vuksic, Robotic process automation: Systematic literature review, ResearchGate. Published August 2019. Accessed March 18, 2021
4. N. Nawaz, Robotic process automation for recruitment process. ResearchGate. Published 2019. Accessed March 18, 2021
5. M. Makadia, How is robotic process automation in contact centers. customerthink. Published April 2019
6. L. Kosonen, O.-E. Lappi, *Impacts, Benefits, and Implementation of RPA and Its Effect on Outsourced Work* (Published November, ResearchGate, 2020)
7. K.V. Rajat, V. Kumar, P. Singh, Automated order management using robotic process automation. SSRN. Published March 2020
8. A. Kannan, How email query processing works in RPA. script. Published March 2020
9. 10xDS Team, Top 10 Benefits of Robotic Process automation, January 2019 www.10xds.com
10. F.Y. Lim, RPA use case: Using a bot for resume screening, November 2019. www.fylim.mediam.com

11. J.-A. Min, How will AI impact recruitment, November 2017. www.hrzone.com
12. N. Lindqvist, A smarter labor market-Be creative, September 2020
13. A. Mitra, Sentiment analysis using machine learning approaches (Lexicon based on movie review dataset). J. Ubiquitous Comp. Commun. Technologies (UCCT) **2**(03), 145–152 (2020)

Image Enhancement and Object Recognition for Night Vision Traffic Surveillance

Monal Patel, Arvind Yadav, and Carlos Valderrama

Abstract Image processing is a vital task. Image processing is very useful in analyzing, detecting, surveillance systems, etc. The most challenging task in detection and surveillance systems is capturing and enhancing images during dark light. Nighttime low-illumination image enhancement is desired for outdoor computing vision applications. Nighttime image and video processing are crucial tasks because of the color density, low brightness, low contrast, and in addition to the high amount of noise available in the captured image or video. Hence, to overcome this challenge the Dark Channel Prior (DCP) filter is used in the proposed work. This filter is used to improve the visibility, color contrast, and brightness of the image or video captured during nighttime. Another challenge in real-time applications is the processing speed, which can be improved by implementing the algorithms on a Field-Programmable Gate Array (FPGA). In this study, an Alveo board was used to improve the processing speed. The board was officially introduced by Xilinx.

Keywords Dark channel prior (DCP) · Vitis Xilinx · Alveo board FPGA · Convolutional neural network (CNN) · Image enhancement · Infrared imaging · Object recognition

1 Introduction

Image and video processing is used to enhance images and videos captured from cameras and sensors. Various techniques and methodologies have been developed

M. Patel (✉) · A. Yadav
Department of Electronics and Communication, Parul University, Vadodara 391760, India

A. Yadav
e-mail: arvind.yadav@paruluniversity.ac.in

C. Valderrama
Department of Electronics and Microelectronics (SEMi), University of Mons, 7000 Mons, Belgium
e-mail: valderramasakuyama@umons.ac.be

G. Ranganathan et al. (eds.), *Soft Computing for Security Applications*,
Advances in Intelligent Systems and Computing 1397,
https://doi.org/10.1007/978-981-16-5301-8_52

and used in image processing during the last few decades, among which many approaches have been developed to enhance captured images for object detection in the surveillance, auto and aviation sectors, space exploration, etc. Object detection in real-time scenarios is one of the most crucial and needful tasks. The quality of images from outdoor scenes is normally degraded by turbid media such as haze, fog, and smoke. This phenomenon occurs because of atmospheric changes. In this paper, the main focus is on the enhancement of images captured during night, haze, smoke, and dark environments. The images or videos captured during the abovementioned environments possess low illumination, brightness, contrast, and poor visibility. Various studies have been conducted that contain a number of algorithms to achieve the desired results on poor images. These types of algorithms need to be optimized for object detection in real-time scenarios. Several approaches and methods such as infrared and gamma correction, Discrete Fourier Transform (DFT), and histogram equalization can be implemented to enhance nighttime captures. From many approaches and filters, the Dark Channel Prior (DCP) is used in this work because it is simple and is the fastest algorithm that can be implemented on hardware. To achieve the speed requirements, various hardware resources, such as CPUs, GPUs, FPGAs, and ASICs have been considered in the implementation of the algorithm. ASIC provides the best results in terms of speed and low power consumption for the implementation of the algorithm, but the main limitation in ASIC is to be application-specific. They cannot be remodified or reconfigured according to the requirements of the user. Hence, these boards result in high costs. An alternative to ASICs is FPGAs. Because of the limitations and cost of ASICs, FPGAs are the most favorable and efficient candidates for hardware implementation. FPGAs possess various advantages over CPU/GPU and ASICs. FPGAs provide a user-friendly environment that allows users to create and customize its architectural structure by implementing any desired function and IP blocks and removing unwanted or unnecessary functionality. These reasons lead to the use of less power and higher speed in comparison with other candidates. The main objective of this paper is to focus on accelerating the processing speed and reducing power consumption while using DCP to images and videos captured during nighttime. The schematic diagram for the enhancement of the images captured during a dark environment or night is presented below.

Figure 1, the input frame of the captured image is enhanced by using the DCP algorithm, which is implemented on an FPGA. This hardware module is placed between the camera and monitor. The enhancement algorithm and filters are optimized to be implemented on the Xilinx's ALVEO board, which is targeted by using the Xilinx unified development environment, that is, VITIS unified software tool.

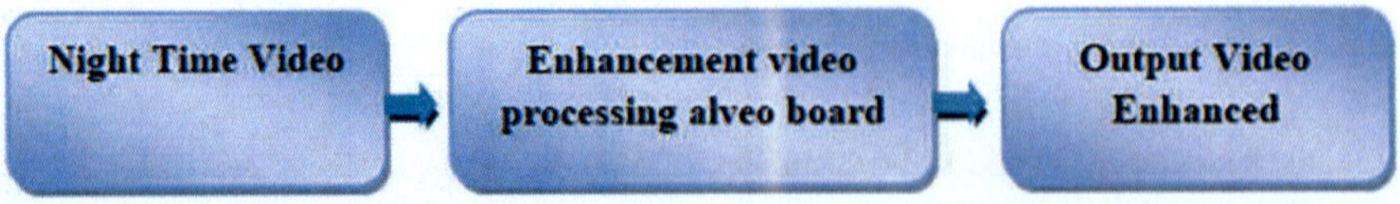

Fig. 1 Enhancement of the nighttime

VITIS is a new tool introduced by Xilinx, which combines all the aspects from Xilinx hardware as well as software in one platform for the development of application-specific programs. VITIS allows the user to develop the hardware board architecture according to the requirements by using an embedded software flow, that is, the Xilinx Software Development Kit (SDK). The Xilinx hardware development environment is very beneficial for users who want to explore more and proceed toward new technologies. The software and hardware are interconnected with each other by the help of a driver named as Xilinx runtime. VITIS also provides a new framework supporting deep and machine learning Python libraries such as Caffe and Tensorflow, which were not available in older versions of Xilinx tools such as VIVADO and SDSoC, which were used for the development of applications on FPGA. This unified development framework makes easy for the user to develop a program as per requirement. In VITIS, the application program is divided into two sections: host and kernel. The language used for hardware acceleration is C++, which is a high-level programming language used in both software and hardware for acceleration. This type of high-level programming language uses various sets of libraries to make the desired development easier for the user. Here, the C++ programming language is used for the development of a program that needs to be used in both the host and kernel.

2 State of Art

Image enhancement is the procedure of improving the quality and information content of original data before processing. A few years ago, a variety of different algorithms were used to enhance nighttime video. Those techniques are described in [1–7] [8]. Infrared is one of the techniques used to improve the quality of dark input images or video. This technique uses a non-contact device that detects infrared energy and converts it into an electronic signal, it can capture only the objects with higher temperature than their surroundings [1]. In addition to being sensitive to sunlight, this technique is expensive compared to other normal cameras. This disadvantage limits the scope of their applications and future development. We can also use the traditional image processing techniques to improve the quality of night input images or video. Histogram equalization and Gamma correction are mostly used for traditional image processing, proposed methods for removal of noise motion adaptive temporal filtering based on the Kalman structured updating. By adaptive adjustment of RGB histograms causes the increment in dynamic range of denoised video. Ultimately, the remaining unwanted factor which is noise can be removed using Non-local means (NLM) denoising. This method exploits Color Filter Array (CFA) raw data to achieve low memory consumption. The final experimental results indicate that this method is highly promising for various real-time applications to consumer digital cameras, especially CCTV and the surveillance video system [2]. Although they reported pleasing results for low lighting images and videos, they still inevitably introduce undesirable halo effects and excessive enhancement phenomena [3]. Bhagya et al. [4] proposed video enhancement using histogram equalization with the JND model. With this technique, besides being used to achieve visually pleasant enhancement effects, the over-enhancement and saturation artifacts are avoided. However, this method is complex, and it requires a lot of hardware resources. R. Peng et al. proposed a contrast stretching method which improves the quality of image by stretching the intensity range, while it stretches the low intensity pixels to lower values and high intensity pixels to higher values. As each value in the input image can have several values in the output image, objects in the original image may misplace their relative brightness values [5]. Choi et al. proposed single scale retinex which is one of the new emerging techniques in the enhancement field. The advantage of this method is the speed of execution. But there is also a limitation with Single retinex, as it deals with dynamic range enhancement and color interpretation, it struggles to do both together [6]. Rahman et al. proposed multiple scale retinex to solve the problem of Single scale retinex, but this method fails to produce good quality enhancement of input videos or images. Indeed, it finds input videos or images having high spectral characteristics in the single band [7]. Xuesong Jiang et al. proposed enhancement of dark input video based on the DCP algorithm. This method is simple and improves the quality of nighttime video or images, but it is quite slower for real-time video processing of ultra-high definition video [8]. Due to the low speed of video processing on CPU, our proposed method is to develop an accelerator for the DCP algorithm, to accelerate the video processing of nighttime, and obtain a continuous time video processing. In this technique, we have split the image into 3 windows, and we provide 3 compute units to process each window. In this manner, in addition to providing an optimal enhancement, we also improve the execution time.

3 Methodology

3.1 Dark Channel Prior

The Dark Channel Prior basically depends on various perceptions of outdoor dimness-free images. In a large portion of the non-sky locations, at least one shading channel possesses few pixels with a low, near zero, intensity. Comparably, the least intensity in such a location is almost zero. To officially portray this perception, we initially characterize the idea of a dark channel. For a self-assertive picture J, its dark channel J^{dark} is given by:

$$J^{dark}(x) = y \in \Omega(x) \min\big(c \in (r, g, b) \min J^{c}(y)\big)$$

Here, J^c is a shading channel of J, and $\Omega(x)$ is a local patch focused at x. A dark channel is the result of two least operators $min_{c\epsilon\{r,\ g,\ b\}}$ performed on every pixel, and $min_{y\epsilon\Omega(x)}$ is a base filter. The least administrators are commutative.

Utilizing the idea of a dark channel, the perception says that if J is an outside dimness-free picture, aside from the sky area, the intensity of J^{dark} is low and will, in general, be zero:

$$J^{dark} \to 0$$

This perception can be considered as a dark channel prior.

The low intensity in this type of dark channel is mostly due to three major factors: the first one is shadows, for example, the shadows of vehicles, structures, and within windows in cityscape pictures, or on the other hand, the shadows of trees, and geological components such as rocks in landscape pictures. The next factor is bright objects or surfaces, for example, any item with low reflectance in any shading channel (e.g., green grass/tree/plant, red or yellow bloom/leaf, and blue water surface) will bring about low qualities in dark channels. The last, but very important factor, is dull articles, or surfaces such as dark tree trunks and stones. As the common open air pictures are typically bright and loaded with shadows, the dark channels of these pictures are truly dim.

3.2 *Mathematical Model of DCP Algorithm*

In this paper, we mention the use of DCP algorithms to enhance the input of images from dark environments. The algorithm, which is used to enhance the dark environment image, enhances the input image based on the minimum pixel value of RGB. Here, we use the McCartney equation to recover the foggy image.

Figure 2 shows that the image captured by the sensor, which is considered as the input, is not as clear as required because of the availability of some components in an aerosol environment (the image was not clearly visible because of the particles in the air). In such a case, the camera absorbs the sunshine while capturing the image. This type of phenomenon is known as an Airlight. Most of the dark channels have

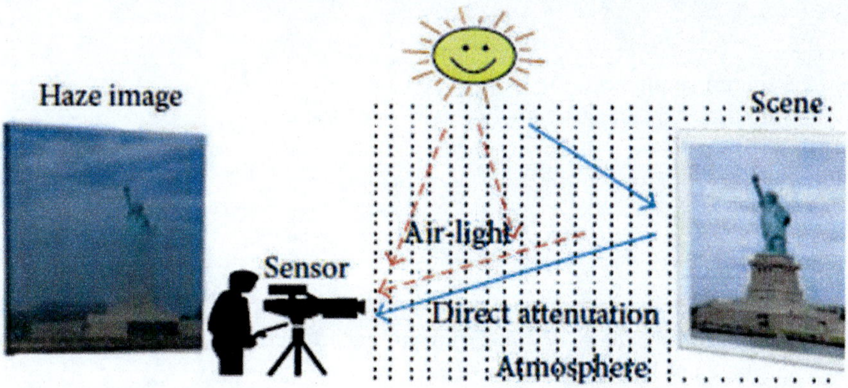

Fig. 2 Atmospheric scattering

very small amounts of intensity, and very few outdoor diness-free images deviate from the prior. Because of the additional Airlight, a hazy image is brighter than its dark version. Because of this, transmission t is very low. Hence, it can be considered that the dark channel of a hazy picture possesses a high intensity value in areas with more haze. Initially, it can be considered that the intensity of the darker channel has a more unclear approximation of the haze thickness.

Using the McCartney Equation in order to recover the enhanced resulting picture frame:

$$I_{\text{inv}}(x) = I_{\text{inv}}(x) + A(1 - t(x))$$

Here,

X is the coordinate of the image $x = (I, j)$.

$I(x)$ depicts the input frame from the dark or night environment,

$I_{\text{inv}(x)}$ is the inversion of the input frame.

$J_{\text{inv}\ (x)}$ is the scene radiance at x.

A is the Global Atmospheric Light.

$t(x)$ is the transmission medium of x.

The main goal of the algorithm is to recover J, A, and t from the inverted input image.

Using Equation in order to recover the input inverted image $J_{\text{inv}}(x)$:

$$J_{\text{inv}}(x) = \frac{I_{\text{inv}}(x) - A(1 - t)}{t(x)}$$

With:

$$t(x) = 1 - \omega I^{c}_{\text{dark}}(x)$$

$$I^{c}_{\text{dark}} = \min_{c \in (r,g,b)} \frac{I^{c}_{\text{inv}}(X)}{A^{c}}$$

Here,

ω: a control parameter.

I^{c}_{dark}: indicates a dark channel prior to inverted frames.

3.3 The Dark Channel Prior Processing Flow

The design flow of the dark channel prior algorithm is discussed here by considering the input image as a foggy or smoky image. Each function used for the hardware implementation of DCP will be described.

As shown in Fig. 3, the first stage of the processing is to provide an input of the impure image in the system. After the insertion of the image, the image is sent to

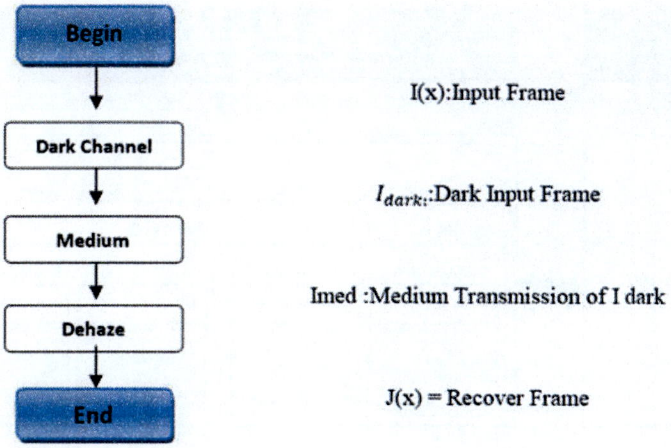

Fig. 3 The dark channel prior processing flow

the dark channel prior (DCP) module, where impurities such as smoke, haze, low or high brightness, low or high saturation, and contrast will be removed or adjusted according to the requirements, and visibility is improved. After processing the image in the DCP module, the resulting image is sent to the medium transmission, which is used to reduce the unwanted noise from an image. The VITIS Vision library plays an important role in determining medium transmission on hardware. Finally, after processing the image in the DCP and Medium Transmission modules, the resulting image is imported to the Dehaze module. The Dehaze function is used to remove the foggy or hazy impurities of the imported image using McCartney's equation.

4 Hardware, Tools and Languages

4.1 VITIS Unified Tool

Xilinx Environment introduced a new development kit known as the VITIS unified platform, which targets various FPGA boards such as Ultrascale, Alveo, and PYNQ. The tool was composed of three parts. The first is the VITIS Library (OpenCV, Ai/ML, Bias Libraries), which is used for the development of programs and kernel programs executing on hardware. It can be useful for various FPGA platforms such as Zynq, Zynq Ultrascale+, and Alveo. The second part is the VITIS Development Kit, which consists of a compiler to convert high-level code into Xilinx Object File linked with the platform for the generation of a binary file (.xclbin), an analyzer that analyzes the work done during the execution of the program, and debugger, which is used to locate the errors, occurs during the compilation of the host and kernel. The last part is Xilinx Run Time (XRT), which is a driver used to maintain communication between

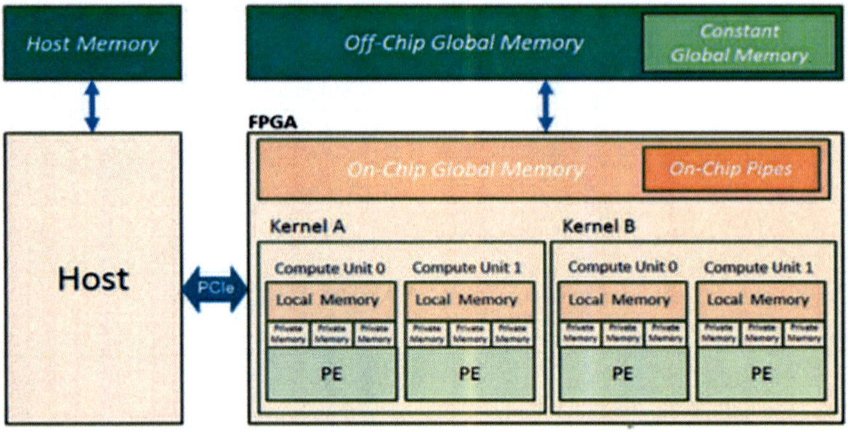

Fig. 4 Global architecture of host and Kernel

the host and kernel. This tool is useful for developing and creating an application program and executing the program on a hardware accelerator. The tool consists of two components: the host and the kernel. The host consists of an application-specified software program, and the kernel consists of a compute unit. In VITIS, APPs can be developed using pure software in which high-level programming languages such as C/C++, OpenCL, and Python are used for the development of AI application-based host code, and C++languages can be used for the development of hardware. The C++ and OpenCL languages were used because of their simplicity. OpenCL API was used for the development of the host code, and the RTL kernel was used for the development of kernel functions. In VITIS, the host and kernel use a peripheral component express for the interconnection. The host code is executed on the CPU and kernel executes in the FPGA. The global architecture of the host and kernel are represented in Fig. 4.

Figure 4 indicates that the host possesses its own memory, which can be accessible only from the host, and the kernel consists of multiple kernels, each having multiple compute units. This compute unit is responsible for the execution of the bitstream on an FPGA. It has a local memory that can only be accessed by the kernel. Global memory is also available, which is known as shared memory and can be accessible by both the host and kernel.

4.2 *OpenCL and C++ Languages*

Open Computing Language (OpenCL) is a framework that is useful for the development of application-specified programs that can be executed on various types of platforms such as CPUs, GPUs, FPGAs, and DSPs. Here, OpenCL is utilized for the development of host code and is also utilized to target the Alveo Board with the help of various commands such as creation of the context, setting the kernel, data writing in shared memory, and the execution of binary files on FPGA.

Because of the simplicity of the C/C++ language, it is very easy to use these languages to develop kernel and RTL kernel functions. In addition, these languages are also fast in comparison with other high-level programming languages, and they also provide several libraries such as OpenCV. C++ was developed in 1979 by a Danish computer scientist named Bjarne Stroustrup in a laboratory in New Jersey. The language was originally an improved version of the C language and named C with classes, but later on, it was modified as C++ in 1983. The language is statistically typed, compiled, general-purpose, case-sensitive, object-oriented, middle-level programming language.

4.3 Alveo Board

The Alveo board is an open-source FPGA board that was introduced by the Xilinx Environment in 2019. The board consists of low-level resources such as gates, flip-flops, SRAMs, and BRAMs. PCIe is used for interconnection of the board with the host. The Alveo board is used for the execution and acceleration of the program developed for enhancement of images captured during shady atmosphere or night-time because the usage of the board is very easy, as it supports the VITIS software tool used for the creation of an application program using an advanced-level languages such as C/C++ and OpenCL. Here, the source code is optimized in order to use it in the creation of the program code in VITIS, which is provided by the Xilinx developing program. Various Alveo boards, such as Alveo u-50, Alveo u-200, Alveo u-250, and Alveo u-280 from Xilinx, and these boards are different on the basis of performance and power consumption. Here, the u-200 board is used for the compilation, implementation, and optimization of the program.

To implement and utilize an Alveo board, there is a requirement for highly compatible hardware resources such as computer systems with high RAM, good graphic card storage, high power supply, multiple modes of connectivity, memory card to write a compatible image, and good Internet connection (Fig. 5).

5 Hardware Software Implementation of Dark Channel Prior Algorithm

5.1 System Architecture

In this study, the Alveo Board was accelerated by the program used to enhance the image or video captured during dark or night environments. The desired system is divided into three parts that is, The host, kernel, and global (shared) memory are

Fig. 5 Alveo board

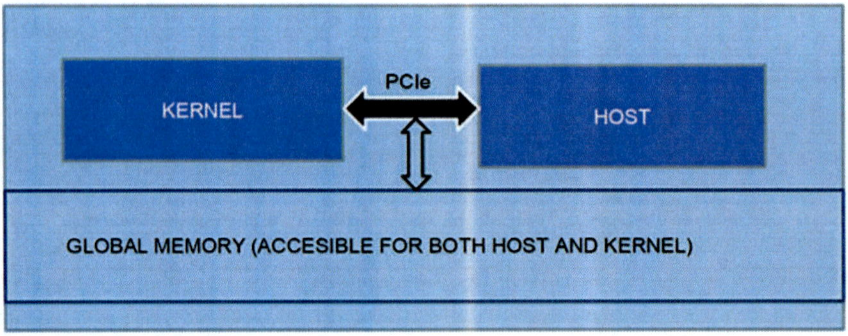

Fig. 6 System architecture

available in the FPGA. The host and kernel parts are internally connected with the help of PCIe (Fig. 6).

Here, a Dell Core i7 CPU, which is configured with 2.7 graphic memory and 16 GB of RAM along with a Linux interface, was used as the host. An Alveo u200 FPGA with 182,240 LUT, 2,364,480 FF, and 6840 BRAM was used as the platform, and the VITIS Tool was used to target and work with the Alveo FPGA. The image below represents the implementation of an Alveo board connected to the host through a peripheral component interconnect. The system was powered with the help of power cables.

5.2 Working of the Global Architecture of Proposed Method

In this global architecture, the execution of the application-based program is the responsibility of the host and kernel. In this case, the CPU acts as the host, and the FPGA acts as the kernel. Here, the input frame is written in the global memory of the connected device through the PCIe interface, and the host application is responsible for this process. After importing the image in the form of data, the host commands the kernel for execution of the Dark Channel Prior algorithm and medium transmission on the input image. After receiving, the kernel follows the command and executes the data with the help of the VITIS vision library, which leads to the outcome of the image and data. The resulting data or images are then written to the shared memory. The host again enters the process, reads the image from shared memory, executes the Dehaze function on the code, and finally writes the output enhancement in the host memory.

5.3 Design Flow of Proposed Method

The host and kernel are two top-level functions that are only and completely accessible by the develop/user. The kernel function utilizes the VITIS to execute the application-based program. The next layer contained the use of tools. These tools are used for the compilation and linking of application-based programs. However, in our scenario, both the host and the kernel were compiled separately. Here, the host uses a GCC compiler. To generate an executable file (.exe) from the high-level (.cpp). On the other hand, the kernel uses the VITIS tools compiling facility to compile the high-level function. The VITIS compiler commands the Vivado HLS tool for conversion of high-level code (.cpp) into the Xilinx object file. The object file from Xilinx is then linked to the platform, which is Alveo u200, using a C++ linker for the generation of byte streams in order to execute it on the FPGA. The finale most or the lowest layer of the program is the platform, the platform is in the hardware that can be utilized for the execution of the file, and the executable file executes on the CPU and the bitstream executes on the FPGA.

Figure 7 above presents the connection of the layers and the flow between the host and kernel in the proposed method. Along with this, there is also the utilization of the pipeline in the kernel to execute the DCP algorithm on hardware.

Figure 8 represents the dataflow of the host and kernel. The input image frames are received at the host, segmented, and divided into a small set of images having 9X9 size and these images are sent to the global memory. The frames from the global/shared memory are fetched by the kernel and the Dark Channel Prior filter, and medium transmission is executed on the data and transmitted to the host through global memory; then, the host reads the resulting image data from the global memory and executes the Dehaze function in order to recover the frame.

6 Results and Evaluation

In this study, the frames of images captured during night or dark environments were enhanced and accelerated on an Alveo board, which acts as an FPGA. The captured

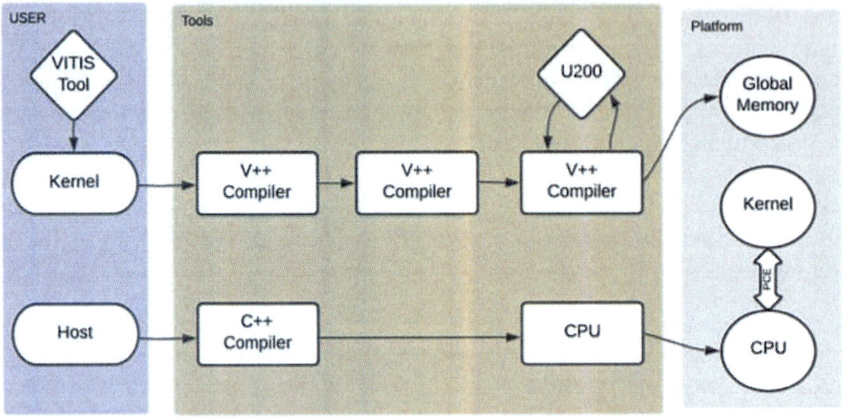

Fig. 7 Connection of the layers and flow between the host and kernel

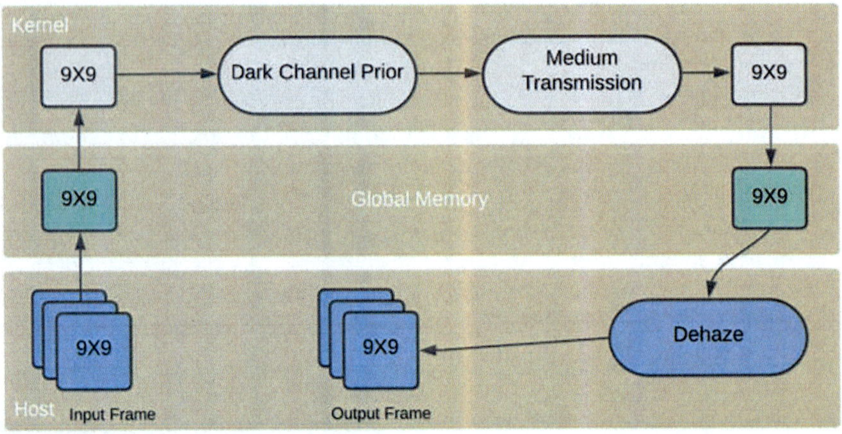

Fig. 8 Dataflow of host and the kernel

frame was enhanced or improved by using the dark channel prior algorithm, medium transmission, and haze filter. The speed of image processing using efficient FPGA resources is the main advantage of this approach. The results of the Dark Channel Prior implemented on FPGA indicate that the process is very fast in comparison with the Dark Channel Prior algorithm implemented on the CPU. The processing speed of the proposed methodology implemented on the FPGA is presented in the figure:

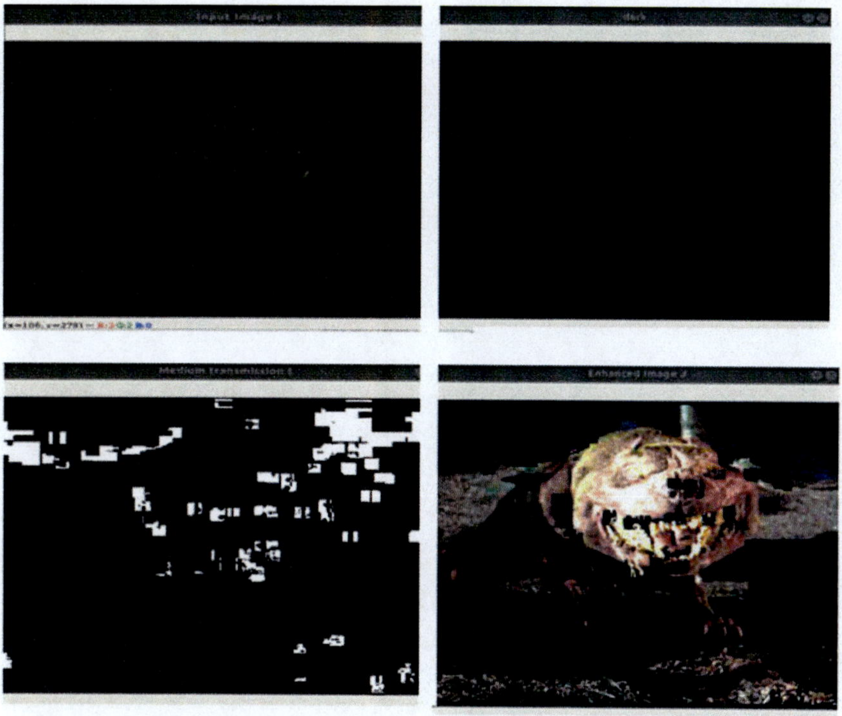

```
+---------+---------+----------+----------+---------+---------+----------+
| Latency (cycles) |  Latency (absolute)  |     Interval      | Pipeline |
|   min   |   max   |    min   |    max   |   min   |   max   |   Type   |
+---------+---------+----------+----------+---------+---------+----------+
|  8303056|  8321548| 27.674 ms | 27.736 ms |  8303044|  8321533| dataflow |
+---------+---------+----------+----------+---------+---------+----------+
```

The proposed methodology utilizes three functions for the execution of data on the computing unit. The functions are the dark channel prior algorithm, medium transmission, and haze function. The table presents the processing time of the implementation of the DCP algorithm on the CPU with different frame sizes.

Frame size	Dark channel function (in sec)	Medium transmission (in sec)	Dehaze function (in sec)
1600 × 900	15.58	15.49	0.14
1200 × 675	5.29	5.49	0.09
640 × 360	0.42	0.44	0.02
320 × 180	0.044	0.046	0.0061

The table presents a comparison of the processing time of the Dark Channel Prior algorithm implemented on the CPU and FPGA (Alveo U200) along with various image resolutions. The comparison shows that the hardware implementation that utilizes the FPGA is the most efficient and fast. Despite the various complications and difficulties that arise in the implementation, the proposed method is the most suitable.

Frame size	CPU (core i7) (in sec)	Alveo board (in ms)
1600 × 900	31.048	4.86
1200 × 675	10.71	2.763
640 × 360	0.865	0.81
320 × 180	0.087	0.23

The targeted clock frequency is 3.33 ns; clock uncertainty is 0.90 ns; and fastest achievable estimated clock frequency is 2.433 ns which belong to the Xilinx Vivado High-Level Synthesis tool. The figure presents the details of the FPGA resources utilized for the proposed methodology:

```
* Summary:
+---------------------+---------+-------+---------+---------+-----+
|         Name        | BRAM_18K| DSP48E|    FF   |   LUT   | URAM|
+---------------------+---------+-------+---------+---------+-----+
|DSP                  |        -|      -|        -|        -|    -|
|Expression           |        -|      -|        0|       46|    -|
|FIFO                 |        0|      -|      183|     1525|    -|
|Instance             |       70|     32|    15770|    47107|    0|
|Memory               |        -|      -|        -|        -|    -|
|Multiplexer          |        -|      -|        -|       72|    -|
|Register             |        -|      -|       18|        -|    -|
+---------------------+---------+-------+---------+---------+-----+
|Total                |       70|     32|    15971|    48750|    0|
+---------------------+---------+-------+---------+---------+-----+
|Available SLR        |     1440|   2280|   788160|   394080|  320|
+---------------------+---------+-------+---------+---------+-----+
|Utilization SLR (%)  |        4|      1|        2|       12|    0|
+---------------------+---------+-------+---------+---------+-----+
|Available            |     4320|   6840|  2364480|  1182240|  960|
+---------------------+---------+-------+---------+---------+-----+
|Utilization (%)      |        1|  0.4  |   0.6   |    4    |    0|
+---------------------+---------+-------+---------+---------+-----+
```

Resource type	Resource used	Total available	Ratio (%)
LUTs	48750	1182240	4
DSPs	32	6840	0.4
FFs	15971	2364480	0.6
URAMs	0	60	0
BRAM_18K	70	4320	1

7 Conclusion and Future Scope

In this paper, it is concluded that the methodologies for the improvement and enhancement of the image frames captured in dark, smoky, and hazy environments have been studied and practiced. The practical has been performed by implementing the three functions, that is, dark channel prior, medium transmission, and haze function on the CPU as well as on the FPGA, which is an Alveo board introduced by Xilinx in 2019. The CPU imposes configurability with Core i7 and 16 GB RAM and evaluates the speed of processing on frames with various resolutions. The speed of video processing captured in a dark environment with 240p, 360p, and 480p can be continuous on the CPU. However, as there is an increase in the size and quality of the input image or video, it will affect the processing speed of the CPU. Hence, the best solution is to use an FPGA as a hardware source for the implementation of the DCP function for the improvement and enhancement of the image, and it can be practiced that the improvement of the input image frame is continuous on FPGA with very little usage of resources in hardware. In addition, the benefit of using an FPGA as a hardware source is that it has low latency and fast processing speed. Here, the input frame is executed on the CPU using the Python programming language and assigned to the hardware. The experimental results show that the proposed methodology and algorithm can attain real-time image and video processing very fast compared to software processing.

Future work of the proposed work is to increase the compatibility and efficiency along with increasing the computing unit in the hardware to improve the design while minimizing the power usage utilized by the FPGA. This is a challenging task; hence, reducing the power consumption can be carried forward in future work.

References

1. R. Kapoor, R. Gupta, R. Kumar, S. Jha, Fog removal in images using improved dark channel prior and contrast limited adaptive histogram equalization. Multimedia Tools Appl. **303**(78), 23281–23307 (2019)
2. Y. Liu, W. Qin, K. Liu, F. Zhang, Z. Xiao, A dual convolution network using dark channel prior for image smoke classification. IEEE Access **7**(29), 60697–60706 (2019)
3. H.M. Khudair, T. Alawsi, A.A. Aldergazly, A.H. Majeed,Design and implementation of aerial vehicle remote sensing and surveillance system, Dehazing technique using modified dark channel prior
4. J. Chen, C.-H. Tan, L.-P. Chau,Haze removal with fusion of local and non-local statistics. IEEE International Symposium on Circuits and Systems (ISCAS) 2020, pp. 1–5
5. Q. Tang, J. Yang, X. He, W. Jia, Q. Zhang, H. Liu,Nighttime image dehazing based on Retinex and dark channel prior using Taylor series expansion. Computer Vision Image Understanding **202**, 103086 (2020)
6. A.S. Parihar, G. Gupta,A study on dark channel Prior based Image enhancement techniques. IEEE 11th International Conference on Computing, Communication and Networking Technologies (ICCCNT, 2020), pp. 1–7
7. N. WangNo, S. Pichai,Hybrid algorithm of dark channel prior and guided filter for single image dehazing. SNRU J. Science Tech. **12**, 182–189 (2020)

8. Q.H. Nguyen, B. Kang, FPGA-based Haze removal architecture using multiple-exposure fusion. J. Korean Society of Info. Tech. **18**, 85–90 (2020)
9. A. Golts, D. Freedman, M. Elad, Unsupervised single image dehazing using dark channel prior loss. IEEE Trans. Image Process. **29**, 2692–2701 (2019)
10. S. Salazar-Colores, J.-M. Ramos-Arreguín, J.-C. Pedraza-Ortega, J. Rodríguez-Reséndiz, Efficient single image dehazing by modifying the dark channel prior. EURASIP J. Image Video Processing, 66 (2019)
11. L. Unzueta, et al., Adaptive multicue background subtraction for robust vehicle counting and classification. IEEE Transactions on Intelligent Transportation Systems **13**(2), 527–540 (2012)
12. S. Sivarama, M.M. Trivedi, Deep Neural Network. IEEE Transactions on Intelligent Transportation Systems **14**(2) (2013)
13. Y. Tang, C. Zhang, L. Peng, B. Yang, Vehicle detection and recognition for intelligent traffic surveillance systems. Multimedia Tools Appl. **76**(4), 5817–5832 (2017)
14. A. Hampapur, L. Brown, J. Connell, A. Ekin, N. Haas, M. Lu, H. Merkl, S. Pankanti, Smart video surveillance: exploring the concept of multiscale spatiotemporal tracking. IEEE Signal Process. Mag. **22**(2), 38–51 (2005)
15. P. Chakraborty, Y.O. Adu-Gyamfi, S. Poddar, V. Ahsani, A. Sharma, S. Sarkar, Traffic congestion detection from camera images using deep convolutional neural networks. Transp. Res. Rec. **2672**(45), 222–231 (2018)
16. K. He, G. Gkioxari, P. Dollár, R. Girshick, In Mask r-CNN. Proceedings of the IEEE international conference on computer vision **2017**, 2961–2969 (2017)
17. V. Mandal, in *Artificial Intelligence Enabled Automatic Traffic Monitoring System* (University of Missouri, Columbia, 2019)

Cyberbullying Detection in Native Languages

Sakshi Shetty, Jesica Dsouza, Ameaza Rodrigues, and Monali Shetty

Abstract Cyberbullying can manifest in various forms like cyber-stalking, identity theft, etc. Teenagers are more prone to be victims of cyberbullying since they spend a significant amount of their time daily, engaging with various social media entities. UNESCO drew attention to this issue by producing case studies from several geographically distant countries. These studies show that the urgency of countering cyberbullying cannot be gainsaid and is a subject of human rights. Similar to bullying, cyberbullying affects a person's mental well-being, often leading to low self-esteem, self-harm, anxiety, depression, suicidal thoughts, etc. Due to the widespread instances of cyberbullying and harassment, combating cyberbullying is a pressing issue. We have completed a thorough literature review of Cyberbullying. The literature study concluded that social media spaces are rife with instances of harassment and bullying. There is a paucity of mere documentation of cyberbullying, let alone actions to prevent it. Of late, researchers are in favor of countering cyberbullying through automation. Machine Learning models have been designed to detect bullying and harassment through text in comments. Nevertheless, these systems are lacking in terms of texts/comments that are available in regional scripts. Training machine learning models with regional languages shall make them more efficient in countries like India, where various regional languages, differing in scripts, are used. Hindi has around 73 million native speakers (1.1% of the world's population). In this paper, we have designed automated systems for Hindi that will significantly impact the safety of online communication spaces.

Keywords Support vector machines (SVM) · Cyberbullying · Social media · Machine learning · Supervised learning · Naive Bayes · Attacks · Random forest · Unsupervised learning · Native languages · Classification · Data preprocessing · Feature selection · Confusion matrix · Accuracy

S. Shetty (✉) · J. Dsouza · A. Rodrigues · M. Shetty
Fr. Conceicao Rodrigues College of Engineering, Bandra, Mumbai, India

M. Shetty
e-mail: monali_shetty@fragnel.edu.in

G. Ranganathan et al. (eds.), *Soft Computing for Security Applications*,
Advances in Intelligent Systems and Computing 1397,
https://doi.org/10.1007/978-981-16-5301-8_53

1 Introduction

Cyberbullying is the act of humiliating, scaring, or causing danger to other people through electronic devices and online networks. Bullies, if left unquestioned of their behavior, can normalize such behavior leading to rampant abuse. Automatic detection and cyberbullying reporting can help keep a check on individuals occupying online spaces and induce a sense of responsibility toward their behavior. Implementation of this method of countering cyberbullying shall require automation of classification of individuals' various activities as offensive or not offensive. Since the most common form of bullying is mean comments, classifying text according to the conveyed sentiments can help bring bullies to book. For this purpose, machine learning (ML) models can be trained and employed.

2 Background

1. **Cyberbullying**
   Cyberbullying prevails in most countries across the world. The paradigm shift within the Internet of Things (IoT) around ten years back resulted in tremendous growth within the range of active social media users. Nowadays this range has exceeded over 3 billion. Cyberbullying takes a toll on a person's mental and emotional health, brought to light in Hinduja et al. [1].
   Traditionally, guidelines and protocols have been drafted, but they involve the victim reporting the abuse to their parents or guardians. This approach of dealing with cyberbullying has not proved very efficient since most victims do not report the incidents. Therefore, automated cyberbullying detection is of immense importance.
2. **Data Availability**
   The increase in the consumption of internet resources in our country might even be attributed to the "Jio effect." Reliance Jio arrived in 2016 with almost negligible cost information plans compared to the already existing plans. The network gained many customers and turned into the most important mobile operator globally with over three hundred million subscribers. Reliance Jio's client base is regarded as massive as the entire population of the USA to place this in context. This also means that the number of engagements people have in social media spaces has increased manifold (Big Data. Era) [2], and many scalable data can be employed to train Machine Learning models.
3. **Machine Learning**
   "ML is the science of getting computers to learn and function like humans do, and enhance their learning over time in an independent fashion, by providing them information and data in the form of observations and real-world interactions" [3]. Computational methodology such as graph mining and relational learning

has been in use for business-centered Data analytics. Reorientation of these methods can be instrumental in mitigating cyberbullying, if not cease it.
Most of the time, the focus while learning Machine Learning is placed on choosing the best algorithm, but equal emphasis should be placed on the data used to train the models.
We tend to propose a possible answer to detect these cyberbullying instances, which contain detrimental and disparaging text or sentences.

3 Related Works

Owing to the importance of this topic, this is a widely covered topic both online and offline. Many research studies are done in psychology, social sciences, and research to detect cyberbullying using current technologies, and testament to the above statement is the abundance of research papers available regarding this topic. This section includes a discussion of the prevalence and seriousness of cyberbullying.

Kontostathis et al. [4] reviewed data from Formspring.me, and manually labeled data using Amazon's Mechanical Turk. They implemented text mining to detect vandalism, spam, internet abuse, and cyberterrorism.

Dinakar et al. [5] at MIT conducted a supervised machine learning approach to develop a model to identify cyberbullying about sexuality, race & culture, intelligence. They collected Youtube comments and manually labeled them, and implemented various binary and multiclass classifiers. Their study revealed that binary classifiers outdid the multiclass classifier. The authors used SVM learner and reached an accuracy of 66.7%.

Yin et al. [6] used supervised learning to develop a model to detect harassment. They employed a bag of words approach based on parameters such as content, sentiment, and contextual attributes of data to train an SVM classifier and reached a recall level of 61.9%

Dadavar et al. [7] analyzed the data from MySpace, a platform that offers an interactive, user-submitted community of friends with blogs, personal profiles, etc., and specifically focused on going down the route of a gender-based approach. To train a gender-based classifier, they used an SVM model. The dataset consists of 381.000 posts. The result obtained by the gender-based approved baseline by 15% in F-measure, 39% in precision, and 6% in the recall.

Nandhini et al. [8] used Naive Bayes machine learning algorithm. They used the dataset from MySpace, and they achieved an accuracy of 91%. They then proposed another model [9], Naive Bayes Classifier, and genetic operation (FuzGen), and they reached 87% accuracy.

Isa et al. [10] retrieved dataset from Kaggle. They employed two classifiers, namely, SVM and Naive Bayes. They scored an accuracy of 92.81% with Naive Bayes, while with the SVM poly kernel; they achieved an accuracy of 97.11%.

Haidar et al. [11] put forward a model to discern cyberbullying by using the Arabic language. They used SVM and achieved 94.1% precision while with Naive Bayes they achieved 90.85% precision.

Chavan et al. [12] implemented support vector machines (SVM) and logistic regression. With logistic regression, they attained 60% recall and 73.76% accuracy, and 64.4% precision. While with SVM, they achieved 70% precision, 77.65% accuracy, and 58% recall.

Eckert et al. [13] presented the work for detection of cyberbullying incidents in social media spaces using DNN-based models.

H. Rosa et al. [14] produced a detailed study of research which focused on automatic detection of cyberbullying in the course of a quantitative systematic review approach. We also inspected this approach with a detailed experiment to examine current implementation by using feature engineering.

Some papers also mentioned deep learning and neural networks as a way to detect cyberbullying. Zhang et al. [15] used innovative pronunciation-based convolution neural networks (PCNN). They used 1300 messages from Formspring.me. and 1313messages from Twitter. They achieved 96% accuracy, a precision of 56%, and 78% recall.

Parime et al. [16] used MySpace as their dataset source and manually marked them. They employed SVM as their classifier.

Capua et al. [17] used GHSOM networks to be well suited for extensive collections of documents that need to be classified. They used a dataset from Kontostathis et al. [4]. They achieved 72% precision. 73% accuracy and 71% on F1 using GHSOM. Using C4.5, they earned 60% precision and 67% recall. Using SVM, they received 67% recall on this dataset. They also used a dataset from Dadavar et al. [7]. Using GHSOM, they achieved 60%precision, 69% accuracy, 94% recall.

Salawu et al. [18] conducted a unique survey under the umbrella of cyberbullying detection.

The studies mentioned above highlighted a significant gap in detecting cyberbullying. Most of them are likely to stick to a dataset of one language, i.e., English.

4 Methodology

In this model, three algorithms are used to accomplish the detection of Cyberbullying. They are as follows: Random Forest, Naive Bayes, and unsupervised machine learning algorithm SVM (Fig. 1).

The proposed model has been trained and tested for various English and Hindi languages datasets to attain the below-mentioned performance.

A. **Data Collection**: We have collected datasets from dhavapotdar/cyberbullying detection, which provided datasets in the English Language, consisting of around 30,000 Facebook comments and kmi-linguistics/track-1 datasets in

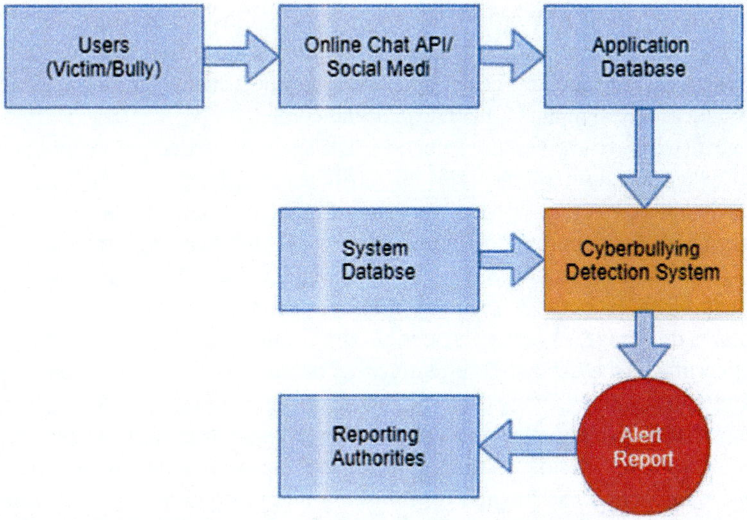

Fig. 1 Overview of the methodology

Hindi Language consisting of approximately 15,000 tweets on github.com. The Hindi dataset consists of data in the Hindi language and Hindi words transliterated using the English language. The datasets consisted of various comments classified as either cyberbullying or not. The classification classes were skewed and distributed such that 12,384 comment entities out of 26,133 fell under the label of cyberbullying, while 13,749 belonged to the other type for one English dataset. The dataset consists of 2 columns. The first column consists of text and the second column consists of labels. The dataset was labeled 1: offensive texts and 0 for non-offensive texts. 80% of the dataset was used for training the model, and the remaining 20% was used for testing the model.

B. **Data Preprocessing**: The preprocessing step is very significant in cyberbullying detection. It consists of cleansing of texts (e.g., removing punctuation marks and stop words) and removing content classified as spam. The planned model has been applied to get rid of and clean irrelevant noise in text detection. In this model, stop words, memorable characters, and removed perennial words. Then, the remaining words were stemmed from their original roots. This preprocessing and the dataset containing clean tweets are created for the planned model to be trained and tested.

C. **Feature Extraction**: Feature extraction is a significant step for text classification in the detection of cyberbullying. The available data is refashioned into an appropriate format viable to feed into machine learning algorithms. Within the planned model, a Count vectorizer is used. Wherein we tend to extract the characteristics of the inputted data and place them in a features list. Vectorization is a standard method of transforming a set of text documents into numerical feature vectors. This particular strategy (tokenization, enumeration, and normalization) is named the Bag of Words or "Bag of n-grams" illustration. The essential plan of the Count vectorizer is that it converts a set of textual files to a matrix of token counts, i.e., the occurrences of tokens in every document. This implementation produces an exemplary illustration of the counts.

D. **Feature Selection***:* Feature selection, as a spatiality reduction method, aiming to select a subset of the relevant attributes and features from the existing attributes and features by getting rid of orthogonal, redundant, or noise-creating attributes and features. It will typically cause better learning performance, i.e., higher learning accuracy, reduced procedure value, and higher model interpretability. Recently, researchers from computer vision and text mining have planned a range of feature selection algorithms, and in terms of theory and experiment, they show the effectiveness of their works. The current paper is aimed toward reviewing state of the art on techniques. Moreover, a radical experiment is conducted to visualize if feature selection will improve learning performance, considering a number of the approaches mentioned within the literature. The experimental results show that unsupervised feature selection algorithms edge machine learning tasks, raising agglomeration performance.

E. **Classification**: In this paper, various classifiers have been used to classify whether a particular tweet is an instance of cyberbullying or not. The different classifiers used are SVM, Naive Bayes, and Random Forest. These classifiers have been discussed in Sect. 5, and the evaluation of their performance is carried out in Sect. 7.

F. **Machine Learning in Cyberbullying Detection**: Machine learning is a data analysis methodology that automates analytical model building. It is a branch of AI that supports the concept where systems will learn from information, determine patterns and build selections with minimum human intervention. Computational methodology such as graph mining and relational learning has been in use for business-centered Data analytics. Reorientation of these methods can be instrumental in mitigating cyberbullying, if not cease it.

5 Algorithms and Techniques

1. **Naive Bayes**: This algorithm is based on Bayes' Theorem with a statement of independence among predictors.
   Bayes theorem formulated by Thomas Bayes calculates the probability of an event occurring based on prior knowledge related to the event. It is based on the formula (1).

$$P(A|B) = P(A) * P(B|A) / P(B) \tag{1}$$

   Multinomial Naïve Bayes (Multinomial NB) is popularly used for Natural Language Processing (NLP).NB was used to detect instances of cyberbullying in many cases. This model assumes that a parametric model produces the text

	label	Text
0	1	Mahmood Ghaznavi Aor ABdali ko bhol gaya ha tu...
1	1	Bhai 60sal pehle desh me kya tha pehle pta kro...
2	1	chutiya friday ko isliye releae krte kyoki wee...
3	1	जय मोदीराज
4	1	UPA walo ne bahot kuch kr diya tha desh k liye

Fig. 2 Annotated dataset in Hindi

and uses training data to determine Bayes-optimal parameter estimates of the model.

2. **Random Forest**: Random Forest is an ensemble of decision trees generally trained via the bagging method. It can be used for both classification and regression. RF gives good accuracy compared to other classification methods, with overfitting not being a concern while using Random Forest.
3. **Support Vector Machine**: SVM may also be referred to as a supervised machine learning algorithm used for classification challenges that provide higher accuracy in the cyberbullying detection project.
   Support Vector Machine is a versatile machine learning model capable of performing linear and non-linear classification.
   The main idea behind SVM is that the two classes are linearly separable. We can think of SVM as fitting the widest possible street between the two classes. SVM classifiers work efficiently even if the classes can be separated linearly. Specific kernels are implemented in the SVM to turn the function space appropriately. Kernelized SVM models help us to achieve this (Figs. 2, 3, 4, 5, and 6).

6 Implementation

7 Experimental Results

Primarily, the model was trained, thus tested, for two different datasets in the English language, each consisting of entries around 30,000 tweets. This resulted in the accuracies of around 91%.

Later, to extricate the drawback of an insufficient dataset, we merged the existing datasets in English (total entries around 60,000 tweets), which realized the accuracies as 92.16% through Naive Bayes, 93.76% through Random Forest, and 94.51% through SVM.

Furthermore, to upgrade the model and outperform the existing systems, the model was trained and tested for a Hindi language dataset (around 15,000 tweets).

```
0        Mahmood Ghaznavi Aor ABdali ko bhol gaya ha tu...
1        Bhai 60sal pehle desh me kya tha pehle pta kro...
2        chutiya friday ko isliye releae krte kyoki wee...
3                                               जय मोदीराज
4           UPA walo ne bahot kuch kr diya tha desh k liye
                                 ...
11995    अगला लोकसभा चुनाव केवल 2 पार्टियाँ लड़ेंगी ।\nए...
11996    Kisi dushre desho ki soldier k sar katke lana ...
11997                                     Kyoki wo Hindu h
11998    mileage ki tension na lo, India me 40 k upar n...
11999    Gulamo Ki baat ka bura nahi mantey Bhai Log......
Name: Text, Length: 12000, dtype: object
```

Fig. 3 Comments from dataset in Hindi

```
print("Accuracy of Naives Bayes : ",mnb.score(x_testcv,y_test))

    Accuracy of Naives Bayes :  0.9007200523674449
```

Fig. 4 Naive Bayes algorithm performance

```
rf = RandomForestClassifier()
rf.fit(x_traincv,y_train)

    RandomForestClassifier(bootstrap=True, ccp_alpha=0.0, class_weight=None,
                           criterion='gini', max_depth=None, max_features='auto',
                           max_leaf_nodes=None, max_samples=None,
                           min_impurity_decrease=0.0, min_impurity_split=None,
                           min_samples_leaf=1, min_samples_split=2,
                           min_weight_fraction_leaf=0.0, n_estimators=100,
                           n_jobs=None, oob_score=False, random_state=None,
                           verbose=0, warm_start=False)
```

Fig. 5 Random forest algorithm performance

```
model = svm.SVC()
model.fit(x_traincv,y_train)

    SVC(C=1.0, break_ties=False, cache_size=200, class_weight=None, coef0=0.0,
        decision_function_shape='ovr', degree=3, gamma='scale', kernel='rbf',
        max_iter=-1, probability=False, random_state=None, shrinking=True,
        tol=0.001, verbose=False)
```

Fig. 6 SVM algorithm performance

Consequently, the following accuracies were achieved: 79.37% through Naive Bayes, 83.45% through SVM, and 85.45% through Random Forest. However, the existing models conferred the accuracy between 50 and 60% only.

Table 1 Performance of the proposed methodology

S. no	Dataset (language & entities)	Accuracy (%)
1	English (30,000)	Around 91
2	More than one English dataset (60,000)	NB—92.16 RF—93.76 SVM—94.51
3	Hindi (15,000)	NB—79.37 RF—85.45 SVM—83.45
4	English + Hindi (70,000)	NB—90.07 RF—91.69 SVM—92.44

In the final analysis, as shown in Table 1, the model was trained and tested for multilingual datasets; wherein, all the datasets as mentioned above were combined, emanating a solid dataset consisting of tweet entries of around 70,000. The accuracy obtained for the same was 90.07% through Naive Bayes, 91.69% through Random Forest, and 92.44% through SVM. It should be noted that the existing models with insufficient datasets provide an accuracy of around 80% for multilingual purposes.

For English language, Table 2 gives the classification report.

Table 3 Shows the confusion matrix for the Hindi language. From the table, true positive values are good, and false positive values are less in SVM than NB and RF

For Hindi and English language, Table 4 gives the classification report.

As shown in Table 5, both the precision and recall for English vocabulary were very high in SVM, leading to a good f1-score and a high accuracy of around 95%.

For more than one English dataset (refer Table 6) had highest accuracy with Linear SVC.

For the Hindi dataset (refer to Table 7), RF had the highest accuracy and lowest for NB.

Table 2 Confusion matrix of various algorithms for English language dataset

Performance criteria	NB	SVM	RF
True positive	457	723	521
True negative	4064	3990	4068
False positive	376	110	312
False negative	60	134	56

Table 3 Confusion matrix of various algorithms for Hindi language dataset

Performance criteria	NB	SVM	RF
True positive	145	215	146
True negative	1760	1762	1907
false positive	309	239	308
False negative	186	184	39

Table 4 Confusion matrix of various algorithms for Hindi and English language datasets

Performance criteria	NB	SVM	RF
True positive	6574	6762	6631
True negative	5827	5875	6002
False positive	650	462	593
False negative	698	650	523

For the Hindi + English dataset (Refer Table 8), RF had better accuracy compared to SVC and NB.

8 Conclusion

Although this is a very well-researched area evident from the abundance of research papers in the field: Detection of cyberbullying, existing research has not attained satisfactory accuracy. The proposed model has utilized various algorithms, namely, supervised machine learning algorithm SVM, Naive Bayes, and Random Forest.

To maintain the efficiency of these systems, one must consider rapidly evolving vocabulary and behavioral patterns. Attention needs to be paid to cultural nuances to classify a particular behavior as offensive or not offensive successfully; former research tested most of the work done in this field for datasets in the English language only. Therefore, we have also tried to expand this research by detecting cyberbullying in other native languages such as Hindi. However, a significant hindrance while doing so was the unavailability of a sufficient dataset. Hence, we have successfully attained utmost accuracies for the combination of datasets in English and Hindi. This model provides 92% for Hindi and English, 86% accuracy for Hindi language and 95% for English language.

Table 5 Performance of algorithms with English dataset

	Algorithm	Accuracy:test	Precision:test	Recall:test	F1 Score:test	Prediction time	Accuracy:train	Precision:train	Recall:train	F1 Score:train	Training:time
0	LinearSVC	0.950777	0.973171	0.967507	0.970331	0.008949	0.999596	0.999697	0.999818	0.999758	0.218552
1	Random foresl classifier	0.926568	0.930009	0.985936	0.957156	1.202811	1.000000	1.000000	1.000000	1.000000	13.404802
2	MultinomialNB	0.912044	0.915315	0.985451	0.949089	0.003016	0.956270	0.959163	0.989573	0.974131	0.007962

Table 6 Performance of algorithms with more than one English dataset

	Algorithm	Accuracy:test	Precision:test	Recall:ttest	FI Score: test	Prediction:time	Accuracy:train	Precision:train	Recall:train	FI Score: train	Training:time
0	LinearSVC	0.948277	0.949141	0.923720	0.936258	0.005054	0.998216	0.997966	0.997582	0.997774	4.452245
1	Random forest classifier	0.937880	0.933290	0.914292	0.923693	6.093843	0.999978	1.000000	0.999945	0.999973	146.85774E
2	MultinomialNB	0.921667	0.888204	0.926077	0.906745	0.010883	0.943850	0.913085	0.950371	0.931355	0.029531

Table 7 Performance of algorithm with Hindi dataset

	Algorithm	Accuracy:test	Precision:test	Recall:test	F1 Score: test	Prediction:time	Accuracy:train	Precision:train	Recall:train	F1 Score: train	Training:time
0	Random forest classifier	0.857083	0.862834	0.979445	0.917449	1 904,548	0.997500	0.997945	0.998972	0.998458	30.410.612
1	LinearSVC	0.823750	0.880180	Q.9Q5961	0.892884	0.000000	0.991146	0.991692	0.997429	0.994552	1.039274
2	MultinomialNB	0.793750	0.850652	0.904419	0.876.712	0.002.185	0.835938	0.885533	0.915927	0.900474	0.010218

Table 8 Performance of dataset with English + Hindi dataset

	Algorithm	Accuracy:test	Precision:test	Recall:test	F1 Score: test	Prediction:time	Accuracy:train	Precision:train	Recall:train	F1 Score: train	Training:time
0	Random forest classifier	0.919703	0.909009	0.923218	0.916058	7.696995	0.999582	0.999348	0.999770	0.999559	293.000310
1	LinearSVC	0.919121	0.927095	0.900383	0.913544	0.007336	0.995982	0.996694	0.994820	0.995756	6.546328
2	MultinomialNB	0.901957	0.899645	0.893027	0.896324	0.021261	0.921867	0.919477	0.915279	0.917373	0.033106

References

1. S. Hinduja, J. Patchin, Offline consequences of online victimization: School violence and delinquency. J. Sch. Violence **6**(3), 89–112 (2007)
2. X. Wu et al., Data mining with big data. IEEE Trans. Knowl. Data Eng. **26**(1), 97–107 (2014)
3. https://emerj.com/ai-glossary-terms/what-is-machine-learning/
4. A. Kontostathis, K. Reynolds, A. Garron, L. Edwards., Detecting cyberbullying: query terms and techniques. Proceedings of the 5th WebSci 2013. ACM, New York (2013)
5. K. Dinakar, R. Reichart, H. Lieberman, Modeling the detection of textual cyberbullying. MIT, International Conference on Weblog and Social Media. Barcelona, Spain (2011)
6. D. Yin, B. D. Davison, Z. Xue, L. Hong, A. Kontostathis, L. Edwards, Detection of Harassment on Web 2.0. Proceedings of CAW2.0 Workshop (2009)
7. M. Dadvar, R.B. Trieschnigg, F.M.G. de Jong, Experts and machines against bullies: A Hybrid approach to detect cyberbullies. 27th Canadian Conference on Artificial Intelligence, University of Waterloo, Montral, Canada (2014)
8. B Nandhini, J.I. Sheeba, Cyberbullying detection and classification using information retrieval algorithms. Proceedings of the 2015 International Conference on Advanced Research in Computer Science Engineering & Technology (ICARCSET 2015), p. 20
9. B Sri Nandhini, J.I. Sheeba, Online social network bullying detection using intelligence techniques. Procedia Computer Sci. **45**, 485–492 (2015)
10. S.M. Isa, L. Ashianti et al., Cyberbullying classification using text mining. In Informatics and Computational Sciences (ICICoS), 2017. 1st International Conference (IEEE, 2017), pp 241–246
11. B. Haidar, M. Chamoun, A. Serhrouchni, A multilingual system for cyberbullying detection: Arabic content detection using machine learning. Adv. Sci. Tech. Engineering Syst. J. **2**(6), 275–284 (2017)
12. V.S. Chavan, S.S. Shylaja, Machine learning approach for detection of cyber-aggressive comments by peers on social media networks. Advances in computing, communications, and informatics (ICACCI), 2015 International Conference (IEEE, 2015), pp. 2354–2358
13. M. Dadvar, K. Eckert, Cyberbullying detection in social networks using deep learning based models; A reproducibility study. Web-based Information Systems and Services, Stuttgart Media University Nebenstrasse 8, 70569 Stuttgart, Germany. Dadvar. @hdm-stuttgart.de, eckert@hdm-stuttgart.de
14. H. Rosaa, N. Pereirac, R. Ribeiroa, P.C. Ferreiraa, J.P. Carvalho, S. Oliveirac, L. Coheura, P. Paulinod, A.M. Veiga Simãoc, I. Trancosoa, Automatic cyberbullying detection: A systematic review. Computers in Human Behavior
15. X. Zhang, J. Tong, N. Vishwamitra, E. Whittaker, J. P Mazer, R. Kowalski, H. Hu, F. Luo, J. Macbeth, E. Dillon, Cyberbullying detection with a pronunciation-based convolutional neural network. 2016 15th IEEE International Conference on Machine Learning and Applications (ICMLA) (IEEE, 2016), pp. 740–745
16. S. Parime, V. Suri. Cyberbullying detection and prevention: Data mining and psychological perspective. In-Circuit, Power and Computing Technologies (ICCPCT). 2014 International Conference (IEEE, 2014), pp. 1541–1547
17. M. Di Capua, E. Di Nardo, A. Petrosino, Unsupervised Cyber Bullying Detection in Social Networks from 2016 23rd International Conference on Pattern Recognition (ICPR) Cancún Center, Cancún, México, December 4–8 (2016)
18. S. Salawu, Y. He, J. Lumsden, Approaches to automated detection of cyberbullying: A survey. IEEE

Intrusion Detection Framework Using an Improved Deep Reinforcement Learning Technique for IoT Network

P. May raju and Govind P. Gupta

Abstract With the exponential increase in usage of the Internet-based services in various smart cities, and smart environment applications, several cyber-attacks are reported in recent years in IoT network deployed for smart monitoring and observation. In order to protect the Internet of Things (IoT) network, there is a need to design a robust and fast intrusion detection system to protect and identify the malicious attacks in time. However, in previous frameworks analysts not used IoT environment and used old datasets which does not contain the IoT traces. This paper proposed an intrusion detection framework using an improved deep reinforcement learning (DRL) model with parameter tuning with dataset which contains the IoT traces, and it is a MedBIoT dataset. Performance of the proposed DRL-based IDS framework is compared with logistic regression and Naive Bayes model in terms of evaluation metrics. Experimental results confirm the superiority of the proposed model with accuracy of 96.99%.

Keywords Intrusion detection · IoT network · Deep reinforcement learning · Machine learning

1 Introduction

In recent years, exponential increase in the deployment of the IoT devices in different applications in smart city, smart grid, smart home environments motivates to consider security and privacy issues.

Due to low-cost and resource limitation of the IoT devices, it is very prone to easily compromise by the malicious attackers. Thus, to protect the IoT-based applications, there is need to design an efficient and fast intrusion detection system (IDS) for IoT

P. May raju (✉) · G. P. Gupta
Department of Information Technology, National Institute of Technology Raipur, Raipur, Chhattisgarh, India

G. P. Gupta
e-mail: gpgupta.it@nitrr.ac.in

G. Ranganathan et al. (eds.), *Soft Computing for Security Applications*, Advances in Intelligent Systems and Computing 1397,
https://doi.org/10.1007/978-981-16-5301-8_54

system to detect and identifies the threats [1–6]. IDS for IoT network can be classified into two groups such as signature-based detection and anomaly-based detection. In signature-based detection, threats are known, and signature of these known threats is maintained in a database, and in order to detect the threats, traffic patterns are searched and matched with the existing threats pattern. In the case of new threat patterns, the signature-based detection scheme is unable to identify. However, anomaly-based detection is opposite of the signature-based detection and is able to detect the new threat patterns, whenever it enters into the IoT network.

Most of the existing IDS framework uses supervised and unsupervised machine learning (ML) model for detection and identification of the threats. In addition, deep learning techniques is also used for detection of intrusions in IoT network. Due to resource constrains and rapid changing nature of IoT devices and IoT gateways, this approach does not provide light weight solutions. In addition, most of the existing IDSs can be able to detect limited type of threats, and these approaches suffer from high computational complexity, require large volume of dataset, and take a lot of time in training the model. Thus, there is a need to design a lightweight IDS system for IoT network which not only easily to deployed at IoT gateways, but also provide efficient detection accuracy.

To provide an intelligent IDS framework for IoT network, this paper proposes a novel IDS framework using reinforcement learning-based technique which can be able to detect various threats automatically by hypertuning of the detection threshold. To validate the efficiency of the proposed framework, a novel IoT Botnet dataset is used, known as MedBIoT dataset [xxx]. Performance of the proposed DRL-based IDS framework is compared with Logistic regression and Naive Bayes model in terms of accuracy, precision, recall, and F1 score. Experimental results confirm the superiority of the proposed model with accuracy of 96.99%.

This research paper is organized as follows: Section 2 discusses the related work. Section 3 presents the proposed DRL-based framework of intrusion detection for IoT network. Section 4 presents data pre-processing and model description and the experimental results. Section 5 presents the conclusion of this paper.

2 Related Work

In the research area, a great number of studies show the evaluation of intrusion detection (ID) using distinct data samples and distinct ML and classification algorithms.

The study also explains the evaluation of the intrusion detection (ID) that applied using reinforcement learning and deep reinforcement learning models performed on the different data samples. First, we will discuss how machine learning algorithms applied for detecting intruders and related works. In Ref. [7] based on the available data, we can say that the ID can be communicated as anomaly detection or as a classification problem. Finding the intrusion detection, most of the ML models used the NSL-KDD dataset.

In Ref. [8], the authors introduced the 5-label prediction model that applied the MLP with three layers and giving an accuracy of 79.9%. In Ref. [9] authors reported a multi-classification problem using the AdaBoost with Naive Bayes as a weak learner and obtained an F1 score of 98% after applying the feature selection. In Ref. [10], a dynamic auto-encoder is presented to perform identification on NSL-KDD using a 5-label multi-classification problem which achieved the accuracy of 80%. In Ref. [11], intrusion detection as anomaly detection, the author proposed an anomaly identifier based on reinforcement learning with a simulated network environment. The attack patterns are introduced in a guided manner and get a positive reward based on the right prediction of the attacks. Although the methods applied in Ref. [12], a Q-table algorithm which is a Q-learning that requires the states and actions, to overcome a crash in the table size. In the present work, we are using the function approximators based on the neural networks, which allow applying any size with continuous and discrete values. In Refs. [13–15], they used the Q-tables with TD learning, same drawback as mentioned above. In Refs. [16, 17], authors introduced the distinct approach to intrusion detection by reinforcement learning in a multi-agent architecture. In this, they formed an architecture in a master–slave manner to detect the illegal actions. The drawback in this model is they used the Q-table for implementing the algorithm.

In some of the research work, the authors applied reinforcement learning with classification. In Ref. [18], author used an actor-critic method to perform classification over different datasets like Iris, Hepatitis. This author increased the state space dimension and with a complex reward process. Because of this, the model impacted complexity and computational performance which is different from the one which we used now. In Ref. [19], authors applied RL for classification of intrusion detection, and they compared RL model with classification models such as support vector machine gained similar proportion without the need of feature engineering. In Ref. [20], the author presented a simulated network environment as cybersecurity with incomplete information is presented. In these, two classifiers are used, i.e. attacker and defender agent; they try to achieve the game. This author used RL model-free method Monte Carlo and Q-learning using the Q-table. In Ref. [21], the author created the attacker and defender environment using the RL. In this, the attacker and defender agent try to adjust the environment periodically. The aim for this focused on the cybersecurity attack.

The results presented are not from the real data. In Ref. [22], authors are used RL for intrusion detection and for to detect the model predicted the correct actions or not they used machine learning for verification technique; the datasets they used are real network data, but the problem is the model is not updated periodically to detect the new patterns and the results presented is the only accuracy on the graph chart. In Ref. [23], authors applied the DRL on cloud infrastructure for detecting intrusion detection, and they performed other ML classifiers for comparison purposes. In this, the accuracy they are getting 83.80% the dataset which authors used is UNSW-NB15 which does not contain the IoT anomalies.

3 Proposed Framework for Intrusion Detection Using an Improved Deep Reinforcement Model

In this section, we discussed the data pre-processing steps and a detail working of the proposed framework as shown in Figs. 1 and 2.

3.1 *Data Pre-processing*

The training and testing dataset contains the different types of numeric and categorical features. It is important to pre-process the dataset to build the model.

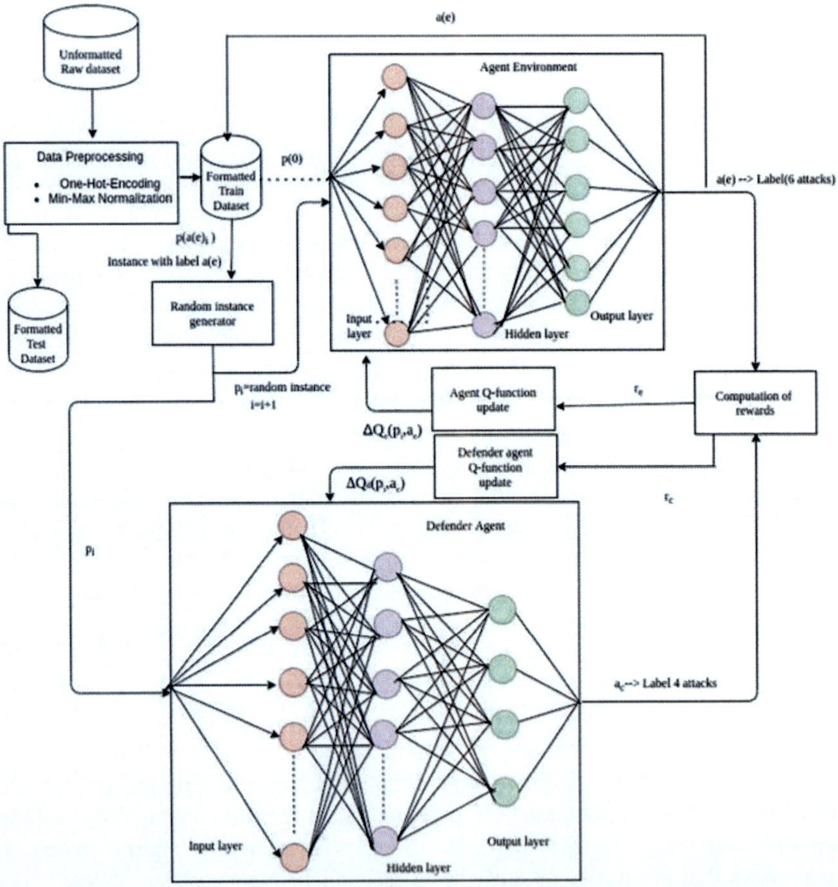

Fig. 1 Deep reinforcement learning model building or training phase

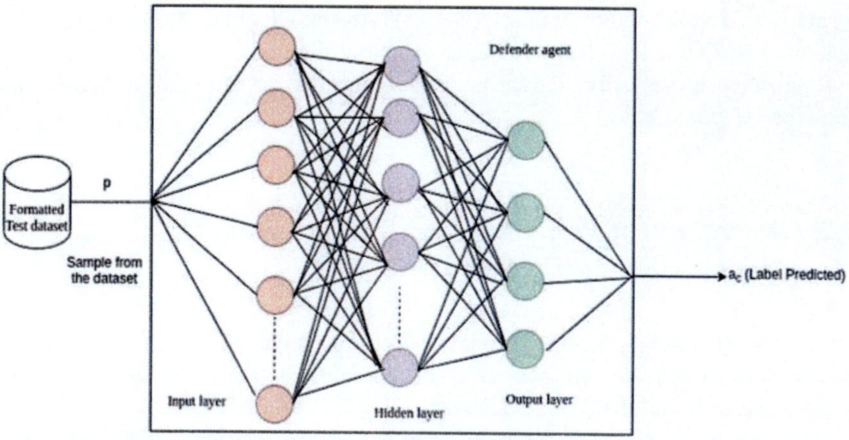

Fig. 2 Deep reinforcement learning model prediction/testing phase

i. **Feature Mapping:** In ML history to train any model, model need all the features will be in numerical form. When the dataset consists of features with categorical values, the model cannot be trained. To handle that situation, a conversion is needed to change the categorical values to numeric values. There are a lot of methods available, but for this framework, we went for the one-hot-encoding (o-h-e) method. The o-h-e will generate the unique columns for the number of unique types a particular categorical column contains. In the dataset, the label contains the six unique types, the o-h-e will produce the six unique columns with their name as a label name, before that the dataset initially have 100 numeric features and 1 categorical column which is a label, after applying o-h-e the dataset transforms into 106 features, The last six features act as a label for the model because it is a multi-classification problem. The classes which are in the training dataset are: (i). Normal, (ii). Mirai-malicious-cc, (iii). Mirai-spread-all, (iv). Torii-malicious, (v). Bash lite-malicious-cc, and (vi). Bash lite-malicious-spread. The classes in the testing dataset are: (i). Normal, (ii). Mirai, (iii). Bash lite, and (iv). Torii.
ii. **Normalization:** In the dataset, the features consist of values of distinct ranges. To maintain the feature with a similar range need of the normalization of the values in the dataset, framework will apply the normalization to features. It is used as a data pre-processing stage in building ML models. In this work, we used Min–Max normalization, which performs the straight modification on the original data, with this the data can range from 0 to 1. The formula for this is as follows:

$$E(\text{Scaled}) = \frac{E - E(\text{smallest})}{E(\text{largest}) - E(\text{smallest})} \tag{1}$$

where E is the value of the column, E(largest) is the maximum value of the particular column, and E(smallest) is the minimum value of the particular column. After the normalization process, the dataset is divided into the formatted train dataset and formatted test dataset.

3.2 *Description of Proposed Deep Reinforcement Learning-Based IDS for IoT Network*

In this, we proposed the working model of DRL-based IDS using RL. In RL models basically, two things will happen: One is concerning the agent responding and the second one with the environment response.

The agent responds from a certain state and takes a particular action that will lead to the other state within the environment. The environment response gives the reward based on the action of the agent, i.e. the chosen action reaches the goal or not if it reaches the goal the environment acknowledges the positive reward, otherwise it gives the negative reward. In RL, solving every problem is based on a Q-function that uses a Q-learning algorithm, i.e. it will find the best state-action set or a value. The Q-function can be computed repetitively using the below equation [23].

$$Q(p_i, a_i) = Q(p_i, a_i) + \alpha[r_{i+1} + \gamma \max_a Q(p_{i+1}, a_{i+1}) - Q(p_i, a_i)] \quad (2)$$

where p_i is the present state, a_i is the present action, $p_{i+1}i$ s the next action, r_{i+1} is the next reward value, α is the learning rate which is 0.1, and γ is the discount factor is 0.01. $Q(p_{i+1}, a_{i+1})$ is Q-value of the next state-action value. The Q-learning works in a small environment because the Q-function consists of a table that is unable to process when the size of the states is increased, to deal with researchers introduced the neural network (N–N) in RL which is called the DQN (Deep-Q-Network), and the variation of the DQN is DDQN which is used. In this, the input to the N–N is the attributes extracted from the labelled dataset. The outcome of the N–N is the Q-function value for the available actions. In this model to approximate the Q-function, a quadratic loss function is used [23]. The equation is as follows:

$$(r_{i+1} + \gamma \max_a (Q(p_{i+1}, a_{i+1}) - Q(p_i, a_i)))^2 \quad (3)$$

In this working model, after pre-processing framework divide the dataset into training and testing. The framework has two agents, the agent environment and the defender agent, who are trained simultaneously. The return of the defender agent will be the intrusion kind prediction for a sample dataset, and the action of the agent environment will be the class of attack, and that is used to generate the next instances from the dataset. In the working model first, we select the initial instance $p(0)$ or p_0 from the sample, and the instance or state is passed through the agent environment. The agent environment processes the instance and selects the action

(a_e). The agent environment consists of N–N is a fully connected N–N, the input to the neural network is the number of available features and the N–N produces Q-function which consists of Q-values of available actions or label values. The model picks an action a_e which consists of the maximum Q-function value. The state p_i or p(i) is an instance, and the action a_e is used for selecting or extracting the next instance p_{i+1} of the sample. In this, the model selects the instances of the sample which belong to the action a_e, by using the random sample model select one instance p_i where $i = i + 1$ that instance is passed to both agent environment and defender agent. The defender agent generates the action a_c by using $p_i(i = i + 1)$ which is the type of attack it belongs to. This agent environment consists of the six types of class/actions, and the defender agent generates the intrusion type which is four types of attacks/actions. The selected actions from the agent environment and the defender agent are used for the reward calculations, and the reward calculation is positive for the agent environment and negative for the defender agent.

In the RL, the environment optimizes needed to optimize the environment model use Q-function; there are two Q-functions, $Q_e(p_i, a_e)$ is for the agent environment, and $Q_d(p_i, a_c)$ is for the defender environment. This function is used for the update of the Q-function values. The framework updates the Q-function values by using Eq. (3), and notation is displayed in Fig. 1 as $\Delta Q_e(p_i, a_e)$ and $\Delta Q_d(p_i, a_c)$. The policy followed is epsilon-greedy which decreased every episode. Firstly, the exploration is very high, and it decreases every episode. An episode is called training the framework using the entire dataset. In this framework, the agent environment and defender agent use their policy, the epsilon value is lower for different episodes. To increase the detection rate, the lower bound of epsilon for the defender agent and the agent environment epsilon value is set by the hyperparameter.

3.3 Algorithm for Proposed DRL Framework

1. Initialize the Q-function values of agent and defender agent to random values.
2. Iteratively do (until the episodes complete).
    1. Pick the initial state p_0 = random function (formatted dataset).
    2. Pass the p_0 to the agent environment, and the agent environment response with action a_{e}.
    3. Perform random sampling, and pick the instance sample p_i that belongs to the action a_{e}.
    4. Iteratively perform (for each $i \to 0$ to terminal state).
        1. Chose the defender action a_c with state p_i using own policy.
        2. Perform step function it will give the p_{i+1}, r_e, r_c.
        3. Rewards are assigned to agent and defender environment r_e, r_c.
        4. p_{i+1} is the next state, and pass that state to the agent environment and it will give the next action a_{e+1}.

5. Change p_{i+1} using random sampling which will it belongs to a_{e+1}.

3. Update the Q-functions of agent and defender environment using Eq. (3).

4 Result Analysis and Discussion

This section presents a performance analysis of the proposed DRL IDS model and its comparison with the latest existing models like logistic regression and Naive Bayes algorithms. The experiment is conducted using the Python programming language, and Scikit-learn Python library is used to implement different machine learning algorithms. The proposed model performance is tested using the MedBIoT dataset. The experiments are performed on Google Colab Engine backend with RAM 12.72 GB disc size with 107.77 GB, and the dataset of training and testing will be 847.19 MB.

4.1 Description of Dataset

The proposed DRL model is evaluated using the actual IoT-based dataset. The MedBIoT is created by the combination of real and emulated IoT devices in the medium-sized network with 83 devices. No other datasets before dealt with the combination of such devices. In this dataset, for generating malicious traffic, actual malware was deployed and provided the real malware traffic data. To create a benign behaviour, an automated execution simulation approach is used. This approach considers the topology of the device, a Python script, and to control the IoT devices, a communication protocol is used for that purpose; the protocol is MQ Telemetry Transport (MQTT). The malicious behaviour is generated by the arrangement of Botnet malware within the environment. The Botnet which is arranged is Mirai, Bash lite, and Torii. The features extracted from the PCA files are used for input to the model.

A total of 100 numerical features are extracted within distinct time slots. The following lines will give the features of the dataset. The collected features are belonging to distinct categories. The first category is Host-Mac and IP, the second category is Channel, the third category is Network Jitter, and the last category is Socket; from these categories, the analyst collected the features.

- **Mirai**: It is defined as malware that uses Linux-based network devices and turns them into a remotely controllable bot. It is used as a Botnet attack in a big-size environment or network.
- **Bash lite**: It is used to trigger the DDOS attack and malware which damages the Linux systems.
- **Torii**: It is a new malware with new technologies, and it will attack every type of system.

Table 1 Classification of malicious and normal samples in training dataset

Attack class	Attack class frequency	Class frequency in (%)
Mirai	131,032	32.758
Normal	112,382	28.0955
Bash lite	85,448	21.362
Torii	71,138	17.7845
Total	400,000	100

Table 2 Classification of malicious and normal samples in testing dataset

Attack class	Attack class frequency	Class frequency in (%)
Mirai	32,758	32.758
Normal	28,096	28.096
Bash lite	21,362	12.362
Torii	17,784	17.784
Total	100,000	100

Tables 1 and 2 show the classification statistics of samples used in training and testing phase. The training dataset contains a count of 400,000 samples. Out of which 131,032 are Mirai samples, 112,382 are Normal samples, 85,448 are Bash lite samples, and 71,138 are Torii samples. The testing dataset contains a count of 1,00,000 samples. In that 32,758 are Mirai samples, 28,096 are Normal samples, 21,362 are Bash lite samples, and 17,784 are Torii samples. The following section extends the evaluation metrics used in the proposed model.

4.2 Description of Evaluation Metrics

To find out the efficiency of the proposed framework, a different evaluation metric has been used. The popular evaluation metrics are Accuracy, Precision, Recall, F1 score, and confusion matrix is used. These metrics are based on the successive four variables:

- **true positive(tp)**: This describes the total count of attack samples in medium-sized IoT network traffic that are perfectly detected as attack activity by the model.
- **true negative(tn)**: This describes the total count of benign samples in medium-sized IoT network traffic that perfectly detected as benign activity by the model.
- **false positive(fp)**: This describes the total count of benign medium-sized IoT network samples that are mistakenly marked as the attack actions by the model.
- **false negative(fn)**: This describes the total count of attack samples in a medium-sized IoT network that is mistakenly marked as normal actions by the model.
- **Accuracy (ACC)**: In ML, to check the efficiency of the model Accuracy is used. It is defined as the sum of right estimations (true positive & true negative) to the

sum of all estimations.

$$\text{ACC} = \frac{\text{tp} + \text{tn}}{\text{tp} + \text{fp} + \text{tn} + \text{fn}}$$

- **Precision (PRE)**: It is defined as the total number of positives activities predicated as positive to the total number of predictions defined as attacks.

$$\text{PRE} = \frac{\text{tp}}{\text{tp} + \text{fp}}$$

- **Recall (REC)**: RE determines the number of correct positive predictions made out of all positive predictions that could have been made.

$$\text{REC} = \frac{\text{tp}}{\text{tp} + \text{fn}}$$

- **F1 score (F1)**: F1 score determines the weighted average of PRE and RE.

$$\text{F1} = 2 * \frac{\text{REC} * \text{PRE}}{\text{REC} + \text{PRE}}$$

- Confusion matrix: A confusion matrix is a table that is often used to describe the performance of a classification model on a set of test data for which true values are studied. It is also known as error matrix.

4.3 *Results Analysis*

In the analysis process, initially an error matrix is used. An error matrix is produced and used to calculate the performance of the framework. Table 3 will represent the error matrix results for the DRL framework. The analytical parameter tp is showed across the diagonal of the error matrix, columns in the error matrix related to the estimated class, while rows in the error matrix are related to the actual classes. Tables 4 and 5 give the error matrix for the logistic regression and Naive Bayes frameworks.

Table 3 Error matrix for the DRL framework

Actual attack classes	Predicted attack classes			
	Mirai	Normal	Bash lite	Torii
Mirai	31,721	0	634	403
Normal	1014	27,054	28	0
Bash lite	385	213	20,764	0
Torii	0	0	330	17,454

Table 4 Error matrix for logistic regression framework

Actual attack classes	Predicted attack classes			
	Mirai	Normal	Bash lite	Torii
Mirai	21,412	16	834	116
Normal	1508	13,574	1929	71
Bash lite	507	899	24,400	548
Torii	64	55	86	13,981

Table 5 Error matrix for Naïve Bayes framework

Actual attack classes	Predicted attack classes			
	Mirai	Normal	Bash lite	Torii
Mirai	7035	451	1333	13,559
Normal	9392	5979	820	891
Bash lite	1777	1872	21,684	1021
Torii	22	78	22	14,064

Figures 3, 4, and 5 represent the normalized confusion matrix of the DRL, logistic regression, and Naive Bayes frameworks.

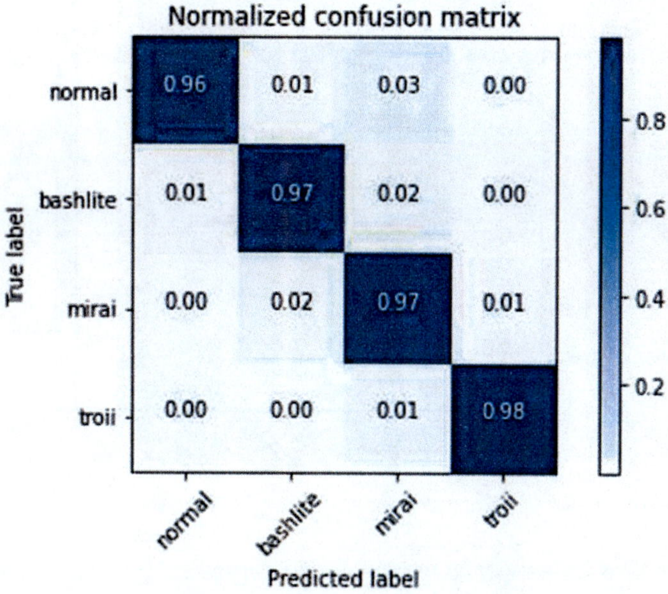

Fig. 3 Normalized confusion matrix for deep reinforcement learning framework

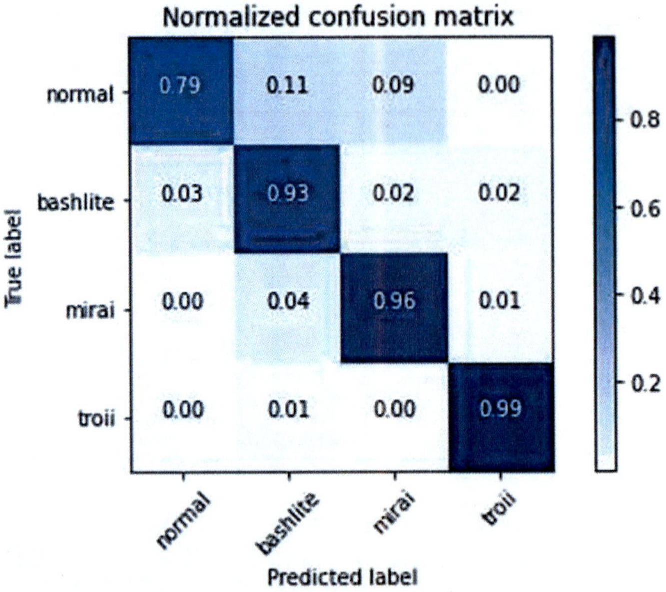

Fig. 4 Normalized confusion matrix for logistic regression framework

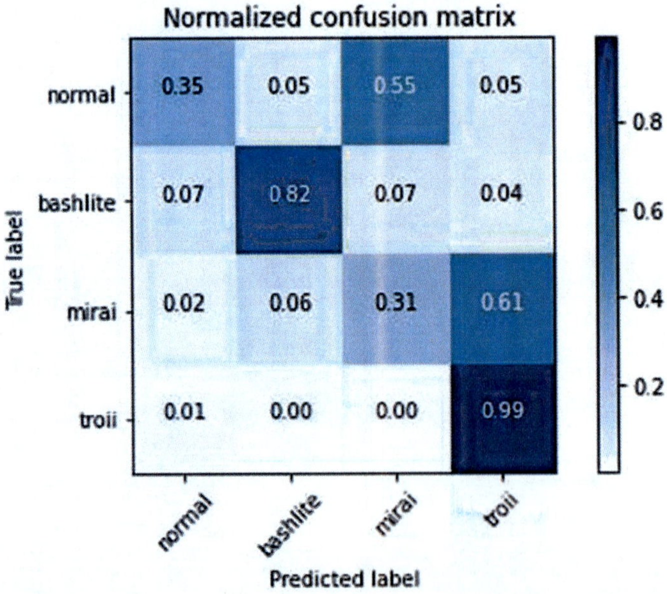

Fig. 5 Normalized confusion matrix for Naïve Bayes framework

Table 6 Performance metrics of the different framework

Frameworks used	Accuracy (%)	Precision (%)	Recall (%)	F1 score (%)
Deep reinforcement learning	96.99	97.02	96.99	97.00
Logistic regression	91.26	91.26	91.51	91.24
Naïve Bayes	60.95	64.41	60.95	59.58

Table 7 Comparison of the parameters used in the previous framework and DRL model

	Periodic iteration	No. of hidden layers	Dataset	Accuracy (%)
Adversarial environment reinforcement learning algorithm for intrusion detection [23]	100	3	NSL-KDD	80.16
Proposed DRL model	150	5	MedBIoT	96.99

Table 6 will show us that the DRL framework shown the highest accuracy of 96.99%, while the other two machine learning frameworks like logistic regression and Naive Bayes framework shown the lowest accuracy. From this, we can say that the deep reinforcement learning framework giving the highest accuracy.

The framework, when compared to the previous model, is somewhat same but what new we introduced in this one is in the last model they used the unbalanced dataset which does not contain the IoT traces but in this, we introduced the IoT traces dataset and the number of periodic iteration and number of layers in N–N also increased. When compared to previous frameworks, the present framework with IoT traces gives the highest accuracy than previous framework which does not contain IoT traces. Table 7 gives the comparison of proposed framework with previous framework.

5 Conclusion

Cyber-attacks are playing a key role in the modern IoT environment. In such cases, an IDS service is implemented in the IoT network. Basically, network intrusion attacks are composite, and network patterns cannot be assigned very easily because of continuous changes and the need to build the novel model every time, which will lead to extra time.

To overcome this type of problem, this work gives a new approach, in the form of a framework called the RL framework to the intrusion detection problem. RL extracted the curiosity to the analyst in distinct areas, and RL techniques are implemented in different areas. The DRL framework combines the supervised and RL models. The framework is implemented under the direction of the RL environment. In a nutshell, the contribution of this paper gives the framework for IoT environment for detecting network intrusion using DRL models. The proposed framework is trained in different periodic iteration with changing the number of hidden layers of neural networks and parameter changing. The framework is developed for the unbalanced samples, the dataset used is MedBIoT which contains the IoT traces, and the framework is compared with the various ML frameworks with the aim of accuracy and framework flexibility. In further work, we plan to use the other policy methods so that we can tried with same RL models.

References

1. A. Shameli-Sendi, M. Cheriet, A. Hamou-Lhadj, Taxonomy of intrusion risk assessment and response system. Comput. Secur. **45**, 1–16 (2014)
2. N. Moustafa, J. Slay, The significant features of the UNSW-NB15 and the KDD99 data sets for network intrusion detection systems. in *2015 4th International Workshop on Building Analysis Datasets and Gathering Experience Returns for Security (BADGERS)* (IEEE, 2015)
3. N. Moustafa, G. Creech, J. Slay, Big data analytics for intrusion detection system: Statistical decision-making using finite Dirichlet mixture models. in *Data Analytics and Decision Support for Cybersecurity* (Springer, Cham, 2017), pp. 127–156
4. N. Moustafa, J. Slay, G. Creech, Novel geometric area analysis technique for anomaly detection using trapezoidal area estimation on large-scale networks. IEEE Trans. Big Data **5**(4), 481–494 (2017)
5. N. Moustafa, G. Creech, J. Slay, Anomaly detection system using beta mixture models and outlier detection. in *Progress in Computing, Analytics, and Networking* (Springer, Singapore, 2018), pp. 125–135
6. N. Moustafa, Hu. Jiankun, J. Slay, A holistic review of network anomaly detection systems: a comprehensive survey. J. Netw. Comput. Appl. **128**, 33–55 (2019)
7. R.R. Guimaraes et al., Intelligent network security monitoring based on optimum-path forest clustering. IEEE Netw. **33**(2), 126–131 (2018)
8. B. Ingre, A. Yadav, Performance analysis of NSL-KDD dataset using ANN. in *2015 International Conference on Signal Processing and Communication Engineering Systems* (IEEE, 2015)
9. Y. Wahba, E. ElSalamouny, G. ElTaweel, Improving the performance of multi-class intrusion detection systems using feature reduction. arXiv preprint arXiv:1507.06692 (2015)
10. M. Lopez-Martin, et al., Conditional variational autoencoder for prediction and feature recovery applied to intrusion detection in IoT. Sensors 17(9), 1967 (2017)
11. A. Servin, D. Kudenko, Multi-agent reinforcement learning for intrusion detection. in *Adaptive Agents and Multi-Agent Systems III. Adaptation and Multi-Agent Learning* (Springer, Berlin, Heidelberg, 2005), pp. 211–223
12. A. Servin, *Towards Traffic Anomaly Detection via Reinforcement Learning and Data Flow* (Department of Computer Science, University of York, UK, 2007)
13. X. Xu, Sequential anomaly detection based on temporal-difference learning: principles, models and case studies. Appl. Soft Comput. **10**(3), 859–867 (2010)
14. A.V. Sukhanov, S.M. Kovalev, V. Stýskala, Advanced temporal-difference learning for intrusion detection. IFAC-PapersOnLine **48**(4), 43–48 (2015)
15. X. Xu, T. Xie, A reinforcement learning approach for host-based intrusion detection using sequences of system calls. in *International Conference on Intelligent Computing* (Springer, Berlin, Heidelberg, 2005)
16. K. Malialis, *Distributed Reinforcement Learning for Network Intrusion Response* (University of York, Diss, 2014)

17. M.A. Wiering, et al., Reinforcement learning algorithms for solving classification problems. in *2011 IEEE Symposium on Adaptive Dynamic Programming and Reinforcement Learning (ADPRL)* (IEEE, 2011)
18. M.G. Lagoudakis, R. Parr, Reinforcement learning as classification: leveraging modern classifiers. in *Proceedings of the 20th International Conference on Machine Learning (ICML-03)* (2003)
19. R. Elderman, et al., Adversarial reinforcement learning in a cyber security simulation. in *ICAART (2)* (2017)
20. M. Zhu, Z. Hu, P. Liu, Reinforcement learning algorithms for adaptive cyber defense against Heartbleed. in *Proceedings of the First ACM Workshop on Moving Target Defense* (2014)
21. R.R. dos Santos, et al., A long-lasting reinforcement learning intrusion detection model. in *International Conference on Advanced Information Networking and Applications* (Springer, Cham, 2020)
22. K. Sethi, et al., Deep reinforcement learning-based intrusion detection system for cloud infrastructure. in 2020 International Conference on COMmunication Systems & NETworkS (COMSNETS) (IEEE, 2020)
23. G. Caminero, M. Lopez-Martin, B. Carro, Adversarial environment reinforcement learning algorithm for intrusion detection. Comput. Netw. **159**, 96–109 (2019)
24. M. Lopez-Martin, B. Carro, A. Sanchez-Esguevillas, Application of deep reinforcement learning to intrusion detection for supervised problems. Expert Syst. Appl. **141**, 112963 (2020)
25. A. Guerra-Manzanares, et al., MedBIoT: generation of an IoT botnet dataset in a medium-sized IoT network. in *ICISSP* (2020)
26. A. Sungheetha, R. Sharma, Service quality assurance in clouddata centers using migration scaling. J. Inf. Technol. **2**(01), 53–63 (2020)
27. A. Sungheetha, R. Sharma, Novel shared key transfer protocol for secure data transmission in distributed wireless networks. J. Trends Comput. Sci. Smart Technol. (TCSST) **2**(2), 98–108 (2020)
28. D. Sivaganesan, Smart contract based industrial data preservation on block chain. J. Ubiquit. Comput. Commun. Technol. (UCCT) **2**(1), 39–47 (2020)

A Survey of Data Mining Techniques in the Field of Cyborg Mining

Nafiza Rahman, Maisha Maimuna, Afroja Begum, Md. Razu Ahmed, and Mohammed Shamsul Arefin

Abstract The growing user population and open nature of social media have attracted a huge number of programs which are automated, also known as bots, which is considered to be a double-edged sword. But, in between human and bot, there is another organism known as cyborg. Cyborg can be defined as bot-assisted human or human-assisted bot. The spread of misinformation in the social media has greatly increased due to social bots as well. We mainly review about different techniques that have been used up until now to detect difference between human, bot or cyborg, fake accounts or spoofing accounts assisted by bot or cyborg with the help of data mining.

Keywords Data mining · Cyborg · Bot

1 Introduction

Data mining is the process which analyzes a huge amount of information to discover trends and patterns. It can be used by corporations to understand what customer prefers. Data mining is also known as knowledge discovery in data (KDD). It is considered as a subfield of computer science and statistics. The main goal of data mining is to extract information from a dataset and convert the information into a better one for further use. As a result of the evolution of data warehousing technology and the evolution of big data, usefulness of data mining techniques has rapidly accelerated over time.

N. Rahman · M. Maimuna · A. Begum · M. S. Arefin (✉)
Department of Computer Science and Engineering, Chittagong University of Engineering and Technology, Chattogram 4349, Bangladesh
e-mail: sarefin@cuet.ac.bd

Md. R. Ahmed
Department of Electronic and Telecommunication Engineering, International Islamic University Chittagong, Chattogram 4318, Bangladesh

G. Ranganathan et al. (eds.), *Soft Computing for Security Applications*,
Advances in Intelligent Systems and Computing 1397,
https://doi.org/10.1007/978-981-16-5301-8_55

Cyborg data mining is a method for collecting data. The data collected are produced by an implantable device that needs to monitor bodily processes for interests in the commercial sectors. An android is a human-like robot, but a cyborg is not. It is an organism whose functioning is completely dependent upon a mechanical or electronic device that relies on some feedback.

In this paper, we mainly reviewed all of the research work related to cyborg mining. In Ref. [1], author researched about chat bots and introduced two component classification systems for detecting chat bots. In Ref. [2], author researched about the classification of human, bot or cyborg accounts in social media. The author in Ref. [3] proposed a dynamic system which is hard for the bot to crack. In Ref. [4], difference between human, bot or cyborg twitter accounts are discussed. Author in Ref. [5] helps to understand which users are most likely to contact with social bots. All of these papers are about how to protect human users from the harmful bots or cyborgs.

In Ref. [6], author provides an approach to detect some content that can help human analysts in their work. In Ref. [7], author identifies real-life influencers and ranks people based on their influence which is decided through data related to Twitter. In Ref. [8], it provides the basic concept and the knowledge about the architecture of cyborg intelligence. Twitter can be considered as one of the most widely used social media platforms. Along with regular users, politicians or celebrities or company representatives also have twitter accounts. Automation can be used as both good and evil in case of Twitter. Legitimate bots can tweet about useful news or blog updates. This can help the twitter to become a widely used information network. On the other hand, malicious bots can spread spam which is harmful. The solution to this problem is given in Ref. [9]. In Ref. [10], the work analyzes overall network simplification measures applied on fake influencer accounts and on legitimate influencer accounts.

In Ref. [11], the author studies about 2016 US election on which social bots had a huge role to play on the public opinion of the candidates. Following spoofing accounts can lead to misinformation. As a result, it is indeed dangerous to follow them on social media. These accounts are spotted in Ref. [12]. We summarize all of the research works related to bots or cyborgs in this paper. Each paper includes their contribution, dataset, implementation and evaluation. The papers reviewed are listed in Table 1. The remaining paper is categorized as: Sect. 2 is about preliminary of cyborg mining. Section 3 contains the detailed information about the reviewed papers. Lastly Sect. 4 holds the conclusion.

2 Cyborg Mining Preliminary

2.1 *Definition of Cyborg Mining*

Cyborg can be defined as a half biological, half mechanical system. It is the enhancement of the biological component and overall a more complicated creation. Cyborg

Table 1 Evolution of data mining in cyborg mining research

2009–2012	2013–2015	2016–2017	2018	2018–2019
Bots in Internet chat [1]	Predicting susceptibility to social bots [5]	Recent progress and future directions in cyborg [13]	Human versus bots reaction to deceptive and trusted news sources [14]	Compromised account detection on Twitter [15]
Bots in Twitter [2]	Adult account detection on Twitter [16]	Twitter Bot detection via warped correlation [9]	Automatic chatting influence in Twitter [10]	Identifying spoofing accounts on Twitter [12]
Dynamic image-based CAPTCA [3]	Detecting Jihadist messages on Twitter [6]	Bot detection and stance classification in Twitter [17]	Twitter influencers or cheated buyers [18]	Analyzing the digital traces of political manipulation [19]
Detecting automation of Twitter accounts [4]	Detecting real-world influence through Twitter [7]	Protecting the privacy of social media users [20]	End-to-End compromised account detection [21]	Interactive refactoring documentation bot [22]
	Labeling automated and non-automated social media accounts [8]	Privacy preserving text mining [23]	Fake account identification in social networks [24]	Twitter Bot detection using neural networks and word embeddings [25]
			Impact of malicious accounts on political tweet sentiment [11]	

data mining is a method for collecting data. The data collected are produced by an implantable device that needs to monitor bodily processes for interests in the commercial sectors. It is a being with completely dependent functionality.

2.2 *Use of Cyborg Mining*

Cyborgs are considered as comprehensive data subjects. And so they can be used as a medium to increase surveillance with the help of optical recording technology. Cyborgs can be used to collect harmless data such as posting pictures to Facebook or recording people's life experiences. Cyborgs can also serve as a means of surveillance on the overall population and gather information about them for further use.

The data mined from cyborg mining technology can help different companies in their research projects which can lead to the development of new technologies.

2.3 Challenges of Cyborg Mining

Cyborg mining comes with its fair share of challenges. It is not an easy task to integrate biological and machine intelligence into a single form of intelligence. It is a very challenging but promising goal. Both biological and artificial method needs to be explored for the enhancement of the performance of the cyborg, which is a very hard task to grasp. Noisy or incomplete or distributed data can also disrupt the system.

3 Review Details

This section focuses on the contribution, dataset, implementation and evaluation of each paper with detailed information. The summary of the reviewed papers is illustrated in Table 2.

3.1 Bots in Internet Chat [1]

Two component classification systems are introduced: (1) a classifier for detecting unknown chat bots which is entropy-based and (2) a classifier for detecting known chat bots which is based on machine learning. Mainly 3 datasets are used: (1) chat logs used by humans from the year 2007 in August (2) chat logs used by bots from the year 2007 in August (3) chat logs used by bots from the year 2007 in November. The total count of human messages to bot messages are 342,696 and 87,049 respectively.

The users that are actually human but are misclassified as bots are given false positive and the ones that are misclassified as human users are given false negative. Cutoff scores are determined by the human training sets which determines a test sample as a human or a bot. Finally, from the chat logs 14 types of chat bots are identified and divided into four more categories.

3.2 Bots in Twitter [2]

This paper classifies cyborg, bot and human accounts on twitter to help human users to understand whom they are sharing information with. The training set includes 1000 users that are cyborg, 1000 users that are bot and 1000 human users thus 3000 samples in total. In a similar way, a test set is created containing 3000 samples. Both sets consist of 8,350,095 tweets posted by sampled users.

Logs that have more than 100 tweets are processed by entropy component. Whether the tweet is a spam or not is determined by the machine learning component

Table 2 Summary of the reviewed cyborg mining papers

Paper Title	Contribution	Dataset	Evaluation
Bots in Internet chat [1]	Provides two types of classification system for detecting unknown or known chat bots	Total 342,696 human messages and 87,049 bot messages are used	Total 14 types of chat bots can be detected through this process
Bots in Twitter [2]	Differentiates between cyborg, bot and human accounts	Both the training set and test set include total 3000 samples	The final result identifies 13.8% users as bot, 37.5% users as cyborg and 48.7% users as human
Dynamic image-based CAPTCA [3]	Proposes a three-tier system that is hard for the bot to crack but easy for human users	Total 200 images are used for this purpose	Neither the image nor the prototype system can be hacked easily
Detecting automation of Twitter accounts [4]	Discusses the true nature of the tweet content and about the account holder	The dataset consists of 500,000 users	The final result identifies 10.5% users as bot, 36.2% users as cyborg and 53.2% users as human
Predicting susceptibility to social bots [5]	Helps to determine some specific feature of the users that share their information with the bots	A dataset with 610 users is used who were messaged by bots	Total number of friends and followers are the specific predictors for contacting with a bot
Adult account detection on Twitter [16]	Proposes an iterative classifier system that can accurately detect accounts containing adult contents	Contains 107 million accounts, 123 million tweets with 174,700 hashtags	Total 563 accounts are labeled as adult account
Detecting Jihadist messages on Twitter [6]	Provides an approach to detect content that can help human analysts in their work	Tweets between 4000 and 7500 are used in total	Weka, a machine learning software is used that performs better than other methods
Detecting real-world influence through Twitter [7]	Identifies real-life influencers and ranks people based on their influence	The training set and testing set consists of 2500 and 4500 users respectively	NLP approach. gives significant results for detecting spammers or influencers
Labeling automated and non-automated social media accounts [8]	Provides an alternative to the traditional system that can speed up the whole process	4,000,000 tweets from at least 13,000 user accounts are used	The results show that a relative amount of precision and recall have been gained

(continued)

Table 2 (continued)

Paper Title	Contribution	Dataset	Evaluation
Recent progress and future directions in cyborg [13]	Simplifies the basic concept and the architecture of cyborg intelligence	Biological and artificial methods are explored for the enhancement of cyborg performance	Some neural decoding and encoding algorithms are presented for the biological and machine component
Twitter bot detection via warped correlation [9]	Develops a finder whose task is to find correlation to determine correlated user accounts in social media	The top clusters containing 9134 bot accounts are picked	The method is evaluated with per-user method and twitter suspension process
Bot detection and stance classification in Twitter [17]	Proposes a framework to help users in identifying real people from bots	3240 tweets are collected from each user	Accurately measures the true number of followers
Protecting the privacy of social media users [20]	Introduces a privacy cyborg which can accurately and privately observe a person's behavior	Accounts that hold tweets over 200 are taken as dataset	Either informs the user about the public opinion or warn the user about the danger of their sensitive attribute
Privacy preserving text mining [23]	Provides a framework for text analysis with privacy policy	The dataset contains a total of 308 tweets	Finds whether a tweet is related to terrorism without revealing owner's sensitive information
Human versus bots reaction to deceptive and trusted news sources [14]	Illustrates the difference between the reactions among human and bot	431,771 tweets from verified accounts are used which consist of 255 news sources	Final result shows that human reacts on verified news channels more than bots and bots take much more time to react than humans
Automatic chatting influence in Twitter [10]	Discusses social bots being turned into influential users	10 re-tweets were used and social bots had 10 likes	Illustrates how much communication skills can be developed by social bots
Twitter influencers or cheated buyers [18]	Shows how to measure fake influencer accounts from the real ones	Dataset consists of 18 real and 18 fake influencers	Compared with the traditional techniques, this paper provides a better and accurate performance

(continued)

Table 2 (continued)

Paper Title	Contribution	Dataset	Evaluation
End-to-End compromised account detection [21]	Provides an account detection framework that can smoothly detect fake accounts or spam	Dataset contains 835 compromised and 7632 non-compromised accounts	The proposed method efficiently detects compromised accounts
Fake account identification in social networks [24]	Proposes a technique for identifying real accounts on social media from the fake ones	9019 tweeter accounts were used to conduct this experiment	Detects efficiently fake accounts through some discovered patterns
Impact of malicious accounts on political tweet sentiment [11]	Mainly determines about the influence of social bots on social media	The dataset consists of total 2,942,808 tweets	This work shows a result where 3 to 19% users are shown as bots
Compromised account detection on Twitter [15]	The proposed model learns through tweet history and embeds the knowledge on a feature representation	There are 5889 users of which 484 are compromised accounts	Detect compromised accounts more efficiently than the traditional ones
Identifying spoofing accounts on Twitter [12]	Proposes a method to detect spoofing accounts using PageRank citation ranking	The authors selected the accounts of 30 people from a popular group	96% real accounts and 75% fake accounts were detected. 27 real accounts and 37 fake accounts were used in this experiment
Analyzing the digital traces of political manipulation [19]	Discusses the rumor of Russia being involved in spreading misinformation in the 2016 election with the help of the bots	A large dataset containing 43.7 million tweets are used	With the help of the large dataset it is found that 4.9% of the liberal users and 6.2% of the conservative users were bots
Interactive refactoring documentation bot [22]	Introduces a documentation bot with the refactoring ability	In this experiment, a survey was conducted using 14 developers	The documentation bots can be used to help the developers to validate the document
Twitter bot detection using neural networks and word embeddings [25]	Focuses on the spam bot detection technique to help human users to understand who they are contacting with	The dataset contains 3474 human accounts and 1455 bot accounts	This is a easier implementation technique in contrast to time consuming feature extraction

based on the text pattern of the tweets. The feature value is considered −1 for tweets with spam and 1 for non-spam. A classification score is given to the users by the decision maker and they are divided into classes based on their score. The final result identifies 13.8% users as bot, 37.5% users as cyborg and 48.7% users as human.

3.3 Dynamic Image-Based CAPTCHA [3]

This paper proposes a three-tier system that is hard for the bot to crack but easy for the human users. Through the help of Google, Bing, etc., total 200 images are collected. For the purpose of unique identification a unique id is given to each image. The source image needs to be filtered and 32 filters are used for this purpose.

Bots cannot attack easily as the system makes a random guess around all the levels. 32 filters are selected at random. As a result, even brute force attack cannot crack the prototype system. Image database cannot be hacked either because of hard maintenance.

3.4 Detecting Automation of Twitter Accounts [4]

This paper is all about identifying the true nature of the tweet content and finding out the true holder of the accounts. The training set consists of 10,000 human users, 10,000 bot users and 10,000 cyborg users and the total dataset consists of 500,000 users.

The periodic or regular timing is detected by the entropy component. Bayesian classification is used to detect patterns of the text of known spam by the spam detection component. To catch bots account-related properties are used by the account properties component. By using random forest algorithm, the decision maker component makes a decision.

3.5 Predicting Susceptibility to Social Bots [5]

This paper researches about the users that are most likely vulnerable to bots. Whether the user will share information with the bot or not can be found through some features and this paper helps to determine those specific features. A dataset with 610 users is used who were messaged by bots, and whether they contacted with the bot or not is seen and the features are determined through this process.

Mainly two experiments are conducted in this research: firstly, the likelihood of a individual to contact with a bot is determined through the ten feature ranker and secondly, building a model to test the prediction. Final result shows that total number of friends and total number of followers are the specific predictors for contacting with a bot.

3.6 Adult Account Detection on Twitter [16]

This paper helps to determine the accounts containing adult contents. An iterative classifier system is proposed that can accurately detect this. As seed accounts, 10 popular adult Twitter accounts are manually selected. Each seed account contains two hop networks. It contains both its friends and followers. Total dataset contains 107 million accounts, 123 million tweets with 174,700 hashtags and 127,000 URLs.

To evaluate the performance 5000 accounts on Twitter are considered as test set. Each account must have more than 5 followers and must have at least 5 tweets. Accounts other than this are mostly fake or inactive and unable to promote adult content. Profile images and background images are checked as well as recently uploaded images for finding adult content. Finally, total 563 accounts are labeled as adult account.

3.7 Detecting Jihadist Messages on Twitter [6]

This paper provides an approach to detect some content that can help human analysts in their work. Whether an account is supporting jihadists group or not can be found through a machine learning approach. There are three different datasets. (1) Tweets are collected based on hashtags and networks related to jihadists. (2) Tweets are collected randomly based on various topics. (3) Tweets are collected that are against ISIS. Tweets between 4000 and 7500 are used in total.

A machine learning software is also known as Weka which is written in java is used in this experiment. Different classification algorithms like SVM, Naive Bayes and Ada-Boost were also used. In this case, Ada-Boost performs better than the rest.

3.8 Detecting Real-World Influence Through Twitter [7]

This paper identifies real-life Influencers and ranks people based on their influence which is decided through data related to Twitter. Several approaches such as machine learning techniques with the help of natural language processing have been used to identify influencers. The training set consists of 2500 users with 796 influencers among them. The testing set also consists of 4500 users with 1563 influencers among them.

The Rep-Lab framework is used to evaluate the estimated rankings. It shows that to detect spammers or influencers classical methods do not give any significant results. The Klout Score does not help the user to detect real-life influence. On the other hand, an NLP approach is considered to have a better performance.

3.9 *Labeling Automated and Non-automated Social Media Accounts [8]*

Traditional bot detection approaches depend on manual observation from which classifiers learn. But this requires a huge amount of manual effort. This paper provides an alternative to this system that can speed up the whole process. Tweets are collected from the Russia–Ukraine conflict consisting of 4,000,000 tweets from at least 13,000 user accounts. Tweets are collected through some selected keywords.

A subset of the dataset is used for training the classifier. A random forest classifier that is based on machine learning is used for data labeling. The results show that a relative amount of precision and recall have been gained. As a result, the classifier can be used in the real world for the purpose of classification with better performance.

3.10 *Recent Progress and Future Directions in Cyborg [13]*

It is not an easy task to integrate biological and machine intelligence into a single form of intelligence. It is a very challenging but promising goal. It simplifies the basic concept and the architecture of cyborg intelligence. Both biological and artificial method needs to be explored for the enhancement of the performance of the cyborg, which is a very hard task to grasp. So, some neural decoding and encoding algorithms were presented for the biological and machine component.

This paper presents an overall concept about cyborg system architecture. This also researched about some functions like sensation, perception and other cyborg skills. Techniques are also developed to understand how animals can be enhanced by AI.

3.11 *Twitter Bot Detection via Warped Correlation [9]*

This paper develops a finder whose task is to find correlation to determine correlated user accounts in social media. This bot detection system works without the need of any labeled data to detect activity correlation. De-Bot ran every 4 h for 16 days and linked all the clusters into one integrated set of clusters using the friend-of-friend technique. The top clusters containing 9134 bot accounts were picked.

It shows that the presence of highly synchronous cross-user activities reveals abnormalities and is a key to detecting automated accounts. To detect bot accounts on Twitter an unsupervised method is developed. The method is evaluated with per-user method and Twitter suspension process. The evaluation shows that Twitter suspends automated accounts with lower rates than this method.

3.12 Bot Detection and Stance Classification in Twitter [17]

This paper mainly focuses on stance detection and bot identification. And proposes a framework to help users in identifying real people from bots and so on. A test case with supporters of both sides of the election is needed. And so 310 supporter of one side and 412 supporter of another side is chosen. The amount of tweets collected from each supporter is 3240 tweets.

For one-item-set and two-item-set hashtags in the tweets, their affinity is evaluated. The final ranking is done based on a threshold. The hashtags with higher affinities are to be used. This method can accurately measure the true number of followers. Thus the real social media popularity can be found.

3.13 Protecting the Privacy of Social Media Users [20]

This paper introduces a privacy cyborg which can accurately and privately observe a person's behavior. It can be used for observing contents that have been published. It can also suggest advice if the person's privacy is at stake. The cyborg needs access to the tweets of its owner and all the other private informations in order to defend them from the attacker. And tweets over 200 are taken as dataset.

The Bayesian model is used for attacking literature. This paper assumes the attacker is very knowledgeable. And thus the attacker can easily determine the index of a sensitive attribute. The cyborg can handle two tasks: Either informing the user about the public opinion or warn the user about the danger about their sensitive attribute.

3.14 Privacy Preserving Text Mining [23]

This paper provides a framework for text analysis with privacy policy. This framework analyzes in a privacy preserving manner. It makes sure that no sensitive information is leaked even to the analyzing engine itself. The dataset contains a total of 308 tweets that are written in Arabic and English language and 236 E-mails. The tweets are labeled as normal or terrorist.

To execute word search in dictionaries encryption techniques are used. Two datasets are used to determine the performance of the proposed system. The system finds whether a tweet is related to terrorism without revealing any sensitive information of the owner.

3.15 Human Versus Bots Reaction to Deceptive and Trusted News Sources [14]

This paper shows the difference between the reactions among human and bot. The responses are labeled as agreement, disagreement, appreciation, humor or negative reaction. Tweets from trusted news sources are used as dataset. So total 431,771 tweets from verified accounts are used which consist of 255 news sources.

This work literally shows how bots and human reacts to different news sources. The number of sources that bots react to and the number of sources that human reacts to be compared. Their speed is compared as well. It is seen that human reacts on verified news channels more than bots and bots take much more time to react than humans.

3.16 Automatic Chatting Influence in Twitter [10]

This paper discusses the influence of the social bot. It shows that social bots can be turned into influential users. A set of accounts were created for the experiment. The social bots were designed to follow some initial accounts. They also needed to have a number of posts. In this experiment, 10 re-tweets were used and social bots had 10 likes.

The goal of this experiment is to see how much communication skills can be developed by the social bots. And surprisingly the follower count increased by 35%, re-tweet rate also increased by 15%. This proves how efficient the social bots are in the communication skill performance.

3.17 Twitter Influencers or Cheated Buyers [18]

Marketing conducted by influencers is more effective than traditional marketing. Fake influencers try to increase their follower count by buying fake accounts from vendors. This paper shows how to measure fake influencer accounts from the real ones. Dataset consists of 18 real and 18 fake influencers. Real influencer accounts consist of politicians, actors, football players, etc. On the other hand, fake accounts consist of at least 50% fake followers.

The main purpose of this paper is to provide a characterization measure to effectively find the fake influencer account from the real one. Compared with the traditional techniques, this also provides a better and accurate performance. The dataset used in this experiment is a unique one.

3.18 *End-To-End Compromised Account Detection [21]*

The goal of this research is to provide an effective account detection framework that can smoothly detect fake accounts or spam. This process uses an Long Short-Term Memory network. Compromised and non-compromised accounts are used for dataset from the year 2015 and 2017. The datasets were achieved through some mentioned keyword. Final dataset contains 835 compromised accounts and 7632 non-compromised accounts.

Stratified sampling is used in this work for splitting training and testing sets of which 80% is training set, 10% validation and 10% test set. Precision and recall can be used to measure the algorithm performance. The proposed method can efficiently detect compromised accounts.

3.19 *Fake Account Identification in Social Networks [24]*

This paper proposes a technique for identifying real accounts on social media from the fake ones. Information is extracted from big data to analyze patterns of fake accounts. One of the most used social media platform is Twitter. And so 9019 tweeter accounts were used to conduct this experiment. Dataset is categorized as real, verified or fake.

The proposed technique shows that it is possible to distinguish between real accounts from the fake ones. Correct information collection can help a lot to understand the pattern of the fake accounts. Thus information is extracted from big data and analyzed thoroughly. Fake accounts cannot fully replicate the real ones and thus it is easy to detect them through some discovered patterns.

3.20 *Impact of Malicious Accounts on Political Tweet Sentiment [11]*

The circulation of misinformation in the social media has greatly increased due to social bots. This study mainly determines about the influence of social bots on social media. The dataset consists of 2,942,808 tweets from 705,381 user accounts. Location or account information had no restriction in this case.

Social media can be used as a election campaign where the social bots can have a essential role. Public opinion on the candidates can be measured with the help of the social bots or cyborgs. Bots can tweet about a specific candidate and influence public opinions. And that's why the bot accounts need to be determined to avoid circulation of misinformation. This work shows a result where 3–19% users are shown as bots.

3.21 Compromised Account Detection on Twitter [15]

Many techniques have been deployed to detect compromised accounts. This paper also presents such a technique. The proposed framework learns through tweet history and embeds the knowledge on a feature representation. Reconstruction errors are used to detect suspicious accounts. Only tweets within the USA are considered. 200 previous tweets of each user are collected. There are 5889 users of which 484 are compromised accounts. Users with less than 150 tweets are removed.

This work uses an unsupervised technique that uses nonlinear encodings. This multiview framework can detect compromised accounts more efficiently than the traditional ones. The final result shows 462 compromised accounts and 5065 normal accounts.

3.22 Identifying Spoofing Accounts on Twitter [12]

Spoofing accounts can easily spread misinformation. It is important to distinguish between real and spoofing accounts. This paper proposes a method to detect spoofing accounts using PageRank citation ranking. For the dataset, the accounts of a Japanese girl group were selected. They have verified accounts as well as many spoofing accounts. The authors selected 30 people from the group who were ranked among the top 20 on vote events in the year of 2013, 2014, 2015.

Famous people are likely to have a lot of spoofing accounts which can circulate misinformation. On the most part, fake accounts always have a lot of follows. But real verified accounts of famous people have a lot of followers but not much follows. After conducting the research 96% of real accounts and 75% fake accounts were detected. 27 real accounts and 37 fake accounts were used in this experiment.

3.23 Analyzing the Digital Traces of Political Manipulation [19]

Russia was accused of spreading biased and misinformation in the 2016 election with the help of the bots. This paper has a detailed explanation of the matter. Whether bots were used as a tool for circulating biased information. In this experiment, a large dataset containing 43.7 million tweets are used. This tweets were posted by at least 5.7 million users in the year of 2016.

Social media can be used as a election campaign where the social bots can have a essential role. Public opinion on the candidates can be measured with the help of the social bots or cyborgs. Thus through bots biased information can be shared. With the help of the large dataset, it is found that 4.9% of the liberal users and 6.2% of the conservative users were bots.

3.24 *Interactive Refactoring Documentation Bot [22]*

To check the documentation and validate the re-factorings developers can use the help from the bot. And such a documentation bot with the re-factoring ability was introduced in this paper. In this experiment, a survey was conducted using 14 developers that were quite active to manually evaluate the document. It also needed to be correct as well.

The documentation bots can be used to help the developers to validate the document. The bot needs to check the file level at first. It needs to ensure if it is really documented by the developer. Then it needs to check the re-factoring. Then the documentation bot needs to recommend a message. Thus the bot works to help the developer.

3.25 *Twitter Bot Detection Using Neural Networks and Word Embeddings [25]*

Spam bots can spread spams which is harmful. To help human users to understand who they are contacting with, this paper focuses on the spam bot detection technique. The dataset contains 3474 human accounts and 1455 bot accounts with 8.4 million tweets and 3 million tweets respectively.

This work proposes a technique that can detect a bot with only the tweet information and does not require specific feature extraction. This method can achieve similar performance compared to the traditional one. The feature extraction technique is a very time consuming technique which is not necessary in the proposed system. This is a easier implementation technique.

4 Conclusion

The social media is a widely used platform through which many users can contact with each other. Social bots or cyborgs take this chance to spread misinformation through various fake accounts. Spammers can spread spams with the help of the cyborgs. Fake influencers can cheat through their fake ids. In this paper, all the works related to cyborgs and bots are reviewed thoroughly. As a result data mining techniques have been used to gather and extract information. To understand the role of cyborgs on social media, we have thoroughly checked the behaviors of humans, bots, and cyborgs on Twitter. The summarized work shows the developing rate of identifying bots or cyborgs with the help of data mining. Total 25 papers are reviewed for cyborg data mining. We outlined the contributions, dataset properties and implementation and evaluation for each research.

References

1. S. Gianvecchio, M. Xie, Z. Wu, H. Wang, Measurement and classification of humans and bots in ınternet chat. in *USENIX Security Symposium* (2008 July), pp. 155–170
2. Z. Chu, S. Gianvecchio, H. Wang, S. Jajodia, Who is tweeting on Twitter: human, bot, or cyborg? in *Proceedings of the 26th Annual Computer Security Applications Conference* (IEEE, 2010 December), pp. 21–30
3. R.U. Rahman, D.S. Tomar, S. Das, Dynamic image based captcha. in *2012 International Conference on Communication Systems and Network Technologies* (IEEE, 2012 May), pp. 90–94
4. Z. Chu, S. Gianvecchio, H. Wang, S. Jajodia, Detecting automation of twitter accounts: are you a human, bot, or cyborg? IEEE Trans. Dependable Secure Comput. **9**(6), 811–824 (2012)
5. R. Wald, T.M. Khoshgoftaar, A. Napolitano, C. Sumner, Predicting susceptibility to social bots on Twitter. in *2013 IEEE 14th International Conference on Information Reuse & Integration (IRI)* (IEEE, 2013 August), pp. 6–13
6. M. Ashcroft, A. Fisher, L. Kaati, E. Omer, N. Prucha, Detecting jihadist messages on Twitter. in *2015 European Intelligence and Security Informatics Conference* (IEEE, 2015 September), pp. 161–164
7. J.V. Cossu, N. Dugué, V. Labatut, Detecting real-world influence through Twitter. in *2015 Second European Network Intelligence Conference* (IEEE, 2015 September), pp. 83–90
8. C. Teljstedt, M. Rosell, F. Johansson, A semi-automatic approach for labeling large amounts of automated and non-automated social media user accounts. in *2015 Second European Network Intelligence Conference* (IEEE, 2015 September), pp. 155–159
9. N. Chavoshi, H. Hamooni, A. Mueen, Debot: Twitter bot detection via warped correlation. in *ICDM* (2016 December), pp. 817–822
10. A. Savvopoulos, P. Vikatos, F. Benevenuto, Socialbots' first words: can automatic chatting improve influence in Twitter? in *2018 IEEE/ACM International Conference on Advances in Social Networks Analysis and Mining (ASONAM)* (IEEE, 2018 August), pp. 190–193
11. B. Heredia, J.D. Prusa, T.M. Khoshgoftaar, The impact of malicious accounts on political tweet sentiment. in *2018 IEEE 4th International Conference on Collaboration and Internet Computing (CIC)* (IEEE, 2018 October), pp. 197–202
12. K. Nakasai, M. Tsunoda, H. Hata, K. Matsumoto, Identifying spoofing accounts on Twitter based on relationships of accounts. in *2018 IEEE International Conference on Big Data, Cloud Computing, Data Science & Engineering (BCD)* (IEEE, 2018 July), pp. 85–90
13. Z. Wu, Y. Zhou, Z. Shi, C. Zhang, G. Li, X. Zheng, N. Zheng, G. Pan, Cyborg intelligence: recent progress and future directions. IEEE Intell. Syst. **31**(6), 44–50 (2016)
14. M. Glenski, T. Weninger, S. Volkova, How humans versus bots react to deceptive and trusted news sources: a case study of active users. in *2018 IEEE/ACM International Conference on Advances in Social Networks Analysis and Mining (ASONAM)* (IEEE, 2018 August), pp. 654–661
15. C. VanDam, P.N. Tan, J. Tang, H. Karimi, Cadet: a multi-view learning framework for compromised account detection on Twitter. in *2018 IEEE/ACM International Conference on Advances in Social Networks Analysis and Mining (ASONAM)* (IEEE, 2018 August), pp. 471–478
16. H. Cheng, X. Xing, X. Liu, Q. Lv, ISC: An iterative social based classifier for adult account detection on twitter. IEEE Trans. Knowl. Data Eng. **27**(4), 1045–1056 (2014)
17. S. Sadiq, Y. Yan, A. Taylor, M.L. Shyu, S.C. Chen, D. Feaster, Aafa: Associative affinity factor analysis for bot detection and stance classification in *Twitter. in 2017 IEEE International Conference on Information Reuse and Integration (IRI)* (IEEE, 2017 August), pp. 356–365
18. S. Zenonos, A. Tsirtsis, N. Tsapatsoulis, Twitter influencers or cheated buyers? in *2018 IEEE 16th International Conference on Dependable, Autonomic and Secure Computing, 16th International Conference on Pervasive Intelligence and Computing, 4th International Conference on Big Data Intelligence and Computing and Cyber Science and Technology Congress (DASC/PiCom/DataCom/CyberSciTech)* (IEEE, 2018 August), pp. 236–242

19. A. Badawy, E. Ferrara, K. Lerman, Analyzing the digital traces of political manipulation: the 2016 Russian interference Twitter campaign. in *2018 IEEE/ACM İnternational Conference on Advances in Social Networks Analysis and Mining (ASONAM)* (IEEE, 2018 August), pp. 258–265
20. T. Georgiou, A. El Abbadi, X. Yan, Privacy cyborg: towards protecting the privacy of social media users. in *2017 IEEE 33rd International Conference on Data Engineering (ICDE)* (IEEE, 2017 April), pp. 1395–1396
21. H. Karimi, C. VanDam, L. Ye, J. Tang, End-to-end compromised account detection. in *2018 IEEE/ACM International Conference on Advances in Social Networks Analysis and Mining (ASONAM)* (IEEE, 2018 August), pp. 314–321
22. S. Rebai, O.B. Sghaier, V. Alizadeh, M. Kessentini, M. Chater, Interactive refactoring documentation bot. in *2019 19th International Working Conference on Source Code Analysis and Manipulation (SCAM)* (IEEE, 2019 September), pp. 152–162
23. G. Costantino, A. La Marra, F. Martinelli, A. Saracino, M. Sheikhalishahi, Privacy-preserving text mining as a service. in *2017 IEEE Symposium on Computers and Communications (ISCC)* (IEEE, 2017 July), pp. 890–897
24. L. Caruccio, D. Desiato, G. Polese, Fake account identification in social networks. in *2018 IEEE International Conference on Big Data (Big Data)* (IEEE, 2018 December), pp. 5078–5085
25. F. Wei, U.T. Nguyen, Twitter bot detection using bidirectional long short-term memory neural networks and word embeddings. in *2019 First IEEE International Conference on Trust, Privacy and Security in Intelligent Systems and Applications (TPS-ISA)* (IEEE, 2019 December), pp. 101–109

BDFN: A Bilingual Model to Detect Online Fake News Using Machine Learning Technique

Fahmida Liza Piya, Rezaul Karim, and Mohammad Shamsul Arefin

Abstract The dissemination of online misinformation is causing increasing concern around the world. Government officials and other responsible agencies are deeply worried about the situation and are working tirelessly to change it, but it seems that some yellow journalists are only interested in making money by selling news with clickbait headlines and fake facts inside the news. Some shady online news sources, both in English and Bengali, are actively working to spread false information in order to create a stir. While there are several systems that can detect fake news from English news corpus, we are unaware of any system that detects fake news from both Bengali and English languages at the same time. So, here we propose a bilingual fake news detection model in this paper that employs TF-IDF and N-Gram Analysis for feature extraction in order to detect fake news from a bilingual perspective. In addition, we compare the results of six separate machine learning algorithms for detecting false news. The model employs a supervised method of operation. Among these, we have acquired the highest performance with Linear Support Vector Classification (Linear SVC) algorithm where the accuracy is 93.29% and the F1 score is 0.93.

Keywords Fake news · NLP · Bilingual · Text analysis · TF-IDF · Text analysis · N-gram models · Online news portals

F. L. Piya · M. S. Arefin (✉)
Department of Computer Science and Engineering, Chittagong University of Engineering and Technology, Chattogram 4349, Bangladesh
e-mail: sarefin@cuet.ac.bd

F. L. Piya
e-mail: u1504107@student.cuet.ac.bd

R. Karim
Department of Computer Science and Engineering, University of Chittagong, Chattogram 4331, Bangladesh
e-mail: rezaul.cse@cu.ac.bd

G. Ranganathan et al. (eds.), *Soft Computing for Security Applications*, Advances in Intelligent Systems and Computing 1397,
https://doi.org/10.1007/978-981-16-5301-8_56

1 Introduction

Fake news is described as false stories that appear to be news and are widely disseminated on the internet or by other media, typically to sway political or other opinions or as a joke. Because of its potential to have harmful consequences, the spread of fake news and its dissemination on various platforms has become a major concern.

Prior to the introduction of the Internet, journalists were charged with checking reports and sources and thoroughly checking the details [1]. Therefore, the public's access to fake news was more limited. Today, social media enables the distribution of unverified and inaccurate knowledge to a larger extent, thereby deeply impacting the perception of the readers and interpretation of the events [2].

As a normal human action, authors [3] stressed the lack of interest in checking the authenticity of much of the knowledge found in blogs and virtual encyclopedias and the inherent propensity to trust in the shared content.

The proliferation of fake news has intensified political polarization, reduced confidence in public institutions, and weakened democracy [4]. In the last decade, there has been a dramatic rise in the dissemination of fake news, most notably during the 2016 US elections [5]. Such a flood of posting stories online that do not adhere to reality has resulted in a slew of issues, not only in politics, but in sports, health, and science as well [6]. The financial markets [7] are one of the sectors affected by false news, where a rumor can have negative effects and even bring the economy to a halt.

Even though fake news is not a new phenomenon, identifying it is thought to be a difficult task due to humans' proclivity for believing false information and the lack of control over the dissemination of fake material. Also detecting fake news presents a number of fresh and difficult study issues. In this paper, we present a machine learning-based model that can detect fake news from a bilingual corpus which means a corpus which contains both Bengali and English languages together.

To accomplish this novel task we have chosen the text analysis technique so that we can extract and classify information from the contents of news. Here, we have analyzed both the headline and the main text body of specific news in both Bengali and English. Logistic Regression, Linear SVC, Decision Tree Classifier, Random Forest Classifier, Multinomial Naive Bayes, and Passive Aggressive Classifier these six machine learning models were compared and used for performing classification to recommend the best model for our fake news detection system.

2 Motivation

Several content based solutions have been proposed in recent years based on text mining approach but there is a problem that denotes most of them were confined to one specific language only. Till now, most of the work done on fake news detection revolves around the language English like the work in Ref. [8], also there is a work done in both Chinese and English language [9], another one in Spanish [10] and in

Bengali [11]. However, fake news detection from online news is still in the early stages of growth, and many difficult problems still need to be investigated further. And previously these works have been related to fake news detection specifically in English or Bengali but very few were done in bilingual approach. According to our research, not a single work has been done that has the ability to detect fake news from a context that can combine both Bengali and English languages. A novel model like this can help in advanced implementations where fake news can be identified regardless of whether the text in the news is in Bengali or English.

3 Contribution of Our Work

The aim of a thesis or research project is to achieve a particular set of objectives, such as defining a new approach or improving an existing one. A thesis or research project should add to the advancement of human knowledge by providing new information. In this thesis of ours, the main focus was to develop e framework for detecting fake news from a bilingual perspective.

- Feature extraction from bilingual dataset using TF-IDF along with N-Grams
- To identify fake news and authentic news from different news articles of online news portals from a bilingual perspective.
- Apply text analysis technique and compare six different machine learning techniques to compare performance.

4 Related Literature Review

A straightforward strategy was suggested to detect fake news in social media networks using certain data from articles such as title, use of title numbers, and capital letters in title, on-site user behavior, and occurrence of title phrases in content and title keywords [12]. A survey was performed by authors [13] analyzing different fake news assessment methods based on linguistic methods and network analysis approach.

Authors [14] suggested that machine learning techniques can be used to detect fake news. Three common approaches are used in their experiments: Naive Bayes, Neural Networks, and Support Vector Machines. The normalization process, according to this study, is an essential step in cleaning data before using the machine learning method to classify it.

In an article [15], the fake news problem is explored by authors by conducting a two-phase analysis of current literature: characterization and detection. During the characterization process, they presented the basic concepts and principles of fake news in both traditional and social media channels. During the detection process, they looked at existing fake news detection approaches from a data mining perspective, including feature extraction and model formation. They have also discussed the datasets, evaluation matrices, and promising future directions in this specific field.

Authors [16] have identified clickbaits and installed browser extensions to warn and prevent users from accessing this type of news with clickbait headlines. To identify false or true claims in social media, websites, and other news sources authors have taken the interaction between the language of the claim and the web sources reliability jointly in a work [17]. In Ref. [18], dishonest opinions were identified by using applied stylometric tools, i.e., lexical and syntactic; Sequential Minimal Optimization (SMO), Naive Bayes and Support Vector Machine (SVM).

N-Gram analysis for a fake news detection model has been used and machine learning techniques have also been used [19]. In this paper, they have got the highest accuracy of 92% using Linear Support Vector Machine (LSVM) which is a classifier and along with this TF-IDF has been used as the feature extractor. Wang [20] took a slightly different approach. For training various ML models, the author used textual features and metadata. The author concentrated on convolutional neural networks (CNN). To capture the dependency between the metadata vectors, a convolutional layer is used, followed by a bidirectional LSTM layer.

In another machine learning-based framework, it uses term frequency-inverse document frequency (TF-IDF) of bag-of-words and the feature extraction technique and their classifier is Support Vector Machine (SVM) [21] for detecting fake news.

Taking into account the content of online news stories, the authors in their work [22] suggested a fake news identification method and investigate two machine learning algorithms using word and character n-grams analysis. Their tests with character n-grams and the Term Frequency-Inverted Document Frequency (TF-IDF) and Gradient Boosting Classifier produce better performance, with a 96% accuracy [22].

Another machine learning-based fake news identification model employs an ensemble approach [23]. Multiple learning algorithms are used in ensemble methods in statistics and machine learning to achieve greater predictive precision than any of the constituent learning algorithms alone. The research looks into various textual properties that can be used to tell the difference between fake and real content. The authors used various ensemble methods to train a variety of machine learning algorithms, then evaluated their output on four real-world datasets. In contrast to individual learners, the proposed ensemble learner method performed better in experiments.

In a recent work [24] on Bangla fake news detection system authors have analyzed their recently released Bangla dataset and created a benchmark framework using cutting-edge NLP techniques to detect Bangla fake news. Zobaer Hossain et al. [24] have used conventional linguistic features as well as neural network-based approaches to develop this framework.

5 Diagram/Overview of Framework

In Fig. 1, the overall mechanism for identifying fake news from a bilingual corpus is addressed. It depicts the step-by-step process for constructing the final model to identify fake news. The first step towards creating our bilingual model is to preprocess the dataset. We have two different datasets initially. The preprocessing step is conducted at first. The steps are mentioned in Fig. 2.

First, we have tokenized them using NLTK, conducted stemming using porter stemmer then removed stopwords. We uploaded a list of Bengali stopwords in our python environment while removing stopwords from the Bengali newspaper dataset. And, the English stop words have been removed using NLTK. Then punctuations

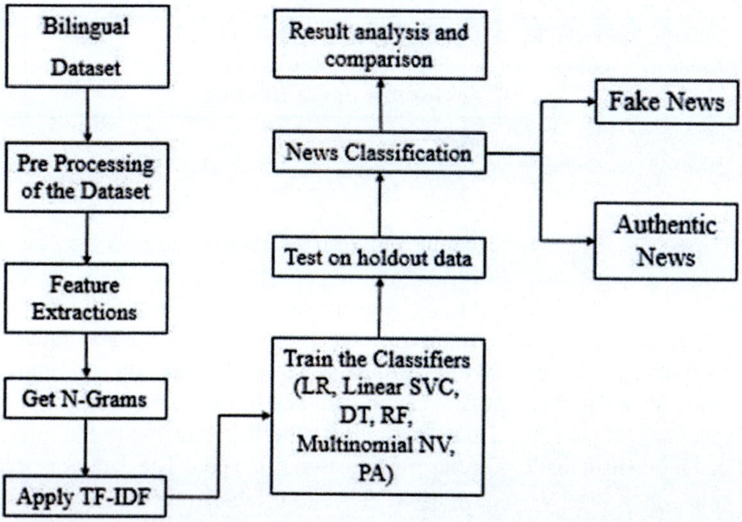

Fig. 1 Framework of Bilingual model for fake news detection

Fig. 2 Preprocessing of Bilingual dataset for fake news detection

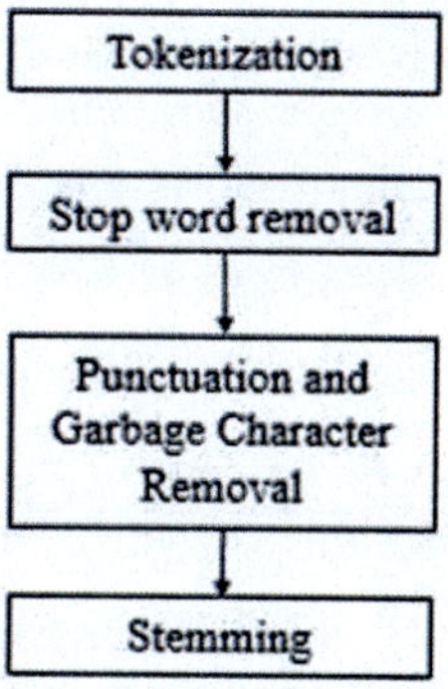

Title	Text	Category	Label
হঠাৎ আফগান ক্রিকেট বোর্ড প্রধানের পদত্যাগ	ক্রিকেট বিশ্বের নতুন চমকের নাম আফগানিস্তান। কয়েক বছরে তাদের পারফরম্যান্স ...	Sports	1
বিশ্বের ১৭৩ জনের মধ্যে ৩য় সৎ সরকার প্রধান শেখ হাসিনা	পিপলস অ্যান্ড পলিটিক্স বিশ্বের পাঁচজন সরকার রাষ্ট্রপ্রধানকে চিহ্নিত করেছেন, যাদের দুর্নীতি স্পর্শ করেনি, বিদেশে কোনো ব্যাংক অ্যাকাউন্ট নেই....	National	0
Muslims in Asia protest against Trump's Jerusalem plan	Thousands of protesters in Muslim-majority countries in Asia rallied on Friday to condemn the U.S. decision to recognize	worldnews	1
Donald Trump Sends Out an Embarrassing Message!	Donald Trump Sends Out Embarrassing New Years Eve Message; This is Disturbing	News	0

Fig. 3 Sample of Bilingual dataset

and garbage characters have been removed from the dataset. Figure 3 shows a sample example of our Bilingual Dataset.

After cleaning the text corpus we implemented the N-Gram model. We have integrated this model in our classification technique as it accelerates the accuracy greatly. Bigrams, Trigrams, and even four grams are extracted in this process to compare and evaluate the accuracy of the fake news detection model. Now taking these sequences of words TF-IDF is applied. The last step is the training of the classifier. Our dataset that has been pre-processed is labeled. The dataset is split into 70:30 ratio which means we have used 70% of our labeled dataset for training our classifier so that later on it can identify any news as true or fake based on its previous training experience using N-Gram features and TF-IDF scores.

Figure 4 depicts, during training our bilingual model, feature extractor (Tf-Idf and N-Gram model) is used to convert each input news content into a feature matrix. Combinations of feature sets and the labels of the news contents are fed into machine learning algorithms. In the testing or classification part, while performing classification, the exact same feature extractors are used to identify feature sets. Features are fed into the machine learning models to perform classification of the news articles.

5.1 Dataset Preparing

For our novel model, we needed a dataset that included both English and Bengali texts. So, after some significant processing, we merged two separate datasets. For the

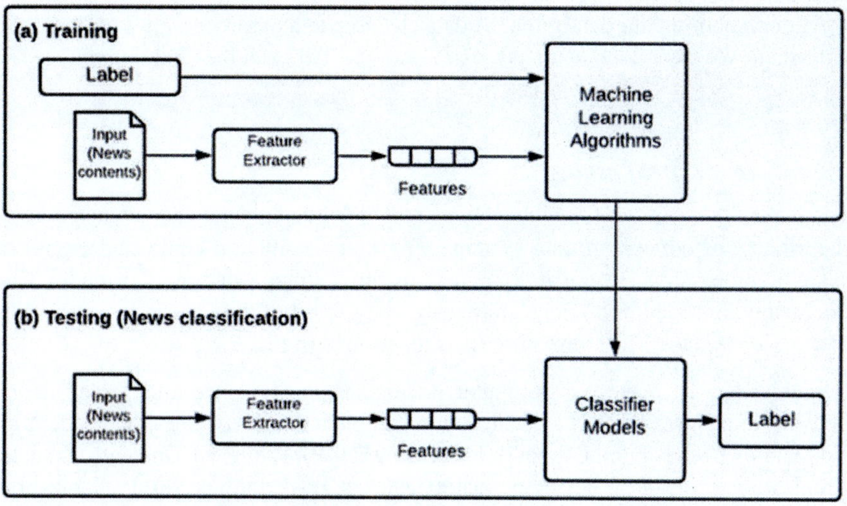

Fig. 4 Fake news classification technique

Bengali texts we used a recently available public dataset [24]. Here, the authors have proposed a dataset for fake news detection containing 50K annotated data. There are 12 categories of data. We have used 7k authentic news and 1k fake news from this specific dataset to build our model. The data that has been selected is labeled as we are approaching building a supervised model. The information columns that the dataset contains are- Label, domain Name, Published Time, Category, Source, Headline-article relation, Headline, Article. We have to drop some of the columns to match with our English Dataset.

In addition, for our English news content, we used a dataset that the authors published in Ref. [19]. This specific dataset is known as ISOT Fake news dataset. There are two types of articles-True and Fake. There are more than 12,600 articles collected from Reuters which are true and also more than 12,600 fake articles collected from various sources. Each article contains the following information: article title, text, type, and date. And to match with the Bengali dataset we labeled both the true and fake articles. The true articles are labeled as '1' and the fake articles are labeled as '0'. To combine these two datasets in different languages we had to drop some of the columns and rename them. Then, we worked with Title, Text, Category, Date, and Label.

- **Title**: Short title text that seeks to grab readers' attention and defines the article's main subject.
- **Text**: This is the main text which elucidates the news story's information.
- **Category**: It shows the category of the news article. Ex: Sports, Politics, etc.
- **Date**: Denotes when the story was published.
- **Label**: Denotes if the news story is true or fake.

After combining our dataframe the data cleaning and preprocessing step has been performed. We have done text analysis of both the Title and the Text column.

5.2 *Data Preprocessing*

The process of converting data to something that a computer could understand is called preprocessing. Before applying any feature extractor or classifier model data processing is needed to be done so that we can get rid of unnecessary information existing in the data. The entire process is mentioned in Fig. 2.

Tokenization: To make our computer understand the text, we must break down the word in a way so that our machine can understand. Here, comes the concept of tokenization from Natural Language Processing (NLP). While working with text data tokenization is one of the key or mandatory aspects. If we don't perform tokenization we will not be able to work with text data. In our bilingual dataframe, we have simply conducted splitting for further cleaning and preprocessing tasks. We have tokenized the texts of both the columns naming Headline and Text. Example of tokenized text is shown in Figs. 5 and 6.

Stop Word Removal: Information extraction is a result of text data mining where unnecessary words that are very repetitive and common which is actually called the Stop word can create noise. Some of the much generalized examples of stopwords are—a, an, the, on, at, which, etc. But there is no universal single list for stop words. These stopwords can take a lot of space and also a large amount of processing time. We can conveniently delete them by saving a list of words. For our English data, we have used the Natural Language Toolkit shortly known as NLTK. In python, we simply needed to improve the nltk and the stopwords for English are already specified in the library. Then from our Headings and Text, we can filter out the stopwords. And for removing the Bengali stopwords we provided a list of custom stopwords in order to filter out these from the Text and Headlines of the Bengali news. Some of the stopwords that have been filtered out are shown in Fig. 7.

Text	Tokenized Text
ক্রিকেট বিশ্বের নতুন চমকের নাম আফগানিস্তান	ক্রিকেট, বিশ্বের, নতুন, চমকের, নাম,আফগানিস্তান
এশিয়া কাপের ষষ্ঠ ম্যাচে বাংলাদেশ দলের বিপক্ষে	এশিয়া, কাপের, ষষ্ঠ, ম্যাচে, বাংলাদেশ, দলের, বিপক্ষে
রাজধানীতে মাদক বিরোধী বিশেষ অভিযান পরিচালনা	রাজধানীতে, মাদক, বিরোধী, বিশেষ, অভিযান, পরিচালনা

Fig. 5 Bengali tokenized text

Text	Tokenized Text
Pope Francis Just Called Out Donald Trump During His Christmas Speech	Pope, Francis, Just, Called, Out, Donald, Trump, During, His, Christmas, Speech.
10 Ways America Is Preparing for World War 3	10, Ways, America, Is, Preparing, for, World, War, 3.
Coronavirus can be spread through contact with food or food packaging	Coronavirus, can, be, spread, through, contact, with, food, or, food, packaging.

Fig. 6 English tokenized text

অতএব, অথচ, অথবা, অনুযায়ী, অনেক, অনেকে, অনেকেই, অন্তত, অন্য, অবধি, অবশ্য, অর্থাৎ, আই, আগামী, আগে, আগেই, আছে, আজ, আদ্যভাগে, আপনার, আপনি, ই, ইত্যাদি, ইহা, উচিত, উত্তর, এ, এদের, কিন্তু, কী, কে, জানায়, জানিয়ে, জানিয়েছ

Fig. 7 List of Bengali stopwords

Punctuation and Garbage character Removal: Cleaning up data means getting rid of the noise in the dataset. In text cleaning, we may have to get rid of all identifiers, HTML instances, punctuation, non-alphabets, and all other characters that might not be part of the language. Some of the punctuation that has been considered while filtering them out is shown in Fig. 8. As we have considered only news articles from different news portals there were no emoticons or characters (which are very much common while working with social media data) found upon manually searching. So, we didn't need to perform any procedure to filter out them. These punctuation and garbage characters can take up a lot of space and increase processing time and as in our model, we are considering the text of our news we filtered out these garbage characters from our dataset to make this a clearer one.

[,(,), $, %, ^, &, *, +, =, {, }, \, [\], :, \, ", |, \'\, ~, `, <, >, /, ¦, !, ?, ½, £, ¶, ¼, ©,
⅐, ⅑, ⅒, ⅓, ⅔, ⅕, ⅖, ⅗, ⅘, ⅙, ⅚, ⅛, ⅜, ⅝, ⅞, 1/0/3, ¤, ¿, °, ;, -,], +....

Fig. 8 Example of punctuation and garbage characters

Stemming: Stemming is the act of reducing a word to its word stem which is appended to suffixes and prefixes or to the roots of words such as lemmas. Stemming is important in understanding and processing the natural language. Stemming can be done by defining some rules. In our preprocessing step we have used the Porter Stemming Algorithm. This is a method of deleting the commoner morphological and inflexional endings from English words. The main use of this is a part of a process of standardization of terms usually performed when setting up information recovery systems. This specific part of preprocessing has helped us to have higher accuracies in some of the models.

5.3 Feature Extraction

Having a huge amount of data it is very important to get rid of unimportant and redundant information and to prepare the dataset that can be applicable to any other machine learning models or applications or algorithms. Due to the huge amount of text data produced in many ways, such as social networks, hospital reports, news media, the NLP has gained so much attention in recent years. It is impossible for humans to go through all of the text data and locate useful information and coordinate vast volumes of data. To have an overview of the method of translation and representation of these data involves measuring word frequencies from the text and in all of the data set. Hence, extracting the correct information is necessary. The mapping of real valued vectors from textual data is called feature extraction. The widely used method of extraction of the feature is TF-IDF. TF-IDF technique is improved for extraction of features for greater performance of a model [25, 26]. We have used TF-IDF alongside the N-Gram model.

N-Gram: It is merely a sequence of n items from a text corpus or a document. These items can be words, letters, phonemes, syllables, or base pairs as needed. This is a very popular approach in the field of Natural Language Processing (NLP). Speech or text corpus is used to get the N-Grams. These have been used in some of the most successful language models. $N = 1, 2, 3$ denotes the unigram, bigram, trigram respectively. A bigram ($n = 2$) is a series of two adjacent components that are usually letters, syllables, or words from a set of tokens. The frequency distribution of each bigram in a string is widely used in many text analyses including computational linguistics, cryptography, voice recognition, etc. A trigram ($n = 3$) is a series of two adjacent components. These are also used in the interpretation of natural language for the mathematical study of texts and for the detection and usage of ciphers and codes in cryptography.

In our work, we have used word-based N-Gram features to see how unigram, bigram, and trigram impact the overall performance of the model and this was one of the feature extractors. And, as we have observed while we used bigrams in our six different models we started getting better accuracy with an enriched model but after using the trigrams and four grams the model performance declined though in a very small range. The N-grams along with the TF-IDF extracted the features for our detection model. Our model is trained with these and later on, based on this training experience the model identified fake and authentic news quite successfully. In Fig. 9 a basic example of N-Gram feature extraction from Bengali sentence and in Fig. 10 a basic example of N-Gram feature extraction from English sentence is shown.

TF-IDF: It stands for Term frequency-Inverse Document Frequency. The value of the TF-IDF increases proportionally to the number of times a term appears in the text and decreases with the number of documents containing the word in the corpus.

Sentence = 'হঠাৎ আফগান ক্রিকেট বোর্ড প্রধানের পদত্যাগ'	n=2	('হঠাৎ', 'আফগান') ('আফগান', 'ক্রিকেট') ('ক্রিকেট', 'বোর্ড') ('বোর্ড', 'প্রধানের') ('প্রধানের', 'পদত্যাগ')
	n=3	('হঠাৎ', 'আফগান', 'ক্রিকেট') ('আফগান', 'ক্রিকেট', 'বোর্ড') ('ক্রিকেট', 'বোর্ড', 'প্রধানের') ('বোর্ড', 'প্রধানের', 'পদত্যাগ')

Fig. 9 Example of N-Gram feature extraction from Bengali

Sentence = 'On Christmas day, Donald Trump announced'	n=2	('On', 'Christmas') ('Christmas', 'day,') ('day', 'Donald') ('Donald', 'Trump') ('Trump', 'announced')
	n=3	('On', 'Christmas', 'day,') ('Christmas', 'day,', 'Donald') ('day', 'Donald', 'Trump') ('Donald', 'Trump', 'announced')

Fig. 10 Example of N-Gram feature extraction from English

$$\text{tf(wd, doc)} = \text{fdoc(wd)} : \text{Frequency of wd in document doc}$$
$$\text{idf(wd, DOC)} = \log 1 + |\text{DOC}|1 + \text{df(doc, wd)}$$
$$\text{tfidf(wd, doc, DOC)} = \text{tf(wd, doc)idf(wd, DOC)}$$

It consists of two subparts. These are TF and IDF. Actually, in simple words, TF-IDF is a product of TF and IDF.

Term Frequency (TF): Term frequency determines how often a word (wd) appears in the text or document (doc) as a whole. It measures the number of times a word exists in a text document with respect to the overall number of words in the following text document.

Inverse Document Frequency (IDF): The inverse frequency of a text is an indicator of whether a word (wd) in the whole text corpus is unusual or common across the records. The symbol DOC is the representation of the entire text corpus means the number of documents existing there in the text corpus in total. And, df (doc, wd) represents the number of documents in which the word appears. Then as a result of idf (wd, DOC) a value is achieved which is logarithmically scaled and it denotes the number of documents in the corpus divided by the number of times word wd appears in the entire corpus. A word which is quite prevalent in the entire text corpus will have a lesser TF-IDF value because of the denominator of idf.

The explanation for combining N-Grams and TF-IDF is that we found that using bigrams and trigrams in different machine learning models improved efficiency when detecting bilingual fake news. TF-IDF, despite being one of the most common feature extraction techniques in the text analysis area, could not outperform this. Similar results were discussed in another study in which the authors [19] only dealt with English news.

We have also applied Bag-of-words which is one of the most common methods in text analysis and specifically in feature extraction, it didn't provide us expected performance as it captures some issues which are not so prominent in the entire dataset. It doesn't care how many times a word appears or the order in which the words appear, all that matters is if the word is in a list.

6 Experimental Setup

6.1 Dataset

In our final bilingual dataset, there are 42,324 data in total. The bilingual data frame contained mainly 5 columns in total which we have chosen for training and testing our model. The data which were authentic has been labeled with 1 and the data which were fake has been labeled with 0 (Table 1).

Table 1 Bilingual dataset

Column name	Number of contents
Title	42,324 non-null object
Text	42,324 non-null object
Category	42,324 non-null object
Date	42,324 non-null object
Label	42,324 non-null object

6.2 Experiments Procedure

We have split our dataset using train-test-split from scikit learn. While performing a machine learning experiment it's a very foolish mistake if the same dataset is used for both the training and testing data because it will give us the perfect accuracy score but will fail to perform well later on any unseen data. That's why holding out a part of the dataset is very important. We have used 70 percent of our dataset for training purposes and 30 percent of our dataset for testing purposes. As we wanted to shuffle the dataset before splitting we set the parameter Shuffle = True. Then we used the different classification models, trained them with our bilingual data, and later on tested them with the holdout spread.

6.3 Evaluation Metrics

Various evaluation criteria have been used to measure the efficiency of algorithms for the fake news detection problem. These metrics are widely used in the machine learning community and enable us to assess a classifier's output from various angles.

Accuracy is a metric that compares the resemblance between expected and actual false news. Precision is a metric that calculates the percentage of all detected false news that is annotated as fake news, resolving the critical issue of determining which news is fake. Due to the distorted nature of fake news datasets, a high level of accuracy may be easily accomplished by making less optimistic forecasts. As a result, recall is used to gauge sensitivity or the percentage of annotated fake news stories that are predicted to be fake news. F1 is used to combine precision and recall, resulting in a prediction value for detecting false news. The higher the value for Precision, Recall, F1, and Accuracy, the better the results.

- **True Positive(TP)**: When the news article which was predicted as fake news is eventually labeled as fake news
- **True Negative(TN)**: When the news article which was predicted as true news is eventually labeled as true news
- **False Negative(FN)**: When the news article which was predicted as true is eventually labeled as fake news

- **False Positive (FP)**: When the news article which was predicted as false is eventually labeled as true news

Accuracy: It is the most common indicator of performance. And this denotes the ratio of correctly predicted observations and total observations. Only higher accuracy doesn't always prove that the model performs really well. But, if the dataset is well balanced then definitely accuracy is the best choice. We have measured the accuracy of each of our models along with the other performance metrics.

$$\text{Accuracy} = \frac{|\text{TP}| + |\text{TN}|}{|\text{TP}| + |\text{TN}| + |\text{FP}| + |\text{FN}|}$$

Precision: It is also called the positive predicted value. Precision is the ratio of accurately forecast positive outcomes to overall expected positive results. A perfect 1.0 accuracy score means any search result was relevant.

$$\text{Precision} = \frac{|\text{TP}|}{|\text{TP}| + |\text{FP}|}$$

Recall: It is also known as sensitivity. Recall is the percentage of correctly expected positives to the real class findings. But, Precision is more important than recall you choose to get less False Positives off in exchange to get more False Negatives.

$$\text{Recall} = \frac{|\text{TP}|}{|\text{TP}| + |\text{FN}|}$$

F1 Score: In most cases, F1 Score is more useful than accuracy when the dataset is not that balanced. Basically, it's the weighted average of precision and recall. It takes both the false positives and false negatives into account which makes it a bit more complicated than accuracy. An F1 score of 1.0 is considered perfect.

$$\text{F1 Score} = 2 \cdot \frac{\text{Precision} \cdot \text{Recall}}{\text{Precision} + \text{Recall}}$$

7 Results and Analysis

For our bilingual fake news detection model, we used six separate machine learning algorithms. They are as follows: (i) Logistic Regression, (ii) Linear Support Vector Classification, (iii) Decision Tree, (iv) Random Forest Classifier, (v) Naive Bayes Multinomial, (vi) Passive Aggressive. We evaluated the output of our six machine learning models using not only accuracy but also using precision, recall, and F1 score

Table 2 Experiments results of six different machine learning models using the main text of the news articles

Machine learning model	N-Grams	Precision	Recall	F1 score	Accuracy (%)
Logistic regression	Unigram	0.92	0.92	0.92	92.23
	Bigram	0.92	0.92	0.92	92.47
	Trigram	0.92	0.92	0.92	91.99
Linear SVC (support vector classifier)	Unigram	0.93	0.93	0.93	92.83
	Bigram	0.89	0.89	0.89	93.29
	Trigram	0.93	0.93	0.93	92.83
Decision tree classifier	Unigram	0.92	0.92	0.92	91.53
	Bigram	0.89	0.89	0.89	91.68
	Trigram	0.91	0.92	0.92	91.61
Random forest classifier	Unigram	0.91	0.91	0.91	91.27
	Bigram	0.91	0.91	0.91	90.79
	Trigram	0.90	0.90	0.90	90.15
Multinomial Naïve Bayes	Unigram	0.84	0.84	0.84	83.52
	Bigram	0.81	0.81	0.81	81.11
	Trigram	0.82	0.81	0.80	80.7
Passive aggressive classifier	Unigram	0.92	0.92	0.92	92.38
	Bigram	0.89	0.89	0.89	93.09
	Trigram	0.93	0.93	0.93	93.11

because we don't have equivalent data for Bengali and English, and our dataset is imbalanced based on the labels.

Table 2 represents the overall data analysis of the text data of news articles. The model was also trained on the headlines of the news articles but as there is less number of words in the headlines it performed poorly. So, the results of models that have been trained on the text body are represented only. In most of the cases, the model performed really well as the precision, recall, F1 score is almost perfect sometimes.

Linear SVC performed best for our data. If we use bigrams it gives the highest accuracy and if we use trigrams for feature extraction it gives us the highest F1 score. Logistic Regression showed better accuracy with bigrams and it has a great performance too.

Decision Tree performs very poorly while it is trained on the headline of the news contents with an F1 score of 0.71 only. And, it gives us a great performance using both unigrams and trigrams while training on the main text. Random forest performed very poorly using the data of the Headline column. But, it showed a great F1 score while analyzing Text data. The reason behind this is, while training on the headlines there were less words to consider than the texts of the news. On the other hand, Random Forest Classifier showed pretty good results while training on the headlines too.

The Multinomial Naive Bayes comparatively showed poor results on both the headlines and the texts. The highest accuracy we have got is 87.43% with bigrams that too using the headlines. The Passive Aggressive Classifier showed performance almost as good as Linear SVC both on headings and texts as shown in Table 2. The F1 score denotes the model performs really satisfactorily while using trigrams.

Six of the model employs supervised method of operation. Among these, we have acquired the highest performance with Linear Support Vector Classifier (Linear SVC) and Passive Aggressive Classifier where the highest accuracy is 93.29% & 93.11% respectively and the F1 score is 0.93 for both. The Linear Support Vector Classifier, however, is the best of the six since it gives us a decent F1 score using both unigrams and trigrams. In the case of the passive aggressive classifier, we must go all the way to trigram to get the maximum performance.

8 Conclusion

In this paper, we developed a model to detect fake news from a bilingual dataset consisting Bengali and English language articles. There are several works in which authors detected fake news from various perspectives, but they often concentrated on a single language. Some of the works addressed the bilingual approach, but to the best of our knowledge, no work has been done to identify fake news from a corpus that includes both Bengali and English.

In addition, we compared and evaluated the output of six different machine learning techniques using various feature extraction techniques. This novel model of ours can be used further to create apps that can filter out fake news from a news portal, regardless of whether the language is Bengali or English.

There are numerous fact-checking websites available these days, most of which actually focus on Bengali news fact-checking in order to improve the internet, but the majority of them are still carried out manually. If an automated mechanism can identify and filter out fake news, it will have a significant impact on the world of online news portals.

After analyzing the performance of six different machine learning models exploring with the N-Gram feature extractor in our work we have got the highest performance with Linear Support Vector Classifier. If we consider accuracy as the performance metric we acquired the highest accuracy using Bigrams and if we take F1-score as our performance metric we have acquired the highest accuracy while using Trigrams. As our dataset is not balanced F1-Score will be the best evaluation metric for our model. In the future, we will be exploring more features and aspects of fake news with an enriched set of data.

There are also some limitations in our proposed work. Data shortage is a major concern when conducting this type of task. As Bengali is a low-resource language, and manually annotating data in Bengali is a difficult job we worked with limited amount of data. However, the more data we collect, the better our model becomes at detecting false news. In the future we can work with vast amount of annotated data to train our model in a better way. Also, since it was trained on a small set of data, it may not work very well on real-time data. We must continuously train our model with a large amount of data in order to achieve real-time efficiency. In the future, once we have a sufficient amount of data, we will experiment with various deep learning models.

References

1. V.L. Rubin, N. Conroy, Discerning truth from deception: human judgments and automation efforts. First Monday **17**, (2012). https://doi.org/10.5210/fm.v17i3.3933
2. A. Zubiaga, A. Aker, K. Bontcheva, M. Liakata, R. Procter, Detection and resolution of rumours in social media: a survey. ACM Comput. Surv. **51**, 32 (2017). https://doi.org/10.1145/3161603
3. C. Shao, G.L. Ciampaglia, O. Varol, K.C. Yang, A. Flammini, F. Menczer, The spread of low-credibility content by social bots. Nat. Commun. **9**, 1–9 (2018). https://doi.org/10.1038/s41467-018-06930-7
4. M. Granik, V. Mesyura, Fake news detection using Naive Bayes classifier. in *2017 IEEE 1st Ukraine Conference on Electrical and Computer Engineering, UKRCON 2017—Proceedings* (Institute of Electrical and Electronics Engineers Inc., 2017), pp. 900–903. https://doi.org/10.1109/UKRCON.2017.8100379
5. PolitiFact—2016 Lie of the Year: Fake news. https://www.politifact.com/article/2016/dec/13/2016-lie-year-fake-news/. Last accessed 14 April 2021
6. D.M.J. Lazer, M.A. Baum, Y. Benkler, A.J. Berinsky, K.M. Greenhill, F. Menczer, M.J. Metzger, B. Nyhan, G. Pennycook, D. Rothschild, M. Schudson, S.A. Sloman, C.R. Sunstein, E.A. Thorson, D.J. Watts, Z.L. Zittrain, The science of fake news: addressing fake news requires a multidisciplinary effort. Science **80**(359), 1094–1096 (2018). https://doi.org/10.1126/science.aao2998
7. S. Kogan, T.J. Moskowitz, M. Niessner, Fake news: evidence from financial markets. SSRN Electron. J. (2018). https://doi.org/10.2139/ssrn.3237763
8. H. Rashkin, E. Choi, J.Y. Jang, S. Volkova, Y. Choi, Truth of varying shades: analyzing language in fake news and political fact-checking. in *Proceedings of the 2017 Conference on Empirical Methods in Natural Language Processing* (Association for Computational Linguistics, Stroudsburg, PA, USA, 2017), pp. 2931–2937. https://doi.org/10.18653/v1/D17-1317
9. Y. Zhu, X. Gao, W. Zhang, S. Liu, Y. Zhang, A Bi-Directional LSTM-CNN model with attention for aspect-level text classification. Future Internet **10**, 116 (2018). https://doi.org/10.3390/fi10120116
10. M. P. del Salas-Zárate, M.A. Paredes-Valverde, M.A. Rodriguez-Garćıa, R. Valencia-Garćıa, G. Alor-Hernández, Automatic detection of satire in Twitter: a psycholinguistic-based approach. Knowl.-Based Syst. **128**, 20–33 (2017). https://doi.org/10.1016/j.knosys.2017.04.009
11. F. Islam, M.M. Alam, S.M. Shahadat Hossain, A. Motaleb, S. Yeasmin, M. Hasan, R.M. Rahman, Bengali fake news detection. in *2020 IEEE 10th International Conference on Intelligent Systems, IS 2020—Proceedings* (Institute of Electrical and Electronics Engineers Inc., 2020), pp. 281–287. https://doi.org/10.1109/IS48319.2020.9199931
12. M. Aldwairi, A. Alwahedi, Detecting fake news in social media networks. Procedia Comput. Sci. 215–222 (2018). https://doi.org/10.1016/j.procs.2018.10.171
13. N.K. Conroy, V.L. Rubin, Y. Chen, Automatic deception detection: methods for finding fake news. Proc. Assoc. Inf. Sci. Technol. **52**, 1–4 (2015). https://doi.org/10.1002/pra2.2015.145052010082
14. S. Aphiwongsophon, P. Chongstitvatana, Detecting fake news with machine learning method. in *ECTI-CON 2018—15th International Conference on Electrical Engineering/Electronics, Computer, Telecommunications and Information Technology* (Institute of Electrical and

Electronics Engineers Inc., 2019), pp. 528–531. https://doi.org/10.1109/ECTICon.2018.8620051

15. K. Shu, A. Sliva, S. Wang, J. Tang, H. Liu, Fake news detection on social media: a data mining perspective (2017) https://doi.org/10.1145/3137597.3137600
16. A. Chakraborty, B. Paranjape, S. Kakarla, N. Ganguly, Stop clickbait: detecting and preventing clickbaits in online news media. in *Proceedings of the 2016 IEEE/ACM International Conference on Advances in Social Networks Analysis and Mining, ASONAM 2016* (Institute of Electrical and Electronics Engineers Inc., 2016), pp. 9–16. https://doi.org/10.1109/ASONAM.2016.7752207
17. K. Popat, S. Mukherjee, J. Stro¨tgen, G. Weikum, Credibility assessment of textual claims on the web. in *International Conference on Information and Knowledge Management, Proceedings* (Association for Computing Machinery, New York, NY, USA, 2016), pp. 2173–2178. https://doi.org/10.1145/2983323.2983661
18. S. Shojaee, M.A.A. Murad, A. Azman, N.M.A. Bin Sharef, S. Nadali, Detecting deceptive reviews using lexical and syntactic features. in *International Conference on Intelligent Systems Design and Applications, ISDA* (IEEE Computer Society, 2014), pp. 53–58. https://doi.org/10.1109/ISDA.2013.6920707
19. I. Ahmad, M. Yousaf, S. Yousaf, M.O. Ahmad, Fake news detection using machine learning ensemble methods. Complexity 2020, (2020). https://doi.org/10.1155/2020/8885861
20. W.Y. Wang, "Liar, Liar pants on fire": a new benchmark dataset for fake news detection. in *Proceedings of the 55th Annual Meeting of the Association for Computational Linguistics*, vol. 2: Short Papers (Association for Computational Linguistics, Stroudsburg, PA, USA, 2017), pp. 422–426. https://doi.org/10.18653/v1/P17-2067
21. N.F. Baarir, A. Djeffal, Fake News detection Using Machine Learning. Presented at March 19 2021. https://doi.org/10.1109/ihsh51661.2021.9378748
22. H.E. Wynne, Z.Z. Wint, Content based fake news detection using N-gram models. in *ACM International Conference Proceeding Series* (Association for Computing Machinery, New York, NY, USA, 2019), pp. 669–673. https://doi.org/10.1145/3366030.3366116
23. I. Ahmad, M. Yousaf, S. Yousaf, M.O. Ahmad, Fake news detection using machine learning ensemble methods. Complexity (2020). https://doi.org/10.1155/2020/8885861
24. M. Zobaer Hossain, M. Ashraful Rahman, M. Saiful Islam, S. Kar, *BanFakeNews: A Dataset for Detecting Fake News in Bangla* (2020)
25. L.H. Patil, M. Atique, A novel approach for feature selection method TF-IDF in document clustering. in *Proceedings of the 2013 3rd IEEE International Advance Computing Conference, IACC 2013* (2013), pp. 858–862. https://doi.org/10.1109/IAdCC.2013.6514339
26. L.P. Jing, H.K. Huang, H.B. Shi, Improved feature selection approach TFIDF in text mining. in *Proceedings of 2002 International Conference on Machine Learning and Cybernetics* (2002), pp. 944–946. https://doi.org/10.1109/icmlc.2002.1174522

Deep Learning Framework for Lung Disease Prognosis Using X-Ray Image

Shruti Patil and Bhavana Tiple

Abstract Lung inconsistency is one of the ordinary infections in individuals of all ages on social occasions and this disease may arise on account of various reasons. Lately, lung defilement on account of SARS-CoV-2 has impacted a greater human neighbourhood, and because of its speed, the World Health Organization (WHO) reported it as a pandemic ailment. The COVID-19 ailment successfully influences the respiratory framework, and the defilement earnestness can be perceived using a picked imaging system. In the proposed research work, the COVID-19 is distinguished using move acquiring from X-ray images. The model is ordered into four classes as COVID-19, normal, pneumonia, and tuberculosis. A two-stage recognition model is proposed to improve the revelation exactness and the frameworks are according to the accompanying information increase and information pre-processing. Phase2-COVID-19 recognition utilizing pre-prepared CNN model. This work has thought about the notable pre-prepared designs, for instance, VGG-16, Xception, InceptionV3, MobileNet, and MobileNetV2 for the preliminary appraisal. In this work, 70% of pictures are considered to set up the training and 30% of pictures are considered to support the testing dataset. The presentation of the considered designs is assessed by figuring the regular exhibition measures. The evaluation parameters considered in this work are accuracy, precision, F1-score, and recall.

Keywords COVID-19 · Pneumonia · Tuberculosis · Transfer learning · CNN

1 Introduction

Around the world, lung-related diseases are common and may harm a person to any extent. There are various ways to categorize lung disease are airway diseases, circulation diseases, and tissue disease [1]. Airway diseases are common and do not have specific symptoms. If we consider airway disease, it causes the interference of the different gases and oxygen into the lungs through the respiratory

S. Patil (✉) · B. Tiple
MIT World Peace University, Pune, India

G. Ranganathan et al. (eds.), *Soft Computing for Security Applications*,
Advances in Intelligent Systems and Computing 1397,
https://doi.org/10.1007/978-981-16-5301-8_57

system. The different airway diseases are asthma, chronic obstructive pulmonary disease (COPD), and bronchitis, as well as emphysema and bronchiectasis. Circulatory disease affects the lungs and blood vessels. These diseases are caused by clots or pain in the blood vessels. It harms the lungs for the intake of oxygen and affects the functioning of the heart. The different circulatory diseases are pulmonary hypertension etc. Tissue diseases are when are person faces shortness of breathing. It becomes difficult for a person to inhale and exhale. Different types of tissue diseases are pulmonary fibrosis and sarcoidosis. Cellular breakdown in lungs pneumonia and other respiratory diseases are the disease that harms the lungs.

One of the contagious dangerous diseases named COVID-19 has spread rapidly all over the world harming people to an extreme. This disease has harmed people of every age and gender. The disease directly affects the respiratory system of the infected patient and forms the layer over the lungs which harms the normal working of lungs. This disease was first found in Wuhan China in December 2019. By February 2020, International Committee on Taxonomy of Viruses (ICTV) reported the name as Severe Acute Respiratory Syndrome Coronavirus2 (SARS-CoV-2) and WHO named it as COVID-19 [2, 3]. The common symptoms of this disease found in people are cough, cold, fever, loss of taste, loss of smell, etc. The second wave that hit India badly showed newer symptoms of COVID-19. When the infected person comes in a direct contact with the healthy person, he may also get infected. When the infected person speaks, sneezes, or coughs through the droplets, the transmission takes place. The humans who easily get infected by this disease are those who are suffering from diabetes, different respiratory diseases, or having low immunity. During the first wave, the old people were majorly affected by this disease. During the second wave, the younger people were harmed by the disease. The experts and researchers are working to find the solution for this disease [4].

The testing units for COVID-19 are categorized in two ways as: Antigen test to find if the person is infected with the disease or not, and antibody test is to find the antibodies in the patient who has recently got infected with the disease. The antigen test known as polymerase chain reaction (PCR) is used to find whether the person is infected with the disease or not. The steps followed in the PCR test are nasopharyngeal swabs, oropharyngeal swabs. These samples are sent to the laboratory for detection of COVID-19 disease. Generally, PCR tests take a longer time for the results. So, the option is the rapid test which gives the results faster but may not be that accurate. Due to the increase in the number of cases, government has faced challenges to reduce the number. But due to a lack of vaccine and drugs and insufficient treatment, we are not able to control the cases of COVID-19 [5].

As PCR test takes a long time to detect whether the person is infected to Covid-19, in these circumstances, clinical images such as CT scan and X-ray images have discovered that gives quick detection of COVID-19 [6, 7]. The CT scan images and the X-ray images are utilized to get the feature from the chest for the detection of COVID-19. COVID-19 has proved to be a various dangerous disease that has caused a severe threat to human civilization. The COVID-19 diagnosis uses specific imaging methods like CT images or X-RAY that are viewed using high-quality methods. Since all the systems use different data sizes for the experiment, the performances provided by the systems cannot be rationally compared. X-ray-based systems resulted in far better diagnosis in comparison to the CT-based systems in terms of image modalities comparison. Most of the framework based on X-ray used benchmark data while the real-time data from the hospital environment is considered [8]. It is contemplated the system which was brought by CT samples were applicable for real-time testing but when it came to X-ray-based schemes, real-time testing was needed with target people before application.

The proposed research work is design to detect lung diseases using X-ray images. This research is divided into two phases—Data augmentation and lung disease prognosis. The available dataset has a limited amount of X-ray images. So, to overcome this, the images are augmented to increase the data size. As the images are less, the problem of overfitting may occur. So, it is necessary to increase the data size. In phase two, the X-ray images are divided into four classes COVID-19, normal, pneumonia, and tuberculosis [9, 10]. The transfer learning methods such as VGG-16, Xception, InceptionV3, MobileNet, and MobileNetV2. The commitment of the proposed work can be summed up as the proposed philosophy with the novel information increase on the restricted dataset is utilized to arrange the X-beam information into four classes, i.e., COVID-19 and normal, pneumonia, and tuberculosis. The exhibition of five pre-prepared exchange learning models is contrasted with address the issue of COVID-19 identification through X-beam with the restricted dataset.

The point of the paper is to analyse the evolution of the deep learning-based coronavirus prognosis framework dependent on information gathered from medical images. Scientific classification is introduced that arranges the inspected framework dependent on a pre-prepared model with deep transfer learning. The rest of the paper incorporates Sect. 2 proposed system, Sect. 3 incorporates architecture, Sect. 4 experiment, Sect. 5 result, Sect. 6 discussion, and Sect. 7 conclusion.

2 Proposed System

Deep learning is a subfield of machine learning stressed over calculations brought about by the designs and elements of the mind called artificial neural networks. The field of artificial intelligence (AI) is fundamentally when machines can take care of undertakings that by and large requires human understanding. Deep learning grants the machine to deal with complex issues regardless, while using a dataset that is very varying, unstructured, and interconnected. The more deep learning computations become acquainted with, the better they perform. Deep learning procedures are utilized in clinical frameworks like brilliant medical care, biomedicine, drug revelation, clinical picture examination, and so forth.

Deep learning-based structures are utilized for forecast of COVID-19 which incorporates a couple of stages, for instance, information gathering, information creation, extricating of components, classification of information and assessment. The virus prognosis framework relies upon deep learning as demonstrated in Fig. 1. In the phase of information assembling, the patients from the clinic climate are considered

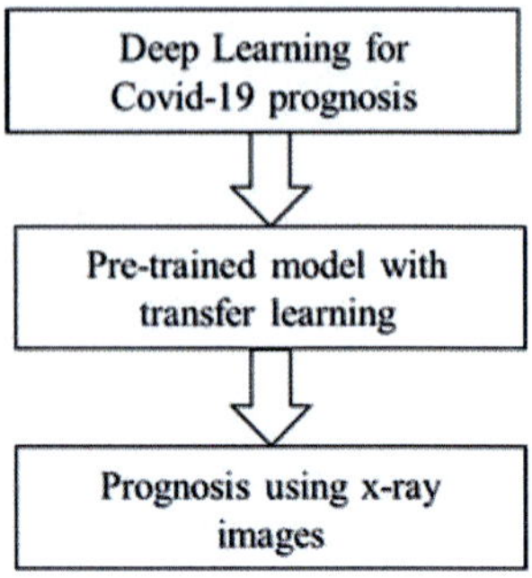

Fig. 1 Deep learning prognosis using X-ray images

as a part. Albeit the information may have various constructions anyway for the visualization of COVID-19, CT, and X-beam tests are taken. The subsequent stage is the data preparation which changes over the information into a proper configuration. Here, data pre-processing incorporates tasks like resizing, normalization, rotation of images, horizontal flip. In this progression, the information is part into preparing, approval and testing set for the test. For data partitioning, commonly the cross-validation technique is used. So, the dataset is divided into training and testing as 70 and 30%. And these images are further classified into four classes as COVID-19, normal, pneumonia, and tuberculosis. So, the primitive step is the feature extraction from the X-ray images for the prognosis of the virus. So, by using deep learning techniques like VGG-16, Xception, InceptionV3, MobileNet, MobileNetV2 by training the model, it accordingly extracts the necessary features from the X-ray images and execute various operations. Based on class labels, the classification is concluded. The next main stage is the prognosis of the disease using different transfer learning algorithms as VGG-16, Xception, InceptionV3, MobileNet, and MobileNetV2. The performance evaluation of the proposed work is done with the help of accuracy, sensitivity, specificity, precision, F1-score, and confusion matrix.

In the Phase I, the data collection and data augmentation is done. The data is collected from the infected patients from COVID-19, normal, pneumonia, and tuberculosis. The data is in the form of X-ray images. As the X-ray images of infected COVID-19 are smaller in number so using Keras, data augmentation is done. Using rotation, shear, horizontal flip the data is augmented and have increased the size of data. The data is pre-processed using the resize of images and normalization of images. In deep learning, it automatically extracts the features from the given dataset of an X-ray image by performing several different operations. The pre-processed dataset is classified in 70% training of images and 30% testing of images. In Phase II, the COVID-19 and other lung disease prognosis are done by the model. For the prognosis, the model uses Xception, InceptionV3, VGG-16, MobileNet, and MobileNetV2 (Fig. 2).

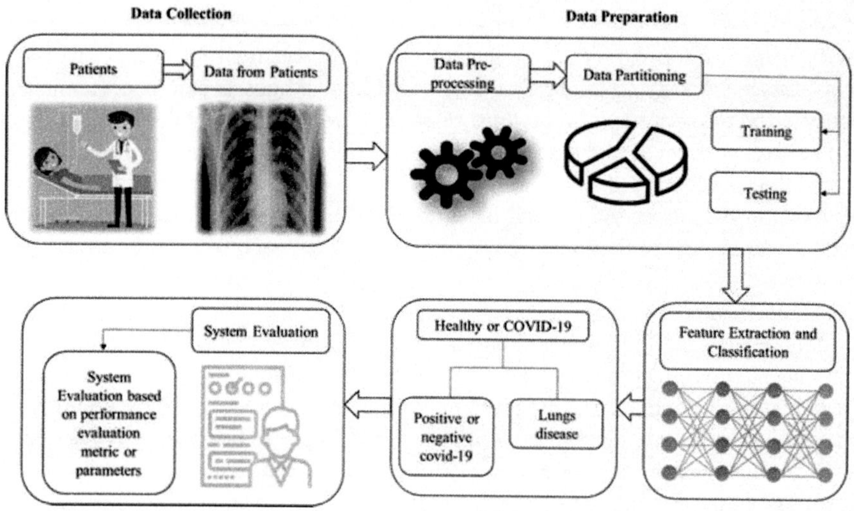

Fig. 2 Proposed methodology

3 Architecture

A pre-trained model is created to solve a particular problem anyone can use it to solve a similar problem. Pre-trained models on image datasets are available for several image classification using convolutional neural network (CNN). ImageNet is one of the models which is a large-scale, accurate, and diverse image database. WordNet hierarchy is used to arrange the images in this database. For this study, four models pre-trained on image dataset have been used, architectures of these modules are as below.

There are sure benefits of move learning-based CNN models. (i) The dataset requires limit pre-handling, (ii) quicker learning measure, (iii) by diminishing, the various boundaries time intricacy can be changed, and (iv) functions admirably on the restricted dataset, accordingly proper for the clinical picture grouping task. In this given work, for parallel order of chest X-ray images into COVID-19, pneumonia, normal, tuberculosis move learning instrument is utilized. Five diverse pre-prepared exchange learning-based CNN models are utilized, for example, VGG-16, Xception, InceptionV3, MobileNet, and MobileNetV2. From the picture net dataset, the CNN models are prepared to order into 1000 article categories. For the double grouping information of chest X-ray images, the pre-prepared exchange learning models are retrained by utilizing four classes.

A model prepared on one issue is used to anticipate names associated to a subsequent issue; this is an interaction utilized in move learning. The fundamental benefit in the decision of move learning is addressed by the diminishing time for the preparation task for a neural organization model. Also, it can bring about more substandard speculation mistakes.

InceptionV3: Inception v3 or GoogleNet was a 22-layer network is a convolutional neural organization for aiding picture investigation and item location. It won the 2014 ImageNet test and has 93.3% top-5 exactness and got its commencement as a module for GoogleNet. It is the third release of Google's Inception Convolutional Neural Network presented during the ImageNet Recognition Challenge. Following adaptations were tended to as Inception VN where N represents rendition number so inceptionV1, inceptionV2, and inceptionV3 [9]. A few even and awry structure blocks were found in the Inception V3 network where each square has a few parts of convolutions, normal pooling, max pooling, linked, dropouts, and completely associated layers.

MobileNetV2: MobileNetV2 contains a convolutional neural organization structure that attempts to perform well on cell phones. It was proposed by Sandler et al. for machines with limited figuring power, as cell phones. The MobileNet accomplishes memory utilization reduction as a critical benefit by presenting the upset residuals-with-straight bottleneck squares of 19 leftover layers and diminishing the quantity of learning boundaries. Plus, numerous famous profound learning structures have the pre-prepared execution of MobileNetV2 is broadly accessible.

VGG-16: There are 16 layers inside a convolutional neural organization which is known as VGG-16. It is feasible to stack a pre-prepared variant of the VGG-16 organization prepared on the ImageNet dataset, which is an informational index of more than 14 million pictures: the pre-prepared organization can characterize pictures into 1,0 0 unique classes, for example, as a feline, pen, and other comparable articles. This model gets 92.7% top-5 test precision in the ImageNet dataset.

MobileNet: As the name applied, the MobileNet model is intended to be utilized in portable applications, and it is TensorFlow's first versatile computer vision model. MobileNet utilizes depthwise divisible convolutions. It essentially decreases the quantity of boundaries when contrasted with the organization with normal convolutions with similar profundity in the nets [11]. This outcomes in lightweight profound neural organizations. MobileNet is a class of CNN that was publicly released by Google, and along these lines, this gives us a phenomenal beginning stage for preparing our classifiers that are madly little and madly quick.

Xception: It is a convolutional neural organization that is 71 layers profound. You can stack a pre-trained adaptation of the organization prepared on more than 1,000,000 pictures from the ImageNet information base [12]. The pre-trained organization can characterize pictures into 1000 item classes, like console, mouse, pencil, and numerous creatures.

4 Experimental Analysis

4.1 Dataset

In a public dataset like GitHub [13] and Kaggle repository [14], 3000 images of X-ray were collected. The dataset ratio is divided into 70% for training images and 30% for testing images For training 2100 images and 900 images for testing. The collected images were classified into four classes COVID-19, normal, pneumonia, and tuberculosis. The samples used by these classes are 750 images for COVID-19 in this 525 images are considered for training and 225 images for testing, 750 samples of normal patients here 525 images are considered for training and 225 images for testing, 750 samples of pneumonia patients here 525 images are considered for training and 225 images for testing, 750 samples of the patient suffering from tuberculosis so the images are classified on the basis of 525 images for training and 225 images for testing. As the COVID-19 images were not in large number, data is augmented to increase the number of images.

4.2 Experimental Setup

The implementation of these five architectures like VGG16, Xception, InceptionV3, MobileNet, and MobileNetV2. Backend as TensorFlow with Keras framework is used. From the ImageNet database, Keras provides pre-trained weights on the pre-trained model despite the fact that the images in the ImageNet dataset on which these models are prepared may not be like pictures gathered for concentrate however may help by moving information figured out how to make the planned errand more productive. Likewise, pre-prepared loads diminish the prerequisite of an enormous volume of dataset for training. Online platform Google Collab was used to execute the Python code. The GPU known as Tesla P4 provided by Google Collab was used for faster processing. To train the networks, Adam optimizer and categorical cross-entropy were used to train the model. The batches size was set to 10 for training and epochs were set.

5 Results

The below-mentioned figures are the confusion matrices of the five models on the testing dataset, i.e., Xception, VGG-16, MobileNet, MobileNetV2, and InceptionV3. Table 1 outlines the accuracy of the five models. The highest accuracy is obtained by MobileNetV2 92%. The MobileNet obtains 96% accuracy. VGG-16 obtains 95% accuracy. The lowest accuracy is obtained by InceptionV3 by 92%. Table 2 outlines the evaluation parameters as precision, recall, F1-score differently of all the four

Table 1 Method versus accuracy for proposed method

Method	Accuracy
InceptionV3	0.92
MobileNet	0.96
MobileNetV2	0.97
VGG-16	0.95
Xception	0.93

Table 2 Evaluation metrics used in detection of various disease

Algorithm	Evaluation parameters	COVID-19	Normal	Pneumonia	Tuberculosis
InceptionV3	Precision	0.93	0.96	0.90	0.91
	Recall	0.92	0.84	0.97	0.96
	F1-score	0.92	0.90	0.93	0.93
MobileNet	Precision	0.94	0.98	0.91	0.99
	Recall	0.99	0.86	0.99	0.99
	F1-score	0.97	0.92	0.95	0.99
MobileNetV2	Precision	0.99	0.98	0.92	0.99
	Recall	0.98	0.91	0.99	1.00
	F1-score	0.98	0.94	0.95	0.99
VGG-16	Precision	0.97	0.99	0.88	0.99
	Recall	0.99	0.83	0.99	1.00
	F1-score	0.98	0.90	0.93	0.99
Xception	Precision	0.93	0.96	0.88	0.95
	Recall	0.94	0.86	0.97	0.93
	F1-score	0.93	0.91	0.92	0.94

classes accordingly. The highest precision is obtained by MobileNetV2. MobileNet gives nearby accuracy but cannot be consider as robust model has precision less than 0.5. Accuracy of all these five models is dependable on the prognosis of the disease. Figures 3, 4, 5, 6, and 7 are the confusion matrix of the implemented method.

6 Discussion

In this section, we review the research performed on lung disease using chest X-ray images of the patient using transfer learning methods.

Hemdan et al. [15] proposed the system in which VGG-19, DenseNet201, ResNetV2, Xception, InceptionV3, InceptionResNetV2, MobileNetV2 methods were used. All these methods have given near about the lowest 60% and highest

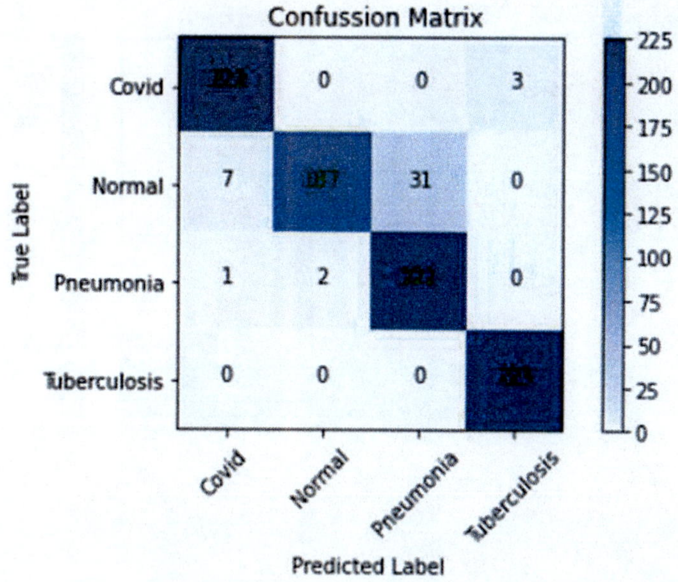

Fig. 3 Confusion matrix of VGG-16

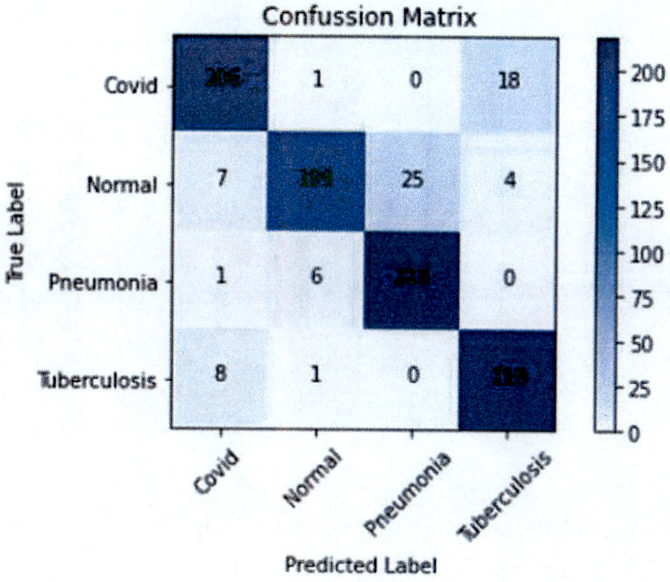

Fig. 4 Confusion matrix for InceptionV3

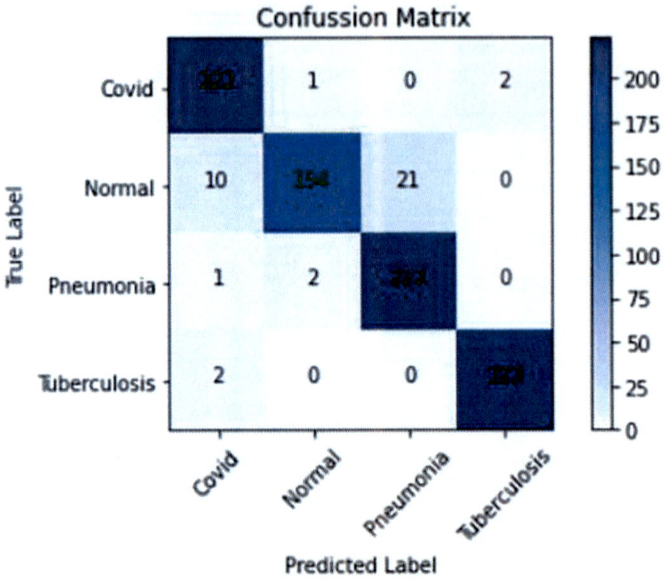

Fig. 5 Confusion matrix for MobileNet

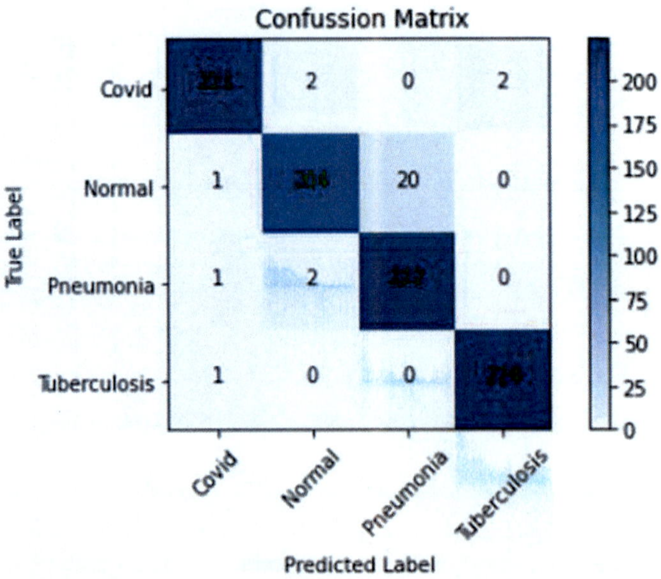

Fig. 6 Confusion matrix for MobileNetV2

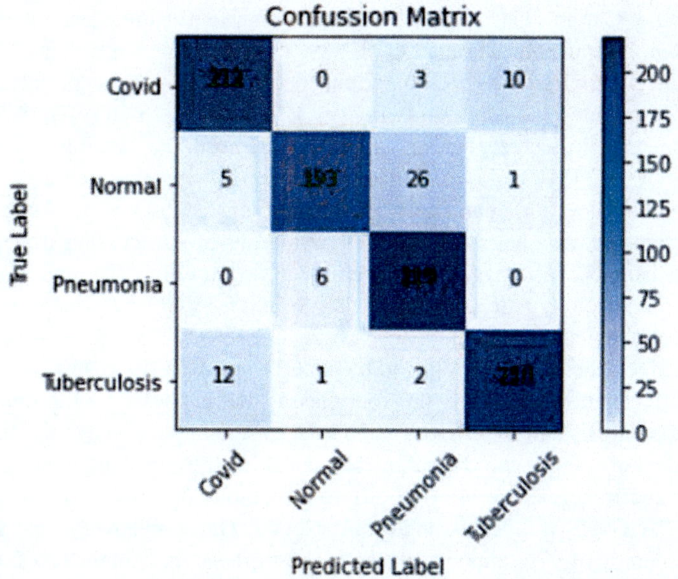

Fig. 7 Confusion matrix of Xception

90% of accuracy. The dataset includes 50 images 25 of COVID-19 and 25-non-COVID patients. The proposed system uses the COVIDXnet framework including deep learning algorithms. The accuracy obtained by VGG-19 is 90%, DenseNet201 is 90%, ResNetV2 is 70%, Xception is 80%, Inceptionv3 is 50%, InceptionResNetV2 is 80%, MobileNetV2 is 60%. For the evaluation parameters, other factors consider are precision, recall, and F1-score.

Sethi et al. [5] proposed a system that uses four deep learning methods as InceptionV3, Xception, MobileNet, and ResNet50. The MobileNet gives the highest accuracy and the ResNet gives the lowest accuracy. The dataset includes 6249 X-ray images of the patient suffering from 320 COVID-19 and 5928 non-COVID-19. The evaluation parameters considered are accuracy, precision, F1-score, specificity, and recall.

Apostolopoulos and Bessiana [16] proposed system for automatic detection of COVID-19 X-ray images using CNN and transfer learning. The pre-trained models considered are VGG19, Inception, Inception ResNetV2, MobileNet, and Exception. In the proposed system, 1427 X-ray images were considered and are classified as 224-COVID-19, 700-pneumonia, 504-normal X-ray images. The performance evaluation is done based on accuracy, specificity, and sensitivity. The model shows the accuracy by VGG-19 is 93.48%, Inception is 92.85%, InceptionResNetV2 is 92.85%, MobileNet is 92.85%, and Xception is 92.85% which is achieved for detection of COVID-19.

Moutounet-Cartan [17] proposed a diagnosis system using a deep convolution neural network for pneumonia and COVID-19. The system uses different CNN architectures like VGG-16, VGG-19, Xception, InceptionV3, and InceptionResNetV2. The dataset includes 327 total X-ray images 152-heathy, 125-COVID-19, 50-other types of pneumonia. The evaluation parameters are accuracy, sensitivity, and recall.

Makris et al. [18] proposed a diagnosis system using the technique of deep learning and convolution neural network; because of the absence of a large number of image samples, data augmentation with numerous pre-processing strategies was performed using Keras Image data generator during training. The models considered for the discovery of COVID-19 are VGG16, VGG-19, Xception, InceptionV3, MobileNetV2, InceptionResNetV2, DenseNet201, Res-Net152V2, and NASNetLarge. The accuracy obtained by VGG16 and VGG19 is 95%.

The current purposed system has four classes that is COVID-19, normal, pneumonia, tuberculosis and a dataset of 3000 images so comparatively, this system has more amount of images and the parameters considered for evaluation are precision, F1-score, and recall. There are five different methodology used are InceptionV3, Xception, VGG-16, MobileNet, and MobileNetV2. The MobileNetV2 has given the highest accuracy of 97% and as well the highest precision. MobileNetV2 turns out to give better accuracy results (Table 3).

7 Conclusion

The second wave has hit India, and the cases are increasing to another extent. This deep learning prognosis system may help the patient with higher risk and were the radiological expertise is low. This proposed technique classifies the lungs X-ray images into COVID-19 and normal, pneumonia, and tuberculosis class. The dataset collected is resized accordingly for implementation. Data augmentation using random rotation, shear operation, and horizontal flip is applied to increase the data size. The two-stage classification prognosis is performed using five different deep transfer learning methods that are VGG-16, Xception, InceptionV3, MobileNet, and MobileNetV2, and their performances are verified. These models are pre-trained on the ImageNet database and have pre-trained weights. The model is classified into four classes as COVID-19, normal, pneumonia, and tuberculosis. After experimenting, the highest accuracy is obtained by MobileNetV2 that is 97%. The highest precision is obtained by MobileNetV2. The lowest accuracy is obtained by inceptionV3. The CNN model gives better results in the medical domain of X-ray and CT images. Transfer learning algorithms play a major role in improving the accuracy. Different pre-trained models may be explored to develop a prognosis system. As a future scope, we can predict different lung disease such as lung cancer and other lungs disease. As future scope, different CNN architectures can be developed for prognosis.

Table 3 Comparison of different evaluation parameters for different models evaluated by various author

Author	Algorithm	Accuracy (%)	Precision	Specificity	Sensitivity/Recall	F1-score
Sethi et al. [5]	InceptionV3	98.1	85.9%	99.2%	81.1%	83%
	ResNet50	82.5	95.6%	99.7%	24.2%	38.6%
	MobileNet	98.6	87.8%	99.3%	87.8%	87.8%
	Xception	97.4	88.9%	99.5%	62.2%	73.2%
El-Din Hemdan et al. [15]	VGG-19	90	–	–	–	–
	DensNet201	90	–	–	–	–
	ResNetV2	70	–	–	–	–
	InceptionV3	50	–	–	–	–
	InceptionRes-NetV2	80	–	–	–	–
	Xception	80	–	–	–	–
	MobileNet-V2	60	–	–	–	–
Apostolopoulos et al. [16]	VGG-19	93.48	–	98.75%	92.85%	–
	Inception	92.85	–	99.7%	12.94%	–
	InceptionRes-NetV2	92.85	–	99.83%	0.01%	–
	MobileNet	92.85	–	97.09%	99.10%	–
	Xception	92.85	–	99.99%	0.08%	–
Moutoun et Cartan [17]	VGG-16	84.1	–	–	87.7%	–
	VGG-19	82.0	–	–	86.1%	–
	InceptionRes-NetV2	61.2	–	–	71.0%	–
	InceptionV3	69.1	–	–	78.0%	–
	Xception	61.4	–	–	75.1%	–
Makris et al. [18]	VGG-16	0.9588	–	–	–	–
	VGG-19	0.9503	–	–	–	–
	MobileNetV2	0.3970	–	–	–	–
	InceptionV3	0.5441	–	–	–	–
	Xception	0.7647	–	–	–	–
	InceptionRes-NetV2	0.5588	–	–	–	–
	DenseNet201	0.3823	–	–	–	–
	ResNet152V2	0.6176	–	–	–	–

References

1. https://medlineplus.gov/ency/article
2. D. Cucinotta, M. Vanelli, WHO declares COVID-19 a pandemic. Acta Bio Medica: Atenei Parmensis **91**(1), 157 (2020)
3. J.P. Cohen, P. Morrison, L. Dao, K. Roth, T.Q. Duong, M. Ghassemi, Covid-19 image data collection: prospective predictions are the future. arXiv preprint arXiv:2006.11988 (2020)

4. R. Sethi, M. Mehrotra, D. Sethi, Deep learning based diagnosis recommendation for COVID-19 using chest x-rays images. in *2020 Second International Conference on Inventive Research in Computing Applications (ICIRCA)* (IEEE, 2020), pp. 1–4
5. M. Phankokkruad, COVID-19 pneumonia detection in chest x-ray images using transfer learning of convolutional neural networks. in *Proceedings of the 3rd International Conference on Data Science and Information Technology*, (2020), pp. 147–152
6. Md. M. Islam, F. Karray, R. Alhajj, J. Zeng, A review on deep learning techniques for the diagnosis of novel coronavirus (covid-19). IEEE Access **9**, 30551–30572 (2021)
7. M. Anthimopoulos, S. Christodoulidis, L. Ebner, A. Christe, S. Mougiakakou, Lung pattern classification for interstitial lung diseases using a deep convolutional neural network. IEEE Trans. Med. Imaging **35**(5), 1207–1216 (2016)
8. C. Szegedy, S. Ioffe, V. Vanhoucke, A. Alemi, Inception-v4, inception-resnet and the impact of residual connections on learning. in *Proceedings of the AAAI Conference on Artificial Intelligence*, vol. 31, no. 1 (2017)
9. A.G. Howard, M. Zhu, B. Chen, D. Kalenichenko, W. Wang, T. Weyand, M. Andreetto, H. Adam, Mobilenets: efficient convolutional neural networks for mobile vision applications. arXiv preprint arXiv:1704.04861 (2017)
10. E. Luz, P.L. Silva, R. Silva, L. Silva, G. Moreira, D. Menotti, Towards an effective and efficient deep learning model for COVID-19 patterns detection in X-ray images. arXiv preprint arXiv: 2004.05717 (2020)
11. F. Chollet, Xception: deep learning with depthwise separable convolutions. in *Proceedings of the IEEE Conference on Computer Vision and Pattern Recognition*, (2017), pp 1251–1258
12. E. El-Din Hemdan, M.A. Shouman, M. Esmail Karar, Covidx-net: a framework of deep learning classifiers to diagnose covid-19 in x-ray images. arXiv preprint arXiv:2003.11055 (2020)
13. https://github.com/ieee8023/covid-chestxray-dataset/
14. https://www.kaggle.com/tawsifurrahman/tuberculosis-tb-chest-xray-dataset
15. I.D. Apostolopoulos, T. Bessiana (2020), Covid-19: automatic detection from x-ray images utilizing transfer learning with convolutional neural networks. arXiv preprint arXiv:2003.11617 (2020)
16. L. Brunese, F. Mercaldo, A. Reginelli, A. Santone, Explainable deep learning for pulmonary disease and coronavirus COVID-19 detection from x-rays. Comput. Methods Programs Biomed. **196**, 105608 (2020)
17. A. Makris, I. Kontopoulos, K. Tserpes, COVID-19 detection from chest x-ray images using deep learning and convolutional neural networks. in *11th Hellenic Conference on Artificial Intelligence*, (2020), pp 60–66
18. A. Kumar, P.K. Gupta, A. Srivastava, A review of modern technologies for tackling COVID-19 pandemic. Diab. Metab. Syndr.: Clin. Res. Rev. **14**(4), 569–573 (2020)

Combined PCA–Segmentation Method: An Efficient Technique for Covered Face Recognition

Md. Mahbubul Alam, Md. Ashikur Rahman Khan, Zayed Us Salehin, Razib Chandra Das, Main Uddin, and Sultana Jahan Soheli

Abstract Face is the most common parameter which defines any person uniquely. Face recognition refers to identify the identity of an individual using their face. Principal component analysis is the very common algorithm for face recognition. In facial recognition, there are some complexities that degraded the efficiency of it for covered face and covered eyes. Iris recognition can remove the covered face complexities, but when someone wears sunglasses which covered eyes, then iris recognition will not work properly. To overcome all these complexities which include covered face or eyes, our proposed efficient model is combined PCA–segmentation face recognition. In this work, principal component analysis and segmentation process will take place combined. The motto of this work is to remove the complexities of beard, mustache, hijab and sunglasses at facial recognition to increase the overall accuracy for improving the performance.

Keywords Face recognition · Image segmentation · PCA · Covered face · Covered eyes · Sunglasses and scarf

1 Introduction

Face recognition is the process to recognize individual uniquely by evaluating face. It is a biometric identification process by which one can uniquely identify and verify individuals by analyzing the contours of faces [1]. Facial recognition is widely used in various security and academic aspects. It is used in automatic attendance system as well as in national security, i.e., identifying terrorist and identifying invaders for home security.

Md. M. Alam (✉) · Md. A. R. Khan · Z. U. Salehin · R. C. Das · M. Uddin · S. J. Soheli
Department of Information and Communication Engineering, Noakhali Science and Technology University, Noakhali 3814, Bangladesh
e-mail: ashik@nstu.edu.bd

Md. M. Alam
Department of of Computer Science and Engineering, Feni University, Feni 3900, Bangladesh

G. Ranganathan et al. (eds.), *Soft Computing for Security Applications*,
Advances in Intelligent Systems and Computing 1397,
https://doi.org/10.1007/978-981-16-5301-8_58

They have various statistical and neural network approaches for recognizing faces in computer system. Statistical approaches [2] include

- Principal component analysis (PCA)
- Discrete cosine transform (DCT)
- Linear discriminant analysis (LDA)
- Locality preserving projections (LPP)
- Gabor wavelet (GW)
- Independent component analysis (ICA)
- Kernel PCA.

They also have some machine learning approaches for facial recognition, i.e.,

- Neural networks with Gabor filters
- Hidden Markov models and neural networks
- Fuzzy neural networks (FNN).

Principal component analysis [3–6] is a very efficient algorithm for recognizing face. PCA method uses eigen faces and eigen vectors to differentiate different faces from one another. A person with covered face is a situation when PCA algorithm will not work efficiently. Many women commonly wear hijab, and many people wear masks for various reasons; in these covered face situations, PCA is not a perfect method because of less accuracy. When a person wears sunglass or somehow a person with covered eyes, iris recognition [7, 8] will not work properly. It is a big limitation of iris recognition method to work with covered or closed eyes.

To overcome these complexities, we have applied combined PCA and segmentation method in order to recognize the face. This is one of the efficient techniques that recognizes a person even after the person wears sunglass, mask or increased beard–mustache. By using segmentation process [9–11], an image can be splitted out into several separate images. After segmentation, the PCA algorithm will be applied on the segmented images. The segmentation process will be work on upper region of face and lower region of face.

We have used MATLAB because it is a reliable platform to work to implement the technique.

This paper proposes PCA–segmentation combined approach which overcomes a number of difficulties of the covered face and eyes.

2 Related Works

Some related previous works using various methods are mentioned as follows, although these approaches are not efficient enough for covered face and eyes.

In 2017, Singh et al. [12] introduced a deep learning framework that disguised face identification by detecting fourteen key points of face and two annotated facial key points data sets. In 2018, Gupta et al. [13] proposed an approach of deep neural network for face recognition. Their approach provided only the extracted facial features but not raw pixel values as input. It provided accuracy of 97.05% and decreased the complexity on Yale data set.

Gondole and Salunkhe [14] in 2018 proposed an approach that face area is divided into small parts, and local binary pattern histograms are extracted from these parts and added into feature histogram. In 2017, Ayi, M. et al. introduced a method for face detection using Arduino and MATLAB [15]. They developed a real-time application that is essential in several fields.

In 2015, Parvaneh and Hadi combined PCA, eigenface and SVM for face recognition using N-fold method [16].

In 2009, Faruqe and Hasan [17] proposed a method of face recognition using PCA, feature extraction and the SVMs, which controls face recognition problem. They illustrated the efficiency of SVMs by using the Cambridge ORL face database.

In 2020, Vijayakumar and Vinothkanna [18] have proposed a cognitive classification model to retrieve information efficiently from multimedia data storage.

In 2021, Adam [19] has implemented and evaluated the fingerprint liveness detection by machine learning approach.

Therefore, we proposed a combined PCA and segmentation technique using ORL and YALE and real image data set for face recognition, which recovered the problems of covered face and eyes.

3 Methodology

In our work, PCA and segmentation methods are being combined. We used three data sets of different images to evaluate the performance and accuracy of the proposed technique for partially covered face recognition.

3.1 Data Set

In the experiment, following three face databases are used:

- ORL face database
- YALE face database
- Real face database.

RL face database [20–22] is mostly used face database of different forty persons and ten different images for each person. Various expressions are in this database.

Yale face database [23–25] has 165 pictures in grayscale of 15 different persons. Each person has eleven images with different expression like happy, sad and surprised, etc. [26].

The most interesting thing is that in this work, they used real face data set with 300 images of sixty different persons with five images for each. The faces of this

Fig. 1 Real face data set (partial)

data set have various expressions like closed or open eyes and faces with sunglass, hijab, beard and mustache. All these data sets were used to perform our technique efficiently (Fig. 1).

3.2 Method

The working procedure of combined PCA–segmentation technique is included preprocessing part, image training part, image testing part and final (Result) part. In these parts, many operations executed like loading, conversion, segmentation, reshaping and resizing along with PCA algorithm.

The process of this work of combined PCA–segmentation is depicted step by step by flowchart in Fig. 2.

Preprocessing part: In preprocessing part to train the image, data sets are being ready to load.

- Load image data: Loading image data is the first step to start the work for the purpose of training.
- Convert RGB to grayscale images: In this step, the RGB images will be converted into grayscale images.

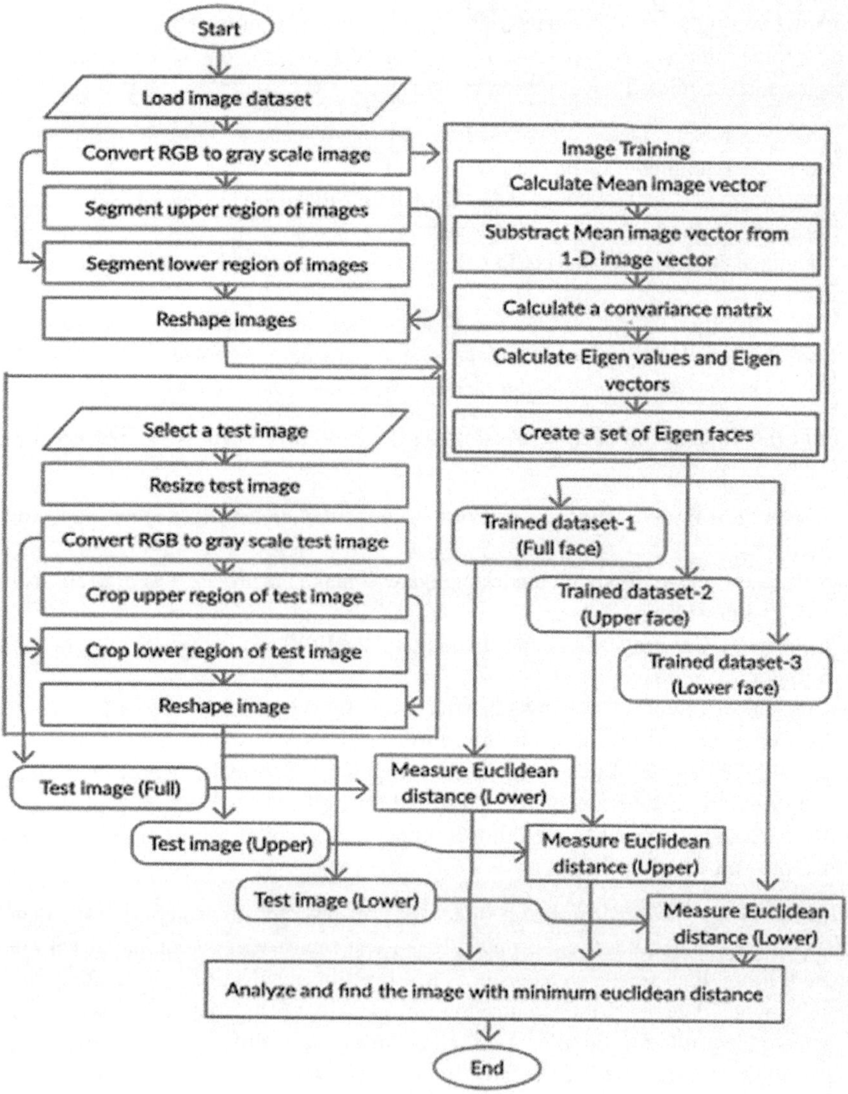

Fig. 2 Flowchart of combined PCA–segmentation technique

- Segment upper region of images: After segmenting of upper region of images, the upper segmented images are stored in a new folder for training.
- Segment lower region of images: After segmenting of lower region of images, the lower segmented images are stored in another folder for training.
- Reshape images: Reshape the segmented upper and lower region images to bring back the previous original shape.

Image training part: In the data training part, the total training process takes place on the three folders which contains original full face images, segmented upper region images and segmented lower region images.

- Calculate mean image vector: Calculate the image vector and find the mean image vector by the number of images.
- Subtract mean image vector: After calculating mean image vector, subtract the mean image vector from 1D image vector.
- Calculate convariance matrix: To calculate eigen values and vectors, calculate the matrix.
- Calculate eigen values and vectors: Calculate eigen values using the convariance and identity matrix and calculate eigen vectors from eigen values.
- Create eigen faces: Create eigen faces set from the eigen values and vectors.

Image testing part: In the image testing part, the test image will be ready for testing with the training data set.

- Select a test image: Selection of a test image is the starting step of image testing part.
- Resize test image: Resizing the test image will make the process easier to compare with training images.
- Convert RGB to gray test images: In this step, the RGB test image will be converted into grayscale image.
- Crop upper region of test image: After cropping the upper region of test image, the upper cropped test image is stored for further steps.
- Crop lower region of test image: After cropping the lower region of test image, the lower cropped test image is also stored for further steps.
- Reshape image: Reshape the cropped upper and lower test images to bring back the previous original shape.

Final part: In the final part, the three test images such as full face test image, cropped upper test image and cropped lower test image will be compared with the three training data sets such as original full face images data set, segmented upper region images data set and segmented lower region images data set, respectively. Then calculate the minimum Euclidean distances for the three different testing iterations. After all these steps, analyze the three Euclidean distances and names of images and find the best matching image with least euclidean distance.

4 Result and Discussion

Combined PCA–segmentation method provides better performance than the solo PCA algorithm. Several epochs were performed with test images to evaluate the performance.

4.1 Result

Performance of recognition for ORL data set is calculated on random choice test images. Table 1 represents performance of proposed technique along with PCA for ORL.

From Table 1, recognition accuracy using PCA algorithm on ORL data set is less than the combined PCA–segmentation method.

Table 2 shows performance of proposed technique along with PCA for Yale.

From Table 2, face recognition accuracy on Yale data set is better for combined PCA–segmentation.

For real face data set, we choose random images from data set to find the least Euclidean distance image from trained data set. The face recognition performance on real face data set is shown in Table 3.

From Table 3, recognition accuracy using PCA algorithm on real data set is less than the combined method.

4.2 Discussion

Combined PCA–segmentation approach is more efficient than PCA method. Proposed method is applied on three image data sets like ORL, Yale and real face data sets which contains real images along with PCA algorithm. This combined method is efficient over PCA algorithm for covered face and eyes situations. It is also good

Table 1 Accuracy for ORL data set

Methods	Accuracy (ORL data set) (%)
Using PCA	72.63
Using combined PCA–segmentation	99.47

Table 2 Accuracy for Yale data set

Methods	Accuracy (Yale data set) (%)
Using PCA	86.67
Using combined PCA–segmentation	97.78

Table 3 Accuracy for real face data set

Methods	Accuracy (Real face data set) (%)
Using PCA	69.57
Using combined PCA–segmentation	95.65

for the sunglass and scarf user. Result part shows that this research methodology works well. So the combined PCA–segmentation technique is fruitful for removing the complexities of covered eyes and faces recognition.

5 Conclusions

Combined PCA–segmentation system provides better recognition result than PCA method for partially covered face recognition. Proposed method can recognize a person even when the person wears sunglass, hijab and scarf. It can detect a person with long beard and mustache. In future, we wish to use more realistic complex image data set and dynamic algorithms along with advance AI to improve the performance.

References

1. What is facial recognition?—Definition from Techopedia. Techopedia.com. https://www.techopedia.com/definition/32071/facial-recognition. Accessed 03 Mar 2020
2. I. Marqués, M. Granã, Face recognition algorithms, 16 June 2010
3. X. Zhang, X. Ren, Two dimensional principal component analysis based independent component analysis for face recognition, in *2011 International Conference on Multimedia Technology* (Hangzhou, 2011), pp. 934–936
4. J. Shermina, Illumination invariant face recognition using discrete cosine transform and principal component analysis, in *2011 International Conference on Emerging Trends in Electrical and Computer Technology* (Nagercoil, 2011), pp. 826–830
5. X. Nie, Z. Tan, Face recognition using bidirectional principal component analysis and wavelet transform, in *2011 International Conference on Electric Information and Control Engineering* (Wuhan, 2011), pp. 4203–4206
6. W.Y. Min, E. Romanova, Y. Lisovec, A.M. San, Application of statistical data processing for solving the problem of face recognition by using principal components analysis method, in *2019 IEEE Conference of Russian Young Researchers in Electrical and Electronic Engineering (EIConRus)* (Saint Petersburg and Moscow, Russia, 2019), pp. 2208–2212
7. S. Lim, K. Lee, O. Byeon, T. Kim, Efficient iris recognition through improvement of feature vector and classifier (2001)
8. R. Wildes, J. Asmuth, G. Green, S. Hsu, R. Kolczynski, J. Matey, S. McBride, A system for automated iris recognition (1994)
9. Y. Song, H. Yan, Image segmentation techniques overview, in *2017 Asia Modelling Symposium (AMS)* (2017)
10. A. Taneja, P. Ranjan, A. Ujjlayan, A performance study of image segmentation techniques, in *2015 4th International Conference on Reliability, Infocom Technologies and Optimization (ICRITO) (Trends and Future Directions)* (2015)
11. G. Scheleyer, C. Cubillos, G. Lefranc, R. Osorio-Comparan, G. Millan, A new colour image segmentation, in *2016 6th International Conference on Computers Communications and Control (ICCCC)* (2016)
12. A. Singh, D. Patil, M. Reddy, S.N. Omkar, Disguised face identification (DFI) with Facial KeyPoints using spatial fusion convolutional network, in *2017 IEEE International Conference on Computer Vision Workshops (ICCVW)* (2017)
13. Dr. P. Gupta, N. Saxena, M. Sharma, J. Tripathi, Deep neural network for human face recognition. Int. J. Eng. Manuf. **1**, 63–71 (2018)

14. D. Gondole, Prof. P.A. Salunkhe, Face recognition based on local binary pattern (2018)
15. M. Ayi, A.K. Ganti, M. Adimulam, B. Karthik, Interfacing of MATLAB with Arduino for face detection and tracking algorithm using serial communication, in *2017 International Conference on Inventive Computing and Informatics (ICICI)* (2017)
16. P. Shayghan Gharamaleki, H. Seyedarabi, Face recognition using Eigen faces, PCA and support vector machines. Eur. J. Appl. Eng. Sci. Res. (2015)
17. M.O. Faruqe, M.A.M. Hasan, Face recognition using PCA and SVM, in *3rd International Conference on Anti-Counterfeiting, Security, and Identification in Communication* (2009)
18. T. Vijayakumar, R. Vinothkanna. Retrieval of complex images using visual saliency guided cognitive classification. J. Innov. Image Process (JIIP) **2**(02), 102–109 (2020)
19. E.E.B. Adam, Evaluation of fingerprint liveness detection by machine learning approach—A systematic view. J. ISMAC **3**(01), 16–30 (2021)
20. Z. Lihong, S. Ying, Z. Yushi, Z. Cheng, Z. Xili, Face recognition based on image transformation, in *2009 WRI Global Congress on Intelligent Systems* (Xiamen, 2009)
21. P. Dinkova, P. Georgieva, A. Manolova, M. Milanova, Face recognition based on subject dependent hidden Markov models, in *2016 IEEE International Black Sea Conference on Communications and Networking (BlackSeaCom)* (Varna, 2016), pp. 1–5
22. M. Agarwal, H. Agrawal, N. Jain, M. Kumar, Face recognition using principle component analysis, eigenface and neural network, in *2010 International Conference on Signal Acquisition and Processing* (Bangalore, 2010), pp. 310–314
23. G. Sun, L. Zhang, H. Sun, Face recognition based on symmetrical weighted PCA, in *2011 International Conference on Computer Science and Service System (CSSS)* (Nanjing, 2011), pp. 2249–2252
24. K. Shailaja, B. Anuradha, Effective face recognition using deep learning based linear discriminant classification, in *2016 IEEE International Conference on Computational Intelligence and Computing Research (ICCIC)* (Chennai, 2016), pp. 1–6
25. M.A. Abuzneid, A. Mahmood, Enhanced human face recognition using LBPH descriptor, multi-KNN, and back-propagation neural network. IEEE Access **6**, 20641–20651 (2018)
26. UCSD computer vision, in *Yale Face Database*. http://vision.ucsd.edu/content/yale-face-database. Accessed 07 Mar 2020

Recent Advances in Transform and Hybrid Domain Digital Watermarking Techniques—A Survey

M. Sajeer, Ashutosh Mishra, and P. S. Sathidevi

Abstract In the older days, watermarking has been extensively used in different forms in diverse fields, mainly to recognize an entity's ownership. At the early stages, its analog rendition has been utilized, like stamps or currencies, perceptible to the users. In the present scenario, the situation changed from analog-to-digital watermarking; mostly, the watermarks are imperceptible to the users, requiring some processing steps to figure out the entity's possession. The processing steps may vary from the type or the application area of the watermark that has been utilized. This paper aims to review the basic idea of digital watermarking techniques, current trends, and future scopes that have been occurred in this area during recent years.

Keywords Digital watermarking · Transform domain schemes · DFT · DCT · DWT · SVD · Hybrid domain schemes

1 Introduction

In the present situation, mostly the data is stored and communicated in digital form. So authenticity has a vital role in identifying the source of data. Digital watermarking resolves this problem [1], which is the technique of inserting data (called watermark) into the original cover or carrier image, and it can extract to verify the authorized creator or provider of the data [2]. It can be perceivable or hidden to the users. A general watermarking approach consists of three processes, watermark creation, watermark implanting, and watermark separation. Watermark creation is the technique of creating a watermark based upon a generation algorithm. The method of hiding a watermark into a host image using an embedding algorithm may define as watermark implanting, and the corresponding image is called a watermarked image. The watermark separation extracts a watermark using an extraction algorithm [3], as shown in Fig. 1. Recently, the watermarking techniques have many potential applications such as ownership authentication, copyright protection, telemedicine,

M. Sajeer (✉) · A. Mishra · P. S. Sathidevi
National Institute of Technology, Calicut, Kerala, India
e-mail: sajeer@sctce.ac.in

G. Ranganathan et al. (eds.), *Soft Computing for Security Applications*,
Advances in Intelligent Systems and Computing 1397,
https://doi.org/10.1007/978-981-16-5301-8_59

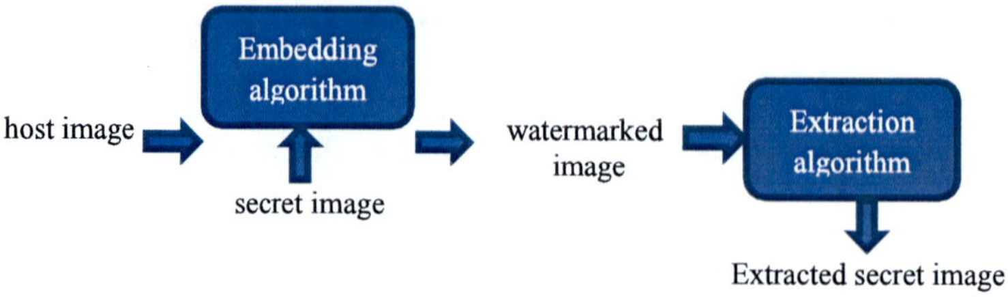

Fig. 1 Watermarking process

electronic governance, media file archiving, web publishing, media forensics, digital libraries, artificial intelligence, military, secure E-voting, different licenses, real-time audio/video, and robotics [4–7].

1.1 Characteristics of a General Watermarking System

There are different characteristics essential for a general watermarking system [8]. Fidelity is an important characteristic, which defines the analogy between the carrier image and the watermarked image [9, 10]. Robustness is the ability to resist different geometric operations and image processing attacks. Data payload (capacity) may be defined as the amount of information present in a watermark [11]. Security is a vital characteristic to identify genuine authors to withstand intentional attacks [12]. Computational complexity explains the total time required for a system to finish watermark embedding and watermark separation processes [13]. Perceptibility is the measurement of distortion in a watermarked image, or it should be invisible to the users [8]. Reversibility outlines the ability to restore the watermark and original image, which is essential for medical applications [14]. Reliability can identify the genuine person to receive the message without data degradation [15].

1.2 Categorization of Digital Watermarking Techniques

Digital watermarking techniques can be categorized as follows [16]. The first classification consists of document, hiding domain, human discrimination, and reversibility [8]. The document type may be image, text, video, or audio. Based on the hiding domain, the watermarking technique is classified into a spatial or transform style. In the spatial method, watermarking is done directly on the host image's pixel values, but in the frequency domain, watermarking performs on the host image's transform coefficients [17]. Invisible (the watermark is invisible to human), visible (the watermark is perceivable), and dual (a combination of visual and hidden) watermarking techniques are coming under human perception [18]. The secret watermarking scheme may be fragile, robust, semi-fragile, or hybrid approaches based on withstanding image processing attacks and geometrical operations [8]. Reversible (lossless watermarking) and non-reversible (lossy watermarking) techniques are another classification of the watermarking scheme [14].

1.3 Types of Watermarking Attacks

Watermarking attacks are the processes that make alterations to the contents of the watermark by unauthorized users. It may also cause the revealing of the hidden data in a watermark—they include.

Signal Processing Attacks They are also called non-geometrical attacks. It may be due to the addition of noise, and some non-geometrical operations such as histogram equalization and gamma correction.

Geometrical Attacks It includes some geometrical operations such as scaling, rotation, translation, shearing, and cropping.

Cryptographic Attacks Cryptographic attacks usually affect the secret information in the watermark.

Protocol Attacks These attacks encompass the copy and protocol attacks, where the interferer will try to change the watermarks in the data.

1.4 Performance Measurement in Watermarking System

There are different matrices to assess the potential of a watermarking system. They compare the cover image and the watermarked image. They include:

Mean Square Error or MSE [19]—It is the average squared error between the original image and the watermarked image. Its value should be deficient. Ideally, its value is zero.

Peak Signal-to-Ratio or PSNR [20]—The ratio of signal-to-noise measures the watermarked image's visual quality. Usually, its value will be high (generally more than 30 dB), which is the measure of similarity between the watermarked and host image.

Normalized Coefficient or NC [20]—It finds the compatibility between embedded and extracted watermark. Its value comes between zero and one. Ideally, it is one.

Structural Similarity Index or SSIM [20]—This shows the closeness of the original and watermarked image. The value of SSIM varies from -1 to $+1$. Ideally, it is $+1$.

Bit Error Ratio or BER [21]—The ratio of the number of received bits as the error to the binary sequence's total length in a watermark. Its value should be low, and ideally, it is zero.

UACI and NPCR [22]—UACI is the unified averaged changed intensity, NPCR is the pixel changing rate used to investigate encryption algorithms' performance against different attacks. Ideally, their value is very high.

The rest of the paper is described as follows: Section 2 explains different domain schemes used in digital watermarking, mainly various transform domains and different hybrid domain techniques; challenges in watermarking domains are presented in Sect. 3. The conclusion summarizes the survey.

2 Survey of Different Watermarking Schemes

Based on the domain, the watermarking technique may be spatial or transform type. The spatial domain schemes have advantages like a high payload, less computational complexity, and less processing time, but they have less imperceptibility and less robustness against watermarking attacks. LSB or least significant bit technique, correlation-based methods, patchwork methods, spread spectrum methods, etc., are coming under this scheme. Transform domain techniques such as discrete Fourier transform (DFT), discrete cosine transform (DCT), discrete wavelet transform (DWT), singular value decomposition (SVD) are having advantages like high imperceptibility and very high robustness against attacks. Disadvantages of these methods are less watermarking capacity, high computational complexity, and increased processing time. Most of the researchers are currently using hybrid approaches, which combine the above transform domain methods based on the applications [22].

2.1 Transform Domain Watermarking Schemes

There are distinct transform domain schemes introduced by various authors like [23–35], and their observations are summarized in Table 1.

Andalibi and Chandler [23] proposed a DWT domain invisible watermarking process using adaptive logo texturization. Initially, the carrier image is separated into a sufficiently textured area and a poorly textured region using the ALT-MARK algorithm. Then modified the watermark into a perceptibly similar structure as that of a carrier image using the methods like Arnold transform, lossless rotation. For every blocks, inserted the modified watermark into that region using the DWT technique. An affine parameter is utilized to withstand geometrical transformations. The process measured the PSNR for checking the imperceptibility and obtained values varying

Table 1 Summary of transform domain watermarking techniques

Refs	Techniques/algorithms performed	Cover image/watermark specifications	Dataset used	Attacks handled	Area of application
[23]	DWT, ALT-MARK algorithm, Arnold transform	512 × 512 (gray), 64 × 64 (gray)	CSIQ	AGN, JPEG, HE, BR, RO, SC, CR, CMB	Logo watermarking
[24]	IWT	265 × 190 (gray), --	CT, MRI and PET images	AGN, S&P, JPEG, LPF	Tamper detection in ROI of medical images
[25]	DCT, Visual cryptography	--, 64 × 64 text image	CASIA V4,UBIRIS V1	AGN, MF, HE, JPEG, S&P, CR	Iris biometric protection
[26]	Hessenberg decomposition, Arnold transform	512 × 512 (color), 32 × 32 (color)	CVG-UGR	AGN, JPEG, S&P, MF, BLPF, SC, BR, RO	Copyright protection of color image
[27]	DCT	512 × 512 (gray), –	–	AGN, JPEG, SC, CLC	Image watermarking
[28]	DCT, Fractal encoding	1024 × 1024 (gray), 256 × 256 (gray)	–	WN, JPEG, GF	Image watermarking
[29]	DCT	512 × 512 (gray), 32 × 32 (binary)	–	AGN, MF, WF, AF, GLP, HE, JPEG, S&P, SN, PN, CR, SC, SH, CMB	Copyright protection
[30]	Ripplet-II transform, DFT	512 × 512 (Gray), 256 × 256 (binary)	USC-SIPI	AGN, MF, GLP, HE, JPEG, S&P, SN, CR, SH, RO, TR	Copyright protection

(continued)

Table 1 (continued)

Refs	Techniques/algorithms performed	Cover image/watermark specifications	Dataset used	Attacks handled	Area of application
[31]	DCT	512 × 512 (color), 32 × 32 (binary)	CVG–UGR, USC-SIPI	AGN, JPEG, MF, S&P, AF, GLP, SC, CR	Logo watermarking
[32]	DCT, additive watermarking	256 × 256 (gray), 32 × 32 (binary)	–	MF, GF, PN, SN, S&P, JPEG, SH, BR	Logo watermarking
[33]	DCT, Bees algorithm	512 × 512 (gray), 32 × 32 (binary)	–	AGN, MF, AF, GLP, HE, JPEG, S&P, CR, SC, SH, GC, RO	Logo watermarking
[34]	Histogram, HFCM	Gray	–	AGN, JPEG, S&P, MF, SC, SH, RO, CR, BE, JT	Image watermarking
[35]	DCT, QIM-DM	1024 × 728 (color), 64 × 64 (color)	USC-SIPI	AGN, JPEG, AF, MF	Color image watermarking

*Non-geometrical attacks: Additive Gaussian Noise (AGN) [23], JPEG compression (JPEG) [24], Histogram Equalization (HE) [25], Motion Blurring (BR) [26], White noise (WN) [28], Poisson Noise (PN) [32], Salt and Pepper Noise (S&P) [30], Gaussian Filter (GF) [32], Median Filter (MF) [31], Constant Luminance Change (CLC) [27], Weiner Filter (WF) [29], Speckle Noise (SN) [30], Average Filter (AF) [31], Gamma Correction (GC) [33]; Geometrical attacks: Rotation (RO) [33], Scaling (SC) [34], Cropping (CR) [34], Translation (TR) [30], Shearing (SH) [33], Bending (BE) [34], Jittering (JT) [34]; Combinational Attacks (CMB) [29]

*Techniques or algorithms used: Discrete Wavelet Transform (DWT) [23], Integer Wavelet Transform (IWT) [24], Discrete Fourier Transform (DFT) [30], Discrete Cosine Transform (DCT) [31], High-Frequency Component Modification (HFCM [34]), Quantization Index Modulation–Dither Modulation (QIM-DM) [35]

from 29.68 to 44.21 dB for different watermarked images. Robustness is evaluated by applying some non-geometrical attacks like JPEG compression, blurring, and Gaussian noise; geometrical attacks like rotation, scaling, and cropping; and combinational attacks. The method obtained good robustness results for the consecutive attacks (Gaussian noise of density 0.1% + counterclockwise direction of 10° + cropping of 40 pixels from the top and bottom of the image) and significant values for

JPEG compression, and cropping. The scheme has to improve the robustness against attacks like histogram equalization, motion blurring, and rotation.

Easwaraiah et al. [24] developed a novel watermarking method for medical images based on an integer wavelet transform (IWT). Medical images consist of region of non-interest (RONI) and region of interest (ROI) regions. In this method, the watermark, which consists of patient data, the hash value of ROI, and recovery data of ROI, is inserted in the RONI region using Cohen–Daubechies–Fauraue IWT. When an attacker tampers the ROI, the receiver can reconstruct it using the RONI watermark. In this method, a rectangle-shaped ROI of size 3×3 is used, marked by the physician. The process has experimented with CT, MRI, Ultrasound, PET images, and 20 images in each modality. The method's performance is measured by using PSNR, weighted PSNR (WPSNR), mean SSIM (MSSIM), and total perceptual error (TPE). The average values obtained for PSNR and WPSNR are above 40 dB, desirable for medical images. The scheme's robustness is evaluated by applying attacks like noise addition, JPEG compression, filtering, and finding the normalized correlation (NC) and obtained significant results. ROI's size should be < 20% of the whole image's size and is the main limitation of the above method.

Abdullah et al. [25] used fusion of visual cryptography and watermarking for the protection of iris image and its templates. It used DCT coefficients of the host image's middle band frequencies to implant text watermark, which holds personal data. Template protection is carried out using visual cryptography technique. Different matrices are employed for measuring the system's performance; PSNR, BER values obtained for the watermarked images are 38.47 dB and 0.22%, and that for separated text are 84.63 dB and 0.023%, respectively. The system's robustness is evaluated by applying some attacks like JPEG compression, the addition of Gaussian noise; image manipulations such as histogram equalization, median filtering, and cropping, which obtained significant values for all the attacks. A strength constant is used to improve withstanding capability against different attacks; the images are distorted for less strength constant values by median filtering and compression attacks. It can enhance them by using high-strength constant values. Another limitation is that the watermark's text length is more dependent on the size of the iris image.

Su [26] designed a watermarking technology using Hessenberg decomposition, which decomposes the original color carrier image into a matrix of size 4×4; inserted the watermark on the specified columns in the orthogonal matrix, obtained by the host image decomposition. The watermark pixel position is shuffled using an Arnold transform. The extraction process did not require either the carrier image or watermark image. The similarity measurement of the scheme is evaluated using PSNR and SSIM, obtained values above 35 dB and 0.9616, respectively, under no attack. The watermark's robustness is calculated using NC values, against attacks like JPEG 2000 compression, the addition of noise, filtering, scaling operation, blurring, and rotation. The method resulted in NC values of more than 0.9 in almost all attacks and deals ranges from 0.8 to 0.9 for JPEG 2000 compression and Gaussian noise attacks, which has to be improved.

Zong et al. [27] introduced a rank-based watermarking scheme. This method divided the cover image into blocks: applied 2D-DCT to obtain DCT coefficients in each block. A confidential key is used to randomly selecting some DCT values for the watermark inserting. A rank-based embedding rule is employed to implant the watermark into the image structure by altering some DCT coefficients. PSNR and BER values are measured to evaluate the watermarked image's perceptual quality and robustness against attacks. The average values obtained for the PSNR are above 40 dB, which is a significant one. The proposed method's robustness is evaluated by applying JPEG compression, constant luminance change, amplitude scaling, and noise addition attacks. The scheme calculated the BER for different episodes with different 'K' values, where 'K' is the number of DCT coefficients selected for embedding the watermarks in each block. They obtained 0 deal for BER without any attacks, amplitude scaling, and constant luminance change attacks. The BER values obtained for JPEG compression and Gaussian noise addition attacks are low, desirable for a watermarking scheme. The inserting capacity also is high since the approach used 2 DCT values to implant a single watermark bit in each block.

Liu et al. [28] developed a digital watermarking technique based on fractal encoding and DCT. First, the host image is encoded by fractal encoding and used encoded parameters in the DCT domain, which enabled a two-level encryption scheme. The watermark is inserted in the middle- and lower-frequency coefficients, resisting more JPEG compression with extensive embedded data. The technique measured the watermarked image's perceptual quality using PSNR and obtained a value of above 40 dB. Different attacks like JPEG compression, noise, and Gaussian filter are employed for the robustness measurement of the proposed scheme and obtained different mutual correlation coefficient (MCC) values of more than 0.9 in all the cases. But the technique handled only a small number of attacks to verify its robustness.

Ernawan and Kabir [29] presented an optimal DCT psycho-visual threshold watermarking method for copyright protection. The watermark is disarranged before inserting, which provides additional security. The carrier image is split into blocks of size 8 $\times$ 8 and computed for each block's modified entropy. Applied DCT to some lowest entropy-valued blocks and selected some coefficients for watermark embedding based on a psycho-visual threshold in lower frequencies. The selected coefficients are modified to insert the shuffled watermark, to form a watermarked image. The proposed scheme's imperceptibility is measured using PSNR, absolute reconstruction error (ARE), and SSIM and obtained a value of above 44 dB for PSNR; the average values of ARE and SSIM are 0.317 and 0.994. Robustness is evaluated using attacks like filtering, noise addition, JPEG compression, histogram equalization, cropping, scaling. The method obtained better robustness results for the attacks like Gaussian noise, median filtering, histogram equalization; weak robustness to the rotation, translation, and Poisson noise, which needs to be improved.

Shahdoosti and Salehi [30] designed a transform-based watermarking technique for copyright protection, used the ripplet-II transform. Here, the ripplet-II transform's rotation-invariant property is used, which provides good robustness to rotation attacks. The cover image is transformed to matrix coefficients using ripplet-II transform, DFT is employed to rows of the resulted matrix to obtain the Fourier matrix. The Fourier matrix consists of a phase matrix and an amplitude matrix, the former saved for inverse transform and the latter for watermarking. A cost function improves the invisibility of the scheme in the ripplet-II transform. Then added a ridge regularization constant to the cost function, which avoids the singularity problem. The weights of the transform are adjusted by varying the cost function. The scheme's watermark invisibility is measured using the matrices PSNR and SSIM and obtained outstanding results a value of above 73 dB for PSNR and 1 for SSIM. The robustness is checked using BER by applying attacks like filtering, noise addition, JPEG compression, histogram equalization, edge sharpening, rotation, translation, and cropping. Significant results are obtained for the attacks like rotation and cropping. The scheme results in fewer values for translation attacks.

Byun et al. [31] proposed a novel watermarking scheme using DCT. Initially, a host image's DCT coefficients for a particular location are calculated and modified the coefficient according to a variation value based on inserting bits and quantization levels. The watermark bits are inserted by changing some pixel merits in DCT, which reduces computational complexity. The method obtained an embedding time of less than 1 ms because of less computational complexity. PSNR and SSIM matrices are employed to evaluate the proposed scheme's invisibility and brought an average value of 42 dB for PSNR and 0.97 for SSIM. Different attacks like noise, filtering, scaling, JPEG compression, and cropping are applied and calculated BER and NC values. The results implied that improvements to be needed for the JPEG attack.

Srivastava et al. [32] introduced a joint imperceptible image watermarking with JPEG compression using DCT. An additive watermarking is performed by adding a scaled watermark to the DCT coefficient of the cover image. The carrier image is subdivided into blocks of size 8×8, applied DCT on each block, embedded watermark using additive watermarking before quantization and after quantization of blocks resulting in a compressed watermarked image. Extraction of the watermark is done using the reverse process of embedding. The scheme's performance is compared in terms of PSNR, correlation, compression ratio, different quantization matrices with various Q factor, and DCT block sizes. The results indicated that the watermarking after quantization performed well compared to watermarking after DCT and obtained the values of 36.24 and 36 dB for PSNR in the above two cases. The scheme got a significant number of 0.83 for the compression ratio. Different attacks like sharpening, blurring, median filtering, Gaussian filtering, Poisson and speckle attacks, salt-and-pepper noise are applied to measure the robustness resulted in moderate values for blurring, Poisson, and speckle attacks. The main limitation is the less robustness against sharpening, median filtering, and Gaussian filtering attacks.

Abdelhakim et al. [33] designed a DCT-based watermarking technique that uses Bees algorithm for optimization. The carrier image is subdivided into blocks, applied DCT to each block, and selected some coefficients in the high-frequency region for the watermark embedding. Fitness functions are used to adjust the optimization between imperceptibility and robustness against different attacks. The method resulted in a maximum PSNR of 46.02 dB with the implemented fitness function and BCR values of more than 0.9 to attacks. The scheme used JPEG compression, filtering, gamma correction, the addition of different noises, histogram equalization, and various geometrical attacks for measuring robustness. The technique has to enhance the robustness against noise attacks.

Zong et al. [34] proposed a gray-level watermarking scheme to increase the robustness to cropping and random bending attacks. The cover image is preprocessed using a Gaussian low-pass filter, which removes high-frequency components. The filtered image's histogram is obtained and selected different gray levels from the histogram using a secret key. An index based on the shape of the histogram is used for choosing a pixel for watermark embedding and a high-frequency component modification to nullify the effect of low-pass filtering. At the receiver end, decoded the watermark using the secret key. The method obtained a PSNR value of above 40 dB and an SSIM of more than 0.9. Robustness is measured to different signal processing, and geometrical attacks and brought significant results. The method has to enhance the robustness against JPEG compression.

Muñoz et al. [35] designed a color watermark embedding scheme using DCT. First, the carrier color image is separated into R, G, B images, further converted each sub-image into non-overlapping blocks of size 8×8. Then performed DCT to each block and arranged the DC coefficients to form a new matrix. DCT is applied to this new matrix, embedded subsampled binary values of watermarks to the new coefficients with the help of the quantization modulation technique. Then performed inverse DCT to form the watermarked image. For watermark extraction, the reverse of the inserting process is done. The technique obtained a maximum PSNR of 44.82 dB and an SSIM value of 0.99. Different attacks are applied to evaluate the performance and resulted in significant results. The method needs to enhance the performance against the average filtering.

2.2 *Hybrid Domain Watermarking Schemes*

There are different hybrid watermarking schemes proposed by various authors like [36–42], whose inferences are demonstrated in Table 2.

Muhammad and Bibi [36] presented a hybrid technique using discrete wavelet transform and singular value decomposition. Initially, the carrier image is fragmented using DWT, SVD, PPLU algorithm, which factorizes the watermark into permutation and triangular matrices. The permutation matrix is the key for the authentication. A weightage-based differential evaluation algorithm is performed to obtain better robustness against attacks. Measurement matrices like PSNR, structure similarity index (SSIM), and correlation coefficient (CC) are utilized to estimate the invisibility of the scheme. The average value obtained for PSNR is above 40 dB; CC index is nearly equal to 1. It has a high embedding capacity of 20,888 bits. Different fragile watermarking operations like noise, rescaling, blurring, histogram, contrast, and gamma corrections are performed to assess the method's robustness and also tried a JPEG compression with a compression ratio between 70–90% for evaluating

Table 2 Summary of hybrid domain watermarking techniques

Refs	Techniques/algorithms performed	Cover image/watermark specifications	Dataset used	Attacks handled	Area of application
[36]	DWT,SVD,PPLU algorithm, Weightage-based differential algorithm	512 × 512 (gray), 256 × 256 (binary)	SIPI	NS, JPEG, SC, BR, HE, CT, GC	Image watermarking
[37]	DWT, QR, FA algorithm	512 × 512 (gray), 64 × 64 (binary)	–	JPEG, SN, GLP, GC, RO, SC	Image watermarking
[38]	DWT, DCT, SVD	512 × 512 (gray), 256 × 256 (binary), Text (50 characters)	–	JPEG, MF, GLP, HE, S&P, SC	Iris biometric protection
[39]	FDCuT, DCT	1024 × 1024 (Gray), 128 × 128 (binary)	Med Pix	AGN, MF, AF, GLP, JPEG, S&P, SN, CR, SH, RO, FP, BR	Medical image watermarking
[40]	NSCT,RDWT, SVD, Logistic map chaotic encryption algorithm	512 × 512 (gray), 256 × 256 (gray), 128 × 128(text)	–	AGN, MF, GLP, JPEG, S&P, SC, RO, HE	Multiple medical image watermarking
[41]	DWT, DCT, SVD, Chaotic encryption algorithm	512 × 512 (gray), 256 × 256 (gray)	–	AGN, MF, GLP, JPEG, S&P, SC, RO, HE, CR, SH	Medical image watermarking
[42]	DWT,SVD, hash function	1024 × 1024 (gray), 1024 × 1024 (gray) (equal size for cover image and watermark)	–	AGN, MF, AF, S&P, SN, SC, RO, HE, CR, GC	Medical and non-medical image watermarking

*Non-geometrical attacks: Additive Gaussian Noise (AGN) [39], JPEG compression (JPEG) [36], Histogram Equalization (HE [38]), Motion Blurring (BR) [39], Noise (NS) [36], Speckle Noise (SN) [42], Salt and Pepper Noise (S&P) [40], Average Filter (AF), Median Filter (MF) [31], Gaussian Low Pass Filter (GLP) [37], Gamma Correction (GC) [37], Contrast (CT) [36]; Geometrical attacks: Rotation (RO) [42], Scaling (SC) [42], Cropping (CR) [41], Shearing (SH) [39], Flipping (FP) [39]
*Techniques or algorithms used: Discrete Wavelet Transform (DWT) [36], Discrete Cosine Transform (DCT) [38], Singular Value Decomposition (SVD) [41], Partial Pivoting Lower–Upper triangular decomposition algorithm (PPLU) [36], Firefly Algorithm (FA) [37], Fast Discrete Curvelet Transform (FDCuT) [39], Non-Subsampled Counterlet Transform (NSCT) [40], Redundant Discrete Wavelet Transform (RDWT) [40]

the robustness and obtained CC values ranging from 0.9090 to 0.9849. The scheme successfully withstands almost all attacks but handled less number of attacks.

Guo et al. [37] designed an optimized blind watermarking approach for images, which uses the firefly algorithm in the DWT-QR domain. Firefly algorithm is an optimization algorithm with advantages like local attraction and automatic regrouping. DWT transform domain methods have properties like high energy compression, high robustness against watermarking attacks, and multi-resolution representation. The main disadvantage is the low withstanding capability to some geometric attacks, which is avoided by the matrix factorization method- QR decomposition. It provides high robustness against geometric attacks. FA is used to optimize embedding strength, which improved the scheme's performance. At the initial stage, the carrier image is scrambled and disintegrated into sub-bands using 1 level DWT, divided LL sub-band into blocks, and applied QR factorization to all blocks. The R-matrix is employed for binary watermark insertion. The watermarked image's invisibility is checked by using performance matrices PSNR, SSIM, correlation, and NC. The method obtained a value of above 35 dB for PSNR and above 0.93 for all other matrices. Here, applied various attacks like rescaling, rotation, JPEG compression, gamma correction, filtering, and speckle noise for robustness measurement. Matrices like SSIM, BER, and NC are utilized for the quality checking of the process. The method resulted in outstanding robustness to gamma correction and comparable performance to other attacks.

Singh et al. [38] proposed multiple watermarking methods using DWT, DCT, and SVD and applied for medical images. At first carrier medical image is disintegrated to the two-level of coefficients using DWT. The resulting LL band is altered using DCT and SVD; the same transformation is practiced to the watermark image also. The embedding is accomplished using the singular values. Text watermark or personal data, embedded at the two-level of the high-frequency band, HH of the carrier image after encoding. Different gain factors like 0.05, 0.5, and 1 are utilized for the experimentation. The method used PSNR, NC, and BER as the measurement matrices. PSNR values of 35.84 dB, BER = 0, and NC = 0.9802 are obtained at a gain factor of 0.01 with encryption. The scheme's robustness is calculated using different attacks like median filtering, noise addition, JPEG compression, scaling, Gaussian filtering, and histogram equalization; significant values are obtained for compression and median filtering. It has low robustness to attacks like histogram equalization, noise, and scaling.

Thanki et al. [39] introduced a blind watermarking process for medical images utilizing DCT and FDCuT, which represents an image as to edges. FDCuT is implemented to cover images to procure different low-, medium-, and high-frequency sub-bands and performed DCT to the HF curvelet sub-band. Gaussian Noise sequences are inserted into mid-band frequency coefficients, to acquire the embedded image. Using the correlation among the Gaussian noise sequence and watermarked data, the watermark separation is done. The scheme's performance is assessed using matrices like NC, PSNR. An average value of PSNR = 49.36 dB, NC = 0.9618, is obtained at a gain factor = 2. Various attacks—noise addition, JPEG compression, cropping, rotation, flipping, filtering, blurring, and sharpening–are used to validate the scheme's robustness and obtained medium values. The method has limitations like noise in the separated watermark and difficulty of embedding the text watermark.

Thakur et al. [40] designed a watermarking method that used Contourlet transform–discrete wavelet transform–SVD techniques. It used multiple watermarking—a medical image and personal data. Initially, the original carrier image is disintegrated into subsampled components using the sub-sampling method and applied Contourlet transform to maximum entropy components. Then applied redundant DWT to the HF sub-bands of Contourlet transform decomposed image and performed SVD to the selected sub-bands of DWT transformed images. Similar strategy is employed for both the images and embedded the watermark utilizing modified SVD coefficients with a pre-defined gain factor. The personal data watermark is separated using the logistic map chaotic encryption algorithm. The scheme's performance is measured by employing PSNR, NC, number of changing pixel rate (NPCR), unified averaged change intensity (UACI) and obtained maximum values 54.49 dB, 0.9993, 0.99, and 0.34. The method's robustness is evaluated by applying salt-and-pepper noise, Gaussian noise, JPEG compression, rotation, Gaussian LPF, scaling, median filtering, and histogram equalization and obtained comparable results.

Thakur et al. [41] introduced a watermarking process, which uses DWT-DCT-SVD with chaotic encryption for tele-health applications. The original carrier image is transformed using DWT-DCT-SVD domain; watermark image is using DCT-SVD domain. Singular values resulted are used for the creation of the final watermarked image and are encrypted by a chaotic encryption algorithm. The recovery process is done using the reverse embedding process. The system's performance is evaluated using PSNR, NC, SSIM, NPCR, and UACI and obtained the maximum value of 74.60 dB, 0.9999, 1, 0.9962, and 0.4835. The scheme's robustness is measured using different attacks like JPEG compression, salt-and-pepper noise, Gaussian noise, cropping, rotation, scaling, Gaussian low-pass filter, sharpening, median filter, histogram equalization and obtained medium values. The technique needs to enhance the robustness against attacks like rotation, Gaussian LPF, and histogram equalization.

Araghi and Manaf [42] developed a hybrid watermarking scheme using DWT-2 level SVD methods, which can apply for both medical and non-medical images. Initially, the watermark is preprocessed by duplicating it to the host image's size. Then applied DWT to the original cover image, separated the high-frequency sub-band into 16×16 non-overlapping blocks. After that, performed 1 level SVD to all the blocks, selected one singular value from each block to form a creative matrix. Two-level SVD is employed in this matrix. Singular value coefficients of this decomposed matrix are employed for the watermark inserting. The same procedure is applied for the watermark after preprocessing. Two-level security is performed using the hash function for authenticity. The scheme obtained an average value of 46.94 dB for PSNR, measured the robustness to distinct attacks filtering, noise addition, scaling, cropping, rotation, histogram equalization, gamma correction and resulted in nearly 0.99 of NC for almost all the attacks. The scheme does not handle any compression attack.

3 Discussion on Research Challenges and Opportunities in the Area of Digital Watermarking

There are different characteristics for a watermarking system: imperceptibility, robustness, embedding capacity, and data security are very important. Many authors presented various techniques to handle the situation. But only a less number of approaches are accomplished in it. So it will be a challenging one to deal with them effectively. Watermarking with cryptographic methods can be employed to enhance security issues, requiring more attention in this field. Multiple watermarking has prominent applications, where users can insert different watermarks, which provide more authenticity to the system. While implementing a watermarking approach for medical application, it should be reversible; otherwise, it may cause the disease's misdiagnosis due to data loss. Nowadays, watermarking has immense applications in biometric security, where different biometric templates like fingerprint and iris can be stored or transmitted with cover images. Here efficient encryption algorithms can be effectively used in order to prevent it from unauthorized users. Another issue is developing real-time watermarking systems, where execution time should be crucial with all other characteristics. Mostly the watermarking schemes are software-based digital signal processing (DSP) boards, field-programmable gate arrays (FPGAs), graphical processing units (GPUs), etc., can be employed for hardware implementation [11]. In the present scenario, we used the tele-biometric system and the tele-medical healthcare system in the medical field. The blockchain–Internet of Things (IoT) combination techniques may come into the picture, where data security happens to be a significant issue. It can use digital watermarking with suitable compression schemes to resolve this problem, protect the data, and be a challenging opportunity for current researchers [43, 44]. Machine learning and deep learning methods can also incorporate into the watermarking, so that we can use the marking technique effectively.

4 Conclusion

Digital watermarking, the process of embedding a secret image into a host image, has immense applications in almost all areas and has been popularly used especially for authenticity and data protection. This survey paper discussed the basics of digital watermarking process and recently used various transform and hybrid domain techniques. Transform domain watermarking techniques such as DFT, DCT, and DWT have advantages like high imperceptibility and very high robustness against attacks. Disadvantages of these methods are less watermarking capacity, high computational complexity, and increased processing time. Hybrid domain approaches combine the transform domain techniques, so that we can utilize the advantages of the above methods. We prepared summary tables with observations, which highlight the attributes of each method. Challenges and research opportunities, which may be beneficial for current researchers working in this relevant field, are explained. New techniques should be introduced, which can efficiently handle all the issues related to conventional schemes.

References

1. F. Shih, *Digital Watermarking and Steganography, Fundamentals and Techniques* (CRC Press, USA, 2008)
2. S.P. Mohanty, Watermarking of digital images. M.S. thesis (Indian Institute of Science, India, 1999)
3. H. Nyeem, W. Boles, C. Boyd, A review of medical image watermarking requirements for teleradiology. J. Digit. Imaging **26**, 326–343 (2013)
4. N. Morimoto, Digital watermarking technology with practical applications, Inf. Sci. Spec. Issue Multimedia Inf. Technol. Part 1 **2**(4), 107–111 (1999)
5. F. Hartung, F. Ramme, Digital rights management and watermarking of multimedia content for m-commerce applications. IEEE Commun. Mag. **38**((11), 78–84 (2000)
6. B.L. Gunjal, S.N. Mali, *Applications of Digital İmage Watermarking in Industries* (CSI Communications, 2012), pp. 5–7
7. A.K. Singh, B. Kumar, M. Dave, S.P. Ghrera, A. Mohan, Digital image watermarking: techniques and emerging applications. In: *Handbook of Research on Modern Cryptographic Solutions for Computer and Cyber Security* (IGI Global, USA, 2016), pp. 246–272
8. S.M. Mousavi, A. Naghsh, S. Abu-Bakar, Watermarking techniques used in medical images: a survey. J. Digit. Imaging **27**, 714–729 (2014)
9. C. Fung, A. Gortan, W.G. Junior, A review study on image digital watermarking, in *The Tenth İnternational Conference on Networks* (2011), pp. 24–28
10. I.J. Cox, M.L. Miller, J.A. Bloom, J. Fridrich, T. Kalker, *Digital Watermarking and Steganography*, 2nd edn. (Morgan Kaufmann, Burlington, 2008), pp. 425–467
11. Mohanty, S.P., Sengupta, A., Guturu, P., E. Kougianos, Everything you want to know about watermarking (IEEE Consumer Electronics Magazine, 2017), pp. 83–91
12. L.O.M. Kobayashi, S.S. Furuie, P.S.L.M. Barreto, Providing integrity and authenticity in DICOM images: a novel approach. IEEE Trans. Inf. Technol. Biomed. **13**, 582–589 (2009)
13. J.D. Gordy, L.T. Bruton, Performance evaluation of digital audio watermarking algorithms, in *Proceedings of the 43rd IEEE Midwest Symposium on Circuits and Systems* (2000), pp. 456–459
14. N. Thilagavathi, D. Saravanan, S. Kumarakrishnan, S. Punniakodi, J. Amudhavel, U. Prabu, A survey of reversible watermarking techniques, application and attacks, in *Proceedings of the 2015 International Conference on Advanced Research in Computer Science Engineering & Technology* (ICARCSET, 2015), p. 37
15. Z. Fan, Z. Hongbin, Capacity and reliability of digital watermarking, in *Proceedings of the IEEE International Conference on the Business of Electronic Product Reliability and Liability* (2004)
16. A.F. Qasim, F. Meziane, R. Aspin, Digital watermarking: applicability for developing trust in medical imaging workflows state of the art review. Comput. Sci. Rev. **27**, 45–60 (2018)
17. A.-N. Yahya, H.A. Jalab, A. Wahid, R.M. Noor, Robust watermarking algorithm for digital images using discrete wavelet and probabilistic neural network. J. King Saud Univ.-Comput. Inf. Sci. **27**, 393–401 (2015)
18. A. Hassani Allaf, M. Ait K bir, A review of digital watermarking applications for medical images exchange security. in*Lecture Notes in İntelligent Transportation and İnfrastructure Book Series* (Springer, 2019)
19. A. Mohanarathinam,·S. Kamalraj, G.K.D. Prasanna Venkatesan, R.V.Ravi, C.S. Manikandababu, Digital watermarking techniques for image security: a review, J. Ambient Intell. Hum. Comput. **11**, 3221–3229 (2020)

20. F.A.P. Petitcolas, R.J. Anderson, M.G. Kuhn, Information hiding—A survey. IEEE **87**(7), 1062–1078 (1999)
21. A.K. Singh, B. Kumar, M. Dave, A. Mohan, Robust and imperceptible dual watermarking for telemedicine applications. Wirel Pers Commun **80**(4), 1415–1433 (2014)
22. A. Anand, A.K. Singh, Watermarking techniques for medical data authentication: a survey, in *Multimedia Tools and Applications* (Springer, 2020), pp. 1–33
23. M. Andalibi, D.M. Chandler, Digital ımage watermarking via adaptive logo texturization. IEEE Trans. Image Process. **24**(12) (2015)
24. R. Easwaraiah, E. Sreenivasa Reddy, Robust medical image watermarking technique for accurate detection of tampers inside region of interest and recovering original region of interest. IET Image Process. **9**(8), 615–625 (2015)
25. M.A.M. Abdullah, S.S. Dlay, W.L. Woo, J.A. Chambers, A framework for iris biometrics protection: a marriage between watermarking and visual cryptography. IEEE Access **4**, 10180–10193 (2016)
26. Q. Su, Novel blind colour image watermarking technique using Hessenberg decomposition. IET Image Process. **10**(11), 817–829 (2016)
27. T. Zong, Y. Xiang, S. Guo, Y. Rong, Rank-based image watermarking method with high embedding capacity and robustness. IEEE Access **4**, 1689–1699 (2016)
28. S. Liu, Z. Pan, H. Song, Digital image watermarking method based on DCT and fractal encoding. IET Image Process. **11**(10), 815–821 (2017)
29. F. Ernawan, M.N. Kabir, A robust image watermarking technique with an optimal DCT-psychovisual threshold. IEEE Access **6**, 20464–20480 (2018)
30. H.R. Shahdoosti, M. Salehi, Transform-based watermarking algorithm maintaining perceptual transparency. IET Image Process. **12**(5), 751–759 (2018)
31. S.-W. Byun, H.-S. Son, S.-P. Lee, Fast and robust watermarking method based on DCT specific location. IEEE Access **6**, 20464–20480 (2018)
32. R. Srivastava, B. Kumar, A.K. Singh, A. Mohan, Computationally efficient joint imperceptible image watermarking and JPEG compression: a green computing approach, in *Multimedia Tools and Applications* (Springer, 2017), pp. 1–13
33. A.M. Abdelhakim, H.I. Saleh, A.M. Nassar, Quality metric based fitness function for robust watermarking optimisation with Bees algorithm. IET Image Process. **10**(3), 247–252 (2016)
34. T. Zong, Y. Xiang, I. Natgunanathan, S. Guo, W. Zhou, G. Beliakov, Robust histogram shape-based method for ımage watermarking. IEEE Trans. Circ. Syst. Video Technol. **25**(5) (2015)
35. D. Muñoz, V. Ponomaryov, R. Reyes-Reyes, C. Cruz-Ramos, S. Sadovnychiy, Embedding a color watermark into DC coefficients of DCT from digital ımages. IEEE Latin Am. Trans. **17**(8) (2019)
36. N. Muhammad, N. Bibi, Digital image watermarking using partial pivoting lower and upper triangular wavelet decomposition into the wavelet domain. IET Image Proces. **9**(9), 795–803 (2015)
37. Y. Guo, B.-Z. Li, N. Goel, Optimised blind image watermarking method based on firefly algorithm in DWT-QR transform domain. IET Image Process. **11**(6), 406–415 (2017)
38. A.K. Singh, M. Dave, A. Mohan, Hybrid technique for robust and imperceptible multiple watermarking using medical images, in *Multimedia Tools and Applications* (Springer, 2015), pp. 1–21
39. R. Thanki, S. Borra, V. Dwivedi, K. Borisagar, An efficient medical image watermarking scheme based on FDCuT-DCT. Eng. Sci. Technol. Int. J. **20**, 1366–1379 (2017) (Elsevier)
40. S. Thakur, A.K. Singh, S.P. Ghrera, A Mohan, Chaotic based secure watermarking approach for medical images, in *Multimedia Tools and Applications* (Springer, 2018), pp. 1–14
41. S. Thakur, A. K. Singh, S.P. Ghrera, M. Elhoseny, Multi-layer security of medical data through watermarking and chaotic encryption for tele-health applications, in *Multimedia Tools and Applications* (Springer, 2018), pp. 1–14
42. T.K. Araghi, A.A.Manaf, An enhanced hybrid image watermarking scheme for security of medical and non-medical images based on DWT and 2-DSVD.Future Gener. Comput. Syst. **101**, 1223—1246 (2019)

43. A. Castiglione, K.-K. Raymond Choo, M. Nappi, F. Narducci, Biometrics in the cloud: challenges and research opportunities. IEEE Cloud Comput. Mag. (2017)
44. H.-N. Dai, M. Imran, N. Haider, Blockchain-enabled internet of medical things to combat covid-19. IEEE Internet Things Mag. (2020)

Improved Genetic Algorithm Using Machine Learning Approaches to Feature Modelled for Microarray Gene Data

Revathi Durgam, Nagaraju Devarakonda, Anand Nayyar, and Ramakrishna Eluri

Abstract One of the most controversial and emerging applications of microarray-based cancer research is the identification of the pathology that are canons and genetic pathway alterations. Though microarray-based classifications of cancer diagnosis continue to be a concern, cancer orthodox has not advanced very far. The primary explanation for this is the vast number of available genes for training proportionately to the number of samples accessible. In this paper, we present a hybrid machine learning approach scenario on genetic algorithms as an improved characteristic set of data sets such as lymphoma, lung, MLL, breast and colon cancers utilized to provide characteristic features on each algorithm proposed. The classification methods such as XG-boost, RFR and extra tree classifications have been implemented with GA and without GA, providing the list of genes until it has reduced the number of features to the machine learning classifier which are trained and tested. These experimental results show that the hybrid model on GA in a few cases with ML approaches increases the accuracy and in others reduces the number of successful levels while providing higher classification accuracy.

Keywords Genetic algorithm (GA) · X gradient boosting (XG-boost) · Random forest regression (RFR) · Decision tree (DT)

1 Introduction

Hybridization method is used to improvise the accuracy level when the combination of ML algorithm is in combination with genetic optimization algorithm. In this paper using two basic macromolecules which is made up of only two kind of molecules i.e; protein and DNA (cell). There are a mere billions of yeast cells on the Earth, while

R. Durgam (✉) · N. Devarakonda · R. Eluri
School of Computer Science & Engineering, VIT-AP University, Amaravati, A.P, India
e-mail: revathi.20phd7015@vitap.ac.in

A. Nayyar
Faculty of Information Technology, Duy Tan University, Da Nang, Vietnam

G. Ranganathan et al. (eds.), *Soft Computing for Security Applications*,
Advances in Intelligent Systems and Computing 1397,
https://doi.org/10.1007/978-981-16-5301-8_60

a human population consists of trillions of them. Every human being has trillions of membrane-long micro- and macromolecular components as well as microglia that together make up every human, which have tailored micro–functions, the nuclei, which serve as the cellular reserve, comprises much of the cell's genetic material in a compact form [1]. In general, a bioinformatic method may be useful for the study of complicated diseases, such as cancer since it is shown to be effective for the classification of such diseases.

Expanding our results using a microarray features (the most likely cancer-associated features) methods (including phasing, hunt-and-and-and-peck and FACS sequencing) on our big gene expression microarray data thin plates with which binds oligonucleotides suspended in a glass bead caging and collected probe sets of thousands of microscopic beads with oligonucleotide nucleotides on them, gene expression and protein synthesis are responsible for the two processes of protein production. The first thing DNA does is it goes through an intermediate stage of being transformed to mRNA, and the second thing is going through a process of being a protein. Genes originate from genes being activated, and proteins are the blueprints for the traits of all living species [2]. With several combinations of protein molecules and tiny beads that can be seen at the same time by an electron microscope, it is known as the functions as a primary tool in the classifying [the function of] gene expression and has changed due to new capabilities that have become available in the modern times This is how it all comes to an end. We process the materials like glass, plastic and silicon chips as the outcome of this. The technique of microarray in formatting and use of microarray technology are for the purposes of identifying and determining genetic individualization and diagnosing and identifying genetic causes.

In addition, if a cell undergoes mutation or undergoes a change in the genetic content, the cell is referred to as being "exposed to cancer". Many different forms of cancer are widely accepted to be very lethal, so there is a strong need for these investigations. When using machine learning strategies including feature selection and classification, we can accurately predict the outcome of the data's performance. The main goal of the present study is to increase the precision of the genetic profile by using a variety of genes and patients.

There is more study being done on colon tumour diagnosis and cancer classification and prediction which looks at the relationships between these data and individual characteristics and possible preexisting conditions and the individual characteristics.

To get the most likely and most significant cancer-associated genes or features through SNR and T-statistics, to the issue, such as this we can readily identify which features are best [3]. These features like decision tree, K-nearest neighbours, etc., when applied can identify which features are the best.

2 Literature Survey

Basic optimization process is a random progression-based algorithm on a population, taking different variables into consideration and adequate sociopsychological principles are taken to find solution for engineering problems.

The advent of DNA microarray technologies enables scientists to track and examine vast numbers of genes at various points in time [3]. The expression levels of two related genes vary in concert with different environmental stimuli. Soft clustering is essential to detect patterns belonging to multiple clusters simultaneously; therefore, this paper suggests the use of a new clustering technique based on weight scale, rather than the standard one which employs the Euclidean distance. Here the issue of microarray analysis is minimized while at the same time is maximized in the whole group. Finally, the latest fuzzy-based algorithm (GWO) is weighed against some current clustering strategies.

Four artificial data sets and three real GEO gene expression data sets have been studied to determine the validity of the current fuzzy WD clustering method in both models [3]. Experimental research and results assessment indicate that fuzzy WDO outperforms the other clustering methods such as PSO-FCM, FA-FCM, FCM and GWO-FCM.

Gene expression has been significantly increased with the use of DNA microarray technologies. This method of identifying the most important genes is often known as "microarray research". A major use of the microarray is the classification of cancers. Due to these factors, gene expression profiles are difficult to identify. The main method for overcoming these problems is feature (gene) selection. This article provides a study and evaluation of the newest approaches that employ bio-inspired approaches as wrappers. Biological information systems, using a range of sophisticated and non-traditional machine learning approaches, are used in diagnosing, managing, even preventing and probing for disease mechanisms [4]. Darwin theory has inspired evolution of genetic algorithm, and these algorithms are adaptive heuristic algorithms.

In addition, it is a field that increases computational capacity and analytical capabilities, particularly in processes that use expression computation on microarray data ("Introducing Biomedical Machine Learning Into RNA-Based Metadata Extraction, Interpretation, and Discovery") section which focuses on these methodologies, which are current and available machine learning methods used in the fields of bioinformatics for the investigation of and production of gene expression and their adoption in these methodologies in that region.

A gene microarray assay is done on the level of several different biological samples with the understanding of being able to provide millions of pieces of knowledge about the levels of expression [5]. With the emergence of high expansion of the bioinformatics, this new technique has emerged as a kind of investigation of gene network inference. Several theoretical approaches have been applied to target gene-based gene expression relationships that influence transcription factor inference models, including (those including) evolutionary, computational and non-evolutionary-based approaches [6]. A key use of these approaches is to aid researchers in the development of rational and effective medications to fight illness, and the biological literature is compatible with these claims. A literature review examines the various gene regulation strategies as well as their capability to discover gene regulatory networks, which will provide readers with a better understanding of the information contained in expression results.

3 Existing Model

3.1 Correlation with PSO

SmypParticle swarm optimization process is a random progression-based algorithm on a population, taking different variables into consideration, ample sociopsychological principles are taken to solve engineering problems like strength data point values and the correlation coefficient from the given feature values to calculate the correlation coefficient, which serves as the correlation criterion (CFS). It does away with the shortcoming of univariate filter methods in that it integrates both features [i.e. it resolves the problem of interaction among features]. It is important to know the recognition capacity of each attribute, since it pertains to a certain subset of attributes. This multivariate approach works well in discovering the association between the various variables [the various traits in the data set], the coefficient of Pearson correlation has a greater impact on the existence of outliers and noise, and their interaction with one another (genes) is linked to a calculation (i.e. means they can be calculated through correlation). Based on the correlation coefficient or Pearson product moment coefficient (PCC), it better explains the relationship between two variables, which correlates or approximates the tendency of two variables (PPMC) [7]. PSO allows particles to thrive profits based on finding the latest inventions and previous knowledge. For applications based on PSO, each particle that flies through the search space represents a candidate solution.

3.2 Extreme Learning Classifier

In contrast to a linear or direct neural network, a feedforward one uses help vector system and has a machine which is compounded when neural networks are implemented in feedforward configurations. The two main explanations are the iterative adjustment of network training of all the parameters using machine learning techniques such as these and the use of learning algorithms that use nonlinearly times to train neural networks rather. Binary classifiers are mostly used for binary classification issues, but, because of their excellent classifier capacity, it is used to expand the boundaries of this functionality. Quadratic kernels have been employed for text generation tasks because of their superior ability to represent nonlinear concepts (LS-SVM) [4, 8]. A massive number of layers are highly interconnected in a single space to enable the propagation of massive parallel information across the network (SHFFN). This network does not need any extra input: the additional, secret layer does not need to be expanded. Single-type and multi-type radial basis, multiple secret, non-layer networks and other concepts, including these, may be used in deep learning systems, like polynomial functions, support vector machines (SVMs) and traditional one-based systems. The special aspect that ELM differs from the other approaches is that ELM just requires the networks' configuration and neurons to be predefined, which are randomly generated [4, 9]. Furthermore, the fact that the goal functions and training and testing data set are not based on the secret parameters makes the methods in this class of learning model special. The theoretical value for the ELM is calculated, instead manually, which is said to save time and money.

In practice, to improve the productivity of the real-world applications, various methods can be taken, such as iterative or non-iterative; some of these approaches are gradual, and others are no incremental. The reconditioning parameters have the property of universality, but do not possess uniform approximation and divergence. These values may be learned as an alternative since the secret nodes are not only different from the training collection, but also alternative to the other values on their relations. In the term secret nodes, the non-selectable nodes and their mapping/relations are called ELM random nodes or ELM neurons or EL relationships. Using the method described above, the training data set can be reduced to see how accurate the classifiers are by determining the threshold weights and avoiding the need to store them, making it feasible to use a smaller training collection of independent variables. ELM deviates from traditional classifiers in which it does not need awareness of the training data ahead of time. However, it is to be noted that in contrast to EOMs, ELMs may be problematic. To put it another way, to optimize classifier ELM consistency, researchers are currently working on a new ensemble classifier.

4 Proposed Design Using Hybrid Approach

Our design performance improvement factor is on the optimization tuning scenario for each case of cancer data set on microarray gene. On the contrary, we implicate the design factor on the different classes observed from each data set as mentioned below in the section design procedure. This design provides the different features of the classes where data modelling and encoding of each type of classification are modelled in data acquisition block as per the CSV file chosen.

According to procedure of Algorithm in different models that helps to visualize in different scenario for improvising the features that is implemented as per the requirement.

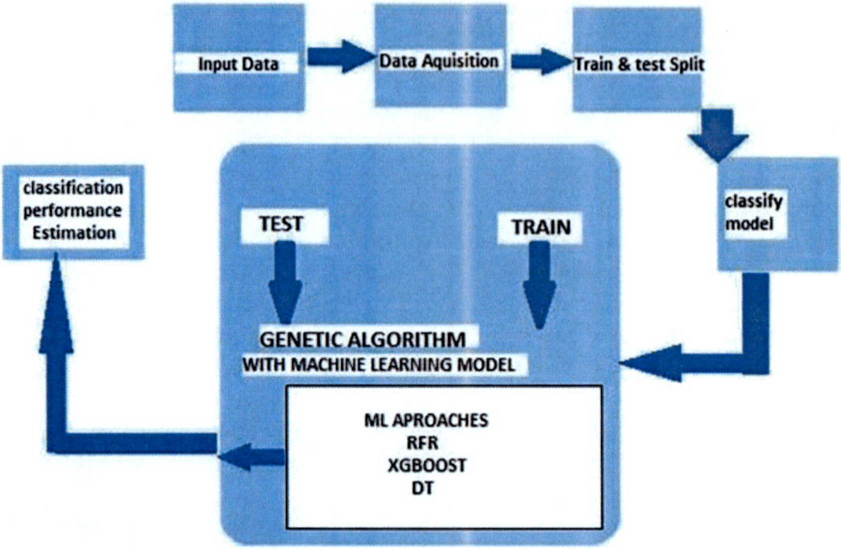

Fig. 1 Gene expression data modelling using hybrid optimization using GA and ML

4.1 Design Procedure

- Initiate data reading from the different gene data sets and analysis for data cleaning.
- Improvise a machine learning approach and to tune different optimization parameters using mathematical formulations.
- Finally apply ac hybrid optimization scheme which improvises the design performance parameters with different scale by tuning (Fig. 1).

4.2 Algorithms

4.2.1 Decision Tree

Directed tree routing includes a function that recursively uses the forecast which is used to manage bagging. From each vertex on the DT explored are classifications and other nodes, under which decision tree analysis is given with hierarchical class probabilities which are considered. The values for each of the forecast classes are the source of error for the total likelihood of the break.

$$E(S) = \sum_{i=1}^{N} p_i \log_2 p_i \tag{1}$$

In our proposed with DT case, we have applied Eq. (1) with the current features model with all the probabilities estimated as mentioned below:

This design improvising all the features with random probabilities and the functional formulation for Xg-boost, DT and RF are calculated using the formulas (1–3).

In the analysis on each algorithm with corresponding formulations, we have represented one of solutions of DT changes which requires probability varying from 0 to 1 with Fig. 2a and log of data values on each feature gene array data in Fig. 2c. Similarly, Eqs. 1–3 are estimated with X_final in Fig. 2b resulting in values utilized with classifier based on the sklearn libraries classification model.

a

```
In [45]: X

Out[45]: array([[-0.299,  0.093, -0.215, ...,  0.206, -0.013,  0.  ],
               [-0.081,  0.009, -0.091, ..., -0.037, -0.211,  1.  ],
               [-0.125,  0.07 , -0.006, ...,  0.034,  0.075,  0.  ],
               ...,
               [-0.167, -0.047, -0.146, ..., -0.075, -0.205,  0.  ],
               [-0.195, -0.097, -0.153, ...,  0.011, -0.03 ,  1.  ],
               [ 0.182, -0.081, -0.133, ..., -0.011,  0.274,  0.  ]])
```

b

```
In [19]: import random

         x = random.random()
         # Random float number
         for i in range(16384):
             print(random.random())

         0.40880310975411216
         0.5646477322444222
         0.061179962453922876
```

c

```
In [44]: X_final[:, 0:16384]

Out[44]: array([[ 0.05970739, -0.01857119,  0.04293341, ..., -0.04113619,
                  0.00259597,  0.        ],
                [ 0.01617491, -0.00179721,  0.01817181, ...,  0.00738854,
                  0.04213465, -0.19969026],
                [ 0.02496128, -0.01397832,  0.00119814, ..., -0.00678947,
                 -0.01497677,  0.        ],
                ...,
                [ 0.03334827,  0.00938544,  0.02915478, ...,  0.01497677,
                  0.0409365 ,  0.        ],
                [ 0.0389396 ,  0.01936996,  0.03055261, ..., -0.00219659,
                  0.00599071, -0.19969026],
                [-0.03634363,  0.01617491,  0.0265588 , ...,  0.00219659,
                 -0.05471513,  0.        ]])
```

d

```
In [14]: dataset.columns

Out[14]: Index(['Contig45645_RC', 'Contig44916_RC', 'D25272', 'J00129',
                'Contig29982_RC', 'Contig26811', 'D25274', 'Contig36292', 'Contig42854',
                'Contig34839',
                ...
                'AK002117', 'AK002118', 'Contig24843_RC', 'Contig44257_RC',
                'Contig36238_RC', 'Contig48760_RC', 'Contig35884_RC', 'Contig54569_RC',
                'Contig62588_RC', 'class'],
               dtype='object', length=16384)
```

Fig. 2 **a** Representing the *X* (original data set) and *x* for random probabilities for each column's length. **b** Representing the final weight probabilities multiplied for each columns in the data set. **c** Representing breast cancer data set columns of length 16,384

4.2.2 XG-Boost

We must provide a design algorithm that includes both supervised and unsupervised methods, and each one of the experimental measurements has an ensemble algorithm where the structure weights are progressively applied to the model, followed by application to the multi-class scenario. Parametric variables are incorporated into the gradient boosting process for that purpose.

Features

- Allows you to choose the level of tree expansion.
- Implemented as memory allocators.
- Computes parallel processing.
- Additional specification requirements are to be cross-validated with the user specifications.
- Pre-supplies almost all the analysis of the investigation on predilection.

Formulation of XG-BOOST

$$\text{SCORE} = \frac{\sum_{i=1}^{N} R^2}{N + \delta} \tag{2}$$

Here R is the residue of each class utilized to form the design scenario.

4.2.3 Random Forest

Random forest (RF) is a classification and regression technique that is built on bootstrap bagging and is dependent on the ensemble approach [5]. Boosting and bagging are two well-known and widely used techniques for tree classification. The random forest is a selection of trees that could also forecast the class mark for a series of data points based on the categorical-dependent variable [7]. A random subset of the original features is used to train each decision tree. A new sample is used to classify the mode of each tree's outputs to determine the sample form. This method has been used in a broad range of speech and language processing applications.

$$\text{RFR}_i = \sum \frac{F_{\text{norm}}(i, j)}{\sum j \in F(i, j), k \in T * F_{\text{norm}}(i)} \tag{3}$$

4.2.4 Genetic Algorithm

One of the current scenarios on evolutionary algorithms which are applied based on the adaptive heuristic approach is known to be genetic algorithms. The solution analysis and its features, best features, identification as a part of exploiting random searches ensure the different data perspectives. Our problem on the improved aspect of the different disease provides a larger source of medical data ensuring that the different procedures and features are represented based on micro-gene array features of five different data sets as mentioned below.

Here the design on GA would improvise a natural selection process ensuring the data to adapt different changes in the features or other database modelling to survive and reproduce the next best features according to the concept of "survival for best".

i. Algorithm

1. We initiate the different stages of the GA with feature acquisition from training, testing of each feature modelled according to the number of iterations.
2. The offsprings are acquired from the best solution of each data feature utilized to initiate the GA.
3. Forecasting of the current data to the next and upcoming offspring are to be fitted with best solution.
4. Successive regression of each feature is represented and estimated for each set of population size.

4.2.5 Design Methodology

Figure 1 describes on which algorithms we could provide a novel formulation changes, providing a drastic effect on the accuracy and classifying the objectives. The GA and ML approaches have two architectural differences which makes them crucial on justifying which formulation has led to better classifying the data. One of the important features is heuristic searching and pattern identification, with the one as GA provides a heuristic model on optimization on evolving data either manually or randomly. In this case to operate such evolving data, GA represents four operational characteristics features which ensures selection crossover, mutation and finally fitness of each corresponding data initialization. With machine learning, each pattern identified is provisioned with different training scenarios or cases which effects the designed network model consisting of multiple algorithms. Here our objective is to identify the different set of class objects and with its features depending upon the data set chosen. These data sets are considered from the website link http://csse.szu.edu.cn/staff/zhuzx/Datasets.html. Considering the formulations from above Eqs. 1–3 with machine learning model improvising the different set of values observed for each gene array observed, data for the type of cancer is represented. With same data, we initiate the improved GA model to calculate the fitness function for the trained data improvising with GA to provide better accuracy and classify the different classes based on performance factors (Fig. 3).

ii. Flow graph
See Fig. 4.

```
import pandas as pd
import numpy as np
import matplotlib.pyplot as plt
from sklearn.decomposition import PCA
from sklearn.cluster import AgglomerativeClustering
from sklearn.preprocessing import StandardScaler, normalize
from sklearn.metrics import silhouette_score, confusion_matrix, classification_report
from sklearn.model_selection import cross_val_score
from sklearn.tree import DecisionTreeClassifier
dataset = pd.read_csv('Lymphoma.csv')
print(len(dataset.columns))
from sklearn import preprocessing
import seaborn as sns
# label_encoder object knows how to understand word labels.
label_encoder = preprocessing.LabelEncoder()

# Encode labels in column 'species'.
dataset['class']= label_encoder.fit_transform(dataset['class'])

X = dataset.iloc[:, 0:4027].values
y = dataset.iloc[:, 4026].values
import random
from sklearn.model_selection import train_test_split

X_train, X_test, y_train, y_test = train_test_split(X, y, test_size=0.2, random_state=0)
from sklearn.preprocessing import StandardScaler

sc = StandardScaler()
X_train = sc.fit_transform(X_train)
X_test = sc.transform(X_test)
classifier = DecisionTreeClassifier(random_state=0)
classifier.fit(X_train, y_train)
y_pred = classifier.predict(X_test)
```

Fig. 3 Pseudo code for hybrid GA-ML algorithm

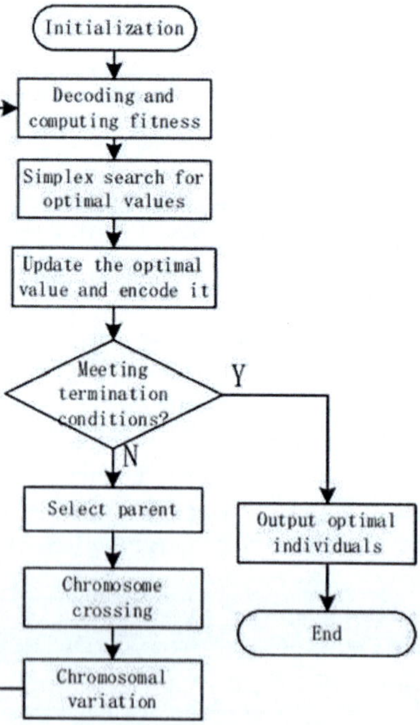

Fig. 4 Representation of flow diagram for proposed work

iii. Computational Complexity of GA with ML.

The design with features values of m is no. of individual values for GA applied, g generations, n population sizes, and we have the complexity as $O(g(mn + f_n))$. Since F_n is fitness function for the proposed design which will improvise the classifier outcome values, overall we initiate the classifier complexity for DT, GA and ML with $O_c(m)$, where m is type of classifier.

If $m = 1$, we have decision tree.

The overall computation for proposed DT is based on Eq. (1) we have

$$F_n = O_c(m = 1) = O_c(n\log(n))$$

As per the design for the overall hybrid GA with DT, we have $O_g(mn + n\log n)$.

Similarly, for the overall hybrid GA with RF, we have $O_g(mn + n^2)$, and for the overall hybrid GA with DT, we have $O_g(mn + nkd)$.

5 Results and Discussions

For each case, we have represented the different classes observed and predicted based on the three different classifiers using machine learning algorithm. This graph provides the values on each section of class variables estimated based on the type mentioned. The proposed result is compared with existing DT and existing RFR which resulted in improvising the scope of efficiency. The results of XGB classifier with GA approach are more efficient when compared to other classifiers (Fig. 5).

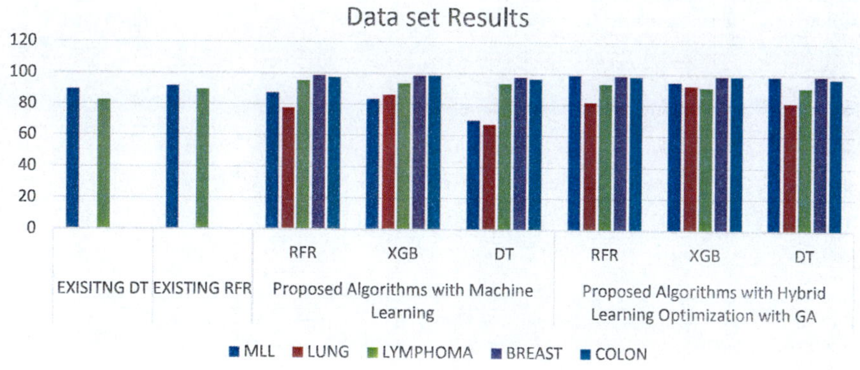

Fig. 5 Representation of data classes with three classifiers with machine learning algorithm and genetic algorithm

6 Accuracy

The classifier XGB with GA in Table 1 provides better accuracy when compared to other algorithms (RFR and DT) for five different data sets as mentioned below. Since existing models did not utilize the lung, breast and colon data sets, we have introduced a scenario where all the five data sets are modelled using ML and hybrid GA as a hybrid approach for better optimization of accuracy. The average accuracy of the XGB classifier is about 95.29% for all the databases included, while for RFR and DT are 93.49 and 86.62% (Table 2).

• Data classes for each databases

For each case, we have represented the different classes observed and predicted based on the different classifier using genetic algorithm. This Table 3 provides the values on each section of class variables estimated based on the type mentioned. In case of MLL, we have represented 5,5,6 as the values for each class of 0,1,2 encoded for each label type on all classifiers.

Table 1 Representing the accuracy tabulations with ML and hybrid logic (GA)

Data sets	Existing DT	Existing RFR	Proposed algorithms with machine learning accuracy			Proposed algorithms with hybrid learning optimization with GA accuracy		
			RFR	XGB	DT	RFR	XGB	DT
MLL	65.3	80.6	81.25	81.25	68.75	81.25	97.56	81.25
Lung	0	0	97.27	93.18	90.24	97.27	90.09	97.5
Lymphoma	78.9	89.5	86.67	85.71	85.71	93.33	92.85	92.85
Breast	0	0	98.6	98.7	98.6	98.3	97.53	98.3
Colon	0	0	98.12	98.5	97.23	97.33	98.46	61.538

Table 2 Representing the test and train sample for microarray gene data

database	RFR		DT		XGB	
	Train	Test	Train	Test	Train	Test
MLL	9902	2680	10,066	2516	9902	2680
Lung	9916	2614	10,080	2520	9916	2614
Colon	1574	426	1600	400	1574	426
Breast	12,864	3490	13,107	3277	12,864	3490
Lymphoma	3311	896	3222	985	3311	896

Table 3 Representation of the data class and its accuracy of prediction of data class using databases of five types with three different classifiers

Database microarray gene	Algorithms	Data class support actual	Data class support predicted after GA
MLL	RFC	16(5,5,6)	13(5,5,6)
MLL	DT	16(5,5,6)	13(5,5,6)
MLL	XGB	16(5,5,6)	16(5,5,6)
Breast	RFR	21(8,13)	21(8,13)
Breast	DT	20(6,14)	20(6,14)
Breast	XGB	20(8,12)	20(8,12)
Lymphoma	RFR	15(1,11,3)	14(1,11,3)
Lymphoma	DT	14(2,8,4)	13(2,8,4)
Lymphoma	XGB	14(2,8,4)	13(2,8,4)
Colon	RFR	13(1,12)	13(1,12)
Colon	DT	14(6,8)	14(6,8)
Colon	XGB	13(6,7)	13(6,7)
Lung	RFR	44(33,3,1,6,1)	43(33,3,1,6,1)
Lung	DT	41(29,4,2,3,3)	40(29,4,2,3,3)
Lung	XGB	44(33,3,1,6,1)	40(33,3,1,6,1)

7 Conclusion

In this paper, we have modelled the different data analytic structures with machine learning algorithms improvising the different accuracies and its importance of classification of tumour and non-tumour. In this paper, the cases all the three classifiers are estimated with their accuracy of each database are considered. Furthermore, the XGB classification will be modelled based on user-defined function and will be utilized to provide the features of each section of the designed databases. As per the existing model, we have seen that there is systematic change in most of the databases using a combination approach as ML + GA which results in XGB + GA providing the highest accuracy on average. We can improvise different scenarios on PSO algorithm, GA algorithm and firefly algorithm based on machine learning techniques to ensure the correct classification and prediction model for each database.

References

1. A. Achom, R. Das, P. Pakray, S. Saha, Classification of microarray gene expression data using weighted grey wolf optimizer based fuzzy clustering, in *TENCON 2019 - 2019 IEEE Region 10 Conference (TENCON)*. Electronic ISSN 2159-3450
2. R. Das, S. Saha, Gene expression classification using a fuzzy point symmetry based PSO clustering technique, in *2015 Second International Conference on Soft Computing and Machine*

Intelligence (ISCMI) (2015), pp. 69–73
3. S. Saha, S. Acharya, K.K. Miriyala, Simultaneous clustering and feature weighting using multiobjective optimization for identifying functionally similar mirnas. IEEE J. Biomed. Health Inform. **22**(5), 1684–1690 (2018)
4. A. Paul, J. Sil, Identification of differentially expressed genes to establish new biomarker for cancer prediction. IEEE/ACM Trans. Comput. Biol. Bioinform **6**(6), 1557–9964 (2019)
5. B. Peter, G. Prager, B. Streubel, Molecular pathology of cancer: How to communicate with disease. ESMO Open **1**(5) (2016)
6. R.K. Saidala, N. Devarakonda, Northern Bald Ibis optimization algorithm: theory and application, in *2018 IEEE 17th International Conference on Cognitive Informatics & Cognitive Computing (ICCI*CC)* (2018), pp. 541–551. https://doi.org/10.1109/ICCI-CC.2018.8482088
7. S. Dudoit, J. Fridlyand, T. Speed, Comparison of discrimination methods for the classification of tumors using gene expression data. J. Amer. Stat. Assoc. **97**(457), 77–87 (2002)
8. M. Sun, T. Ding, X.-Q. Tang, K. Yu, An efficient mixed-model for screening differentially expressed genes of breast cancer based on LR-RF. IEEE/ACM Trans. Comput. Biol. Bioinform. **16**(1) (2019). Electronic ISSN: 1557-9964.
9. C. Grace, E.P. Nacheva, Significance analysis of microarrays (SAM) offers clues to differences between the genomes of adult philadelphia positive all and the lymphoid blast transformation of CML. Cancer Inf. **11**, 173–183 (2012)
10. H. Abusamra, A comparative study of feature selection and classification methods for gene expression data (2013)
11. M.H. Bailey, C. Tokheim, E. Porta-Pardo, S. Sengupta, D. Bertrand, A. Weerasinghe, A. Colaprico, M.C. Wendl, J. Kim, B. Reardon et al., Comprehensive characterization of cancer driver genes and mutations. Cell **173**(2), 371–385 (2018)
12. L.A. Torre et al., Global cancer statistics 2012. CA Cancer J. Clin. **65**(2), 87–108 (2015)
13. S. Perinel, M. Launay, É. Botelho-Nevers, Towards optimization of hydroxychloroquine dosing in intensive care unit COVID-19 patients. Clin. Infect. Dis. **71**(16), 2227–2229 (2020)

Improved Image Enhancement of Natural Images with Median Mean-Based Sub-Image Clipped Histogram Equalization

Rangu Srikanth, Koppisetti Lakshmi Sowmya, Sika Anjana, Gajarajula Vamshi, and Annem Ram Mohan Reddy

Abstract This paper introduces a powerful image contrast enhancement algorithm that is developed based on the energy curve equalization technique. Instead of the histogram, an energy curve is used for considering spatial contextual information in the image. This is Improved Image Enhancement of natural images with median mean-based sub-image clipped histogram equalization. The algorithm consists of the following steps, first, obtaining the energy curve, secondly, calculating median and mean intensity values of the image, third, the energy curve is clipped using a threshold level which is its mean occupancy, and fourth, the clipped energy curve is divided into two halves based on median and then further partitioned into four subparts based on mean intensity values, all these four portions are equalized and then combined to form an enhanced image. This promotes natural enhancement and levies control over the rate of enhancement. The simulation conveys that the proposed method shows supremacy over the other previously existing methods. This shows an increase in entropy which is the information content held in the resultant image.

Keywords Image enhancement · Energy curve · Entropy · Equalization

1 Introduction

Enhancing the contrast of an image is considered to be the most important pre-processing step for many applications which improves visual perception of the image. In the past, numerous image refinement methodologies were proposed to conserve the quality of the image and are most significant in the present-day multimedia requirements. Due to the lack of proficiency of the device operator, malfunction of the acquisition device, or due to surrounding conditions while acquisition may lead to acquiring an image having lower contrast. Images with lower contrast can lead to the degradation of visual perception. This setback can be ruled out by contrast enhancement by expanding the range of information content of the image. Obtaining

R. Srikanth (✉) · K. L. Sowmya · S. Anjana · G. Vamshi · A. R. M. Reddy
Department of ECE, Kakatiya Institute of Technology and Science, Warangal, India
e-mail: rs.ece@kitsw.ac.in

G. Ranganathan et al. (eds.), *Soft Computing for Security Applications*,
Advances in Intelligent Systems and Computing 1397,
https://doi.org/10.1007/978-981-16-5301-8_61

an image consisting of more information and an appealing view is the primary object of the contrast enhancement technique. In the literature, there are many methods that adjust the image's dynamic range to get an enhanced image.

The most simple technique used for contrast enhancement is histogram equalization [1]. This technique is more frequently used because it is easy to implement. Histogram equalization is one of the spacial—domain techniques. It is also a grayscale image enhancement technique that extends the dynamic range of the gray levels in an image and improvizes image contrast [2]. Using the cumulative distribution function the global histogram equalization technique maps the gray levels of the image to another which flattens the probability distribution function. This method cannot be universally used for all applications because sometimes it may give over-enhanced and can even produce unwanted artifacts. The brightness levels of the image may drastically change. To overcome this setback numerous methodologies were suggested in the literature. First of all, is the BBHE (Brightness preserving bi-histogram equalization) which was proposed by Kim [2] in the year 1997. This method divides the original histogram into two parts with the mean intensity of the image as a boundary limit. Those two histograms are equalized separated to form a better image. The next method which was proposed was DSIHE (Dualistic sub-image histogram equalization) by Wan et al. in 1999 [3]. This method was proposed for brightness preservation and entropy conservation. It is similar to BBHE except that the histogram is divided based on the median level intensity of the image. The two histograms have the nearly same number of pixels. In extension to BBHE, a method was presented by Chen and Ramli which is named MMBEBHE (minimum—mean—brightness error bi-histogram equalization) [4]. This method calculates the AMBE (Absolute mean brightness error) repetitively in the range of 0 to $L - 1$ Gy level. When a minimum brightness error is obtained the corresponding mean intensity level Xm is taken as the threshold to bisect the image histogram. Apart from these methods, some iterative procedures were also proposed. RMSHE [5] (Recursive—mean-separate histogram equalization) technique iteratively divides the image, that is depending upon the mean value histogram is dived into two parts and further mean values are obtained for these two and are again divided this process continues for several times based on the order. The order defines the number of iterations. We can also consider this method to be the recursive approach of the BBHE technique which was developed by Chen and Ramli. Similarly, if the same process is carried out based on median intensity then we call it to be RSIHE (Recursive—sub-image histogram equalization) proposed by Sim et al. [6]. These iterative methods give excellent results but the problem is that the optimized solution is not recognized properly and due to its recursive nature the algorithm becomes more and more complex for implementation. A similar technique is presented by Kim and Chung [7] named RSWHE (Recursively—Separated and Weighted Histogram Equalization). It is also an iterative weighting process where a normalized power-law function is used to modify the histogram. However, all the techniques don't have a provision to adjust the rate of enhancement. To overcome this limitation some techniques were proposed which include clipping of histogram [8–10]. This histogram clipping helps to control the enhancement rate by keeping a check of the maximum level of the histogram. To clip the histogram we need a threshold. Many techniques are proposed to determine this threshold. DHE (Dynamic histogram equalization) [11] is presented by Abdullah-Al-Wadud et al. In this technique, the histogram is partitioned depending upon the local minima and then allocate a different dynamic range to the histogram. But DHE shows a huge change in the mean brightness level. Further Brightness preserving version of DHE is proposed by Ibrahim and Kong [12] BPDHE (Brightness preserving-dynamic histogram equalization) in which before the histogram portioning it uses the Gaussian filter to smoothen the image. Then the histogram partitioning is carried out based on local maxima instead of local minima. A small variant of this method was proposed called BPDFHE (Brightness Preserving—Dynamic Fuzzy Histogram Equalization) [13]. In BPDFHE smoothen is done using Fuzzy histogram computation. This method claims to be eminent in brightness conservation and lower computational time. In extension to these Ooi and Isa [14] proposed QDHE (Quadrants Dynamic Histogram Equalization) in which histogram is divided into four parts based on median intensity levels and the histogram is clipped depending upon the mean occurrence. Separate dynamic ranges are allocated to the four sub histograms, which are then equalized to form an enhanced image.

Kuldeep Singha and Rajiv Kapoor proposed a hybrid method named MMSICHE (Median mean-based sub-image clipped histogram equalization) [15–18]. Initially, the histogram is cleaved into two segments with median intensity as bound and these two segments are further separated to form four sub histograms based on their corresponding mean intensity values. The histogram is then clipped considering median occurrence as the threshold. The histograms are individually equalized and united to form an MMSICH equalized image. This technique comprises the features such as brightness conservation, enhancement control, and realistic enhancement.

The present method being proposed is an advancement of MMSICHE in which we obtain the energy curve of the image and split it into four parts firstly based on median and secondly based on the means of the respective parts. Calculations of the thresholds and equalization are all carried out on the energy cure of the image. The main difference of the proposed method is using an Energy curve instead of the histogram to improve the efficiency of image enhancement. Using the energy curve instead of the histogram increased the enhancement rate and provided good visual context to the images.

2 Proposed Method

The algorithm along with the equations required to carry out this image enhancement technique is presented in this section. The basic idea is to divide the energy curve of the image, equalize them separately and join them to get an enhanced image. Improved Image Enhancement of natural images with median mean-based sub-image clipped histogram equalization process is done as follows.

- Obtaining energy curve.

- Calculation of median and mean values.
- Clipping of energy curve
- Partitioning, equalization, and merging of image

A detailed description of each process is explained below.

2.1 Obtaining Energy Curve

Instead of using the histogram of the image, to find the threshold values we use the energy curve to get better results out of it. This helps us to get optimized thresholds and increases efficiency [16].

Assume the image as a function of its special coordinates i and j, here $i = 1, 2, \ldots, N$ and $j = 1, 2, \ldots, M$ (i.e., image is $I = x(i, j)$). The total number of pixels in the image is $T = N \times M$. Consider a neighborhood N which is of order d and the image consisting of the spatial coordinates as (i, j), its spatial correlation with the neighboring pixels of the image can be defined as $N_{ij}^{d} = \{(i+u, j+v), (u, v) \in N^{d}\}$.

To generate the energy curve of the image, we measure second-order neighborhood systems, i.e., $(u, v) \in \{(\pm 1, 0), (0, \pm 1), (1, \pm 1), (-1, \pm 1)\}$. Initially, we have to generate the binary matrix of the image which is the energy of every pixel intensity of the grayscale image.

$B_x = \{b_{ij}, 1 \le i \le M, i \le j \le N\}$ here, if $x_{ij} > xb_{ij} = 1$, else $b_{ij} = -1$. Consider one more matrix $C = \{c_{ij}, 1 \le i \le M, 1 \le j \le N\}$ here, $\forall (i, j), c_{ij} = 1$. E_x is the energy of each pixel x in the image I is defined below.

$$E_x = -\sum_{i=1}^{M}\sum_{j=1}^{N}\sum_{pq \in N_{ij}^{2}} b_{ij}.b_{pq} + \sum_{i=1}^{M}\sum_{j=1}^{N}\sum_{pq \in N_{ij}^{2}} c_{ij}.c_{pq} \tag{1}$$

The last term of the above equation should be constant and $E_x \ge 0$, i.e., the energy of every pixel of the image should be a positive value. From Eq. (1) it is evident that when every element of B_x is either -1 or 1 then, that particular gray level has zero energy. In other words, we can say that in the image $I(i, j)$ all the pixels with gray levels are less than x or greater than x, if not the pixel value x will have a positive energy value.

Properties of the energy curve.

There are some appealing properties for the energy curve which is generated from the equation defined previously. All the objects in the image will have certain gray level ranges for example let us consider a gray level range $[t_1, t_2]$ about an image, the elements of the matrix B_x corresponding to that particular object in the image will be 1 at $x = t_1$. Some of the elements of B_x will change to -1 as x increases. The elements of the matrix B_x about the pixels of the object will turn to become -1 at$x = t_2$. So, for the gray level range $[t_1, t_2]$ of the object, the energy curve looks like a bell shape. Identification of the objects in the image is possible with the help of peaks and valleys seen in the energy plot of the image.

2.2 *Calculation of Median and Mean Values*

The number of samples in the image is denoted by N.

$$N = \sum_{k=0}^{L-1} E_k \tag{2}$$

E_k is the energy curve which is obtained previously. $z(k)$ is an intermediate variable which we consider to calculate the median level of the image

$$z(k) = z(k-1) + E_k, \quad k = 0, 1, \ldots, L-1 \tag{3}$$

The initial value is considered as, $z(0) = E_0$. The values of $z(k)$ are calculated iteratively until the condition $z(k) \geq \frac{N}{2}$ satisfies, now the median value is given as $X_e = k$ (Fig. 1).

After dividing the energy curve into two parts based on the median value previously calculated we get two curves, first one is from gray levels 0 to $X_e - 1$ and the second one is from X_e to $L - 1$. For these two histograms, we find mean values X_{ml} and X_{mul} respectively. These values are obtained by Eq. 4 given below.

Fig. 1 Energy curve partitions

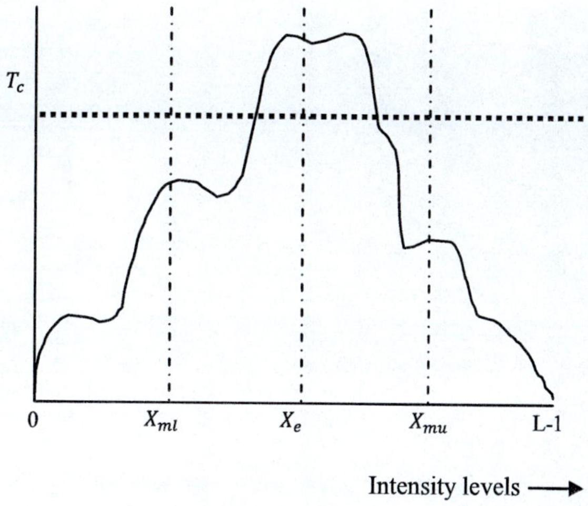

$$X_{\mathrm{ml}} = \sum_{k=0}^{X_{\mathrm{e}}-1} P_{\mathrm{l}}(k) \times k$$

$$X_{\mathrm{mu}} = \sum_{k=X_{\mathrm{e}}}^{L-1} P_{\mathrm{u}}(k) \times k \tag{4}$$

$P_{\mathrm{l}}(k)$ and $P_{\mathrm{u}}(k)$ are the probability distribution functions of the two sub energy curves from 0 to $X_{\mathrm{e}} - 1$ and X_e to $L - 1$ gray levels respectively and are calculated by Eq. 5 (Fig. 2).

$$P_{\mathrm{l}}(k) = \frac{E_k}{N_{\mathrm{l}}}, \quad \text{for} \quad k = 0, 1, \ldots, X_{\mathrm{e}} - 1$$

$$P_{\mathrm{u}}(k) = \frac{E_k}{N_E}, \quad \text{for} \quad k = X_{\mathrm{e}}, X_{\mathrm{e}} + 1, \ldots, L - 1 \tag{5}$$

N_{l} is the number of pixels present in the lower part of the energy curve and N_{u} is the number of pixels present in the upper part of the energy curve. By this we obtain four

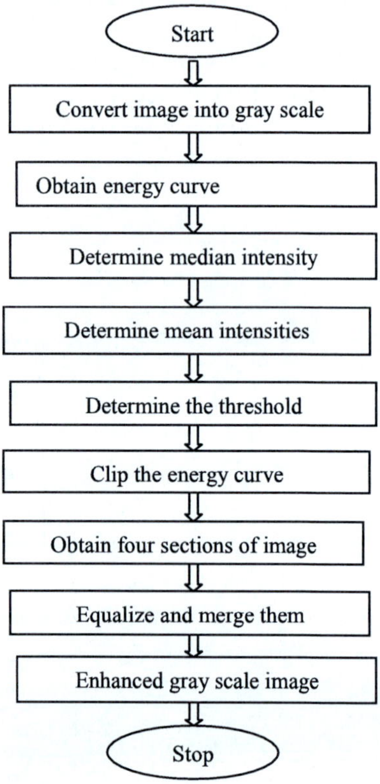

Fig. 2 Flowchart of the proposed method

different energy curves having a range from 0 to X_{ml}, $X_{\mathrm{ml}}+1$ to X_{e}, $X_{\mathrm{e}}+1$ to X_{mu} and $X_{\mathrm{mu}}+1$ to $L-1$.

2.3 *Clipping of Energy Curve*

Some part of the energy curve is clipped before dividing it. This process is done so that the enhancement rate is controlled. By this, we can get an image that appears to be realistic. The threshold level for clipping is nothing but the mean intensity occurrence. All the intensity levels which are higher than the threshold level are leveled to the threshold. Other intensity levels, that are lower than the threshold remain as they are. The threshold level is written as,

$$T_{\mathrm{c}} = \mathrm{mean}[\mathrm{E_k}] \tag{6}$$

Clipped energy curve

$$\begin{aligned} E_c(k) &= T_c \quad \text{for} \quad E_k \geq T_c \\ E_c(k) &= E_k \quad \text{for} \quad E_k < T_c \end{aligned}$$

Here E_k is actual energy curve and $E_{\mathrm{c}}(k)$ is the clipped energy curve, generally, we will consider histogram.

2.4 *Partitioning, Equalization, and Merging of Image*

The given image is divided into four sub-images W_{Ll}, W_{Lu}, W_{Ul}, W_{Uu} these are formed by steps discussed earlier. $P_{\mathrm{Ll}}(k)$, $P_{\mathrm{Lu}}(k)$, $P_{\mathrm{Ul}}(k)$, $P_{\mathrm{Uu}}(k)$ are the probability distribution functions of the image energy curve which can be calculated as follows

$$\begin{aligned} P_{\mathrm{Ll}} &= \frac{E_{\mathrm{c}}(k)}{N_{\mathrm{Ll}}}, \quad 0 \leq k \leq X_{\mathrm{ml}}\frac{1}{2} \\ P_{\mathrm{Lu}} &= \frac{E_{\mathrm{c}}(k)}{N_{\mathrm{Lu}}}, \quad X_{\mathrm{ml}}+1 \leq k \leq X_e \\ P_{\mathrm{Ul}} &= \frac{E_{\mathrm{c}}(k)}{N_{\mathrm{Ul}}}, \quad X_{\mathrm{e}}+1 \leq k \leq X_{\mathrm{mu}} \\ P_{\mathrm{Uu}} &= \frac{E_{\mathrm{c}}(k)}{N_{\mathrm{Uu}}}, \quad X_{\mathrm{mu}}+1 \leq k \leq L-1 \end{aligned} \tag{7}$$

N_{Ll}, N_{Lu}, N_{Ul}, N_{Uu} are the number of pixels in the sub-images. The cumulative distribution function is the cumulative sum of the probability distribution functions up to the desired number of pixels. The CDFs of the four sub-images are

$C_{\mathrm{Ll}}(k), C_{\mathrm{Lu}}(k), C_{\mathrm{Ul}}(k), C_{\mathrm{Uu}}(k)$ respectively.

$$C_{\mathrm{Ll}}(k) = \sum_{k=0}^{X_{\mathrm{ml}}} P_{\mathrm{Ll}}(k) \quad C_{\mathrm{Ll}}(\mathrm{k}) = \sum_{k=X_{\mathrm{ml}}+1}^{X_{\mathrm{e}}} P_{\mathrm{Lu}}(k) \tag{8}$$

$$C_{\mathrm{Ul}}(k) = \sum_{k=X_{\mathrm{e}}+1}^{X_{\mathrm{mu}}} P_{\mathrm{Ul}}(k) \quad C_{\mathrm{Uu}}(k) = \sum_{k=X_{\mathrm{mu}}+1}^{L-1} P_{\mathrm{Uu}}(k) \tag{9}$$

Further, all the four parts are to be separately equalized for this we need the transfer functions of these sub-image energy curves. The transfer functions $F_{\mathrm{Ll}}, F_{\mathrm{Lu}}, F_{\mathrm{Ul}}, F_{\mathrm{Uu}}$ of these sub-images are given by

$$\begin{aligned} F_{\mathrm{Ll}} &= X_{\mathrm{m_l}} \times C_{\mathrm{Ll}} \\ F_{\mathrm{Lu}} &= (X_{\mathrm{ml}} + 1) + (X_{\mathrm{e}} - X_{\mathrm{ml}} + 1) \times C_{\mathrm{Lu}} \end{aligned} \tag{10}$$

$$\begin{aligned} F_{\mathrm{Ul}} &= (X_{\mathrm{e}} + 1) + (X_{\mathrm{mu}} - X_e + 1) \times C_{\mathrm{Ul}} \\ F_{\mathrm{Uu}} &= (X_{\mathrm{mu}} + 1) + (L - X_{\mathrm{mu}} + 1) \times C_{\mathrm{Uu}} \end{aligned} \tag{11}$$

Finally, the Improved Image Enhancement of natural images with median mean-based sub-image clipped histogram equalized output is the amalgamation or union of the above transfer functions.

3 Simulation Results

This section presents the comparative analysis of the simulated results. The proposed method is compared with the existing methods such as HE (Histogram Equalization), BBHE (Brightness-preserving bi-histogram equalization), DSIHE (Dualistic sub-image histogram equalization), RMSHE (Recursive mean-separate histogram equalization), and MMSICHE (Median mean-based sub-image clipped histogram equalization). All the above-mentioned methods are applied on five different images, which are girl, top-view, butterfly, hand, and women.

The quality measure which speaks about the extent of enhancement is entropy. Entropy is the average information content of the image. The primary objective is to obtain an image which imbibed with more information. Even minute details of the image should be visible clearly. A greater entropy value indicates the presence of more information in the image. Entropy is defined as follows.

$$E = -\sum_{k=0}^{L-1} P(k) \log P(k) \tag{12}$$

Table 1 Entropy values different image enhancement techniques

Images	Original	HE	BBHE	DSIHE	RMS HE	MMSICHE	Proposed method
Butterfly	5.0074	4.7161	4.7966	4.8081	4.7236	4.8526	4.9710
Hand	5.0082	4.8509	4.9556	4.9597	4.8531	4.9372	4.9427
Women	5.3518	4.8213	5.1294	5.1841	5.2608	5.2167	5.3266
Top-view	5.6693	5.3761	5.5559	5.5621	5.5598	5.5796	5.6167
Girl	6.6215	5.8622	6.2239	6.2227	6.5493	6.5628	6.586

3.1 *Quantitative Assessment*

Table 1 presents the details of the various methods and their entropy values of different images. Each row corresponds to different images and each column represents various methods that are compared. The last column shows greater entropy values. From Table 1, it is evident that the proposed method shows better entropy values as compared to other methods. If the entropy values are near to the entropy values of the original image it implies that natural enhancement is achieved. The entropy obtained by the proposed method is much closer to the entropy of the original ones. This designates that the proposed method performs natural enhancement.

3.2 *Qualitative Assessment*

Apart from quantitative assessment qualitative assessment is also important to judge the image enhancement. The qualitative analysis deals with the visual perception and relies on the fact that how appealing the enhanced image looks.

Five different pictures and their results for various enhancement techniques are depicted here. By observing the images it is clear that the proposed method giving good results in terms of appearance. We can also notice that all the enhanced images of the proposed method look realistic.

Butterfly images in Fig. 3, indicate that HE, BBHE, and DISHE methods over-enhanced the image especially the background of the butterfly. RMSHE and MMSICHE nearly degraded the appearance of the butterfly. Image enhanced Proposed method contains all the details of butterfly along with the branch of the plant. In the hand image Fig. 4, BBHE and DSIHE produced over-enhanced the image. In RMSHE and MMSICHE hand is not clearly seen. The proposed method the enhanced image is a little dark but the hand looks good. The above image is the graphical representation of the results. The vertical axis denotes the entropy and the horizontal axis denotes various methods used in comparative analysis. Of all the techniques the proposed method shows much higher entropy values are nearer to that of the original image. This indicates the superiority of the method promoting natural enhancement.

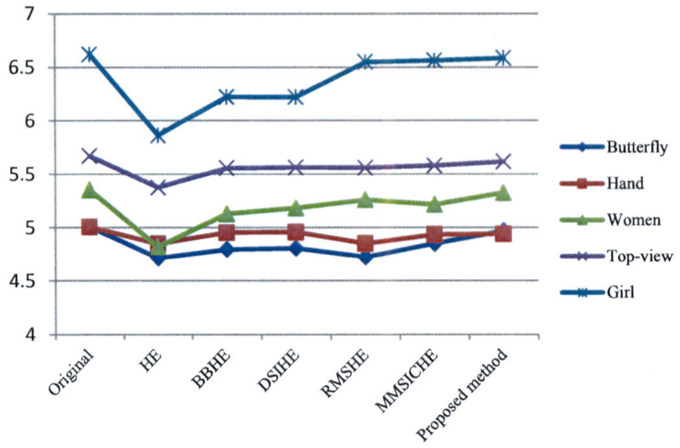

Fig. 3 Graph of entropy

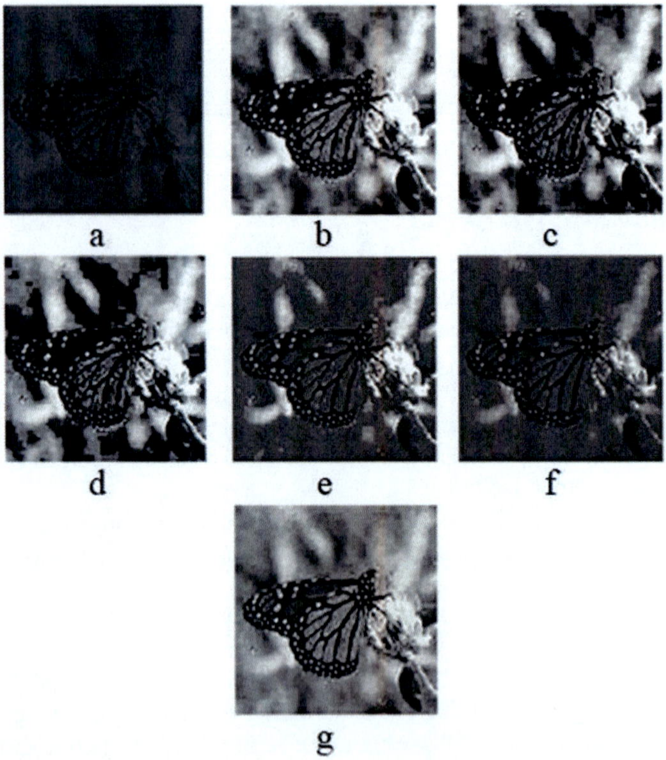

Fig. 4 Butterfly images. **a** Original image, **b** HE, **c** BBHE, **d** DSIHE, **e** RMSHE, **f** MMSICHE, **g** Proposed method

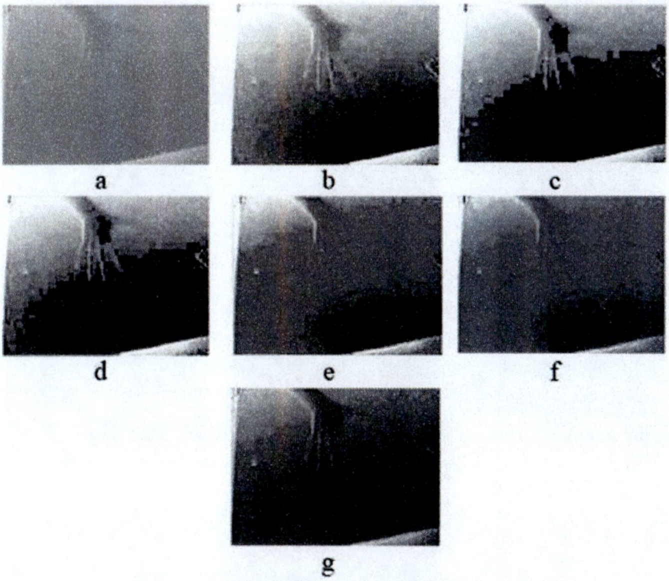

Fig. 5 Hand images. **a** Original image, **b** HE, **c** BBHE, **d** DSIHE, **e** RMSHE, **f** MMSICHE, **g** proposed method

The women image, enhanced by the proposed method in Fig. 5, seems to be natural and has good quality. HE, BBHE, and DSIHE images look blurred and others are unrealistic. The Top-view of the river is shown in Fig. 6, RMSHE and MMSICHE images we cannot see the bridge across the river. Proposed techniques showed better results, while other methods produced artifacts and highlighted unimportant details at the bottom left corner of the image. In Fig. 7, it is clear that the image produced by the proposed technique shows a good result as the features of the girl and the design of threads on the hat looks natural, whereas RMSHE and MMSICHE showed brighter image. As shown in Fig. 7, the proposed method can be applied to satellite images and can also apply to medical images (Fig. 8).

Overall the techniques Improved Image Enhancement of Natural Images with median mean-based sub-image clipped histogram equalization yield superior quality image with needed enhancement rate and noise immunity.

4 Conclusion

This paper presented a contrast enhancement technique which is Improved Image Enhancement of Natural Images with median mean-based sub-image clipped histogram equalization. The algorithm is devised on the energy curve of the image which is obtained based on the difference between the pixels in the neighborhood.

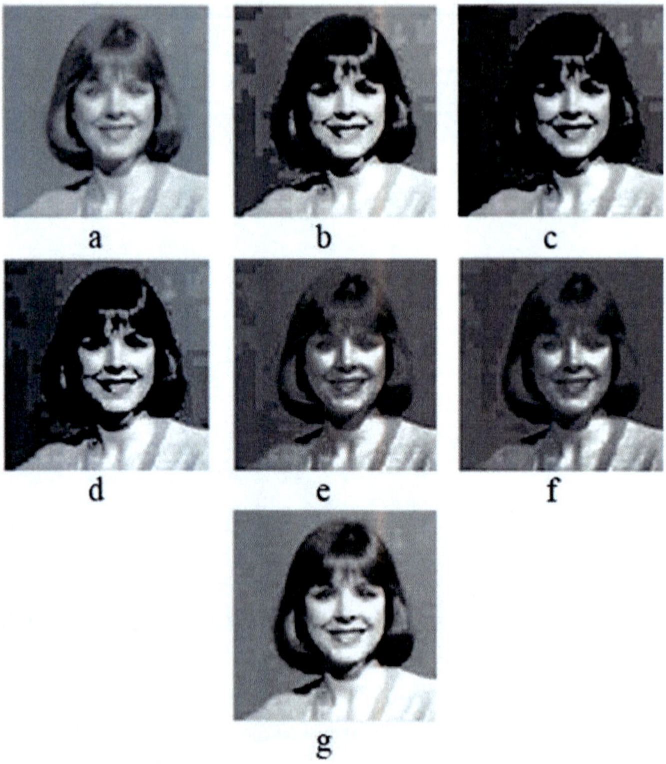

Fig. 6 Women images. **a** Original image, **b** HE, **c** BBHE, **d** DSIHE, **e** RMSHE, **f** MMSICHE, **g** proposed method

This technique boosts the appearance of the image and promotes natural enhancement. The simulated results convey that this algorithm naturally enhances the image without any significant artifacts. As the equalization of the energy curve is done locally, it enlightens the realistic features of the image. The comparative analysis reflects the dominance of the algorithm over other techniques.

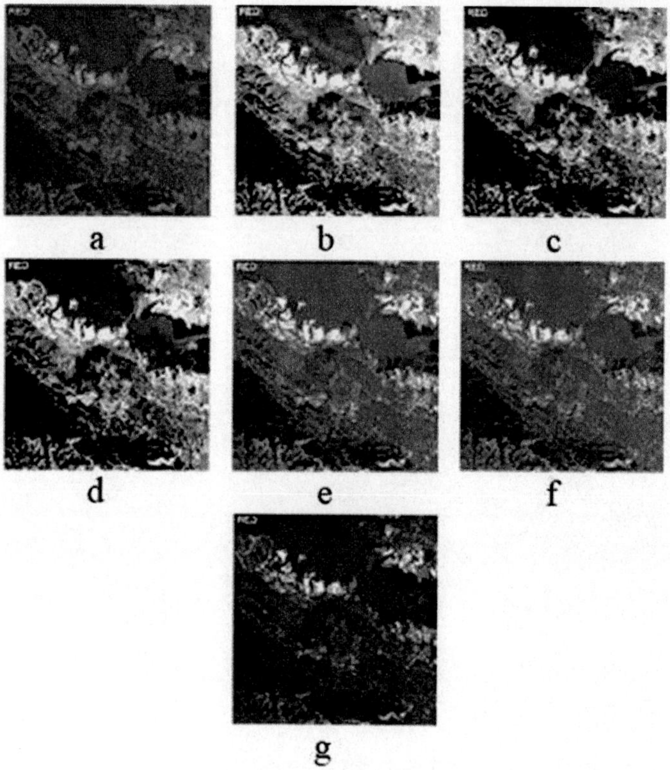

Fig. 7 Top-view images. **a** Original image, **b** HE, **c** BBHE, **d** DSIHE, **e** RMSHE, **f** MMSICHE, **g** proposed method

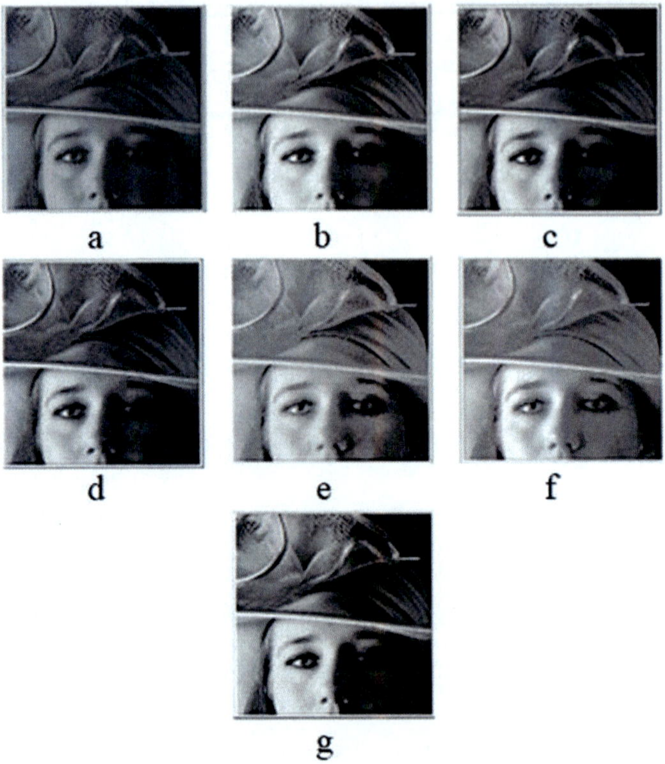

Fig. 8 Girl images. **a** Original image, **b** HE, **c** BBHE, **d**. DSIHE, **e**. RMSHE, **f** MMSICHE, **g** proposed method

References

1. R.C. Gonzalez, R.E. Woods, *Digital Image Processing*, 2nd edn. (Prentice-Hall, Englewood Cliffs, NJ, 2002)
2. Y.T. Kim, Contrast enhancement using brightness preserving bi-histogram equalization. IEEE Trans. Consum. Electron. **43**, 1–8 (1997)
3. Y. Wan, Q. Chen, B.M. Zhang, Image enhancement based on equal area dualistic sub-image histogram equalization method. IEEE Trans. Consum. Electron. **45**, 68–75 (1999)
4. S.D. Chen, A.R. Ramli, Minimum mean brightness error bi-histogram equalization in contrast enhancement. IEEE Trans. Consum. Electron. **49**, 1310–1319 (2003)
5. S.D. Chen, A.R. Ramli, Contrast enhancement using recursive mean-separate histogram equalization for scalable brightness preservation. IEEE Trans. Consum. Electron. **49**, 1301–1309 (2003)
6. K.S. Sim, C.P. Tso, Y.Y. Tan, Recursive sub-image histogram equalization applied to grayscale images. Pattern Recogn. Lett. **28**, 1209–1221 (2007)
7. M. Kim, M.G. Chung, Recursively separated and weighted histogram equalization for brightness preservation and contrast enhancement. IEEE Trans. Consum. Electron. **54**, 1389–1397 (2008)
8. Q. Wang, R.K. Ward, Fastimage/video contrast enhancement based on weighted thresholded histogram equalization. IEEE Trans. Consum. Electron. **53**, 757–764 (2007)

9. T. Kim, J. Paik, Adaptive contrast enhancement using gain-controllable clipped histogram equalization. IEEE Trans. Consum. Electron. **54**, 1803–1810 (2008)
10. C.H. Ooi, N.S.P. Kong, H. Ibrahim, Bi-histogram with a plateau limit for digital image enhancement. IEEE Trans. Consum. Electron. **55**, 2072–2080 (2009)
11. M. Abdullah-Al-Wadud et al., A dynamic histogram equalization for image contrast enhancement. IEEE Trans. Consum. Electron. **53**, 593–600 (2007)
12. H. Ibrahim, N.S.P. Kong, Brightness preserving dynamic histogram equalization for image contrast enhancement. IEEE Trans. Consum. Electron. **53**, 1752–1758 (2007)
13. D. Sheet, H. Garud, A. Suveer, M. Mahadevappa, J. Chatterjee, Brightness preserving dynamic fuzzy histogram equalization. IEEE Trans. Consum. Electron. **56**, 2475–2480 (2010)
14. C.H. Ooi, N.A.M. Isa, Quadrants dynamic histogram equalization for contrast enhancement. IEEE Trans. Consum. Electron. **56**, 2552–2559 (2010)
15. K. Singh, R. Kapoor, Image enhancement via median-mean based sub-ımage-clipped histogram equalization. Optik **125**(17), 4646–4651 (2014)
16. S. Patra, R. Gautam, A. Singla, A novel context-sensitive multilevel thresholding for image segmentation. Appl Soft Comput J **23**, 122–127 (2014). https://doi.org/10.1016/j.asoc.2014.06.016
17. R. Dhaya, Improved image processing techniques for user immersion problem alleviation in virtual reality environments. J. Innov. Image Process. (JIIP) **2**(02), 77–84 (2020)
18. S. Dutta, A. Banerjee, Highly precise modified blue whale method framed by blending bat and local search algorithm for the optimality of image fusion algorithm. J. Soft Comput. ing Paradigm (JSCP) **2**(04), 195–208 (2020)

Detection of DDoS Attack Using Ensemble Machine Learning Techniques

M. Beulah and B. Pitchai Manickam

Abstract In recent years, the number of Internet users has increased and the attackers are also establishing more attacks and one among them is distributed denial-of-service (DDoS) attacks. The DDoS attacks are still the major attacks in which the attacker controls the bots by taking advantage of vulnerable devices in order to generate a variety of attacks, which are affecting the organizations as well as corporations. This paper proposes an ensemble learning classification with support vector machine and logistic regression to detect the attacks which occur in the network. The voting classifier is used for combining the classifiers that help to attain high accuracy rate, low false positives and enhance the overall detection performance. It will dynamically identify the attack by analyzing the information available in the KDD CUP dataset, which contains 41 features. The evaluation of the proposed work shows that, the ensemble learning has achieved an accuracy rate of 99.2% and error rate of 0.8% by using an anaconda Spyder IDE.

Keywords Distributed denial of service (DDoS) · Network security · Machine learning · Classification algorithms · Probing · Anomaly detection · Support vector machine · Logistic regression · UDP and smurf attacks

1 Introduction

Since network services are vulnerable to intruder attacks, network security has become critical in all businesses, banking, social media, and e-mail services. As a result, attackers and defenders are always battling it out to take advantage of and safeguard the system. The attacker is the one who generates the attack in the victim or targeted server, service, or system to access the user information without their knowledge by using different kinds of strategies accompanied by the malware exploitation. There are different kinds of security issues; they are malware, phishing attacks, cryptojacking, and distributed denial of services (DDoS). The attacks affect the servers,

M. Beulah (✉) · B. Pitchai Manickam
Department of Computer Science and Engineering, Kalasalingam Academy of Research and Education, Krishnankoil, Tamil Nadu, India

G. Ranganathan et al. (eds.), *Soft Computing for Security Applications*,
Advances in Intelligent Systems and Computing 1397,
https://doi.org/10.1007/978-981-16-5301-8_62

software systems, network routers, and end user PCs. The victim is affected by the incoming traffic flooding that is originated from potentially different sources. The distributed denial of service attacks the targeted service or system in order to crash the server by flooding them with more traffic.

The KDD CUP dataset with 41 features has been used. The ensemble with SVM and logistic regression algorithm is used to identify and detect whether it is a normal or malicious attack. The support vector machine and logistic regression algorithm both are supervised machine learning algorithms, which are used for solving the classification challenges. From SVM, support vector classifier (SVC) is used. Since both are the same algorithm, the implementation is just different, wherein SVC supports the linear kernel.

The DDoS attack consists of three types; they are volume-based (UDP, ICMP, spoofed packet floods), protocol (SYN, ping of death, fragmented packet attacks, smurf attack), and application attacks (GET/POST floods, low and slow attacks). The proposed work gives low-false positives and low-false negatives. The primary objective is to find the results effectively, reduce the sparsity problems and increase the accuracy rate, error rate, and enhance the overall classification results by using an ensemble learning classification algorithm.

The contribution of this paper is as follows: The complexity of existing mechanisms and methodologies is high and expensive, when it comes to the implementation to overcome all the challenges, this research work has proposed the ensemble with support vector machine and logistic regression algorithm that is implemented in the IDE spyder. The ensemble learning improves the combined contributing models output, which gives accurate results to achieve the improved performance and robustness and reduce the dispersion and model performance. Classifiers must be simple so that they do not overfit. The proposed method helps to increase the accuracy, precision, recall, $F1$ score and reduce the false positive rate when compared to the existing methods that are discussed in the related works.

The paper is organized as follows: In Sect. 2, related works are mentioned. Section 3 discusses about the proposed work of ensemble with SVM (SVC) and logistic regression algorithm for increasing the accuracy rate and enhancing the overall performance. Section 4 discusses about the implementation and outcomes. Section 5 offers the conclusion and future research directions.

2 Related Work

Author	Year	Methodology	Detection result	Limitations
Kumar et al. [1]	2020	The author proposed a novel misuse-based IDS using UNSW-NB15 and RTNITP18 dataset	The accuracy of the IDS method gives 83.3%	Increase the detection rate in real-time dataset and not have the ability to detect the current unknown attacks
Bhuvaneswari et al. [2]	2021	Class-center-based triangle area vector method	This detection method achieves the accuracy of 92.062% to 98.243%	Computational complexity is high
Selvaraj et al. [3]	2016	Ant-based technique that using virtual honeypots	It reduces the false positive of 20% and increases the throughput by 0.8	Honeypot unaware of the activities unless it is directly attacked
Mehmood et al. [4]	2018	Naïve-Bayesian classification in IDS	The work gives better performance and a high accuracy rate of 98%	Rarely implementing in real time
Yuan et al. [5]	2017	Deep defense method using deep learning	It evaluates the results using ISCX2012 dataset and the deep defense reduces the error rate of 7.517–2.103%	Using big data RF model that decreases the performance
Wang et al. [6]	2017	Sky-shield defense system uses bloom filters and two sketches	The bloom and CAPTCHA technique mitigate the app DdoS attacks	Designed for lightweight DDoS attacks and have to use the real-time Web data for advanced detection
Jain et al. [7]	2019	Honeypot with ML-based framework	To lure the attacker and recording the information as in the log files that are given as an input to predict the abnormal activities	İmplementing in real-time scenario and take on a cloud server to deal with the extreme constrained devices of Internet of things
Elejla et al. [8]	2018	Flow-based IDS that detects the ICMPv6-based DDoS attacks	Using weka that gives an accuracy of 85.83% and low false-positive rates of 30.8%	Features are identified by the assumption and to enchrich the accuracy rate have to extract the extra features of the detected flows

(continued)

(continued)

Author	Year	Methodology	Detection result	Limitations
Jaber et al. [9]	2020	FCM-SVM-based IDS detection	The NSL-KDD dataset is used and implemented in the weka platfom that gives an accuracy of 99.1% and false classification rate of 1.35%	It restricts the overall classification accuracy based on the size of the dataset
Kesavamoorthy et al. [10]	2019	An autonomus multiagent system that uses the swarm intelligence in the cloud platform	The proposed method gives an accuracy of 98%	Have to increase the optimization in the attack detection phase
Wang et al. [11]	2018	LFA defense system	It is implemented in the .net framework. The threshold of bandwidth is 93Mbps, packet loss is 4%	Expensive and defense response time is slow
Ravi et al. [12]	2020	BALANCE approach which detects and mitigates the LFA attack via hybrid SDN	It detects the link flooded attack bots at a precision of 97.64%	Improve the identification of the link flooded attack bots and prevent the congestion in the network
Kalkan et al. [13]	2016	Survey on filtering-based defense mechanism and intrusion detection	For filtering techniques classification approaches are recommended for the DDoS attacks compared to the individual mechanisms	The study covers the filtering mechanisms and the criticality of the ICT systems and the challenges of the fundamental operations
Gilberto et al. [14]	2019	Discussed about the survey on network anomaly detection, hybrid method, detection techniques, and attack categories	The paper helps to understand the theoretically problem of anomaly detection with different aspects	Have to discuss about the open issues of the unknown attacks and to improve the effectiveness of the anomaly detection

(continued)

(continued)

Author	Year	Methodology	Detection result	Limitations
Sharma et al. [15]	2018	Honeydos and data mining approach that tested in three interfaces	The packet analyzer captures the network packets using SVM and within the 2 min the number of packets are captured by each interfaces	Unprotected machine can be compromised in the network. Fringerprinted honeypot that attacker can create spoofed attacks to distract them
Smys [16]	2019	Convolutional neural network and support vector machine (CNSVM)	The DDoS attack can be detected in the telecommunication network that shows the accuracy of 89% compared to NN and SVM	There will exist some loophole in the method and cannot use the statistical features

From the above survey, the ensemble with support vector classifier and logistic regression is not utilized for the DDoS detection. The existing algorithms for DDoS attack provide high-prediction error rate and low accuracy rate. The authors have proposed this technique for finding the effective solution for the DDoS attack. This is the first attempt to use a support vector classifier the same as the support vector machine but the implementation process is different in ensemble learning with logistic regression algorithm is proposed for the DDoS attack detection that it reduces the prediction error and gives a correct prediction rate and high accuracy rate. The SVM and logistic regression algorithms are evaluated separately and that is compared with the ensemble classifier that uses the voting classifier to combine the two or more classifiers to improve the accuracy rate of the attack detection. The ensemble's key benefits are that it improves the average prediction performance is the frequent reduction in the variance component of prediction errors created by the contributing models and it gives robustness and enhances the overall performance. This is the novelty of the paper.

3 Proposed Work

SVM and logistic regression algorithm both are supervised learning algorithms used for classification and regression. The support vector machine uses different kernel functions like linear kernel, nonlinear kernel, sigmoid, radial basis function kernel, polynomial, and exponential. During this work, a linear kernel function is employed. The SVM gives hyperplane output (decision boundary) for the separation. The logistic regression uses the statistical approach that is used for prediction using independent variables. Ensemble learning that uses multiple models (two or more) which are politically generated and combined to unravel a specific problem that improves the high accuracy for classification, prediction rate, and assurance for the safety on the network.

The machine learning algorithm of the voting classifier is used that has the estimator and voting for combining the classifier models that help to improve the results and performance. It trains the ensemble of numerous models and predicts the output based on the high votes. The classification model construction is over, and the data values are inserted into the algorithm that makes a decision based on the variable inputs. The voting hyperparameter is either hard or soft. Most of them recommend soft voting. It is used to fit the independent variables with the dependent variables of the dataset after fitting it makes the prediction and gives an improved accuracy. The support vector classifier and logistic algorithm can be combined by taking the advantage of the algorithm to improve the classification results and performance using the voting classifier. The SVM uses the linear SVC that has more flexibility, and it can also solve the regression problems so it is called a support vector regression. The dataset has four categories of attacks and that has 24 types of attacks (Table 1).

The Neptune, teardrop, smurf are the dos attacks that the Neptune makes a victim not able to use the memory resources by sending a TCP packet requesting. In teardrop,

Table 1 Features of TCP connection and traffic features of the dataset

Features	Type	Features	Type
Duration	Continuous	is_guest_login	Discrete
Protocol_type	Discrete	Count	Continuous
Service	Discrete	serror_rate	Continuous
Src_bytes	Continuous	rerror_rate	Continuous
dst_bytes	Continuous	same_srv_rate	Continuous
Flag	Discrete	diff_srv_rate	Continuous
Land	Discrete	srv_count	Continuous
Wrong_fragment	Continuous	srv_serror_rate	Continuous
Urgent	Continuous	srv_rerror_rate	Continuous
Hot	Continuous	srv_diff_host_rate	Continuous
Num_failed_logins	Continuous	dst_host_count	Continuous
Logged_in	Discrete	dst_host_srv_count	Continuous
Num_compromised	Continuous	dst_host_same_srv_rate	Continuous
Root_shell	Discrete	dst_host_diff_srv_rate	Continuous
su_attempted	Discrete	dst_host_same_src_port_rate	Continuous
num_root	Continuous	dst_host_srv_diff_host_rate	Continuous
num_file_creations	Continuous	dst_host_serror_rate	Continuous
num_shells	Continuous	dst_host_srv_serror_rate	Continuous
num_access_files	Continuous	dst_host_rerror_rate	Continuous
num_outbound_cmds	Continuous	dat_host_srv_rerror_rate	Continuous
is_hot_login	Discrete		

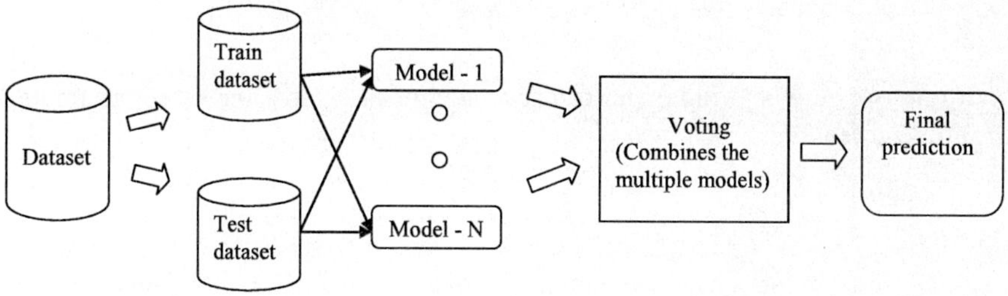

Fig. 1 Ensemble learning

the targeted device can be flooded with intentionally fragmented packets that occurs error when the device tries to rebuild the packets. The smurf causes the victim that it can be flooded by exploiting the ICMP. Warezclient and warezmaster are the r2l that the system can be unauthorized from a remote machine or system. U2r is that the root can be accessed by the unauthorized person, and the Nmap, satan, port sweep, and ip sweep are the probing such as port scanning. The satan is a security administer tool used to search and test for vulnerabilities.

The ensemble learning uses multiple models, these are called base models, and then combines them to produce a more accurate rate than the single model that gives improved results and performance, and it can be applied for regression and classification. After importing the dataset, it can be trained and tested by using a train test split. It uses a voting technique that is used for the classification is a machine learning model that trains an ensemble of various models and predicts an output based on their highest probability. Each classifier's findings are passed into the voting classifier that predicts the output based on the majority of voting for each output class. There are two types of voting such as hard and soft voting. Soft voting is used for the result that is based on the probability of votes (Fig. 1).

Algorithm

Step 1: Import the dataset and divide into training and testing set $D = (a_1, b_1), \ldots (a_n, b_n)$.

Step 2: Feature set $F = \{f_1, f_2 f_3, \ldots, f_n\}$, where f represents the features of the dataset and.

Where $C_i = \{C_1, \ldots C_n\}$ C_i represents the class (i.e., attack or normal).

Step 3: Sort the features f_i in order to determine the output class C_i according to its importance.

Step 4: Checking the null values NaN in the dataset.

Step 5: Encoding categorical data for regression N-1.

Assign value from 1 to N, where the N is the number of categories.

Step 6: Feature extraction

$$\{f_1, \ldots f_n\} \rightarrow \{fi_1, \ldots fi_n\}$$

It improves the algorithms performance significantly, and this area is to finding the efficient search.

$$\{g_1(f_1, \ldots f_n), \ldots g_j(f_1, \ldots f_n), \ldots g_m(f_1, \ldots f_n)\}$$

To construct an effective model, optimized feature extraction is the important key that it creates a new features from the existing ones such that it reduces the number of features in the dataset. This set of features is able to sum up the most of the information that hold in the original features that identify the DDoS attack based on the ensemble method.

Step 7: SVM classifier that uses the hyperplane separates between the normal and attack.

$w.x + b = 0$, where b is called bias.

Step 8: Ensemble classifier combining the classifier 1 and the classifier 2 using voting.

Where the classifier 1 is the SVC from SVM classifier and classifier 2 is the logistic regression that uses the linear model.

Step 9: Construct the dataset for prediction to classifier.

Step 10: Classify the normal or attacked.

Parameters Used in the Algorithm

The parameter that is used in the algorithm is the features of the dataset that can be represented as F, and the class label of C is the classification of the dataset that is whether normal or attacked. The NaN denotes the null values that exist in the dataset, if it exists in the dataset that will solve by the imputation process. The $N - 1$ is the number of categories that the one hot encoder will solve order issues. Each category value is transformed into the column assigned as 1 or 0 to the column. After done this thing in the data preprocessing, the feature extraction is done that it reduces the features by creating new in the dataset. After selecting the features, it can be extracted that help to improve the performance significantly. In the formulation, f denotes the concept or target function and g represents the predictor space or function. In support vector machine uses the hyperplane to represent the classification between the normal or attacks and w and b denote the weight and bias. The voting classifier assists the different models in combines and constructed for predictive and gives better accuracy.

The proposed work will increase performance and the accuracy of the classification that results by classifying the data based on the attacks and others using an ensemble learning algorithm with a support vector and logistic regression machine learning algorithm which is used for the detection of distributed denial of service (DDoS) using the KDD CUP dataset that contains the 41 features and contains four different types of attacks such as Dos, R2L, U2L, and probing and 24 training attack types (Fig. 2).

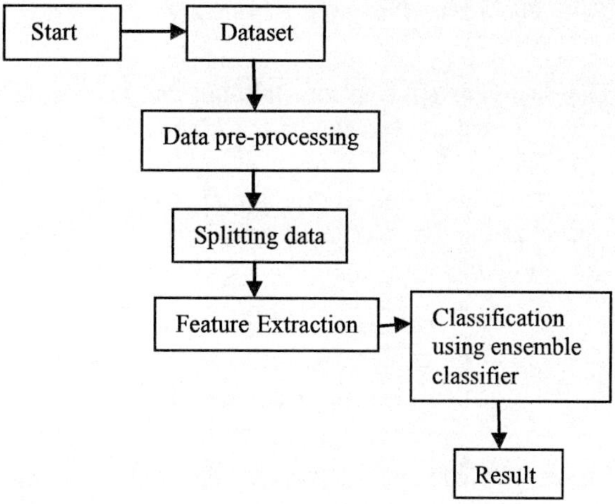

Fig. 2 Flow diagram for proposed method

3.1 Data Selection

The KDD CUP dataset is used for the detection that contains the information about the duration, flag, service, src_bytes, dest_bytes, and class labels. This dataset from which the replica instances have been detached to get rid of the biased classification results and the dataset has the wide variety of versions available, out of which 20% of the training data is used which is identified as train+ , 20% with a total number of 25,192 instances. The test data is recognized by the name of test+ and has a total of 22,544 instances.

3.2 Data Preprocessing

The preprocessing means modification applied to the data before supply it to the algorithm that converts the raw into a clean dataset. In the data preprocessing, two things can be that are missing null values and the other one is encoding categorical data. Here, the missing null values mean, if we consider a large dataset then there will exist some null values term as a NaN or NA (missing data values) so this will lead to a corrupted result and observation. The missing data values are removed by using an imputer (imputation) library that means the null values can be cut out by the mean of the data values of a certain data column. The other one encoding categorical data, here the categorical data are converted into numerical values such as 0s and 1s. For example, if consider a categorical data of men and women then by converting them into a numerical value, then men as 0 and female as 1 by using the label encoder and one hot encoder can be used that it takes a column (which has a categorical data) that has been label encoded to convert them into an integer data.

3.3 Splitting Data and Feature Extraction

The dataset is split up into training and testing data, this train-test split plan of action is used to evaluate the machine learning algorithms performance that make a prediction. Otherwise this plan of action should not be used if you have a small dataset but additional configuration is required. The feature extraction is used to normalize the data within a particular range. It helps to speed up the calculation in an algorithm.

4 Implementation and Results

The ensemble learning algorithm combines the multiple learning algorithms that are constructed to solve a particular problem and that helps to improve the machine learning results and better predictive accuracy performance. The support vector classifier from the support vector machine algorithm uses the linear kernel, and both are the same algorithm that is a supervised machine learning algorithm but used for different implementation. The support vector classifier returns a hyperplane (best fit) that categorizes the data in a higher dimension and to see the predicted class we can give some features to the classifier. The logistic regression is used for classification that returns the probability value by transforming the output using logistic sigmoid.

The proposed work is implemented in Python using anaconda of spyder IDE. The two algorithms are implemented separately and both algorithms are combined by using an ensemble learning algorithm to get predictive accuracy results and enhance performance. After checking the null values, categorical data can be encoded by using a label encoder and one hot encoder. The non-attack and attack can be considered as 0 (non-attack) and 1 (attack). The non-attack is 51.55% and attack is 48.45% of the dataset. Before performing the ensemble classification, we have to check the equally distributed classes; if it not equally distributed, then the sampling can be performed to get an equally distributed class. The result shows that there is no null value after done the data preprocessing.

The imbalance correlation matrix shows the linear relation between them which is shown in the correlation matrix in a table which exhibits the correlation shown in Fig. 3. The imbalance is referred to as unequal distributed classes that exist in the dataset that gives a biased output shown in Fig. 3. It is one of the common issues that can be solved by using the undersampling which is shown in Fig. 3 that represents the before and after sampling. The Python packages used in the implementation are Numpy, matplotlib, pandas, and seaborn library. Those are used for data cleaning, analysis, and transforming the data. The transformed data are visualized by using the seaborn. The heat map is used to represent the imbalanced dataset in the correlation matrix, and the sampling correlation matrix is shown in Figs. 3 and 4 that demonstrate the linear relationship and the data can be displayed as a scatterplot. Any dataset that has an unequal distribution class is said to be an imbalance.

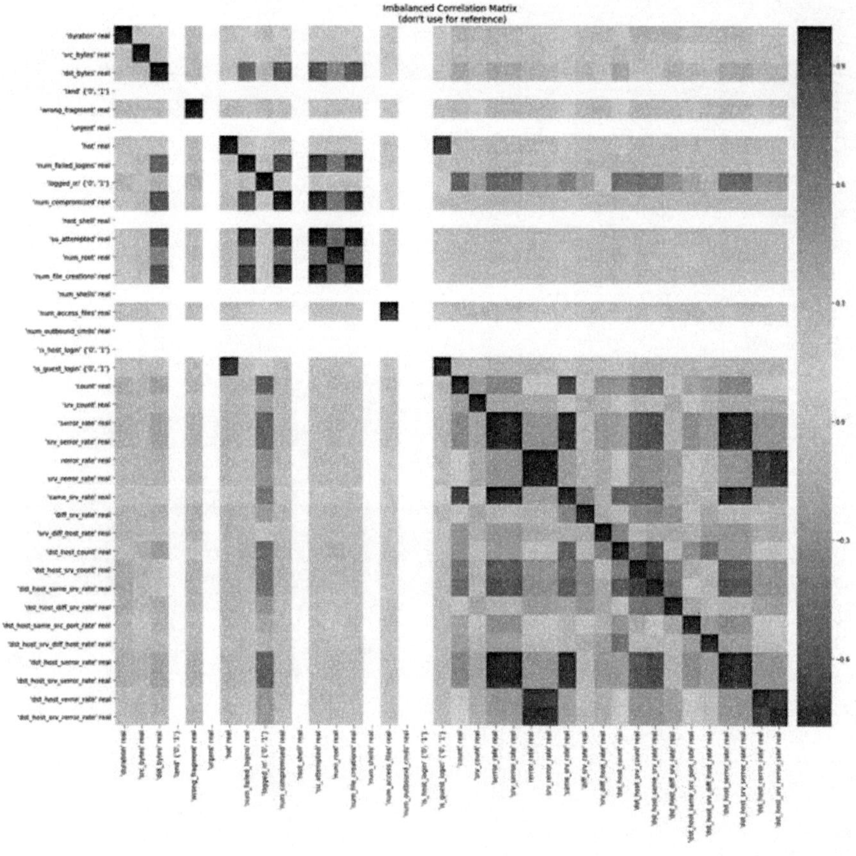

Fig. 3 Imbalance correlation matrix

The subsample correlation matrix is that the imbalanced data can be balanced by using a sampling process which is a statistical analysis that takes a large number of observations depends on the analysis type being performed. The correlation matrix can be created by collecting the data and the data frame is created by using pandas and the visualizing data is represented by using the seaborn and matplotlib.

Distribution of the classes in the subsample dataset.

0	0.504098
1	0.495902

The support vector classifier and logistic algorithm give the accuracy rate of 99.2% and 99%. The ensemble classifier with the combination of SVC and logistic algorithm gives the accuracy of 99.2%. The proposed approach performance is evaluated using some measures such as accuracy, precision, recall, and $F1$ score.

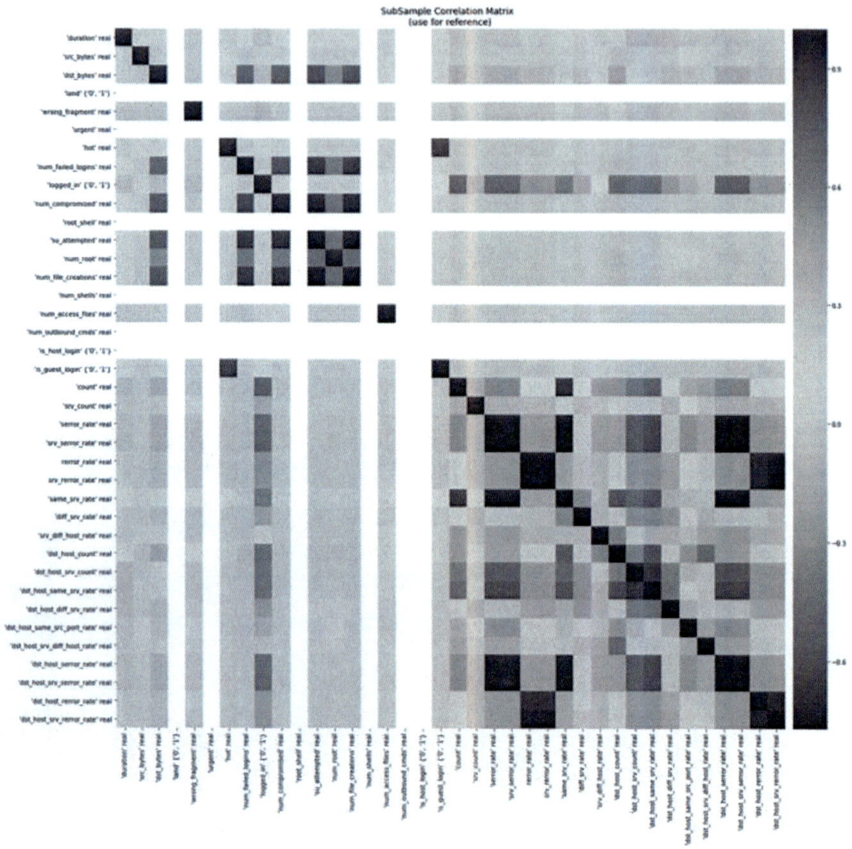

Fig. 4 Subsample correlation matrix

$$\text{Accuracy} = (\text{TP} + \text{TN})/(\text{TP} + \text{TN} + \text{FP} + \text{FN}) \tag{1}$$

where

TP True positives
TN True negatives
FP False positives
FN False negatives.

Precision is the correctly predicted positive to the total predicted positive observation.

$$\text{Precision} = \text{TP}/(\text{TP} + \text{FP}) \tag{2}$$

Recall (is called as sensitivity) is the number of positive class predictions made out of the all positive.

Table 2 Parameters

S. No.	Parameters	Percentage (%)
1	True positive	99.2
2	False positive	0.8
3	False negative	0.8
4	True negative	99.2
5	Accuracy	99.2
6	Error rate	0.8
7	Precision	99.2
8	Recall	99.2
9	$F1$ score	99.2

$$\text{Recall} = \text{TP}/(\text{TP} + \text{FN}) \tag{3}$$

$F1$ score or $F1$ measure is the weighted mean of the precision and recall (Table 2).

$$F1 = 2\text{TP}/(2\text{TP} + \text{FP} + \text{FN}) \tag{4}$$

Figure 5 shows the detection accuracy and compared with the machine learning algorithms of support vector machine, logistic regression, KNN and the proposed method of using ensemble shows the improved precision, recall, $F1$ score, and high accuracy rate of 99.2%. The support vector machine uses the linear kernel of support vector classifier that provides the clear separation and effective and logistic regression help to describe and explain the relationship. For correct classification, ensemble

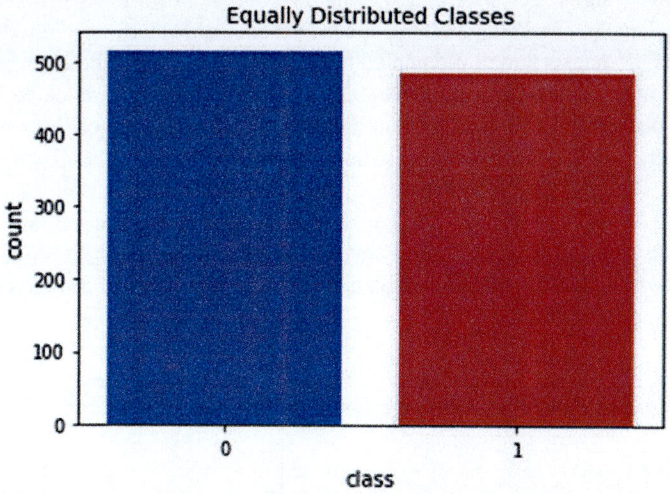

Fig. 5 Distribution of the classes in the subsample

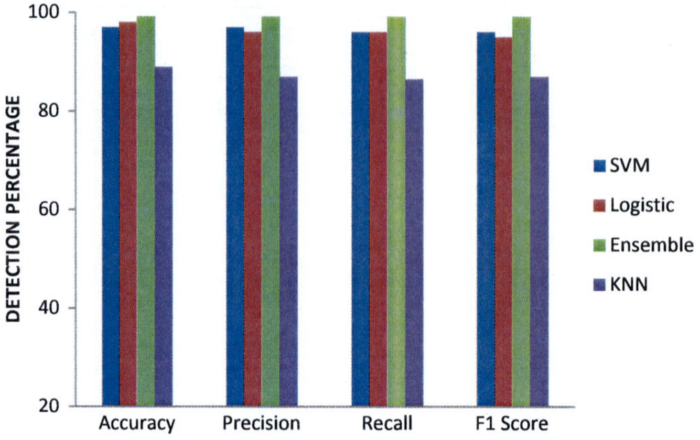

Fig. 6 DDoS attack detection using machine learning

works better than the KNN algorithm and the results show that the ensemble method is recommended regarding the excellent classification technique used for the detection of distributed denial-of-service attacks and enhance the overall performance better (Fig. 6).

5 Conclusion and Future Work

The DDoS attacks destroy most of the organization and corporation information, and serious damages are caused as the attacks are increasing. Most of the existing mechanisms of signature-based and filtering methods are not efficient in detecting and preventing attacks. In this paper, we proposed the ensemble learning method that uses the KDD CUP dataset and implemented in the spyder IDE using the voting classifier (that used in the coding for the ensemble to combine the multiple classifiers) that combines both support vector classifier and logistic regression algorithm and it gives the accuracy of 99.2% and enhances the performance of overall classification and prediction. It alleviates the sparsity problem and reduces information loss. Systems are moving toward analytics and data network to find the prediction through machine learning or deep learning. In future work, we try to analyze and investigate recent and unknown DDoS attacks and implement the classification of the data by using deep learning in SDN.

References

1. V. Kumar, D. Sinha, A.K. Das, S.C. Pandey, R.T. Goswami, An integrated rule based intrusion detection system: analysis on UNSW-NB15 data set and the real time online dataset. Clust. Comput. **23**(2), 1397–1418 (2020)
2. N.B. Amma, S. Selvakumar, A statistical class center based triangle area vector method for detection of denial of service attacks. Clust. Comput. **24**(1), 393–415 (2021)
3. R. Selvaraj, V.M. Kuthadi, T. Marwala, Ant-based distributed denial of service detection technique using roaming virtual honeypots. IET Commun. **10**(8), 929–935 (2016)
4. A. Mehmood, M. Mukherjee, S.H. Ahmed, H. Song, K.M. Malik, NBC-MAIDS: Naïve Bayesian classification technique in multi-agent system-enriched IDS for securing IoT against DDoS attacks. J. Supercomput. 74(10), 5156–5170 (2018)
5. X. Yuan, C. Li, X. Li, DeepDefense: identifying DDoS attack via deep learning, in*IEEE International Conference on Smart Computing (SMARTCOMP)* (2017), pp. 1–8
6. C. Wang, T.T. Miu, X. Luo, J. Wang, SkyShield: a sketch-based defense system against application layer DDoS attacks. IEEE Trans. Inf. Forensics Secur. **13**(3), 559–573 (2017)
7. R. Vishwakarma, A.K. Jain, A honeypot with machine learning based detection framework for defending IoT based botnet DDoS attacks, in 3rd *International Conference on Trends in Electronics and Informatics (ICOEI)*. (IEEE, 2019), pp. 1019–1024
8. O.E. Elejla, M. Anbar, B. Belaton, B.O. Alijla, Flow-based IDS for ICMPv6-based DDoS attacks detection. Arab. J. Sci. Eng. **43**(12), 7757–7775 (2018)
9. A.N. Jaber, S.U. Rehman, FCM–SVM based intrusion detection system for cloud computing environment. Cluster Comput. 1–11 (2020)
10. R. Kesavamoorthy, K.R. Soundar, Swarm intelligence based autonomous DDoS attack detection and defense using multi agent system. Clust. Comput. **22**(4), 9469–9476 (2019)
11. J. Wang, R. Wen, J. Li, F. Yan, B. Zhao, F. Yu, Detecting and mitigating target link-flooding attacks using SDN. IEEE Trans. Depend. Secure Comput. **16**(6), 944–956 (2018)
12. N. Ravi, S.M. Shalinie, D.D.J. Theres, BALANCE: link flooding attack detection and mitigation via hybrid-SDN. IEEE Trans. Netw. Serv. Manage. **17**(3), 1715–1729 (2020)
13. K. Kalkan, G. Gür, F. Alagöz, Filtering-based defense mechanisms against DDoS attacks: a survey. IEEE Syst. J. **11**(4), 2761–2773 (2016)
14. G. Fernandes, J.J. Rodrigues, L.F. Carvalho, J.F. Al-Muhtadi, M.L. Proença, A comprehensive survey on network anomaly detection. Telecommun. Syst. **70**(3), 447–489 (2019)
15. P. Sharma, B. Nagpal, HONEYDOS: a hybrid approach using data mining and honeypot to counter denial of service attack and malicious packets. Int. J. Inf. Technol. 1–10 (2018)
16. S. Smys, DDoS Attack detection in telecommunication network using machine learning. J. Ubiquit. Comput. Commun. Technol. (UCCT) **1**(01), 33–44 (2019)

Cyberbullying Detection and Prevention Using Artificial Intelligence

Ronak Philip Jacob, Kevin Manoj, Divya Mohan, Sneha Issac, and Dhanya Sudarsan

Abstract With the surge of more social media platforms (SMPs), the problems associated with it have also increased. Cyberbullying is one of the major issues in SMPs. It can be defined as an intentional and rude act which is carried out by a person against another person by making use of electronic forms of contact such as a social media platform. It is the need of the hour to restrict such activities as it leads to severe depressions and even suicide attempts. Detecting online bullying on social media is a challenging task as it takes various forms. In this paper, we have given a summary of the literature survey that we conducted to understand the technologies and challenges in cyberbullying detection. After analysis, we have proposed a BLSTM-based neural architecture to detect and prevent cyberbullying. Based on our observations from the study conducted, a scope for future enhancement is also suggested.

Keywords Deep learning (DL) · Artificial intelligence (AI) · Bidirectional long short-term memory (BLSTM) · Recurrent neural network (RNN) · Social media platform (SMP) · Term frequency-inverse document frequency (TFIDF) · Natural language processing (NLP) · Attention mechanism

1 Introduction

Latest developments in Internet connectivity and social networking have led to people having an easy way to communicate and express themselves. However with the advent of new technologies, new concerns arise. One of the greatest challenges we face today is online bullying on social media. A form of bullying that happens on digital devices such as mobile phones, computers and tablets is called cyberbullying. It can happen in different forms on social media platforms, discussion boards, where

R. P. Jacob (✉) · K. Manoj · D. Mohan · S. Issac · D. Sudarsan
Department of Computer Science and Engineering, Muthoot Institute of Technology and Science, Kochi, Kerala, India

D. Sudarsan
e-mail: dhanyasudarsan@mgits.ac.in

G. Ranganathan et al. (eds.), *Soft Computing for Security Applications*,
Advances in Intelligent Systems and Computing 1397,
https://doi.org/10.1007/978-981-16-5301-8_63

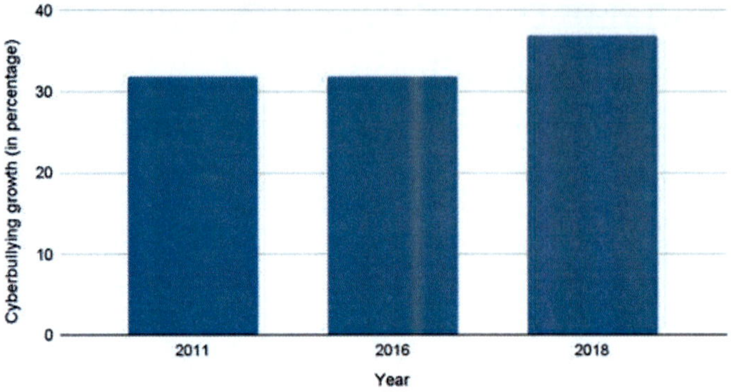

Fig. 1 Growth of cyberbullying over years

people can see, post and share content. Cyberbullying includes various activities like sending, posting or sharing negative, harmful or fake content about another person. It takes various forms across various platforms (sexist, racist, etc.). The victims of cyberbullying have been found to experience stress, anxiety, depression and physical and mental imbalances. So it has become the need of the hour to tackle this issue.

Cyberbullying is a major issue faced by people in this digital era. Figure 1 shows the increase in cyberbullying in India over the years. As our life is becoming more and more centered on social media applications, the impact this is creating in the lives of people is a big concern.

Section 2 summarizes our findings through the literature survey that we conducted. In this, the previous works done in the domain of cyberbullying detection and different natural language processing tasks are analyzed. Section 3 explains a BLSTM-based deep learning model for cyberbullying detection and prevention. Section 4 discusses the future scope in cyberbullying detection.

2 Literature Survey

Ali et al. [1] in their paper defines cyberbullying as a threatening or harassing act done by a person toward another person using an electronic media. In order to identify bullying keywords within a corpus, they made use of machine learning methods. Bullying can be detected on the basis of the content and sentiment of the content. But a single feature is not enough to detect cyberbullying with good accuracy. Integrating both the features will give higher precision. But the major limitation of cyberbullying detection is that there is no standard definition for cyberbullying. In addition, they found out that implementing cyberbullying detection using keywords solely as a feature is not at all reliable because of the dynamic nature of language used by adolescents. Thus, supplementary factors would be required to prove that a message contains bullying content.

Dang et al. [2] found that deep learning solutions have more accuracy for sentiment analysis compared to traditional models. In traditional machine learning methodologies, manual or feature selection methods were used to define and extract features, whereas in deep learning models, features are learned and extracted automatically which would ultimately result in an improvement in precision and performance. For performing sentiment analysis, text preprocessing techniques are initially applied on the input data. The main preprocessing techniques used were tokenization, cleaning and normalization. After preprocessing, we choose the sentiment classifier that best fits our needs and requirements. The first technique to be applied for sentiment analysis was lexicon-based approach. This method uses a dictionary of terms such as SentiWordNet and WordNet to perform sentiment classification. Here the text data is compared with the dictionary and those matched words are given a sentiment score. We have several neural networks such as CNN and RNN for sentiment analysis. For text data, RNN performs better. The advantage of RNN model over traditional neural networks is that RNN model can remember the previous computation of information and apply it to the next item in the input sequence. In their work, they performed sentiment analysis on a set of Twitter data and movie reviews. From their observations, it is seen that RNN combined with word embedding yields better results in most cases on Twitter datasets and IMDB reviews, compared to other approaches. The major drawback of RNN model is that it has processing time which is 10 times longer than that of the CNN model. RNN model has the highest degree of accuracy when used with word embedding.

Galassi et al. [3] talks about attention mechanisms. They pointed out the fact that in order to obtain a compact representation of the data as well as to highlight the relevant information, attention mechanism can be applied to different input parts or different features. The selection process is carried out by a distribution function, which can look at the locality in different dimensions such as space, time or even semantics. This mechanism may be used to perform comparison between input data and query element based on similarity or significance metrics. It also has the capability to learn relevant information independently, by creating a representation which encodes what the significant data should be similar to. The incorporation of attention in neural architectures can therefore generate significant performance gain.

It also looks at the role of attention in solving the basic problems of AI. They have shown how attention can be employed to inject knowledge into the neural model which aids in representing specific features, or to make use of previously acquired knowledge, as in transfer learning settings. Recent studies also indicate that attention could improve the performance in the case of unsupervised learning architectures, by guiding the training process.

Ray et al. [4] present deep bidirectional LSTM architecture for text recognition. It is an approach which uses connectionist temporal classification (CTC) for learning the sequence. This has given favorable results, and it also indicates that deep networks do not need as much data as in the case of shallow networks. They also observed the fact that by increasing the number of hidden LSTM layers, the error rate shows a significant reduction in each epoch proving that deep networks are more effective than their shallow counterparts.

Al-garadi et al. [5] deal with steps and challenges in creating a cyberbullying prediction system. The author discusses two techniques for cyberbullying prediction. One is a lexicon-based approach, and the other is by using machine learning algorithms. Lexicon-based approach is hardly used for cyberbullying prediction because of the unstructured appearance of texts on social media Web sites. It makes cyberbullying detection based only on lexicons difficult. So in order to implement the construction of a cyberbullying prediction model, we have to perform three steps. Primary step is data collection which involves the collection of data from these social media sites and all. One method used by the authors to categorize data from these Web sites is by using keywords and hashtags. Next is feature engineering, which is the selection of features from the given data for the ML classifier. In most models of cyberbullying, a lexicon of words considered as bullying was developed to denote bullying, and those words were used as input features to machine learning algorithms. Then a suitable ML algorithm is selected. There is no ideal machine learning algorithm that works best for every problem. As a result, the majority of researchers selected and compared several supervised classifiers to determine which classifiers worked significantly well for their problem. Here they had discussed a set of generally used ML algorithms like support vector machine where we create a hyperplane to separate the classes, Naive Bayes classifier, which is based on Bayes theorem, logistic regression, etc.

Table 1 summarizes the different methods used and their advantages and disadvantages.

3 BLSTM-Based Deep Learning Model

The proposed solution is an application based on deep learning that could detect bullying content on social media platforms. When a user posts something on the social media platform, our application will take that text as input and check whether it contains abusive content or not based on the context of the post. The text will be analyzed by the neural model, and an output value will be generated showing whether the text is abusive or not. If a post or a message produces a value greater than the threshold value, it will restrict the user from posting it and shows a yes or no output but not both simultaneously. Depending on the diversity of language used, the threshold value changes. The threshold value has to be set by the programmer, whereas the score value will be the output obtained from the process itself. This application will make use of bidirectional long short-term memory (BLSTM) with attention mechanism for detecting the cyberbullying content in an efficient and accurate manner. Deploying this deep learning model over cloud services enables integration with new applications seamless and easy. It could be accessed via API calls.

Table 1 Summary of literature survey

Method	Dataset	Advantages	Disadvantages	Accuracy (%)
Lexicon-based approach	Product review	High performance when context is not taken into account	Dictionary is necessary. There is a need for strong language resources that are not always available	84.6
Support vector machine	Amazon customer review data	Large dimension of input space Irrelevant features are few Sparse document vectors	Requires large amounts of training set Long-term data collection	83.4
Naive Bayes	Twitter dataset	A simple, intuitive approach It brings together efficiency and reasonable precision	Primarily used when the training dataset is smaller in size. It presupposes conditional independence among linguistic characteristics	76.8
LSTM	Twitter dataset	Selectively remembers the important details Solved the problem of vanishing gradient in RNNs	Though it was supposed to handle long-term dependencies, with large sequences, LSTMs failed to attain good performance	85
BLSTM	Formspring and Twitter	It developed a better understanding of context as there were two sequences of LSTMs, one with normal input and other with inverted input	It requires more complex computing The output of the encoder–decoder model depended on a single output vector from the encoder. It is logically not possible to encode all necessary details onto a single vector	87.1

3.1 System Architecture

The raw textual data that is received as input will initially undergo preprocessing. It involves various steps like tokenization, cleaning and normalization. The output from this stage will be word tokens. Before this input is given to the neural network, it needs to be converted into numerical representations that the system will be able to understand. The success of neural networks was largely because of the advent of word embeddings where each word is converted into a real-valued vector. In this word embedding phase, the neural network automatically captures certain features from the text. In the case of machine learning, this was done manually, but in the case of deep learning, this is done autonomously. The output vectors are fed to the neural architecture, i.e., BLSTM with attention mechanism where it analyzes the text and classifies it as abusive or not. The attention mechanism was a breakthrough in natural language processing where the network weighs different words based on its importance in the meaning conveyed. It was developed as an improvement over the traditional encoder–decoder model where the output generated in the decoder phase depended only on the output from the encoder's last hidden state. The output from the Softmax classifier of neural architecture would be a value between 0 and 1. Based on a previously fixed threshold, these would be classified as cyberbullying or not (Fig. 2).

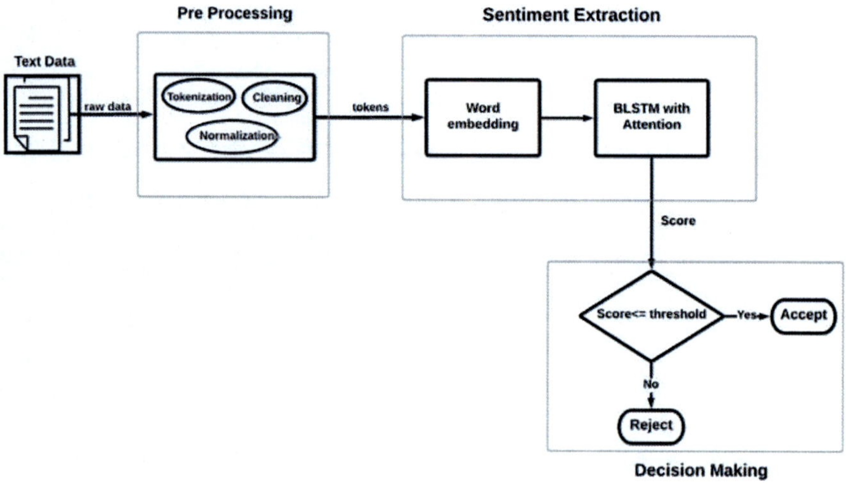

Fig. 2 System architecture

3.2 Preprocessing

Data preprocessing is the process by which the text to be classified is cleaned and prepared. Online texts usually have a considerable amount of noise and non-informative chunks. Preprocessing lowers the noise in the text and enhances the performance of the ML model and accelerates the classification process. So the main text preprocessing methods are as follows:

1. Tokenization
2. Text cleaning
3. Normalization

Tokenization—The action of breaking the given text into smaller units called tokens is known as tokenization. It is simply breaking the raw text into smaller parts called tokens.

Text cleaning—It is the process of removing different symbols and words which do not convey meaning to the model while training.

Normalization—It is the process of transforming non-standard words into their standard format.

3.3 Sentiment Extraction

Various methods can be used to extract sentiment from preprocessed data. Preprocessed tokens are fed as input to this phase. These tokens need to be converted into numerical representations, and these vectors are fed into a sentiment classifier for sentiment analysis.

This phase includes

1. Word embeddings
2. BLSTM

Word Embeddings—Word embedding is a recently developed technique which has largely contributed to the success of natural language processing projects. It maps words and phrases into real-valued vectors. It is capable of understanding the semantics and relationship between words in a text document.

BLSTM—The real-valued vectors that we get as output from the word embedding are fed as input to bidirectional LSTMs. These are advanced versions of LSTM networks. In this architecture, we will have two LSTM sequences working in opposite directions. One will be having the normal text sequence as input, while the other will have an inverted or reversed text sequence. Analyzing the text in two different orders, we will have a better understanding of the context. The output that is produced at each time step will be a combination of the outputs from each of these sequences. They are combined to produce a final output.

3.4 Decision Making

A threshold value needs to be previously fixed in order to classify text. If the score generated from the sentiment extraction phase is greater than threshold value, then it will be classified as cyberbullying. But, this threshold has no standard value. One of the major limitations of all the projects that we have reviewed is that there is no such standard value to classify bullying contents. Because, the way we use language is changing rapidly. We know that the words which are considered as extreme bullying content have now become a part of our normal language. This threshold value depends on the language under consideration and common form of language in conversations. So this threshold value is an assumption that we have to take in order to classify text.

4 Analysis and Observations

From the previous works, it has been found that bidirectional long short-term memory neural tends to bring the best results for natural language processing tasks like text classification.

Some of the main challenges in cyberbullying detection are as follows.

4.1 Definition

This is one of the major challenges in creating a prediction model. There is no proper definition to cyberbullying so that we can easily determine whether a text is bullying or not. Classification of manually labeled data as cyberbullying or non-cyberbullying is difficult due to different definitions.

4.2 Effect of Rapid Changes in Culture

What used to be classified as cyberbullying is perhaps no longer classified as cyberbullying. Online social networks have globalized the culture. So continuous training of machine learning algorithms is required. From introducing new words to new meanings for old words to changing the way we communicate, social media makes a difference.

Table 2 Result obtained on toxic comment dataset

Epoch number	Accuracy obtained
1	88.76
2	90.1
3	91.2

4.3 Language Dynamics

The language is changing rapidly, especially among the youngsters. The new slang is incorporated into the language culture on a regular basis. Also the definition of existing words keeps on changing. So we need to regularly train machine learning algorithms by using newly introduced words. Since social media has changed how we as a society communicate in this modern world, it is difficult to find the semantic of a word in its appropriate context. Therefore, the state of language is very important to social media platforms.

All existing works for cyberbullying detection have at least one of the following three problems. First, most of these projects focus only on one social media platform. The kind of cyberbullying present in different platforms is totally different, and it is important for us to understand these different forms. Second, they address just one form of cyberbullying like racism, sexism, etc. Thirdly, all these models depend on carefully handcrafted features of the data.

Addition of BLSTM with attention mechanism in this neural architecture improves the performance in cyberbullying detection drastically. Also attention mechanism has shown to give more weightage to important words and thus helps in understanding the text better. The accuracy was also increased as the number of hidden layers was increased. Using trained word embedding also tends to improve accuracy.

The developed system was trained on the toxic comment dataset. The performance of the model was improved by a pre-trained word embedding. The model produced an accuracy of nearly 91% on the validation set. It was also found that with more epochs of training, the accuracy attained would increase. Table 2 summarizes the result our model produced with the toxic comment dataset.

5 Future Scope

One of the major limitations to such a project is lack of large amounts of proper data. Though currently we have access to a good amount of data, data from different sources containing different forms of cyberbullying will help to improve the overall efficiency of cyberbullying detection. Newer models like BERT and GPT-3 have shown excellent results in various NLP tasks. So deploying these cyberbullying detection using those models would definitely elevate the efficiency of cyberbullying detection and prevention.

6 Conclusion

Existing models for cyberbullying detection mostly depend on ML models. The basic steps for constructing a cyberbullying prediction model are data collection and preprocessing, feature engineering and choosing the best machine learning algorithm. Most of the existing models use manual feature extraction. Deep neural network architecture has been found to significantly outperform the ML models in NLP tasks. Word embeddings when combined with LSTMs led to best accuracy in NLP tasks. Bidirectional long short-term memory (BLSTM) model helps in understanding the context better. It is observed that error can be reduced by increasing the number of hidden BLSTM layers. In each epoch, the error is significantly reduced proving that deep networks are more efficient than their shallow counterparts. Attention model has become ubiquitous in NLP applications. Attention autonomously learns what should be considered pertinent, by creating a representation encoding what the significant data should be similar to. Incorporating attention in neural architecture thus yields a considerable performance gain. Thus, BLSTM with attention mechanism will help in cyberbullying detection. Transfer learning enables us to understand the different forms of cyberbullying across social media platforms.

References

1. W.N.H.W. Ali, M. Mohd, F. Fauzi, Cyberbullying detection: an overview. IEEE (2018)
2. N.C. Dang, M.N. Moreno-García, F. De la Prieta, Sentiment analysis based on deep learning: a comparative study. IEEE (2020)
3. A. Galassi, M. Lippi, P. Torroni, Attention in natural language processing. IEEE (2018)
4. A. Ray, S. Rajeshwar, S. Chaudhary, Text recognition using deep BLSTM networks. IEEE (2015)
5. M.A. Al-garadi, M.R. Hussain, N. Khan, G. Murtaza, H.F. Nweke, I. Ali, G. Mujtaba, Predicting cyberbullying on social media in the big data era using machine learning algorithms: review of literature and open challenges. IEEE (2019)

Location Prediction on Twitter Using Hybrid Model

Chetali Surti, Pradnya Rane, and Vaishali Jadhav

Abstract Online social networks have gained a lot of popularity over the last decade, especially Twitter. The great use of Twitter leads to a massive collection of data all around the world. Present location shared by the users frequently through tweets is called geo-tagged tweet. On another side of the coin, several users tend to hide their current location. Location prediction serves several platters like marketing, rescue procedures during crisis, suggestions, etc., location prediction from the manuscript of a tweet is a grim task as it contains wrong spelling and unknown abbreviations. Hence, location prediction on Twitter has attracted many researchers nowadays. The proposed system aims to use two hybrid models taking advantage of convolutional neural network (CNN) and long short-term memory (LSTM) for location prediction on Twitter using the text of tweet as input. Convolutional neural network (CNN) is useful in extracting local features from text, and long short-term memory (LSTM) preserves contextual information. The experimental outcomes show that the proposed hybrid model outperforms standard convolutional neural network (CNN) and long short-term memory (LSTM) in terms of accuracy and other parameters.

Keywords Twitter · Location prediction · Word embedding · Machine learning · Neural network · fastText · LSTM · CNN · Hybrid model

1 Introduction

In these days, microblogging and social networking are the typical means of interaction within the whole world. They include daily used social networks such

C. Surti (✉) · P. Rane
Department of Computer Engineering, St. Francis Institute of Technology, Mumbai, India

P. Rane
e-mail: pradnyarane@sfit.ac.in

V. Jadhav
Department of Information Technology, St. Francis Institute of Technology, Mumbai, India
e-mail: vaishalijadhav@sfit.ac.in

G. Ranganathan et al. (eds.), *Soft Computing for Security Applications*,
Advances in Intelligent Systems and Computing 1397,
https://doi.org/10.1007/978-981-16-5301-8_64

as Facebook, Twitter, photo-sharing Web sites for instance Instagram and some career-oriented social networks such as LinkedIn.

Twitter is American microblogging and has been ranked as the second-largest social networking service on which users publish and interact with messages of up to 280 characters well-known as "tweets." The data includes (i) short texts (tweets) sent by Twitter users, (ii) the social network between the users, as well as (iii) user profile information. The availability of these three types of data makes it possible to use Twitter for geolocation prediction. Tweets play a great role in the context of geolocation prediction for several reasons. First and most importantly, tweets can consist of location information in different forms. The location information might appear as location names in hashtags or as pure words or in other media. Additionally, users can attach a location to their tweets. This action is called tweet geotagging or simply geotagging. The location mentioned in tweets can be used in the location-based recommendation, identifying and managing the crisis, sensing location-centric groups, demographics analysis and targeted advertising [1].

Regardless of the fame of Twitter and a huge number of tweets, only some of the tweets (under 1%) are geotagged that they were published from as a result confining the usefulness of many tweets for location-based facilities and investigations. On other hand, there are many challenges in using the text of the tweets such as spelling mistakes, non-standard language, differences in the spelling of location, for example, Mumbai and Bombay. From a machine learning perception, such a complicated work would need more efficient feature creation than just linking surface features. Using the deep learning approach will benefit to solve this problem. Hence, proposing a hybrid neural network model for location prediction of Twitter users in four metropolitan cities of India.

2 Related Work

Chi et al. [1] In the proposed method, location indicative words and several textual features are used to train multinomial Naive Bayes classifier to guess the location of Twitter users from tweets.

Indira et al. [2] proposed a framework where they have predicted the location of the users from the tweet text taking advantage of machine learning methods that is support vector machine, Naïve Bayes and decision tree. They gather more than 1000 tweets of specific keywords, namely "Chennai, Mumbai and Kerala." Their experiment study concludes that the decision tree is appropriate for the study of tweet text and location prediction.

Ouaret et al. [3] proposed the usage of a semi-supervised method to use unlabeled information, which is often ample at the location of an emergency incident, with less identified information. Their study uses the Random Forest model to train data with coordinates and labels as input collected during an earthquake.

Rehman et al. [4] proposed a latest set of linguistic features to determine locations from the texts of the tweet more efficiently and used five well-known machine learning models which are random forest, logistic regression, support vector machine, classification and regression trees and neural network. In all the experiments done, the random forest classifier has outperformed.

Hemamalini et al. [5] aimed at inferring the location of Twitter users, from the location information of his/her friends, through K-mean clustering. Luke S. Snyder et al. [6] improved DeepGeo by using word2vec embedding.

Thomas et al. [7] introduced a study for Twitter geolocation prediction using the effect of text and metadata and text-derived contextual features. The text of tweets and metadata fields comprising texts are transformed to word embedding, which are then sent to a long short-term memory (LSTM) unit.

Singh et al. [8] proposed a model to predict the subsequent probable location of Twitter user called as sentiments-based labeled LDA model (SLLDA) which used sentiment analysis method and topic modeling.

Lourentzou et al. [9] studied the usage of neural networks to resolve the challenge of geolocation prediction and experimented with several methods to optimize neural networks for geolocation detection through text.

Inkpen et al. [10] proposed an approach, consisting of two phases. In the first phase, trained conditional random fields (CRF) models were able to recognize the locations stated in the posts. They trained three classifiers for provinces/states, cities and countries, with a number of sets of features. The second phase resolved the leftover uncertainties, like the post where it consists of more than a single location with a similar name.

Table 1 consists of summarization of existing work for location prediction on Twitter based on problem category whether next location of user or any time location which is the location of user at any given time [11], dataset, features extracted and techniques used, model trained for predicting the location.

3 Proposed Method

This unit gives the details of the proposed method which consists of data collection, preprocessing of raw tweets, feature extraction from preprocessed tweets using word embedding and two types of hybrid model that uses LSTM and CNN (Fig. 1).

3.1 Data Collection

Live stream tweets are collected using Twitter API. To collect live stream tweets, we require authenticating our Twitter account by using consumer_key, access_token, consumer_secret and access_token_secret. When Twitter API is called, it gives JSON of each live tweet as output. We have analyzed the different data present in a tweet JSON object with a special focus on the data related to the tweeted text, the user who authored the tweet, locations in the tweet text. The tweets are converted to the

Table 1 Summarization of existing work for location prediction on Twitter

	Problem category	Dataset	Feature extracted and techniques	Model
[1]	Any time location	Twitter API	Location indicative words, city/country names, #hashtags, @mentions and a combination of all, Bag of features	Multinomial Naive Bayes classifier
[2]	Any time location	Live stream of Twitter data	tweetid, screen_name, name, tweet_text, TweetLocation, HomeLocation, MentionedLocation	SVM, Naïve Bayes and Decision tree
[3]	Any time location	Tweets collected from Barcelonette Earthquake on April 7, 2014	Bounding box in terms of latitude and longitude for all cities and states location names using GeoNames API	Random forest prediction which is an ensemble method
[4]	Any time location	Making sense of Micro-posts 2013 (MSM2013) collection and the Ritter	Textual/linguistic features using Linguistic Inquiry and Word Count (LIWC) lexicon	Random forest, logistic regression, SVM, neural network and CART
[5]	Any time location	Twitter API	GeoCoding to convert the obtained locations into latitudes and longitudes	K-means to cluster the followers and center of the cluster will be the predicted location for user
[6]	Any time location	Tweets collected from Philadelphia, PA, Chicago, IL, New York, NY	(1) Tweet text; (2) tweet creation time; (3) user UTC offset; (4) user time zone; (5) user-listed profile location (text); (6) user account creation time	Deep learning model integrated with word2vec embedding
[7]	Any time location	Second workshop on noisy user-generated text (WNUT'16)	Simple whitespace tokenizer then converted text of tweets and metadata fields containing texts to word embedding	Long short-term memory (LSTM) unit and classification are done using a softmax activation function

(continued)

Table 1 (continued)

	Problem category	Dataset	Feature extracted and techniques	Model
[8]	Next location	Tweets were collected of the users from New York City	Geo-coordinates and location indicative words	Sentiment-based labeled LDA model (SLLDA) is based on LLDA
[9]	Any time location	GeoText, TWUS, TWWORLD	Bag-of-words text representation and extract TF-IDF features	Neural network and batch normalization
[10]	Any time location	Twitter API from June 2013 to November 2013	Bag-of-words, part-of-speech, Gazetteer	Conditional random fields (CRF) models

required format, and the labeling of each tweet is done based on the location present. We have collected more than 1000 tweets for India's four metropolitan cities, i.e., Mumbai, Chennai, Delhi and Kolkata.

3.2 *Preprocessing*

We need to preprocess the tweets as it contains many abnormalities like different letter casing, punctuation marks, special character, etc.

3.2.1 Letter Casing

Letter casing converts all letters in the tweeted text to lower case to remove different letter casing.

3.2.2 Noise Removal

The noise removal technique eliminates unwanted characters, punctuation marks, special characters, white spaces, etc. We have used regular expressions to remove noise from tweets.

3.2.3 Tokenizing

By using tokenization, we have converted the tweets into tokens. Tokens are words parted by spaces in a text. We have used nltk word_tokenize to tokenize the tweets.

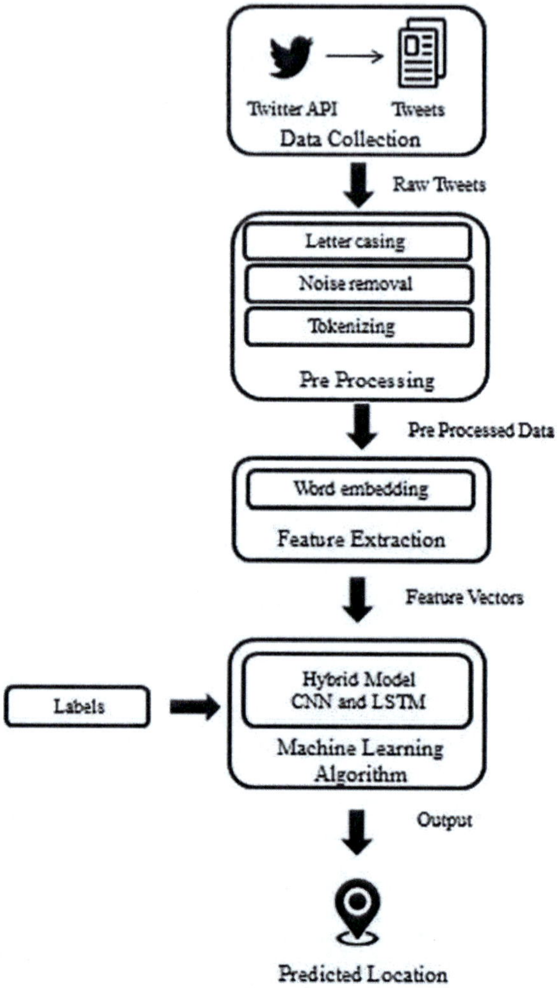

Fig. 1 Proposed system workflow of location prediction on Twitter using hybrid model

3.3 Feature Extraction

We have used the word embedding method for feature extraction. Word embedding is the mathematical representation of words as vectors. They are generated by analyzing a body of text and representing each word, phrase or whole document as a vector in a high-dimensional space. If two words or documents have analogous embedding, they are semantically similar. Word embedding will benefit to maintain the semantics of text corpus. Generally, one hot encoding or word2vec technique is used to convert words to vectors. But both have certain disadvantages like vectors formed using one hot encoding are large and do not capture the semantics of text, whereas word2vec is unsuccessful to offer any vector depiction for words which are not in the model vocabulary. fastText is a word embedding technique which is an enhancement to the word2vec model. fastText denotes every word as an n-gram of letters rather than a learning vector for the words. So, let us say, the word, "institute" with $n = 3$, the fastText illustrates this word as < in, ins, nst, sti, tit, itu, tut, ute, te > , where the angular brackets signify the beginning and end of the word. This enables embedding to identify suffixes and prefixes and also captures the significance of smaller words. If a word was not present during the training, it can be fragmented into n-grams to get its embedding. This will help in working with words that were not present in tweets used while training and get better accuracy [12].

3.4 *Machine Learning Algorithm*

In our hybrid model, we have used convolutional neural network layer (CNN), long short-term memory networks layer (LSTM) and dense layer below are the detailed steps in each network.

3.4.1 Convolution Layer

The convolutional neural network (CNN) is a particular kind of neural network model intended to work with the 2D image, even if they can be used with 1D and 3D data. The CNN constructs a feature map through a chain of convolutions and pooling using a convolutional layer, and a maximum pooling layer as shown, words are fragmented into features and inputted to a convolutional layer. The outputs of the convolution are "pooled" or aggregated to a specific number.

Steps in convolution neural network layer are as follows:

- Step 1: Convolution layer

Move the filter map across the word embedding created from the text of tweets and compute the dot product to get convolved feature map.

$$V_K = X_k \cdot U + bV_k \tag{1}$$

V_k Current feature window,
U Convolution kernel,
X_k Window vector,
b Bias.

This step will help us to extract local features from the text of tweets.

- Step 2: ReLu Activation Function

The ReLu stands for the rectified linear unit. Once the feature matrix or map is obtained, the next phase is to calculate the ReLu activation function wherein an element wise computation is done and assigns all the negative value to 0.

$$f(z) = \max\{0, z\} \tag{2}$$

This step will eliminate the negative value from the feature maps reducing our computation time in further steps.

- Step 3: Max Pooling Layer

Pooling is an operation that lessens the dimensionally of feature maps. The feature map or matrix from the above step at this instant is given to a pooling layer to produce a pooled feature maps or matrix.

$$\lambda_t = \max(V_{k-m+1}, V_{k-m+2}, \ldots, V_k) \tag{3}$$

λ_t Word representation feature,
V_k Current feature window,
m Size of sliding window.

This step helps in reducing the dimensionality of the feature map.

3.4.2 LSTM Layer

Long short-term memory networks typically just called "LSTMs" are a unique type of RNN and are able to learn long-term dependencies. LSTMs were created to deal with the vanishing gradient issue that came across when training traditional RNNs. All recurrent neural networks have the ordering of a chain of repeating units of neural network. LSTMs also have a similar chain structure, but the repetition units have a different structure. In place of having one neural network layer, there are four, interrelated in a very unique way [13].

- Memory cell keeps trail of the dependencies among the elements in the input vector using Eqs. (6) and (7).
- Input gate manages the level to which a new value moves to the cell using Eq. (5)
- Forget gate manages the level to which a value stays in the cell using Eq. (4).
- Output gate manages the level to which a value in the cell is used to calculate the outcome activation of the LSTM unit using Eqs. (8) and (9).

The input to LSTM for the sequential model will be the text feature vector from CNN layer and for the parallel model, and it will be word embedding's feature vector.

Below is the calculation formula used for each unit of LSTM:

$$f_t = \sigma\left(W_f.\left[h_{t-1}, x_t\right] + b_f\right) \tag{4}$$

$$i_t = \sigma\left(W_i \cdot \left[h_{t-1}, x_t\right] + b_i\right) \tag{5}$$

$$\acute{Ç}_t = tanh(W_C.[h_{t-1}, x_t] + b_C) \tag{6}$$

$$C_t = f_t{*}C_{t-1}{+}i_t{*}\acute{Ç}_t \tag{7}$$

$$o_t = \sigma\left(W_o \cdot \left[h_{t-1}, x_t\right] + b_o\right) \tag{8}$$

$$h_t = o_t\ ^*\ \tanh(C_t) \tag{9}$$

x_t The input at the current time,
σ The logistic function,
h_t The final output of LSTM unit,

W_f, W_i, W_C and W_o respectively, represent weight matrix of h_t corresponding to forget gate, input gate, memory gate and output gate.

tanh The activation function.

3.4.3 Dense Layer

The dense layer is a fully connected layer with activation function as softmax to find the probability distribution over labels to classify the output of LSTM in case of the sequential model and output of concatenated layer in case of the parallel model into location. The city with maximum probability is the predicted location of user.

The probability of classifying z is

$$\sigma(z)_j = \frac{e^{z_j}}{\sum_{k=1}^{K} e^{z_k}} \tag{10}$$

z vector of the inputs to the output layer.
J indexes of the output units, so $j = 1, 2, \ldots, 4$.

After getting the probability distribution, categorical cross-entropy loss is computed using (11) and location is predicted using (12)

$$\text{Loss} = \log\left(\sigma(z)_j\right) \tag{11}$$

$$P = \text{argmax}\left(\sigma(z)_j\right) \tag{12}$$

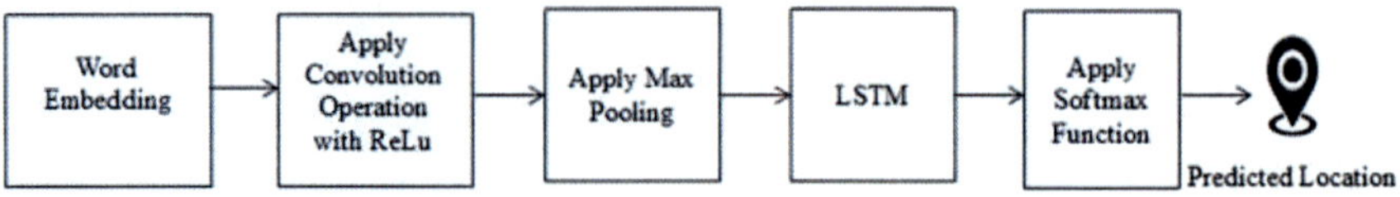

Fig. 2 Sequential model for location prediction on Twitter

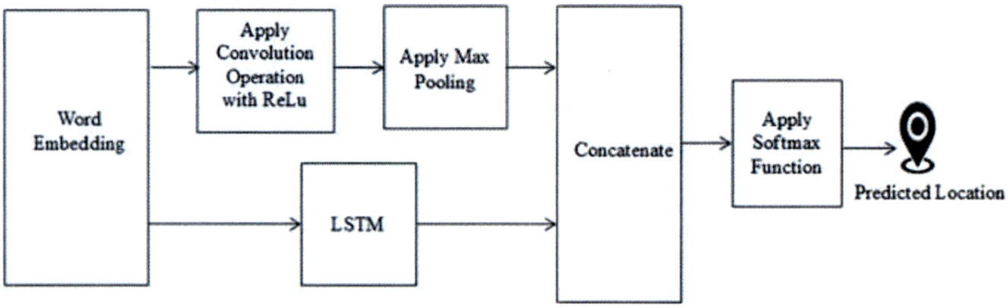

Fig. 3 Parallel model for location prediction on Twitter

3.4.4 Hybrid Model

Sequential Model

Output from word embedding will be given to convolution layer which includes convolution operation with ReLu activation function and max pooling where local context information will be extracted; its output will be passed to LSTM layer which will help to track long-term dependencies. Then output from LSTM will be passed to dense layer which uses softmax function for classification (Fig. 2).

Parallel Model

Output from word embedding layer will be given separately to convolution layer which comprises of convolution operation with ReLu activation function and max pooling and LSTM layer. At that moment, output from both layers will be concatenated and passed to dense layer which uses softmax function for classification (Fig. 3).

4 Experimental Results

4.1 Dataset Description

We collected openly accessible Twitter data using Twitter API. The dataset comprises tweets of four metropolitan cities of India. Around 6411 totals were collected in the period. As Twitter API gives us tweet JSON for each tweet that contains much information of a tweet, we extract only important data for our experiment (Table 2).

4.2 Performance Metrics

We have used three performance metrics to analyze the quality of our hybrid model with LSTM and CNN individual model. Below are performance metrics and its formulas:

4.2.1 Classification Accuracy

$$\text{Accuracy} = \frac{\text{TruePositive} + \text{TrueNegative}}{\text{TotalSample}}.$$

4.2.2 Precision

$$\text{Precision} = \frac{\text{TruePositive}}{\text{TruePositive} + \text{FalsePositive}}$$

4.2.3 Recall

$$\text{Recall} = \frac{\text{TruePositive}}{\text{TruePositive} + \text{FalseNegative}}$$

4.3 Results

For comparison, we have used our hybrid models which consist of two models: one sequential training model and another parallel training model, standard LSTM and CNN. We have examined the effects of changing activation function of convolution layer by using ReLu, tanh and sigmoid, dropout probability with 0.2, 0.3, 0.4, 0.5 and

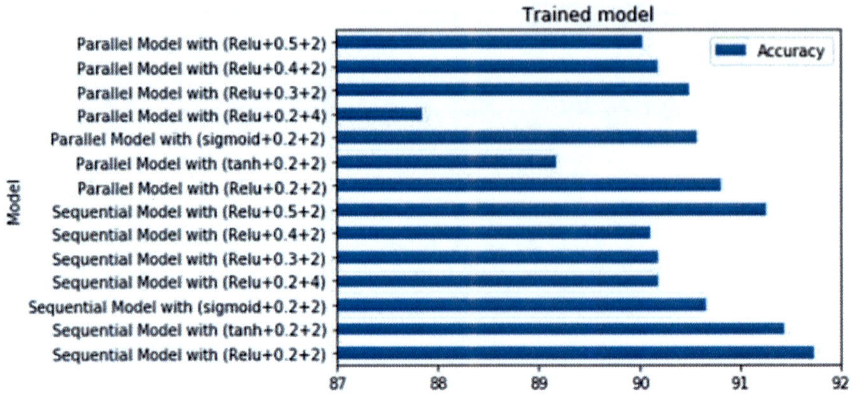

Fig. 4 Accuracy analysis based on different parameter for hybrid model

pool size with 2 and 4 of max pooling layer parameters (Fig. 4) on the accuracy of the sequential model and parallel model. The sequential model is built with embedding size of 100 units, convolution layer consists of filters size of 128, and LSTM unit as 128, dropout probability as 0.2 and dense layer with four units as we are predicting location of four metropolitan cities of India gave the optimal accuracy. Using the same parameters, we trained our parallel model as the parameters selected for sequential model were giving optimal accuracy for parallel as well, where the output from the embedded layer was given to the convolution layer and LSTM layer parallel and then concatenates both outputs. In our analysis, we found that hybrid model gives better accuracy than standard CNN and LSTM (Table 3). The sequential training model gave 91.73% accuracy while the parallel training model gave 90.80% accuracy. Hence, sequential training model gave better accuracy than the parallel training model as local features are extracted from word embedding using CNN, and then by using this local features, LSTM extracts the temporal features and preserves contextual information which helps to attain better accuracy rather than concatenating both local and temporal features.

Table 2 Total count of tweets for each city

City	Count
Delhi	1976
Mumbai	1659
Chennai	1599
Kolkata	1177

Table 3 Performance metric and models

Model	Accuracy	Precision	Recall
Sequential	91.73	94.4	88.6
Parallel	90.8	94.6	91.6
LSTM	88.5	94.6	85.6
CNN	87.5	94.1	79.7

5 Conclusion and Future Work

In this paper, we have proposed two CNN-LSTM hybrid models for location prediction on Twitter which are the sequential model and the parallel model. On one hand, the model takes advantage of LSTM to learn long-term dependencies and resolve vanishing gradient problem and on another CNN extracts the local feature from the tweets. Preprocessing of data was not done, and one hot encoding was used for researches where standard LSTM and CNN were used which would have increased accuracy of model. In our experiment, the sequential model gave better accuracy than the parallel model as a first local feature from tweets is extracted and then passed to LSTM to preserve contextual information, while in parallel model local feature extracted by CNN and contextual feature extracted by LSTM are concatenated and then given to dense layer for classification. For experimental purposes, we have taken only four cities of India, but in the future, this model can be used for location prediction in all cities of India and also can be applied to multiple domains.

References

1. L. Chi, K.H. Lim, N. Alam, C.J. Butler, Geolocation prediction in Twitter using location indicative words and textual features, in *Proceedings of the 2nd Workshop on Noisy User-generated Text (WNUT)* (Osaka, Japan, December 2016), pp. 227–234. http://aclweb.org/anthology/W16-3930
2. K. Indira, E. Brumancia, P.S. Kumar, S.P.T. Reddy, Location prediction on Twitter using machine learning Techniques, in *2019 3rd International Conference on Trends in Electronics and Informatics (ICOEI)* (Tirunelveli, India, 2019), pp. 700–703. https://doi.org/10.1109/ICOEI.2019.8862768
3. R. Ouaret, B. Birregah, E. Soulier, S. Auclair, F. Boulahya, Random forest location prediction from social networks during disaster events, in *2019 Sixth International Conference on Social Networks Analysis, Management and Security (SNAMS)* (Granada, Spain, 2019), pp. 535–540. https://doi.org/10.1109/SNAMS.2019.8931863
4. F. Rehman, M. Malik, M.S. Faisal, Significance of linguistic indicators for location prediction, in *2019 22nd International Multitopic Conference (INMIC)* (Islamabad, Pakistan, 2019), pp. 1–-6. https://doi.org/10.1109/INMIC48123.2019.9022735
5. S. Hemamalini, K. Kannan, S. Pradeepa, Location prediction of twitter user based on friends and followers. Int. J. Pure Appl. Math **118**, 2817–2824 (2018)
6. L.S. Snyder, M. Karimzadeh, R. Chen, D.S. Ebert, City-level geolocation of tweets for real-time visual analytics, in *Proceedings of the 3rd ACM SIGSPATIAL International Workshop on AI*

for Geographic Knowledge Discovery (GeoAI 2019). Association for Computing Machinery (New York, NY, USA, 2019), pp. 85–88. https://doi.org/10.1145/3356471.3365243
7. P. Thomas, L. Hennig, Twitter geolocation prediction using neural networks, in *Language Technologies for the Challenges of the Digital Age. GSCL 2017*. Lecture Notes in Computer Science, ed. by G. Rehm, T. Declerck, vol 10713 (Springers, Cham, 2018). https://doi.org/10.1007/978-3-319-73706-5_21
8. R. Singh, D. Toshniwal, Location prediction using sentiments of twitter users, in *Big Data Analytics and Knowledge Discovery. DaWaK 2018*. Lecture Notes in Computer Science, ed by C. Ordonez , L. Bellatreche, vol 11031 (Springer, Cham, 2018). https://doi.org/10.1007/978-3-319-98539-8_8
9. I. Lourentzou, A. Morales, C. Zhai, Text-based geolocation prediction of social media users with neural networks, in *2017 IEEE International Conference on Big Data (Big Data)* (Boston, MA, 2017), pp. 696–705. https://doi.org/10.1109/BigData.2017.8257985
10. D. Inkpen, J. Liu, A. Farzindar et al., Location detection and disambiguation from twitter messages. J. Intell. Inf. Syst. **49**, 237–253 (2017). https://doi.org/10.1007/s10844-017-0458-3
11. S. Xu, X. Fu, J. Cao et al., Survey on user location prediction based on geo-social networking data. World Wide Web **23**, 1621–1664 (2020). https://doi.org/10.1007/s11280-019-00777-8
12. https://cai.tools.sap/blog/glove-and-fasttext-two-popular-word-vector-models-in-nlp/#:~:text=fastText%20is%20another%20word%20embedding,an%20n%2Dgram%20of%20characters.&text=This%20helps%20capture%20the%20meaning,to%20understand%20suffixes%20and%20prefixes.
13. https://colah.github.io/posts/2015-08-Understanding-LSTMs/#:~:text=Long%20Short%20Term%20Memory%20networks,of%20learning%20long%2Dterm%20dependencies.&text=In%20standard%20RNNs%2C%20this%20repeating,RNN%20contains%20a%20single%20layer

Investigation of SAC-OCDMA FSO System Performance Under Fog Attenuation Conditions Using DW-ZCC Code

S. Boukricha, A. Bouzidi, E.-M. AR Reyouchi, K. Ghoumid, R. Yahiaoui, and O. Elmazria

Abstract The present paper investigates a Free Space Optical (FSO) communication system based on Spectral Amplitude Coding—Optical Code Division Multiple Access (SAC-OCDMA). The main advantage of SAC-OCDMA technology is the ability to provide a unique signature code for each user, which differentiates it from other existing users and thus reduces and eliminates Multiple Access Interference (MAI). Several codes with minimal or zero cross-correlation value are exploited in this regard. Furthermore, free space transmission performance is affected by certain factors, notably attenuation caused by unfavorable weather conditions. This work could be considered as an improvement over previous work that has been published in the literature. This is due to the fact that an appropriate study of the performance of a SAC-OCDMA FSO system has been performed using a specific called Double Weight-Zero Cross-Correlation (DW-ZCC). This code represents the main interest of the present study as it is distinguished by a zero cross-correlation value which is better than that of other codes already proposed in previous works related to SAC-OCDMA FSO systems. The present paper investigates the performance of the proposed system using DW-ZCC code under fog attenuation conditions, by evaluating the bit error rate (BER) and the signal-to-noise ratio (SNR) against many parameters such

Supported by CNRST Morocco Under an Excellence Research Grant

S. Boukricha (✉) · A. Bouzidi · K. Ghoumid
Department of Electronics, Informatics and Telecommunications, ENSAO, Mohamed Premier University, Oujda, Morocco
e-mail: s.boukricha@ump.ac.ma

K. Ghoumid
e-mail: k.ghoumid@ump.ac.ma

E.-M. A. Reyouchi
Department of Telecommunication and Computer Science, Abdelmalek Essaadi University, Tétouan, Morocco

R. Yahiaoui
NanoMedicine Lab, Therapeutic Imagery, Franche-Comte University, Besançon, France

O. Elmazria
Université de Lorraine, CNRS, IJL 54000, Nancy, France

G. Ranganathan et al. (eds.), *Soft Computing for Security Applications*,
Advances in Intelligent Systems and Computing 1397,
https://doi.org/10.1007/978-981-16-5301-8_65

as: transmitted power, data rate and FSO link range. The simulation results, being improved over those already achieved in previous works, prove that the transmitted signal quality is significantly dependent on the attenuation conditions, so that the proposed SAC-OCDMA FSO system provides a maximum range of 2.12 km in clear weather, 1.42 km in light fog, 0.41 km in moderate fog and 0.073 km in heavy fog, at a bit rate of 2.5 Gbits/s and a transmitted power of 14 dBm.

Keywords Free space optics (FSO) · Spectral amplitude code—optical code division multiple access (SAC-OCDMA) · Zero cross-correlation · Double weight-zero cross-correlation (DW-ZCC) code · Fog attenuation · Bit error rate (BER) · Signal-to-noise ratio (SNR)

1 Introduction

Currently, FSO technology is attracting particular attention because of its ability to transmit data with large bandwidth and high data rates [4]. It is an optical technology that ensures the transmission of different types of data, including data, voice and video, at a speed comparable to that of optical fibers while profiting from the advantages of wireless technologies, mainly flexibility of installation and deployment, lack of license requirement, resistance to interferences and high security [14, 20, 27]. An FSO link is based on the propagation of a beam of light generally located in the visible or infrared range, in free space between two points communicating with each other. In particular, a significant security is ensured as the intensity of the beam propagated from the transmitter to the receiver is precisely concentrated in the direction of the receiver, which makes it quite difficult for unauthorized individuals or units to gather the transmitted information [2, 6, 16]. However, FSO links are subject to atmospheric attenuation caused by atmospheric composition and adverse weather conditions (fog, rain, snow...) [25, 26].

Spectral Amplitude Code-Optical Code Division Multiple Access (SAC-OCDMA) has attracted the attention of many researchers in the telecommunications field due to its increased capacity, high security, all-optical equipment as well as asynchronous and simultaneous processing [10, 19, 21].

SAC-OCDMA supports several subscribers sharing the same transmission channel bandwidth, each user being assigned a unique code that constitutes its own signature. This fact makes SAC-OCDMA one of the most appropriate techniques to eliminate Multiple Access Interference (MAI) presenting the main issue that affects the signal performance and resulting from the overlap between interfering users [5, 23].

Codes with low or zero cross-correlation value have been developed by many researchers to encode existing multiple users and to facilitate the identification of the desired user among other interfering users. Among these codes, we can cite Enhanced Double Weight (EDW) [12], Modified Double Weight (MDW) [3], Khazani Syed (KS)[13], Random Diagonal (RD) [9], Zero Cross-Correlation (ZCC) [24] and Double Weight-Zero Cross-Correlation (DW-ZCC) [18].

The objective of this work is the implementation of a code with advantageous characteristics and particularly a perfectly zero cross-correlation value allowing to eliminate as much as possible the multiple access interference between the different users constituting the SAC-OCDMA system. Thus, the use of such a code allows the transmission of signals from different users propagating in free space with a minimum of losses and over longer transmission ranges.

For this purpose, we used the code called DW-ZCC which allowed to achieve better results in comparison with other results already achieved by other researchers using other types of codes, namely the EDW code [1], the MDW code [3] and the ZCC code [17].

Furthermore, the present paper focuses on the performance investigation of a SAC-OCDMA FSO system under fog effect using the DW-ZCC code with optisystem 17 software. An appropriate study has been carried out by investigating the BER and SNR performance of the studied system as a function of several parameters, namely: transmitted power, data rate, distance and attenuation.

Finally, the FSO link and fog attenuation are introduced in Sect. 2. The construction of the DW-ZCC code for a system of 7 users is developed in Sect. 3. The SAC-OCDMA FSO system design is described in Sect. 4. The performance evaluation of the studied system is demonstrated in Sect. 5. Last, the conclusion and perspectives are provided in Sect. 6.

2 FSO Link and Fog Attenuation

An FSO link is mainly composed of a transmitter and a receiver that must obey the line-of-sight (LOS) condition, so that one is placed in front of the other in order to ensure a reliable and secure transmission of information [6].

As depicted in Fig. 1, the transmitter emits a light beam that propagates in the free space constituting the transmission channel and is then received by the opposite receiver. Thanks to free space propagation, FSO systems have the advantage of a high transmission speed compared to fiber optic systems. However, the propagating signal undergoes many attenuations mainly due to scintillation and weather conditions, including: fog, haze, clouds, rain and snow [15].

Atmospheric attenuation varies with visibility, which depends on weather conditions and thus on the particles contained in the atmosphere [7]. The present paper focuses on the fog case and investigates its effect on the performance of the proposed SAC-OCDMA FSO system.

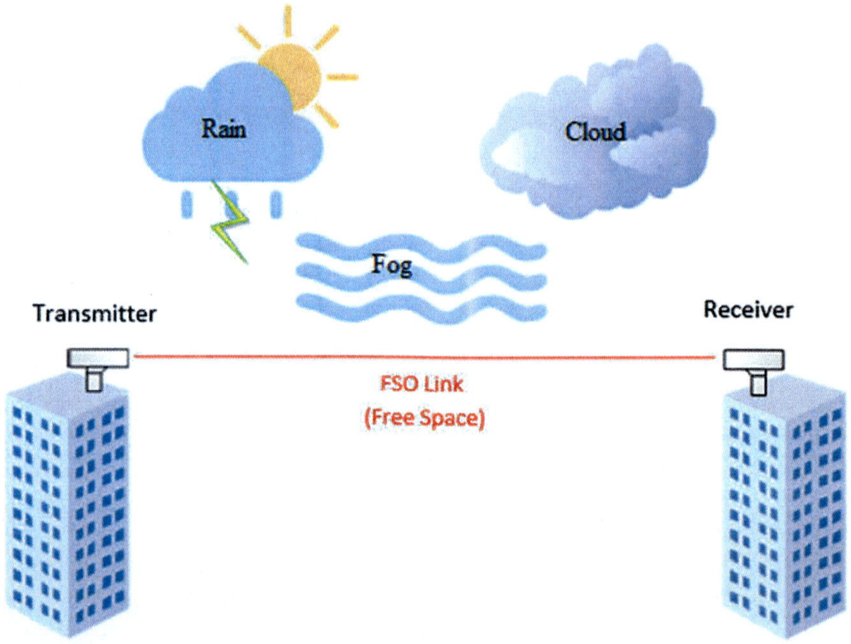

Fig. 1 Point-to-Point FSO link model

2.1 Fog Attenuation

Fog has a considerable attenuating effect on the optical light beam propagation in the atmosphere.

The atmospheric attenuation coefficient depends on several factors, namely: visibility, wavelength and particles size in the air. It can be determined analytically by the Kim and Kruse model, and it is given by [8]:

$$\alpha_{\text{fog}}(\lambda) = \frac{3.91}{V}\left(\frac{\lambda}{550}\text{nm}\right)^{-q} \tag{1}$$

where V is the visibility (km), λ is the operating wavelength (nm) and q is the particles size distribution.

The particles size distribution q, the visibility V and the attenuation values are given in Table 1 [6].

Table 1 Fog attenuation conditions

Weather condition	Visibility (km)	q	Attenuation (dB/km)
Clear weather	18.1	1.3	0.2440
Light fog	2.8	0.788	2.6806
Moderate fog	0.5	0	33.9618
Heavy fog	0.05	0	330.6183

3 DW-ZCC Code Constructions

Codes with low or zero cross-correlation value are largely used in SAC-OCDMA systems because of their effect on reducing existing MAI. They offer the advantage that each user has its own code sequence, and there are no common bits "1" between users. The cross-correlation value between the code sequences of each 2 users is therefore zero.

Double Weight-Zero Cross-Correlation code (DW-ZCC) was proposed by M.Moghaddasi and al. in [18] as a new code structure combining the advantages of two codes MDW (Modified Double Weight) and ZCC (Zero Cross-Correlation).

The DW-ZCC code characteristics are $[L, K, W, \lambda_c]$ where L is code length, K is users number, W is code weight and λ_c is cross-correlation value that is equals to 0. The code length is calculated as $L = K \times W$ and the code matrix as $K \times L$.

The code construction is done as follows [18]:

The location of each pair of adjacent bits "1" can be determined by the following formula:

$$P''(i) = P(i), P(i) + 1 \tag{2}$$

where i = 1,2,3...W/2 and defines the number of existing pairs in code sequences.

For each user k = [1, 2,...k], the ith pair location of adjacent bits "1" is given as:

$$P''_k(i) = \begin{cases} (k \times 2) - 1, k \times 2 & \text{for } i = 1 \\ P''_k(i-1) + [2(K-1) + 2] & \text{for } i > 1 \end{cases} \tag{3}$$

On the other side, the empty spaces between pairs are occupied by zeros whose number is calculated as:

$$Z = 2(K - 1) \tag{4}$$

In this paper, we construct the DW-ZCC code matrix for a number of users K = 7 and a weight W = 4 following the steps and formulas mentioned above.

For K = 7 and W = 4, we get $k = \{1, 2, 3, 4, 5, 6, 7\}$ and $i = \{1, 2\}$. The locations of both pairs in each user's code sequence are determined as follows:

For user 1: $\begin{cases} P''_1(1) = (1 \times 2) - 1, (1 \times 2) = (1, 2) \\ P''_1(2) = P''_1(1) + [2(7-1) + 2] = (15, 16) \end{cases}$

For user 2: $\begin{cases} P_2''(1) = (2 \times 2) - 1, (2 \times 2) = (3, 4) \\ P_2''(2) = P_2''(1) + [2(7-1) + 2] = (17, 18) \end{cases}$

For user 3: $\begin{cases} P_3''(1) = (3 \times 2) - 1, (3 \times 2) = (5, 6) \\ P_3''(2) = P_3''(1) + [2(7-1) + 2] = (19, 20) \end{cases}$

For user 4: $\begin{cases} P_4''(1) = (4 \times 2) - 1, (4 \times 2) = (7, 8) \\ P_4''(2) = P_4''(1) + [2(7-1) + 2] = (21, 22) \end{cases}$

For user 5: $\begin{cases} P_5''(1) = (5 \times 2) - 1, (5 \times 2) = (9, 10) \\ P_5''(2) = P_5''(1) + [2(7-1) + 2] = (23, 24) \end{cases}$

For user 6: $\begin{cases} P_6''(1) = (6 \times 2) - 1, (6 \times 2) = (11, 12) \\ P_6''(2) = P_6''(1) + [2(7-1) + 2] = (25, 26) \end{cases}$

For user 7: $\begin{cases} P_7''(1) = (7 \times 2) - 1, (7 \times 2) = (13, 14) \\ P_7''(2) = P_7''(1) + [2(7-1) + 2] = (27, 28) \end{cases}$

The number of zeros between the pairs is calculated by (4) and is equal to 12. Therefore, the resulting matrix is as follows:

$$C_7 = \begin{bmatrix} 1\,1\,0\,0\,0\,0\,0\,0\,0\,0\,0\,0\,0\,0\,1\,1\,0\,0\,0\,0\,0\,0\,0\,0\,0\,0\,0\,0 \\ 0\,0\,1\,1\,0\,0\,0\,0\,0\,0\,0\,0\,0\,0\,0\,0\,1\,1\,0\,0\,0\,0\,0\,0\,0\,0\,0\,0 \\ 0\,0\,0\,0\,1\,1\,0\,0\,0\,0\,0\,0\,0\,0\,0\,0\,0\,0\,1\,1\,0\,0\,0\,0\,0\,0\,0\,0 \\ 0\,0\,0\,0\,0\,0\,1\,1\,0\,0\,0\,0\,0\,0\,0\,0\,0\,0\,0\,0\,1\,1\,0\,0\,0\,0\,0\,0 \\ 0\,0\,0\,0\,0\,0\,0\,0\,1\,1\,0\,0\,0\,0\,0\,0\,0\,0\,0\,0\,0\,0\,1\,1\,0\,0\,0\,0 \\ 0\,0\,0\,0\,0\,0\,0\,0\,0\,0\,1\,1\,0\,0\,0\,0\,0\,0\,0\,0\,0\,0\,0\,0\,1\,1\,0\,0 \\ 0\,0\,0\,0\,0\,0\,0\,0\,0\,0\,0\,0\,1\,1\,0\,0\,0\,0\,0\,0\,0\,0\,0\,0\,0\,0\,1\,1 \end{bmatrix}$$

The code sequences and assigned wavelengths of each user are presented on Table 2.

Table 2 Code sequences of DW-ZCC code with $K = 7$ and $W = 4$

User	Code sequence	Assigned wavelengths
1	1100000000000011000000000000	$\lambda_1, \lambda_2, \lambda_{15}, \lambda_{16}$
2	0011000000000000110000000000	$\lambda_3, \lambda_4, \lambda_{17}, \lambda_{18}$
3	0000110000000000001100000000	$\lambda_5, \lambda_6, \lambda_{19}, \lambda_{20}$
4	0000001100000000000011000000	$\lambda_7, \lambda_8, \lambda_{21}, \lambda_{22}$
5	0000000011000000000000110000	$\lambda_9, \lambda_{10}, \lambda_{23}, \lambda_{24}$
6	0000000000110000000000001100	$\lambda_{11}, \lambda_{12}, \lambda_{25}, \lambda_{26}$
7	0000000000001100000000000011	$\lambda_{13}, \lambda_{14}, \lambda_{27}, \lambda_{28}$

4 SAC-OCDMA FSO System Description

Figure 2 describes the design of the proposed SAC-OCDMA FSO system. A CW laser array emits 28 wavelength values corresponding to the DW-ZCC code length ($L = K \times W$ for $K = 7$ and $W = 4$). They are gathered using a multiplexer (28×1) and then demultiplexed using a demultiplexer (1×28) before the encoding process occurs. The encoding process is carried out using an encoder with the purpose of assigning a unique code sequence to each user. The encoder for each user consists of a 4 inputs WDM multiplexer where each input corresponds to a specific wavelength value constituting the code sequence corresponding to the user concerned. It is worth noting that the insertion losses are taken into consideration and are of 2 dB. Next, the signal is modulated using a Mach–Zehnder modulator with the binary data generated by a Pseudo-Random Bit sequence generator. An optical power combiner is then utilized to combine the coded and modulated signals to be transmitted through the FSO channel under fog conditions.

At the receiver, the encoded signals are divided using an optical power splitter and the signal corresponding to each user is decoded using a series of Fiber Bragg gratings (FBGs) with the same wavelength values as the encoder. Afterward, the decoded signals are detected by a PIN photodetector allowing the optical-electrical conversion and then filtered with a low-pass Bessel filter to remove undesired signals. Finally, a BER analyzer is introduced at the output of the low-pass Bessel filter in order to visualize the performance of the studied system by displaying the BER value.

5 Performance Analysis and Simulation Results

The simulation is carried out on SAC-OCDMA FSO system employing DW-ZCC code with seven users and weight $W = 4$. The FSO transmitter and receiver aperture diameter are set to 5 cm and 30 cm, respectively. The transmitter divergence angle is about 1 mrad. The insertion losses of the FSO channel, and the multiplexer and demultiplexer are, respectively, 1 dB and 2 dB.

The performance analysis is performed by evaluating the effect of many parameters, mainly: transmitted power, data rate, FSO range and fog attenuation.

Figure 3 presents the signal and noise power variation versus the FSO range for different transmitted power values. The primary purpose of this simulation is to examine the impact of transmitted power and link range on the acquired signal power and generated noise power after crossing the free space. It is clearly indicated that increased transmitted power leads to improved system performance in terms of received signal power. In contrast, increasing the distance of the FSO link decreases the received signal power because of atmospheric attenuation due to fog. The variation of the noise power corresponding to all transmitted power values is almost stable around −94 dBm for distances from 1 km to 3 km. At 1 km, the signal power for a transmitted

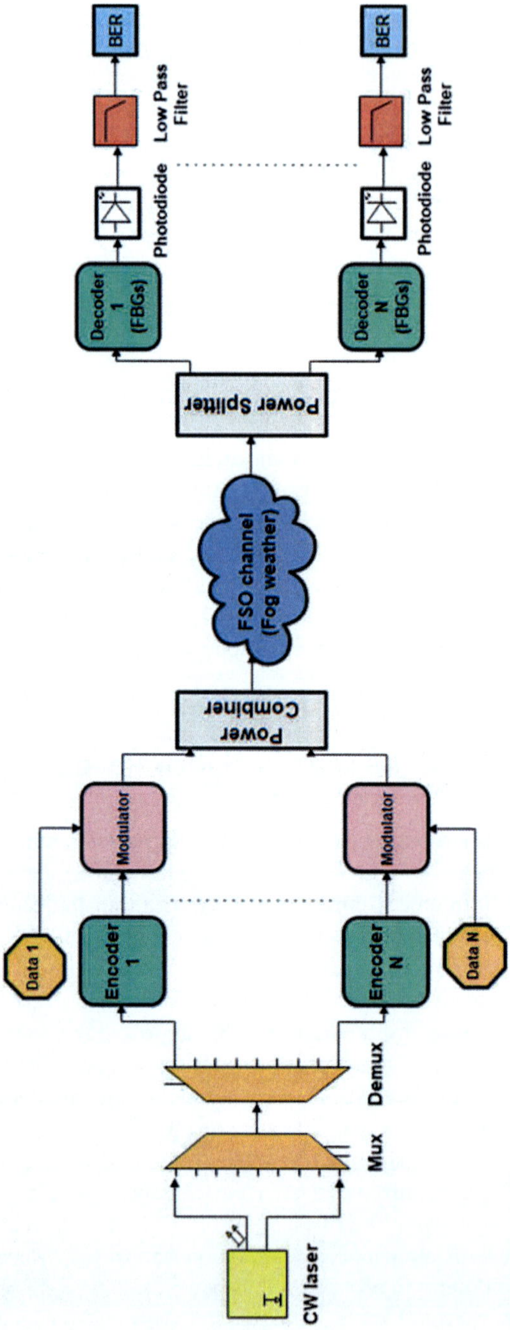

Fig. 2 SAC-OCDMA FSO system under fog conditions using DW-ZCC code

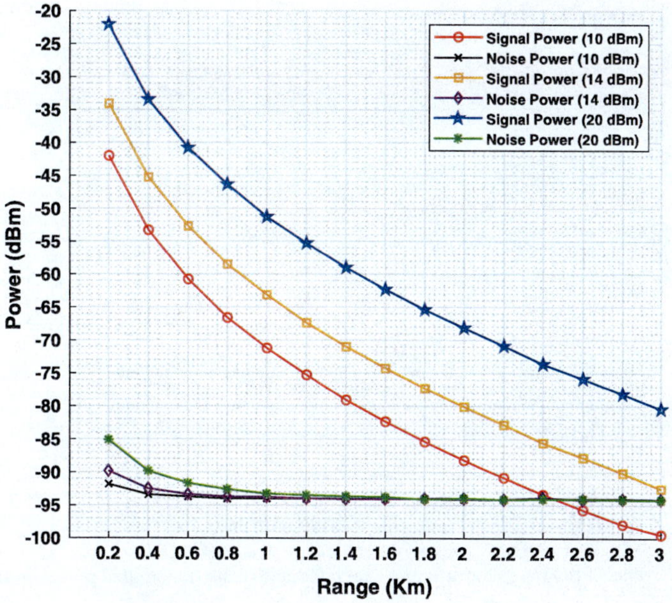

Fig. 3 Signal and noise power versus FSO range for different transmitted power values under light fog (attenuation = 2.6806 dB/km)

power of 10 dBm, 14 dBm and 20 dBm is −71.22 dBm, −63.18 dBm and −51.28 dBm, respectively.

Figure 4 shows the BER variation as a function of FSO range under light fog effect for three data rate values: 2.5 Gbps, 5 Gbps and 10 Gbps. The transmitted power and attenuation are set to 20 dBm and 2.6806 dB/km, respectively. It is clearly observed that as the data rate and the distance traveled by the signal increase, the signal performance deteriorates due to the attenuation occurring in free space.

The maximum distance that could be reached at 2.5 Gbps, 5 Gbps and 10 Gbps is 2.27 km, 2.04 km and 1.8 km, respectively. The corresponding BER values are, respectively, 1.39E− 09, 1.91E−09 and 1.44E−09.

These results show that the system performs better at 2.5 Gbps than at 5 Gbps and 10 Gbps. This is due to the minimization of the pulse width and thus the overlap between bits 1 and 0 that generates signal dispersion at high data rates [11, 22].

Figures 5 and 6 illustrate BER performance and SNR variation against the FSO transmission distance for different fog attenuation conditions, respectively. The attenuation corresponding to each fog condition is set according to that presented in Table 1. The data rate and transmitted power are set to 2.5 Gbps and 14 dBm, respectively.

The obtained results shown in Fig. 5 indicate that the maximum transmission distance achieved with an acceptable BER depends on fog conditions, so that in clear

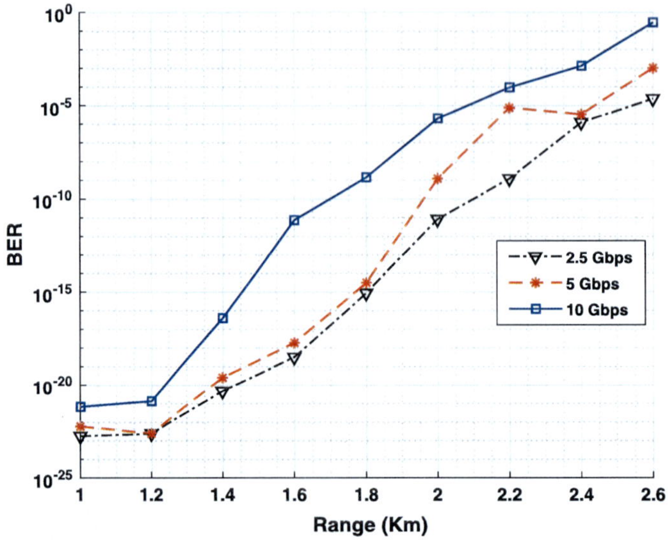

Fig. 4 BER versus FSO range for different data rates at 20 dBm transmitted power under light fog

weather, the SAC-OCDMA FSO system performs well for up to 2.12 km with BER= 5.52E−09, while in light fog, moderate fog and heavy fog, the maximum distance is decreased to 1.42 km, 0.41 km and 0.073 km with corresponding BERs of 3.78E−09, 1.82E−09 and 1.04E−09, respectively. This fact ensures that the system efficiency is strongly dependent on atmospheric attenuation conditions.

Similarly, Fig. 6 shows that the SNR itself depends on atmospheric conditions and degrades with transmission distance. This is because the signal propagating in the air undergoes attenuation and dispersion that are significant at longer distances and higher attenuation. At 0.4 km, the SNR is approximately 48.96 dBm and 23.64 dBm under clear weather and moderate fog, respectively. While, at 1 km the SNR is about 35.32 dBm under clear weather and −5.96 dBm under moderate fog.

Moreover, the SNR corresponding to the maximum distances deduced from Fig. 5 is 22.44 dBm for clear weather at 2.12 km, 22.6 dBm for light fog at 1.42 km, 22.66 dBm for moderate fog at 0.41 km and 23.62 dBm for heavy fog at 0.073 km.

6 Conclusion

In the present study, the performance of a SAC-OCDMA FSO system using DW-ZCC code with zero cross-correlation value has been effectively investigated under fog weather conditions. The results of performed simulations have shown that the system performance with regard to BER and SNR depends on many parameters such as data rate, transmitted power and fog attenuation. An increase in transmitted power

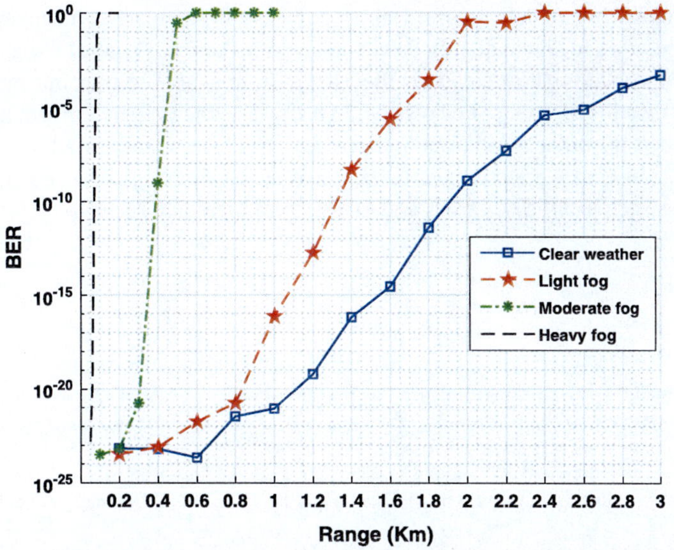

Fig. 5 BER versus FSO range at 2.5 Gbps and 14 dBm under different fog attenuation conditions

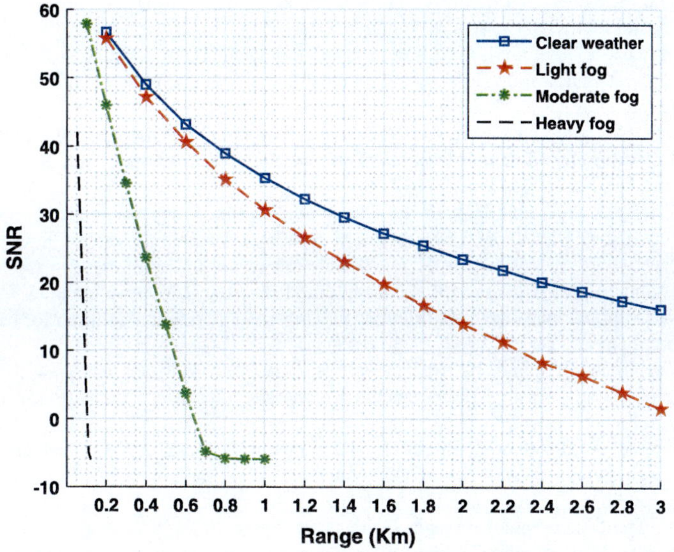

Fig. 6 SNR versus FSO range at 2.5 Gbps and 14 dBm under different fog attenuation conditions

leads to an improvement in received signal power, while an increase in data rate leads to a deterioration in signal quality due to dispersion. Analysis against fog attenuation conditions showed that in clear weather; the system is able to transmit over longer distances than in moderate or heavy fog. In the future, the system performance could be examined for other weather conditions, i.e., rain, snow and turbulence. It could also be improved to support longer transmission distances using advanced modulation techniques such as DPSK, QPSK and QAM, with high data rates.

References

1. A. Aissaoui, L. Hacini, Performance comparison of different SAC OCDMA-FSO detection techniques in presence of atmospheric losses, in *2019 6th International Conference on Image and Signal Processing and their Applications (ISPA)* (IEEE, 2019), pp. 1–5
2. E.-M. Ar reyouchi and K. Ghoumid. Technical accuracy based on efficiency algorithm for measuring standing wave ratio in wireless sensor network. Int. J. Commun. Antenna Propag. **9**(2), 137–143 (April 2019)
3. S. Aljunid, M. Ismail, A. Ramli, B.M. Ali, M.K. Abdullah, A new family of optical code sequences for spectral-amplitude-coding optical CDMA systems. IEEE Photonics Technol. Lett. **16**(10), 2383–2385 (2004)
4. S. Bloom, E. Korevaar, J. Schuster, H. Willebrand, Understanding the performance of free-space optics. J. Opt. Netw. **2**(6), 178–200 (2003)
5. S. Boukricha, K. Ghoumid, S. Mekaoui, E. Ar-Reyouchi, H. Bourouina, R. Yahiaoui, SAC-OCDMA system performance using narrowband Bragg filter encoders and decoders. SN Appl. Sci. **2**(6), 1–9 (2020)
6. Y. Chatea, M. Hammouti, E.-M. Ar reyouchi, K. Ghoumid, Downlink and uplink message size impact on round trip time metric in multi-hop wireless mesh networks. Int. J. Adv. Comput. Sci. Appl. **8**(3), 223–229 (March 2017)
7. M.K. El-Nayal, M.M. Aly, H.A. Fayed, R.A. AbdelRassoul, Adaptive free space optic system based on visibility detector to overcome atmospheric attenuation. Results Phys. **14**, 102392 (2019)
8. H.A. Fadhil, A. Amphawan, H.A. Shamsuddin, T.H. Abd, H.M. Al-Khafaji, S. Aljunid, N. Ahmed, Optimization of free space optics parameters: an optimum solution for bad weather conditions. Optik **124**(19), 3969–3973 (2013)
9. H.A. Fadhil, S.A. Aljunid, R. Ahmad, Performance of random diagonal code for OCDMA systems using new spectral direct detection technique. Opt. Fiber Technol. **15**(3), 283–289 (2009)
10. K. Ghoumid, A. Ghadban, S. Boukricha, R. Yahiaoui, S. Mekaoui, M. Raschetti, C. Lepers et al., Spectral coded phase bipolar OCDMA technological implementation thanks to low index modulation filters. Telecommun. Syst. **73**(3), 433–441 (2020)
11. D. Goff, /textitFiber Optic Reference Guide (CRC Press, 2002)
12. S. Kumawat, M.R. Kumar, Generalized optical code construction for enhanced and modified double weight like codes without mapping for SAC-OCDMA systems. Opt. Fiber Technol. **30**, 72–80 (2016)
13. S. Kumawat, R.K. Maddila, Development of odd weight code construction for ks code with ex-or detection for sac-ocdma system, in *2019 IEEE International Conference on Advanced Networks and Telecommunications Systems (ANTS)* (IEEE, 2019), pp. 1–5
14. A.K. Majumdar, Free-space optical (FSO) platforms: Unmanned aerial vehicle (UAV) and mobile, in *Advanced free space optics (FSO)* (Springer, Berlin, 2015), pp. 203–225
15. A. Malik, P. Singh, Free space optics: current applications and future challenges. Int. J. Opt. **2015** (2015)

16. H. Manor, S. Arnon, Performance of an optical wireless communication system as a function of wavelength. Appl. Opt. **42**(21), 4285–4294 (2003)
17. L. Mesri, A. Djebbari, Performance limits of FSO based SAC-OCDMA system under weather conditions. J. Opt. Commun. **1**(ahead-of-print) (2018)
18. M. Moghaddasi, S. Seyedzadeh, I. Glesk, G. Lakshminarayana, S.B.A. Anas, Dw-zcc code based on SAC-OCDMA deploying multi-wavelength laser source for wireless optical networks. Opt. Quantum Electron. **49**(12), 1–14 (2017)
19. S. Mostafa, A.E.N.A. Mohamed, F.E Abd El-Samie, A.N.Z. Rashed, Performance evaluation of SAC-OCDMA system in free space optics and optical fiber system based on different types of codes. Wirel. Personal Commun. **96**(2), 2843–2861 (2017)
20. J. Perez, S. Zvanovec, Z. Ghassemlooy, W.O. Popoola, Experimental characterization and mitigation of turbulence induced signal fades within an ad hoc FSO network. Opt. Exp. **22**(3), 3208–3218 (2014)
21. P.R. Prucnal, *Optical Code Division Multiple Access: Fundamentals and Applications* (CRC Press, 2018)
22. R.K.Z. Sahbudin, M. Kamarulzaman, S. Hitam, M. Mokhtar, S.B.A. Anas, Performance of sac OCDMA-FSO communication systems. Optik-Int. J. Light Electron Opt. **124**(17), 2868–2870 (2013)
23. R. Sahbudin, M. Abdullah, M. Mokhtar, Performance improvement of hybrid subcarrier multiplexing optical spectrum code division multiplexing system using spectral direct decoding detection technique. Op. Fiber Technol. **15**(3), 266–273 (2009)
24. S. Singh, D. Singh, B. Utreja, Performance enhancement of a FSO-based hybrid SAC-OCDMA system using zero cross correlation code. IJECT **5**(2), 108–112 (2014)
25. B.R. Strickland, M.J. Lavan, E. Woodbridge, V. Chan, Effects of fog on the bit-error rate of a free-space laser communication system. Appl. Opt. **38**(3), 424–431 (1999)
26. Y. Tan, J.Z. Guo, Study of channel model of free-space laser communications system, in *Free-Space Laser Communications* (vol. 5892, International Society for Optics and Photonics, 2005), p. 58921S
27. H. Willebrand, B.S. Ghuman, *Free space optics: enabling optical connectivity in today's networks* (SAMS Publishing, 2002)

An Approach to Predict the Location of Crime Using Machine Learning

Siddhartha Banerjee, Surajit Giri, Debayan Das, and Pramit Kumar Mandal

Abstract In order to prevent crimes, it is very important to recognize the patterns of the criminal activities. If the crime patterns of different geological points of a city are known to the police and detective agencies then they can work more efficiently in order to solve the problem. The proposed methodology provides an automated technique to predict the geological location of a crime depending upon the date, time, and type of the crime from past crime behavior. Linear regression and support vector regression algorithm are used for this purpose. The proposed methodology is tested on a dataset containing the crime records of Indore city in the month of February and March 2018. The result obtained using linear regression and support vector regression algorithms is compared, and it is found that support vector regression provides better result.

Keywords Crime prediction · Machine learning · Linear regression · Support vector regression · RMS error

1 Introduction

Socioeconomic conditions as well as the life style of people get affected by the criminal activities that are very familiar to every part of the world today. The criminal activities include different types of events like pick pocketing, robberies, rapes, murders, smuggling, etc. The life of common people is at risk due to these criminal activities.

People feel very insecure with all the antisocial activities happening around them. The works of the criminals are very organizing now, and they hardly left any clue behind to trace them. It is also observed that in some cases the criminals have link up with national and international agencies. Thus, dealing with such criminal activities becomes a big issue to the government.

S. Banerjee (✉) · S. Giri · D. Das · P. K. Mandal
Ramakrishna Mission Residential College (Autonomous), Narendrapur, Kolkata, West Bengal, India

G. Ranganathan et al. (eds.), *Soft Computing for Security Applications*,
Advances in Intelligent Systems and Computing 1397,
https://doi.org/10.1007/978-981-16-5301-8_66

The scientific study for understanding the behaviour of crime, nature of crime, and deriving some anticrime strategies by indentifying the characteristics of crime is known as criminology. Crime analysis is a sub-branch of criminology. It studies the pattern of the crimes and tries to find indicators of the events. But presence of huge amount of data and due to increasing crime rate make it impossible for security authority and their personnel to manually analyze these data and find the hidden secrets within these data [1–4]. Application of data science technology appeared as promising and reliable solution in last few years to handle business strategies, smart decision making, marketing and weather and environment forecasting [5, 6]. The fact is that every incident holds a valuable information that can be used for forecasting the occurrence of the incident in future, i.e., past is the true providence of future [7]. The crime is also predictable since human nature is not reversible [8]. Human nature generally evokes periodically to perform the same event. With the advancement of big data and easy to implement algorithms for analysis of data, prediction of crime is a growing field of study.

In this paper, our focus is to study the crime patterns and behaviors of the crime and predict the geological location of the crime. The following types of crimes are considered in our study:

- Robbery (Act 379)
- Gambling (Act 13)
- Accident (Act 279)
- Violence (Act 323)
- Murder (Act 302)
- Kidnapping (Act 363)

To predict the geological location of the crime, it is necessary to provide the date, time, and type of crime in the proposed model. The proposed methodology used linear regression and support vector regression algorithms to predict future crime pattern from past crime behavior. Testing has been done on a dataset containing the crime records of Indore city in the month of February and March 2018. The result obtained using linear regression and support vector regression algorithms is compared, and it is found that support vector regression provides better result.

2 Related Work

Many researches have been carried out to predict the occurrence of crime before its commencement. To deal with criminal activities, the prediction of crime is essential. The government across the globe provides priority for crime prediction. To identify the hotspot of crimes, different projects are carried out by different government, non-government, and academic organizations. Researches have also been carried out to study the relationship between socioeconomic variables (e.g., education level, income level, and unemployment) and criminal activities.

Bogomolov et al. [9] analyze demographic information and the usage of mobile network infrastructure to predict the hotspot area of criminal activities. Decision tree and Naive Bayesian classifier are used by Rizwan [10] on two datasets for prediction of crime category. The study achieves 83.95% accuracy. Different machine learning algorithms have been applied by Shojaee et al. [11] to predict the crime category. Among these techniques, K-nearest neighbor algorithm achieves best performance with 89.50% accuracy. The feature selection process is improved by using chi-square method. The crime pattern is studied by Wang et al. [12] by proposing a machine learning agent known as 'Series Finder,' which tries to find crime committed by same offender or groups of offenders. Yadav et al. [13] use clustering method to study the patterns of criminal behavior and geographic criminal history. The roles of social networks have also been studied to predict the criminal activities. Sentiment analysis has been carried out on Twitter data by Chen et al. [14] to predict crime in real time. The concentration of crime occurrences and large-scale hotspots locations are also identified by analyzing these Twitter data.

Most of these researches focused on classification of different type of crimes and prediction of crime events. But these predictions do not include any location information where these predicted crimes may occur. Thus in this research, our target is not just to predict the occurrence of crime events but also to predict the location of the crime.

3 Material and Methods

To analyze the crime pattern and to predict the location of crime, we use the crime records of Indore city which is available publicly on the Web site of the Indore city police. It provides information of crime incidents reported in Indore in the months of February and March of 2018. This dataset contains 2091 crimes information with different attributes as shown in Table 1. Before performing the analysis and prediction on crime data, a preprocessing step is performed. In this step, the Timestamp attribute is processed to obtain ten different attributes, i.e., 'year,' 'month,' 'day,' 'hour,' 'day of year,' 'week,' 'week of year,' 'day of week,' 'weekday,' and 'quarter.' The location of crime (i.e., latitude and longitude of the crime) is predicted from these ten different attributes and the type of crime. Linear regression [15] and support vector regression [16] are used as prediction model.

Linear regression is the simplest approach for statistical learning. In this model, the variable which is to be estimated is called the dependent variable and the other variables are called independent variable. There are two types of linear regression model, namely single variable and multivariable linear regression. In this experiment, multivariable linear regression model is used where dependent variable Y is estimated from a set of independent variables $X_1, X_2, \ldots, X_p$ using Eq. (1).

$$Y = \beta_0 + \beta_1 X_1 + \beta_2 X_2 + \cdots + \beta_p X_p \tag{1}$$

Table 1 Dataset attributes description

Attributes	Description
Timestamp	A timestamp when the given crime occurred
ACT 379	Signifies if the crime is under act 379, i.e., robbery
ACT 13	Signifies if the crime is under act 13, i.e., gambling
ACT 279	Signifies if the crime is under act 279, i.e., accident
ACT 323	Signifies if the crime is under act 323, i.e., violence
ACT 363	Signifies if the crime is under act 363, i.e., murder
ACT 202	Signifies if the crime is under act 202, i.e., kidnapping
Latitude	It signifies the latitude of the location of the crime
Longitude	It signifies the longitude of the location of the crime

where β_i indicates the coefficient of independent variable X_i.

Support vector regression (SVR) algorithm works with continuous values. In simple regression model, error rate is minimized. But in SVR, the error is fitted within a certain threshold limit. As shown in Fig. 1, SVR considers the points that are within a boundary line (shown in red color line). In simple terms, the SVR takes only those points which have least error rate. Thus, it gives a better fitting model.

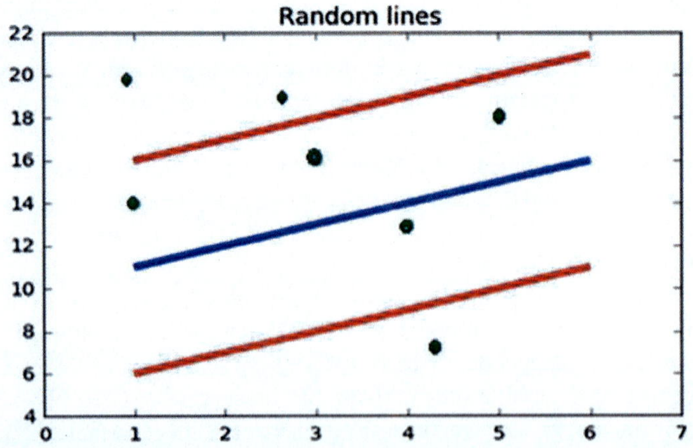

Fig. 1 SVR boundary with data points

In this two models, latitude and longitude of the crime are considered as dependent variable whereas ten attributes which are obtained from Timestamp and type of crime are considered as independent variable. Thus, to predict location of a particular type of crime, the ten attributes derived from Timestamp and type of the crime should be given in these models. These models will predict the location of crime by providing lattitude and longitude as output.

4 Results and Discussion

4.1 Crime Pattern Analysis

In this study, we analyzed the total number of crimes in different time of the day for the months of February and March of 2018. In this plot (Fig. 2), y-axis shows the total number of crimes and x-axis on the graph shown the hour of the day in 24 h format. It is clear from the plot that the most of the crime occurs after evening (after 6 p.m. onward). From this information, it can be concluded that, the prediction of crime events is very necessary after 6 p.m.

We also study the rate of different kinds of crimes occurs in these two months. The plot is shown in Fig. 3. In the plot, x-axis represents different types of crimes and y-axis shows the total number of crimes during these two months. From this plot, we can conclude that the violence, accident, and robbery occur with high frequency but kidnapping event occurs with low rate.

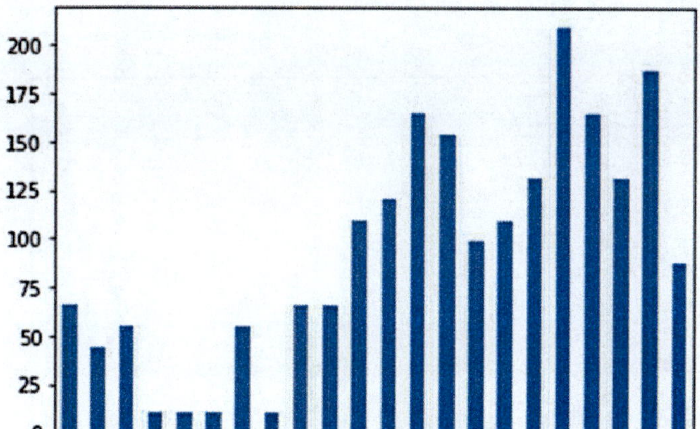

Fig. 2 Criminal activities occur in the different hour of the day

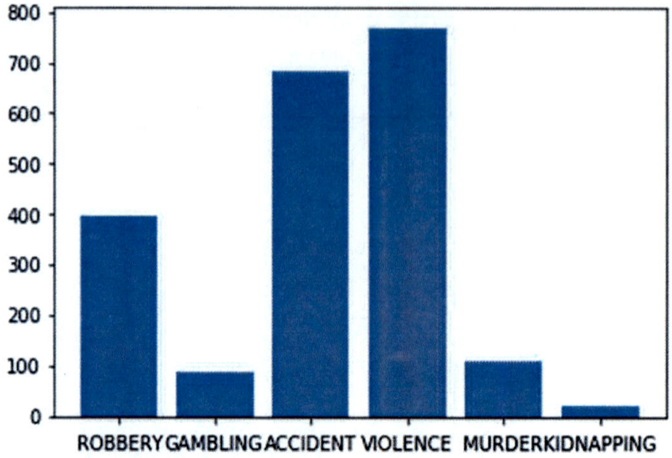

Fig. 3 Rate of diiferent types of criminal activities

4.2 Crime Location Prediction

The linear regression and support vector regression models are used to predict the geological locations, i.e., latitude and longitude of the crimes. The locations of the crime in a particular month of a year at a specific time are computed for different types of crime using linear regression and support vector regression. Figure 4 shows the predicted locations for different types of crime for the month of March, 2018 at 22 h using linear regression as a representative. The same investigation is also done using support vector regression and shown in Fig. 5.

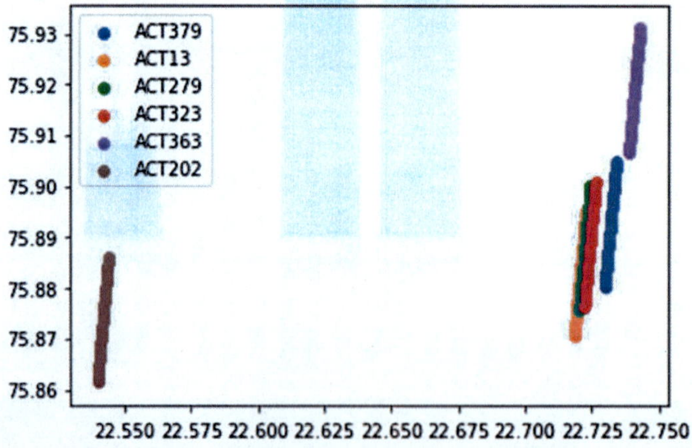

Fig. 4 Crime area prediction by linear regression

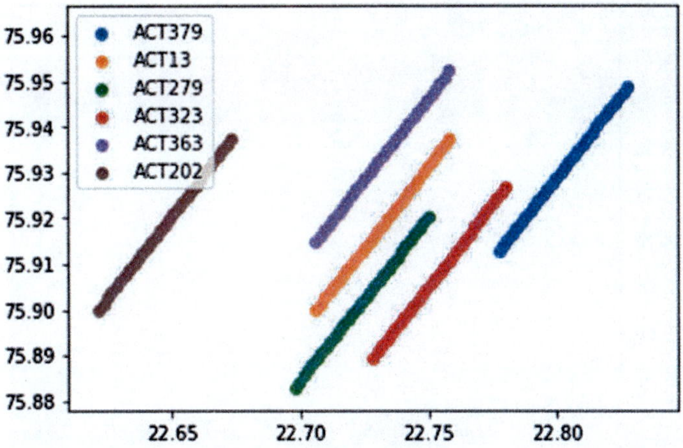

Fig. 5 Crime area prediction by support vector regression

Accuracy of the prediction is measured by dividing the dataset into two parts, namely the training dataset and the testing dataset. The records with the features as well as target level constitute the training whether testing dataset contains only the features. Machine learning model is used to predict the target label. The test dataset contains 20% of the original dataset, and rest is used for training purpose. In testing phase, the location for a particular crime is predicted using linear regression and support vector regression and compared with original information. Figure 6 shows the predicted location and original location for robbery (ACT 379) as an example. Similarly, Fig. 7 shows the same information using support vector regression.

Error estimation in prediction of location of crime (i.e., latitude and longitude) is done for the testing dataset for these two models. The root mean square error (RMS)

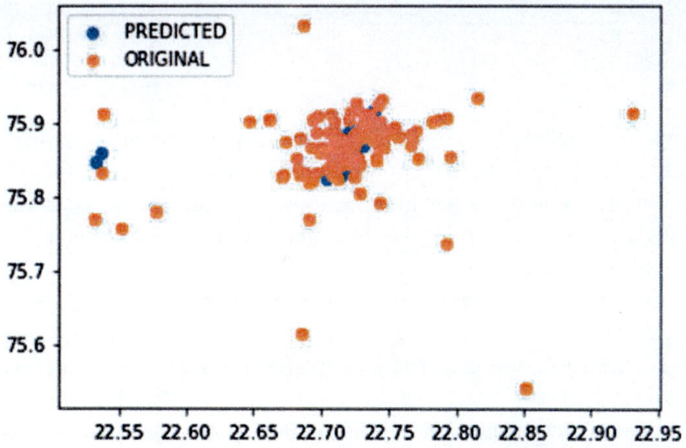

Fig. 6 Original and predicted location for robbery (ACT 379) using linear regression

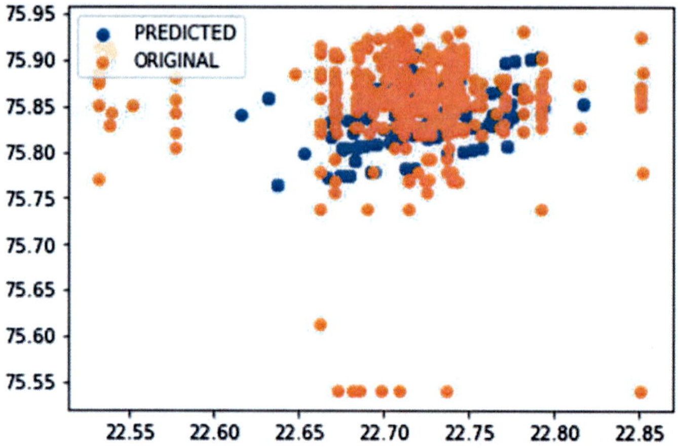

Fig. 7 Original and predicted location for robbery (ACT 379) using support vector regression

Table 2 Error estimation for latitude and longitude

Model	Latitude		Longitude	
	RMS error	MAE error	RMS error	MAE error
Linear regression	0.0024	0.0493	0.0367	0.1917
Support vector regression	0.0007	0.0274	0.0004	0.0207

and mean absolute error (MAE) are calculated using predicted and original values on latitude and longitude. The error estimation is shown in Table 2. It can be observed from the table that support vector regression better estimates the crime location over linear regression model.

5 Conclusions

Data mining and machine learning have become a vital part of the crime detection and prevention nowadays. This paper proposed an automated methodology to predict the location of the crime using machine learning algorithm from past crime records. The effectiveness and accuracy of two machine learning algorithms, namely linear regression and support vector machines are evaluated using dataset containing the crime records of Indore city. From the experimental results, it can be concluded that support vector regression works well and has more accuracy if the data points lie within the clustering region as it has low RMS error value. But if the data points are located near the boundaries, linear regression model also provides results with high accuracy. This paper predicts the location of crimes by providing latitude and longitude of the location from the past crime records, i.e., from Timestamp and type of past crimes events. This work can be exended by considering other attributes like unemployment rate of the area, literacy rate of the area, etc.

References

1. H. Hassani, X. Huang, E.S. Silva, M. Ghodsi, A review of data mining applications in crime. Stat. Anal. Data Min. ASA Data Sci. J. **9**(3), 139–154 (2016)
2. E. Ahishakiye, I. Niyonzima, R. Wario, A performance analysis of business intelligence techniques on crime prediction. Int. J. Comput. Inf. Technol. **6**(2), 84–90 (2017)
3. S. Shojaee, A. Mustapha, F. Sidi, M.A. Jabar, A Study on classification learning algorithms to predict crime status. Int. J. Digit. Content Technol. its Appl. **7**(9), 361–369 (2013)
4. A.G. Ferguson, Big data and predictive reasonable suspicision. Univ. PA. Law Rev. **163**(2), 329–346 (2012)
5. U. Saeed, M. Sarim, A. Usmani, A. Mukhtar, A.B. Shaikh, S.K. Raffat, Application of machine learning algorithms in crime classification and classification rule mining. Res. J. Recent Sci. Res. J. Recent Sci. **4**(3), 106–114 (2015)
6. D. Usha, K. Rameshkumar, A Complete Survey on application of frequent pattern mining and association rule mining on crime pattern mining. Int. J. Adv. Comput. Sci. Technol. **3**(4), 264–275 (2014)
7. S. Schneider, Predicting crime: a review of the research, School of Justice Studies, Ryerson University, Department of Justice (2002)
8. W.L. Perry, B. McInnes, C.C. Price, S.C. Smith, J.S. Hollywood, Predictive policing: the role of crime forecasting in law enforcement operations, National Institute of Justice (2013)
9. A. Bogomolov, B. Lepri, J. Staiano, N. Oliver, F. Pianesi, A. Pentland, Once upon a crime: towards crime prediction from demographics and mobile data, in *Proceedings of the 16th İnternational Conference on Multimodal İnteraction* (ACM, 2014)
10. I. Rizwan, M. Masrah, A. Mustapha, P. Panahy, K. Nasim, An experimental study of classification algorithms for crime prediction. Indian J. Sci. Technol. **6**, 4219–4225 (2013)
11. S. Shojaee, A. Mustapha, F. Sidi, M.A. Jabar, A study on classification learning algorithms to predict crime status. Int. J. Digital Content Technol. Its Appl. **7**(9) (2013)
12. T. Wang, C. Rudin, D. Wagner, R. Sevieri: Learning to detect patterns of crime: machine learning and knowledge discovery in databases, in *European Conference, ECML PKDD* (Prague, Czech Republic, 2013)
13. R. Yadav, S.K. Savita: Analysis of criminal behavior through clustering approach. Int. J. Comput. Sci. Eng. **6** (2018)
14. X. Chen, Y. Cho, S.Y. Jang, Crime prediction using Twitter sentiment and weather in *Systems and Information Engineering Design Symposium* (2015)
15. D.C. Montgomery, E.A. Peck, G.G. Vining, *Introduction to Linear Regression Analysis*, 5th edn (Wiley, 2012)
16. M. Awad, R. Khanna: *Support Vector Regression, Efficient Learning Machines* (Springer, Berlin, 2015)

Privacy-Preserving in Fingerprint Templates Security Using Combinatorial Logic Circuits

R. Ashoka Rajan and P. Kumaran

Abstract Biometric recognition is one of the most popular person authentication schemes than any other recognition approaches like password or token-based authentication method. Biometrics uses either physiological or behavioral characteristics of a person which is unique in nature. Although biometrics provides lot of advantages, protecting its template database information is highly a challenging task. Fingerprint trait is one of the compact and convenient traits embedded in many digital devices for pattern recognition. In this paper, a novel fingerprint-based template transformation technique is proposed which consist of various important phases such as finding distance between minutiae points, customizing distance matrix, formulating combinatorial logic gate circuit, and disk scheduling that eventually provides multilevel template security. The experimental analysis clearly shows that the proposed method provides better security and improved equal error rate (EER) that is 2.3% when compared with other state-of-art methods.

Keywords Biometric security · Fingerprint template security · Combinatorial logic circuit · Biometric template transformation · Template protection · Fingerprint image transformation · Distance matrix

1 Introduction

The biometric user need not to remember the trait information as like password where one has to remember everything. Biometric identities cannot be shared with anyone. This is the simplicity and convenience of biometrics. Features like palm print, iris, and face recognition are used in many application [1, 2]. However, fingerprint scanners are very compact and available everywhere even in smartphones and laptops.

R. Ashoka Rajan (✉)
School of Computer Science and Engineering, Vellore Institute of Technology, Chennai, India

P. Kumaran
Department of Computer Science and Engineering, SRM University-AP, Andhra Pradesh, India
e-mail: kumaran.p@srmap.edu.in

G. Ranganathan et al. (eds.), *Soft Computing for Security Applications*, Advances in Intelligent Systems and Computing 1397,
https://doi.org/10.1007/978-981-16-5301-8_67

Hence, the proposed method utilized the fingerprint as input and transformed the original features before storing the information into the template database [3]. Person authentication is an act of identifying or verifying user's unique identity to access the system. Authentication system plays an imperative role to keep information private and secure from unauthorized parties. The evolution of user authentication starts with the concept of personal identification number (PIN), passwords, tokens, etc. In password authentication, security is entirely based on the confidentiality and the strength of the password. In order to reinforce the password, different alphabets, numbers, and symbol combinations were used for various applications [4]. A password has a relatively a greater number of combinations since the lengthier the password, the more difficult it is to remember [5]. Similarly, in token-based authentication, tokens may be shared or misused by the malicious users. Biometrics authentication system serves as a better option than conventional security systems.

Although biometrics provide more advantages to its user, there are some challenging issues in biometric system. Synthetic fingerprint can be generated by the hacker. To avoid this problem sensor, detect the liveness of fingerprints. Feature vectors may be overridden so that old or modified data can be replied to the actual user. Hacker can attack the matcher so that the authentication process could be completely disrupted [6, 7]. It is possible to steal or change the original features from the biometric template database. The proposed method attempts to enhance the security of template database without compromising the fingerprint matching performance.

2 Literature Survey

Komeili et al. [8] proposed a method where fingerprint is used along with ECG. This method found the authenticity of biometrics against fake traits by sensing the liveness of modalities. Also, this method updates the stored template information periodically. But still, the increased time complexity of this method reduced the matching performance. Similar work is proposed by Hammad et al. [9] where a neural network scheme is utilized by combining the fingerprint as well as ECG physiological traits. However, scanning ECG traits need more cost compare to other biometric features. Adithya et al. [1] proposed a method where cloud storage is safeguarded. Similarly, Haoxiang et al. [10] proposed memory efficient-based cloud security system.

Othman and Arun [11] have explained the work of mixing of two dissimilar fingerprints in 'On Mixing Fingerprints' related to two different fingers, to create a new fingerprint. For mixing two fingerprints, it is separated into two different components. One is the continuous component, and the other is the spiral components. Then continuous components of one fingerprint are combined with the spiral component of the other fingerprint. Li et al. [11] have proposed the idea to overcome the disadvantages of reconstructing the entire fingerprint from the stored template. Lu et al. [12] proposed a novel method in which breath biometrics is used as traits. This method provides a better matching performance. However, intra-class matching and inter-class matching pave the way for the hacker to access the template database using synthetic breath biometrics. S. Li and A. C. Kot have proposed a system of privacy protection in the fingerprint authentication system [13]. Then during authentication process, the user information is extracted from the templates which are already stored, to check the genuineness of the person whose fingerprint is provided. A ridge-based co-ordinated system is proposed to find new ridges apart from the conventional ridge finding algorithms [10]. It is difficult for a hacker to differentiate a genuine fingerprint from a template of minutiae. If combined minutiae template is stolen, then the entire minutiae of separate fingerprint is unsatisfied.

Based on the above literature survey, some methods provide better security but they take more time to authenticate the users. Similarly, other methods focused on provide better authentication with poor security improvements. The challenge has to be addressed here is the security of the template database that has to be improved without affecting the conventional fingerprint recognition performance.

3 Proposed System

3.1 Problem Statement

A non-invertible biometric approach is used in this paper. That is, even if the intruder is able to acquire the stored biometric template information, it is not possible to reconstruct the original fingerprint trait because the proposed method uses one way encryption technique. Thus, the reliability of biometric security system is completely depending on how securely the biometric template is stored in the database. But in some cases, intra- and inter-class variation of fingerprint features could reveal the original identity of the biometric user. Therefore, proposed work focuses on transforming the trait information and improving security.

3.2 Proposed Work

The proposed method is categorized into two major sections such as enrollment and verification.

Enrollment: Initially, the fingerprint is obtained from the user. Instead of extracting the entire fingerprint information, features extracted from a small portion of finger which consist of large minutiae features. This is called as region of interest. The unique features from the fingerprint such as ridges and bifurcation are extracted. The minutiae of these features and the Euclidian distance between all pair of minutiae points is calculated, and it is used to generate an n*n matrix. Two values at a time are taken from the matrix sequentially, and binary logic techniques are applied to it.

The manipulated values are given as input to the combinatorial logic circuit, and two values are obtained as output. One of the outputs is stored as the matrix value, while the other is fed as an input for further computations. Based on the produced matrix, a unique graph will be generated using different disk scheduling approaches. This graph is stored in the template database.

Verification: The process mentioned above is repeated during the verification phase, and the generated matrix is compared against the corresponding stored matrix in the database during the verification phase. Figure 1 shows the architecture diagram of the proposed work. The Euclidean distance between every pair of minutiae point is calculated, and n*n matrix is generated as shown in Table 1.

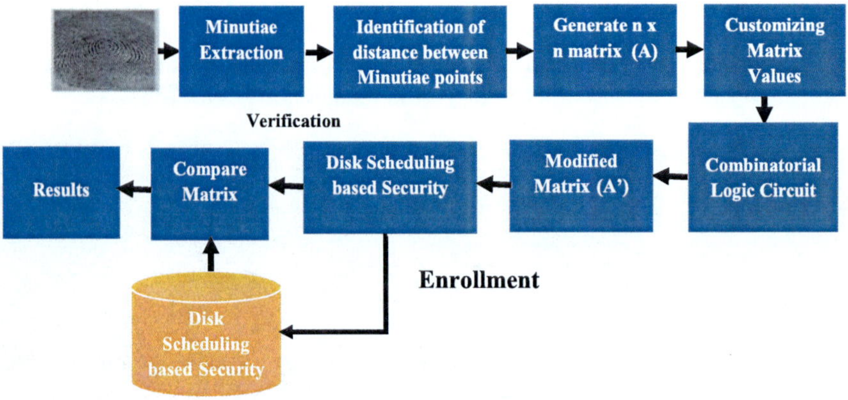

Fig. 1 Proposed system architecture diagram

Table 1 $n * n$ distance matrix-A

	1	2	3	4	5	6	…	n
1	75	175	96	100	44	206	…	…
2	4	76	18	189	23	58	…	…
3	67	34	9	234	192	98	…	…
4	190	16	55	23	155	79	…	…
5	14	73	97	137	189	33	…	…
6	92	87	3	210	178	28	…	…
…	…	…	…	…	…	…	…	…
n	…	…	…	…	…	…	…	…

Algorithm 1: Finding the n × n Distance Matrix

Input	: Finger print
Output	: n x n matrix (A)

Step 1	:	Take Fingerprint as input
Step 2	:	Convert it into binary image of zeros and ones, mat [][]
Step 3	:	**for** every 3X3 matrix
Step 3.1	:	**if** (mat [2][2] ==1)
Step 3.2:		**if** the matrix has 2 ones around mat [2][2] **Store** the (x, y) point in [ridge_ x, ridge_ y]
Step 3.3	:	**end if**
Step 3.4	:	**end if**
Step 3.5	:	**end for**
Step 4	:	**for** every 3X3 matrix
Step 4.1	:	**if** (mat [2][2] ==1)
Step 4.2	:	**if** the matrix has 3 ones around mat [2][2]
Step 4.3	:	**Store** (x, y) point in (bifurcation_ x, bifurcation_ y)
Step 4.4	:	**end if**
Step 4.5	:	**end if**
Step 4.6	:	**end for**
Step 5	:	concatenate ridge and bifurcation matrices
Step 6	:	**for** every minutiae point (x1, y2) n
Step 6.1	:	**Calculate** its distance to remaining n-1 points (x2, y2)
Step 6.2	:	$Dt=SQRT ((x2 - x1)^2 + (y2 - y1)^2)$ /* The Euclidean distance between two points A(x1, x2) and B (y1,y2) */
Step 7	:	**Store** the value of **dt** in n x n matrix (A)

Two values (say *A*, *B*) are taken sequentially and manipulated as depicted in the Fig. 2.

Step 1: ***Bits Swapping***

The numbers A and B are split up into two nibbles *A* (*a*, *b*) and *B* (*c*, *d*). The first nibble of *C* (*c*) is swapped with second nibble of the *B* (*b*).

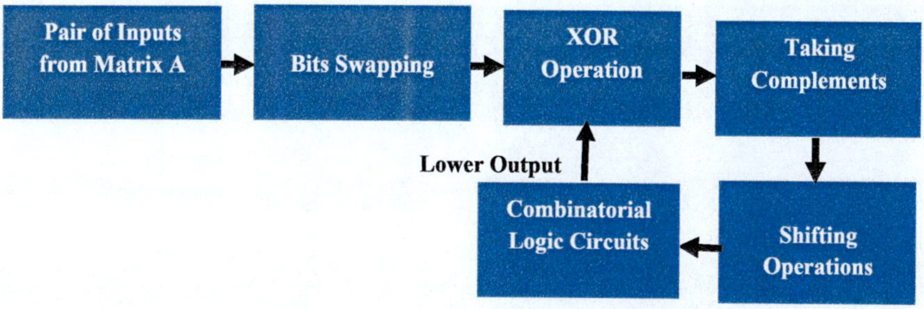

Fig. 2 Customizing input module

Take two pair of inputs	**A=75**	**B=175**
	A=0100 1011	**B=1010 1111**
Bits Swapping	**A=0100 1111**	**B=1010 1011**

Step 2: ***XOR Operation***

The number B is XOR-ed with 1's complement of the previous output from logic circuit. The number C is XOR-ed directly with the previous output from the logic circuit. Initially, a user-defined is used as the previous output.

XOR	**A=0100 1111**	**B=1010 1011**
	^	**^**
	Last lower output	**Last lower output**
	(Initially with user defined 8 bit)	**(Initially with user defined 8 bit)**
	Let O/P: 1010 1101	**1001 1110**

Step 3: ***Taking Complements***

The 1's complement of $A(a)$ and 2's complement of $A(b)$ are XOR-ed to give the first nibble of $A(a)$. The 2's complement of $A(a)$ and 1's complement of $A(b)$ are XOR-ed to give the second nibble of $A(b)$. The same operation is performed on number B.

Complement of **(^) 2's Complement of b**	**Complement of b** **(^) 2's Complement of b**
------------------------------	------------------------------
0101 **(^) 0011**	**0110** **(^) 0010**
------------------------------	------------------------------
0110 **2's Complement of a** **(^) Complement of b**	**0100** **2's Complement of c** **(^) Complement of d**
------------------------------	------------------------------
0110 **(^) 0010**	**0111** **(^) 0001**
------------------------------	------------------------------
0100 **O/P: 0110 0100**	**0110** **O/P: 0100 0110**

Step 4: ***Shifting Operations***

The number A is circular left shifted twice, and the number B is circular right shifted twice.

(Left shift by 2)	**(Right shift by 2)**
1001 0001	**1001 1001**
(A_0 A_1...A_7)	**(B_0 B_1...B_7)**

3.3 Customizing Inputs Module

3.4 Combinatorial Logic Circuit

The manipulated values are given as input to the combinatorial logic circuit which is composed of a number of logic gates. The circuit outputs two values *C* and *D*. *C* is stored in the resulting manipulated matrix, and *D* is used as the output from the logic circuit to perform the XOR operation in step 2. The logic circuit is shown in the Fig. 3.

Fig. 3 Logic circuit design

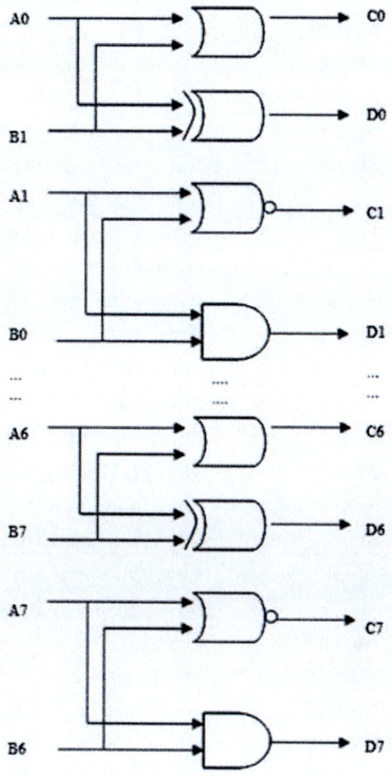

Algorithm 2: Customizing Matrix Values

Input : original n x n matrix
Output : customized n x n matrix (A')

Step 1 : **for** i = 0 to n-1 *// i – rows*
Step 2 : **for** j = 0 to n-1 *// j – columns*
Step 3 : Let A = M [i , j] , B = M [i, j+1] *// Bits Swapping*
Step 3.1 : **end for**
Step 3.2 : **end for**
Step 4 : **Swap** last 4 - bits of both A and B together. *// XOR Operation*
Step 5 : **if** j = 0
Step 6 : A and B are XOR ed with user defined 8-bits value
Step 7 : **else**
Step 8 : **do** A = A ^ D, B = B ^ D *// complementing values*
Step 9 : **Let** a = first four bit value of A and b = second four bit value of A,
Let c = first four-bit value of B and d = second four-bit value of B.
Step 10 : **Now** first four bit value of A = (complement of a) ^ (2's complement of b) **Then** second four bit value of A = (2's complement of a)^(complement of b)
Step 11 : **Now** first four bit value of B = (complement of c) ^ (2's complement of d) **Then** second four bit value of B = (2's complement of c)^(complement of d)
Step 12 : By combining these we get new values of A and B // **Shifting**
Step 13 : **Do** Left Circular Shift of A by 2(A<<2)
Step 14 : **Do** Right Circular Shift of B by 2(B>>2)
/* The 8bits Value of A and B are given as input to the Combinatorial Logic Circuit , to get the values C and D. */
// Step 15 to 19 is used to get even bits of C and D
Step 15 : **Let** x=0
Step 16 : **While** x < 8
Step 17 : C[x] = A[x] | B[x+1] *//OR Operation*
Step 18 : D[x] = A[x] ^ B[x+1] *// XOR Operation*
Step 19 : x = x + 2
Step 19.1: **End while**
//step 20 to 25 is used to get even bits of C and D
Step 20 : **Let** x=1
Step 21 : **While** x < 8
Step 22 : C[x] = !(A[x] | B[x-1]) *// NOR Operation*
Step 23 : D[x] = A[x] & B[x-1] *// AND Operation*
Step 24 : x = x + 2

Step 24.1: **End while**
Step 25 : **Finally** we get the 8-bits values of C and D
Step 26 : **Now,** Store the value of C in the n x n matrix.
Step 27 : **Store** the value of D(lower input) to be used next XOR operation in the customizing matrix values algorithm.
Step 28 : j = j + 1 ***// Repeat from step 3***
Step 29 : End

3.5 Disk Scheduling Based Security

The resulting customized matrix values are used to generate a graph using various disk scheduling algorithms. In the customized matrix, every row uses a different disk scheduling approaches to generate the graph which improves the security of the fingerprint template. From the disk scheduling graph, the difference between two successive points, i.e., larger number–smaller number, is again stored in the customized matrix, and for the last value, the difference between initial and final points is taken.

For example, for an 8×8 matrix, if the input sequence is 95, 180, 34, 119, 11, 123, 62, 64 with the initial point at the 95 and the end point is at 255, then the graph is drawn as given in Fig. 4.

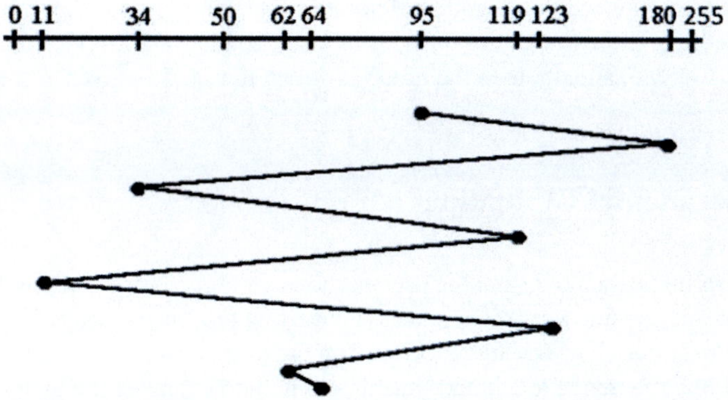

Input Sequence : 95, 180, 34, 119, 11, 123, 62, 64
Output Sequence : 85, 146, 8, 108, 12, 61, 2, 31

Fig. 4 Transformation using disk scheduling approaches

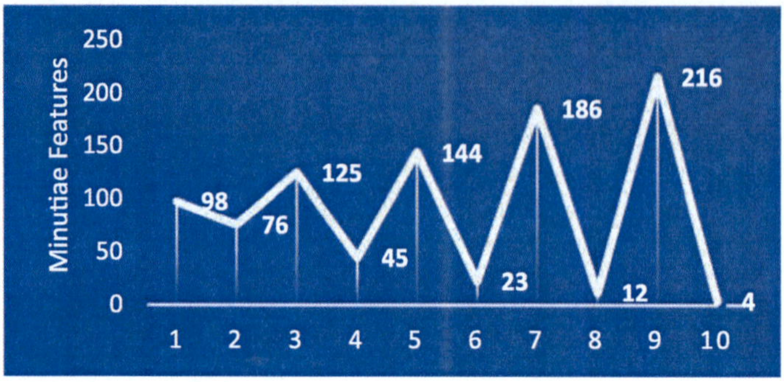

Input Sequence	: 12, 76, 125, 216, 4, 98, 23, 144, 186, 45
Output Sequence	: 22, 49, 80, 99, 121, 163, 174, 204, 212, 94

Fig. 5 User-defined disk scheduling approach

3.5.1 User-Defined Disk Scheduling Algorithm

Besides using the default disk scheduling algorithms, another user-defined disk scheduling algorithm is proposed for every column of customized matrix. According to this algorithm, the column values are sorted, and the middle value is taken as initial point. Then it services the lower and higher value requests sequentially in order, such that it traverse all the points. Now again the differences between two successive points are stored in the respective column of the customized matrix, and for the last value, the differences between initial and final points are taken.

For example, if the first column values of 10×10 matrixes are 12, 76, 125, 216, 4, 98, 23, 144,186, and 45, then the graph is drawn in Fig. 5 as given.

4 Experimental Evaluation

The fingerprint is obtained from the user as shown in Fig. 6a for enrollment process. Various pre-processing operations such as converting to a binary image (Fig. 6b) and thinning the image (Fig. 6c) are performed on the fingerprint.

Converting the image to a binary image will help to visualize the image in terms of zeros and ones. Thinning the image will help to identify the ridges and bifurcation points efficiently. The ridge endings and bifurcation points are the features that are capable of identifying the unique characteristics of an individual. Since the fingerprint is converted into a binary image, identification of these features becomes easier. From the binary value of the image, segments of 3 * 3 window size are taken which is known as region of interest (Fig. 6d). If the center cell is 1 and if it is surrounded by three 1's, the point is identified as a bifurcation (*b*) point. This way, ridge endings and bifurcation (*b*) points are identified and plotted as shown in Fig. 6e.

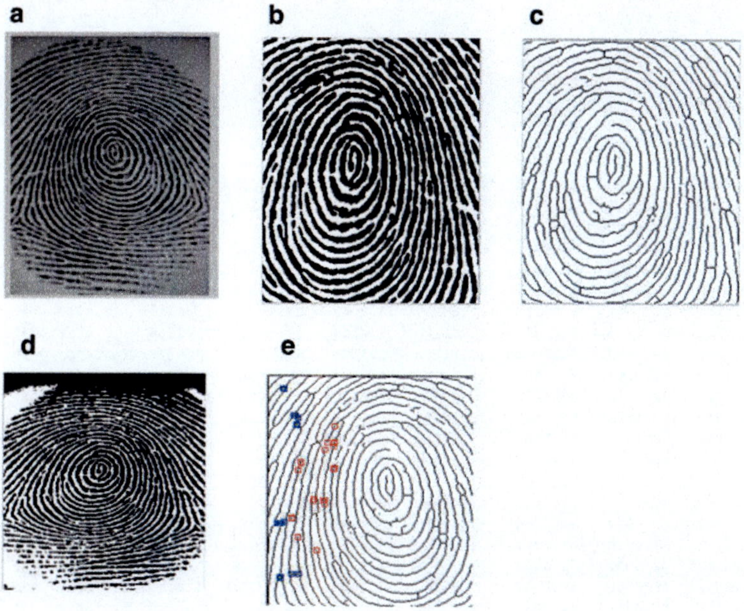

Fig. 6 **a** Input fingerprint. **b** Binary image. **c** Thinned image. **d** Region of interest. **e** Minutiae extraction

The values of these minutiae (*m*) points are stored. Then the Euclidean distance (*d*) between every pair of minutiae (*m*) points are calculated, and $n * n$ matrix is generated. Figure 7 shows the $n * n$ matrix for few pairs of minutiae (*m*) points.

Two values (*A*, *B*) from the obtained distance matrix are taken sequentially, and bit operation techniques are applied.

4.1 Bits Swapping

The number is split into two nibbles. The first nibble of A becomes the second nibble of B, and the second nibble of B becomes the first of A.

4.2 XOR Operation

Then XOR operation is performed. The number *A* is XOR-ed with the 1's complement of previous output from the logic circuit. The number B is XOR-ed directly with the previous output from the logic circuit. Initially, a user-defined number is used.

Columns 1 through 9

0	1.0000	58.4209	25.0000	68.7314	66.7533	31.2410	33.9706	48.8262
1.0000	0	59.4138	24.0416	69.7209	67.7422	31.8904	34.6554	48.0104
58.4209	59.4138	0	82.0000	10.4403	8.5440	41.6293	39.3573	100.2247
25.0000	24.0416	82.0000	0	92.0489	90.0500	47.1699	50.3289	26.4008
68.7314	69.7209	10.4403	92.0489	0	2.0000	50.0000	47.4342	109.4897
66.7533	67.7422	8.5440	90.0500	2.0000	0	48.0833	45.5412	107.5174
31.2410	31.8904	41.6293	47.1699	50.0000	48.0833	0	3.1623	60.1332
33.9706	34.6554	39.3573	50.3289	47.4342	45.5412	3.1623	0	63.0714
48.8262	48.0104	100.2247	26.4008	109.4897	107.5174	60.1332	63.0714	0
40.8167	41.3401	46.4004	53.0000	53.2353	51.4782	11.0454	10.1980	61.4003
90.8735	91.7878	39.0512	111.1261	30.8869	31.9061	64.3273	61.1882	123.3288
42.5441	43.0465	47.6340	54.0833	54.2033	52.4786	13.0384	12.1655	61.6604
40.7185	41.1461	50.8035	50.8035	57.7062	55.9464	13.3417	13.4164	57.7062
100.3195	101.2373	47.3814	120.6027	38.4187	39.6989	73.7564	70.6116	132.5443
98.8838	99.7647	49.4065	117.9873	41.5933	42.5441	70.8378	67.6757	128.4134
75.4321	76.2168	40.0125	92.1358	38.0789	37.6563	45.2769	42.1900	100.8067
121.8524	122.7721	67.8012	142.0458	58.1378	59.6657	95.0789	91.9239	153.3101
103.7690	104.6566	53.2635	123.0325	44.9444	46.0435	75.8947	72.7324	133.5066
102.8834	103.7690	52.6308	122.0901	44.4185	45.4863	74.9466	71.7844	132.5179
77.6209	78.4092	41.1096	94.3716	38.6394	38.3275	47.5079	44.4185	102.9612
0	2.0000	3.0000	67.0671	4.0000	66.1211	67.1193	165.1091	165.1484
2.0000	0	1.0000	67.0075	2.0000	66.0303	67.0298	165.0485	165.0757
3.0000	1.0000	0	67.0000	1.0000	66.0076	67.0075	165.0273	165.0485
67.0671	67.0075	67.0000	0	67.0075	1.4142	1.0000	232.0194	232.0345
4.0000	2.0000	1.0000	67.0075	0	66.0000	67.0000	165.0121	165.0273
66.1211	66.0303	66.0076	1.4142	66.0000	0	1.0000	231.0087	231.0195
67.1193	67.0298	67.0075	1.0000	67.0000	1.0000	0	232.0086	232.0194
165.1091	165.0485	165.0273	232.0194	165.0121	231.0087	232.0086	0	1.0000
165.1484	165.0757	165.0485	232.0345	165.0273	231.0195	232.0194	1.0000	0
164.1493	164.0762	164.0488	231.0346	164.0274	230.0196	231.0195	1.4142	1.0000
8.2462	6.3246	5.3852	69.1809	4.4721	68.1175	69.1158	163.0123	163.0031

Fig. 7 $n * n$ distance (d) matrix

4.3 Taking Complements

Next the complementing technique is used. The 1's complement of the first nibble of A and 2's complement of the second nibble of *A* are XOR-ed and stored as the first nibble of *A*. The 2's complement of the first nibble of *A* and 1's complement of the second nibble of *A* are XOR-ed and stored as the second nibble of *A*. The same operation is performed on the number *B*.

4.4 Shifting Operation

Finally, it comes the shifting operation. Number *A* is circular left shifted once, and the number *B* is circular right shifted once is shown in Fig. 8.

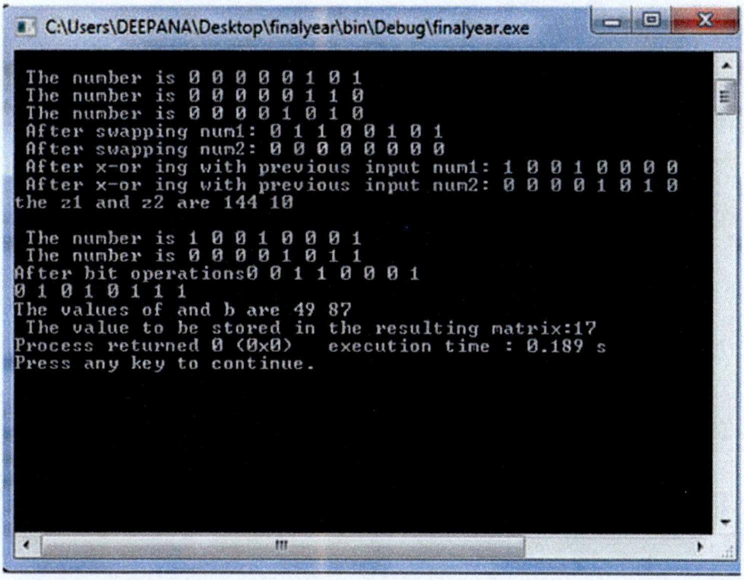

Fig. 8 Customized values output

4.5 *Combinatorial Logic Circuit*

The manipulated values are given as input to the combinatorial logic circuit which is composed of a number of logic gates. The circuit outputs two values C and D. C is stored in the resulting manipulated matrix, and D is used as the output from the logic circuit to perform the XOR operation in step 2. An example of the operation performed by the logic circuit is shown in the Fig. 9.

4.6 *User-Defined Disk Scheduling Algorithm*

The modified matrix after the combinational logic circuit is again modified row-wise using existing disk scheduling algorithms. Then the user-defined disk scheduling algorithm is applied for all the columns of the matrix. An example of the user-defined disk scheduling operation is performed as shown in Fig. 10.

The proposed method is tested against different fingerprint databases such as finger verification competition 2002, 2004 databases. Equal error rate (EER) is measured based on the ratio of wrongly accepted users and wrongly rejected users. Based on EER parameter, the accuracy of the biometric recognition method is measured. The works proposed by komeili et al. and Abdullahi et al. produced the EER as 3.1% and 2.8%, respectively, against FVC 2002 database. However, the proposed method

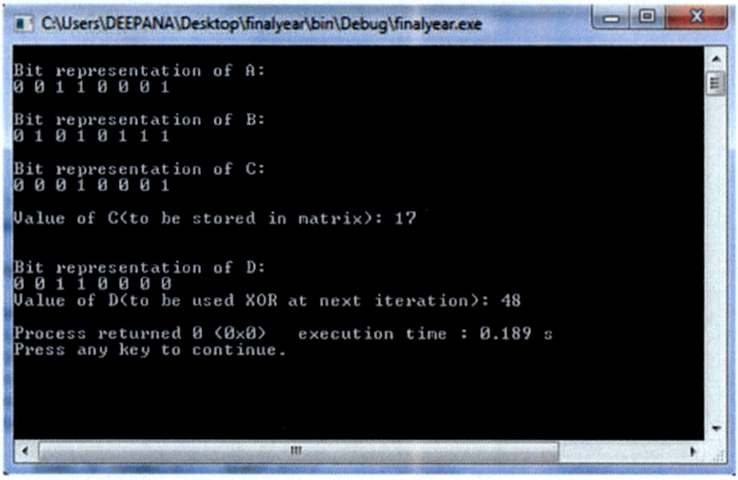

Fig. 9 Logic circuit output

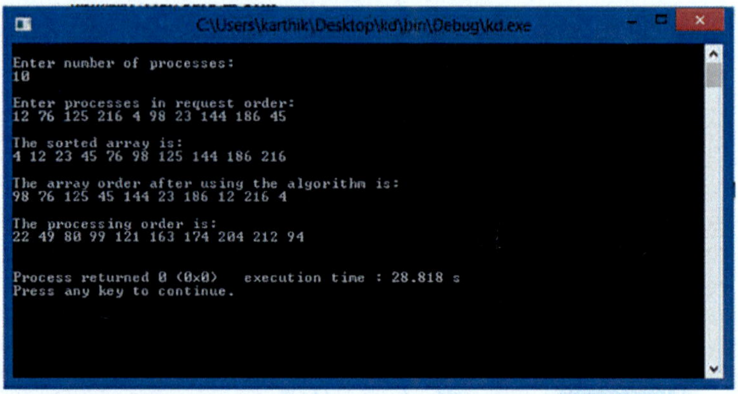

Fig. 10 User-defined disk scheduling output

takes reduced EER of 2.3%. This shows the efficiency of the proposed combinatorial logic gate approach. The detailed comparison is plotted in Fig. 11.

Similarly, Fig. 16 represents the comparison of matching accuracy against FVC 2004 database. Komeili and Abdullahi provided EER as 5.3%, 4.9% respectively. Even in FVC 2004 database also, the proposed method provides EER as 4.3% which is definitely better than the existing methods. The pictorial comparison of matching performance is given in Fig. 12.

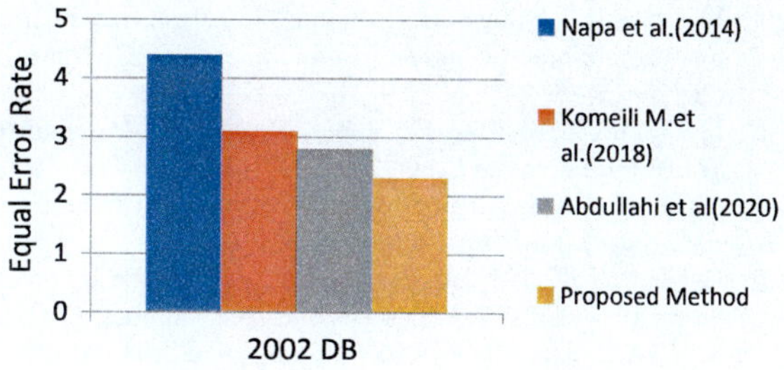

Fig. 11 Proposed method vs existing method using FVC 2002 database

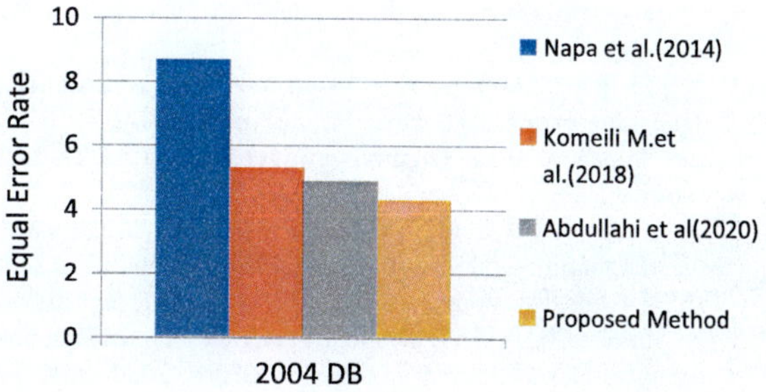

Fig. 12 Proposed method vs existing method using FVC 2004 database

5 Conclusion

The proposed work has been carried out as minutiae extraction, distance calculation, performing bit operations and obtaining a customized matrix by inputting the manipulated values to the user-defined combinatorial logic circuit and a user-defined disk scheduling algorithm. The multilayer feature transformation technique ensures the security of the proposed method. Since most of the modules utilized the bit based logical operations, the efficiency of the biometric matching performance is improved when compared with other state-of-art techniques. Experimental analysis clearly depicts the efficiency as well as the improved security of the proposed method. In future, instead of using uni-modal biometric traits, multi-modal biometrics can be used to improve the performance further.

References

1. M. Adithya, P.G. Scholar, B. Shanthini, Security analysis and preserving block-level data de-duplication in cloud storage services. J. Trends Comput. Sci. Smart Technol. (TCSST) **2**(02), 120–126 (2020)
2. A. Nagar, K. Nandakumar, A.K. Jain, Multi-biometric cryptosystems based on feature-level fusion. IEEE Trans. Inf. Forensics Secur. **7**(1), 255–268 (2011)
3. R. Cappelli, M. Ferrara, D. Maltoni, Fingerprint indexing based on minutia cylinder-code. IEEE Trans. Pattern Anal. Mach. Intell. **33**(5), 1051–1057 (2010)

4. K. Simoens, J. Bringer, H. Chabanne, S. Seys, A framework for analyzing template security and privacy in biometric authentication systems. IEEE Trans. Inf. Forensics Secur. **7**(2), 833–841 (2012)
5. M. Ferrara, D. Maltoni, R. Cappelli, Noninvertible minutia cylinder-code representation. IEEE Trans. Inf. Forensics Secur. **7**(6), 1727–1737 (2012)
6. S. Li, A.C. Kot, An improved scheme for full fingerprint reconstruction. IEEE Trans. Inf. Forensics Secur. **7**(6), 1906–1912 (2012)
7. L. Lu, L. Liu, M.J. Hussain, Y. Liu. I sense you by breath: Speaker recognition via breath biometrics." IEEE Transact. Depend. Sec. Comput. **17**(2) (2017): 306–319
8. S. Li, A.C. Kot, Privacy protection of fingerprint database. IEEE Signal Process. Lett. **18**(2), 115–118 (2010)
9. M. Hammad, Y. Liu, K. Wang, Multimodal biometric authentication systems using convolution neural network based on different level fusion of ECG and fingerprint. IEEE Access **7**, 26527–26542 (2018)
10. N. Sae-Bae, N. Memon, K. Isbister, K. Ahmed, Multitouch gesture-based authentication. IEEE Trans. Inf. Forensics Secur. **9**(4), 568–582 (2014)
11. A. Othman, A. Ross, On mixing fingerprints. IEEE Trans. Inf. Forensics Secur. **8**(1), 260–267 (2012)
12. M.H. Lim, A.B.J. Teoh, An analytic performance estimation framework for multibit biometric discretization based on equal-probable quantization and linearly separable subcode encoding. IEEE Trans. Inf. Forensics Secur. **7**(4), 1242–1254 (2012)
13. K. Simoens, P. Tuyls, B. Preneel, Privacy weaknesses in biometric sketches, in 2009 30th IEEE Symposium on Security and Privacy (IEEE, 2009 May), pp. 188–203

Author Index

G. Ranganathan et al. (eds.), *Soft Computing for Security Applications*, Advances in Intelligent Systems and Computing 1397,
https://doi.org/10.1007/978-981-16-5301-8